Introductory M[illegible]tics & Statisti[illegible] Business

4th Edition

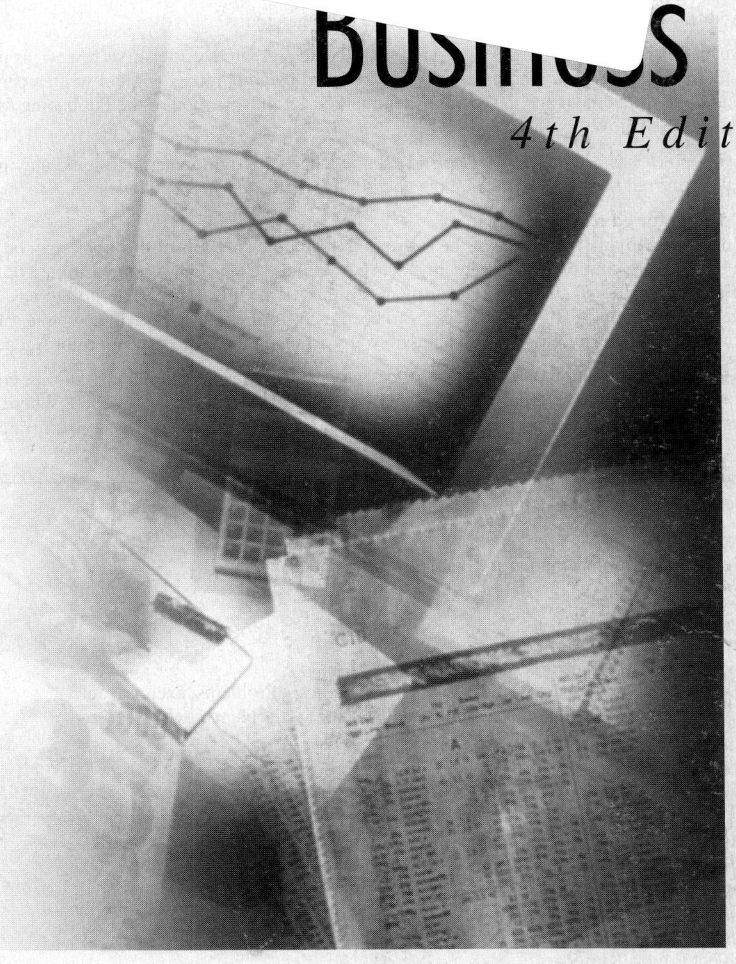

The McGraw-Hill Companies, Inc.

Beijing Bogotà Boston Burr Ridge IL Caracas Dubuque IA Lisbon London Madison WI Madrid Mexico City Milan Montreal New Delhi New York San Francisco Santiago Seoul Singapore St Louis Sydney Taipei Toronto

John S. Croucher

BA(Hons) MSc PhD MAustMS

Professor of Statistics

Macquarie University

McGraw-Hill/Irwin

A Division of The ***McGraw·Hill*** *Companies*

National Library of Australia Cataloguing-in-Publication data:

Croucher, John S.
Introductory mathematics and statistics for business.

4th ed.
Includes index.
ISBN 0 074 71042 7.

1. Business mathematics. 2. Business mathematics – Problems, exercises, etc.
3. Commercial statistics. 4. Commercial statistics – Problems, exercises, etc.
I. Title.

650.01513

Published in Australia by
McGraw-Hill Australia Pty Limited
4 Barcoo Street, Roseville NSW 2069, Australia
Sponsoring Editor: Michael Tully
Project Editor: Kim Ross
Editor: Caroline Hunter, Burrumundi Partnership
Proofreader: Tim Learner
Designer: Kate Barry
Typesetter: Jobs On Mac, Hazelbrook, NSW
Permissions Editor: Susan Gentle
Indexer: Glenda Browne
Marketing Manager: Sara Dougall
Printed by Prowell Productions Limited, Hong Hong

In loving memory of darling Kay
(1949–2001)

If of thy mortal goods thou art bereft
And of thy meagre store
Two loaves alone to thee are left
Sell one, and with the dole
Buy hyacinths to feed thy soul
(The Gulistan of Saadi, 1270)

Contents

STATISTICS

S3 MEASURES OF CENTRAL TENDENCY S.75

S4 MEASURES OF DISPERSION S.109

S5 SAMPLING S.139

Preface

Like many other elementary texts, this book aims to give a non-mathematical person a grounding in business mathematics and statistics. However, unlike many others, it aims to develop a genuine understanding of mathematical and statistical methods, rather than simply to provide a set of formal procedures. This book presents the general principles, but also supports them by many worked examples. It is important that the reader views working with actual numbers as an essential part of the learning process; no matter how clearly principles are emphasised, they will never be fully appreciated unless put into practice.

The book has been specially directed to the needs of those working in business, management and commerce. In particular, most examples have been designed to demonstrate the relevance of mathematics and statistics to making decisions in day-to-day commercial contexts. Since the need is for an Australian book that covers both mathematics and statistics, many examples relate to the Australian economy and business world.

This is the fourth edition of the text, and the changes made from the popular previous three editions have resulted from both student and teacher feedback. I have also been mindful of the information provided by instructors and the content of syllabuses, which vary from state to state. In doing so I have endeavoured to capture those topics considered essential for many courses, without writing a book that would be of gigantic proportions if all possible topics were included.

This edition is still of manageable size and content, although there may be several topics that individual instructors may choose to omit without detracting from the rest of the course. Chapters have been revised and updated with modern data and detailed explanation of difficult concepts.

Major changes from the third edition include:

1. The number of quick quiz questions at the end of every chapter has been increased from eight to ten.
2. The number of multiple-choice questions at the end of every chapter has been increased from six to ten.
3. The list of learning objectives at the beginning of each chapter has been expanded.
4. Updated attractive and appealing visual displays are used to maintain student interest.
5. A new integrated statistics chapter on the analysis of frequency data, including a discussion of the Chi-square test statistic.
6. New sections on the statistical analysis of two-sample data.
7. New sections to incorporate the GST and changes to taxation rates.
8. Updated rates and methods of depreciation.
9. The glossary of key terms has been updated and expanded. Terms contained in the glossary are set in bold type in the chapters.
10. Three sample examination papers based on the material in the text (one at the end of the maths section and two at the end of the statistics section).

11. Updated data have been used in many exercises and examples, including the latest information from the Australian Bureau of Statistics.
12. An appendix on computer packages includes modern software relevant to the syllabus. Packages covered include the statistical packages Minitab® and SPSS®, and the spreadsheet Microsoft Excel®. The capabilities of each are detailed and a brief description of elementary commands is provided.
13. The number of worked examples within chapters has been increased to over 400.
14. The number of exercises has been increased to over 1400, including 220 quick quiz and 220 multiple choice questions. This enables instructors to have a wide selection of problems to set for homework, as well as providing enough exercises for students to check their own answers.
15. Answers have been provided for 40% of the exercises at the end of each chapter.
16. A glossary of key terms. Terms contained in the glossary are set in bold type in the chapters.

For the instructor there is once again a separate manual containing the full solutions to all problems in the text, including the quick quiz and multiple choice questions. The testbank in this manual has been expanded to include over 200 additional exercises with solutions, which instructors can set for extra work. The case studies include eight practical situations that involve writing a report. These case studies relate to a discussion of the problem as an employer might propose it, perhaps as part of a job. The answers require the student to suggest possible methods of approaching the problem and to criticise how others have attempted to solve the problem. Suggested answers are provided to these case studies.

The appendixes in the text include an explanation of summation notation and essential financial and statistical tables. There is also a complete updated summary of all mathematical formulae used in the text at the end of the maths section and all statistical formulae used in the text at the end of the statistics section. In each case, these formulae are referenced by their equation number.

This book is the result of many years of experience in teaching mathematics and statistics both in Australia and overseas. However, it could not have been written without the special help and understanding of friends and colleagues that is so essential in the writing of any work. In this regard my appreciation goes to Catriona Lavermicocca for her assistance in preparing the sections on depreciation and taxation.

I would particularly like to thank the editors Michael Tully and Caroline Hunter for their splendid efforts in making this text into the finished product and Kim Ross for her assistance. Their enthusiasm, encouragement and easygoing manner were very much appreciated. Thanks also go to Kj Byun who provided valuable advice and checking.

To my family goes a special mark of appreciation for their patience and good nature in understanding the long hours of work that necessitated my being locked away in the study for endless hours, night after night and on weekends, working on the manuscript.

This book is also specially dedicated to my two daughters Joanne and Amy, who have now embarked on their own tertiary education that, perhaps unfortunately for them, includes a healthy dose of statistics.

GUIDED TOUR

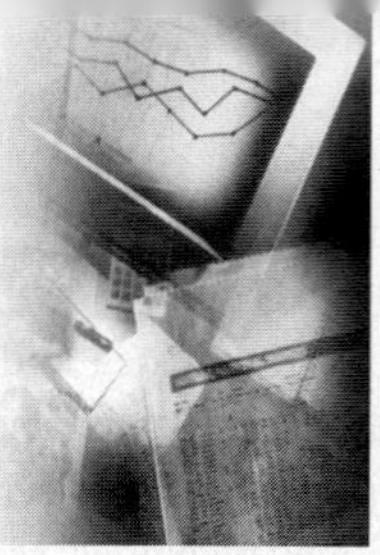

Clear agenda

Identify your goals. Each chapter starts by clearly defining learning objectives. Use this section to identify the difference between the chapters.

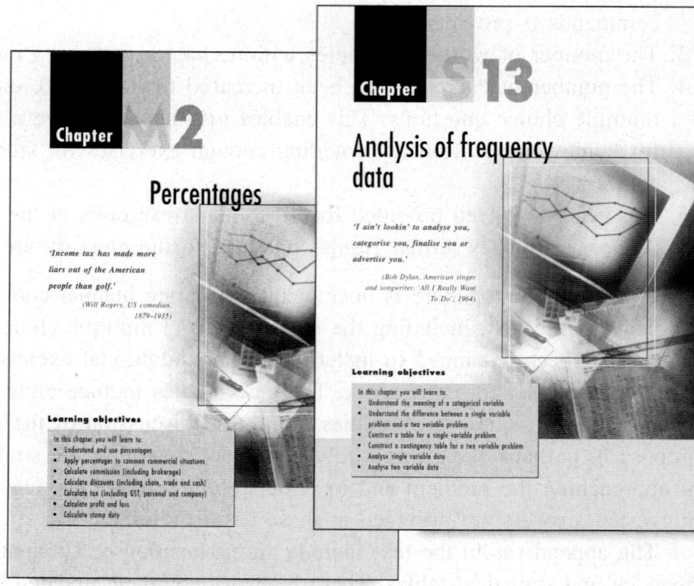

Chapters marked with the letter 'M' relate to the Mathematics section of the book.

Chapters marked with the letter 'S' relate to the Statistics section of the book.

Living applications

Real examples test your knowledge as you go, ensuring you develop the skills to successfully complete each learning objective.

A selection of worked solutions are provided to build your problem-solving skills.

Example M8.2

Using the information in Example M8.1, construct a depreciation schedule for the full 5 years of the effective life of the carpet.

Solution

From Example M8.1(a), the accumulated depreciation *increases* by $1100 each year and the book value *decreases* by $1100 each year. Note that in the schedule in Table M8.4 the first row corresponds to 'end of year 0', which represents the time of the original purchase.

Using the prime cost method, note that the book value at the end of the effective life is always *zero*.

Table M8.4 Depreciation schedule for Example M8.2

End of year	Annual depreciation ($)	Accumulated depreciation ($)	Book value ($)
0	0	0	5500
1	1100	1100	4400
2	1100	2200	3300
3	1100	3300	2200
4	1100	4400	1100
5	1100	5500	0

Introductory Mathematics and Statistics for Business, 4E, has more applications to real situations and data than any other book on the market.

'M' indicates this figure relates to the Mathematics section of the book.

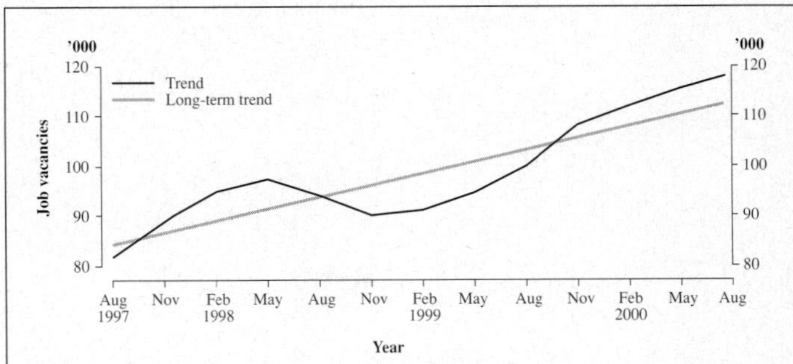

Figure M9.10 The changing value of the number of job vacancies in Australia between August 1997 and August 2000

Source: Australian Bureau of Statistics, *Australian Economic Indicators*, December 2000, Cat. No. 1350.0

Practice makes perfect

Over 1400 **exercises** throughout the book help you understand each topic. Questions are designed to highlight key concepts from each chapter. **Solutions** have been provided to 40% of the exercises in the book.

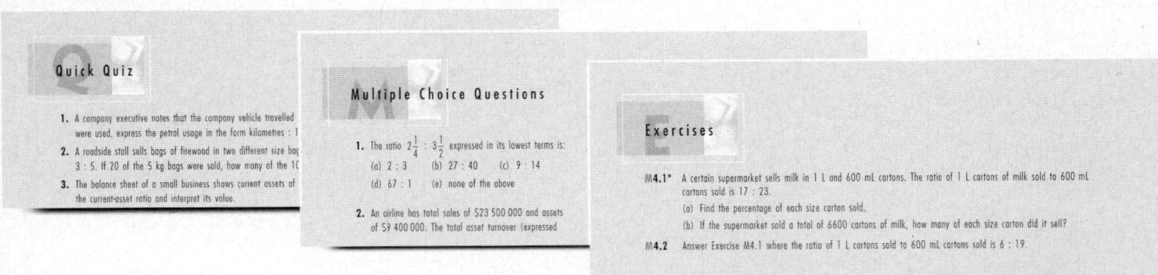

Quick Quiz

1. A company executive notes that the company vehicle travelled ... were used, express the petrol usage in the form kilometres : 1
2. A roadside stall sells bags of firewood in two different size bag... 3 : 5. If 20 of the 5 kg bags were sold, how many of the 1(
3. The balance sheet of a small business shows current assets of ... the current-asset ratio and interpret its value.

Multiple Choice Questions

1. The ratio $2\frac{1}{4} : 3\frac{1}{2}$ expressed in its lowest terms is:
(a) 2 : 3 (b) 27 : 40 (c) 9 : 14
(d) 67 : 1 (e) none of the above
2. An airline has total sales of $23 500 000 and assets of $9 400 000. The total asset turnover (expressed

Exercises

M4.1* A certain supermarket sells milk in 1 L and 600 mL cartons. The ratio of 1 L cartons of milk sold to 600 mL cartons sold is 17 : 23.
(a) Find the percentage of each size carton sold.
(b) If the supermarket sold a total of 6600 cartons of milk, how many of each size carton did it sell?

M4.2 Answer Exercise M4.1 where the ratio of 1 L cartons sold to 600 mL cartons sold is 6 : 19.

Exams: are you ready?

Take the exams on pages M.212 and S.448 to test your learning before you get to your final exams.

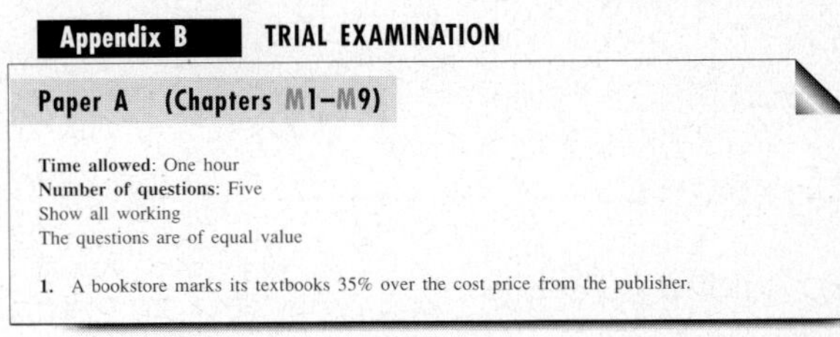

Appendix B TRIAL EXAMINATION

Paper A (Chapters M1–M9)

Time allowed: One hour
Number of questions: Five
Show all working
The questions are of equal value

1. A bookstore marks its textbooks 35% over the cost price from the publisher.

Added value

To get the most out of your text, access the appendixes, tables and summary of useful mathematical formulae at the end of the mathematics part along with the appendixes, tables, summary of useful statistical formulae and complete glossary at the end of the statistics part.

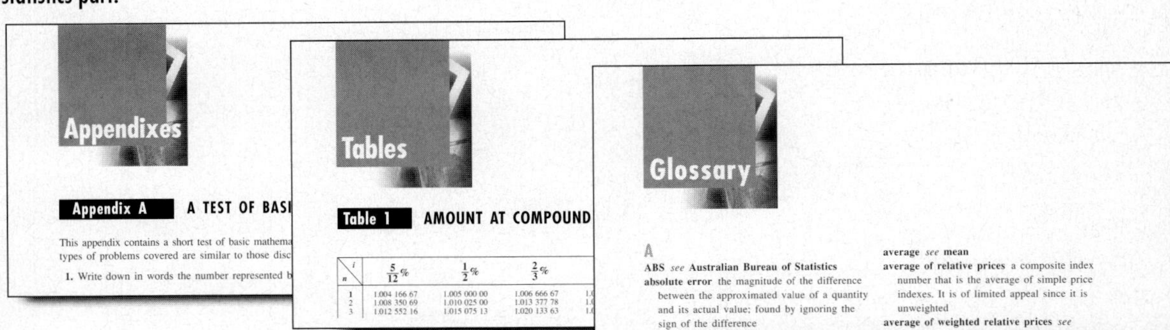

Appendixes

Appendix A A TEST OF BASI

This appendix contains a short test of basic mathema... types of problems covered are similar to those disc...

1. Write down in words the number represented b...

Tables

Table 1 AMOUNT AT COMPOUND

n \ i	$\frac{5}{12}$%	$\frac{1}{2}$%	$\frac{2}{3}$%
1	1.004 166 67	1.005 000 00	1.006 666 67
2	1.008 350 69	1.010 025 00	1.013 377 78
3	1.012 552 16	1.015 075 13	1.020 133 63

Glossary

A

ABS *see* **Australian Bureau of Statistics**

absolute error the magnitude of the difference between the approximated value of a quantity and its actual value; found by ignoring the sign of the difference

average *see* **mean**

average of relative prices a composite index number that is the average of simple price indexes. It is of limited appeal since it is unweighted

average of weighted relative prices *see*

Mathematics

Chapter M1

Basic mathematics

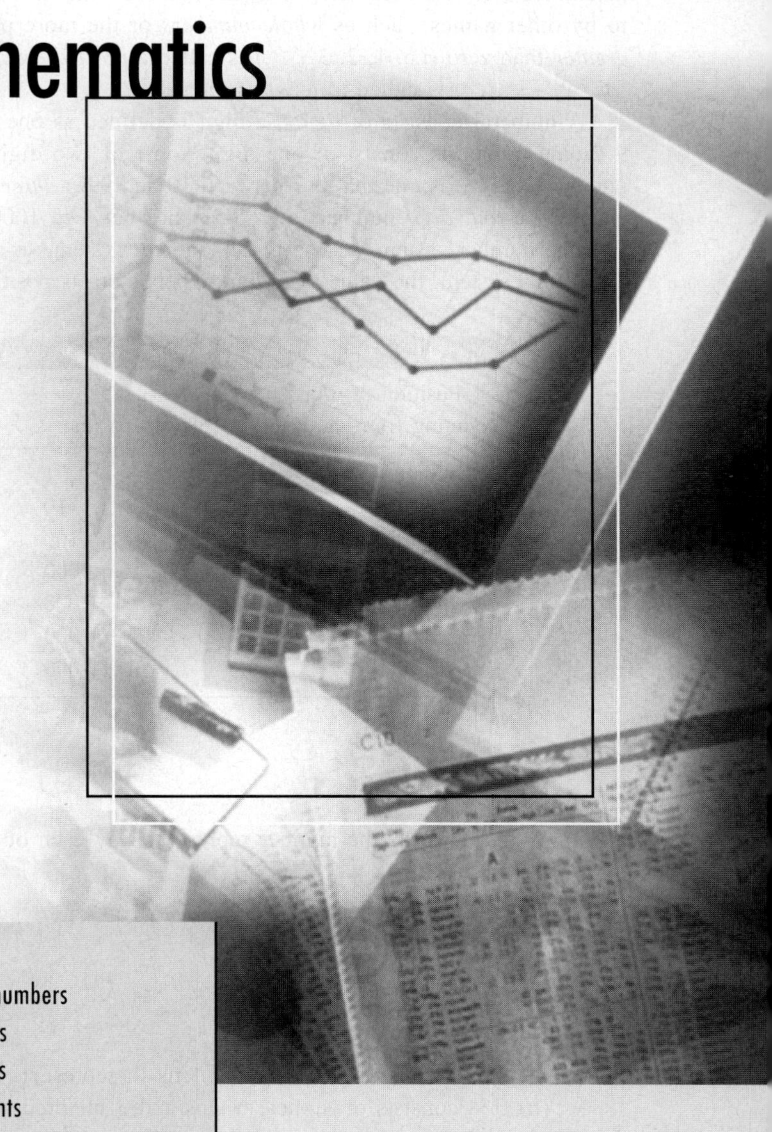

'Mathematics may be defined as the subject in which we never know what we are talking about, nor whether what we are saying is true.'

(Bertrand Russell, British philosopher and mathematician, Mysticism and Logic, *1918)*

Learning objectives

In this chapter you will learn to:

- Carry out calculations involving whole numbers
- Carry out calculations involving fractions
- Carry out calculations involving decimals
- Carry out calculations involving exponents
- Use and understand scientific notation
- Use and understand logarithms

M1.1 WHOLE NUMBERS

The simplest numbers in our number system are those represented by the symbols 0, 1, 2, 3, etc. To be precise, the actual symbols are *numerals* that represent *numbers*. However, in general practice, 'numbers' is often used to mean 'numerals'. ('Numerals' is usually used only when there is a possibility of confusion.)

These numerals represent the *natural numbers*, which are the numbers usually used to count whole objects rather than some fraction of them. (Some people do not include the numeral 0 as a natural number; however, there is no hard and fast rule about this.) Natural numbers are often referred to by other names, such as *whole numbers* or the more popular term **integers**. An integer that is *greater* than zero (i.e. 1, 2, 3, . . .) is called a *positive integer*, while one that is *less* than zero (i.e. –1, –2, –3, . . .) is called a *negative integer*.

A numeral consists of *digits*; a digit is defined as one of the numerals 0, 1, 2, 3, 4, 5, 6, 7, 8, 9. Some numerals consist of one digit, some of two digits, others of three or more. A *two-digit* number (e.g. 43) is one that lies between 10 and 99; a *three-digit* number (e.g. 841) lies between 100 and 999; a *four-digit* number (e.g. 2853) lies between 1000 and 9999, and so on.

The number system we use is the *decimal* system, so-called because it is based on the number 10. In this system, the value of a digit depends on its position within the numeral (see Table M1.1).

Table M1.1 Values of digits in the decimal system

Position of digit (counting from the right)	Value	Name
First	1	Ones (or units)
Second	10	Tens
Third	100	Hundreds
Fourth	1 000	Thousands
Fifth	10 000	Tens of thousands
Sixth	100 000	Hundreds of thousands
Seventh	1 000 000	Millions

Example M1.1

Write in words the number represented by each of the following numerals:

(a) 56
(b) 368
(c) 1679
(d) 367 297

Solution

(a) 56 consists of 6 ones and 5 tens. It represents the number 'fifty-six'.

(b) 368 consists of 8 ones, 6 tens and 3 hundreds. It represents the number 'three hundred and sixty-eight'.

(c) 1679 consists of 9 ones, 7 tens, 6 hundreds and 1 thousand. It represents the number 'one thousand, six hundred and seventy-nine'.

(d) 367 297 consists of 7 ones, 9 tens, 2 hundreds, 7 thousands, 6 tens of thousands and 3 hundreds of thousands. It represents the number 'three hundred and sixty-seven thousand, two hundred and ninety-seven'.

There are four basic mathematical operations that can be performed on numbers. They are:

1. multiplication (often represented by $\times$)
2. division (often represented by $\div$ or /)
3. addition (represented by +)
4. subtraction (represented by –)

The *order* in which these operations should be performed in an expression is:

(a) multiplications and divisions first
(b) then additions and subtractions

For example, $3 + 2 \times 5 = 3 + 10 = 13$. However, to avoid any ambiguity, we can use *parentheses* (or brackets), which take precedence over *all* four basic operations.

In the above case we could have written $3 + (2 \times 5)$ to make the expression clearer, although we would still obtain the same answer of 13.

Suppose, however, that we wanted the 3 and the 2 to be added together *before* multiplication by 5. In this case we would write $(3 + 2) \times 5 = 5 \times 5 = 25$.

A brief discussion of each of the operations is given below.

Multiplication

There are several ways of indicating that two numbers are to be multiplied. For example, suppose we want to multiply the number 3 by the number 4. Some common notations for this are:

1. 3×4
2. $3 \,.\, 4$
3. (3)(4)
4. 3(4) or (3)4

All these expressions mean the *product* of 3 and 4.

An expression of the type $3 \times 4 = 12$ is a mathematical statement that the product '3×4' and the number '12' are really the same. This type of expression is called an **equation**, since it equates two quantities.

An important property of multiplication is its *symmetry*. For example, we could reverse the order of 3 and 4 in the above equation and we would get exactly the same result, i.e. $4 \times 3 = 12$.

Also, if the two numbers to be multiplied have the *same* sign, the result will be *positive*; if they have *different* signs, the result will be *negative*.

Remember that when multiplying numbers in expressions that contain parentheses (brackets), you must perform the calculations *within* the parentheses first.

Example M1.2

Calculate:

(a) $3 \times (6 + 7)$

(b) $(5 + 9) \times 4$
(c) $(4 + 3 - 1) \times (8 - 5 + 6)$
(d) 4×-3
(e) -5×6
(f) -7×-9

Solution

(a) $3 \times (6 + 7) = 3 \times 13 = 39$
(b) $(5 + 9) \times 4 = 14 \times 4 = 56$
(c) $(4 + 3 - 1) \times (8 - 5 + 6) = 6 \times 9 = 54$
(d) $4 \times -3 = -12$
(e) $-5 \times 6 = -30$
(f) $-7 \times -9 = 63$

Division

There are several ways of indicating that two numbers are to be divided. For example, suppose we want to divide the number 12 by the number 4. Some common notations for this are:

1. $12 \div 4$
2. $12/4$
3. $\dfrac{12}{4}$

In this case the number to be divided (12) is called the *numerator* or *dividend*, and the number that is to be divided by (4) is called the *denominator* or *divisor*. The answer to the division (in this case 3) is called the *quotient*.

Division does *not* have symmetry. If the *order* in which the numbers appear in a division is changed, the answer is also changed. For example, $\dfrac{12}{4}$ is *not* the same as $\dfrac{4}{12}$.

Also, if the signs of the numerator and denominator are the *same*, the quotient will be *positive*; if the signs are *different*, the quotient will be *negative*.

Example M1.3

Calculate:

(a) $\dfrac{-3}{6}$ (b) $\dfrac{5}{-15}$ (c) $\dfrac{-20}{-4}$

Solution

(a) $\dfrac{-3}{6} = -\dfrac{3}{6} = -\dfrac{1}{2}$

(b) $\dfrac{5}{-15} = -\dfrac{5}{15} = -\dfrac{1}{3}$

(c) $\dfrac{-20}{-4} = \dfrac{20}{4} = 5$

Addition

Addition *does* have symmetry; the order in which the numbers appear does not affect the result. For example, the expressions '2 + 3' and '3 + 2' are the same as they are both equal to 5.

Subtraction

Unlike addition, subtraction does *not* have symmetry; the order in which the numbers appear does affect the result. For example, the expression '8 − 3' is equal to 5, but '3 − 8' is equal to −5, not 5.

When performing additions and subtractions, you should treat a + and − sign immediately next to each other as simply a − sign. Treat two − signs immediately next to each other as a + sign.

Example M1.4

Evaluate:

(a) $8 + (-5)$
(b) $12 - (+3)$
(c) $18 - (-5)$

Solution

(a) $8 + (-5) = 8 - 5 = 3$
(b) $12 - (+3) = 12 - 3 = 9$
(c) $18 - (-5) = 18 + 5 = 23$

M1.2 FRACTIONS

A **fraction** is an expression that appears in the form:

$$\frac{a}{b} \tag{M1.1}$$

As in the previous section, a is called the *numerator* and b is called the *denominator*. The denominator cannot be zero, because if it is, the result is undefined.

A *proper* fraction is one in which the numerator is *less* than the denominator (e.g. $\frac{2}{3}$, $\frac{6}{17}$); an *improper* fraction has the numerator greater than the denominator (e.g. $\frac{5}{2}$, $\frac{25}{3}$).

Addition of fractions with the same denominator

To add fractions with the same denominator, add together the numerators to obtain the new numerator. The denominator remains the same.

Example M1.5

Add the following fractions:

(a) $\frac{1}{5}$ and $\frac{3}{5}$

(b) $\frac{4}{9}$ and $\frac{1}{9}$

(c) $\frac{2}{15}$, $\frac{4}{15}$ and $\frac{7}{15}$

Solution

(a) $\frac{1}{5} + \frac{3}{5} = \frac{1+3}{5}$ or $\frac{4}{5}$

(b) $\frac{4}{9} + \frac{1}{9} = \frac{4+1}{9}$ or $\frac{5}{9}$

(c) $\frac{2}{15} + \frac{4}{15} + \frac{7}{15} = \frac{2+4+7}{15}$ or $\frac{13}{15}$

Addition of fractions with different denominators

We must change fractions with different denominators into fractions with identical denominators before we can add the numerators. The identical denominator used is called the *lowest common denominator* (LCD) and is the smallest number into which the denominators will all divide. This smallest number itself is called the *lowest common multiple* (LCM). It may be found by multiplying together common factors of the numbers.

Example M1.6

Find the lowest common multiple (LCM) of:

(a) 2, 5, 6, 8

(b) 5, 12, 15, 20

Solution

The lowest common multiples are:

(a) $2 \times 5 \times 3 \times 4 = 120$

(b) $5 \times 3 \times 4 = 60$

Example M1.7

Evaluate:

(a) $\frac{1}{3} + \frac{2}{9} + \frac{5}{6}$

(b) $\frac{2}{5} + \frac{4}{15} + \frac{1}{12}$

Solution

(a) The LCM of 3, 9 and 6 is 18, so each fraction must be converted to an equivalent one with a denominator of 18.

$$\frac{1}{3} = \frac{6}{18} \quad \frac{2}{9} = \frac{4}{18} \quad \frac{5}{6} = \frac{15}{18}$$

and so:

$$\frac{1}{3} + \frac{2}{9} + \frac{5}{6} = \frac{6+4+15}{18} = \frac{25}{18}$$

(b) The LCM of 5, 15 and 12 is 60.

$$\frac{2}{5} = \frac{24}{60} \quad \frac{4}{15} = \frac{16}{60} \quad \frac{1}{12} = \frac{5}{60}$$

and so:

$$\frac{2}{5} + \frac{4}{15} + \frac{1}{12} = \frac{24 + 16 + 5}{60} = \frac{45}{60} = \frac{3}{4}$$

Subtraction of fractions

The rules for addition also apply to subtraction. This time the numerators are subtracted.

Example M1.8

Evaluate:

(a) $\frac{11}{15} + \frac{7}{15} - \frac{14}{15}$

(b) $\frac{5}{16} - \frac{1}{8} + \frac{1}{12}$

Solution

(a) The denominators are the same, so:

$$\frac{11}{15} + \frac{7}{15} - \frac{14}{15} = \frac{11 + 7 - 14}{15} = \frac{4}{15}$$

(b) This time the denominators (16, 8 and 12) are different. The LCM of 16, 8 and 12 is 48, so:

$$\begin{aligned} \frac{5}{16} - \frac{1}{8} + \frac{1}{12} &= \frac{15}{48} - \frac{6}{48} + \frac{4}{48} \\ &= \frac{15 - 6 + 4}{48} \\ &= \frac{13}{48} \end{aligned}$$

Multiplication of fractions

To multiply fractions we multiply the numerators to get the new numerator and multiply the denominators to get the new denominator. If there are common factors in the resulting fraction, they should be divided.

Example M1.9

Evaluate:

(a) $\frac{2}{3} \times \frac{5}{6}$

(b) $\frac{4}{5} \times \frac{3}{4} \times \frac{7}{12}$

(c) $3 \times \frac{6}{11} \times \frac{1}{24}$

Solution

(a) $\frac{2}{3} \times \frac{5}{6} = \frac{2 \times 5}{3 \times 6} = \frac{10}{18} = \frac{5}{9}$

(b) $\frac{4}{5} \times \frac{3}{4} \times \frac{7}{12} = \frac{4 \times 3 \times 7}{5 \times 4 \times 12} = \frac{84}{240} = \frac{7}{20}$

(c) $3 \times \frac{6}{11} \times \frac{1}{24} = \frac{3 \times 6 \times 1}{1 \times 11 \times 24} = \frac{18}{264} = \frac{3}{44}$

It is also possible to cancel common factors of the numerators and denominators *before* multiplying. This can often save tedious arithmetic.

Example M1.10

Evaluate:

$$\frac{5}{6} \times \frac{2}{15} \times \frac{7}{8}$$

Solution

$$\frac{5}{6} \times \frac{2}{15} \times \frac{7}{8} = \frac{\cancel{5}^1}{6} \times \frac{\cancel{2}^1}{\cancel{15}_3} \times \frac{7}{\cancel{8}_4} = \frac{7}{72}$$

Division of fractions

To divide one fraction by a second fraction, *invert* (i.e. turn upside down) the second fraction, then multiply it by the first. Every division problem can be changed into one of multiplication.

Example M1.11

Evaluate:

(a) $\frac{1}{2} \div \frac{5}{12}$

(b) $6 \div \frac{2}{3}$

(c) $\frac{5}{14} \times \frac{1}{8} \div \frac{3}{4}$

Solution

(a) $\frac{1}{2} \div \frac{5}{12} = \frac{1}{2} \times \frac{12}{5} = \frac{12}{10} = \frac{6}{5}$

(b) $6 \div \frac{2}{3} = 6 \times \frac{3}{2} = \frac{18}{2} = 9$

(c) $\frac{5}{14} \times \frac{1}{8} \div \frac{3}{4} = \frac{5}{14} \times \frac{1}{8} \times \frac{4}{3} = \frac{20}{336} = \frac{5}{84}$

M1.3 DECIMALS

Since our number system is based on the number 10, it is possible to express any fraction as a *decimal*. For example, we use a decimal system in our currency and in metric measurement. A decimal is really a fraction in which the numerator has been divided by the denominator to yield an equivalent decimal expression.

A decimal consists of three components: an integer, followed by a *decimal point*, followed by another integer. The values of the digits that occur *after* the decimal point depend on how many positions they are after the point (see Table Ml.2).

Table M1.2 Values of digits after the decimal point

Position of digit after decimal point	Value	Name
First	0.1	Tenths
Second	0.01	Hundredths
Third	0.001	Thousandths
Fourth	0.000 1	Ten-thousandths
Fifth	0.000 01	Hundred-thousandths
Sixth	0.000 001	Millionths

If a number is expressed in decimal form, any zeros on the right-hand end after the decimal point and after the last digit do not change the number's value. For example, the decimals 0.5, 0.50, 0.500 and 0.5000 all represent the same number.

Also, if a number that is expressed in decimal form is less than 1, it is not uncommon for the 0 before the decimal point to be omitted. For example, 0.6 may be written simply as .6.

Example M1.12

Express as fractions:

(a) 0.3 (b) 0.47 (c) 0.763 (d) 0.25

Solution

(a) 0.3 is three-tenths or $\frac{3}{10}$

(b) 0.47 is four-tenths plus seven-hundredths or forty-seven hundredths or $\frac{47}{100}$

(c) 0.763 is seven-tenths plus six-hundredths plus three-thousandths or seven hundred and sixty-three thousandths or $\frac{763}{1000}$

(d) 0.25 is two-tenths plus five-hundredths or twenty-five hundredths or $\frac{25}{100}$ or $\frac{1}{4}$

Example M1.13

Express as decimals:

(a) $\frac{1}{5}$ (b) $\frac{3}{4}$ (c) $\frac{7}{8}$

Solution

These may be found by expressing the numerators in decimal form and performing division as shown.

(a) $\begin{array}{r} 0.2 \\ 5\overline{)1.0} \end{array}$ Thus $\frac{1}{5} = 0.2$

(b) $\begin{array}{r} 0.75 \\ 4\overline{)3.00} \end{array}$ Thus $\frac{3}{4} = 0.75$

(c) $\begin{array}{r} 0.875 \\ 8\overline{)7.000} \end{array}$ Thus $\frac{7}{8} = 0.875$

Addition and subtraction of decimals

The rules for adding and subtracting decimals are similar to those for integers. The main point to remember is to *align* the numbers so that the decimal points are directly under each other. You may fill any unfilled spaces with zeros so that each number has the same number of digits after the decimal point.

Example M1.14

Evaluate:

(a) 2.3 + 0.34 + 1.672
(b) 5.7 − 2.49
(c) 245.879 4 + 356.376 21

Solution

(a) $\begin{array}{l} 2.3000 \;\; + \\ 0.340 \;\; + \\ \underline{1.672} \\ \underline{4.312} \end{array}$

(b) $\begin{array}{l} 5.70 \;\; - \\ \underline{2.49} \\ \underline{3.21} \end{array}$

(c) $\begin{array}{l} 245.879\,40 \;\; + \\ \underline{356.376\,21} \\ \underline{602.255\,61} \end{array}$

Multiplication and division of decimals

When *multiplying* two decimals, follow these steps:

1. Count the number of digits to the right of the decimal point for each number.
2. Add the number of digits found in Step 1 to obtain a number, say x.
3. Multiply the two original decimals, ignoring the decimal points.
4. In the answer obtained in Step 3, mark the decimal point so that there are x digits to the right of the decimal point. This is the answer to the original multiplication.

Example M1.15

Evaluate:

(a) 3.12 × 2.7
(b) 5.542 × 6.78

Solution

(a) Since 3.12 and 2.7 have 2 and 1 digits after their decimal points, respectively, $x = 2 + 1 = 3$.

Perform the multiplication, ignoring the decimal points:

$$\begin{array}{r} 312 \ \times \\ 27 \\ \hline 8424 \\ \hline \end{array}$$

Marking the decimal point in 8424 so that there are 3 digits to the right of the point yields 8.424, which is the answer.

(b) Since 5.542 and 6.78 have 3 and 2 digits after their decimal points, respectively, $x = 3 + 2 = 5$.

Perform the multiplication, ignoring the decimal points:

$$\begin{array}{r} 5\,542 \ \times \\ 678 \\ \hline 3\,757\,476 \\ \hline \end{array}$$

Marking the decimal point in 3 757 476 so that there are 5 digits to the right of the point yields 37.574 76, which is the answer.

When *dividing* decimals, follow these steps:

1. Count the number of digits that are in the divisor to the right of the decimal point. Call this number x.
2. Move the decimal point in the dividend x places to the right (adding zeros as necessary). Do the same to the divisor.
3. Divide the transformed dividend (Step 2) by the transformed divisor (which now has no decimal point). The quotient of this division is the answer.

Example M1.16

Evaluate:

(a) $3.248 \div 0.04$ (b) $82.5 \div 4.125$

Solution

(a) The divisor (0.04) has 2 digits to the right of the decimal point. Thus, $x = 2$. Move the decimal point 2 places to the right in the dividend (3.248) to obtain 324.8. Do the same to 0.04 to obtain 4. We now divide 324.8 by 4:

$$\begin{array}{r} 81.2 \\ 4\overline{)324.8} \end{array}$$

Therefore, $3.248 \div 0.04 = 81.2$

(b) The divisor (4.125) has 3 digits to the right of the decimal point. Thus, $x = 3$. Move the decimal point 3 places to the right in the dividend (82.5) to obtain 82 500. Do the same to 4.125 to obtain 4125. We now divide 82 500 by 4125:

$$\begin{array}{r} 20 \\ 4125\overline{)82\,500} \end{array}$$

Therefore, $82.5 \div 4.125 = 20$

M1.4 EXPONENTS

An **exponent** (or *power*) is a number that is written as a superscript to another number called the **base**. It tells how many times the base is to be multiplied by itself. The position of the superscript is to the right of and a little above the base.

A number that is expressed as a base with an exponent is said to be in *exponential form*. This form is written as a^n where a is the base and n is the exponent (or power).

Example M1.17

Evaluate:

(a) 4^2
(b) 3^5

Solution

(a) In this case, the base is 4 and the exponent is 2. Therefore:

$$4^2 = 4 \times 4 = 16$$

(b) In this case, the base is 3 and the exponent is 5. Therefore:

$$3^5 = 3 \times 3 \times 3 \times 3 \times 3 = 243$$

There are a number of rules that can often simplify calculations involving exponents.

Positive exponents

Suppose that two numbers with the same *base* are expressed in exponential form.

1. The *product* of those two numbers will have the same base with an exponent equal to the *sum* of the two original exponents. That is:

$$a^m \times a^n = a^{m+n} \qquad \textbf{(M1.2)}$$

2. The *quotient* of those two numbers will have the same base with an exponent equal to the *difference* between the original exponents. That is:

$$a^m \div a^n = a^{m-n} \qquad \textbf{(M1.3)}$$

Example M1.18

Evaluate:

(a) $2^3 \times 2^4$ (b) $3^6 \div 3^2$

Solution

(a) In this case, the base is 2 and the exponents are 3 and 4. Using Equation M1.2:

$$2^3 \times 2^4 = 2^{3+4} = 2^7 = 2 \times 2 \times 2 \times 2 \times 2 \times 2 \times 2 = 128$$

(b) In this case, the base is 3 and the exponents are 6 and 2. Using Equation M1.3:

$$3^6 \div 3^2 = 3^{6-2} = 3^4 = 3 \times 3 \times 3 \times 3 = 81$$

If a number already expressed in exponential form is itself raised to *another exponent*, the result is a number with the original base and an exponent equal to the *product* of the two exponents. That is:

$$(a^m)^n = a^{mn} \qquad \textbf{(M1.4)}$$

Example M1.19

Evaluate $(5^3)^2$.

Solution

In this case, the base is 5 and the exponents are 3 and 2. Using Equation M1.4:

$$(5^3)^2 = 5^{3 \times 2} = 5^6 = 5 \times 5 \times 5 \times 5 \times 5 \times 5 = 15\,625$$

Negative exponents

A number expressed in exponential form with a *negative* exponent is equal to the *reciprocal* of the same number with the negative sign removed from the exponent. That is:

$$a^{-n} = \frac{1}{a^n} \qquad \textbf{(M1.5)}$$

Example M1.20

Evaluate:

(a) 2^{-3}
(b) $3^{-1} \times 3^{-2}$
(c) $(4^{-1})^2$
(d) $(2^2)^{-3}$

Solution

(a) Using Equation M1.5:

$$2^{-3} = \frac{1}{2^3} = \frac{1}{8}$$

(b) Using Equations M1.2 and M1.5:

$$3^{-1} \times 3^{-2} = 3^{-1-2} = 3^{-3} = \frac{1}{3^3} = \frac{1}{27}$$

(c) Using Equations M1.4 and M1.5:

$$(4^{-1})^2 = 4^{-2} = \frac{1}{4^2} = \frac{1}{16}$$

(d) Using Equations M1.4 and M1.5:

$$(2^2)^{-3} = 2^{2\times-3} = 2^{-6} = \frac{1}{2^6} = \frac{1}{64}$$

Fractional exponents

The definition of an exponent (a^n) can be extended to include situations where n is a fraction. In particular, if n is of the form $\frac{1}{k}$ (where k is an integer), $a^{\frac{1}{k}}$ is said to be the 'kth root of a'. The kth root of a number is one such that when it is multiplied by itself k times you get that number.

Special kth roots are $k = 2$ (called the *square root*) and $k = 3$ (called the *cube root*). Equations M1.6 and M1.7 illustrate the relationship between fractional exponents and the root of a number.

$$a^{\frac{1}{k}} = \sqrt[k]{a} \qquad \textbf{(M1.6)}$$

$$a^{\frac{m}{n}} = (\sqrt[n]{a})^m \qquad \textbf{(M1.7)}$$

Example M1.21

Simplify:

(a) $9^{\frac{1}{2}}$ (b) $\left(5^{\frac{1}{2}}\right)^4$ (c) $(16)^{\frac{3}{4}}$ (d) $\sqrt[3]{27}$ (e) $(\sqrt{50})(\sqrt[4]{64})$

Solution

(a) $9^{\frac{1}{2}} = (3^2)^{\frac{1}{2}} = 3^{2\times\frac{1}{2}} = 3^1 = 3$

(b) $\left(5^{\frac{1}{2}}\right)^4 = 5^{\frac{1}{2}\times 4} = 5^2 = 25$

(c) $(16)^{\frac{3}{4}} = (2^4)^{\frac{3}{4}} = 2^{4\times\frac{3}{4}} = 2^3 = 8$

(d) $\sqrt[3]{27} = 27^{\frac{1}{3}} = (3^3)^{\frac{1}{3}} = 3^{3\times\frac{1}{3}} = 3^1 = 3$

(e)
$$\begin{aligned}(\sqrt{50})(\sqrt[4]{64}) &= (50)^{\frac{1}{2}}(64)^{\frac{1}{4}} \\ &= (2\times 25)^{\frac{1}{2}}(4\times 16)^{\frac{1}{4}} \\ &= (2)^{\frac{1}{2}}(5^2)^{\frac{1}{2}}(2^2)^{\frac{1}{4}}(2^4)^{\frac{1}{4}} \\ &= (2)^{\frac{1}{2}}(5)^1(2)^{\frac{1}{2}}(2)^1 \\ &= 10(2)^{\frac{1}{2}}(2)^{\frac{1}{2}} \\ &= 10\times 2^1 \\ &= 20\end{aligned}$$

Zero exponent

Under our definitions, the expression a^0 has no meaning. However, using Equation M1.3:

$$\frac{a^n}{a^n} = a^{n-n} = a$$

for all values of n. It therefore seems reasonable to define

$$a^0 = 1 \qquad \textbf{(M1.8)}$$

for all values of a. (The only exception is 0^0, which is undefined.)

Example M1.22

Find the value of 5^0.

Solution

From Equation M1.8, $5^0 = 1$

M1.5 SCIENTIFIC NOTATION

Some measurements, often those occurring in scientific work, involve very large or very small numbers. **Scientific notation** is a shorthand way of writing such numbers in a more convenient form. The procedure for expressing a number in scientific notation involves the **reference position** of the decimal point as defined below.

Definition M1.1

The *reference position* for the decimal point in a number N is immediately to the right of the first non-zero digit in N.

Example M1.23

Find the reference position for the decimal point in each of the following numbers:

(a) 315.68
(b) 2793
(c) 0.456
(d) 0.003 472
(e) 467 000
(f) 50

Solution

The reference position for the decimal point in each case is indicated by an arrow.

(a) $3_{\uparrow}15.68$
(b) $2_{\uparrow}793$
(c) $0.4_{\uparrow}56$
(d) $0.003_{\uparrow}472$

(e) 4↑67 000
(f) 5↑0

The rule for writing a number N in scientific notation is:

$$N = N' \times 10^c$$

where N' = the digit before the reference position

c = the number of digits between the reference position and the decimal point. If the decimal point is to the *right* of the reference position, then the value of c is *positive*. If the decimal point is to the *left* of the reference position, then the value of c is *negative*.

If the number is an integer then the value of c is the number of digits after the reference position.

Example M1.24

Write the numbers in Example M1.23 in scientific notation.

Solution

The values of N', c and the scientific notation are shown in Table M1.3.

Table M1.3 Scientific notation for the numbers in Example M1.23

Number	N'	c	Scientific notation
315.68	3	2	3.1568×10^2
2793	2	3	2.793×10^3
0.456	4	−1	4.56×10^{-1}
0.003 472	3	−3	3.472×10^{-3}
467 000	4	5	4.67×10^5
50	5	1	5×10^1

Most calculators will automatically switch to scientific notation if the answer resulting from a calculation is either too large or too small for normal display.

Example M1.25

Use your calculator to find the values of:

(a) 52 658 × 49 875
(b) 257 ÷ 458 964

Solution

(a) Unless your calculator displays an 'E' (for error) because it cannot handle large numbers, then the display should be (depending on how many digits your calculator displays):

2.62631 09

which is in scientific notation. This corresponds to the number $2.626\ 31 \times 10^9$. There are already 5 digits after the decimal point, so we add 4 zeros after these (since the exponent is 9) to obtain the number 2 626 310 000. Note that this is an approximation of the exact answer, which is 2 626 317 750.

(b) Unless your calculator displays an 'E' (for error) because it cannot handle small numbers, then the display should be (depending on how many digits your calculator displays):

5.59956 −04

which is in scientific notation. This corresponds to the number $5.599\ 56 \times 10^{-4}$. There is already 1 digit (5) before the decimal point, so we add 3 zeros before this digit (since the exponent is −4) to obtain the number 0.000 559 956. Note that this is still an approximation of the exact answer.

M1.6 LOGARITHMS

A very important mathematical function is the **logarithm**, which is closely connected to the theory of exponents. Logarithms play an important part in the physical sciences such as physics, engineering, statistics, chemistry and mathematics. We will look at some of the properties of logarithms and see how they may be used in numerical calculations.

Until the advent of calculators in the mid-1970s, the usual manner in which students divided numbers or raised a number to a power was with the aid of logarithms. This involved tediously looking up various logarithm tables, and it was certainly very easy to make a mistake. Since that time, of course, this technique has become a lost art. Such calculations are now performed by pressing a few calculator buttons. Nevertheless, logarithms do play a prominent role in mathematics since they arise in a wide variety of scientific problems. One such example is in the *growth model* in Chapter S11 of this book.

Studying and using logarithms can also give an insight into their fundamental nature, which is of great use when encountering them in more advanced subjects. This section provides a few simple examples of how logarithms work, along with an understanding of the suffering endured by students before the introduction of calculators!

In practical terms, however, nobody today would perform arithmetic calculations by means of logarithms, since using a calculator or computer would be far more efficient. However, by understanding elementary calculations involving logarithms, the more advanced mathematics involving their use will be made much easier.

The theory of logarithms is based on the laws of exponents. That is, there is a direct relationship between the exponents and logarithms of numbers, as given in Definition M1.2.

Definition M1.2

The *logarithm* of a number N to a base b is the power to which b must be raised to obtain N. The notation is:

$$\log_b N$$

That is, if $x = \log_b N$, then $N = b^x$.

Typical bases are the number 10 and the special number *e* (which stands for *exponential*). In fact, logarithms to the base *e* are called **natural logarithms** or *Napierian logarithms* after the Scotsman John Napier who published his work in 1614.

Example M1.26

Suppose we consider logarithms to the base 10.

$$\text{Since } 10^0 = 1, \text{ then } \log_{10} 1 = 0$$
$$\text{Since } 10^1 = 10, \text{ then } \log_{10} 10 = 1$$
$$\text{Since } 10^2 = 100, \text{ then } \log_{10} 100 = 2$$
$$\text{Since } 10^3 = 1000, \text{ then } \log_{10} 1000 = 3$$
$$\text{Since } 10^4 = 10\,000, \text{ then } \log_{10} 10\,000 = 4$$
$$\text{Since } 10^5 = 100\,000, \text{ then } \log_{10} 100\,000 = 5$$

Example M1.27

Suppose we consider logarithms to the base 2.

$$\text{Since } 2^1 = 2, \text{ then } \log_2 2 = 1$$
$$\text{Since } 2^2 = 4, \text{ then } \log_2 4 = 2$$
$$\text{Since } 2^3 = 8, \text{ then } \log_2 8 = 3$$
$$\text{Since } 2^4 = 16, \text{ then } \log_2 16 = 4$$
$$\text{Since } 2^5 = 32, \text{ then } \log_2 32 = 5$$

In theory we can take the logarithm of a number to *any* base. However, since our decimal system is based on the number 10, the base that is most commonly used is 10. In fact, logarithms to the base 10 are called *common logarithms*. For our purposes, if the base is omitted, then it is assumed that the base is 10.

The definition of logarithms also applies to negative exponents, as in the following example.

Example M1.28

$$\text{Since } 10^{-1} = 0.1, \text{ then } \log_{10} 0.1 = -1$$
$$\text{Since } 10^{-2} = 0.01, \text{ then } \log_{10} 0.01 = -2$$
$$\text{Since } 10^{-3} = 0.001, \text{ then } \log_{10} 0.001 = -3$$
$$\text{Since } 10^{-4} = 0.000\,1, \text{ then } \log_{10} 0.000\,1 = -4$$
$$\text{Since } 10^{-5} = 0.000\,01, \text{ then } \log_{10} 0.000\,01 = -5$$

From Examples M1.26 to M1.28 we can see that the logarithms of numbers *greater than zero but less than 1* are *negative*, the logarithm of *1* itself is *zero* and the logarithms of numbers *greater than 1* are *positive*. The logarithms of *negative numbers* and *zero* are *undefined*.

Characteristic and mantissa

While it is relatively easy to calculate the logarithms of numbers that are powers of 10, the logarithms of other numbers are not immediately obvious. However, it is possible to find logarithms for *all* positive numbers.

Definition M1.3

Suppose that the logarithm of a positive number is expressed as an integer plus a non-negative decimal fraction. Then:

- the integer is called the *characteristic* of the logarithm
- the decimal fraction is called the *mantissa* of the logarithm

The *characteristic* of the common logarithm of a positive number N is numerically equal to the number of digits between the reference position and the decimal point in N.

If the decimal point is to the *right* of the reference position, then the characteristic is *positive*. If the decimal point is to the *left* of the reference position, then the characteristic is *negative*.

Note that if the number is an *integer*, then the characteristic is equal to *one less* than the number of digits in the number.

Example M1.29

Find the characteristics of the following numbers:

(a) 235.89
(b) 12.658
(c) 0.256
(d) 0.045
(e) 3.68
(f) 69

Solution

The reference positions and characteristics are shown in Table M1.4.

Table M1.4 Reference positions and characteristics for Example M1.29

Number	Reference position	Characteristics
235.89	2↑35.89	2
12.658	1↑2.658	1
0.256	0.2↑56	−1
0.045	0.04↑5	−2
3.68	3↑.68	0
69	6↑9	1

To find the mantissa of a common logarithm of a number we must use Table 5 on pages M.236–7. Before using this table it is helpful, first, to write the number in scientific notation. We then use the three following properties of logarithms, which hold for any two positive numbers A and B and for any base.

$$\log (A \times B) = \log A + \log B \qquad \textbf{(M1.9)}$$
$$\log (A \div B) = \log A - \log B \qquad \textbf{(M1.10)}$$
$$\log A^n = n \times \log A \qquad \textbf{(M1.11)}$$

Example M1.30

Write the following numbers in scientific notation, and use Equations M1.9 and M1.11 to help find their logarithms:

(a) 563.2
(b) 0.238

Solution

In scientific notation:

(a) $563.2 = 5.632 \times 10^2$

Therefore,

$$\begin{aligned} \log 563.2 &= \log (5.632 \times 10^2) \\ &= \log 5.632 + \log 10^2 && \text{(using Equation M1.9)} \\ &= \log 5.632 + 2\log 10 && \text{(using Equation M1.11)} \\ &= \log 5.632 + 2 && \text{(since } \log 10 = 1\text{)} \end{aligned}$$

Note that the integer 2 is the characteristic of the required logarithm.

(b) In scientific notation:

$$0.238 = 2.38 \times 10^{-1}$$

Therefore,

$$\begin{aligned} \log 0.238 &= \log (2.38 \times 10^{-1}) \\ &= \log 2.38 + \log 10^{-1} && \text{(using Equation M1.9)} \\ &= \log 2.38 - \log 10 && \text{(using Equation M1.11)} \\ &= \log 2.38 - 1 && \text{(since } \log 10 = 1\text{)} \end{aligned}$$

Note that the -1 is the characteristic of the required logarithm.

In both (a) and (b) we have reduced the logarithm of our number to the sum of the logarithm of a number with only *one* digit before the decimal point plus (or minus) an integer (which is also the characteristic).

It is the logarithms of those numbers with only *one* digit before the decimal point that are given in Table 5.

Example M1.31

Use Table 5 to find the logarithms of the numbers in Example M1.30.

Solution

We have seen in Example M1.30 that log 563.2 = log 5.632 + 2. Note that Table 5 provides the logarithms of numbers where the number consists of *one* digit and up to *two* decimal places. (Note that the first decimal place is down the left-hand side of the table and the second decimal place is across the top.)

In our case we want log 5.632, but the table only provides us with log 5.63 (= 0.7505) and log 5.64 (= 0.7513). The answer we want, therefore, lies between 0.7505 and 0.7513. A process called *linear interpolation* could help approximate a more exact value, but for our purposes we will simply use the closer value of log 5.63 = 0.7505. This is the mantissa of the logarithm of 563.2. We already know that the characteristic is 2, hence we have:

$$\log 563.2 = 2.7505$$

We have also seen in Example M1.30 that log 0.238 = log 2.38 – 1. In our case we want log 2.38, and from Table 5 this value is 0.3766. This is the mantissa of the logarithm of 2.38. We already know that the characteristic is –1, hence we have:

$$\log 0.238 = 0.3766 - 1$$

(Note that this is the usual way to write the logarithms of numbers that have negative characteristics.)

Antilogarithms

The **antilogarithm** is the value of the number that corresponds to a given logarithm. That is, we have the value of the logarithm, and the aim is to find the value of the number that has that logarithm.

Example M1.32

Find the numbers that have antilogarithms of:

(a) 2.8756
(b) 0.5119 – 1

Solution

(a) From Table 5, the mantissa of .8756 corresponds to $N = 7.51$. Also, the characteristic of 2 corresponds to a factor of 10^2.
Hence, the required number is $7.51 \times 10^2 = 751$.
(b) From Table 5, the mantissa of .5119 corresponds to $N = 3.25$. Also, the characteristic of –1 corresponds to a factor of 10^{-1}. Hence, the required number is $3.25 \times 10^{-1} = 0.325$.

Calculations involving logarithms

Logarithms may be used to find the solution to arithmetic problems such as multiplication, division and the raising of numbers to powers. Indeed, before the advent of calculators, this (together with slide rules) was the most common method of performing such calculations.

Example M1.33

Use logarithms to find the value of:

(a) 4.32×5.89

(b) $32.8 \div 4.65$

(c) $\sqrt[3]{48.7}$

(d) $\dfrac{\sqrt{20} \times (2.31)^2}{\sqrt[3]{54} \times 0.86}$

Solution

(a) Using Equation M1.9,

$$\begin{aligned} \log (4.32 \times 5.89) &= \log 4.32 + \log 5.89 \\ &= 0.6355 + 0.7701 \\ &= 1.4056 \end{aligned}$$

Hence, 4.32×5.89 is the antilogarithm of 1.4056. From Table 5 this is approximately $2.54 \times 10^1 = 25.4$. (The exact answer is 25.444 8.)

(b) Using Equation M1.10,

$$\begin{aligned} \log (32.8 \div 4.65) &= \log 32.8 - \log 4.65 \\ &= 1.5159 - 0.6675 \\ &= 0.8484 \end{aligned}$$

Hence, the value of $32.8 \div 4.65$ is the antilogarithm of 0.8484. From Table 5 this is approximately 7.05.

(c) Using Equation M1.11,

$$\begin{aligned} \log (48.7)^{\frac{1}{3}} &= \left(\frac{1}{3}\right) \log 48.7 \\ &= \frac{1}{3} \times 1.6875 \\ &= 0.5625 \end{aligned}$$

Hence, the value of $(48.7)^{\frac{1}{3}}$ is the antilogarithm of 0.5625. From Table 5 this is approximately 3.65.

(d) Using Equations M1.9 to M1.11,

$$\log \frac{\sqrt{20} \times (2.31)^2}{\sqrt[3]{54} \times 0.86} = \log (\sqrt{20} \times (2.31))^2 - \log(\sqrt[3]{54} \times 0.86)$$

$$= \log 20^{\frac{1}{2}} + \log 2.31^2 - \log 54^{\frac{1}{3}} - \log 0.86$$

$$= \left(\frac{1}{2}\right) \log 20 + 2 \log 2.31 - \left(\frac{1}{3}\right) \log 54 - \log 0.86$$

$$= \left(\frac{1}{2}\right)(1.3010) + (2)(0.3636) - \left(\frac{1}{3}\right)(1.7324) - (0.9345 - 1)$$

$$= 0.6505 + 0.7272 - 0.5775 + 0.0655$$

$$= 0.8657$$

Hence, the value of our answer is the antilogarithm of 0.8657. From Table 5 this is 7.34.

M1.7 REMARKS

A thorough knowledge of fractions, decimals and exponents is essential for an understanding of basic mathematical principles. The advent of hand-held calculators in the 1970s, followed by computers, has taken the tedium out of many of the calculations. Arithmetic problems that might have taken many minutes can now be done in seconds by pressing a few buttons.

Nevertheless, you should not be too reliant on modern technology to solve every problem. (A calculator may not always be nearby when you need it!) Even though we may use this equipment as an aid, you are far better prepared if you are also aware of the processes that the calculator is undertaking when performing calculations.

We have discussed how logarithm tables can be used to perform calculations and have provided a table of common logarithms (Table 5). Most calculators have a *log* button that will find the logarithms of numbers more quickly and more accurately than these tables. (Note that your calculator may also have a button labelled *ln*, which refers to logarithms to the base e.) The (common) antilogarithm of a number x can be found from a calculator using the button labelled 10^x.

Quick Quiz

1. Find the lowest common multiple (LCM) of 12, 18, 20, 30.
2. Evaluate:

   $52 + (-12) - (-6) + (18 - 4)$
3. Evaluate:

   $$\frac{24 - 16 + (15 - 5)}{3 + (16 - 4) - 9}$$
4. Evaluate:

   $(\sqrt{27})(\sqrt{72})(\sqrt{32})(\sqrt{3})$
5. Evaluate:

   $$\frac{2}{3} \times \frac{5}{8} - \frac{1}{4}$$
6. Evaluate:

   $3.64 + 7.096 - 4.8$
7. It takes each worker at a manual carwash 5 minutes to wash a car and 4 minutes to dry it. If the carwash has 6 workers who can wash or dry, how many cars can it wash and dry in 3 hours?
8. One bottle of orange soft drink contains 50 g of sugar, 800 mL of water and 100 g of orange syrup. A scout group has available 310 g of sugar, 6 L of water and 590 g of orange syrup. How many bottles of orange soft drink can they make?
9. Write 692.7 in scientific notation.
10. Express 0.875 as a fraction in its lowest terms.

Multiple Choice Questions

1. The value of $(\sqrt[3]{64})(\sqrt[5]{52}) \div 2^3$ is:
(a) 8 (b) 4 (c) 1
(d) 2 (e) none of the above

2. The fraction $\frac{10}{16}$ expressed as a decimal is:
(a) 0.160 (b) 0.600 (c) 0.100
(d) 0.625 (e) none of the above

3. The value of 0.327 − 0.159 + 1.643 is:
(a) 2.129 (b) 1.811 (c) 1.157
(d) −1.475 (e) none of the above

4. The value of $(6^2 \times 3^{-1} \times 2^3) \div (6 \times 2^2)$ is:
(a) 3 (b) 36 (c) 18
(d) 4 (e) none of the above

5. The value of $\frac{3}{16} + \frac{1}{4} - \frac{3}{8} + \frac{5}{12}$ is:
(a) $\frac{1}{2}$ (b) $\frac{29}{24}$ (c) $\frac{5}{12}$
(d) $\frac{3}{20}$ (e) none of the above

6. The value of $\left(\frac{1}{3} + \frac{3}{4}\right) \times \left(2 + \frac{1}{4}\right)$ is:
(a) $\frac{9}{2}$ (b) $\frac{17}{14}$ (c) $\frac{39}{16}$ (d) $\frac{9}{4}$
(e) none of the above

7. The number 0.0372 expressed in scientific notation is:
(a) 3.72×10^{-1} (b) 3.72×10^{-2}
(c) 3.72×10^{-3} (d) 37.2×10^{-2}
(e) none of the above

8. The value of $(64)^{2/3}$ is:
(a) 4 (b) 8 (c) 16
(d) 32 (e) none of the above

9. Which of the following statements is correct?
(a) log 5 = log 12 − log 4 − log 3
(b) log 25 = 2log 10
(c) log 8 = log 6 + log 2
(d) log 20 = log 5 + 2log 2
(e) none of the above

10. It takes a worker exactly 2.5 minutes to make one pie. The maximum number of pies the worker can make in 44 minutes is:
(a) 18 (b) 19 (c) 20
(d) 21 (e) none of the above

Exercises

M1.1* Write down in words the number represented by each of the following numerals:

(a) 23 (b) 59 (c) 346 (d) 782

M1.2 Write down in words the number represented by each of the following numerals:

(a) 1428 (b) 21 347 (c) 66 469 (d) 932 158

(e) 3 457 622

M1.3 Evaluate:

(a) $4 + 3 + 8$ (b) $7 - 2 + 5$ (c) $15 - 6 + 4 - 2 + 1$

(d) $28 + (-13) - (+2) - (-6)$

M1.4* Evaluate:

(a) $2 \times (4 + 3)$ (b) $(2 + 5) \times 3$ (c) $(2 \times 4) + 3$ (d) $2 + (5 \times 3)$

M1.5 Evaluate:

(a) -2×7 (b) 6×-3 (c) -2×-5

(d) $(2 + 5 - 1) \times (8 - 2 + 3)$

M1.6* Evaluate:

(a) $\frac{-2}{8}$ (b) $\frac{2}{-8}$ (c) $\frac{-2}{-8}$

M1.7 Evaluate:

(a) $\frac{2 + 6}{20 - 4}$ (b) $\frac{8 - 20}{6 + 10}$ (c) $\frac{3 - 18}{2 - 17}$

M1.8* Find the lowest common multiple (LCM) of:

(a) 2, 3, 12, 20 (b) 20, 25, 30 (c) 2, 6, 9, 15

M1.9 Find the lowest common multiple (LCM) of:

(a) 24, 60, 144 (b) 7, 14, 35, 49 (c) 100, 150, 250

M1.10* Evaluate:

(a) $\frac{1}{2} + \frac{1}{3} + \frac{1}{6}$ (b) $\frac{1}{5} + \frac{1}{10} + \frac{1}{20}$ (c) $\frac{1}{10} + \frac{1}{30} + \frac{1}{60}$

M1.11* Evaluate:

(a) $\frac{3}{4} - \frac{1}{2} + \frac{1}{8}$ (b) $\frac{5}{6} - \frac{1}{3} + \frac{1}{4}$ (c) $\frac{1}{2} + \frac{1}{4} - \frac{3}{4}$

M1.12 Evaluate:

(a) $\frac{5}{24} - \frac{1}{12} - \frac{3}{8}$ (b) $2 - \frac{2}{3} + \frac{1}{6}$ (c) $\frac{3}{16} - \frac{1}{8} + \frac{1}{4}$

M1.13* Evaluate:

(a) $\frac{3}{4} \times \frac{2}{9}$ (b) $4 \times \frac{1}{8}$ (c) $\frac{5}{6} \times \frac{3}{10}$

M1.14 Evaluate:

(a) $\frac{5}{6} \times \frac{3}{20} \times 4$ (b) $\frac{3}{11} \times \frac{22}{15} \times \frac{5}{18}$ (c) $\frac{1}{12} \times \frac{9}{14} \times \frac{7}{3}$

M1.15* Evaluate:

(a) $\frac{1}{6} \div \frac{5}{6}$ (b) $2 \div \frac{1}{4}$ (c) $\frac{1}{2} \times \frac{1}{3} \div \frac{1}{6}$

M1.16 Evaluate:

(a) $\left(\frac{1}{2} + \frac{1}{4}\right) \div \frac{4}{9}$ (b) $\left(2 + \frac{2}{3}\right) \div (5 - 3)$

(c) $\left(\frac{1}{2} + \frac{1}{8} + \frac{3}{16}\right) \div \left(\frac{5}{8} + \frac{3}{16}\right)$

M1.17* Express as fractions (in simplest terms):

(a) 0.50 (b) 0.075 (c) 0.125

M1.18 Express as decimals:

(a) $\frac{1}{2}$ (b) $\frac{3}{8}$ (c) $\frac{5}{12}$

M1.19* Express as decimals:

(a) $\frac{5}{8}$ (b) $\frac{7}{12}$ (c) $\frac{9}{20}$

M1.20 Evaluate:

(a) 0.8 + 2.7 + 6.5 (b) 6.3 − 2.1 + 1.9 (c) 1.43 + 3.05 − 0.29

M1.21* Evaluate:

(a) 0.4 − 0.18 + 0.326

(b) 4.89 − 3.291 + 1.002

(c) 3.9385 + 4.06 − 1.072

M1.22* Evaluate:

(a) 1.6 × 2.0 (b) 0.8 × 0.4 (c) 0.25 × 0.50

M1.23 Evaluate:

(a) 4.13 × 2.05 (b) 6.8 × 2.125 (c) 1.625 × 0.24

M1.24* Evaluate:

(a) 6.40 ÷ 1.6 (b) 0.625 ÷ 0.125 (c) 40.8 ÷ 0.04

M1.25 Evaluate:

(a) 6.876 ÷ 1.8 (b) 61.8 ÷ 0.16 (c) 9.45 ÷ 1.75

M1.26* Evaluate:

(a) $2^4 \times 2^5$ (b) $3^{12} \div 3^{10}$ (c) $5^2 \times 5 \times 5^{-4}$

M1.27 Evaluate:

(a) 2^{-5} (b) $4^{-1} \times 4^{-2}$ (c) $(2^{-1})^2$

M1.28* Evaluate:

(a) $2^8 \times 2^{-3} \times 2^{-1}$ (b) $(3^2)^2 \times 3^{-5} \times 3^2$

(c) $(5^{-1})^2 \times 5^4 \times (5^2)^{-2}$

M1.29 Evaluate:

(a) $16^{\frac{1}{2}}$ (b) $9^{\frac{3}{2}}$ (c) $8^{\frac{2}{3}}$

M1.30 Evaluate:

(a) $(\sqrt[3]{64})(\sqrt{36})$ (b) $(\sqrt[4]{32})(\sqrt[8]{64})$ (c) $(\sqrt[5]{64})(\sqrt[5]{16})$

M1.31* Find the value of 100^0.

M1.32 Three workmen shared a pizza for lunch. Alan ate one-third, Bill ate one-quarter and Ron ate one-sixth. What fraction of the pizza was *uneaten*?

M1.33* Evaluate:

(a) $(-5 + (-2) + (-4 \times 2)) \times (-5 + 8)$

(b) $[(7 \times 8) - (18 \div 3) + (-2) \times (-4)] \times (6 \div 3)$

M1.34 If 1 L of paint covers 12 m^2, how much paint would be required to cover an area of:

(a) 1.5 m by 3 m?

(b) 0.25 m by 0.75 m?

(c) 8.00 m by 4.5 m?

(*Hint*: 1 L = 1000 mL.)

M1.35* To make 1 cup of porridge requires $\frac{1}{4}$ cup of oats and $\frac{3}{4}$ cup of water.

(a) How much oats and water is required to make 5 cups of porridge?

(b) If you have 6 cups of oats and 15 cups of water, how many cups of porridge can you make?

M1.36* Express the following numbers in scientific notation:

(a) 23.87 (b) 0.359 (c) 0.000 46

M1.37 Express the following numbers in scientific notation:

(a) 0.068 (b) 0.0064 (c) 3564.8

M1.38* Find the common logarithms of the following numbers:

(a) 5.65 (b) 48.2 (c) 0.356

M1.39 Find the common logarithms of the following numbers:

(a) 237.0 (b) 34.5 (c) 0.0451

M1.40* Use logarithms to calculate the value of the following expressions. Show each step. (Check your answers on a calculator and comment on the accuracy of using logarithms.)

(a) 2.56×7.61 (b) $\dfrac{34.7}{2.48}$ (c) $\dfrac{\sqrt[3]{129} \times 1.38}{2.56 \times 1.80}$

M1.41 Evaluate:

(a) $567 - 342 + 678$ (b) $1234 + 3215 - 678$

(c) $-37 + 67 - 54$

M1.42* Evaluate:

(a) $43 - (-14) + (-25)$

(b) $-57 - (-21) - (-17)$

(c) $666 - (+444) + (-555)$

M1.43 Evaluate:

(a) $12 \times (43 - 36)$ (b) $(15 - 9) \times -5$ (c) $(7 - 12) \times (-6 + 8 - 9)$

M1.44 Evaluate:

(a) $(13 - 7 + 16) \times (45 - 52 + 4)$

(b) $(3 - 5 + 9) \times (17 - 19 - 4)$

(c) $(35 + 2 + 6) \times (21 - 39 + 16)$

M1.45 Find the lowest common multiple (LCM) of:

(a) 36, 72, 180 (b) 9, 16, 20, 24 (c) 9, 20, 30

M1.46* Find the lowest common multiple (LCM) of:

(a) 48, 156, 180 (b) 7, 21, 126 (c) 12, 42, 64

M1.47 Evaluate:

(a) $\dfrac{25}{7} \times \dfrac{12}{5} \times \dfrac{28}{3}$ (b) $\dfrac{5}{14} \times \dfrac{36}{25} \div \dfrac{6}{7}$ (c) 1.08×0.03

M1.48 Evaluate:

(a) $3^3 \times 9^{-2} \times 2^2$ (b) $5^{-2} \times 4^{-3} \times 20^2$ (c) $\sqrt[3]{250} \times 10^{\frac{2}{3}} \times \sqrt[6]{256}$

M1.49* Express the following in scientific notation:

(a) 0.0456 (b) 0.000 067 (c) 345.89

M1.50 Express the following in scientific notation:

(a) 256.987 (b) 1.000 05 (c) 17.171 7

M1.51 (a) Use Table 5 to find the *common* logarithms of the following numbers:

(i) 27.58 (ii) 0.567 (iii) 6.99

(b) Check your answers in (a) by using the *log* key on your calculator.

M1.52 Use the *ln* key on your calculator to find the *natural* logarithms of the following numbers:

(a) 444.4 (b) 1.008 (c) 0.006

M1.53* A 2 L bottle of lemonade will provide 2 cups each for 4 people. How many bottles will be required to provide:

(a) 6 cups to each of 6 people?

(b) 4 cups to each of 25 people?

(c) 3 cups to each of 40 people?

M1.54 A restaurant employs casual staff on a Saturday at a pay rate each of \$15 per hour and on a Sunday at a pay rate each of \$20 per hour. Find the total pay if the restaurant employs:

(a) 12 people each for 6 hours on each of Saturday and Sunday

(b) 5 people each for 5 hours on Saturday and 4 people each for 8 hours on Sunday

(c) 7 people each for 7 hours on Saturday and 6 people each for 8 hours on Sunday

M1.55* One small radiator can heat a room of 20 m^3, while one large radiator can heat a room of 30 m^3. How many cubic metres can be heated by a combination of:

(a) 3 small and 2 large radiators?

(b) 5 small and 6 large radiators?

(c) 10 small and 12 large radiators?

M1.56 In Exercise M1.55 you are now told that a small radiator costs \$40 and a large radiator costs \$55. Find the total cost for each combination in (a), (b) and (c).

M1.57 A hairdresser takes 8 minutes to wash a client's hair and 20 minutes to cut it. What would be the total time taken to:

(a) wash 5 clients' hair and cut 7 clients' hair?

(b) wash and cut 8 clients' hair?

(c) cut, but not wash, 9 clients' hair?

M1.58* If the hairdresser in Exercise M1.57 charges \$10 to wash hair and \$25 to cut hair, what would be the total charge for each of the combinations in (a), (b) and (c)?

M1.59 If the hairdresser in Exercise M1.57 decides to wash the hair of 15 clients, how many clients' hair could be cut in the remaining work time if the total amount of work time available is:

(a) 4 hours? (b) 6 hours? (c) 8 hours?

M1.60 A football field consists of 9600 m^2 of lawn. Two employees, Sid and Jack, are given the task of mowing the entire field. Sid can mow 10 m^2 each minute, while Jack can mow 12 m^2 each minute. How long would it take to mow the entire field if:

(a) Sid mowed it by himself?

(b) Jack mowed it by himself?

(c) Sid and Jack mowed half each (at the same time)?

(d) Sid and Jack mowed half each (one after the other)?

(e) Sid mowed for 3 hours and then Jack mowed the rest of the field?

M1.61* Suppose in Exercise M1.60 that Sid is paid \$14 per hour and Jack is paid \$15 per hour. If Sid and Jack commenced mowing at the same time and finished when the field was completely mowed:

(a) how long would it take to finish the job?

(b) what would be the total cost?

Chapter **M2**

Percentages

'Income tax has made more liars out of the American people than golf.'

(Will Rogers, US comedian, 1879–1935)

Learning objectives

In this chapter you will learn to:

- Understand and use percentages
- Apply percentages to common commercial situations
- Calculate commission (including brokerage)
- Calculate discounts (including chain, trade and cash)
- Calculate tax (including GST, personal and company)
- Calculate profit and loss
- Calculate stamp duty

M2.1 CONVERSION TO AND FROM PERCENTAGES

One of the most important concepts in business is that of the **percentage**. To express one number as a percentage of another is to give its *relative* value expressed as the numerator of the equivalent fraction in which 100 is the denominator. The symbol used for a percentage is %.

To convert a *fraction to a percentage*, multiply the fraction by 100 and write the percentage symbol after the answer.

Example M2.1

Convert to percentages:

(a) $\frac{1}{2}$ (b) $\frac{3}{5}$ (c) $\frac{5}{8}$ (d) $\frac{2}{3}$

Solution

(a) $\frac{1}{2} = 100(\frac{1}{2})\% = \frac{100}{2}\% = 50\%$

(b) $\frac{3}{5} = 100(\frac{3}{5})\% = \frac{300}{5}\% = 60\%$

(c) $\frac{5}{8} = 100(\frac{5}{8})\% = \frac{500}{8}\% = 62\frac{1}{2}\%$

(d) $\frac{2}{3} = 100(\frac{2}{3})\% = \frac{200}{3}\% = 66\frac{2}{3}\%$

Note that these results agree with the definition of a percentage given above. That is:

$$\frac{1}{2} = \frac{50}{100} \quad \frac{3}{5} = \frac{60}{100} \quad \frac{5}{8} = \frac{62\frac{1}{2}}{100} \quad \frac{2}{3} = \frac{66\frac{2}{3}}{100}$$

To convert a *decimal to a percentage*, we also multiply it by 100. This is done by moving the decimal point two digits to the right and writing the percentage symbol after the answer. (Add zeros as necessary.)

Example M2.2

Convert to percentages:

(a) 0.25 (b) 0.872 (c) 0.4 (d) 2.35

Solution

(a) 0.25 = 25% (b) 0.872 = 87.2%
(c) 0.4 = 40% (d) 2.35 = 235%

To convert a *percentage to a fraction*, divide the percentage by 100 and remove the percentage symbol. Simplify the fraction if possible.

Example M2.3

Convert to fractions:

(a) 37.5% (b) 72% (c) $33\frac{1}{3}\%$ (d) 250%

Solution

(a) $37.5\% = \dfrac{37.5}{100} = \dfrac{3}{8}$

(b) $72\% = \dfrac{72}{100} = \dfrac{18}{25}$

(c) $33\frac{1}{3}\% = \dfrac{33\frac{1}{3}}{100} = \dfrac{1}{3}$

(d) $250\% = \dfrac{250}{100} = \dfrac{5}{2}$

To convert a *percentage to a decimal*, divide the percentage by 100. That is, place a decimal point two digits to the left of the percentage symbol (or two digits to the left of the decimal point if the percentage contains a decimal point) and remove the symbol.

Example M2.4

Convert to decimals:

(a) 80% (b) 25% (c) 45.78% (d) 150%

(e) $\frac{1}{10}$ (f) $\frac{1}{5}$ (g) $\frac{1}{4}$ (h) $\frac{1}{3}$

Solution

(a) 80% = 0.80 or 0.8 (b) 25% = 0.25 (c) 45.78% = 0.4578

(d) 150% = 1.50 or 1.5 (e) $\frac{1}{10} = 0.10$ (f) $\frac{1}{5} = 0.20$

(g) $\frac{1}{4} = 0.25$ (h) $\frac{1}{3} = 0.3\dot{3}$

(Note that the 'dot' above the second 3 in solution (h) indicates that this is a repeating decimal. That is, the remaining figures after this 3 are an infinite sequence of 3s. That is, 0.333333 ....)

There are many instances where economic and business statistics are expressed as percentages, along with reports from surveys in a variety of areas that appear in all sections of the media as well as in scientific journals. To fully comprehend such articles it is necessary to understand the meaning of percentages. Some examples are:

1. The population of the Earth reportedly consists of 57% Asians, 21% Europeans, 14% North and South Americans and 8% Africans. Of these, 51% are female and 49% are male, while 70% are non-white and 30% are white. The number of people who are unable to read is 70%. (*Source*: *Sunday Telegraph*, 20 April 1997.)

2. In an academic report it was estimated that 50% of teenage pregnancies in Australia end in abortion. (*Source*: *What is Happening with Teenage Pregnancies*?, Research Paper No. 011/96, School of Economic and Financial Studies, Macquarie University, p. 5.)
3. In a *Newspoll* survey on the way in which adults in Australia view the behaviour of teenagers, 79% thought that teenagers take too much for granted, while 65% felt that teenagers rarely show any courtesy such as giving up a seat on a bus or a train. In addition, 68% felt that they did not show enough respect to their parents and 59% were of the view that teenagers would benefit by compulsory national service. To make matters even worse, 76% felt that teenagers were too materialistic. (*Source*: *Daily Telegraph*, 10 January 2001.)
4. A survey reported that in about 65% of occupations women had a lower rate of participation than men in the over-30s age group. In addition, women managers earned 39% less than men, women TAFE teachers an average 11% less and women typists an average 17% less. In the 15–24 age group, men were paid 10% more than women across the board. In welfare and the performing arts, women earned about 5% more than their male colleagues, two of the very few professions in which women came out ahead. (*Source*: *Sunday Telegraph*, 29 June 1997, p. 47.)
5. In 1983, the percentage of convictions for every car stolen in New South Wales was 3.2%. By the beginning of 2001 this had fallen to just 2.1%. (*Source*: *Sun-Herald*, 10 December 2000.)
6. A report in 2001 regarding car crashes revealed that 25% of all car crashes resulted from crashes in car parks or when a driver hit a parked vehicle. (*Source*: *Daily Telegraph*, 17 January 2001, p. 14.)
7. In a national survey of the level of interest in Australian Rules football, a survey in October (just after the grand final) revealed that 51% of people had a positive interest in the game, 33% had been to watch a game and 94% watched it on television, while only 5% actually played it. (*Source*: *Sydney Morning Herald*, 14 October 1996, p. 25.)
8. A report on the habits of Australian shoppers revealed that 54% of people prefer to shop alone. The largest shopping group (33%) is aged 55 years or older and the smallest (9%) is aged between 16 and 24 years. The report found that 80% of all shoppers are female, while this rose to 86% when it came to main grocery shopping. Single-person households make up 14% of shoppers, with 31% living in two-person households, 18% in three-person households and 37% in households of four or more. (*Source*: Radio Marketing Bureau.)
9. In a survey of the television-viewing habits of Australians, it was discovered that 79.4% watched news and current affairs programs, 45.6% watched movies, 40.5% watched documentaries, 38.2% watched sports and 22.6% watched series or serials. There was a significant difference between the sexes when it came to sports, with 54.9% of males saying they watched sports compared to only 22.2% of females. (*Source*: Bureau of Transport and Communications economics survey.)
10. A study of 256 homicides in New South Wales over a three-year period found that 50.4% of the offenders were sentenced for murder, while 49.2% were sentenced for manslaughter. The remaining 0.4% (one case) was for infanticide. Of the 56 cases involving the killing of a sexual partner, 83.9% of the offenders were male, with 80% of these being aged 35 years or less. When facing court, 24.2% of offenders raised the defence of provocation or diminished responsibility. (*Source*: *Report of the Judicial Commission of NSW*, December 1995.)

M2.2 COMMISSION

There are many types of goods and services that are usually sold through an agent who is paid a percentage of the total amount received from the sale. This remuneration is referred to as a **commission**. It may be the only payment received by the agent, or it may be combined with a fixed amount of payment. Examples of such goods and services are manufactured goods, insurance, real estate, vehicles, brokers' services, media promotion, modelling and services of consultants in many fields.

The amount of commission will, of course, vary from activity to activity and will depend on the individual involved. However, the commission is often expressed as a percentage of the amount of the gross sale, or perhaps of the amount of capital that the agent is saving or investing on behalf of the client. On the other hand, the commission may simply be or include a fixed amount (sometimes called a *bonus* or *retainer*) that is independent of the amount of the sale.

One of the difficulties that may be encountered in business is the problem of determining what kind of commission is appropriate. For example, suppose that you are about to launch your career as a new-car salesperson and are trying to negotiate your weekly wage with your prospective employer. Suppose you are offered the following options, of which you must choose one:

1. a fixed wage of $570 per week; or
2. a fixed wage of $400 per week, plus 0.5% of the gross sales that you make; or
3. no fixed wage—only a straight commission of 1.5% of the gross sales that you make.

In order to decide which option is the best for you financially, you need to obtain some information (such as likely sales) and make some calculations.

Suppose that we define the quantities:

S = amount of sales
R = rate of commission per sale (often expressed as a percentage, but must be expressed as a decimal in the equations)
F = fixed amount paid (irrespective of sales)
C = commission earned

Then the commission earned is:

$$C = F + (S \times R) \qquad \textbf{(M2.1)}$$

In the case in which there is no fixed amount but only a straight commission, $F = 0$ and Equation M2.1 becomes:

$$C = S \times R \qquad \textbf{(M2.2)}$$

Example M2.5

An insurance agent works on a straight commission basis, being paid 2% commission per sum insured sold. In a particular month, the agent made the following sales:

- one policy for $45 000
- two policies for $25 000 each
- three policies for $5000 each

What was the total commission for that month?

Solution

The total amount of sales was:

$$\begin{aligned} S &= \$45\,000 + (2 \times \$25\,000) + (3 \times \$5000) \\ &= \$110\,000 \end{aligned}$$

The commission rate is 2%. That is, $R = 0.02$.

From Equation M2.2, the total commission for that month was:

$$\begin{aligned} C &= S \times R \\ &= \$110\,000 \times 0.02 \\ &= \$2200 \end{aligned}$$

Thus, the agent's commission for the month was $2200.

Example M2.6

A travelling salesman earns a straight 2.5% on the gross sales value of his products. How much would he have to sell in order to earn $600?

Solution

This time we have a commission rate of 2.5% or $R = 0.025$. The desired commission is $C = \$600$. From Equation M2.2:

$$\begin{aligned} C &= S \times R \\ \$600 &= S \times 0.025 \end{aligned}$$

which becomes:

$$\begin{aligned} S &= \frac{\$600}{0.025} \\ &= \$24\,000 \end{aligned}$$

Thus, the travelling salesman would have to sell $24 000 worth of products to earn a commission of $600.

Example M2.7

A bookseller received $225 commission.on sales of $7500 worth of books. Assuming that a straight commission was paid, what was the rate?

Solution

In this case we have $C = \$225$ and $S = \$7500$. From Equation M2.2:

$$\begin{aligned} C &= S \times \mathrm{R} \\ \$225 &= \$7500 \times R \end{aligned}$$

which becomes:

$$\begin{aligned} R &= \frac{\$225}{\$7500} \\ &= 0.03 \end{aligned}$$

Thus, a commission rate of 3% was paid.

Example M2.8

Consider the problem of the car salesperson's wage on page M.37. Suppose that you estimate that you will have an average of \$36 000 worth of sales each week. Which type of commission would you choose?

Solution

Each option should be considered in turn.
In option 1, the weekly wage is fixed at \$570 irrespective of sales.
In option 2, F = \$400, S = \$36 000 and R = 0.005. Using Equation M2.1:

$$\begin{aligned} C &= F + (S \times R) \\ &= \$400 + (\$36\,000 \times 0.005) \\ &= \$580 \end{aligned}$$

In option 3, S = \$36 000 and R = 0.015. Using Equation M2.2:

$$\begin{aligned} C &= S \times R \\ &= \$36\,000 \times 0.015 \\ &= \$540 \end{aligned}$$

Option 2 yields the largest average weekly wage, although the amounts for all options are about equal. This decision is very much dependent on the estimated weekly sales of \$36 000. Perhaps as you become more proficient at selling cars, option 3 could well be better!

Example M2.9

The amount of commission on property sold that can be charged by real estate agents in New South Wales was deregulated in May 1993. In the interests of competition, since that time they have been able to set their own commission rates. However, prior to that time their commission rate was set at the following amounts:

5.0% on the first \$15 000 of real estate sold
3.0% on the next \$45 000
2.5% on the next \$40 000
2.0% on the amount exceeding \$100 000

(There was also a minimum charge of \$200 in case the commission did not reach \$200.)

Using the above rates, calculate the commission payable to an agent on real estate worth the following amounts:

(a) \$3000
(b) \$18 000
(c) \$65 000
(d) \$165 000

Solution

(a) The commission payable on \$3000 is \$3000 × 0.05 = \$150. Since this does not reach the minimum amount of \$200, the agent would receive \$200.

(b) The commission payable on the first \$15 000 is \$15 000 × 0.05 = \$750.
The commission on the next \$3000 is \$3000 × 0.03 = \$90.
Hence, the total commission payable is \$750 + \$90 = \$840.

(c) The commission payable on the first \$15 000 is \$15 000 × 0.05 = \$750.
The commission on the next \$45 000 is \$45 000 × 0.03 = \$1350.
The commission on the next \$5000 is \$5000 × 0.025 = \$125.
Hence, the total commission payable is \$750 + \$1350 + \$125 = \$2225.

(d) The commission payable on the first \$15 000 is \$15 000 × 0.05 = \$750.
The commission on the next \$45 000 is \$45 000 × 0.03 = \$1350.
The commission on the next \$40 000 is \$40 000 × 0.025 = \$1000.
The commission on the next \$65 000 is \$65 000 × 0.02 = \$1300.
Hence, the total commission payable is \$750 + \$1350 + \$1000 + \$1300 = \$4400.

Brokerage

A (stock) broker is an agent who acts on your behalf and consequently charges you commission. This commission is called **brokerage**. When the broker has bought or sold on your behalf, he or she sends you a *contract note*, which sets out the price of the shares, the brokerage and the government stamp duty.

Brokerage rates vary according to the type of transaction, but general rules that apply are:

1. When *selling*, brokerage is subtracted from the proceeds of the sale.
2. When *buying*, brokerage is added to the amount that you must pay for the stock or shares.

Example M2.10

A broker charges customers a fixed fee of \$50 plus 2% of the amount of all investments made. If a customer invests a total of \$1200 through the broker, what is the total brokerage payable?

Solution

In this case, F = \$50, S = \$1200 and R = 0.02. Using Equation M2.1:

$$\begin{aligned} C &= F + (S \times R) \\ &= \$50 + (\$1200 \times 0.02) \\ &= \$74 \end{aligned}$$

Thus, a total brokerage of \$74 is payable.

Example M2.11

Suppose that the broker in Example M2.10 raises the fixed fee to \$75. What would the additional percentage have to be so that the total brokerage payable on \$1000 is \$110?

Solution

In this case, F = \$75, S = \$1000 and C = \$110. Using Equation M2.1:

$$C = F + (S \times R)$$
$$\$110 = \$75 + (\$1000 \times R)$$

which becomes:

$$R = \frac{\$35}{\$1000}$$
$$= 0.035$$

Thus, an additional brokerage of 3.5% would have to be charged.

M2.3 DISCOUNTS

When goods and services are sold there is often an inducement for the customer to make the purchase. These inducements can take many forms, such as two for the price of one, an advertised special with an expiry date, a fixed amount off the price, or a reduction in the unit price if large quantities are purchased. A reduction in price is referred to as a **discount** and is often expressed in the form of a percentage.

On some occasions there may be more than one discount on an item. Examples include where the first discount has still failed to attract a purchaser and a second (further) discount is given, or where a trade discount (see next section) is given on top of an already discounted item. These multiple discounts are called **chain discounts**.

To calculate the amount of discount, the following definitions are required:

L = list price
D = amount of discount
R = rate of discount (expressed as a fraction)
DP = discount price or net price

The *list price* is the manufacturer's recommended retail price or catalogue price.

The *rate of discount* is:

$$R = \frac{D}{L} \qquad \text{(M2.3)}$$

To express the rate of discount as a percentage, multiply R by 100.

The *amount of discount* may be found by rewriting Equation M2.3 as:

$$D = R \times L \qquad \text{(M2.4)}$$

The *discount price* (or *net price*) is:

$$DP = L - D \qquad \text{(M2.5)}$$

Example M2.12

A store is now selling for \$699 a videocassette recorder that it normally sells for \$799. What percentage discount is it offering?

Solution

In this case, D = \$799 – \$699 = \$100. Also, L = \$799. From Equation M2.3:

$$R = \frac{D}{L} = \frac{\$100}{\$799} = 0.125$$

Hence, the percentage discount is 100 × R = 12.5% off the list price.

Example M2.13

A hairdressing salon usually charges men \$18 for a hair cut and style. It is prepared to offer a 20% discount for customers who come on Tuesdays. What would the discount price be?

Solution

In this case, L = \$18 and R = 0.20. From Equation M2.4, the amount of discount is:

$$\begin{aligned} D &= R \times L \\ &= 0.20 \times \$18 \\ &= \$3.60 \end{aligned}$$

Hence, from Equation M2.5, the discount price is \$18.00 – \$3.60 = \$14.40.

Example M2.14

A store is selling a vacuum cleaner for \$120, which it claims represents a discount of 25%. What is the usual list price?

Solution

In this case, R = 0.25 and so

$$\begin{aligned} D &= R \times L \\ &= 0.25\,L \end{aligned}$$

Since DP = \$120, using Equation M2.5:

$$\begin{aligned} DP &= L - D \\ \$120 &= L - 0.25\,L \\ &= 0.75\,L \end{aligned}$$

Therefore, \$120 represents 75% of the list price. This may be written:

$$L = \frac{\$120}{0.75} = \$160$$

Hence, the usual list price is \$160.

Trade discounts

When purchases of goods or services are made from one business by another business, there is usually a special discount known as a **trade discount**. A typical trade discount might involve a builder purchasing timber from a timber yard, a service station obtaining tyres from a manufacturer or an electrician purchasing cables and switches from an electrical supplier.

A typical trade discount would range between 10% and 25%, depending on the trade and the item. If a number of discounts are given, for example a trade discount combined with a further discount for a cash payment, the discount is called a *chain discount*. The method for calculating trade discounts is the same as in the previous examples.

Example M2.15

A plumber wishes to purchase 20 m of copper pipe, which has a list price of $3.50 per m. The plumber is entitled to a trade discount of 15% off the list price.

(a) What total amount must the plumber pay?

(b) If a further discount of 10% is offered for cash, what price must the plumber pay now if he pays cash?

Solution

(a) The total list price of 20 m at $3.50 per metre is:

$$20 \times \$3.50 = \$70$$

From Equation M2.4, the amount of discount is:

$$\begin{aligned} D &= R \times L \\ &= 0.15 \times \$70 \\ &= \$10.50 \end{aligned}$$

From Equation M2.5, the discount price is:

$$\begin{aligned} DP &= L - D \\ &= \$70 - \$10.50 \\ &= \$59.50 \end{aligned}$$

This is the total amount that the plumber must pay.

(b) In part (a), the trade discount of 15% meant that the plumber must pay a total of $59.50.

The store is now offering a second discount (or chain discount) of a further 10% for cash payment. If the plumber takes up this offer, we now have the following:

$$\begin{aligned} \text{Trade discount price} &= \$59.50 \\ \text{Cash discount} &= 10\% \text{ of } \$59.50 \\ &= 0.10 \times \$59.50 \\ &= \$5.95 \end{aligned}$$

New discount price = trade discount price − cash discount
= \$59.50 − \$5.95
= \$53.55

Hence, if the plumber takes advantage of the cash discount, the total price will now be \$53.55.

Cash discounts

It is not uncommon for a seller to deduct a certain percentage of the bill provided that the purchaser pays within a specified period of time. Alternatively, this offer may also extend to purchasers who pay in the form of cash or a cheque. Such reductions are referred to as **cash discounts**, and the percentage of the discount may depend on how quickly the bill is paid.

Example M2.16

A supplier of electrical goods informs retailers that the following discounts are available for early payment of purchases:

- Within 7 days 10.0%
- Within 14 days 7.5%
- Within 28 days 2.5%

An electrical store has purchased \$30 000 worth of goods from this supplier. How much will it pay for these goods if payment is made after:

(a) 5 days?
(b) 10 days?
(c) 21 days?
(d) 35 days?

Solution

(a) After 5 days the discount is 10%. Using Equation M2.4, the amount of discount is:

$$\begin{aligned} D &= R \times L \\ &= 0.10 \times \$30\,000 \\ &= \$3000 \end{aligned}$$

From Equation M2.5, the discount price is:

$$\begin{aligned} DP &= L - D \\ &= \$30\,000 - \$3000 \\ &= \$27\,000 \end{aligned}$$

This is the amount that must be paid.

(b) After 10 days the discount is 7.5%. Using Equation M2.4, the amount of discount is:

$$\begin{aligned} D &= R \times L \\ &= 0.075 \times \$30\,000 \\ &= \$2250 \end{aligned}$$

From Equation M2.5, the discount price is:

$$\begin{aligned} DP &= R - D \\ &= \$30\,000 - \$2250 \\ &= \$27\,750 \end{aligned}$$

This is the amount that must be paid.

(c) After 21 days the discount is 2.5%. Using Equation M2.4, the amount of discount is:

$$\begin{aligned} D &= R \times L \\ &= 0.025 \times \$30\,000 \\ &= \$750 \end{aligned}$$

From Equation M2.5, the discount price is:

$$\begin{aligned} DP &= L - D \\ &= \$30\,000 - \$750 \\ &= \$29\,250 \end{aligned}$$

This is the amount that must be paid.

(d) After 35 days no discount is available. The store must pay the entire bill of $30 000.

M2.4 GST (GOODS AND SERVICES TAX)

On 1 July 2000 a **GST** (goods and services tax) was introduced into Australia. This is a broad-based tax of 10% on most supplies of goods and services consumed in Australia. The GST replaced a number of other taxes, including wholesale sales tax, that were applied at varying rates to a range of products.

The GST does not apply to all supplies. Examples where the GST is not applicable are:

- specified food (including meat, fruit and vegetables)
- health
- education
- residential rent
- child care
- nursing home care
- cars for the disabled
- religious services
- exports and various other supplies

However, some food and beverages items *do* have GST included. Examples include prepared food, takeaway food, restaurant meals, confectionery, ice-cream, snack foods, alcoholic beverages, soft drinks and bakery products (excluding bread).

The amount of GST payable can be calculated easily by working out 10% of the cost of the goods or service.

Example M2.17

Jenny manages a hairdressing salon and sets a price (before GST) of $60 for a shampoo, cut and blow dry. How much GST will the customer have to pay?

Solution

In this case, the GST is 10% of \$60, that is 0.10 × \$60 = \$6. This means that the customer will have to pay a total of \$66, comprising \$60 for Jenny and \$6 for the GST.

Consumers bear the cost of the GST, not the providers of the goods or services. However, the *liability* to pay the GST to the Australian Taxation Office (ATO) does rest with the providers of the goods and services, not with their customers. Most providers, even if they do not include GST in their prices, are still liable to pay the GST to the ATO if the GST is applicable to the transaction.

To calculate its GST liability, a business can use a *total invoice* approach by totalling the prices of all taxable items, where the GST is 1/11 of the price, on an invoice. It can then make a single 'divide by 11' calculation of the gross taxable amount to arrive at the total GST. This total will then be rounded to the nearest whole cent. The rounding rule is 'round down to the nearest whole cent below half a cent; round up to the nearest whole cent from half a cent upwards'.

Most businesses prefer the above option to the more laborious *line item* approach, where the GST is calculated on each taxable item or invoice and then added up to arrive at the total GST for an invoice or for the total GST liability. This is because the line item method requires the recording of the GST on each taxable item or line of items to as many decimal points as a business' system will allow and the total GST for the invoice is then found by summing the individual line item amounts of the GST.

While the GST is added to each sale in the production and distribution chain, businesses claim a full credit for the GST they pay on goods and services they use in their business. This means that for each product or service, the GST is effectively paid only once; that is, in the final price.

If a business is registered for the GST, it is entitled to claim **input tax credits** for the GST included in the price paid for items acquired and the GST paid on importations if used in the enterprise. If the business is *not* registered for the GST, then input tax credits cannot be claimed. An *input tax credit* is the amount allowed to offset GST included in the price paid for an acquisition or the GST paid on importations used in the enterprise.

An understanding of how the GST works can be seen in Example M2.18, which is based on the ATO publication, *Guide to GST*.

Example M2.18

Julian, a timber merchant, sells a quantity of timber for \$220 (including \$20 GST) to Andrea, who is a furniture manufacturer.

Andrea uses the timber to make a table and sells it to Don, a furniture retailer, for \$440 (including \$40 GST).

Don sells the table to Lyn, a consumer, for \$550 (including \$50 GST).

The GST and input tax credit situation for each of these people is as follows:

- Julian pays \$20 GST to the ATO.
- Andrea is entitled to an input tax credit for the \$20 GST included in the price paid to Julian. She offsets this \$20 against the \$40 GST payable on the supply of the table to Don and pays \$20 to the ATO.
- Don is entitled to an input tax credit for the \$40 GST included in the price paid to Andrea. He offsets this \$40 against the \$50 GST payable on the supply of the table to Lyn and pays \$10 to the ATO.

- Lyn, the consumer who buys the table, bears the GST since consumers cannot claim input tax credits.

For the purpose of the examples that involve the GST in this chapter, the issue of input tax credits will *not* be considered.

Customers may easily calculate the amount of GST they have paid for a good or service. This is done by dividing the final price (including GST) by 11.

Example M2.19

(a) Edna bought a watch for a GST-inclusive price of \$52.80. How much of this price was GST?

(b) Janice paid a plumber a GST-inclusive amount of \$187 to repair some leaking pipes in her bathroom. How much of this price was GST?

Solution

(a) To obtain the amount of GST paid, divide the final price by 11.

GST = \$52.80/11 = \$4.80

Therefore, Edna paid a GST of \$4.80 in the price of her watch.

(b) To obtain the amount of GST paid, divide the final price by 11.

GST = \$187/11 = \$17.00

Therefore, Janice paid a GST of \$17.00 in the price of her plumbing.

With the introduction of the GST on 1 July 2000, consumers found that some items rose in price and others decreased in price, while some stayed much the same. For example, clothing had previously attracted no sales tax but rose in price since it became subject to the GST, while the cost of some items such as televisions, videos, sound systems and cameras fell significantly since they previously attracted a sales tax of 32%.

M2.5 PERSONAL INCOME TAX

In Australia, individuals who receive income are liable for personal income tax. *Income* generally excludes windfall gains (e.g. lottery and gambling wins), inheritances and gifts (except those gained from employment), but includes profits from property bought for profit-making purposes.

The definitions below refer to the *Income Tax Assessment Act 1997* (as amended) and to regulations made under that Act.

Definition M2.1

Gross income = total amounts received or accrued
Assessable income = gross income *less* exempt income
Allowable deductions = costs of producing income and certain concessional deductions
Taxable income = assessable income *less* allowable deductions
Tax payable = tax according to the table on taxable income *less* rebates

While many individuals fill out their own personal taxation forms, many people use a *registered tax agent* to assist in this process. However, responsibility for the correctness of the information entered on the form still rests with the individual.

For the *taxation year* ending 30 June 2002 (also called the 2001/02 taxation year), *personal income tax rates* in Australia are as shown in Table M2.1. The third column shows the **marginal tax rate**, which is the percentage tax that must be paid on each dollar in that income range.

Note that the *tax-free threshold* for most resident individuals is \$6000. This means that they do not pay tax unless their taxable income exceeds \$6000. There may also be a special rebate of up to \$150 for low-income earners.

As well as liability for personal income tax, individuals must also pay a **Medicare levy**, currently set at 1.5% of taxable income. This Medicare levy is payable in addition to income tax payable and cannot be reduced by any rebates you may be entitled to.

However, there are special Medicare rates for low-income earners:

- if your taxable income is \$13 550 or less, you are not liable to pay the Medicare levy
- if your taxable income is more than \$13 550 but less than \$14 649, you are liable to pay 20 cents for every dollar above \$13 550 but below \$14 649.

Table M2.1 Personal income tax rate in Australia for the year ending 30 June 2002

Taxable income 2001/02 (\$)	Marginal tax rate (%)
0–6 000	0
6 001–20 000	17
20 001–50 000	30
50 001–60 000	42
60 001 and over	47

Using the information in Table M2.1, Table M2.2 shows how much tax is payable for specific income levels.

Table M2.2 Calculation table for personal income tax Australia for the year ending 30 June 2002

Taxable income (\$)	Tax payable (%)
6 000	0
20 000	2 380
50 000	11 380
60 000	15 580

Example M2.20

Use the information in Tables M2.1 and M2.2 to find the personal tax payable in the 2001/02 taxation year for the following taxable incomes:

(a) \$4560
(b) \$17 543

(c) $28 765
(d) $72 600

Solution

(a) Since $4560 is less than $6000, no tax is payable.
(b) Since $17 543 is more than $6000 but less than $20 000, the tax payable is:

$$\$0 + 17\% \text{ of } (\$17\,543 - \$6000) = (0.17 \times \$11\,543) = \$1962.31$$

(c) Since $28 765 is more than $20 000 but less than $50 000, the tax payable is:

$$\begin{aligned} \$2380 + 30\% \text{ of } (\$28\,765 - \$20\,000) &= \$2380 + (0.30 \times \$8765) \\ &= \$2380 + \$2629.50 \\ &= \$5009.50 \end{aligned}$$

(d) Since $72 600 is above $60 000, the tax payable is:

$$\begin{aligned} \$15\,580 + 47\% \text{ of } (\$72\,600 - \$60\,000) &= \$15\,580 + (0.47 \times \$12\,600) \\ &= \$15\,580 + \$5922 \\ &= \$21\,502.00 \end{aligned}$$

M2.6 COMPANY TAX

A *company* is a distinct legal entity with its own income tax liability, separate to your personal income tax. A company must lodge a company tax return (Form C), which shows the income and deductions of the company and is used to calculate the income tax that the company should pay.

As with individuals, the income tax of companies is calculated on taxable income, which is the income earned by the company less any allowable deductions. Deductions are the expenses in running the company, such as wages, the cost of stock, rent, bad debts and previous year losses.

A company's income tax is calculated as a percentage of the taxable income earned by the company during the financial year. The amount of tax to be paid on lodgement of the tax return is reduced by any *PAYG* (*Pay As You Go*) instalments paid during the year.

The general rate of tax payable by companies on 2000/01 income was 34% and in the 2001/02 financial year this rate dropped to 30%. Parliament decides when the company tax rates change. (Refer to the *Income Tax Rates Act 1986*, Section 23(2).) Special tax rates apply to pooled development funds and various components of life assurance companies and registered organisations.

Example M2.21

Calculate the tax payable if a company had a taxable income of $124 500 in 2001/02.

Solution

The company tax rate in that year was 30 cents for every dollar of taxable income. Therefore, the tax payable on $124 500 is:

$$\$124\,500 \times 0.30 = \$37\,350$$

M2.7 PROFIT AND LOSS

One of the main aims of embarking on a business venture is to make a profit. In other words, the selling price of goods or services must exceed the total cost of obtaining the goods or providing the services. We define the *actual profit* to the seller on an item as the difference between the price at which the item is sold (the *selling price*) and the cost of obtaining that item (the *cost price*). The profit is also known as a *profit margin* or *mark-up*. Note that the value of any applicable GST is *not* included in the selling price for these calculations unless specified.

In some situations an item may be sold for an amount that is *less* than the cost price. In this case, the profit is *negative* and is referred to as a *loss*.

The actual cost of an item is often difficult to calculate, since it involves not only the cost of obtaining the item from the supplier but other costs as well. These include the general costs of running a business, such as wages, insurance, taxes, stationery, equipment, electricity, rent and other overhead expenses. For this reason, it is easier to express a profit or loss as a percentage of the selling price than of the cost price. (Retail discounts are usually expressed as a percentage of the selling price.) However, this tends to *understate* the profit and *overstate* any loss.

The following definitions will be useful:

SP = selling price (not including any applicable GST)

CP = cost price

P = actual profit (when the value of SP exceeds the value of CP)

L = actual loss (when the value of CP exceeds the value of SP)

P_S = profit rate (or mark-up rate) expressed as a fraction of *selling price*

P_C = profit rate (or mark-up rate) expressed as a fraction of *cost price*

(The profit rates expressed as percentages are $100 \times P_S$ and $100 \times P_C$, respectively.)

L_S = loss rate (or mark-down rate) expressed as a fraction of *selling price*

L_C = loss rate (or mark-down rate) expressed as a fraction of *cost price*

(The loss rates expressed as percentages are $100 \times L_S$ and $100 \times L_C$, respectively.)

The following relationships hold:

$$P = SP - CP \tag{M2.6}$$

$$L = CP - SP \tag{M2.7}$$

$$P_S = \frac{P}{SP} \tag{M2.8}$$

$$P_C = \frac{P}{CP} \tag{M2.9}$$

$$L_S = \frac{L}{SP} \tag{M2.10}$$

$$L_C = \frac{L}{CP} \tag{M2.11}$$

Note that the selling price *including* GST would be (SP + 10% of SP) = $1.1SP$. That is, the selling price *including* GST is 11/10 of the selling price *not including* GST.

Example M2.22

A store buys lawnmowers for \$150 from the supplier and sells them for \$225 (not including GST). Calculate the profit as a percentage of:

(a) the selling price
(b) the cost price

Solution

In this case, $CP = \$150$ and $SP = \$225$. From Equation M2.6, actual profit is:

$$P = SP - CP$$
$$P = \$225 - \$150$$
$$= \$75$$

(a) From Equation M2.8:

$$P_S = \frac{P}{SP}$$
$$= \frac{\$75}{\$225}$$
$$= \frac{1}{3}$$

Hence, the profit as a percentage of the *selling price* is $100 \times \frac{1}{3} = 33\frac{1}{3}\%$.

(b) From Equation M2.9:

$$P_C = \frac{P}{CP}$$
$$= \frac{\$75}{\$150}$$
$$= \frac{1}{2}$$

Hence, the profit as a percentage of the *cost price* is $100 \times \frac{1}{2} = 50\%$.

Example M2.23

A furniture store purchased a lounge suite from a supplier for \$1200, but has not been able to sell it. To make a quick sale, the manager has decided to mark it down to \$1100 (including GST). Calculate the profit or loss as a percentage of:

(a) the selling price
(b) the cost price

Solution

In this case, $CP = \$1200$. The GST component of the selling price is \$1100/11 = \$100. Therefore, the value of the selling price *not including* GST is \$1100 – \$100 = \$1000, and so $SP = \$1000$.

Since CP is greater than SP, the actual loss is found from Equation M2.7 as:

$$\begin{aligned} L &= CP - SP \\ &= \$1200 - \$1000 \\ &= \$200 \end{aligned}$$

(a) From Equation M2.10:

$$\begin{aligned} L_S &= \frac{L}{SP} \\ &= \frac{\$200}{\$1000} \\ &= \frac{1}{5} \end{aligned}$$

Hence, the loss as a percentage of the *selling price* is $100 \times \frac{1}{5} = 20\%$.

(b) From Equation M2.11:

$$\begin{aligned} L_C &= \frac{L}{CP} \\ &= \frac{\$200}{\$1200} \\ &= \frac{1}{6} \end{aligned}$$

Hence, the loss as a percentage of the *cost price* is $100 \times \frac{1}{6} = 16\frac{2}{3}\%$.

Example M2.24

A bookstore marks up its books $33\frac{1}{3}\%$ over the price that it pays the supplier (i.e. the cost price).

(a) If the store pays the supplier \$12 for a book, for what price will the book retail?
(b) If the store sells a book for \$39.60 (including GST), what price did it pay the supplier?

Solution

In this case, $P_C = \frac{1}{3}$.

(a) Since $CP = \$12$, from Equation M2.9:

$$P_C = \frac{P}{CP}$$

or $$\frac{1}{3} = \frac{P}{\$12}$$

which becomes:

$$\begin{aligned} P &= \$12 \times \frac{1}{3} \\ &= \$4 \end{aligned}$$

From Equation M2.6:

$$P = SP - CP$$

and so:

$$\$4 = SP - \$12$$

$$\text{or} \quad SP = \$12 + \$4 = \$16$$

Hence, the book will retail for \$16 plus a GST of \$1.60 for a total of \$17.60.

(b) From Equation M2.9:

$$P_C = \frac{P}{CP}$$

$$\text{or} \quad \frac{1}{3} = \frac{P}{CP}$$

which may be written as:

$$P = \frac{1}{3} \times CP$$

From Equation M2.6:

$$P = SP - CP$$

and so

$$\frac{1}{3} \times CP = SP - CP$$

In this case, the selling price (including GST) is \$39.60. The GST component of this price is \$39.60/11 = \$3.60 and so SP = \$36. Therefore:

$$\frac{1}{3} \times CP = \$36 - CP$$

Hence:

$$\left(\frac{1}{3} \times CP\right) + CP = \$36$$

$$\text{or} \quad \frac{4}{3} \times CP = \$36$$

and so:

$$CP = \$36 \div \frac{4}{3} = \$36 \times \frac{3}{4} = \$27$$

Hence, the bookstore paid the supplier \$27 for the book.

Example M2.25

The manager of a used-car yard has a customer who requested a 1962 green Hillman Minx sedan (the type of car he used to own as a young man). The manager did not have one in stock, but promised the customer that when one was found she would sell it to him at cost plus 20%.

Eventually, one of the salespeople in the yard found such a car (at a motor wreckers, but still driveable), was told that it could be had for $300 and the manager agreed on this price. However, the manager also had the following charges:

- delivery $75
- detailing $55
- commission for salesperson 15% of cost price

Keeping her promise in mind, what price should the manager charge the customer for the car?

Solution

The answer to this question depends on how the term 'cost' is interpreted. Suppose that the manager takes the view that 'cost' means only the cost from the wrecker. This cost was $300, and a profit of 20% on this amount would yield an actual profit P, using Equation M2.9, of:

$$\begin{aligned} P_C &= \frac{P}{CP} \\ 0.20 &= \frac{P}{\$300} \\ P &= 0.20 \times \$300 \\ &= \$60 \end{aligned}$$

From Equation M2.6:

$$\begin{aligned} P &= SP - CP \\ \$60 &= SP - \$300 \\ \text{or } SP &= \$300 + \$60 \\ &= \$360 \end{aligned}$$

Hence, the manager should charge the customer $360. (No GST is payable by the purchaser of a used car.)

Suppose, however, the manager takes the view that 'cost' includes *all* costs of supplying the car to the customer. Additional costs are now $75 (delivery), $55 (detailing) and $0.15 \times \$300 = \$45$ commission, giving a total additional cost of $175. The cost to the manager is then $\$300 + \$175 = \$475$. A 20% profit on this figure would yield an actual profit of $0.20 \times \$475 = \$95$ for a selling price of $\$475 + \$95 = \$570$.

Whether the manager should charge the customer $360 or $570 is a matter of conjecture (and of how her original promise should be interpreted). If she charges $360, she will actually *lose* $\$475 - \$360 = \$115$ on the deal.

The point is that there are many types of cost, and it is necessary to be quite clear which type is meant when discussing mark-ups, discounts, profits and losses.

M2.8 STAMP DUTY

Each state in Australia levies a charge, known as **stamp duty**, on certain legal and commercial transactions, such as property transactions and the transfer of registration of a motor vehicle. In each case stamp duty is paid by the purchaser.

For motor vehicles, as a general rule purchasers must pay stamp duty on the market value of a vehicle or the price they paid for it, whichever is the greater. In New South Wales, in addition to stamp duty there is a transfer fee of $21. Since 1 July 1997, stamp duty on passenger motor vehicles up to $45 000 is $3 for every $100 or part of $100. For vehicles over $45 000, the stamp duty on the amount over $45 000 is $5 for every $100 or part of $100.

If a purchaser does not apply for transfer of registration within 14 days of purchasing a vehicle, a late application fee of at least $74 is also charged.

Example M2.26

(a) If a car with a market value of $52 995 is sold, what stamp duty is payable?

(b) What can you say about the market value of a vehicle that attracted a stamp duty of $621.00?

Solution

(a) The market value of $52 995 consists of $45 000 + $7995.

The stamp duty on the $45 000 is $\$3 \times 450 = \$1350$.

The remaining $7995 consists of 79 lots of $100 plus $95. The stamp duty payable on 79 lots of $100 is:

$79 \times \$5.00 = \$395.00$

Since stamp duty is payable also on parts of $100, the $95 itself attracts a $5.00 stamp duty, so that the stamp duty payable on the $7995 portion is:

$\$395.00 + \$5.00 = \$400.00$

Therefore, the total stamp duty payable on the car is $\$1350 + \$400 = \$1750$.

(b) If the stamp duty payable was $621.00, the number of multiples (or part thereof) of $100 to make up the market value was:

$$\frac{\$621}{\$3} = 207$$

Hence, the market value of the car was somewhere between $20 601 and $20 700.

Stamp duty is payable on the purchase price of property (or valuation in certain circumstances). The duty must be paid within three months of the date of exchanging contracts, unless buying 'off the plan' or a house and land package.

The amount of stamp duty payable for various property prices is shown in Table M2.3.

Table M2.3 Stamp duty payable on the purchase price of a property

Purchase price ($)	Stamp duty rate
Up to 14 000	$1.25 per $100 ($10 min.)
14 001–30 000	$175 plus $1.50 per $100 over $14 000
30 001–80 000	$415 plus $1.75 per $100 over $30 000
80 001–300 000	$1290 plus $3.50 per $100 over $80 000
300 001–1 000 000	$8990 plus $4.50 per $100 over $300 000
Over 1 000 000	$40 490 plus $5.50 per $100 over $1 000 000

Example M2.27

What is the stamp duty payable on a property with a purchase price of:

(a) $217 000?
(b) $592 000?

Solution

(a) From Table M2.3, the amount of stamp duty payable on a property with a purchase price of $217 000 is:

$$\begin{aligned}\$1290 + \$3.50 \times (2170 - 800) &= \$1290 + (\$3.50 \times 1370) \\ &= \$6085\end{aligned}$$

(b) From Table M2.3, the amount of stamp duty payable on a property with a purchase price of $592 000 is:

$$\begin{aligned}\$8990 + \$4.50 \times (5920 - 3000) &= \$8990 + (\$4.50 \times 2920) \\ &= \$22\,130\end{aligned}$$

M2.9 REMARKS

The application of percentages in modern business practice is widespread and this chapter has presented some of the more common examples. The introduction of the GST into Australia meant significant tax reform including substantial personal income tax cuts and the removal of a number of indirect taxes.

In using the taxation tables and related information, it is important to be aware that rates charged may vary from year to year and are determined by the federal government. Further information regarding taxation in Australia can be found from the ATO Web site at **http://www.ato.gov.au**.

Quick Quiz

1. Convert the fraction $\frac{17}{25}$ to a percentage.
2. A salesperson earns a straight commission of 6.5% on the gross value of products sold. For what gross value of products would the salesperson earn $250?
3. A store normally sells a certain brand of fridge for $700 but is currently offering it for sale at $577.50. What percentage discount is the store offering on the fridge?
4. The cost price of a chest of drawers is $170. Taking the value of the GST into account, calculate the retail price that would give the dealer a profit of $43.
5. A plumber gets a trade discount of 15% on all plumbing supplies purchased. If the plumber's total bill was $276.25 (excluding GST), what was the total list price of the goods that the plumber purchased?
6. A sculptor pays $995 (excluding GST) for supplies after receiving a trade discount of 15% as well as a further 2.5% (after the trade discount) for cash payment. What was the list price of the supplies?
7. A television set usually sells for $660 (including GST). A store is advertising a discount of 15% if it is purchased on a Sunday and a further 10% discount off that price if cash is used. What is the cash price (including GST) for a Sunday purchase?
8. A store sells cordless telephones for the usual list price of $176 (including GST). On an Easter weekend it advertises a 20% discount off this price. What percentage mark-up must it add to the discount price to get back to the original list price?
9. An electrician makes repairs to a company switchboard and presents a final bill for $260.70. How much is the GST component of the bill?
10. Sally has a taxable income of $52 284 for the year ending 30 June 2001. Calculate how much personal tax she is liable for, along with her Medicare levy.

Multiple Choice Questions

1. The fraction $\frac{5}{16}$ converted to a percentage is:
(a) 32% (b) 5% (c) 3.125%
(d) 31.25% (e) none of the above

2. A sales representative earns a straight 7.50% commission on gross sales value. How much would have to be sold in order to earn $487.50 commission?
(a) $3656.00 (b) $65.00 (c) $650.00
(d) $6500.00 (e) none of the above

3. A broker charges customers a fixed fee of $75 plus 3.50% of the amount of all investments made. If a customer invests $15 000 through the broker, the total brokerage payable is:
(a) $600 (b) $5325 (c) $480
(d) $127.50 (e) none of the above

4. A store is selling a massage chair for $1163.50, which it claims represents a discount of 35%. The usual list price is:
(a) $1204.22 (b) $3324.29 (c) $1790.00
(d) $1570.73 (e) none of the above

5. An electronics store marks up its hand-held computer games by 45% over the price it pays the supplier. If the store sells these games for $36.25 (*not* including GST), the price it pays the supplier is:
(a) $16.31 (b) $19.94 (c) $34.69
(d) $25.00 (e) none of the above

6. A service station buys brake pads for $36 from the supplier and sells them for $41.40 (*not* including GST). The profit as a percentage of cost price is:
(a) 15.0% (b) 5.4% (c) 13.0%
(d) 54% (e) none of the above

7. A coffee machine sells for $61.60 in a department store. The GST component of this price is:
(a) $6.16 (b) $1.60 (c) $6.60
(d) $56.00 (e) none of the above

8. A dress has a selling price of $88.00 (*not* including GST). The amount of GST payable by the customer is:
(a) $8.80 (b) $8.00 (c) $80.00
(d) $0.80 (e) none of the above

9. Linda purchases a car for $16 000 with a market value of $17 500. The amount of stamp duty payable by Linda is:
(a) $480 (b) $526.50 (c) $528.00
(d) $175.50 (e) none of the above

10. A ripe mango sells for $4.00. It is then marked up 25% in price. The new price is then reduced by 25%. The resulting price of the mango is:
(a) $3.00 (b) $3.75 (c) $4.00
(d) $5.00 (e) none of the above

Exercises

M2.1* Convert to percentages:

(a) $\frac{1}{4}$ (b) $\frac{1}{3}$ (c) $\frac{1}{8}$

M2.2 Convert to percentages:

(a) $\frac{5}{12}$ (b) $\frac{3}{8}$ (c) $\frac{3}{4}$

M2.3* Convert to percentages:

(a) 0.72 (b) 0.505 (c) 0.8273

M2.4* Convert to fractions (expressed in their lowest terms):

(a) 22.5% (b) $66\frac{2}{3}\%$ (c) 87.5%

M2.5 Convert to decimals:

(a) 70% (b) 30.72% (c) $16\frac{2}{3}\%$

M2.6* A used-car salesperson works on a straight commission of 2.5% of the sales price of all vehicles sold. Calculate the commission payable if the total sales price of all vehicles sold is:

(a) \$25 000 (b) \$126 000 (c) \$290 000

M2.7* The salesperson in Exercise M2.6 has had the basis of commission changed to a fixed amount of \$200 per week plus 1.5% of the sales price of all vehicles sold. Calculate the commission payable in those weeks where the total sales price of all vehicles sold was:

(a) \$30 000 (b) \$150 000 (c) \$245 000

M2.8 Suppose that you were the salesperson described in Exercises M2.6 and M2.7.

(a) Which of the two types of commission would you prefer?

(b) What factors influenced your answer in (a)?

(c) If the total sales price of the vehicles you sold each week had an average of \$15 000, which of the two types of commission would you prefer? Justify your answer.

M2.9* An insurance salesperson has sold policies with a total value of \$250 000. Assuming that a straight commission was paid on this value, what was the rate of commission (expressed as a percentage) if the commission earned was:

(a) \$2500? (b) \$1875? (c) \$8125?

M2.10 A broker charges a fixed fee of \$75 plus 1.75% on the amount of all investments made. If a customer invests a total of \$2500 through the broker, what is the total brokerage payable?

M2.11 Suppose that the broker in Exercise M2.10 raises the fixed fee to \$100. What must the percentage commission be if the total brokerage payable on \$1500 is \$142.75?

M2.12 A real estate agent advertises that he will charge the following commission (based on the selling price) on any property that he sells:

5.0% on the first \$50 000

3.0% on the next \$75 000

2.5% on the next \$75 000

2.0% on the amount exceeding \$200 000

Calculate the commission that will be charged on property that sells for:

(a) \$120 000 (b) \$165 000 (c) \$195 000 (d) \$475 000

M2.13* The real estate agent in Exercise M2.12 is asked to value a property for a prospective seller.

(a) If the agent wants to receive a commission of \$20 000, for how much should the property sell?

(b) If the agent knows that any prospective purchaser will ask for a 10% discount off the advertised price, for how much should the property be advertised for the agent to receive a \$20 000 commission?

M2.14 A store normally has its electric can-openers marked for sale at \$29.99, but currently has them on special for \$25.50.

(a) What approximate percentage discount is the store offering?

(b) If the store wished to offer a 17.5% discount off the normal price of the can-openers, for what price should it sell them?

M2.15* A store is selling jars of jam for \$1.70 each, which it claims represents a discount of $33\frac{1}{3}\%$. What is the usual list price of these jars?

M2.16 A builder purchases 30 m of timber that has a list price of \$12.50 per m. The builder is entitled to a trade discount of 12.5% off the list price. What total amount must the builder pay for the timber?

M2.17 A supplier of goods to a university has informed the university that the following discounts are available for early payment for purchases:

- within 10 days 5.0%
- within 20 days 2.5%
- within 30 days 1.5%

The university has purchased \$25 000 worth of goods from this supplier. How much will it pay for these goods if payment is made after:

(a) 8 days? (b) 15 days? (c) 27 days? (d) 90 days?

M2.18* A bedside table has a retail price of \$159.50 including GST. Calculate the retail price *excluding* GST.

M2.19 A camera retails for \$325.60, including GST and a dealer profit of \$65. What was the cost price?

M2.20* An electronics store marks up its voice-activated tape-recorders by 40% over the cost price from the supplier.

(a) If the store pays the supplier \$65 for such a recorder, for what price will it retail (including GST)?

(b) If the store sells a recorder for \$107.80 (including GST), what was the cost price from the supplier?

M2.21 Answer the questions in Exercise M2.20 if the store marks up its recorders by:

(a) 100% (b) 150%

M2.22* A painter, who was entitled to a trade discount of 20% on all materials purchased, was sent a bill for \$350. What was the list price of the goods that the painter purchased?

M2.23 The number of males and females aged 18 years and over who participate in various organised sports in Australia is shown in the table below.

Sport	Number of males ('000)	Number of females ('000)	Total ('000)
Basketball	219.7	91.4	311.1
Golf	1059.2	265.6	1324.8
Lawn bowls	243.5	124.5	368.0
Swimming	885.3	1026.3	1911.6
Tennis	508.1	512.4	1020.5

Source: Australian Bureau of Statistics, *Participation in Sport and Physical Activities, 1999–2000*, Cat. No. 4177.0.

(a) For each sport calculate the percentage of participants who are male and the percentage who are female.

(b) Find the totals for each of the three columns. Hence, calculate the overall percentage of participants who are male and the percentage who are female.

M2.24 At 30 June 2000 the male population of Queensland was 1 783 744. The population in various age groups is shown in the table below.

Age group (years)	Number of males
0–9	255 400
10–19	265 866
20–39	534 324
40–64	544 425
65 and over	183 729

Source: Australian Bureau of Statistics, *Australian Demographic Statistics*, June Quarter 2000, p. 17, Cat. No. 3101.0.

Calculate (to two decimal places):

(a) the percentage of the male population in each age group

(b) the percentage of the male population aged 40 or over

(c) the percentage of the male population aged 19 or under

M2.25 The population of Australia on 30 June 1995 was 18 054 000 and on 30 June 2000 was 19 157 000. In 1995 the Australian Bureau of Statistics made predictions of what Australia's population might be in future years at 10-year intervals, as shown below.

Year	Projected population ('000s)
2005	20 154
2015	21 909
2025	23 459
2035	24 716
2045	25 623

Source: Australian Bureau of Statistics, *Year Book Australia 1997*, p. 74.

(a) For each of the above five years, find the percentage increase (to two decimal places) in the projected population from that of 30 June 1995.

(b) For each year in the table, find the percentage increase from the figure for 10 years previously.

(c) What do your answers in (b) tell you about the predicted value of Australia's population growth?

M2.26 Use the information in Section M2.8 to find the amount of stamp duty (plus transfer fee) that must be paid on the transfer of registration in New South Wales of a vehicle with a market value of:

(a) \$3650 (b) \$21 786 (c) \$60 450

M2.27* Use the information in Section M2.8 to find the amount of stamp duty (plus transfer fee) that must be paid on the transfer of registration in New South Wales of a vehicle with a market value of:

(a) \$9870 (b) \$32 990 (c) \$15 400

M2.28 A store sells washing machines for a retail price of \$1000 (including GST). To clear out stock in preparation for the new model, it marks these down by 25%. A customer comes into the store and says he will buy a discounted machine, but wants a further discount of 10% if he pays cash. How much in cash is the customer willing to pay for the machine?

M2.29* The store in Exercise M2.28 says that it will not give a further discount of 10% for cash, but will give the customer a total discount of 35% off the original retail price. Is the customer better or worse off with this deal than with his own offer? Explain.

M2.30 The local hardware store is having a sale on brush-cutters and has them marked at '15% off the recommended retail price' at \$102. A customer (who wants to buy a brush-cutter) has a voucher from the newspaper entitling him to '\$15 off the recommended retail price of any item in the store: not to be used in conjunction with any sale price'. He therefore has the option of using his voucher or taking advantage of the discount. Which option should the customer take? Explain.

M2.31 Use the information in Tables M2.1 and M2.2 to find the personal income tax payable on taxable incomes of:

(a) \$6790 (b) \$25 410 (c) \$125 600

M2.32* Use the information in Tables M2.1 and M2.2 to find the personal income tax payable on taxable incomes of:

(a) \$12 543 (b) \$37 520 (c) \$87 540

M2.33 Use the information in the section on company tax in this chapter to find the company tax payable in 2001/02 on a taxable income of:

(a) \$56 450 (b) \$15 490 (c) \$327 900

M2.34* Use the information in Section M2.6 to find the company tax payable in 2001/02 on a taxable income of:

(a) \$16 250 (b) \$75 320 (c) \$534 860

M2.35 A woman purchased a car in New South Wales with a market value of \$24 750. How much transfer fee and stamp duty did she pay?

M2.36* A customer went to the bank manager to borrow \$540 for the payment of stamp duty (not including the transfer fee) on the transfer of registration in New South Wales of his new car. What possible market values could the car have?

M2.37 A store purchases a washing machine from the manufacturer and marks it up 40% on the cost price for sale. After virtually no customer interest for 3 months, the manager decides to sell it to a client for a 10% discount off the marked-up price. If the client pays \$252, find:

(a) the original cost price

(b) the marked-up price

M2.38 Answer the questions in Exercise M2.37 if it is now revealed that the manager in fact gave the client a 40% discount (*not* a 10% discount) off the marked-up price.

M2.39 For the following incomes in 2001/02, determine whether the amount of tax payable would be larger if it were company tax or personal income tax. In each case find the tax payable for both types of tax.

(a) \$70 000 (b) \$80 000 (c) \$90 000

M2.40 For the taxation year 2001/02, find the income (to the nearest dollar) where the company tax payable on that income would equal the personal tax payable on that income. (*Hint*: Your answers to Exercise M2.39 may give you a clue on where to start looking.)

M2.41 For the taxation year 2001/02, find the income that would correspond to a personal tax of:

(a) \$7861.00 (b) \$27 113.80

M2.42* For the taxation year 2000/01, find the income that would correspond to a company tax of:

(a) \$42 732.90 (b) \$282 939.00

M2.43* A tiling company gives retailers a 5% discount if their accounts are paid within 5 days and a 3% discount if they are paid within 10 days. Otherwise, the retailer pays the full amount. Find the amount that a retailer who receives an account of \$22 500 must pay if the account is paid after:

(a) 4 days (b) 9 days (c) 26 days

M2.44 In Exercise M2.43, a retailer who paid his account after 7 days sent a cheque for \$1406.50. How much was the original account?

M2.45 A store advertises a 4% discount off the retail price of its refrigerators for customers who pay cash. It then offers a further 2% discount off that discounted price to customers who also arrange for their own delivery of the refrigerator.

Find the amount that a customer who pays cash and arranges for their own delivery must pay for a refrigerator that retails for:

(a) \$1560 (b) \$2400 (c) \$3600

M2.46* In Exercise M2.45, a refrigerator for a customer who paid cash and arranged for their own delivery had a final price of \$1128.96 (including GST). What was the original retail price?

M2.47 Answer Exercise M2.46 if the final price was \$2352.00.

M2.48 In Exercise M2.45, a refrigerator retails for \$1800. If a customer paid cash and arranged for their own delivery, find:

(a) the amount the customer paid

(b) the percentage discount the customer received off the original retail price

M2.49 Do you think that the answer to Exercise M2.48(b) is the same for *any* retail price of the refrigerator? Explain your reasoning.

M2.50 Suppose that you are offered a job selling nature books from door to door, and that the company offers you a choice in wages of either:

(i) \$350 per week, *or*

(ii) \$200 per week plus 2.0% of the value of all sales, *or*

(iii) \$80 per week plus 4.0% of the value of all sales

The choice is a difficult decision since you are unsure as to the value of the sales you are likely to make. Calculate the best choice, along with the weekly wage, if the value of the sales made each week is:

(a) \$5000 (b) \$8000 (c) \$10 000

M2.51 Suppose that in Exercise M2.50 you select choice (i), but in fact make \$9000 worth of sales each week. Over a 52-week year, how much extra could you have made in wages by selecting the best choice?

M2.52 In Exercise M2.50, find the value of sales each week such that the wages received under:

(a) choices (i) and (ii) are equal

(b) choices (i) and (iii) are equal

M2.53* By the end of the weekend 10–11 February 2001, the film *What Women Want* starring Mel Gibson had total earnings of \$20 170 000 at the Australian box office. (*Source*: Motion Picture Distributors Association of Australia.) If its earnings in the week ending 26 January 2001 alone were \$3 010 000, express these earnings as a percentage (to two decimal places) of the total earnings by the end of the weekend 10–11 February 2001.

M2.54 By the end of the weekend of 10–11 February 2001, the two films *Cast Away* and *Meet the Parents* had combined total earnings of \$36.3 million at the Australian box office. Of this, \$15.81 million was due to *Cast Away* alone. (*Source*: Motion Picture Distributors Association of Australia.) By the end of the weekend of 27–28 January 2001, *Cast Away* had total earnings of \$4.54 million and *Meet the Parents* had total earnings of \$17.46 million.

Find:

(a) the percentage of combined total earnings attributable to each film by the end of the weekend of 27–28 January 2001

(b) the percentage of combined total earnings attributable to each film by the end of the weekend of 10–11 February 2001

(c) the percentage of the total combined earnings attributable to each film for the period 29 January–11 February 2001

M2.55* A broker charges customers a fixed fee of \$90 plus 2.5% of the amount of all investments made. Find the total brokerage payable if a client invests through the broker an amount of:

(a) \$25 000 (b) \$56 000 (c) \$125 000

M2.56 In Exercise M2.55, find the amount of the investment made by a client who had a total brokerage payable of:

(a) \$1890 (b) \$2340 (c) \$3990

M2.57 What is the stamp duty payable in New South Wales on a property with a purchase price of:

(a) \$147 000? (b) \$492 000? (c) \$1 234 980?

Chapter 3

Algebra

'What is algebra exactly? Is it one of those three-cornered things?'

(Sir James Barrie, British novelist, 1860–1937, getting confused with triangles)

Learning objectives

In this chapter you will learn to:

- Understand and use algebraic terms
- Manipulate algebraic expressions
- Solve simple linear equations (including transposition)
- Solve simultaneous linear equations (including the graphical technique)
- Solve business problems using simple algebra

M3.1 DEFINITIONS

The *Macquarie Dictionary* (3rd edition, 1997, published by Macquarie Library Pty Ltd) defines *algebra* as 'the mathematical art of reasoning about (quantitative) relations by means of a systematised notation including letters and other symbols'. In other words, algebra is all about applying arithmetic concepts where the unknown quantities are represented by symbols. For our purposes, the symbols used will be letters of the alphabet, that is a, b, c, . . . , x, y, z, although combinations of letters (or even words) may be used to represent a single quantity.

Before commencing the application of algebraic techniques, it is necessary to have some normal definitions of commonly used terms.

A **constant** is a term whose value does not change. For example, 12, $\frac{1}{2}$, -45, 5.8, 0, -234.589 and π (pi) are all constants because their values do not change. In algebra it is sometimes convenient to represent a constant by a letter of the alphabet. Although strictly speaking any letter will suffice, it is usual to use letters near the beginning of the alphabet, that is a, b, c, d.

A **variable** is a term that represents a quantity that may have a number of different values. Variables also are usually represented by letters of the alphabet. While *any* letter will suffice, letters near the end of the alphabet are often used, that is x, y, z. However, if a variable represents a particular quantity, it is also common for, say, the *first* letter of the name of the quantity to be used as the variable name.

Example M3.1

Suppose that in various problems we are trying to find the values of certain unknown quantities. Although any variable names will do, it is easier to use symbols that relate to the quantities themselves. However, in the same problem, the same variable name should not be used to represent *different* quantities. Some examples of variable names that could be used for the listed quantities are as follows:

Quantity	*Variable name*
Number of apples to buy	a
Length of string to purchase	l
Time to leave for work	t
Price to charge for an item	p
Cost of running a car each week	c

An (*algebraic*) **expression** is a combination of constants and variables by the use of arithmetic operations (such as addition, subtraction, multiplication and division). Examples of expressions are:

2	$x^3 + 4xy + y^4$	$\dfrac{y - 6}{x + 7}$
$3x$	$5(24m - 17n)$	
$3a + 5b - 9$	$3x^2 + 2x - 4$	

A **term** is a part of an expression that is connected to another term by a + or a − sign. For example, in the expression $3x + 2y - z$ the terms are $3x$, $2y$ and $-z$. If there is no sign written in front of a term, it is always understood that the sign is +.

A (numerical) **coefficient** is a factor by which the rest of a term is multiplied. For example, the term $5xy$ has a coefficient of 5 and the term $-15z$ has a coefficient of -15.

The **degree of an expression** is the highest exponent of any variable contained in the expression. For example, the expression $3x + 7$ is a *linear* (or *first-degree*) expression, $5x^2 - 4x + 5$ is *quadratic* (or *second-degree*) and $21y^3 + 4y^2 - 2y + 8$ is *cubic* (or *third-degree*). A **linear equation** is therefore one in which the largest value of the exponents is 1.

In an expression, terms that differ only in their coefficients are said to be *like terms*. For example, in the expression $15x - 3y + 21 - 4x + 6xy + y + x^2$, the terms $15x$ and $-4x$ are like terms. Also, the terms $-3y$ and y are like terms.

An **equation** is a statement that two expressions are equal. The two expressions are shown to be equal by means of an equals (=) sign.

Example M3.2

Suppose that the two expressions $3x + 6$ and $5x - 2$ are equal. The resulting equation is:

$$3x + 6 = 5x - 2$$

The process of finding the value of a variable that will make the equation a true statement is called *solving the equation*. The value of the variable that satisfies the equation is called the *solution* to the equation. In particular, the value $x = 4$ is the solution to the equation in Example M3.2; $x = 4$ is said to *satisfy* the equation.

Simultaneous equations are ones that must be solved simultaneously (i.e. at the same time) in order to find the values of the variables that will solve *both* equations. An example of simultaneous equations in the variables x and y is:

$$4x + 3y = 11$$
$$5x - 2y = 8$$

A **formula** is a rule or principle that is expressed in terms of algebraic symbols. An example would be a formula to find the area of a rectangle, expressed as:

$$A = L \times B$$

where A = area of rectangle
L = length of rectangle
B = breadth of rectangle

To find the area (A), the values of L and B are substituted into the formula.

On some occasions we may wish to use the same formula to find different quantities. For example, we may wish to find the length of a rectangle for which we know the area and the breadth. In this case, we can rewrite the formula making L the subject by dividing both sides of the equation by B. Hence, we have:

$$L = \frac{A}{B}$$

Formulae may be rewritten using the techniques described later in this chapter.

M3.2 ALGEBRAIC EXPRESSIONS

In order to find the solution to an equation it is important to understand some of the basic rules that apply to algebraic expressions.

RULE If an expression contains **like terms**, these terms may be combined into a single term. Like terms are those that differ only in their numerical coefficient. Constants may also be combined into a single constant.

Example M3.3

The expression $5x - 2y + 4z + 7 - 3x + 7y - 6z + 4$ contains like terms:

$5x$ and $-3x$ are like terms.
$-2y$ and $7y$ are like terms.
$4z$ and $-6z$ are like terms.
7 and 4 are constants.

$5x$ and $-3x$ may be combined to form a single term:

$$5x - 3x = 2x$$

$-2y$ and $7y$ may be combined to form a single term:

$$-2y + 7y = 5y$$

$4z$ and $-6z$ may be combined to form a single term:

$$4z - 6z = -2z$$

7 and 4 may be combined to form a constant:

$$7 + 4 = 11$$

Hence, the expression may be simplified to:

$$2x + 5y - 2z + 11$$

RULE When an expression is contained in parentheses (brackets), each term within the parentheses is multiplied by any coefficient written outside the parentheses.

Example M3.4

Consider the expression:

$$2(3x + 4y - 1)$$

To remove the parentheses we must multiply each term within the parentheses by the coefficient 2. Thus, the expression may be written:

$$2(3x) + 2(4y) + 2(-1) = 6x + 8y - 2$$

RULE To multiply one expression by another, multiply each term of one expression by each term of the other expression. The resulting expression (which should be simplified by collecting any like terms) is said to be the *product* of the two expressions.

Example M3.5

Consider the expression:

$$5x(2x - y + 4)$$

To remove the parentheses we must multiply each term within the parentheses by the coefficient $5x$. Thus, the expression may be written:

$$5x(2x) + 5x(-y) + 5x(4) = 10x^2 - 5xy + 20x$$

Example M3.6

Find the product of the two expressions $3x + 2$ and $2x - 1$.

Solution

We are trying to find the value of $(3x + 2)(2x - 1)$. This may be written as:

$$\begin{aligned}(3x + 2)(2x - 1) &= 3x(2x - 1) + 2(2x - 1)\\ &= 3x(2x) + 3x(-1) + 2(2x) + 2(-1)\\ &= 6x^2 - 3x + 4x - 2\\ &= 6x^2 + x - 2\end{aligned}$$

M3.3 SOLVING LINEAR EQUATIONS

To find the value of a variable that will satisfy a linear equation, follow these basic steps:

1. Place like terms on the same side of the equation.
2. Any arithmetic operation that you perform on one side of the equation you must also perform on the other side.

When terms are moved from one side of an equation, they are said to be *transposed* (or transferred). This process is called **transposition.**

In particular, when solving an equation that involves only *one* variable, follow these steps:

1. Place all the terms involving the variable on the *left-hand side* of the equation and the constant terms on the *right-hand side*.
2. Collect the like terms involving the variable, and collect the constant terms.
3. Divide both sides of the equation by the coefficient of the variable.

The left-hand side of the equation should now consist of the variable only. The right-hand side of the equation is the solution.

Example M3.7

Solve the following equation for x:

$$9x - 27 = 4x + 3$$

Solution

We are asked to find the value of x that makes the equation a true statement. Suppose we follow the steps above. First, transpose the term involving $4x$ to the left-hand side of the equation. To do this, we subtract $4x$ from both sides:

$$9x - 27 - 4x = 4x + 3 - 4x$$

After collecting like terms in x, the equation is:

$$5x - 27 = 3$$

In this case, the terms in x disappear from the right-hand side, which now consists of the constant 3 only. However, the effect of doing this procedure is simply to transpose the $4x$ from the right-hand side to the left-hand side where it becomes the term $-4x$, i.e. the same term but *with the sign changed*. This illustrates the following rule involving the transposition of terms from one side of an equation to the other.

> **RULE** Any term may be transposed from one side of an equation to the other. However, when the transposition is made, the *sign* of the term must change from that which it originally had. That is, a '+' term becomes a '–' term and a '–' term becomes a '+' term.

This rule may now be applied to the constant terms in the equation. That is, we can transpose the term –27 from the left-hand side to the right-hand side, where it becomes +27:

$$5x = 3 + 27$$
$$5x = 30$$

Now divide both sides by the coefficient of x (i.e. 5). This yields:

$$\frac{5x}{5} = \frac{30}{5}$$
$$x = 6$$

That is, the solution to the equation is $x = 6$.

(Check this solution by substituting the value of $x = 6$ into the equation and verifying that it is a true statement. We get:

$$9(6) - 27 = 4(6) + 3$$
$$54 - 27 = 24 + 3$$
$$27 = 27$$

Hence, our solution appears to be correct.)

Example M3.8

Solve the following equation for y:

$$12 - 4y = 6y + 37$$

Solution

Transpose the term $6y$ to the left-hand side, where it becomes $-6y$:

$$12 - 4y - 6y = 37$$

Transpose the constant term 12 to the right-hand side, where it becomes -12:

$$-4y - 6y = 37 - 12$$

Collect like terms:

$$-10y = 25$$

Divide both sides by -10:

$$\frac{-10y}{-10} = \frac{25}{-10}$$

$$y = -2.5$$

Therefore, the solution to the equation is $y = -2.5$.

(To verify our solution, substitute $y = -2.5$ into the equation. We get:

$$12 - 4(-2.5) = 6(-2.5) + 37$$

$$12 + 10 = -15 + 37$$

$$22 = 22$$

Hence, our solution appears to be correct.)

Example M3.9

Solve the following equation for x:

$$5(3x - 4) = 2(6 - 2x)$$

Solution

First, remove the parentheses on both sides of the equation:

$$15x - 20 = 12 - 4x$$

Transpose the term $-4x$ to the left-hand side:

$$15x + 4x - 20 = 12$$

Transpose the constant term -20 to the right-hand side:

$$15x + 4x = 12 + 20$$

Collect like terms:

$$19x = 32$$

Divide both sides by 19:

$$x = \frac{32}{19}$$

Therefore, the solution to the equation is $x = \frac{32}{19}$.

(To verify our solution, substitute $x = \frac{32}{19}$ into the equation. We get:

$$5\left(\frac{96}{19} - 4\right) = 2\left(6 - \frac{64}{19}\right)$$

$$5\left(\frac{20}{19}\right) = 2\left(\frac{50}{19}\right)$$

$$\frac{100}{19} = \frac{100}{19}$$

Hence, our solution appears to be correct.)

Example M3.10

Solve the following equation for z:

$$3(2z + 5) = \frac{3z}{4} + 36$$

Solution

First, remove the parentheses on the left-hand side to get:

$$6z + 15 = \frac{3z}{4} + 36$$

Transpose the term $\frac{3z}{4}$ to the left-hand side:

$$6z + 15 - \frac{3z}{4} = 36$$

Transpose the constant term 15 to the right-hand side:

$$6z - \frac{3z}{4} = 36 - 15$$

Collect like terms:

$$\frac{21z}{4} = 21$$

Multiply both sides by $\frac{4}{21}$:

$$\frac{4}{21} \times \frac{21z}{4} = 21 \times \frac{4}{21}$$

$$z = 4$$

Therefore, the solution to the equation is $z = 4$.

(To verify our solution, substitute $z = 4$ into the equation. We get:

$$3(8 + 5) = \frac{3(4)}{4} + 36$$
$$3(13) = 3 + 36$$
$$39 = 39$$

Hence, our solution appears to be correct.)

M3.4 SOLVING SIMULTANEOUS LINEAR EQUATIONS

The problems in Section M3.3 involved finding the solution to a single equation that contained only *one* variable. In this section we consider the situation in which we have to find the values of *two* variables. To do this we must have *two* equations involving the variables. In this case, there will usually be unique values of the variables that will satisfy *both* equations. Such equations are referred to as **simultaneous equations**.

Suppose, for example, that we have two simultaneous equations involving two variables. The steps involved in the solution are as follows:

1. Make the coefficient of either of the variables in one equation equal to its coefficient in the other equation. (It does not matter which variable is used for this purpose.) This may be achieved by multiplying both sides of one equation (or possibly both equations) by a constant. (If both equations are involved, a different constant will be used for each equation.)

2. Eliminate the variable that now has the same coefficient in both equations. How you do this depends on the signs (+, –) before the coefficients:

(a) If the signs are the *same*, *subtract* one equation from the other. (It does not matter which equation is subtracted.)

(b) If the signs are *different*, *add* the equations.

Write the equations under one another for this step, making sure that like terms are directly under each other.

3. After the subtraction or addition is made, you should be left with an equation in only *one* variable, namely the variable that was *not* eliminated in Step 2. Solve this equation using the techniques in Section M3.3.

4. Find the value of the eliminated variable by substituting the value of the variable found in Step 3 into *either* of the two original equations. You will have an equation involving only the eliminated variable, which you can solve using the techniques in Section M3.3.

5. Verify your solution by substituting the values of the two variables into *each* of the two original equations and checking that they are true statements.

Example M3.11

Solve the following simultaneous equations for x and y:

$$3x + 4y = 33 \quad (1)$$
$$2x - 3y = 5 \quad (2)$$

Solution

Suppose that we first eliminate the variable x from both equations. At the moment, x has a coefficient of 3 in the first equation and 2 in the second. We have to make this coefficient the same in both equations (Step 1).

We can achieve this by multiplying both sides of Equation (1) by 2 and both sides of Equation (2) by 3. This will make the coefficient of x to be 6 in the two resultant equations. That is:

$$\begin{aligned} 6x + 8y &= 66 \quad (3) \\ 6x - 9y &= 15 \quad (4) \end{aligned}$$

Note that Equations (3) and (4) are really the same as Equations (1) and (2), respectively.

Now, because the signs are the *same*, *subtract* one equation from the other (Step 2). Suppose that we subtract Equation (4) from Equation (3):

$$\begin{array}{rl} & 6x + 8y = 66 \\ \text{minus} & 6x - 9y = 15 \\ \hline \text{equals} & 8y - (-9y) = 51 \end{array}$$

We are left with an equation that involves only y, which we solve (Step 3):

$$\begin{aligned} 8y - (-9y) &= 51 \\ 17y &= 51 \\ \frac{17y}{17} &= \frac{51}{17} \\ y &= 3 \end{aligned}$$

We have now obtained the solution for y. To obtain the solution for x (Step 4), we substitute the value $y = 3$ into, say, Equation (1). This yields:

$$\begin{aligned} 3x + 4(3) &= 33 \\ 3x + 12 &= 33 \\ 3x &= 33 - 12 \\ 3x &= 21 \\ \frac{3x}{3} &= \frac{21}{3} \\ x &= 7 \end{aligned}$$

Therefore, the solution to the simultaneous equations is $x = 7$ and $y = 3$.

Step 5 involves verification of the solution. Substituting these values into Equations (1) and (2) yields:

$$\begin{aligned} 3(7) + 4(3) &= 33 \\ 2(7) - 3(3) &= 5 \end{aligned}$$

You can verify that both of these statements are true, and so our solution appears to be correct.

Example M3.12

Solve the simultaneous equations in Example M3.11 by first eliminating the variable y.

Solution

In this case, we note that the *signs* of the coefficients of the variable to be eliminated are different in each equation, that is +4 in Equation (1) and –3 in Equation (2). If we ignore the signs, we can make the coefficients of y the same by multiplying both sides of Equation (1) by 3 and both sides of Equation (2) by 4. This yields:

$$\begin{aligned} 9x + 12y &= 99 \qquad (5) \\ 8x - 12y &= 20 \qquad (6) \end{aligned}$$

We can now eliminate the variable y by *adding* Equations (5) and (6):

$$\begin{aligned} & 9x + 12y = 99 \\ \text{plus} \quad & 8x - 12y = 20 \\ \hline \text{equals} \quad & 17x = 119 \end{aligned}$$

From this:

$$\begin{aligned} \frac{17x}{17} &= \frac{119}{17} \\ x &= 7 \end{aligned}$$

Now substitute $x = 7$ into, say, Equation (1):

$$\begin{aligned} 3(7) + 4y &= 33 \\ 21 + 4y &= 33 \\ 4y &= 33 - 21 \\ 4y &= 12 \\ \frac{4y}{4} &= \frac{12}{4} \\ y &= 3 \end{aligned}$$

Therefore, the solution obtained is $x = 7$ and $y = 3$. As expected, this is the same as that in Example M3.11.

M3.5 APPLICATIONS

There are many problems in which the value of some unknown quantity is to be found. In many cases, we can represent the unknown quantity by some variable name and construct an equation involving that variable. We can then solve the equation using the techniques in Section M3.3. Where there are two unknown quantities, we define two variables, construct two simultaneous equations, and find the values as in Section M3.4.

Since there is a vast array of such problems, this section can only provide a sample of just a few types and indicate the manner in which they may be solved.

Example M3.13

A customer left his car for service and received a combined bill (for parts and labour) of \$228. If the labour cost twice as much as the parts, find the amount that parts and labour each cost.

Solution

Let the amount charged for parts (in dollars) = x. Then the amount charged for labour (in dollars) = $2x$. Since:

$$\text{Parts charge} + \text{labour charge} = \$228$$

$$\begin{aligned} \text{Then: } x + 2x &= 228 \\ 3x &= 228 \\ \frac{3x}{3} &= \frac{228}{3} \\ x &= 76 \end{aligned}$$

Therefore, cost of parts = $\$x$ = \$76 and cost of labour = $\$2x$ = \$2(76) = \$152.

Example M3.14

A family consisting of 2 adults and a number of children went to the cinema. Each adult paid \$9 and each child paid \$3.50. If the total bill was \$39, how many children went?

Solution

Let the number of children = x. Then:

$$\begin{aligned} \text{Cost of 2 adults at \$9 each} &= 2 \times \$9 = \$18 \\ \text{Cost of } x \text{ children at \$3.50 each} &= \$3.5x \\ \text{Total bill} &= \$39 \\ \text{Cost of adults} + \text{cost of children} &= \$39 \\ 18 + 3.5x &= 39 \\ 3.5x &= 39 - 18 \\ 3.5x &= 21 \\ \frac{3.5x}{3.5} &= \frac{21}{3.5} \\ x &= 6 \end{aligned}$$

Therefore, 6 children went to the cinema.

Example M3.15

A bookstore purchased 200 copies of a book, some being the softcover version and some the hardcover version. The softcovers sold for \$47.50 each and the hardcovers for \$65.00 each. When all the books were sold, the total amount that the store received from the sale was \$10 725. How many each of the softcover and hardcover versions were sold?

Solution

Let the number of hardcovers sold = x.

Since the total number sold was 200 books:

$$
\begin{aligned}
\text{Number of softcovers sold} &= 200 - x \\
\text{Amount received for hardcovers} &= \$65x \\
\text{Amount received for softcovers} &= \$47.5(200 - x) \\
\text{Total amount received} &= \$10\,725
\end{aligned}
$$

Therefore:

$$
\begin{aligned}
65x + 47.5(200 - x) &= 10\,725 \\
65x + 47.5(200) - 47.5(x) &= 10\,725 \\
65x + 9500 - 47.5x &= 10\,725 \\
65x - 47.5x &= 10\,725 - 9500 \\
17.5x &= 1225 \\
\frac{17.5x}{17.5} &= \frac{1225}{17.5} \\
x &= 70
\end{aligned}
$$

Therefore, the number of hardcover versions sold = 70, and the number of softcover versions sold = 200 − 70 = 130.

Example M3.16

Joanne is 3 years older than Amy. Three times Joanne's age is 15 years more than twice Amy's age. How old are Joanne and Amy?

Solution

Let Amy's age = x. Then:

$$
\begin{aligned}
\text{Joanne's age} &= x + 3 \\
\text{Three times Joanne's age} &= 3(x + 3) \\
\text{Twice Amy's age} &= 2x
\end{aligned}
$$

Therefore:

$$
\begin{aligned}
3(x + 3) &= 2x + 15 \\
3(x + 3) - 2x &= 15 \\
3x + 9 - 2x &= 15 \\
3x - 2x &= 15 - 9 \\
x &= 6
\end{aligned}
$$

Therefore, Amy is 6 years old and Joanne is 6 + 3 = 9 years old.

Example M3.17

Mark and John have been employed to paint a room. Mark estimates that he could paint it in 3 hours if he did it alone, and John estimates that it would take him only 2 hours if he did it alone. If John and Mark painted the room together, how long would it take?

Solution

Let x = time (in hours) it would take them to paint the room together. Then in 1 hour the fraction of the room painted = $\frac{1}{x}$.

In 1 hour Mark could paint $\frac{1}{3}$ of the room and John could paint $\frac{1}{2}$ of the room. Therefore:

$$\frac{1}{2} + \frac{1}{3} = \frac{1}{x}$$
$$\frac{5}{6} = \frac{1}{x}$$

Multiply both sides of the equation by x to get:

$$\frac{5}{6}(x) = \frac{1}{x}(x)$$
$$\frac{5x}{6} = 1$$

Multiply both sides of the equation by $\frac{6}{5}$ to get:

$$\frac{5x}{6}\left(\frac{6}{5}\right) = 1\left(\frac{6}{5}\right)$$
$$x = \frac{6}{5}$$

Therefore, if John and Mark painted the room together, it would take $\frac{6}{5}$ = 1.2 hours or 1 hour 12 minutes.

Example M3.18

A balloon is fully inflated to its maximum capacity of air and hung from the ceiling of a party room. When the party is over, air is let out of the balloon at the rate of 100 cc per second. After 12 seconds the balloon has been reduced to two-thirds of maximum capacity. How much air will be in the balloon when it is at 50% of maximum capacity?

Solution

Let x = maximum capacity of air in the balloon (in cc).

In 12 seconds the amount of air let out of the balloon = 12×100 = 1200 cc. Therefore:

$$\frac{2}{3}x = 1200$$
$$x = 1800$$

Hence, the maximum capacity of air in the balloon is 1800 cc.

Therefore, when it is at 50% of maximum capacity it will contain 0.5×1800 cc = 900 cc of air.

Example M3.19

A customer buys 5 apples and 6 pears at a fruit stall and is charged \$3.10. Another customer buys 2 apples and 5 pears and is charged \$2.15. How much each are the apples and pears?

Solution

In this case, we have *two* unknowns (i.e. the prices of the apples and pears), so two variables are required.

Let A = price (in dollars) of 1 apple.
Let P = price (in dollars) of 1 pear.

Then:

$$\begin{aligned} 5A + 6P &= 3.10 \qquad (1) \\ 2A + 5P &= 2.15 \qquad (2) \end{aligned}$$

We now have two simultaneous equations in the variables A and P. Suppose we eliminate A from the equations. To do this we first multiply Equation (1) by 2 and Equation (2) by 5:

$$\begin{aligned} 10A + 12P &= 6.20 \qquad (3) \\ 10A + 25P &= 10.75 \qquad (4) \end{aligned}$$

If we then subtract Equation (4) from Equation (3) we have:

$$\begin{aligned} & 10A + 12P = 6.20 \\ \text{minus} \quad & \underline{10A + 25P = 10.75} \\ \text{equals} \quad & \quad -13P = -4.55 \end{aligned}$$

Therefore:

$$\begin{aligned} \frac{-13P}{-13} &= \frac{-4.55}{-13} \\ P &= 0.35 \end{aligned}$$

Substituting $P = 0.35$ into Equation (1):

$$\begin{aligned} 5A + 6(0.35) &= 3.10 \\ 5A + 2.10 &= 3.10 \\ 5A &= 3.10 - 2.10 \\ 5A &= 1.00 \\ \frac{5A}{5} &= \frac{1.00}{5} \\ A &= 0.20 \end{aligned}$$

Hence, the solution to the equations is $A = 0.20$ and $P = 0.35$. Therefore, each apple is 20 cents and each pear is 35 cents.

(These values should now be verified by substituting them into both Equations (1) and (2).)

M3.6 REMARKS

In this chapter we have seen how to solve linear equations in one degree only. If an equation involves a squared term (e.g. x^2) it is called a **quadratic equation**, and although a formula exists for solving such equations (called the *quadratic formula*), it is beyond the scope of this book.

Similarly, formulae exist for solving **cubic equations** (containing terms such as x^3) and **quartic equations** (containing terms such as x^4). However, no formula exists for finding the solutions to equations containing terms with powers of 5 or higher, a fact uncovered by the French mathematician Évariste Galois (1811–32).

Quick Quiz

1. Simplify the following expression:

   $3x - 2y + 7z - (2x - y) + (4x + 3z) - (z - 2x)$

2. Five apples, 3 oranges and 6 pears cost a total of \$4.55. If 1 apple costs 40 cents and 1 pear costs 25 cents, how much does 1 orange cost?

3. The perimeter of a rectangular field is 110 m. Three times the length is equal to eight times the breadth. Find the area of the field.

4. Three pencils and 2 rulers cost a total of \$6.15. A ruler costs 45 cents more than a pencil. Find the total cost of 2 pencils and 5 rulers.

5. Solve for x and y:

   $3x + 2y = 87$

   $2x + 5y = 135$

6. If Amy were 15 cm taller, she would be 5 cm taller than Joanne. If Joanne were 30% shorter, she would be 29 cm shorter than Amy. Find the *sum* of the heights of Amy and Joanne.

7. The combined cost of a coffee and a dessert at a restaurant is \$7.50. If a dessert costs \$3.50 more than a coffee, how much would 2 coffees plus 3 desserts cost?

8. Jack can paint 2 small rooms and a balcony in 50 hours. If a small room takes twice as long to paint as a balcony, how long will it take him to paint 1 small room plus 3 balconies?

9. Solve the following equation for x.

   $3x - 2 = 16 - 9x$

10. A customer purchases 4 screws and 3 bolts from a hardware store at a total cost of 82 cents. A bolt costs 4 cents more than a screw. How much would 10 bolts and 8 screws cost?

Multiple Choice Questions

1. The expression $(2x - 1)(x + 3) - (x + 1)(2x - 3)$ may be simplified to:
(a) 6 (b) $6x$ (c) $6x - 6$
(d) $4x$ (e) none of the above

2. The solution for x in the equation
$3(2x - 5) + 2(2 - 5x) = x + 14$ is:
(a) -5 (b) 2.5 (c) -2.5
(d) 2 (e) none of the above

3. The cost of 2 toffees and 4 creams is 22 cents. The cost of 3 toffees and 5 creams is 30 cents. Therefore, the cost of 7 toffees and 2 creams is:
(a) 52 cents (b) 58 cents (c) 48 cents
(d) 41 cents (e) none of the above

4. Three adults and 2 children went to a football match. Each adult paid \$9.50 and the total bill was \$36.50. The amount paid for each child was:
(a) \$3.50 (b) \$4.50 (c) \$5.50
(d) \$4.00 (e) none of the above

5. Company A has been in business for 10 years more than its rival company B. Two years ago company B had been in business for only half as long as company A at that time. In 3 years time the number of years that company B will have been in business for will be:
(a) 25 (b) 22 (c) 12
(d) 15 (e) none of the above

6. A rectangular property has a perimeter of 22 km. Fifteen times the length is equal to 40 times the breadth. The area of the property is:
(a) 25 km^2 (b) 20 km^2 (c) 24 km^2
(d) 28 km^2 (e) none of the above

7. Suppose that x and y satisfy the following equations.
$2x + 3y = 20$
$x - y = 5$
Then the value of $x + y$ is:
(a) 9 (b) 10 (c) 12
(d) 15 (e) none of the above

8. A nursery sells large pots for \$25 and small pots for \$15. Total sales of pots came to \$490 one weekend when customers bought three times as many small pots as big pots. The total number of pots sold on the weekend was:
(a) 20 (b) 23 (c) 25
(d) 27 (e) none of the above

9. A balloon inflates at a constant rate of 250 cc per second and is fully inflated after 12 seconds. The volume of the balloon when it is 60% inflated is:
(a) 1600 cc (b) 1800 cc (c) 2000 cc
(d) 2200 cc (e) none of the above

10. A share trades on a Monday for \$20. On that day Jack sells 35 such shares and spends the entire proceeds purchasing beer at a cost of \$2.50 per can. The number of cans that Jack purchased was:
(a) 200 (b) 250 (c) 280
(d) 300 (e) none of the above

Exercises

M3.1* Simplify:

(a) $9x - 5x + 8x - 6x$

(b) $12x - 5y + 7x - 2y - 3x + 3y$

(c) $13 - 3x + 23 - 5x + 4$

M3.2 Simplify:

(a) $(2x + 1)(3x - 4)$

(b) $(5 - 4x)(2x + 3)$

(c) $2x(3 - x) - 7x(3x + 2) + 4(2x + 1)$

M3.3* Solve for x:

(a) $3x - 2 = 8 - 2x$

(b) $4(x - 1) = 6$

(c) $3x(5 - 2x) = -6x(x + 2) + 81$

M3.4* Solve for y:

(a) $(y - 2)(y + 2) = y^2 - 2y$

(b) $2(3y - 6) = 3(1 + y)$

(c) $4(7y + 2) = 7 + 3(10y - 2)$

M3.5 Solve for z:

(a) $\frac{5z}{3} + 2 = \frac{7(z - 1)}{2}$

(b) $\frac{3z}{4} + 5 = 2 - z$

(c) $4\left(\frac{z}{3} + 2\right) = 5\left(3 - \frac{z}{6}\right)$

M3.6* Solve for x and y:

$5x - 2y = 1$

$3x + 2y = 7$

M3.7* Solve for x and y:

$3x - 8y = 39$

$4x + 3y = 11$

M3.8* Solve for x and y:

$4x - 5y = 16$

$8x + 3y = 253$

M3.9 A customer purchased 3 writing pads from a store. She received \$7.90 change from a \$10 note. How much was each writing pad?

M3.10* Three diaries and 4 pens cost a total of \$11.70. If each pen costs \$1.20, how much does each diary cost?

M3.11 Five plain doughnuts and 6 iced doughnuts cost a total of \$9.80. If each iced doughnut costs 50% more than a plain doughnut, find the individual costs of a plain and an iced doughnut.

M3.12* Two adults and 3 children visited an adventure park. The total cost of entrance was \$24.50. Each adult paid \$5 more than each child. Find the entrance cost for each adult and each child.

M3.13 Jennifer is 7 years younger than Melissa. In 4 years time Jennifer will be 1 year older than half of Melissa's age at that time. How old are Jennifer and Melissa now?

M3.14* An apple is 5 cents more expensive than an orange. Six apples and 4 oranges cost a total of \$1.00. How much each is an apple and an orange?

M3.15* The perimeter of a rectangular field is 70 m. Twice the length is equal to five times the breadth. Find:

(a) the length and breadth of the field

(b) the area of the field

M3.16 John and Sid work separately on a production line where they assemble television components. Between them they can currently assemble a total of 480 components in an hour. If John worked at twice his rate and Sid at three times his rate, they could assemble a total of 1260 components in an hour.

Find how many components each John and Sid can currently assemble in 1 minute.

M3.17* The claims department at a certain insurance company received a number of enquiries on the Monday morning following a wet weekend. The manager noted that if the number of enquiries from men had doubled, it still would have been 6 fewer than the number of enquiries from women. The secretary noted that 11 of the enquiries from women were regarding their husbands' vehicles. If these had been counted as enquiries from men instead of from women, the number of enquiries from men and women would have been equal.

How many enquiries were made by men and how many by women?

M3.18* Three friends had a meal at a restaurant and shared the bill. Pamela paid twice as much as Jenny, and Wilma paid 50% more than Pamela. If the total bill was \$36, how much did each person pay?

M3.19 A sales representative, having found himself lost in a country town, asks at the local service station how far it is to the main highway. The response is that the representative is 30% further away than he would be if he had travelled another 24 km towards the highway.

How far away is the main highway from where the representative is currently?

M3.20* A bucket and spade together cost \$2.75. If the bucket costs 35 cents more than the spade, how much would 3 buckets and 2 spades cost?

M3.21* If 3 apples and 4 oranges cost a total of 78 cents, and 5 apples and 2 oranges cost a total of 74 cents, how much will 10 apples and 8 oranges cost?

M3.22 A swimming pool has developed a leak and is losing water at the constant rate of 28 L per minute. After 20 minutes of leaking the pool is at 90% of full capacity. How much water will be in the pool when it is at 70% of full capacity?

M3.23 Simplify:

(a) $15x - 7y + 23 - 5y + 2x - 4$

(b) $(3x - 2)(x + 5)$

(c) $2x(x - 7) - 3x(2x + 3) + x(4x + 5)$

M3.24 Solve for x:

(a) $3x - 8 = 24 - 5x$

(b) $2x + 7 = 9x - 6$

(c) $10x + 58 = 26 - 4x$

M3.25* Solve for x:

(a) $(2x + 1)(4 - x) = 10 - 2x^2$

(b) $2(x + 6) = 5 + (12 - 3x)$

(c) $10 - 2(3x - 7) = 3 + (5x + 9)$

M3.26 Solve for x and y:

$$2x + 5y = 29$$
$$5x + 2y = 20$$

M3.27* Solve for x and y:

$$12x - 7y = 63$$
$$5x + 6y = 53$$

M3.28 Solve for x and y:

$$10x + 15y = 135$$
$$3x - 10y = -3$$

M3.29* Solve for x and y:

$$25x - 12y = 356$$
$$10x - 3y = 164$$

M3.30 Six key rings and 5 plastic wallets cost a total of \$10.20. If a plastic wallet costs 50 cents more than a key ring, find the cost of:

(a) 1 key ring (b) 1 plastic wallet (c) 2 key rings and 3 plastic wallets

M3.31 In Exercise M3.30, if a customer has a total of \$20 to spend and wants to purchase 12 key rings, what is the maximum number of plastic wallets they can purchase?

M3.32 The total cost of 3 ceramic mugs is \$6 more than the total cost of 2 diaries. Each diary costs \$1 more than a mug. Find the total cost of:

(a) 1 diary (b) 1 mug (c) 7 diaries and 4 mugs

M3.33* In Exercise M3.32, if a customer has a total of \$100 to spend and wants to purchase 5 diaries, what is the maximum number of mugs they can also purchase?

M3.34 Two wooden picture frames and 3 brass picture frames cost a total of \$33.75. Three wooden picture frames and 5 brass picture frames cost a total of \$54.60. Find the total cost of:

(a) 1 wooden picture frame

(b) 1 brass picture frame

(c) 4 wooden picture frames and 7 brass picture frames

M3.35* It costs a total of \$30.60 to purchase 5 lipsticks and 6 combs in a store. If the store raises the price of lipsticks by 20% and the price of combs by 10%, the total price of these items is \$35.76. Find the *original* price of:

(a) 1 lipstick (b) 1 comb (c) 3 lipsticks and 7 combs

M3.36 In Exercise M3.35, find the raised total price of:

(a) 1 lipstick (b) 1 comb (c) 3 lipsticks and 7 combs

M3.37 The perimeter of a rectangular field is 160 m. If the length were extended by 10% and the width were reduced by 5 m, the perimeter would remain unchanged. Find:

(a) the original and new length of the field

(b) the original and new width of the field

(c) the original and new area of the field

M3.38* Alicia and Rose working together can do a job in 4.8 hours. If doing the job by herself Alicia would take 12 hours, how long would Rose take to do the job by herself?

M3.39 A football stadium contains 80% of its full capacity of spectators for a Grand Final. At the end of the game, spectators leave the ground through the various exits at the rate of 10 spectators per second. After 40 minutes the stadium contains 60% of its full capacity of spectators.

(a) What is the full capacity of the stadium?

(b) How many spectators will be in the stadium 50 minutes after the end of the game?

M3.40 Consider the following simultaneous equations:

$3x - 5y = 46$

$x - 10y = 3$

Explain why these equations have no solution.

M3.41 Consider the following simultaneous equations:

$3(4x + 8y) = 40$

$2(3x + 6y) = 20$

Explain why these equations cannot be solved in the usual way. Write down one pair of solutions for x and y.

M3.42* Three friends, Kay, Dale and Carmel, shared in a \$150 000 lottery prize. Dale received 20% more than Kay, while Carmel received \$20 000 less than Dale. How much did each person receive?

Chapter M4

Ratios and proportions

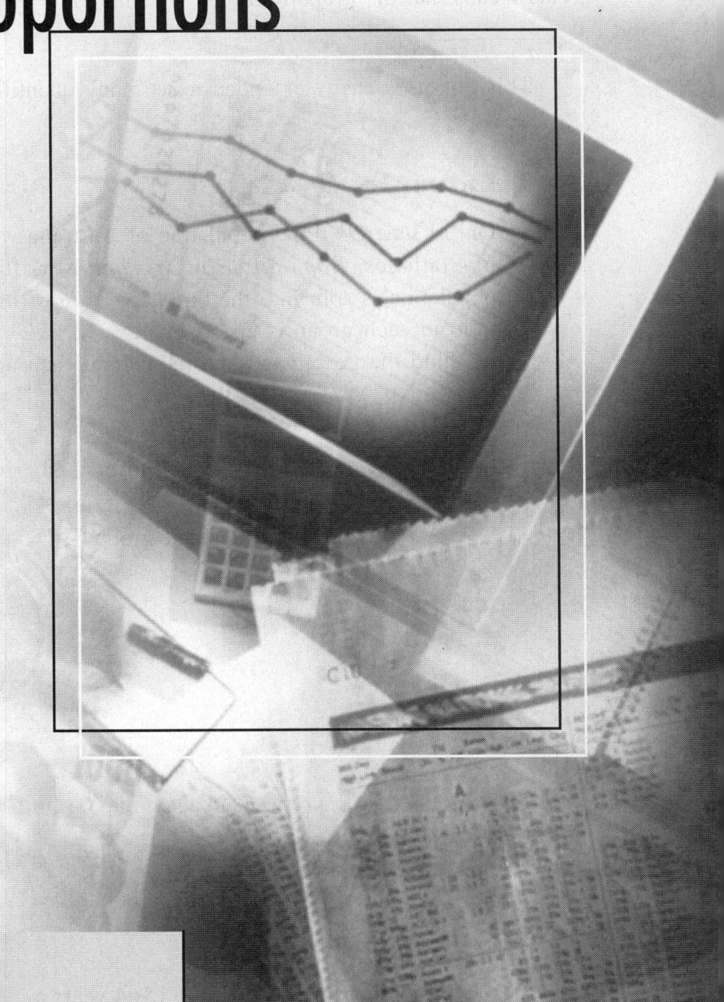

'There is no excellent beauty that hath not some strangeness in the proportion.'

(Francis Bacon, English philosopher and essayist, Of Beauty*, 1625)*

Learning objectives

In this chapter you will learn to:

- Calculate ratios and proportions
- Calculate and apply profit ratios
- Calculate and apply efficiency ratios
- Calculate and apply liquidity ratios

M4.1 RATIOS AND PROPORTIONS

A **ratio** is a method of comparing two or more numbers or rates. A **proportion** represents the *relative* contribution of a quantity to the whole. By definition, the value of a proportion should lie between 0 and 1 (both inclusive).

Ratios are often reduced to proportions. In particular, if two quantities X and Y occur in the ratio $a : b$, respectively, it means that X occurs the proportion $\frac{a}{a + b}$ of the time and Y occurs the proportion $\frac{b}{a + b}$ of the time.

Similarly, suppose that the three quantities X, Y and Z occur in the ratio $a : b : c$, respectively. This means that the proportion that X occurs is $\frac{a}{a + b + c}$, Y occurs is $\frac{b}{a + b + c}$ and Z occurs is $\frac{c}{a + b + c}$.

This notion is easily extended to as many quantities as desired.

Example M4.1

On 30 June 2000 the population of Australia was 19.157 million. (*Source*: Australian Bureau of Statistics, *Population by Age and Sex*, June Quarter 2000, Cat. No. 3201.0.) If this population is split into the three age groups (in years) of 0–14, 15–64 and 65 and over, the ratio in each group is approximately 72 : 224 : 41.

Find the percentage and number in each age group.

Solution

In this case, $a = 72$, $b = 224$ and $c = 41$. Hence:

$$\text{Proportion in age group 0–14} = \frac{a}{a+b+c} = \frac{72}{72+224+41} = \frac{72}{337}$$

Therefore:

$$\text{\% in age group 0–14} = 100 \times \frac{72}{337} = 21.365\%$$

$$\text{Number in age group 0–14} = 19.157 \text{ million} \times 0.21365 = 4.093 \text{ million}$$

$$\text{Proportion in age group 15–64 years} = \frac{b}{a+b+c} = \frac{224}{72+224+41} = \frac{224}{337}$$

Therefore:

$$\% \text{ in age group 15–64} = 100 \times \frac{224}{337}$$
$$= 66.469\%$$
$$\text{Number in age group 15–64} = 19.157 \text{ million} \times 0.66469$$
$$= 12.733 \text{ million}$$
$$\text{Proportion in age group 65 years and over} = \frac{c}{a+b+c}$$
$$= \frac{41}{72+224+41}$$
$$= \frac{41}{337}$$

Therefore:

$$\% \text{ in age group 65 years and over} = 100 \times \frac{41}{337}$$
$$= 12.166\%$$
$$\text{Number in age group 65 years and over} = 19.157 \text{ million} \times 0.12166$$
$$= 2.331 \text{ million}$$

Note that the sum of the three percentages is 21.365 + 66.469 + 12.166 = 100%, as you would expect.

As is the case with fractions, a ratio is unaltered if both sides are multiplied or divided by the same amount. In general, a ratio should be expressed in its *lowest terms*. That is, both numbers in the ratio should be divided by any common factors.

It is also usual for ratios to be expressed such that both sides are **integers**. If this is not the case, it may be desirable to multiply both sides by the same number so that they become integers.

Example M4.2

Express in lowest terms:

(a) 100 : 50
(b) 4 : 6
(c) 5.5 : 2.5
(d) $2\frac{1}{3} : 3\frac{1}{2}$
(e) 3.25 : 1

Solution

(a) The sides have a common factor of 50, so the ratio 100 : 50 is the same as the ratio 2 : 1.
(b) The sides have a common factor of 2, so the ratio 4 : 6 is the same as the ratio 2 : 3.
(c) In this case, neither side is an integer. Both 5.5 and 2.5 can be made integers by multiplying by 2. Hence, the ratio 5.5 : 2.5 is the same as the ratio 11 : 5.

(d) The lowest factor that $2\frac{1}{3}$ and $3\frac{1}{2}$ can be multiplied by to make them both integers is 6. Hence, the ratio $2\frac{1}{3} : 3\frac{1}{2}$ is the same as the ratio $(6 \times 2\frac{1}{3}) : (6 \times 3\frac{1}{2})$, which is 14 : 21 or, in its lowest terms, 2 : 3.

(e) In this case, only 3.25 is a non-integer. It can be made an integer by multiplying by 4. When both sides of the ratio are multiplied by 4, we obtain the equivalent ratio 13 : 4.

Rates

It is not unusual, especially when comparisons involving large numbers are made, for ratios to be used to express the **rate** at which events take place. For example, we speak of the birth rate per 100 000 population and the number of new cases of AIDS per 100 000 population. In some cases, the *second* number in the ratio is a base figure such as a standard unit of measurement. When this happens, the first number is sometimes a non-integer. In other cases, the *first* number is the base figure and the second may be a non-integer. Hence, a rate is really a ratio expressed with a *specified base*.

Example M4.3

A sales representative notes that in a particular week he drove 1350 km and used 120 L of petrol. His sales partner (in another vehicle) drove 940 km and used 88 L of petrol.

Express their petrol usages as a rate in the form *kilometres : 100 L*.

Solution

In this case, the petrol usage has to be expressed as kilometres : 100 L. The first driver has a ratio of 1350 : 120. To convert this ratio to an equivalent one with a second number of 100, we divide both sides by 1.2. This yields a rate $\frac{1350}{1.2} : \frac{120}{1.2}$ or 1125 : 100. That is, the travelling sales representative had a petrol usage rate of 1125 km travelled for every 100 L of petrol used.

The second driver has a ratio of 940 : 88. To convert this ratio we divide both sides by 0.88. This yields a rate of $\frac{940}{0.88} : \frac{88}{0.88}$ or 1068.2 : 100. That is, the second driver had a petrol usage rate of 1068.2 km travelled for every 100 L of petrol used.

Example M4.3 illustrates one of the advantages of expressing rates as ratios in which the second number is common. In this case, per 100 L of petrol, the first driver drove 1125 km and the second driver only 1068.2 km. We can therefore use these ratios as a basis of comparison for fuel economy.

Ratios with common *first* numbers may also be used to compare several items and to indicate the percentage difference between one figure and another. One such example is illustrated with the use of *index numbers* in Chapter S10.

Example M4.4

Three chartered accountants (Roy, Anna and Irene) have decided to pool their resources by forming a partnership and dividing the profits. However, to establish the business a sum of \$60 000 is required. Roy is willing to provide \$30 000, Anna \$10 000 and Irene \$20 000. The partnership agreement states that the profits in the first year will be shared in the same proportion as their contributions.

If the company makes a profit of \$45 000 in the first year, how much will each accountant receive?

Solution

In this case, the profits for Roy, Anna and Irene will be in the ratio 30 000 : 10 000 : 20 000. This ratio is equivalent to 3 : 1 : 2. Hence, we have:

$$\begin{aligned}
\text{Proportion of profit for Roy} &= \frac{3}{3+1+2} \\
&= \frac{1}{2} \\
\text{Amount received by Roy} &= \frac{1}{2} \times \$45\,000 \\
&= \$22\,500 \\
\text{Proportion of profit for Anna} &= \frac{1}{3+1+2} \\
&= \frac{1}{6} \\
\text{Amount received by Anna} &= \frac{1}{6} \times \$45\,000 \\
&= \$7500 \\
\text{Proportion of profit for Irene} &= \frac{2}{3+1+2} \\
&= \frac{1}{3} \\
\text{Amount received by Irene} &= \frac{1}{3} \times \$45\,000 \\
&= \$15\,000
\end{aligned}$$

M4.2 PROFIT RATIOS

By an analysis of past results it is possible to determine whether a company has been profitable. This knowledge is important, since it provides information on investment returns and is an indicator of the risk of bankruptcy. To do this we use **profit ratios**, which express the relationship of profit (gross and/or net) to some financial quantity (e.g. total assets or sales).

The financial position of a company is indicated by a **balance sheet**. A balance sheet lists the resources of value (assets) and liabilities. An important issue is whether the company has the assets to cover its liabilities. Also vital is whether the company has adequate *liquid assets* to pay immediate debts. Liquid assets may include cash, accounts receivable, inventory and prepayments.

The components of a **profit and loss statement** (or income statement) for a business produce the ratio net profit : sales. This is often referred to as the *net profit ratio*. Whether or not the resulting figure is satisfactory depends on the *type* of business considered.

Suppose that the business has a *high turnover*, such as a supermarket that sells *large* volumes of items. It would normally earn a lower profit per item sold than a business with a *low turnover*, such as a company that sells luxury cars, where only a relatively *small* number are sold. In this case, the company would earn a higher profit on each item sold.

Asset turnover

An **asset** is a resource of value controlled by a business (e.g. cash, bank deposits, inventory, receivables). The rate of turnover is measured by the ratio of sales to total assets and is known as the **total asset turnover**. It is defined as:

$$\text{Total asset turnover} = \frac{\text{sales}}{\text{total assets}} \qquad \textbf{(M4.1)}$$

Sales are measured in dollars and are obtained from the profit and loss statement. Total assets are obtained from the balance sheet and are measured in dollars also.

The aim of any asset utilisation is to determine the use being made of assets to an organisation and to uncover any trend in the make-up of the total asset base. A business should only expand its asset base if this will result in an improvement in its profits and sales.

The asset turnover measures the overall effectiveness of a company's current management. In particular, it indicates whether there are sufficient liquid assets on hand to meet current commitments, while at the same time not having an excess that could lead to resources being underutilised.

The total asset turnover is usually expressed as a *percentage*. That is, the ratio in Equation M4.1 is multiplied by 100.

Example M4.5

A company that sells mainframe computers has total sales of \$3 400 000 in a period and assets of \$1 200 000. Find the total asset turnover.

Solution

In this case, sales = \$3 400 000 and total assets = \$1 200 000.

From Equation M4.1:

$$\begin{aligned}\text{Total asset turnover} &= \frac{\text{sales}}{\text{total assets}}\\ &= \frac{\$3\,400\,000}{\$1\,200\,000}\\ &= 2.833 \text{ (to three decimal places)}\end{aligned}$$

In percentage terms, this is $2.833 \times 100 = 283.3\%$. This means that there was a 283.3% turnover in sales in relation to assets. In other words, for every \$1 of total assets the company generated \$2.83 of sales.

A company with a low net profit margin should have a larger total asset turnover than a company with a high net profit margin. Whether the higher turnover compensates for the lower profit margin depends on the overall **return on investment**. The return on investment for a company may be broken down into its profit margin and asset turnover components as follows:

$$\begin{aligned}\text{Return on investment} &= \frac{\text{net profit}}{\text{total assets}} \\ &= \frac{\text{net profit}}{\text{sales}} \times \frac{\text{sales}}{\text{total assets}} \\ &= \frac{\text{net profit}}{\text{sales}} \times \text{total asset turnover}\end{aligned} \qquad \textbf{(M4.2)}$$

To obtain a clearer picture of the performance of a company, we should consider other measures also.

M4.3 EFFICIENCY RATIOS

The *efficiency* of a company can be measured by its ability to achieve the maximum return for the lowest possible level of assets. The optimal level of stock (or inventory) depends on the type of company and the sales volume. For example, if there is a large turnover of stock (such as in a supermarket), more stock must be carried to meet demand. On the other hand, companies with a smaller stock turnover (such as motor vehicle sales) need not have as great a stock level.

The art of formulating inventory policies to meet demand for items can be a complex one that relies on many variables. It is a branch of mathematics known as *inventory control*, which is often taught in courses in the subject area of *operations research*.

A measure of the efficiency of a company's stock control is given by the **stock turnover** (or *inventory turnover*), where:

$$\text{Stock turnover} = \frac{\text{cost of goods sold}}{\text{average stock}} \qquad \textbf{(M4.3)}$$

The *average stock* may be found by averaging the opening and closing stocks from the company's balance sheets. (The closing stock shown in the balance sheet at the end of a period is the opening stock for the next period.) The cost of goods sold is found from the profit and loss statement. That is,

$$\text{Average stock} = \frac{\text{opening stock} + \text{closing stock}}{2} \qquad \textbf{(M4.4)}$$

If a company is involved in an enterprise in which the business is seasonal, it is advisable to use monthly figures in the average.

Whether the stock turnover ratio is satisfactory depends on the type of business concerned. A figure of 30 days would be too large for a manufacturer of, say, wrapped cakes, but might be in order for a manufacturer of whitegoods.

Example M4.6

A store that sells electrical appliances had total sales throughout the year of \$640 000. The opening stock was \$110 000 and the closing stock was \$150 000. Calculate the average stock and the stock turnover ratio.

Solution

From Equation M4.4:

$$\text{Average stock} = \frac{\text{opening stock} + \text{closing stock}}{2}$$

$$= \frac{\$110\ 000 + \$150\ 000}{2}$$

$$= \$130\ 000$$

From Equation M4.3:

$$\text{Stock turnover} = \frac{\text{cost of goods sold}}{\text{average stock}}$$

$$= \frac{\$640\ 000}{\$130\ 000}$$

$$= 4.92 \text{ (to two decimal places)}$$

Hence, the stock turned over 4.92 times during the year, or once every $\frac{365}{4.92} = 74$ days.

It should be noted that in some cases the end-of-year stock level is used instead of the average stock level if it is considered to be more representative of the level of inventory held. The stock turnover ratio can also be distorted by seasonal or cyclical variations (see Chapter S11), since the inventory may be unusually high or low depending on whether it is a peak or low selling season.

If the stock turnover ratio is *low*, it may indicate that the inventory level is too high in relation to sales and that excessive *carrying* (e.g. storage) costs are being incurred. If it is too *high*, it may mean that sales are being lost due to inability to meet customer demand.

Debtor turnover

Another measure of the operating efficiency of the credit policy of a business is the **debtor turnover** (or *accounts receivable turnover*), defined as:

$$\text{Debtor turnover} = \frac{\text{gross credit sales}}{\text{average gross accounts receivable}} \quad \textbf{(M4.5)}$$

The debtor turnover uses gross figures and any bad debts are *not* deducted. The accounts receivable figure may be subject to seasonal or cyclical variations.

The debtor turnover may also be expressed in terms of the *average collection period* (in days) by dividing it into 365. That is:

$$\text{Average collection period} = \frac{365}{\text{debtor turnover}} \quad \textbf{(M4.6)}$$

The debtor turnover measures the effectiveness of *credit control*; that is, how quickly a firm collects its receivables, with a general rule being that the higher the turnover, the better. This ratio also indicates whether a company may be extending its credit facilities beyond an acceptable limit. If the average collection period is much larger than the normal credit period allowed, the result might be extensive bad debts.

Example M4.7

A company has gross credit sales of \$4 300 000 and average gross accounts receivable of \$520 000. Find the debtor turnover and average collection period.

Solution

From Equation M4.5:

$$\begin{aligned}\text{Debtor turnover} &= \frac{\text{gross credit sales}}{\text{average gross accounts receivable}} \\ &= \frac{\$4\ 300\ 000}{\$52\ 000} \\ &= 8.27 \text{ (to two decimal places)}\end{aligned}$$

From Equation M4.6:

$$\begin{aligned}\text{Average collection period} &= \frac{365}{\text{debtor turnover}} \\ &= \frac{365}{8.27} \\ &= 44.1 \text{ days (to one decimal place)}\end{aligned}$$

M4.4 LIQUIDITY RATIOS

A matter of some importance is the ability of a company to pay its immediate debts. That is, some assessment of short-term solvency may be required to measure how well current assets compare with monetary commitments. Two such measures, known as **liquidity ratios**, are discussed below.

Current-asset ratio

One measurement that indicates how well current liabilities are covered by current assets is the **current-asset ratio** (or *working-capital ratio*). It is defined as:

$$\text{Current-asset ratio} = \frac{\text{current assets}}{\text{current liabilities}} \qquad \textbf{(M4.7)}$$

(Both current assets and current liabilities may be obtained from the balance sheet.) If a company has a high current-asset ratio, this might indicate that it is able to meet its obligations. On the other hand, it might indicate a poor stock turnover or a slack debt-collection policy.

Whether a current-asset ratio is acceptable or not depends on how readily stock and accounts receivable can be converted into cash and how quickly cash flows in from sales. This ratio can be used to compare companies of varying sizes as well as measuring the liquidity of the same company from year to year. One problem with the current-asset ratio is that it may be difficult to interpret individual components, such as how quickly current assets can be converted into cash to pay current liabilities or how soon current liabilities are due for payment. An experienced analyst may obtain this detailed information from the balance sheet.

Example M4.8

The balance sheet of a company reveals current assets of \$315 000 and current liabilities of \$420 000. Calculate the current-asset ratio and interpret its value.

Solution

From Equation M4.7:

$$\begin{aligned}\text{Current-asset ratio} &= \frac{\text{current assets}}{\text{current liabilities}}\\ &= \frac{\$315\,000}{\$420\,000}\\ &= 0.75\end{aligned}$$

This indicates that the company has 75c in current assets for every \$1 in current liabilities owed.

Acid-test ratio

The current-asset ratio includes assets that are expected to be turned into cash and current liabilities that are due to be repaid within 12 months. However, liquidity may be more immediately measured by excluding the less liquid current assets and less urgent current liabilities.

For example, current assets that may be excluded include inventory and hire-purchase debtors. Also, current liabilities excluded include bank overdrafts since it is uncommon for banks to demand sudden repayment.

The resulting ratio is therefore a modification of the current-asset ratio and is known as an **acid-test ratio** (or *quick-test ratio*). It is defined as:

$$\text{Acid-test ratio} = \frac{\text{current assets} - \text{inventory}}{\text{current liabilities} - \text{bank overdraft}} \qquad \textbf{(M4.8)}$$

The acid-test ratio gives an indication of a company's ability to use its own liquid assets to meet its immediate financial commitments. It also provides a better indication than the current-asset ratio of the company's ability to pay short-term debts.

Example M4.9

Consider the company in Example M4.8. Suppose that current assets include \$92 000 of inventory and current liabilities include \$145 000 of bank overdraft. Calculate the acid-test ratio.

Solution

From Equation M4.8:

$$\text{Acid-test ratio} = \frac{\text{current assets} - \text{inventory}}{\text{current liabilities} - \text{bank overdraft}}$$

$$= \frac{\$315\ 000 - \$92\ 000}{\$420\ 000 - \$145\ 000}$$

$$= \frac{\$223\ 000}{\$275\ 000}$$

$$= 0.81 \text{ (to two decimal places) or } 81\%$$

Since this is less than 100%, this would be of concern in the short run at least.

M4.5 REMARKS

There are a number of other ratios that reveal changes and trends over a period. They include the following:

1. The *debt/equity ratio* tests the leverage of an entity by measuring the reliance of funds on outside sources compared to internal sources such as shareholders' funds.
2. The *proprietary ratio*, when considered with the previous ratio, indicates an entity's long-term financial stability.
3. *Return on investment* is actually a class of ratios of interest to current and potential stockholders.
4. The *rate of return* ratio indicates that the dollar value of profits is less important than the rate of return that a company can earn.

Whatever ratio, or combination of ratios, is used to test the efficiency of a company, it is important that a systematic approach be adopted and that each ratio be interpreted in the correct context.

Quick Quiz

1. A company executive notes that the company vehicle travelled 580 km during the past week. If 102 L of petrol were used, express the petrol usage in the form *kilometres : 100 L*.
2. A roadside stall sells bags of firewood in two different size bags. The ratio of 10 kg bags sold to 5 kg bags sold is 3 : 5. If 20 of the 5 kg bags were sold, how many of the 10 kg bags were sold?
3. The balance sheet of a small business shows current assets of $80 000 and current liabilities of $60 000. Calculate the current-asset ratio and interpret its value.
4. Suppose that the current assets of the small business in Question 3 include $42 000 of inventory and its current liabilities include $35 000 of bank overdraft. Calculate the acid-test ratio.
5. A cement plant has gross credit sales of $650 000 and average gross accounts receivable of $48 000. Find (to the nearest day) the average collection period.
6. The total sales throughout the year at a discount store amounted to $1 250 000. The opening stock was $350 000 and the closing stock was $520 000. Calculate how often (to the nearest day) the stock turned over.
7. Yolande and Susie are company directors and agree to share the company profits in the following manner: Yolande receives the first $32 000 or 40% of the profit, whichever is the larger, and Susie receives $16 000 or 20% of the profit, whichever is the smaller, while any remaining money is placed in the bank. If the profit is $90 000, how much does each director receive and how much is placed in the bank?
8. A company has an average collection period of 62.5714 days (based on a 365-day year). If gross credit sales were $3.5 million, how much were average gross accounts receivable (to the nearest dollar)?
9. Of a sample of 58 diners at a restaurant, 23 ordered coffee, 21 ordered tea, 10 ordered some other drink and the remainder did not order a drink at all. What percentage of these diners either did not order a drink or ordered tea?
10. Express the following ratio in its lowest terms:

    224 : 480

Multiple Choice Questions

1. The ratio $2\frac{1}{4} : 3\frac{1}{2}$ expressed in its lowest terms is:
(a) 2 : 3 (b) 27 : 40 (c) 9 : 14
(d) 67 : 1 (e) none of the above

2. An airline has total sales of \$23 500 000 and assets of \$9 400 000. The total asset turnover (expressed as a percentage) is:
(a) 25.0% (b) 250.0% (c) 40.0%
(d) 400.0% (e) none of the above

3. A store had total sales throughout the year of \$875 000. The opening stock was \$230 000 and the closing stock was \$105 000. The stock turnover ratio (to two decimal places) is:
(a) 2.61 (b) 5.22 (c) 167.50
(d) 0.38 (e) none of the above

4. A company has gross credit sales of \$3 200 000 and average gross accounts receivable of \$370 000. The average collection period in days (to one decimal place) is:
(a) 8.6 (b) 21.1 (c) 17.3
(d) 42.2 (e) none of the above

5. The local newsagent has total sales of \$524 000 and assets of \$368 000. The total asset turnover (to two decimal places) is:
(a) 0.70 (b) 1.42 (c) 0.71
(d) 0.42 (e) none of the above

6. A manufacturing plant has current assets of \$120 000 of inventory, and current liabilities include \$185 000 of bank overdraft. Current assets in total are \$424 000 and current liabilities in total are \$327 000. The acid-test ratio (to two decimal places) is:
(a) 1.06 (b) 1.30 (c) 2.14
(d) 1.49 (e) none of the above

7. A company has current assets of \$150 000, inventory of \$80 000 and a bank overdraft of \$20 000. Its acid-test ratio is 0.50. Its current liabilities are:
(a) \$50 000 (b) \$70 000 (c) \$160 000
(d) \$190 000 (e) none of the above

8. A company has current assets of \$1.4 million and a current-asset ratio of 0.80. Its current liabilities are:
(a) \$1.12 million (b) \$571 429
(c) \$2.2 million (d) \$1.75 million
(e) none of the above

9. Cartons of milk are sold in three sizes: small, medium and large. The ratio of their sales is 11 : 15 : 10. If 240 000 cartons are sold, the number of these that are either small or large is:
(a) 140 000 (b) 21 000 (c) 210 000
(d) 100 000 (e) none of the above

10. Using the data in Example M4.1, the proportion of the population of Australia who are aged 65 or over is:
(a) 41/224 (b) 41/296 (c) 41/337
(d) 41/100 (e) none of the above

Exercises

M4.1* A certain supermarket sells milk in 1 L and 600 mL cartons. The ratio of 1 L cartons of milk sold to 600 mL cartons sold is 17 : 23.

(a) Find the percentage of each size carton sold.

(b) If the supermarket sold a total of 6600 cartons of milk, how many of each size carton did it sell?

M4.2 Answer Exercise M4.1 where the ratio of 1 L cartons sold to 600 mL cartons sold is 6 : 19.

M4.3* Express in lowest terms:

(a) 16 : 12 (b) 3 : 18 (c) 8.5 : 21.25

M4.4 Express in lowest terms:

(a) 8.125 : 5 (b) $3\frac{1}{4} : 4\frac{1}{3}$ (c) 150.75 : 201

M4.5* A sales representative notes that during the past month she has driven 1650 km and used 125 L of petrol. Express her petrol usage in the form *kilometres : 100 L*.

M4.6 Express the petrol usage of the sales representative in Exercise M4.5 in the form of *metres : litre*.

M4.7* A store has the following sales figures for beach umbrellas during the first 4 months of the year:

January	240
February	203
March	124
April	51

(a) Express each of the sales figures as a ratio compared to February (the base month) where the second number in the ratio is 100.

(b) Use your answers in (a) to calculate the percentage changes in sales in each of the other months compared to February.

M4.8 Answer the questions in Exercise M4.7 using April as the base month.

M4.9* During a sale on men's shorts, a store sold 568 pairs in 22 minutes. Express the sales as:

(a) number of pairs sold per minute

(b) number of pairs sold per hour

(c) number of pairs sold as 1 per so many seconds

M4.10* Three business partners (Sandra, Don and Annette) agree to share the profits from a business venture in the ratio 5 : 8 : 7, respectively. If the venture makes a profit of \$55 000, what is the share of each partner?

M4.11 For the data in Exercise M4.10, calculate the share of each partner on a profit of \$165 000 if they now share the profits in the ratio 3 : 14 : 8.

M4.12* A company that sells musical instruments has total sales of \$385 000 and assets of \$152 000. Find the total asset turnover (to three decimal places).

M4.13* The balance sheet of a football club shows current assets of $132 000 and current liabilities of $120 000. Calculate the current-asset ratio and interpret its value.

M4.14* The balance sheet of a bookstore shows current assets of $52 000 and current liabilities of $65 000. Calculate the current-asset ratio and interpret its value.

M4.15 Suppose that the assets of the football club in Exercise M4.13 include $48 000 of inventory and its current liabilities include $45 000 of bank overdraft. Calculate the acid-test ratio.

M4.16* Suppose that the assets of the bookstore in Exercise M4.14 include $14 000 of inventory and its current liabilities include $21 000 of bank overdraft. Calculate the acid-test ratio.

M4.17 According to the report of a local council to ratepayers, its current-asset ratio has changed from 0.67 in the previous year to 0.58 in the current year. Give an interpretation of this statement.

M4.18 A local council neighbouring that in Exercise M4.17 reports that its current-asset ratio has changed from 0.76 in the previous year to 1.77 in the current year.

(a) Give an interpretation of this statement.

(b) Can you tell which of the two councils is currently in the better financial position? Explain.

M4.19* A discount chain has three stores, which had gross credit sales of $480 000, $340 000 and $730 000, respectively. Combined, the average gross accounts receivable was $176 000. Calculate:

(a) the debtor turnover of the chain

(b) the average collection period

M4.20* Four TAFE lecturers decide to form a consulting business and calculate that they need $50 000 to begin operations. Carmel is willing to provide $12 000, Jack $20 000, Mick $13 000 and Tanya $5000. They agree to divide the profits each year in the same proportion as their contributions. If they make a profit of $2500 in their first year, how much will each of them get?

M4.21 Prior to expansion, a company presents the balance sheet shown below.

Assets	$'000	Equities	$'000
Current assets	43	Current liabilities	26
Fixed assets	54	Long-term liabilities	21
		Owner's equity	50
Total	97	*Total*	97

Calculate the current-asset ratio and interpret its meaning.

M4.22 If the company in Exercise M4.21 has sales of $250 000, what is the company's total asset turnover? Interpret its value.

M4.23* An automotive company that sells engine components has total sales for the financial year of $760 000 and total assets of $320 000.

(a) Calculate the total asset turnover.

(b) Interpret the value in (a).

M4.24 A computer company had stock of $110 000 on 1 July 2001 and closing stock of $230 000 on 30 June 2002. Total sales during the financial year were $625 000.

(a) Calculate the average stock and the stock turnover ratio.

(b) On average, how often does the stock turn over?

M4.25* A furniture company had stock of $650 000 on 1 July 2001 and closing stock of $810 000 on 30 June 2002. Total sales during the financial year were $1 320 000.

(a) Calculate the average stock and the stock turnover ratio.

(b) On average, how often does the stock turn over?

M4.26 Explain why bank overdrafts are usually not included in calculation of the quick-test ratio.

M4.27 Explain why the *quality* of assets is important in the interpretation of a liquidity ratio.

M4.28 Obtain a recent annual report of an Australian public company engaged in the manufacturing industry and listed on the stock exchange. Use the consolidated reports to calculate the following values for the previous 12 months:

(a) current-asset ratio

(b) quick-test ratio

(c) ratio of net profit to sales

(d) total asset turnover

(e) ratio of net profit to shareholders' funds

(f) ordinary earnings per share

M4.29* Donna, Michael and Annie form a partnership in which Donna invests $50 000, Michael $40 000 and Annie $60 000. Donna and Annie are each to receive a salary of $20 000 out of the profit, while the remaining profit is to be divided between the three in the ratio of their investments. Find how much each partner will receive in total if the profit is:

(a) $75 000 (b) $100 000 (c) $160 000

M4.30 In Exercise M4.29, Michael suggests that he be paid a straight salary of $25 000 and not share the profits. Under which of the profits (a), (b) and (c) would Michael be better off under this new arrangement?

M4.31 Vijay and Irene are partners in a clothing business in which Vijay has contributed a total of $4000 over 9 months and Irene a total of $10 000 over 6 months. If profits are to be shared according to average monthly investment, calculate how much each would receive on a profit of:

(a) $20 000 (b) $40 000 (c) $100 000

M4.32 Three brothers, Peter, Alan and David, agree to share their company's profits under the following arrangement:

- Peter receives 35% of the profit or $20 000, whichever is the smaller amount.
- After Peter takes his share, Alan receives 35% of the remaining profit.
- David receives what is left after Peter and Alan take their share.

Calculate how much each brother will receive if the profit is:

(a) $50 000 (b) $100 000

M4.33 Suppose that, under a new arrangement in Exercise M4.32, each brother decides to divide the company profits into equal shares. In each of (a) and (b), calculate how much better off or worse off each brother will be under this new deal.

M4.34* In an Australian Rules football match a total of 128 points was scored. Each goal is worth 6 points and each *behind* is worth 1 point. If the ratio of behinds : goals scored was 9 : 4, calculate how many goals and behinds were scored.

M4.35 Answer Exercise M4.34 if the number of points scored was 92 and the ratio of behinds : goals scored was 5 : 3.

M4.36 In 40% of his innings, a certain cricketer averages a score of 15. In the remainder of his innings he averages a score of 24. What would his average score be if he played a total of:

(a) 10 innings? (b) 20 innings?

M4.37* A bookstore manager notes that the ratio of customers who browse in the store to those who actually buy something is 5 : 3. In a total of 160 customers, find the number in each category.

M4.38 In Exercise M4.37, the manager notes that the ratio for male customers only is 3 : 2. If 50% of the customers are female, what is the ratio for female customers?

M4.39 A real estate agent has two rectangular blocks of land (A and B) for sale. The areas of these blocks are in the ratio 5 : 8 for A and B, respectively. Block A has a width of 30 m and block B has a length of 60 m. The perimeter of block B is 40 m greater than the perimeter of block A. Can you find the area of each block? If so, then find them. If not, explain why it cannot be done.

M4.40 The sum of the diameters of two circles (X and Y) is 70 m. The ratio of the areas of X and Y is 16 : 9. Can you find the radius of each circle? If so, then find them. If not, explain why it cannot be done.

Chapter 5

Simple interest

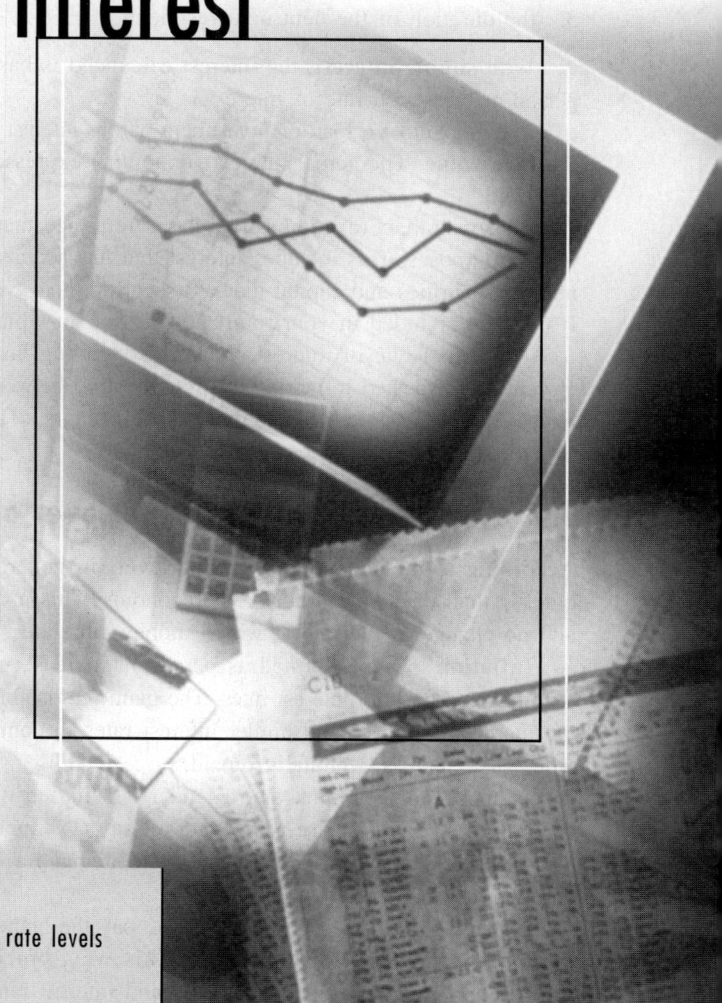

'I'd like to live like a poor man with lots of money.'

(Pablo Picasso, Spanish painter and sculptor, 1881–1973)

Learning objectives

In this chapter you will learn to:

- Understand the effect of inflation on interest rate levels
- Perform calculations involving simple interest
- Manipulate the simple interest formula
- Distinguish between, and calculate, flat and effective rates of interest
- Estimate the effective rate of interest

M5.1 INTRODUCTION

In general terms, the word **interest** denotes any income received from capital that has been invested, or the amount of money paid for the use of borrowed money. The actual amount of money that has been lent or invested, or the current balance of the original amount, is called the **principal** (or *capital*).

The amount of money received as interest depends on three factors:

1. the amount of money lent or invested (i.e. the principal or capital)
2. the *rate of interest*, which is usually expressed as the ratio or percentage of interest earned to the principal over some unit of time
3. the duration of the debt or investment

With **simple interest**, the interest calculated is based on the *original* principal during the entire period at the stated rate of interest.

The amount of a loan or investment before interest is added on (or a discount taken off) is called the **face value**. The actual interest (or return) expressed as a percentage of the price paid is called the **yield**.

The Australian Taxation Office has determined that all interest credited to you must be declared for tax purposes. This includes interest that has accrued each year on discounted securities, deferred interest securities and capital-indexed securities that were originally issued after 16 December 1984. It must be included in your return in the year in which it accrues.

The actual rate of interest charged by lending institutions depends on many factors, including current and expected inflation rates, competition between financial institutions, the size of Australia's budget deficit and often the amount of the principal. (In some cases, a higher rate of interest is paid on larger amounts of money invested.)

The effect of inflation on the level of interest rates

According to monetary theory, price levels in general will rise if the money supply grows more quickly than the total output of the economy. This proposition assumes essentially full employment and no change in the rate at which money circulates.

Inflation may be defined as persistent increase in the general level of prices and is normally accompanied by high interest rates. The main reason for this is that lenders are often able to anticipate the increases and demand higher interest rates to compensate for the reduced value of their money when their loan is eventually repaid.

Suppose, for example, that a lender is satisfied with an 8% yield on the loan in conditions of stable prices. If prices rise at the rate of, say, 2% annually, the 'real' yield on the loan will be only $8 - 2 = 6\%$. In that case, the lender will demand a 10% interest rate: the original 8% plus 2% to compensate for the 2% inflation.

Those who borrow money are able to pay the higher interest rate if inflation enables them to raise their selling price to pay for it. In this way, borrowers themselves are helping to create an **inflationary spiral** in which inflation and interest rates continually push each other higher.

High interest rates also help to promote an inflationary spiral in that they affect the capacity to produce, production costs and the value of money. In each case, there is an upward pressure on prices.

The interest rate that occupies the mind of many Australians is that charged by banks and other lending institutions on *home loans*. Rates fell from the high levels of around 17% in the late 1980s to around 8% to 9% by the late 1990s. By 2001 they had fallen again to between 6.5% and 7.5%.

Interest rates still remain a concern for those wishing to take out a mortgage to purchase a new home or even to upgrade an existing one. On the other hand, those who rely solely on their investments for their income and do not have any loans do not really mind interest rate rises as much as some others might.

One of the factors affecting mortgage rates of interest is the cost of goods and services. If this interest rate rises, the result is that a fixed amount of money today will purchase more than it can in the future. That is, inflation is going to reduce the value of the money and mortgage rates will rise.

One option given by lending institutions to potential customers is to offer an interest rate that is fixed for a number of years (e.g. 1, 3 or 5 years). This rate is usually higher than the current standard home-mortgage rate of interest, but gives purchasers a kind of insurance that, whatever happens to interest rates during this fixed period, they will not be affected. Of course, borrowers are gambling that interest rates will increase to above the level they are locked into paying, while the lending institution is gambling that this will not happen.

Many factors affect interest rates, including high unemployment rates (often leading to a reduction in interest rates), overseas interest rates, balance of payments figures, deregulation of the financial system, the amount held in deposits and especially government policy. As is evidenced by the diverse opinion of economists, the prediction of possible rises or falls in interest rates is an extremely difficult task!

M5.2 CALCULATION OF SIMPLE INTEREST

To find the simple interest on any principal at a given interest rate for a certain number of years, multiply the principal by the rate, then by the number of years. It is useful to define:

P = principal
R = rate of interest (expressed as a fraction, usually as a rate per annum)
T = time (usually in years)
I = total simple interest
S = amount or maturity value of the principal

The *simple interest* (I) can be found from the formula:

$$I = P \times R \times T \qquad \textbf{(M5.1)}$$

The *maturity value* (S) of the principal is:

$$S = P + I \qquad \textbf{(M5.2)}$$

which, when substituting for I from Equation M5.1, may be written:

$$S = P(1 + RT) \qquad \textbf{(M5.3)}$$

In some cases, we may wish to know what amount of principal will give a certain maturity value. In this case, we can write Equation M5.3 as follows:

$$P = \frac{S}{1 + RT} \qquad \textbf{(M5.4)}$$

When P is calculated from S as in Equation M5.4, the value of P is called the *present value* of S or the *discounted value* of S.

Example M5.1

Calculate the simple interest earned on \$15 000 deposited at a rate of 1.5% per annum for a period of:

(a) 2 years
(b) 4 years and 3 months

Solution

In this case, $P = \$15\,000$ and $R = 0.015$.

(a) Here $T = 2$. From Equation M5.1:

$$\begin{aligned} I &= P \times R \times T \\ &= \$15\,000 \times 0.015 \times 2 \\ &= \$450 \end{aligned}$$

Hence, the total simple interest earned is \$450.

(b) Here $T = 4.25$. From Equation M5.1:

$$\begin{aligned} I &= P \times R \times T \\ &= \$15\,000 \times 0.015 \times 4.25 \\ &= \$956.25 \end{aligned}$$

Hence, the total simple interest earned is \$956.25.

Example M5.2

A bank customer invests \$5000 for 18 months at a simple interest rate of 9.5% per annum. Find:

(a) the total interest accumulated
(b) the maturity value of the principal

Solution

In this case, $P = \$5000$, $R = 0.095$ and $T = 1.5$.

(a) From Equation M5.1:

$$\begin{aligned} I &= P \times R \times T \\ &= \$5000 \times 0.095 \times 1.50 \\ &= \$712.50 \end{aligned}$$

Hence, the total interest accumulated is \$712.50.

(b) From Equation M5.3:

$$\begin{aligned} S &= P(1 + RT) \\ &= \$5000\,(1 + 0.095 \times 1.50) \\ &= \$5000\,(1.1425) \\ &= \$5712.50 \end{aligned}$$

Hence, the maturity value of the principal is \$5712.50 (consisting of \$5000 for the original principal plus \$712.50 interest).

Example M5.3

The Acme Loan Company lends a customer \$500 and states that it must be repaid at the end of the month, when \$560 will be due. What annual simple interest rate is the company charging?

Solution

In this case, $P = \$500$, $I = \$60$ and $T = \frac{1}{12}$. This time we use Equation M5.1, where the unknown quantity is R. Hence:

$$I = P \times R \times T$$

$$\$60 = \$500 \times R \times \frac{1}{12}$$

which may be written:

$$R = \frac{\$60}{\$500 \times \frac{1}{12}}$$
$$= 1.44$$

Hence, the annual simple interest rate charged is 144%.

Example M5.4

An investor wishes to have an amount of \$4000 on maturity in 5 years time. He can invest his money at a simple interest rate of 12.0% per annum. How much money should he invest now in order to achieve his aim?

Solution

In this case, $S = \$4000$, $R = 0.12$ and $T = 5$.
From Equation M5.4:

$$P = \frac{S}{1 + RT}$$
$$= \frac{\$4000}{1 + 0.12 \times 5}$$
$$= \frac{\$4000}{1.60}$$
$$= \$2500$$

Hence, he should invest \$2500 now in order to have \$4000 at the end of 5 years.

Example M5.5

How long will it take for an investment of \$16 000 to earn a total interest of \$2860 if it attracts a simple interest rate of 3.25% per annum?

Solution

In this case, P = \$16 000, R = 0.0325 and I = \$2860. This time we use Equation M5.1, where the unknown quantity is T (years). Hence:

$$I = P \times R \times T$$

$$\$2860 = \$16\,000 \times 0.0325 \times T$$

which may be written:

$$T = \frac{\$2860}{\$16\,000 \times 0.0325} = 5.5$$

Hence, it will take 5 years and 6 months to earn a total interest of \$2860.

Minimum monthly balance

Many financial institutions in Australia pay simple interest based on the *minimum monthly balance* in an account each month. That is, if a sum of money is deposited and withdrawn during the same month, it will not attract any interest.

Example M5.6

A customer opens a savings account on 15 October 2001 with a sum of \$1000. Table M5.1 shows the transactions in the account until the end of April 2002.

Table M5.1 Savings account transactions for Example M5.6

Date	Deposits (\$)	Withdrawals (\$)	Balance (\$)
15.10.01	1000		1000
11.11.01	2500		3500
21.12.01		800	2700
03.01.02	750		3450
08.03.02		400	3050
29.03.02	1400		4450
09.04.02		650	3800
22.04.02	500		4300

The bank pays simple interest on the account at a rate of 2.5% per annum based on the minimum monthly balance.

Calculate the interest due each month and the total interest due for the year ending 30 April 2002.

Solution

In this case, it is useful to construct a table of minimum monthly balances as shown in Table M5.2. (Note that the minimum monthly balance for the months from May 2001 to October 2001, both inclusive, is zero.)

The interest for each month can be found using Equation M5.1, where P = minimum monthly balance, R = 0.025 and $T = \frac{1}{12}$.

Table M5.2 Minimum monthly balances and interest for Example M5.6

Month	Minimum balance ($)	Interest ($)
November 01	1 000	2.08
December 01	2 700	5.62
January 02	2 700	5.62
February 02	3 450	7.19
March 02	3 050	6.35
April 02	3 800	7.92
Total	16 700	34.78

For example, the interest for the month of November 2001 is:

$$\begin{aligned} I &= P \times R \times T \\ &= \$1000 \times 0.025 \times \frac{1}{12} \\ &= \$2.08 \text{ (rounded to the nearest cent)} \end{aligned}$$

Hence, the total interest due for the year ending 30 April 2002 is $34.78.

If we were not specifically concerned with the interest paid each month, the *total* interest could have been calculated by finding the minimum monthly balances and finding 1 month's interest on their sum at the rate of 2.5%. That is, we could use Equation M5.1 with P = \$16 700, R = 0.025 and $T = \frac{1}{12}$. Hence:

$$\begin{aligned} I &= P \times R \times T \\ &= \$16\,700 \times 0.025 \times \frac{1}{12} \\ &= \$34.79 \end{aligned}$$

This is the same as the figure obtained in Table M5.2, except for a 1 cent rounding error.

M5.3 EFFECTIVE RATE OF INTEREST

Often, when a short-term or personal loan is made for a special purpose, a financial institution will quote a **flat rate** of interest. This interest rate is always based on the original principal, even though the amount of the loan outstanding is reducing.

For example, suppose that a loan of $10 000 was taken out over a period of 5 years at a flat interest rate of 9.5%. At the beginning of, say, the fifth year the amount of outstanding debt would be much less than $10 000; however, interest for that fifth year would still be calculated on the original $10 000.

In order that borrowers understand precisely the type of interest that they are paying, as well as the flat rate of interest, the loan contract usually also includes the **effective rate** of interest. This represents the rate of interest actually paid by the borrower; that is, as if the borrower had been paying that rate of interest on the remaining balance of the loan each time.

Thus, for every flat rate of interest there will be an equivalent effective rate. The effective rate of interest will be larger than the stated flat rate, and its approximate value can be found from the formula:

$$E = \frac{2N \times R}{N + 1} \tag{M5.5}$$

where E = effective rate of interest
R = flat rate of interest
N = number of repayments in loan

It can be seen from Equation M5.5 that, if the number of repayments (N) in the loan is fairly large (e.g. more than 12), the effective rate of interest will be slightly less than twice the equivalent flat rate.

To find the flat rate of interest that is equivalent to a given effective rate, we can rewrite Equation M5.5 as:

$$R = \frac{(N + 1) \times E}{2N} \tag{M5.6}$$

Example M5.7

A customer takes out a $15 000 personal loan from the bank to be repaid monthly over 5 years. The bank has quoted a flat rate of interest of 9.5%.

Calculate:

(a) the total simple interest on the loan
(b) the equivalent effective rate of interest

Solution

(a) In this case, P = $15 000, R = 0.095 and T = 5.
From Equation M5.1:

$$\begin{aligned} I &= P \times R \times T \\ &= \$15\,000 \times 0.095 \times 5 \\ &= \$7125 \end{aligned}$$

Hence, the total simple interest paid is $7125.

(b) Since the loan is repaid monthly over 5 years, the number of repayments is $N = 5 \times 12 = 60$. From Equation M5.5:

$$\begin{aligned} E &= \frac{2N \times R}{N + 1} \\ &= \frac{2 \times 60 \times 0.095}{60 + 1} \\ &= 0.1869 \end{aligned}$$

Hence, the effective rate of interest is 18.69%. (Note that this is slightly less than twice the value of the flat rate.)

Example M5.8

A credit union is offering home equity loans (with first or second mortgage security) to be repaid over 20 years with monthly repayments and an effective rate of interest of 17.0%. Calculate the equivalent flat rate of interest.

Solution

In this case, $N = 20 \times 12 = 240$ and $E = 0.17$. From Equation M5.6:

$$\begin{aligned} R &= \frac{(N + 1) \times E}{2 \times N} \\ &= \frac{(240 + 1) \times 0.17}{2 \times 240} \\ &= \frac{40.97}{480} \\ &= 0.08535 \end{aligned}$$

Hence, the equivalent flat rate of interest is 8.535%. (Note that this is slightly more than half the effective rate.)

Example M5.9

A builder wants to take out a loan from a lending institution to purchase equipment and materials. The institution manager quotes a flat rate of interest of 6.4% on the loan. The builder enquires as to the equivalent rate in terms of an effective interest rate and is told this would be 12.0%.

If this were true, how many repayments would there be on the loan?

Solution

In this case, $E = 0.120$, $R = 0.064$ and N is the unknown. Rewriting Equation M5.6:

$$\begin{aligned} N &= \frac{E}{2R - E} \\ &= \frac{0.120}{2(0.064) - 0.120} \\ &= 15 \end{aligned}$$

Hence, 15 repayments would have to be made. (Note that this is independent of the amount that the builder wants to borrow.)

M5.4 REMARKS

Calculations involving the formulae in this chapter may also be performed with the aid of a **financial calculator**. This is a special calculator with inbuilt functions that include the use of simple interest formulae.

A computer spreadsheet may be used to perform simple interest calculations quickly. Discussion of one of the more popular spreadsheets, Microsoft Excel for Windows, is provided in Appendix C at the end of the statistics section on page S.436.

Quick Quiz

1. Calculate the amount of simple interest earned on a \$3500 principal at 7.5% per annum over 4 years.
2. Find the rate of simple interest that will result in a principal of \$750 amounting to \$984.38 after 5 years.
3. A finance company lends a customer \$2000 and states that it must be repaid at the end of 2 months, when \$2400 will be due. What annual simple interest rate is the company charging?
4. A store advertises a washing machine for a price of \$800. It states that it can be purchased for 10% deposit and repayments of \$40 per month over 2 years. What is the flat rate of interest?
5. For the data in Question 4, what is the effective rate of interest?
6. A bank is offering personal loans (without security) to be repaid over 10 years with monthly repayments and an effective rate of interest of 15%. Calculate the equivalent flat rate of interest.
7. If a loan has 20 equal repayments, find the flat rate of interest that corresponds to an effective interest rate of 12.5%.
8. A solicitor earned total simple interest of \$864 after 4.5 years of initially investing \$6000 in an account. What annual rate of simple interest was paid on the account?
9. What happens to price levels if money supply grows more rapidly than the total output of the economy?
10. Explain briefly what is meant by the term *inflationary spiral*.

Multiple Choice Questions

1. The simple interest earned on $35 000 deposited at a rate of 6.5% per annum for a period of 5 years 3 months is:
(a) $11 943.75 (b) $2275.00
(c) $12 057.50 (d) $1194.38
(e) none of the above

2. A bank lends a client $10 000, which must be paid at the end of 9 months, when $10 281.25 will be due. The annual simple interest rate the bank is charging is:
(a) 5.75% (b) 3.75% (c) 7.50%
(d) 2.81 % (e) none of the above

3. A newsagent wants to have an amount of $25 000 in 6 years time when she retires. She can invest her money at a simple interest rate of 8.5% per annum. How much money (to the nearest dollar) should she invest now to achieve her aim?
(a) $23 041 (b) $17 482 (c) $15 000
(d) $16 556 (e) none of the above

4. A dentist borrows $27 000 to be repaid monthly over 3 years. There is a flat rate of interest of 7.25%. The total simple interest on the loan is:
(a) $1957.50 (b) $19 575.00 (c) $5000.00
(d) $5872.50 (e) none of the above

5. For the data in Question 4, the effective rate of interest (to one decimal place) is:
(a) 7.0% (b) 14.1% (c) 11.8%
(d) 12.0% (e) none of the above

6. Referring to Equation M5.4, which of the following statements is true?
(a) If the value of *R* increases, the value of *P* increases
(b) If the value of *S* increases, the value of *P* decreases
(c) If the value of *T* decreases, the value of *P* increases
(d) If the values of *R* and *T* *both* increase, the value of *P* increases
(e) If the value of *S* decreases *and* the value of *R* increases, the value of *P* increases

7. The term *face value* of a loan refers to:
(a) the appearance of the loan document
(b) the value of the loan including interest
(c) the tax payable on the loan
(d) the amount of the loan before interest is added on
(e) none of the above

8. If you take out a *fixed interest* home mortgage, you are suggesting that:
(a) mortgage rates will rise in the future
(b) mortgage rates will decrease in the future
(c) mortgage rates will stay about the same
(d) the economy will improve very quickly
(e) none of the above

9. A carpenter opens a savings account on 8 January at a local bank where the simple interest rate is 0.02% paid on the minimum monthly balance. The amount deposited on 8 January is $500. On 15 January he deposited a further $400 and on 25 January he deposited a further $100. The interest paid by the bank on this account for January is:
(a) 0 (b) $2 (c) $10
(d) $20 (e) none of the above

10. The absent-minded professor forgot to pay the amount of $821.21 due on his credit card by the required date. He then had to pay simple interest charges of $19.50 for being 3 months overdue. The annual interest rate charged was (to one decimal place):
(a) 2.4% (b) 9.5% (c) 42.1%
(d) 0.6% (e) none of the above

Exercises

M5.1* Calculate the amount of simple interest earned on:

(a) \$4000 principal at 8% per annum over 4 years

(b) \$12 500 principal at 6% per annum over 5 years

(c) \$25 000 principal at 7% per annum over 3 years

M5.2* Find the annual rate of simple interest that will result in:

(a) a principal of \$1000 amounting to \$1100 after 5 years

(b) a principal of \$5000 amounting to \$6600 after 3 years

(c) a principal of \$2500 amounting to \$2812.50 after 5 years

M5.3 Find the annual rate of simple interest that will result in:

(a) a principal of \$250 amounting to \$282.50 after 4 years

(b) a principal of \$1500 amounting to \$2412 after 8 years

(c) a principal of \$900 amounting to \$1498.50 after 7 years

M5.4* After what period will:

(a) a principal of \$3000 amount to \$3300 at 5% simple interest per annum?

(b) a principal of \$6500 amount to \$7523.75 at 4.5% simple interest per annum?

(c) a principal of \$18 000 amount to \$30 397.50 at 9.5% simple interest per annum?

M5.5 After what period will:

(a) a principal of \$2200 amount to \$2511.67 at 4.25% simple interest per annum?

(b) a principal of \$430 amount to \$650.73 at 8.8% simple interest per annum?

(c) a principal of \$345 000 amount to \$474 375 at 6.25% simple interest per annum?

M5.6* A building society customer invests an amount of \$600 for 15 months at an annual simple interest rate of 7.5%. Find:

(a) the total interest accumulated

(b) the maturity value of the principal

M5.7 Answer the questions in Exercise M5.6 assuming that the customer invests \$2300 for 42 months at an annual simple interest rate of 10.5%.

M5.8* An investor wishes to have an amount of \$3000 in 5 years time. She can invest her money at a simple interest rate of 13.25% per annum. How much money should she invest now in order to achieve her aim?

M5.9 Answer the question in Exercise M5.8 assuming that the investor wants an amount of \$10 000 in 8 years time.

M5.10 A customer opens a savings bank account on 3 March 2001 with a sum of \$50. The table below shows the transactions in the account until the end of June 2001.

Date	Deposits ($)	Withdrawals ($)	Balance ($)
3.03.01	50		50
23.03.01	120		170
15.04.01		35	135
27.05.01	75		210
29.05.01		20	190
8.06.01	200		390

The bank pays simple interest at the rate of 5.5% per annum based on the minimum monthly balance.

(a) Calculate the minimum monthly balance and interest due each month.

(b) Calculate the total interest that will be paid by 30 June 2001.

M5.11 Answer the questions in Exercise M5.10 if the bank pays simple interest at the rate of 9.25% per annum based on the minimum monthly balance.

M5.12* A student borrows $3500 from a bank to be repaid monthly over 3 years. The bank has quoted a flat rate of interest of 6.25%. Calculate:

(a) the total simple interest on the loan

(b) the equivalent effective rate of interest

M5.13 A finance company is offering car loans to be repaid over 15 years with monthly repayments and an effective rate of interest of 14.0%. Calculate the equivalent flat rate of interest.

M5.14* Answer Exercise M5.13 if the effective rate of interest is 18.5%.

M5.15 Answer Exercise M5.13 if payments are made semi-annually.

M5.16* A sewing machine is on sale for $1000. If it is purchased on terms, it will cost 16% deposit and repayments of $38.50 per month over 2 years. Find:

(a) the total cost of the sewing machine to the purchaser

(b) the total interest paid

(c) the flat rate of interest

(d) the estimated effective rate of interest

M5.17 Answer the questions in Exercise M5.16 if the sewing machine were on sale for $500 with terms of no deposit and repayments of $12.25 per month over 4 years.

M5.18* A TAFE student has saved $3500 towards purchasing a car. Assuming that $4750 is required in 4 years time, what is the minimum annual rate of simple interest that the student must get on her money?

M5.19* How long would it take an investment of $25 000 to amount to $40 000 at a flat rate of interest of 9.5% per annum?

M5.20* A building society has quoted a customer an effective interest rate of 15.5% on a loan that is to be repaid in quarterly instalments over 8 years.

(a) What is the equivalent flat rate of interest that the building society is charging the customer?

(b) If the amount of the loan is $20 000, what is the total amount of interest that the customer will have paid after 8 years?

M5.21 How long would it take for an investment of $20 000 to earn interest of $5312.50 at a simple interest rate of 6.25% per annum?

M5.22* The owner of a fruit shop invests $15 000 at a simple interest rate of 7.50% per annum. How long will it take him to double his money?

M5.23 A mother invested $5000 for her daughter for a period of 12 years. It had a maturity value of $7250 at the end. What was the simple interest rate per annum?

M5.24 The director of a company decides to purchase a luxury car that has a price of $80 000. She pays 20% deposit, with the remainder to be paid off monthly as a loan over 6 years at a flat rate of interest of 8.5% per annum.

(a) Calculate the total simple interest she will pay.

(b) Estimate the effective rate of interest.

(c) Calculate the total amount she will pay for the car.

M5.25* An engineer lends a colleague $2000 for a period of 50 days. At that time the colleague repays him $2200. What annual rate of simple interest does the colleague pay? (Assume 365 days in a year.)

M5.26 A solicitor lends a client $4000 for 120 days at a simple interest rate of 5.0% per annum.

(a) How much interest will the solicitor earn?

(b) How much in total will the client pay back to the solicitor?

M5.27* A store advertises a special price of $3600 on its new range of digital televisions. A customer is able to put down a deposit of 10% on 5 January 2002 and agrees to pay off the remainder over the next 37 weeks at a special simple interest rate of 6.5% per annum. (Assume 365 days in a year.)

(a) At what date will the television be paid off?

(b) What total interest will the customer pay?

(c) Calculate the weekly repayments (to the nearest cent).

M5.28 Suppose in Exercise M5.27 that the store offers the customer a deal in which he can take longer to pay off the remainder. The store suggests that he can have until 31 December 2002 and that it will drop the simple interest rate to only 5.0% per annum.

(a) How much will the customer pay to the store each week under this new deal?

(b) What is the total amount that the customer will pay for the television? Is it a deal the customer should accept? Explain.

M5.29* A marketing consultant requires $8000 for a small business venture. She approaches the bank for a 6-month loan, which she will repay on a monthly basis. The bank states that it will require the full amount of interest, quoted as a simple interest rate of 6.5% per annum, in advance.

(a) How much money should she borrow in total?

(b) What will be her monthly repayments?

M5.30 Answer Exercise M5.29 if the consultant wants to borrow $25 000 over 18 months at the same simple interest rate.

M5.31 On 8 January 2002 a mechanic invests $50 000 with a lending institution at a simple interest rate of 7.25% per annum. He wants to invest the money until 8 July 2006, when he will collect the maturity value of the principal.

(a) How much interest will he earn?

(b) What will be the maturity value of the principal?

M5.32* A credit union is offering a loan to be repaid over 15 years with monthly repayments and an effective interest rate of 14.0%. Calculate the equivalent flat rate of interest.

M5.33 A building society is offering a loan to be repaid over 25 years with monthly repayments and an effective interest rate of 17.5%. Calculate the equivalent flat rate of interest.

M5.34* If a promoter borrows $200 000 from a bank to be repaid quarterly over 8 years at a flat interest rate of 6.75%, find:

(a) the total simple interest on the loan

(b) the equivalent effective rate of interest

M5.35 An engineer borrows $75 000 from a credit union, to be repaid half-yearly over 10 years at a flat interest rate of 5.50%. Find:

(a) the total simple interest on the loan

(b) the equivalent effective rate of interest

M5.36* A customer is offered the choice of two different types of deals on a loan of $30 000, to be repaid quarterly over 10 years. The deals are either a flat rate of interest of 4.75% per annum or an effective rate of interest of 9.0%.

(a) Calculate the total amount of interest to be paid under each deal.

(b) Which deal should the customer take? Explain.

M5.37 A hairdresser who wants to borrow $15 000 is offered the choice of two deals, in both of which the loan is to be repaid quarterly over 5 years. The deals are either a flat rate of interest of 6.50% per annum or an effective rate of interest of 12.0%.

(a) Calculate the total amount of interest to be paid under each deal.

(b) Which deal should the hairdresser take? Explain.

M5.38* On 15 January 2001 a schoolteacher opens a savings bank account with a deposit of $150. The table below shows the transactions in the account until the end of October 2001.

Date	Deposits ($)	Withdrawals ($)	Balance ($)
15 January 2001	150		150
12 April 2001	500		650
8 June 2001		325	325
21 September 2001	120		445
15 October 2001		90	355

The bank pays simple interest at a rate of 3.25% per annum based on the minimum monthly balance.

(a) Calculate the minimum monthly balance each month and interest paid each month (to the nearest cent).

(b) Calculate the total interest paid by 31 October 2001.

M5.39 On 24 September 2001 an architect opens a savings bank account with a deposit of $500. The table below shows the transactions in the account until the end of March 2002.

Date	Deposits ($)	Withdrawals ($)	Balance ($)
24 September 2001	500		500
13 December 2001		120	380
23 December 2001		60	320
22 February 2002	250		570
3 March 2002		50	520
17 March 2002	100		620

The bank pays simple interest at a rate of 5.50% per annum based on the minimum monthly balance.

(a) Calculate the minimum monthly balance each month and interest paid each month (to the nearest cent).

(b) Calculate the total interest paid by 31 March 2002.

M5.40* Suppose that a student wants to have $12 000 in 6 years time to purchase a new car. A bank offers her a savings account that pays simple interest at a rate of 5.0% per annum. If she invests $9000 today in the account, will she have enough to purchase the car? Explain.

Chapter M6

Compound interest

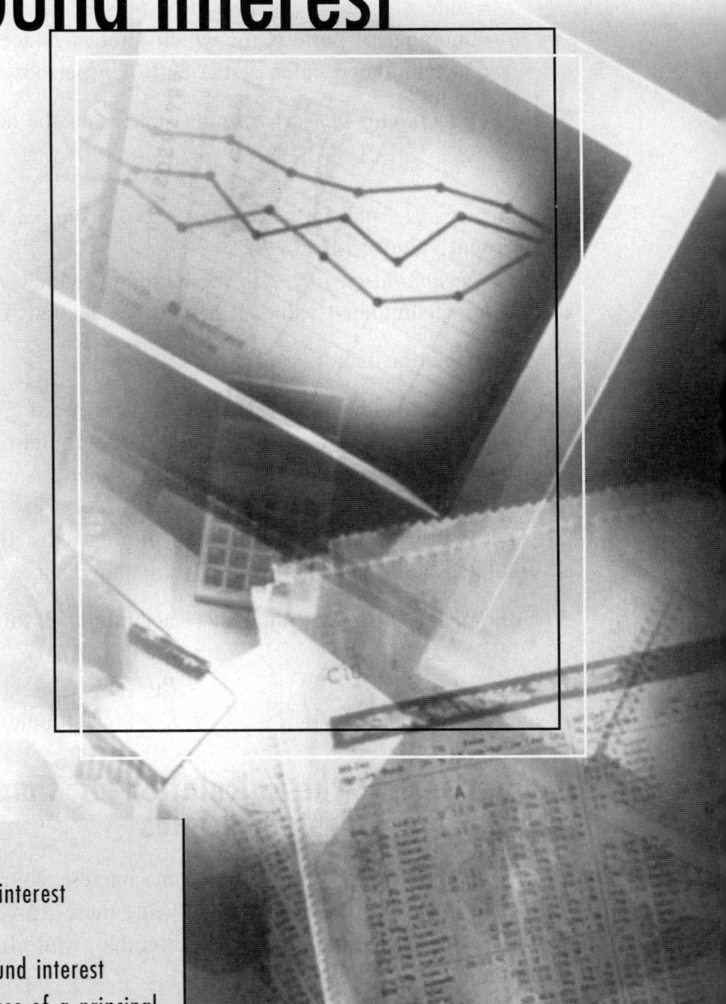

'The darkest hour of any man's life is when he sits down to plan how to get money without earning it.'

(Horace Greely, US newspaper editor and politician, 1811–1872)

Learning objectives

In this chapter you will learn to:

- Distinguish between simple and compound interest
- Calculate compound interest
- Compare calculations of simple and compound interest
- Calculate the present and accumulated values of a principal of money
- Solve problems that involve transposing the compound interest formula

M6.1 INTRODUCTION

Chapter M5 discussed how simple interest is always calculated on the amount of the *original* principal. In this chapter we consider the case in which the interest due is added to the principal at the *end* of each interest period and this interest itself also earns interest from that point onwards. In this case, the interest is said to be *compounded*, and the sum of the original principal plus total interest earned is called the *accumulated value* or *maturity value*. The difference between the accumulated value and original principal is called **compound interest**.

Suppose that we define:

P = principal at the beginning
i = rate of interest per period (expressed as a fraction or decimal)
n = number of periods for which interest is accumulated
S = accumulated value at the end of n periods

Then the *compound interest* formula states that the accumulated value at the end of n periods is:

$$S = P(1 + i)^n \qquad \textbf{(M6.1)}$$

(Note that some financial calculators use the symbol FV (future value) instead of S and the symbol PV (present value) instead of P.)

The *accumulation factor* is the factor by which you multiply the original principal in order to obtain the accumulated value. It may be calculated from:

$$\text{Accumulation factor} = (1 + i)^n \qquad \textbf{(M6.2)}$$

The accumulation factor can be found easily using a calculator equipped with a power key, often labelled y^x, x^y or a^x. However, values of accumulation factors for certain rates of interest are also given in Table 1 on pages M.214–17. Note that the value of the accumulation factor is independent of the value of the beginning principal P.

The actual amount of compound interest earned after n years is the difference between the accumulated value and the original principal. That is:

$$\begin{aligned}\text{Amount of compound interest} &= \text{accumulated value} - \text{original principal}\\ &= P(1 + i)^n - P\\ &= P[(1 + i)^n - 1]\\ &= P \times (\text{accumulation factor} - 1) \qquad \textbf{(M6.3)}\end{aligned}$$

Comparison of the calculation of simple interest and compound interest from first principles

For any given principal P, given the same interest rate i and the same period of an investment or loan, compound interest will always have a value more than simple interest. In this case, from an investor's point of view, compound interest is preferable, while from a borrower's point of view simple interest is preferable.

For example, the amount of *simple interest* earned each year is a constant $P \times i = Pi$. The amount of *compound interest* earned in the *first* year is also Pi. However, the amount of compound interest earned in the *second* year is $Pi(1 + i)$, which is greater than Pi. The amount of compound interest earned in the third year is $Pi(1 + i)^2$, which is also greater than Pi.

This can be illustrated in the general case by using some simple algebra.

$$\text{Amount of compound interest earned in the } k\text{th year} = Pi(1 + i)^{k-1}$$
$$\text{Amount of simple interest earned each year} = Pi$$

Since

$(1 + i)^{k-1} > 1$ for $i > 0$ and $k \geq 1$

it follows that

$Pi\,(1 + i)^{k-1} > Pi$ for $i > 0$ and $k \geq 1$ and $Pi > 0$

which means that the value of compound interest will exceed the value of simple interest for the same values of i and k.

Example M6.1

A market stall owner invests $1000 in an account that pays an interest rate of 8% per annum. The money is to be invested for 3 years.

Find the amount of interest earned each year and the total interest earned after 3 years if the account attracts:

(a) simple interest
(b) compound interest

Solution

(a) Using Equation M5.1, with P = $1000, R = 0.08 and I = 1, the simple interest earned each year is a constant:

$$I = P \times R \times T = \$1000 \times 0.08 \times 1 = \$80$$

Hence, at the end of 3 years, the total *simple* interest earned is:

$$3 \times \$80 = \$240$$

(b) Suppose that compound interest at the annual rate of 8% is added.

At the end of year 1:
Interest added during year 1 = $1000 × 0.08 = $80
New principal at the end of year 1 = $1000 + $80 = $1080

At the end of year 2:
Interest for the second year is calculated on a year-beginning principal of $1080.
Interest added during year 2 = $1080 × 0.08 = $86.40
New principal at the end of year 2 = $1080 + $86.40 = $1166.40

At the end of year 3:
Interest for the third year is calculated on a year-beginning principal of $1166.40.
Interest added during year 3 = $1166.40 × 0.08 = $93.31
New principal at the end of year 3 = $1166.40 + $93.31 = $1259.71

Hence, at the end of 3 years, the total *compound* interest earned is:

$1259.71 − $1000 = $259.71

which, as expected, is larger than the $240 in (a) when simple interest was used.

M6.2 CALCULATION OF COMPOUND INTEREST

Example M6.2

Suppose $8000 is invested at a compound interest rate of 5% per annum. Find the accumulation factor, accumulated value and amount of compound interest earned after:

(a) 3 years
(b) 10 years

Solution

In this case, P = $8000 and $i = 0.05$.

(a) Here $n = 3$. From Equation M6.2:

$$\begin{aligned}\text{Accumulation factor} &= (1 + i)^n \\ &= (1 + 0.05)^3 \\ &= (1.05)(1.05)(1.05) \\ &= 1.157\,625\end{aligned}$$

(Note that this value can be obtained from Table 1 using an interest rate of 5% and $n = 3$.)

From Equation M6.1:

$$\begin{aligned}\text{Accumulated value} = S &= P(1 + i)^n \\ &= \$8000(1.05)^3 \\ &= \$8000 \times 1.157\,625 \\ &= \$9261\end{aligned}$$

From Equation M6.3:

$$\begin{aligned}\text{Amount of compound interest} &= P \times (\text{accumulation factor} - 1) \\ &= \$8000 \times (1.157\,625 - 1) \\ &= \$8000 \times 0.157\,625 \\ &= \$1261\end{aligned}$$

(Note that this could also be found by subtracting the original principal of $8000 from the accumulated value of $9261.)

(b) Here $n = 10$. From Equation M6.2:

$$\begin{aligned}\text{Accumulation factor} &= (1 + i)^n \\ &= (1 + 0.05)^{10} \\ &= (1.05)^{10} \\ &= 1.628\,894\,63\end{aligned}$$

(Note that this value could also be obtained from Table 1 using an interest rate of 5% and $n = 10$.)

From Equation M6.1:

$$\begin{aligned}\text{Accumulated value} = S &= P(1 + i)^n \\ &= \$8000(1.05)^{10} \\ &= \$8000 \times 1.628\,894\,63 \\ &= \$13\,031.16 \text{ (to the nearest cent)}\end{aligned}$$

From Equation M6.3:

$$\begin{aligned}\text{Amount of compound interest} &= P \times (\text{accumulation factor} - 1) \\ &= \$8000 \times (1.628\,894\,63 - 1) \\ &= \$8000 \times 0.628\,894\,63 \\ &= \$5031.16 \text{ (to the nearest cent)}\end{aligned}$$

(Note that this could also be found by subtracting the original principal of \$8000 from the accumulated value of \$13 031.16.)

Example M6.3

Calculate the amount of *simple* interest earned after 10 years on \$8000 at an annual rate of 5%. Compare the answer with that in Example M6.2(b).

Solution

From Equation M5.1, the amount of simple interest is:

$$\begin{aligned}I &= P \times R \times T \\ &= \$8000 \times 0.05 \times 10 \\ &= \$4000\end{aligned}$$

As expected, this is significantly less than the \$5031.16 (found in Example M6.2(b)) that would be earned if the interest were compounded.

In the preceding two examples we have assumed that the interest is compounded *annually*. However, in many instances the interest may be compounded using other time periods, such as semi-annually, monthly, weekly or even daily. This rate, when expressed as a rate per annum, is known as a *nominal rate* of interest.

Specifically, when dealing with nominal rates of interest, we must adjust both the interest rate (i) and number of time periods (n) in our formulae. In particular, the interest rate is divided by the number of periods per year for which the interest is compounded. Also, the number of time periods (n) is now the total number of time periods involved. The technique is demonstrated in Examples M6.4 to M6.7.

Example M6.4

An amount of \$20 000 is invested for 10 years in a bank that pays an interest rate of 6% per annum, compounded semi-annually. Find:

(a) the accumulation factor
(b) the accumulated value after 10 years
(c) the total compound interest earned

Solution

In this case, $P = \$20\,000$. Since the interest is compounded twice per year, the interest rate per 6-month period is:

$$i = \frac{0.06}{2} = 0.03$$

The total number of interest rate periods is:

$$n = 2 \times 10 = 20$$

(a) From Equation M6.2:

$$\begin{aligned}\text{Accumulation factor} &= (1 + i)^n \\ &= (1 + 0.03)^{20} \\ &= (1.03)^{20} \\ &= 1.806\,111\,23\end{aligned}$$

(Note that this value could also be obtained from Table 1 using an interest rate of 3% and $n = 20$.)

(b) From Equation M6.1:

$$\begin{aligned}\text{Accumulated value} = S &= P(1 + i)^n \\ &= \$20\,000(1.03)^{20} \\ &= \$20\,000 \times 1.806\,111\,23 \\ &= \$36\,122.22 \text{ (to the nearest cent)}\end{aligned}$$

(c) From Equation M6.3:

$$\begin{aligned}\text{Total compound interest earned} &= P \times (\text{accumulation factor} - 1) \\ &= \$20\,000 \times (1.806\,111\,23 - 1) \\ &= \$20\,000 \times 0.806\,111\,23 \\ &= \$16\,122.22\end{aligned}$$

(Note that this could also be found by subtracting the original principal of \$20 000 from the accumulated value of \$36 122.22.)

Example M6.5

Solve Example M6.4, but assume that the interest is compounded monthly and that the amount is invested for only 5 years.

Solution

We still have $P = \$20\,000$. Since the interest is compounded monthly (i.e. 12 times per year), the interest rate per monthly period is:

$$i = \frac{0.06}{12} = 0.005$$

The total number of interest rate periods is:

$n = 12 \times 5 = 60$

(a) From Equation M6.2:

$$\begin{aligned}\text{Accumulation factor} &= (1 + i)^n \\ &= (1 + 0.005)^{60} \\ &= (1.005)^{60} \\ &= 1.348\,850\,15\end{aligned}$$

(Note that this value could also be obtained from Table 1 using an interest rate of $\frac{1}{2}$% and $n = 60$.)

(b) From Equation M6.1:

$$\begin{aligned}\text{Accumulated value} = S &= P(1 + i)^n \\ &= \$20\,000(1.005)^{60} \\ &= \$20\,000 \times 1.348\,850\,15 \\ &= \$26\,977.00 \text{ (to the nearest cent)}\end{aligned}$$

(c) From Equation M6.3:

$$\begin{aligned}\text{Total of compound interest} &= P \times (\text{accumulation factor} - 1) \\ &= \$20\,000 \times (1.348\,850\,15 - 1) \\ &= \$20\,000 \times 0.348\,850\,15 \\ &= \$6977.00 \text{ (to the nearest cent)}\end{aligned}$$

(Note that this could also be found by subtracting the original principal of \$20 000 from the accumulated value of \$26 977.00.)

Example M6.6

A company secretary has an investment opportunity in which a lending institution offers her an interest rate of 4.0% compounded quarterly. If she decides to invest an amount of \$6000 under the scheme for 8 years, calculate:

(a) the accumulation factor
(b) the accumulated value after 5 years
(c) the total compound interest earned

Solution

In this case, $P = \$6000$. Since the interest is compounded 4 times per year, the interest rate per 3-month period is:

$$i = \frac{0.04}{4} = 0.01$$

The total number of interest rate periods over 8 years is:

$n = 4 \times 8 = 32$

(a) From Equation M6.2:

$$\begin{aligned} \text{Accumulation factor} &= (1 + i)^n \\ &= (i + 0.01)^{32} \\ &= (1.01)^{32} \\ &= 1.374\,940\,68 \end{aligned}$$

(Note that this value could also be obtained from Table 1 using an interest rate of 1% and $n = 32$.)

(b) From Equation M6.1:

$$\begin{aligned} \text{Accumulated value} = S &= P(1 + i)^n \\ &= \$6000(1.01)^{32} \\ &= \$6000 \times 1.374\,940\,68 \\ &= \$8249.64 \text{ (to the nearest cent)} \end{aligned}$$

(c) From Equation M6.3:

$$\begin{aligned} \text{Total compound interest earned} &= P \times (\text{accumulation factor} - 1) \\ &= \$6000 \times (1.374\,940\,68 - 1) \\ &= \$6000 \times 0.374\,940\,68 \\ &= \$2249.64 \end{aligned}$$

(Note that this could also be found by subtracting the original principal of \$6000 from the accumulated value of \$8249.64.)

Example M6.7

Suppose that the lending institution in Example M6.6 offers the secretary an alternative deal in which she would earn *simple* interest at a rate of 4.6% per annum over the 8 years. Which deal should she accept?

Solution

The better deal will be the one that has the largest accumulated value at the end of 8 years. Under the alternative simple interest deal, with $R = 0.046$, $P = \$6000$ and $T = 8$, the maturity value S using Equation M5.3 is:

$$\begin{aligned} \text{Maturity value} = S &= P(1 + RT) \\ &= \$6000[1 + (0.046 \times 8)] \\ &= \$6000(1.368) \\ &= \$8208.00 \end{aligned}$$

From Example M6.6(b) it was found that under the compound interest deal she would have an accumulated value of \$8249.64. Although there is not much in it, the compound interest deal is slightly better by \$41.64.

M6.3 PRESENT VALUE

In some business situations it is useful to determine what principal P will accumulate to a specified amount S at a given interest rate over n periods. That is, we would like to know how much we must invest today to accumulate a specified amount at some future time. In this case, the unknown quantity is P, which may be found by rewriting Equation M6.1 with P as the subject. That is:

$$P = S(1 + i)^{-n} \qquad \textbf{(M6.4)}$$

(Note that Equation M6.4 is sometimes written as:

$$PV = FV(PVIF\ i,\ n)$$

where PV = present value
FV = future value
$PVIF$ = present value interest factor.)

The value of P is called the *present value* (or *discounted value*) at compound interest of S. The *present value factor* (or *discount factor*) is the amount by which you multiply the specified amount in order to obtain the original principal. It may be calculated from:

$$\text{Present value factor} = (1 + i)^{-n} \qquad \textbf{(M6.5)}$$

The present value factor can also be found easily using a calculator equipped with a power key such as x^y, y^x or a^x. However, values of present value factors for certain rates of interest may be found in Table 2 on pages M.218–21.

Example M6.8

A bank customer wishes to have an amount of \$30 000 at the end of 10 years. The bank pays an interest rate of 8% per annum, compounded annually. How much money will the customer have to invest with the bank now?

Solution

In this case, S = \$30 000, i = 0.08 and n = 10. From Equation M6.5:

$$\begin{aligned}\text{Present value factor} &= (1 + i)^{-n}\\ &= (1 + 0.08)^{-10}\\ &= 0.463\,193\,49\end{aligned}$$

(Note that this value can be obtained from Table 2 using an interest rate of 8% and n = 10.)

From Equation M6.4:

$$\begin{aligned}P &= S(1 + i)^{-n}\\ &= \$30\,000 \times 0.463\,193\,49\\ &= \$13\,895.80 \text{ (to the nearest cent)}\end{aligned}$$

Hence, the customer must invest an amount of \$13 895.80 now to accumulate the specified amount of \$30 000 at the end of 10 years.

Note: You can check your answer using Equation M6.1. That is, you can verify that at the end of 10 years an initial investment of \$13 895.80 will have grown to \$30 000 with an interest rate of 8% per annum, compounded annually.

Example M6.9

Consider the problem in Example M6.8, but now assume that the bank pays an annual interest rate of 8% compounded *monthly* and that the customer wishes to have an amount of \$30 000 at the end of 5 years. How much will the customer have to invest in this case?

Solution

We still have S = \$30 000. Since the interest is compounded monthly (i.e. 12 times per year), the interest rate per monthly period is:

$$i = \frac{0.08}{12} = 0.006\,666\,667 \quad \left(\text{or } \frac{2}{3}\% \text{ per month}\right)$$

The total number of interest rate periods is:

$$n = 12 \times 5 = 60$$

From Equation M6.5:

$$\begin{aligned}\text{Present value factor} &= (1 + \text{i})^{-n} \\ &= (1 + 0.006\,666\,667)^{-60} \\ &= 0.671\,210\,44\end{aligned}$$

(Note that this value can be obtained from Table 2 using an interest rate of $\frac{2}{3}\%$ and n = 60.)

From Equation M6.4:

$$\begin{aligned}P &= S(1 + i)^{-n} \\ &= \$30\,000 \times 0.671\,210\,44 \\ &= \$20\,136.31 \text{ (to the nearest cent)}\end{aligned}$$

Hence, the customer must invest an amount of \$20 136.31 to accumulate the specified amount of \$30 000 at the end of 5 years.

Calculation of the interest rate

The compound interest formula in Equation M6.1 can also be used to find the interest rate charged when a principal of P has accumulated to an amount S after n periods. To do this we must make i the subject of the formula in Equation M6.1, as follows:

$$S = P(1 + i)^n$$

Divide both sides by P and transpose the sides:

$$(1 + i)^n = \frac{S}{P}$$

Take the nth root of both sides (i.e. raise both sides to the power of $\frac{1}{n}$):

$$(1 + i) = \left(\frac{S}{P}\right)^{\frac{1}{n}}$$

Subtract 1 from both sides:

$$i = \left(\frac{S}{P}\right)^{\frac{1}{n}} - 1 \qquad \textbf{(M6.6)}$$

where $\left(\frac{S}{P}\right)^{\frac{1}{n}}$ is the nth root of $\frac{S}{P}$.

The nth root may be found using a calculator equipped with a power key such as x^y, y^x or a^x. Note that the value of i obtained is the interest rate *per period*. To obtain the nominal rate of interest (per annum) we must multiply this value of i by the number of periods in a year. For example, if interest is compounded *quarterly*, the obtained value of i should be multiplied by 4; if *monthly*, it should be multiplied by 12.

Example M6.10

Suppose that \$1000 has amounted to \$1460 in 20 years with interest compounded quarterly. What annual rate of compound interest was used?

Solution

In this case, a period was a quarter and we have:

P = \$1000, S = \$1460 and $n = 20 \times 4 = 80$.

Using Equation M6.6:

$$\begin{aligned} i &= \left(\frac{1460}{1000}\right)^{\frac{1}{80}} - 1 \\ &= (1.46)^{\frac{1}{80}} - 1 \\ &= 1.004\,74 - 1 \\ &= 0.004\,74 \end{aligned}$$

Since the interest was compounded quarterly, we multiply this value of i by 4 to obtain $0.004\,74 \times 4 = 0.018\,96$. This corresponds to an interest rate of 1.896% per annum.

Example M6.11

Suppose that \$200 has amounted to \$310 in 10 years with interest compounded semi-annually. What annual rate of compound interest was used?

Solution

In this case, a period was 6 months and we have:

P = \$200, S = \$310 and $n = 10 \times 2 = 20$.

Using Equation M6.6:

$$\begin{aligned} i &= \left(\frac{310}{200}\right)^{\frac{1}{20}} - 1 \\ &= (1.55)^{\frac{1}{20}} - 1 \\ &= 1.002\ 215 - 1 \\ &= 0.002\ 215 \end{aligned}$$

Since the interest was compounded semi-annually, we multiply this value of i by 2 to obtain $0.002\ 215 \times 2 = 0.004\ 43$. This corresponds to an interest rate of 0.443% per annum.

Calculation of the number of periods

The compound interest formula in Equation M6.1 can also be used to find the number of periods (n) required for an amount of P to accumulate to an amount of S when interest is at a rate of i per period. To do this we must make n the subject of the formula in Equation M6.1, as follows:

$$S = P(1 + i)^n$$

Take logarithms of both sides (any base, e.g. base 10, may be used):

$$\log S = \log P(1 + i)^n$$

Using the properties of logarithms we can write:

$$\log S = \log P + \log (1 + i)^n \qquad \text{(using Equation M1.9)}$$

$$\log S = \log P + n\log (1 + i) \qquad \text{(using Equation M1.11)}$$

Subtract $\log P$ from both sides of the equation (and transpose the sides):

$$n\log (1 + i) = \log S - \log P$$

Using the properties of logarithms:

$$n\log (1 + i) = \log\left(\frac{S}{P}\right) \qquad \text{(using Equation M1.10)}$$

Divide both sides by $\log (1 + i)$:

$$n = \frac{\log\left(\frac{S}{P}\right)}{\log(1 + i)} \qquad \textbf{(M6.7)}$$

The values of the logarithms may be found by using the *log* key on your calculator.

Care must be taken with the value of i when using Equation M6.7. That is, the value of i must be written as the interest rate *per period*. Note also that the value of n obtained will not always be a whole number and so we can only approximate the actual number of periods required. To obtain the number of *years* required, we divide the value of n by the number of periods in a year.

Example M6.12

Find how long it will take for \$800 to amount to \$1500 if interest is at 5% per annum compounded quarterly.

Solution

In this case, a period is a quarter and we have:

$$P = \$800,\ S = \$1500 \text{ and } i = \frac{0.05}{4} = 0.0125.$$

Using Equation M6.7:

$$\begin{aligned} n &= \frac{\log\left(\frac{1500}{800}\right)}{\log\ (1 + 0.0125)} \\ &= \frac{\log\ 1.875}{\log\ 1.0125} \\ &= \frac{0.2730}{0.0054} \\ &= 50.56 \end{aligned}$$

Hence, the number of periods required is 50.56. Since in this case a period is a quarter, the amount of time is approximately equal to $\frac{50.56}{4} = 12.64$ years.

Example M6.13

A TAFE graduate is anxious to purchase a used car, for which he estimates that he will need a total of \$8000. He currently has \$6200 and is offered two different investment plans, A and B. Under Plan A he would be paid simple interest at an annual rate of 5.4%, and under Plan B he would be paid an annual interest rate of 4.8% compounded monthly.

Under each plan, how long would it be before he reached his target of \$8000?

Solution

Under Plan A he has $P = \$6200$, $R = 0.054$ and $S = \$8000$. Using Equation M5.3 (where T = the unknown time in years):

$$\begin{aligned} S &= P(1 + RT) \\ \$8000 &= \$6200(1 + 0.054T) \\ \frac{\$8000}{\$6200} &= 1 + 0.054T \\ 1 + 0.054T &= 1.290\ 322\ 58 \\ T &= \frac{0.290\ 322\ 58}{0.054} \\ &= 5.376 \text{ years} \end{aligned}$$

Alternatively, you could rewrite Equation M5.3 with T as the subject. This yields:

$$T = \frac{S - P}{PR}$$

$$= \frac{\$8000 - \$6200}{\$6200 \times 0.054}$$

$$= 5.376 \text{ years (as before)}$$

Under Plan B he has $P = \$6200$, $S = \$8000$ and $i = \frac{0.048}{12} = 0.004$.
The unknown time is n (in months).
Using Equation M6.7:

$$n = \frac{\log\left(\frac{800}{6200}\right)}{\log\ (1 + 0.004)}$$

$$= \frac{\log\ 1.290\ 322\ 58}{\log\ 1.004}$$

$$= \frac{0.110\ 698\ 297}{0.001\ 733\ 713}$$

$$= 63.32 \text{ months}$$

$$= 5.32 \text{ years}$$

Hence, there is very little between the two plans (0.056 years or about 20 days), although Plan B would technically be the quicker.

M6.4 REMARKS

As is the case with simple interest, calculations involving compound interest formulae can also be performed with the aid of **financial calculators**, which often have inbuilt functions for this purpose. Alternatively, if you have a programmable calculator you may be able to program the formulae yourself to save on some of the tedious calculations.

Computer *spreadsheets* can also be used to perform compound interest calculations, and to consider quantities such as the effects of different rates of interest and make a comparison with simple interest values. There are a number of such spreadsheets available. One of the more popular is Microsoft Excel for Windows, which is discussed in Appendix C at the end of the statistics section on page S.436.

Quick Quiz

1. An amount of \$8000 is invested at an interest rate of 5% per annum, compounded annually. Find the amount of compound interest earned after 6 years.
2. An amount of \$50 000 is invested for 5 years in a bank that pays an interest rate of 7% per annum, compounded semi-annually. Find the accumulation factor and the accumulated value after 5 years.
3. How much should be deposited now into an 8% per annum interest-bearing account, compounded quarterly, so that an amount of \$25 000 will be accumulated in 6 years?
4. A customer wishes to have an amount of \$5000 at the end of 5 years. The bank pays an interest rate of 6% per annum, compounded annually. How much money must the customer invest in the bank now?
5. How long will it take for \$2500 to amount to \$3500 if interest is at 5% per annum, compounded quarterly?
6. Suppose that \$20 000 amounts to \$44 161 in 10 years with interest compounded quarterly. What annual rate of compound interest has been used?
7. Calculate (to two decimal places) the annual interest rate required for an investment of \$1000 to have an accumulated value of \$1600 in 5 years if interest is compounded quarterly.
8. A gambler who won \$1 million at a casino wants to invest the winnings for 5 years. The two choices available are an account (*X*) that pays simple interest at an annual rate of 5.5%, and an account (*Y*) that pays interest compounded monthly at an annual rate of 5.0%. Which account should the gambler choose to maximise the amount of interest paid?
9. A pilot wants to invest \$5000 for a period of 3 years. Plan A offers simple interest at an annual rate of 8.0%, while Plan B offers an interest rate of 2.7% compounded annually. Which plan should the pilot accept? Explain.
10. Explain why your answer to Question 9 would be the same for *any* amount of money that the pilot wants to invest.

Multiple Choice Questions

1. A lawyer invests \$13 000 at an interest rate of 4% per annum, compounded annually. The accumulated value (to the nearest \$) earned after 5 years is:
(a) \$15 260 (b) \$15 500 (c) \$15 816
(d) \$15 600 (e) none of the above

2. In Question 1, the amount of compound interest earned (to the nearest dollar) after 5 years is:
(a) \$2260 (b) \$2816 (c) \$2500
(d) \$2600 (e) none of the above

3. An amount of \$75 000 is invested for 8 years with a bank that pays an interest rate of 8% per annum, compounded semi-annually. The accumulated value (to the nearest dollar) after 6 years is:
(a) \$120 077 (b) \$123 000 (c) \$140 474
(d) \$111 000 (e) none of the above

4. If a university lecturer wants an amount of \$100 000 at the end of 20 years, and the bank pays an annual interest rate of 6% compounded monthly, how much (to the nearest dollar) must the lecturer invest with the bank now?
(a) \$47 169 (b) \$90 506 (c) \$30 209
(d) \$50 000 (e) none of the above

5. A garage mechanic is retiring in 5 years and wants to have \$12 000 put aside to take his family on a round-the-world cruise. The manager of a financial institution offers him an account that will pay interest compounded quarterly, and says that the mechanic must pay \$10 081 now in order to have \$12 000 at the end of 5 years. The annual interest rate the manager is offering is:
(a) 2.00% (b) 1.75% (c) 3.00%
(d) 3.50% (e) none of the above

6. A schoolteacher invests \$750 in an interest-bearing account that pays 5% per annum, compounded monthly. To the nearest month, how many months will it take for the \$750 to amount to \$1500?
(a) 150 (b) 167 (c) 133
(d) 142 (e) none of the above

7. The *accumulated value* of an investment is:
(a) the value of the principal at the beginning of the loan
(b) the number of periods over which the investment is made
(c) the sum of the original principal plus total interest earned
(d) the accumulated interest over the period of the loan
(e) none of the above

8. The *present value* of an investment is:
(a) the value of the investment on maturity
(b) the value of the investment on maturity plus any interest
(c) the total interest earned to the present time
(d) the current value of the investment divided by the interest rate
(e) none of the above

9. Annie has \$80 to invest in a savings account for 4 years at an annual compounded rate of 8%. Her accumulated value at the end of 4 years is:
(a) \$105.60 (b) \$832.00 (c) \$108.33
(d) \$352.00 (e) none of the above

10. Xanthe deposits \$8000 in her savings account that pays 6% interest, compounded quarterly. The total amount of interest that she will earn after 5 years is:
(a) \$2775.20 (b) \$2400.00 (c) \$2705.80
(d) \$960.00 (e) none of the above

Exercises

M6.1* An amount of \$5000 is invested at an interest rate of 4% per annum, compounded annually. Find the accumulation factor, the accumulated value and total amount of compound interest earned after:

(a) 2 years (b) 5 years

M6.2 Answer the questions in Exercise M6.1 if the amount invested is \$12 000 and the interest rate is 6% per annum, compounded annually.

M6.3* An amount of \$16 000 is invested for 5 years in a credit union that pays an annual interest rate of 5%, compounded semi-annually. Find:

(a) the accumulation factor

(b) the accumulated value after 5 years

(c) the total compound interest earned

M6.4 Answer the questions in Exercise M6.3 if an amount of \$25 000 is invested for 4 years.

M6.5* Answer the questions in Exercise M6.3 assuming that the interest is compounded *monthly.*

M6.6 A building society investor wishes to have an amount of \$8000 at the end of 6 years. The building society pays an interest rate of 9% per annum, compounded annually. How much money must the customer invest with the building society now?

M6.7* Answer the question in Exercise M6.6 assuming that the interest is compounded *monthly.*

M6.8 Answer the question in Exercise M6.6 assuming that the interest is compounded *quarterly.*

M6.9* A retired police officer invests \$5000 in a credit union that pays an interest rate of 6% per annum, compounded quarterly. Exactly 2 years later she invests a further \$6000 with the credit union under the same conditions. How much total interest will she have earned 5 years after the original investment?

M6.10 Answer Exercise M6.9 assuming that interest is compounded monthly.

M6.11* The postmaster wants to retire on his 65th birthday with \$100 000 from an investment he is about to make in his bank on his 45th birthday. If the bank pays an interest rate of 6% per annum, compounded semi-annually, how much must the postmaster invest now?

M6.12* How long will it take for \$3000 to amount to \$5000 if interest is at 4% per annum, compounded quarterly?

M6.13 Answer Exercise M6.12 assuming that interest is compounded annually.

M6.14* Suppose that \$5000 amounts to \$6352 in 12 years with interest compounded quarterly. What annual rate of compound interest has been used?

M6.15 Suppose that \$2500 amounts to \$3750 in 6 years with interest compounded monthly. What annual rate of compound interest has been used?

M6.16 Suppose that you had the choice of the following two investments on \$3000 over the next 5 years:

(a) a simple interest rate of 8% per annum

(b) an interest rate compounded quarterly at the rate of 7% per annum.

Which of the two investments would you choose? Give reasons.

M6.17* Susan wants to buy a horse costing $1400 on her 15th birthday, 3 years and 6 months from now. Her parents have only $800 to invest on her behalf in an account that will pay interest, compounded semi-annually. What annual interest rate is required for Susan to be able to buy her horse?

M6.18 In Exercise M6.17, suppose that interest is compounded monthly. What annual interest rate is required now?

M6.19* Ray is going to inherit $25 000 in 10 years time. What is the present value of the money if 5% interest (compounded annually) is added to achieve this figure?

M6.20 On 1 July 2001, Jason invested $30 000 with the Dodgy Investment Company, which promised him interest compounded quarterly at an annual rate of 4.5%. When Jason goes to close his account on 31 December 2005, how much should he receive?

M6.21* When Jennifer was born her father put aside $3000 in an account for her to receive on her 16th birthday. If the account pays 6% per annum, compounded semi-annually, how much should Jennifer receive?

M6.22 Find the present value of the following accumulated values:

(a) $6000 for 4 years at 6% per annum interest, compounded monthly

(b) $3000 for 2 years at 2% per annum interest, compounded semi-annually

(c) $10 000 for 20 years at 9% per annum, compounded annually

M6.23 What amount of principal would be required as an investment over 3 years to receive:

(a) $25 000 at 5.0% annual interest compounded quarterly?

(b) $100 000 at 3.5% annual interest compounded monthly?

M6.24* A grocery store manager has $50 000 to invest for a period of 5 years, and her bank offers a choice of two investment schemes, Plan A and Plan B. Under Plan A she would receive an annual interest rate of 8.30% compounded quarterly, while under Plan B she would receive an annual interest rate of 8.25% compounded monthly.

(a) Calculate the amount of interest she would earn under each plan.

(b) Which plan should she choose? How much extra interest would she earn over the other plan?

M6.25 Suppose that the bank in Exercise M6.24 offers the manager a new option of Plan C, in which she would receive an interest rate of 8.60% compounded annually. Should she choose this plan instead of your choice in Exercise M6.24(b)? Explain.

M6.26 Suppose that a rival bank offers the manager in Exercise M6.24 a scheme in which she would receive simple interest at a flat rate of 10.25% per annum. Is this deal attractive when compared to the three compound-interest plans offered in Exercises M6.24 and M6.25? Give reasons.

M6.27* Although he does not get paid his weekly salary until 4.00 p.m. on Friday, Rupert has a gambling debt of $400 that must be paid at 10.00 a.m. on the day before. His friend Harry agrees to lend Rupert the money at 10.00 a.m. on Thursday at an interest rate of 25% per day (24-hour period), compounded hourly, until 4.00 p.m. the following day when it will be repaid.

(a) How much interest will Rupert have to pay Harry for such a deal?

(b) If Rupert is paid an after-tax weekly salary of $550 in cash, is he going to have trouble paying Harry at 4.00 p.m. on Friday? Explain.

M6.28 In Exercise M6.27, as an alternative to using Harry, Rupert could borrow the $400 from his neighbour Mary, who would charge him a simple interest rate of 1.20% *per hour*. Is Mary's offer more attractive than Harry's? Explain.

M6.29* Calculate the annual interest rate (to one decimal place) required for:

(a) an investment of $3000 to have an accumulated value of $3500 in 3 years, if interest is compounded *monthly*

(b) an investment of $10 000 to have an accumulated value of $15 000 in 6 years, if interest is compounded *quarterly*

(c) an investment of $200 000 to have an accumulated value of $240 000 in 5 years, if interest is compounded *annually*

M6.30 Vivian wants to invest some money in a special account at her local bank so that she will have an accumulated value of $5000 on 1 August 2004, just in time for her trip to the Athens 2004 Olympic Games. The bank is offering a deal in which she will be paid interest at the rate of 6% per annum, compounded monthly. If Vivian makes her investment on 1 February 2002, what amount must she invest to reach her target?

M6.31* Glenn invested $300 in an account that paid interest compounded semi-annually. After 5 years the investment had an accumulated value of $390. What annual rate of interest was Glenn paid?

M6.32 In Exercise M6.31, Glenn discovers that his interest was in fact compounded monthly, and not semi-annually as he first thought. In this case, what annual rate of interest was Glenn paid?

M6.33* The Shady Bank is offering 'unbeatable deals' on its savings accounts, where it will pay investors interest that is compounded quarterly at an annual rate of 4%. Shannon wants to save $10 000 to purchase a car, but has only $8000 to invest in the savings account on 1 June 2001. Under the bank's deal, on approximately what date will Shannon reach her target?

M6.34 Answer Exercise M6.33 if the bank pays Shannon interest at an annual rate of 6% compounded monthly.

M6.35 Rank the following interest-rate deals in order of merit (best to worst) for a customer who wants to invest $10 000 for a 5-year period:

(a) interest compounded annually at an annual rate of 4.85%

(b) interest compounded quarterly at an annual rate of 4.6%

(c) interest compounded semi-annually at an annual rate of 4.8%

(d) interest compounded monthly at an annual rate of 4.5%

(e) simple interest of 5.1% per annum

M6.36 In Exercise M6.35, would your rankings change if the customer had $1 million to invest? Explain.

M6.37 George and Jim were brothers who were each left a will inheritance of $20 000. George invested his money in an account for 5 years at an annual interest rate of 6.0% compounded quarterly. Jim invested his money for 5 years at an annual simple interest rate of 6.75%. At the end of that time they agreed to pool their accumulated principals into one account for a further 5 years with an annual interest rate of 5.0% compounded monthly.

(a) How much accumulated principal did George and Jim each have after 5 years? Hence, determine how much money they had to pool.

(b) What was the accumulated principal of their pooled money at the end of the 10-year period?

M6.38 In Exercise M6.37, when George and Jim first received their inheritance, the executor of the will suggested that they pool their resources immediately for 10 years and invest in an account that paid an annual interest rate of 6.75%, compounded monthly. In view of your answer to Exercise M6.37(b), should they have taken this advice? Explain.

M6.39* Mr S. L. Eaze opens a shopfront business where he will lend money to anyone (without security) if they are willing to pay his interest rate of 2% per week compounded daily. The first customer borrows \$5000 to buy a used car, but is unable to pay Mr Eaze back for 12 weeks. How much will the customer have to pay back at this time?

M6.40 In Exercise M6.39, assuming that there are 365 days in a year, what annual simple interest rate is equivalent to that charged by Mr Eaze?

M6.41* Jacqui invests \$2000 for 8 years in an account that for 5 years pays interest at a rate of 5%, compounded semi-annually, and then for the final 3 years pays interest at a rate of 6%, compounded quarterly. What is the accumulated value after 8 years?

M6.42 In Exercise M6.41, Jacqui had the option of investing her \$2000 for the entire 8 years at an interest rate of 5.6%, compounded annually. Should she have chosen this option? Explain.

Chapter M7

Annuities

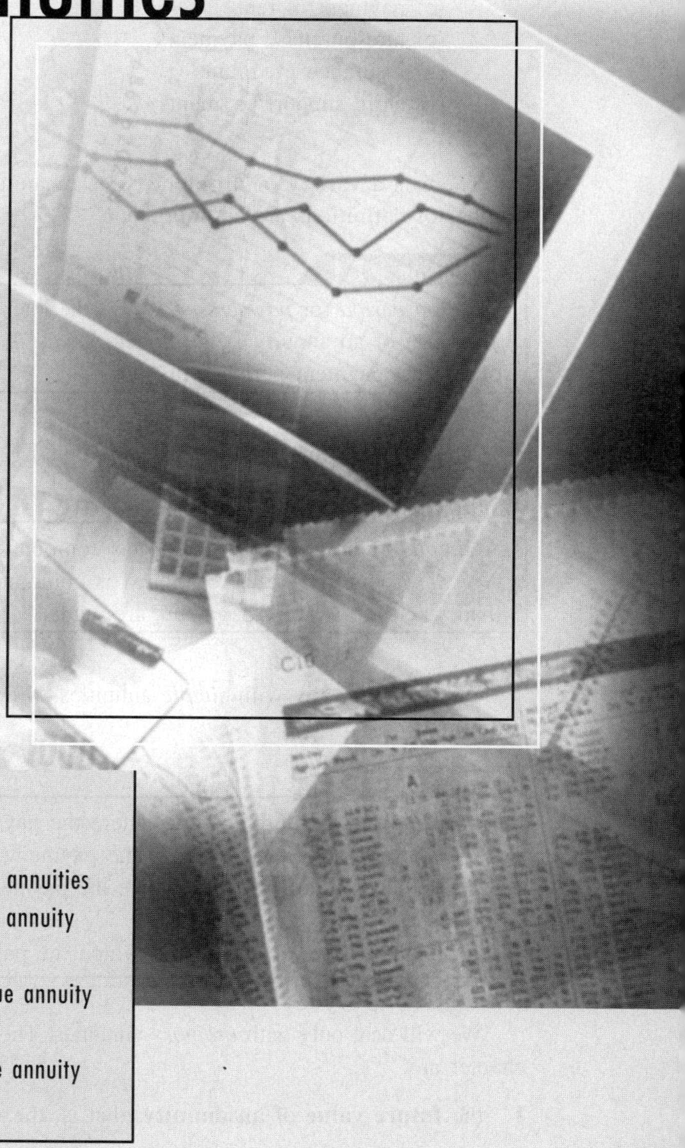

'I don't like money actually,

but it quiets my nerves.'

(Joe Louis, American boxer, 1914–1981)

Learning objectives

In this chapter you will learn to:

- Understand and apply annuities
- Distinguish between future and present value of annuities
- Solve problems involving the future value of an annuity
- Calculate the present value of an annuity
- Calculate the periodic payment of a present value annuity (amortisation)
- Calculate the periodic payment of a future value annuity (sinking fund)

M7.1 INTRODUCTION

An **annuity** is a series of (usually equal) payments, made at equal intervals of time. In everyday life many Australians are involved in either the payment or receipt of annuities at intervals of time, which may be weekly, fortnightly, monthly, semi-annually or annually.

Example M7.1

Some examples of annuity payments are:

(a) mortgage payments
(b) any loan repayment
(c) payment of rent
(d) Austudy allowance
(e) hire-purchase payments
(f) dividend payments
(g) insurance premiums
(h) interest payments on bonds
(i) child support payments
(j) superannuation payments

There is a variety of different types of annuities, and before we begin discussion of them a number of definitions are appropriate.

Definition M7.1

The *rent period* (or *payment period*) is the interval of time between the payments of an annuity. The *term* of an annuity is the time from the beginning of the first payment period to the end of the last payment period.

The two types of annuities are *simple* and *general*.

Definition M7.2

A *simple* annuity is where interest is compounded at the same times as the annuity payments. A *general* annuity is where interest is compounded at times that are either greater or smaller than when the annuity payments are made.

We will deal only with *simple* annuities, which themselves can be classified into four types, as follows.

Definition M7.3

1. An **ordinary annuity** is one where the payments are made at the *end* of each period.
2. An **annuity due** is one where the payments are made at the *beginning* of each period.
3. A **deferred annuity** is one where the payments do not commence until a period of time has elapsed.
4. A **perpetuity** is an annuity in which the payments continue indefinitely.

We will deal only with *ordinary* annuities. The aspects of these annuities that are discussed in this chapter are:

1. the **future value of an annuity**; that is, the value of the annuity at the end of its term

2. the **present value of an annuity**; that is, the value of the annuity at its beginning
3. **amortisation**; that is, the making of periodic payments to repay a debt (including the principal and interest)
4. **sinking fund**; that is, the periodic payments necessary to realise a given sum of money in the future

Each of the above aspects will be discussed in some detail. You will see that there are a number of formulae that require calculations involving numbers that are to be raised to powers (some of which are negative). These calculations can be performed with the aid of a calculator that has a power key x^y, y^x or a^x. Alternatively, you will be shown how to use the tables on pages M.213–M.237 that will make the arithmetic in these problems considerably easier.

You will also notice that some of the values in these calculations involve numbers that are taken to eight decimal places in the tables on pages M.213–M.237. This may seem an excessive number of decimal places, but in some cases these numbers are multiplied by large numbers and using a small number of decimal places could lead to rounding errors.

M7.2 FUTURE VALUE OF AN ANNUITY

The *future value* (or *accumulated value*) of an annuity is the amount due at the end of the term. That is, it is the sum of all the periodic payments made and interest accrued up to and including the final payment period. The future value of an annuity may be calculated by repeated application of the compound interest formula given in Equation M6.1. The idea is illustrated in Example M7.2.

Example M7.2

Consider an annuity in which 4 annual payments each of $200 are made into an account where the compound interest rate is 5% per annum. If the payments are made at the end of each year, find the future value of the annuity. (That is, find the balance in the account at the end of the fourth year.)

Solution

The calculations may be made as shown in Table M7.1.

Table M7.1 Calculations for Example M7.2

Year	Balance at beginning of year ($)	Interest on balance ($)	Annual payment ($)	Total at end of year ($)
1	0.00	0.00	200.00	200.00
2	200.00	10.00	200.00	410.00
3	410.00	20.50	200.00	630.50
4	630.50	31.53	200.00	862.03

The 'interest on balance' column is found by calculating 1 year's interest on this balance at the beginning of the year at the given interest rate of 5%. For example, at the beginning of year 3 the balance was $410. Thus, we had P = $410, i = 0.05 and n = 1 (since we are considering only 1 year at a time). The interest for that year was then $\$410 \times 0.05 \times 1 = \$20.50$. This is then added to the original balance ($410) and the payment for year 3 ($200), to obtain the total at the end of year 3 of $630.50.

From Table M7.1 we see that the future value of this annuity at the end of the fourth year is $862.03.

Calculations such as those in Table M7.1 can become very tedious if a large number of payment periods are involved. For this reason, the future value of an annuity can be more easily found with the aid of a formula. Suppose we now define:

R = amount of the annuity payment made per period
n = total number of payments (i.e. compounding periods)
i = interest rate per payment period (i.e. the periodic interest rate)
S = future value of the annuity (i.e. total payments plus accrued interest)

It can be shown that the future value of an annuity is:

$$S = R \times \frac{(1 + i)^n - 1}{i} \qquad \textbf{(M7.1)}$$

Because of its importance, part of Equation M7.1 is given a special notation. It is:

$$s_{\overline{n}|i} = \frac{(1 + i)^n - 1}{i} \qquad \textbf{(M7.2)}$$

The value of $s_{\overline{n}|i}$ (or s_n at i%) may be thought of as the future value at an interest rate of i% of $1 paid at the end of each period for n periods. By combining Equations M7.1 and M7.2, the future value of an annuity may be found from:

$$S = R \times s_{\overline{n}|i} \qquad \textbf{(M7.3)}$$

(Note that Equation M7.3 is sometimes written as:

$FVA = R(FVIFA\ i, n)$

where FVA = future value of an annuity
$FVIFA$ = future value interest factor.)

The quantity $s_{\overline{n}|i}$ may be found using a calculator equipped with a power key function of x^y, y^x or a^x. However, values of $s_{\overline{n}|i}$ for certain interest rates are also in Table 3 on pages M.222–8.

Example M7.3

Solve the problem in Example M7.2 with the aid of Table 3.

Solution

In this case, $R = \$200$, $i = 0.05$ and $n = 4$. From Table 3, using an interest rate of 5% and $n = 4$:

$$s_{\overline{n}|i} = s_{\overline{4}|0.05} = \frac{(1 + 0.05)^4 - 1}{0.05}$$

$$= 4.310\,125\,00$$

Hence, from Equation M7.3, the future value of the annuity is:

$$\begin{aligned} S &= R \times s_{\overline{n}|i} \\ &= 200 \times s_{\overline{4}|0.05} \\ &= \$200 \times 4.310\,125\,00 \\ &= \$862.03 \end{aligned}$$

which agrees with the answer obtained in Example M7.2.

If an interest rate is quoted as a rate *per annum* but the interest is *not* compounded annually, then the values of n and i in Equations M7.1 to M7.3 need special attention. Suppose that interest is compounded k times per annum. Then we let:

$$n = \text{number of interest periods} = k \times (\text{number of years})$$

$$i = \text{interest rate per period} = \frac{\text{interest rate per annum}}{k}$$

Example M7.4

A customer deposits \$250 every 3 months into a building society account that pays interest at a rate of 8% per annum convertible quarterly (i.e. compounded every 3 months). How much money will be in the account at the end of 10 years?

Solution

In this case, R = \$250. Since the interest is compounded *quarterly* (i.e. k = 4 times per year), the interest rate per quarterly period is:

$$i = \frac{0.08}{4} = 0.02 \quad \text{(or 2.0\% per quarter)}$$

The total number of interest periods is:

$$n = 4 \times 10 = 40$$

From Table 3, using an interest rate of 2% and n = 40:

$$\begin{aligned} s_{\overline{40}|0.02} &= \frac{(1 + 0.02)^{40} - 1}{0.02} \\ &= 60.401\,983\,18 \end{aligned}$$

Hence, from Equation M7.3, the future value of the annuity is:

$$\begin{aligned} S &= R \times s_{\overline{n}|i} \\ &= \$250 \times s_{\overline{40}|0.02} \\ &= \$250 \times 60.401\,983\,18 \\ &= \$15\,100.50 \quad \text{(to the nearest cent)} \end{aligned}$$

Hence, the customer will have \$15 100.50 in the account at the end of 10 years.

M7.3 PRESENT VALUE OF AN ANNUITY

The *present value* (or *discounted value*) of an annuity is its value at the beginning of the initial rent period. That is, it is the sum of the compound present values of all the payments one period before the initial payment.

Example M7.5

A business owns a contract by which it will receive \$5000 at the end of each year for the next 4 years, and this money will earn 6% per annum compounded annually. Find the present value of this annuity at the beginning of the first rent period, that is 1 year before the first payment is made to the business.

Solution

The rent period in this case is 1 year. The present value may be found by calculating the sum of the individual compound present values of all the payments 1 year before the first payment. In this case, R = \$5000 and $i = 0.06$, so Equation M6.4 can be used (with $S = R$ = \$5000) to calculate the present values (P) of each payment for $n = 1, 2, 3$ and 4. These calculations are shown in Table M7.2.

Table M7.2 Present value calculations for Example M7.5

Payment number	Present value of payment at beginning of first rent period (\$)
1	$5000(1 + 0.06)^{-1} = 4\,716.98$
2	$5000(1 + 0.06)^{-2} = 4\,449.98$
3	$5000(1 + 0.06)^{-3} = 4\,198.10$
4	$5000(1 + 0.06)^{-4} = 3\,960.47$
	Total 17 325.53

From Table M7.2, the present value of this annuity (i.e. 1 year before the first payment is made) is \$17 325.53.

As before, calculations such as those in Table M7.2 can become quite tedious if a large number of payments are involved. Consequently, the present value (A) of an annuity can be found more easily with the aid of the formula in Equation M7.4:

$$A = S(1 + i)^{-n} \tag{M7.4}$$

Combining Equations M7.1 and M7.4, it can also be shown that:

$$A = R \times \frac{[1 - (1 + i)^{-n}]}{i} \tag{M7.5}$$

Because of its importance, part of Equation M7.5 is given a special notation:

$$a_{\overline{n|}\,i} = \frac{1 - (1 + i)^{-n}}{i} \tag{M7.6}$$

The value of $a_{\overline{n}|i}$ (or a_n at $i\%$) may be thought of as the present value at an interest rate of $i\%$ per period of \$1 paid at the end of each period for n periods. In other words, the value of $a_{\overline{n}|i}$ may be considered as the present value of an n-year annuity, when payments are made at the end of intervals, of \$1 per annum discounted at a compound annual interest rate of $i\%$.

By combining Equations M7.5 and M7.6, the present value (A) of an annuity may be found from:

$$A = R \times a_{\overline{n}|i} \qquad \textbf{(M7.7)}$$

The quantity $a_{\overline{n}|i}$ may be found using a calculator equipped with the power key x^y, y^x or a^x. However, values of $a_{\overline{n}|i}$ for certain rates of interest are also given in Table 4 on pages M.229–35.

Example M7.6

Solve the problem in Example M7.5 with the aid of Table 4.

Solution

In this case, R = \$5000, $i = 0.06$ and $n = 4$. From Table 4, using an interest rate of 6% and $n = 4$, we have:

$$\begin{aligned} a_{\overline{4}|0.06} &= \frac{1 - (1 + 0.06)^{-4}}{0.06} \\ &= 3.465\,105\,61 \end{aligned}$$

Hence, from Equation M7.7, the present value of the annuity is:

$$\begin{aligned} A &= R \times a_{\overline{n}|i} \\ &= \$5000 \times a_{\overline{4}|0.06} \\ &= \$5000 \times 3.465\,105\,61 \\ &= \$17\,325.53 \quad \text{(to the nearest cent)} \end{aligned}$$

which agrees with the answer in Example M7.5.

Example M7.7

A student took out a loan from a credit union in order to purchase a home computer. She was to repay the loan in monthly instalments of \$120 for 5 years. Calculate the present value of these repayments if the interest rate was 9%, convertible monthly (i.e. compounded every month).

Solution

In this case, R = \$120. Since the interest is compounded monthly (i.e. 12 times per year), the interest rate per monthly period is:

$$i = \frac{0.09}{12} = 0.0075 \quad \left(\text{or } \frac{3}{4}\% \text{ per month}\right)$$

The total number of interest periods is:

$$n = 12 \times 5 = 60$$

From Table 4, using an interest rate of $\frac{3}{4}\%$ and $n = 60$:

$$a_{\overline{60}|0.0075} = 48.173\,373\,52$$

Hence, from Equation M7.7, the present value of the annuity is:

$$\begin{aligned} A &= R \times a_{\overline{n}|i} \\ &= \$120 \times a_{\overline{60}|0.0075} \\ &= \$120 \times 48.173\,373\,52 \\ &= \$5780.80 \quad \text{(to the nearest cent)} \end{aligned}$$

That is, the repayments are worth \$5780.80 at the beginning of the loan.

Example M7.8

A family purchases a new home and agrees to place an immediate deposit of \$25 000 and make monthly repayments of \$750 over the next 20 years. If the interest rate is 9%, compounded monthly, what is the equivalent cash price of the home?

Solution

First, we must find the present value of the monthly annuity of \$750 made over 20 years at an interest rate of 9% compounded monthly.

In this case, $R = \$750$ and the interest rate per monthly period is:

$$i = \frac{0.09}{12} = 0.0075 \quad \left(\text{or } \frac{3}{4}\% \text{ per month}\right)$$

The total number of interest periods is:

$$n = 12 \times 20 = 240$$

We therefore require the value of $a_{\overline{240}|0.0075}$. Since Table 4 only caters for up to $n = 120$ periods, we must go directly to Equation M7.6 and use a calculator (or equivalent) equipped with the exponential function x^y to calculate this value.

$$\begin{aligned} a_{\overline{240}|0.0075} &= \frac{1 - (1 + 0.0075)^{-240}}{0.0075} \\ &= \frac{1 - 0.166\,413}{0.0075} \\ &= \frac{0.833\,587}{0.0075} \\ &= 111.144\,954 \end{aligned}$$

From Equation M7.7, the present value of the annuity is:

$$\begin{aligned} A &= R \times a_{\overline{n}|i} \\ &= \$750 \times a_{\overline{240}|0.0075} \\ &= \$750 \times 111.144\,954 \\ &= \$83\,358.72 \quad \text{(to the nearest cent)} \end{aligned}$$

Since the family has already paid a deposit of \$25 000, the equivalent cash price of the home is the sum of this deposit and the present value of the annuity. That is:

$$\begin{aligned}\text{Equivalent cash price} &= \$25\,000 + \$83\,358.72\\ &= \$108\,358.72\end{aligned}$$

Example M7.9

An annuity pays \$1000 per annum for 4 years and then \$2000 per annum for 5 years. Find the value of this annuity 1 year before the first payment if the interest rate is 10% per annum, compounded annually.

Solution

This problem involves a combination of two annuities and may be divided into two parts.

For the first annuity, we have R = \$1000, $i = 0.10$ and $n = 4$. Hence:

$$\begin{aligned}a_{\overline{4}|0.10} &= \frac{1-(1+0.10)^{-4}}{0.10}\\ &= 3.169\,865\,45\end{aligned}$$

From Equation M7.7:

$$\begin{aligned}\text{Present value of 1st annuity} &= R \times a_{\overline{n}|i}\\ &= \$1000 \times a_{\overline{4}|0.10}\\ &= \$1000 \times 3.169\,865\,45\\ &= \$3169.87 \quad \text{(to the nearest cent)}\end{aligned}$$

To find the present value of the second annuity is a little more complex and two steps are involved, namely:

(a) Find the value of the annuity 1 year before *its* first payment.
(b) Discount the value in (a) back to 1 year before the payment of the *first* annuity.

To find the value of the second annuity 1 year before its payment, we now have R = \$2000, $i = 0.10$ and $n = 5$.

From Equation M7.7:

$$\begin{aligned}&\text{Value of 2nd annuity}\\ &\text{1 year before its payment} = R \times a_{\overline{n}|i}\\ &= \$2000 \times a_{\overline{5}|0.10}\\ &= \$2000 \times \frac{1-(1+0.10)^{-5}}{0.10}\\ &= \$2000 \times 3.790\,786\,77\\ &= \$7581.57 \quad \text{(to the nearest cent)}\end{aligned}$$

Using Equation M7.4, the value of this amount 1 year before payment of the first annuity is:

$$\begin{aligned} \text{Present value of 2nd annuity} &= S(1 + i)^{-n} \\ &= \$7581.57(1 + 0.10)^{-4} \\ &= \$5178.32 \text{ (to the nearest cent)} \end{aligned}$$

Hence, the value of the total annuity 1 year before the first payment is the sum of the present values of the two annuities, that is \$3169.87 + \$5178.32 = \$8348.19.

Example M7.10

A company wishes to donate a prize of \$2000 to the top student in each offering of the first-year business mathematics course at a certain college. The course is offered in each of the two semesters per year. Hence, the prize is to be awarded every 6 months (at the end of June and at end of December). At this stage, the company wishes to guarantee the prize for a total of 5 years.

In situations such as this, the college usually requests that the company donate all monies at the beginning of the first year that the prize is to be offered. The amount donated should be such that there will be a guarantee that the prize will be covered for the 5-year period.

The college is able to invest any money it receives at an interest rate of 7% per annum compounded semi-annually (i.e. every 6 months).

How much money should the college request that the company donate at the beginning of the first year of the prize in order to cover prizes for the 5-year period?

Solution

In this case, R = \$2000. Since interest is compounded twice per year, the interest rate per 6-month period is:

$$i = \frac{0.07}{2} = 0.035 \quad \left(\text{or } 3\tfrac{1}{2}\% \text{ every 6 months}\right)$$

The total number of interest periods is:

$$n = 2 \times 5 = 10$$

From Table 4, using an interest rate of $3\frac{1}{2}\%$ and $n = 10$:

$$a_{\overline{10}|0.035} = 8.316\,605\,32$$

Hence, from Equation M7.7, the present value of the annuity is:

$$\begin{aligned} \text{A} &= R \times a_{\overline{n}|i} \\ &= \$2000 \times a_{\overline{10}|0.035} \\ &= \$2000 \times \frac{1 - (1 + 0.35)^{-10}}{0.035} \\ &= \$2000 \times 8.316\,605\,32 \text{ (From Table 4)} \\ &= \$16\,633.21 \quad \text{(to the nearest cent)} \end{aligned}$$

Therefore, the company should donate \$16 633.21 to the college at the beginning of the first year that the prize is offered in order to guarantee the prize for 5 years.

M7.4 AMORTISATION

When an individual or business pays a debt (including interest) by making periodic payments at regular intervals, the debt is said to be *amortised*. An amortisation problem involves finding the periodic payment that will discharge a debt.

In particular, as each periodic payment is made, this amount covers part of the principal and interest on the balance of the principal. In turn, there is a reduction in the remaining principal such that at the end of the term of the annuity (when the last payment is made) the debt is *extinguished*; that is, the balance is zero.

In amortisation problems we are usually trying to find the periodic payments (or annuity) required to pay off a debt, or the debt that remains after a specified number of repayments have been made. To do this we can use Table 4 on pages M.229–35.

Example M7.11

A family undertakes a mortgage of $40 000 from a bank in order to buy its new home. The bank charges interest at a rate of 12% per annum, compounded quarterly over 20 years. What quarterly payments will the family have to make on this loan?

Solution

In this case, we know that the present value of the annuity is A = $40 000. Since the interest is compounded quarterly, the interest rate per quarterly period is:

$$i = \frac{0.12}{4} = 0.03 \quad \text{(or 3\% per quarter)}$$

The total number of interest periods is:

$$n = 4 \times 20 = 80$$

The unknown quantity is the value of R, the amount of the annuity to be paid each quarter. From Table 4, using an interest rate of 3% and n = 80 we have:

$$a_{\overline{80}|0.03} = 30.200\,763\,45$$

From Equation M7.7:

$$\begin{aligned} A &= R \times a_{\overline{n}|i} \\ \$40\,000 &= R \times a_{\overline{80}|0.03} \\ &= R \times 30.200\,763\,45 \\ &= R \times \frac{\$40\,000}{30.200\,763\,45} \\ &= \$1324.47 \quad \text{(to the nearest cent)} \end{aligned}$$

Hence, the family must repay the loan at the rate of $1324.47 per quarter.
(Note that the family will, in fact, be paying back to the bank a total of $1324.47 × 80 = $105 957.60, considerably more than the original $40 000 loan!)

Example M7.12

Find the amount of the outstanding balance for the loan in Example M7.11 at the end of the fifth year.

Solution

At the end of the fifth year there are still 15 years remaining in the loan. We can once again use Equation M7.7, except that now the unknown quantity is the present value of the loan at that time.

Since there are 15 years remaining in the loan, the number of repayment periods left is:

$$n = 4 \times 15 = 60$$

From Table 4, still using an interest rate of 3% and now $n = 60$:

$$a_{\overline{60}|0.03} = 27.675\,563\,67$$

From Equation M7.7, the present value of the annuity is:

$$\begin{aligned} A &= R \times a_{\overline{n}|i} \\ &= \$1324.47 \times 27.675\,563\,67 \\ &= \$36\,655.45 \quad \text{(to the nearest cent)} \end{aligned}$$

Hence, at the end of the fifth year the family will still owe $36 655.45 to the bank.

(At the end of the fifth year the family will have made $4 \times 5 = 20$ payments each of $1324.47; that is, a total of $26 489.40. However, the principal will have been reduced by only $40 000 − $36 655.45 = $3344.55. Therefore, their payment of $26 489.40 will have consisted of:

$$\$26\,489.40 - \$3344.55 = \$23\,144.85$$

of interest repayments. This constitutes:

$$\frac{\$23\,144.85}{\$26\,489.40} \times 100 = 87.4\%$$

of the total repayments made to that time. This illustrates the fact that, in the early years of repaying a loan, much of the repayments goes towards the interest payments and relatively little reduction is made to the principal. This situation is reversed towards the end of the loan period.)

Example M7.13

In order to finance his earthmoving equipment business, Bob Catt borrowed $100 000 over 20 years at 10% interest per annum, compounded quarterly. However, with interest rates falling, the bank has announced that after 5 years of repayment the interest rate on Mr Catt's loan will drop to 8% per annum.

(a) Find the quarterly repayments that Mr Catt originally paid on his loan.
(b) At the end of 5 years, find the new amount that Mr Catt will repay each quarter if he still wants to keep the total term of the loan at 20 years.

Solution

(a) The present value of the annuity is A = \$100 000. Since the interest is compounded quarterly, the interest rate per quarter is:

$$i = \frac{0.10}{4} = 0.025 \quad \text{(or 2.5\% per quarter)}$$

The total number of interest periods $= n = 4 \times 20 = 80$.

To find the value of the quarterly payments (R) we use Equation M7.7:

$$\begin{aligned} \text{A} &= R \times a_{\overline{n|}i} \\ \$100\,000 &= R \times a_{\overline{80|}0.025} \\ &= R \times 34.451\,817 \quad \text{(from Table 4)} \\ R &= \frac{\$100\,000}{34.451\,817} \\ &= \$2902.60 \quad \text{(to the nearest cent)} \end{aligned}$$

which are the quarterly payments that Mr Catt paid on his loan.

(b) Now we find the amount owing on the loan (A) after 5 years repayments. In this case, we have:

$$\begin{aligned} R &= \$2902.60,\ n = 4 \times 15 = 60, \\ i &= \frac{0.10}{4} = 0.025 \quad \text{(or 2.5\%)} \end{aligned}$$

Using Equation M7.7:

$$\begin{aligned} A &= R \times a_{\overline{n|}i} \\ &= \$2902.60 \times a_{\overline{60|}0.025} \\ &= \$2902.60 \times 30.908\,656\,49 \quad \text{(from Table 4)} \\ &= \$89\,715.47 \end{aligned}$$

Now we calculate the quarterly payments (R) required for the remaining 15 years, when the interest rate is 8% per annum, compounded quarterly. In this case we have:

$$A = \$89\,715.47,\ n = 60,\ i = \frac{0.08}{4} = 0.02 \quad \text{(or 2\%)}$$

Using Equation M7.7:

$$\begin{aligned} A &= R \times a_{\overline{n|}i} \\ \$89\,715.47 &= R \times a_{\overline{60|}0.02} \\ &= R \times 34.760\,886\,68 \quad \text{(from Table 4)} \\ R &= \frac{\$80\,715.47}{34.760\,886\,68} \\ &= \$2580.93 \end{aligned}$$

Therefore, Mr Catt's quarterly repayments will fall from \$2902.60 to \$2580.93.

M7.5 SINKING FUNDS

A *sinking fund* is a fund into which periodic payments are made so as to accumulate to a nominated amount of money at the end of a specified period. An example would be when an amount of money is required for a new car or an overseas holiday, and you would like to know the amount that you must deposit (or sink into the fund) each period in order to accumulate the desired amount when the time comes. It is assumed that each deposit earns compound interest until the end of the period.

The actual periodic payment required can be calculated using Equation M7.1, where the unknown is now R (the amount of the annuity payment made per period), and S (the future value of the annuity) is now known. Thus, Equation M7.1 may be written as:

$$R = \frac{S}{\dfrac{(1+i)^n - 1}{i}} \qquad \textbf{(M7.8)}$$

The denominator of Equation M7.8 is $s_{\overline{n|}i}$ as defined in Equation M7.2. Since many of the values of $s_{\overline{n|}i}$ are given in Table 3, the values of the required periodic payments R may usually be found more easily using Equation M7.9:

$$R = \frac{S}{s_{\overline{n|}i}} \qquad \textbf{(M7.9)}$$

Example M7.14

Connie is planning to spend her Christmas vacation in 2 years time in the United States. She estimates that she will need an amount of \$6000 to pay for her air fares and other expenses. She is going to save her money in an account that attracts 10% interest per annum, compounded quarterly.

How much will Connie have to deposit into her account at the end of each quarter, to have the desired amount at the end of 2 years?

Solution

In this case, $S = \$6000$ and $i = \dfrac{0.10}{4} = 0.025$ (or 2.5%). The total number of interest periods is $n = 2 \times 4 = 8$.

From Table 3:

$$s_{\overline{8|}0.025} = \frac{(1+0.025)^8 - 1}{0.025} = 8.736\,115\,90$$

Using Equation M7.9, the quarterly deposit required is:

$$R = \frac{\$6000}{8.736\,115\,90} = \$686.80$$

That is, if Connie deposits \$686.80 every 3 months over the 2 years, she will have the required \$6000 at the end of that time.

Note that the total amount paid in by Connie will be \$686.80 × 8 = \$5494.40, so the total interest earned will be \$6000 − \$5494.40 = \$505.60.

Example M7.15

Kylie is just starting university to undertake a 3-year degree. At the end of this time she wants to have \$8000 to buy a used car. The bank is offering an interest-bearing account that attracts 5% interest per annum, compounded monthly.

How much should Kylie deposit into such an account each month to have \$8000 at the end of the 3 years? Also find how much interest she will earn during that time.

Solution

In this case, $S = \$8000$ and $i = \dfrac{0.05}{12} = 0.041\,666\,7$ (or $\frac{5}{12}\%$). The total number of interest periods is $n = 3 \times 12 = 36$.

From Table 3:

$$s_{\overline{36}|0.041\,666\,7} = 38.753\,335\,52$$

Using Equation M7.9, the monthly deposit required is:

$$\begin{aligned} R &= \frac{\$8000}{38.753\,335\,52} \\ &= \$206.43 \end{aligned}$$

That is, if Kylie deposits \$206.43 into the account every month over 3 years, she will have the required \$8000 at the end of that time.

The total amount paid in by Kylie will be \$206.43 × 36 = \$7341.48, so the total interest earned will be \$8000 − \$7341.48 = \$568.52.

Example M7.16

A block of strata title units has a body corporate that has estimated that a sinking fund of \$30 000 will be required in 4 years time. The sinking fund currently has no funds in it, and the unit owners must deposit money into an account that attracts 7% interest per annum, compounded semi-annually.

How much money must be deposited at the end of every 6 months, and how much interest will be earned by the sinking fund during that time?

Solution

In this case, $S = \$30\,000$ and $i = \dfrac{0.07}{2} = 0.035$ (or 3.5%). The total number of interest periods is $n = 4 \times 2 = 8$.

From Table 3:

$$s_{\overline{8}|0.035} = 9.051\,686\,77$$

Using Equation M7.9, the semi-annual deposit required is:

$$R = \frac{\$30\,000}{9.051\,686\,77}$$
$$= \$3314.30$$

That is, if the unit owners deposit \$3314.30 every 6 months over the 4 years, they will have the required \$30 000 at the end of that time.

The total amount paid in by the unit owners will be \$3314.30 × 8 = \$26 514.40, so the total interest earned will be \$30 000 − \$26 514.40 = \$3485.60.

Example M7.17

Sandra is an overworked and underpaid administrative officer who is celebrating her 43rd birthday. Her wish is to retire on her 60th birthday with a lump sum of \$100 000. Enquiries reveal that she can place her money into an interest-bearing account that attracts an interest rate of 13% per annum, compounded quarterly.

How much should Sandra deposit into the account at the end of each quarter in order to realise her wish? How much interest will she earn?

Solution

In this case, $S = \$100\,000$ and $i = \frac{0.13}{4} = 0.0325$ (or 3.25%). The total number of interest periods is $n = 17 \times 4 = 68$.

The value of $s_{\overline{68}|0.0325}$ is not given in Table 3. In this case, its value must be found directly from Equation M7.2. That is:

$$s_{\overline{68}|0.0325} = \frac{(1 + 0.0325)^{68} - 1}{0.0325}$$
$$= 240.025\,831\,7$$

Using Equation M7.9, the quarterly deposit required is:

$$R = \frac{\$100\,000}{240.025\,831\,7}$$
$$= \$416.62$$

That is, if Sandra deposits \$416.62 every 3 months over 17 years, she will have the required \$100 000 at the end of that time on her 60th birthday.

The total amount paid in by Sandra will be \$416.62 × 68 = \$28 330.16, so the total interest earned will be \$100 000 − \$28 330.16 = \$71 669.84.

M7.6 REMARKS

Annuities are an important issue for many Australians, particularly those paying off a home loan or an investment property. When undertaking a mortgage from a lending institution, some of the more typical questions from prospective customers include, for example:

1. If we can afford to pay $300 per week off our loan for, say, 20 years, how much can we afford to borrow if interest rates remain much the same as they are now?
2. If we take out a mortgage of $200 000 over 25 years on our new home, what will the monthly repayments be?

Fortunately, most institutions have their own specially written computer programs that can give the customer detailed information on the scenario that they are requesting. In fact, they are usually only too happy to enter your details into the program, which will then provide a year-by-year output of repayments, interest paid and current balance for each year of the loan. In this way you can also see exactly what total interest you will be up for by the time the loan is paid off!

As is the case for simple and compound interest, some *financial calculators* have specially programmed buttons that will compute some of the quantities in this chapter without all the tedious calculations.

Quick Quiz

1. Consider an annuity in which 4 annual payments each of \$600 are made with a compound interest rate of 6% per annum. If the payments are made at the end of the year, find the value of the annuity at the end of the fourth year.
2. A customer deposits \$150 every 3 months into a bank account that pays interest at an annual rate of 4%, compounded every 3 months. How much money will be in the account at the end of 5 years?
3. A chemist took out a loan in order to purchase an airconditioning unit for the pharmacy. He was required to repay the loan in monthly instalments of \$75 for 3 years. Calculate the present value of the loan if the interest rate was 5%, compounded every month.
4. A schoolteacher borrows \$25 000 from a credit union to remodel the lounge room and kitchen of her home. The credit union charges interest at a rate of 8% per annum, compounded semi-annually over 10 years. What repayments will she have to make every 6 months?
5. Margot wants to have \$8000 in 5 years time to purchase a used car. The bank is offering her an account that attracts interest at an annual rate of 3.5%, compounded annually. How much money must Margot deposit into the account every year to have \$8000 in 5 years time?
6. An investment firm owns a contract by which it will receive \$10 000 at the end of each year for the next 3 years, and this money will earn 5% per annum, compounded annually. Find the present value of the annuity at the beginning of the first rent period, that is 1 year before the first payment.
7. Henry wants to have \$25 000 in an account on 31 December 2004. If he invests \$21 000 in a new account on 31 December 2001 where the annual interest rate is 6.0% compounded monthly, is he going to achieve his goal?
8. Find (to two decimal places) the equivalent annual simple interest rate to an interest rate of 6% that is compounded quarterly over 5 years.
9. Explain the difference between a *deferred annuity* and a *perpetuity annuity*.
10. Explain the difference between a *simple annuity* and a *general annuity*.

Multiple Choice Questions

1. A business manager deposits $400 every month into an account that pays interest at a rate of 5% per annum, compounded monthly. The amount of money (to the nearest dollar) in the account at the end of 3 years will be:
(a) $14 400 (b) $15 120 (c) $15 501
(d) $15 132 (e) none of the above

2. If Tony borrows $11 500 to buy a motorbike and is charged interest at an annual rate of 6% compounded quarterly, the quarterly payment (to the nearest cent) he will have to make to repay the loan in 3 years is:
(a) $990.32 (b) $1015.83 (c) $958.33
(d) $1054.32 (e) none of the above

3. A barrister is repaying a loan in annual instalments of $3500 for 8 years. If the interest rate is 7% per annum compounded annually, the present value of the loan (to the nearest dollar) is:
(a) $20 500 (b) $21 900 (c) $26 215
(d) $22 100 (e) none of the above

4. Suppose that you want to have $10 000 in your bank account in 4 years time and that the bank is offering an account that attracts interest at an annual rate of 5% per annum, compounded semi-annually. The amount of money you must deposit (to the nearest dollar) into the account every 6 months is:
(a) $1145 (b) $1200 (c) $1250
(d) $1312 (e) none of the above

5. An investor deposits $600 every 3 months into an account that pays 7% compound interest, convertible quarterly. The amount of money in the account (to the nearest dollar) at the end of 6 years will be:
(a) $17 707 (b) $14 400 (c) $15 408
(d) $10 800 (e) none of the above

6. An engineer borrows $60 000 from a bank that charges interest at a rate of 9% per annum, compounded quarterly over 15 years. The quarterly payments the engineer must make (to the nearest dollar) on this loan are:
(a) $1832 (b) $1941 (c) $2053
(d) $2351 (e) none of the above

7. The *rent period* of an annuity is:
(a) the total period that an annuity is paid
(b) only applicable to simple annuities
(c) the time between payments of an annuity
(d) the rent paid by the holder of the annuity
(e) none of the above

8. The *future value* of an annuity is:
(a) the future amount of the series of payments plus interest
(b) the future amount of the series of payments excluding interest
(c) the total interest paid at the end of the annuity
(d) the value of the annuity at the beginning of the initial rent period
(e) none of the above

9. *Amortisation* is:
(a) the total value of an annuity
(b) the process of paying back a loan by equal periodic payments
(c) the life expectancy of the borrower of a loan
(d) the process of creating a sinking fund for the payments
(e) none of the above

10. The *discounted value* of an annuity is:
(a) the future amount of the series of payments plus interest
(b) the total interest paid at the end of the annuity
(c) the same as the present value of an annuity
(d) the interest rate applicable to the annuity
(e) none of the above

Exercises

M7.1 Apart from those given in Example M7.1, give three other examples of annuity payments.

M7.2* A customer deposits \$100 every month into a bank account that pays interest at the rate of 6% per annum, compounded monthly. How much will be in the account after 5 years?

M7.3 A customer deposits \$250 every month into a bank account that pays interest at the rate of 9% per annum, compounded monthly. How much will be in the account after 3 years?

M7.4* Frank deposits \$500 every 3 months into an interest-bearing deposit account that attracts compound interest at the rate of 4% per annum, compounded quarterly. How much will Frank have in the account at the end of 6 years?

M7.5 A musician has an agreement with the owners of a nightclub: for the next 3 years the owners will deposit \$3000 at the end of every 6 months into an account that attracts compound interest at an annual rate of 6%, compounded semi-annually. Find the present value of this annuity at the beginning of the first rent period (i.e. 6 months before the first deposit is made).

M7.6 Answer the question in Exercise M7.5, assuming that the nightclub owners have agreed to deposit \$5000 over 5 years on the same terms.

M7.7* Sophie is planning a holiday in the United Kingdom and wants to be able to withdraw \$250 every month for 2 years from an account that earns 5% per annum, compounded monthly. How much money should Sophie deposit in the account before she leaves?

M7.8 Answer the question in Exercise M7.7 assuming that Sophie needs \$500 every month for 3 years.

M7.9* A newlywed couple are saving to buy a block of land. They deposit \$500 every 3 months into a bank account that pays an annual interest rate of 10%, convertible quarterly. How much money will be in their account at the end of 5 years?

M7.10 A solicitor owns a contract by which he will receive \$4500 at the end of each of the next 5 years; the money earns an annual interest rate of 8%, compounded annually. Find the present value of this annuity at the beginning of the first rent period (i.e. 1 year before the first payment).

M7.11* Answer the question in Exercise M7.10 assuming that the amount received is \$7500 at the end of each year and that the annual interest rate is 3.5%, compounded annually.

M7.12 A carpenter has taken out a bank loan to purchase more modern tools. The loan is to be repaid in monthly instalments of \$45 over 4 years. Calculate the present value of these repayments if the annual interest rate is 5%, convertible monthly.

M7.13* A college lecturer purchases a new car, placing an immediate deposit of \$2000 and agreeing to make monthly repayments of \$240 over the next 5 years. If the annual interest rate is 6% compounded monthly, what will be the equivalent cash price of the car?

M7.14 Answer the question in Exercise M7.13 assuming that the monthly repayments will be \$500 instead of \$240.

M7.15* The manager of a new business undertakes a \$40 000 mortgage from a bank to secure new premises. The bank charges an annual interest rate of 6%, compounded quarterly over 15 years. What quarterly payments will the manager have to make on the loan?

M7.16 Answer the question in Exercise M7.15 assuming that the mortgage is \$65 000 and that the annual interest rate is 11%, compounded quarterly over 12 years.

M7.17* For the loan in Exercise M7.15, find the amount of the outstanding balance at the end of the tenth year.

M7.18 For the loan in Exercise M7.16, find:

(a) the amount of the outstanding balance at the end of the sixth year

(b) the total amount that the manager will have paid the bank by the end of the twelfth year

(c) the percentage of the original principal that will have been paid by the end of the fourth year

M7.19 Derive Equation M7.5 from Equations M7.1 and M7.4.

M7.20* Consider the problem of Mr Catt in Example M7.13 on page M.152. How much interest in total did Mr Catt save when the bank dropped his annual interest rate from 10% to 8% after 5 years?

M7.21 A husband and wife took out a \$90 000 mortgage on their new house, for which the annual interest rate was 9.5%, compounded semi-annually over 25 years. At the end of 10 years the bank raises its interest rate to 11.0% per annum, compounded semi-annually. Find:

(a) the original payments that the husband and wife paid off their loan

(b) the new amount at the end of 10 years that will have to be repaid every 6 months

M7.22 Answer the questions in Exercise M7.21 assuming that the original mortgage was \$150 000 over 20 years.

M7.23* The parents of a 10-year-old girl have promised her a stereo wide-screen television set for her own bedroom in 3 years time if she does well at school in the meantime. The parents estimate that the set will cost \$1500, so they have decided to invest some money in an account that will accumulate the \$1500 at the end of 3 years. The account they are going to use attracts 8% interest per annum, compounded monthly.

(a) How much money must the parents deposit into the account each month?

(b) How much interest will the account earn during the 3 years?

M7.24 The manager of a travel agency wishes to retire at the end of another 15 years with an amount of \$200 000. The bank is currently offering an interest-bearing account in which the interest rate is 6% per annum, compounded semi-annually.

(a) How much money will the manager have to deposit into the account every 6 months to accumulate \$200 000 at the end of 15years?

(b) How much interest will the account earn during the 15 years?

M7.25* Karen deposits \$500 every 6 months into a credit union account that pays an annual interest rate of 6%, compounded semi-annually. How much will Karen have in her account at the end of 3 years?

M7.26 Answer Exercise M7.24 assuming that the interest is compounded quarterly.

M7.27* Phyllis is an architect who plans to retire in 25 years time with \$250 000. To do this she plans to deposit an amount of money now into an account that pays an annual interest rate of 6%, compounded semi-annually. How much must Phyllis deposit into her account now?

M7.28* An annuity pays \$500 per annum for 3 years and then \$1000 per annum for 2 years. Find the value of this annuity 1 year before the first payment if the interest rate is 8%, compounded annually.

M7.29 The manager of a small business borrows \$15 000 from a bank in order to purchase a network of computers for the office. The bank charges an interest rate of 3% per annum, compounded quarterly over 10 years.

(a) What quarterly payments will the manager have to make on this loan?

(b) What will be the total amount paid back to the bank?

M7.30* Jack is a fisherman who would like to have \$9000 in 4 years time to purchase a new boat. He has found an account that will pay 8% interest per annum, compounded monthly.

(a) How much money must Jack deposit each month into the account to have \$9000 at the desired time?

(b) How much total interest will Jack earn on his account?

M7.31 Suppose in Exercise M7.30 that Jack discovers that interest is actually compounded *quarterly*.

(a) How much money must Jack deposit each quarter into the account to have \$9000 at the desired time?

(b) How much interest will Jack earn on his account?

(c) Do you think that Jack would prefer to have the interest compounded monthly or quarterly? Give reasons.

M7.32* (a) How much should be deposited monthly into a 9% per annum interest-bearing account, with interest compounded monthly, if the aim is to accumulate \$40 000 at the end of 6 years?

(b) How much interest will be earned during that period?

M7.33 On 1 July 2002 Darcy retires with \$24 000 and begins investing \$2000 per quarter into an account. The investment account used attracts an annual interest rate of 5%, compounded quarterly. How much will Darcy have in the account on:

(a) 1 July 2004? (b) 1 January 2006?

M7.34 In Exercise M7.33, suppose that Darcy had decided to invest the entire \$24 000 into the account on 1 July 2002. How much would Darcy have in the account on:

(a) 1 July 2004? (b) 1 January 2006?

M7.35* Jeannie borrows \$200 000 from her bank to start a small clothing business. The bank charges an annual interest rate of 4.5%, compounded quarterly over 12 years. What quarterly payments will Jeannie have to make on the loan?

M7.36 In Exercise M7.35, suppose that the bank had decided to compound the interest *monthly* over 12 years.

(a) What monthly payments would Jeannie have to make?

(b) How much more would Jeannie have to pay back to the bank in total under this scheme compared to that in Exercise M7.35?

M7.37 Two parents want to have \$20 000 available in 5 years time when their daughter starts university studies.

(a) How much should they deposit each month into an account that pays 6% interest per annum, compounded monthly?

(b) How much interest will be earned during the 5 years?

M7.38 Answer Exercise M7.37 if interest is compounded *quarterly* and deposits are made each *quarter*.

M7.39* A builder takes a loan from a credit union to purchase materials for an upcoming job. The loan is to be repaid with monthly instalments of \$80 for 2 years. Calculate the present value of these repayments if the interest rate is 6%, convertible monthly.

M7.40 Answer Exercise M7.39 if the loan is to be repaid quarterly with an interest rate of 6%, convertible quarterly.

M7.41* Spiro is the recipient of an annuity in which 5 annual payments of \$500 are deposited into his account, where the annual compound interest rate is 4%. If the payments are made at the end of each year, find the value of the annuity at the end of each year.

M7.42 An annuity pays \$2000 per annum for 2 years and then \$3000 per annum for 3 years. Find the value of the annuity 1 year before the first payment if the interest rate is 6% per annum, compounded annually.

M7.43* A cricket team decide to purchase a house where they can hold their functions near their oval. They take out a mortgage of \$120 000 from the Ruptcy Bank, which charges an annual interest rate of 5%, compounded semi-annually over 10 years.

(a) What payments will the team have to make every 6 months on this loan?

(b) What is the total amount of interest they will pay?

M7.44 In Exercise M7.43, the rival Fleece Finance offers the cricket team a loan for which it will charge an annual interest rate of 4.8%, compounded monthly over 10 years.

(a) What payments will the team have to make every month on this loan?

(b) What is the total amount of interest they will pay?

M7.45 In view of your answers to Exercises M7.43 and M7.44, from which lending institution should the cricketers take their loan?

Chapter 8

Depreciation

'The value of life lies not in the length of days but in the use you make of them.'

(Michel Eyquem de Montaigne, French novelist and essayist, Essais, *1533–1592)*

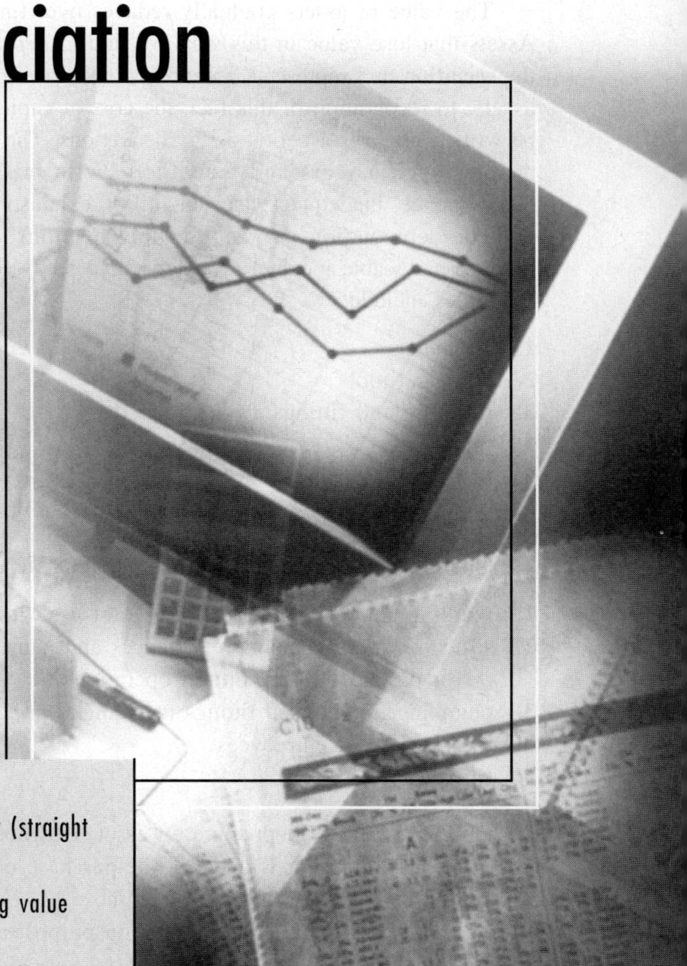

Learning objectives

In this chapter you will learn to:

- Calculate depreciation rates using the prime cost (straight line) method
- Calculate depreciation rates using the diminishing value (reducing balance) method
- Calculate depreciation rates using the units-of-production method
- Prepare a depreciation schedule
- Calculate the current written down value (book value) of an asset
- Learn how to keep depreciation records

M8.1 INTRODUCTION

Depreciable assets

Under Australian income tax law, you are allowed to claim deductions for expenses incurred in earning assessable income. Examples of these include recurring expenses such as rent, wages and electricity, while other expenses such as the cost of acquiring capital assets are not allowable. *Capital assets* (or *fixed assets* or *non-current assets*) are those that provide a benefit over a number of years, such as buildings, real estate, office equipment, motor cars and machinery.

The value of assets gradually reduces over time as they approach the end of their useful lives. Assets that lose value in this way are said to *depreciate*, and the amount of this lost value is called **depreciation** and represents a *depreciation expense*. The assets themselves are known as **depreciable assets**. In recognition of this fact, the cost of capital assets used in producing assessable income can be written off over a period as tax deductions. The term 'depreciation' is usually applied to plant and articles such as cars and machinery, with possible causes including physical deterioration, obsolescence, inadequacy, lack of utility, exhaustion or simply the passage of time.

The depreciation provisions apply to **plant**, which means articles or assets that are used to produce assessable income, or are installed ready for use to produce income and are held in reserve. Examples include:

1. computers
2. electrical tools
3. furniture and fittings
4. furnishings
5. manufacturing machinery
6. motor vehicles

Other assets that are regarded as depreciable under Australian income tax law include:

1. working animals (e.g. racehorses) not used in primary production
2. structural improvements (e.g. fences and dams) on land used for a primary production business or for pastoral or agricultural operations
3. plumbing fixtures and fittings in premises principally for the use of employees or for the care of the children of employees
4. software

Only owners can depreciate an asset. However, this does not extend to an asset that is being leased or hired unless it is under a hire-purchase or similar arrangement, in which case you have the right to purchase the asset at a later time.

The cost of an item for depreciation purposes includes:

1. the original purchase price or construction cost
2. transport costs
3. installation costs
4. customs duty
5. relocation costs

Depreciation methods

It is not always easy to determine at any one time the exact amount of depreciation. Under Australian income tax law, the two alternative methods of working out depreciation deductions are the **prime cost method** (also known as the **straight line method**) and the **diminishing value method** (also known as the **reducing balance method**). You must choose which method to use on an item-by-item basis at the time you purchase an item. Once a particular method of depreciation for an item is adopted, it is not permitted to change to the other method.

Other methods of depreciation include the **units-of-production method** and the **sum-of-the-years'-digits method**. Although neither of these is used extensively in Australia, the units-of-production method is discussed in Section M8.4 to demonstrate other methods of depreciation.

Depreciation rates

Rates of depreciation are fixed by reference to the **effective life** of each asset, where the effective life of a particular item is the period for which the item can reasonably be expected to be used in producing assessable income if it is maintained in good condition. The longer the effective life of an item, the lower its depreciation rate. The **salvage value** (or **scrap value**) of an asset is its estimated value, if any, at the end of its effective life. (This may be the value that a business feels it can get by disposing of the asset.)

Methods of depreciation vary according to when the asset was acquired. The three important dates in this respect are:

- assets acquired up to 26 February 1992
- assets acquired after 26 February 1992 but before 11.45 a.m. on 21 September 1999
- assets acquired since 11.45 a.m. on 21 September 1999

Each of these is discussed below.

Assets acquired up to 26 February 1992

Taxation Ruling IT 2685 contains specific depreciation rates for assets in this category. The basic rates applied in individual cases can be obtained from the taxpayer's historical records and these rates should continue to apply.

Assets acquired after 26 February 1992 but before 11.45 a.m. on 21 September 1999

During this period the actual depreciation rate was higher than the basic rate and was achieved by adding a 20% loading to the basic rate. The rates were then categorised into one of seven common rates, thereby providing *accelerated rates* of deduction for depreciation. These accelerated rates were calculated according to various *asset categories* and continued to apply to assets acquired until 11.45 a.m. on 21 September 1999. These rates are shown in Table M8.1, but do not include cars, motorcycles, artworks, software and employee amenities, for which special rates apply. These rates applied to small business taxpayers until 30 June 2001, from which time the *Simplified Tax System* applied.

Table M8.1 Rates of depreciation according to years of effective life of items in general		
Effective life	Prime cost method (%)	Diminishing value method (%)
Less than 3 years (or cost $300 or less)	100	100
3 years to less than 5 years	40	60
5 years to less than $6\frac{2}{3}$ years	27	40
$6\frac{2}{3}$ years years to less than 10 years	20	30
10 years to less than 13 years	17	25
13 years to less than 30 years	13	20
30 years or more	7	10

Source: Australian Taxation Office, *Guide to Depreciation*, 1999–2000, p. 6.

Taxpayers can make their own estimate of the effective life for assets acquired during this time period and use the depreciation rates in Table M8.1 according to the method of depreciation selected. However, great care needs to be taken when making your own estimate of effective life and you have to be able to demonstrate that you took all relevant information into account and that your conclusions were reasonable.

Table M8.2 (from Taxation Ruling IT 2685) shows the effective lives of some commonly used assets. The third and fourth columns of this table show the percentage reduction in the value of the asset according to the depreciation method adopted. Note that the diminishing value percentage is approximately 1.5 times the value of the prime cost percentage. The percentages shown for each method in Table M8.2 could also be found from Table M8.1 once the effective life is known.

Table M8.2 Examples of the effective life and depreciation rates as determined by the ATO for items acquired after 26 February 1992 but before 11.45 a.m. on 21 September 1999			
Item	Effective life (years)	Prime cost method (%)	Diminishing value method (%)
Calculators	10	17	25
Photocopiers	10	17	25
Cars (other than taxis)	7	15	22.5
Furniture	15	13	20
Carpets, business	5	27	40
Curtains and drapes	7	20	30
Radio	10	17	25
Libraries	10	17	25
Washing machines	7	20	30

Source: Australian Taxation Office, *Guide to Depreciation*, 1999–2000, pp. 16–17.

Assets acquired since 11.45 a.m. on 21 September 1999

A number of changes were made to the depreciation system for assets that were acquired or commenced to be constructed after this time. The most important change was that depreciation rates were fixed by reference to the effective life of each asset. This was the basic rate free of any factors that might secure a quicker write-off.

Another important change during this period was that taxpayers had the option of re-estimating the effective life periods of their depreciable assets. This could occur if you felt that the effective life that you had been using was no longer accurate because of changed circumstances, such as changes in technology or the use of the asset turned out to be more or less rigorous than expected.

From 1 January 2001, the ATO revised the effective life of some items (Taxation Ruling TR 2000/18), as illustrated by the examples in Table M8.3. The effective life of these items is now determined at the time the item is first used, or is installed ready for use, for assessable income-producing purposes. Taxpayers still have the choice of using these figures or opting for their own estimate of effective life.

Table M8.3 Effective life of several items according to the ATO as from 1 January 2001

Item	Effective life (years)	Item	Effective life (years)
Electric fence	20	Crates	4
Concrete silo	100	Curtains	$6\frac{2}{3}$
Alarms	20	Portable safe	40
Artworks	100	Furniture	$13\frac{1}{3}$
Binoculars	10	Photocopier	5
Bulldozers	$6\frac{2}{3}$	Projectors	10
Concrete mixers	10	Signs	20
Television set	10	Car phone	$6\frac{2}{3}$

Depreciated value

Once an asset has been depreciated, it is then said to have a certain **book value**, which is its original value minus all the depreciation deductions to date. In equation form:

$$\text{Book value} = \text{original cost of asset} - \text{depreciation to date} \qquad \textbf{(M8.1)}$$

Book value is also referred to as *(current) written down value* (WDV) or *depreciated value*. Note that the book value of an asset will only by coincidence be the same as its market value. The book value of an asset simply represents the amount that remains of the original cost of the asset that has not been charged as an expense.

For each of the depreciation methods discussed in this chapter we can construct a *depreciation schedule* showing the depreciation expense each year, the accumulated depreciation to date and the book value at the end of each year. This is also known as a *fixed asset register* and should be maintained by a business with fixed assets. This register is important to keep insurance policies up to date and for recording the written-down value, so that when an asset is sold or scrapped, a proper adjustment to the depreciation expenses can be ascertained.

M8.2 PRIME COST METHOD (STRAIGHT LINE METHOD)

In this case, the *prime cost method* assumes that the value of an asset depreciates by a *fixed sum* each year over the life of the asset. It is also called *straight line depreciation* because the cumulative depreciation plotted against time yields a straight line (see Section M9.3). The annual depreciation is given by:

$$\text{Annual depreciation} = \frac{\text{original cost price}}{\text{effective life in years}} \qquad \textbf{(M8.2)}$$

The *annual rate* of depreciation (as a percentage) is given by:

$$\text{Annual rate of depreciation} = \frac{\text{annual depreciation}}{\text{original cost price}} \times 100\% \qquad \textbf{(M8.3)}$$

In accordance with Equation M8.1, the book value of an asset at the end of k years is:

$$\text{Book value} = \text{original cost} - (k \times \text{annual depreciation}) \qquad \textbf{(M8.4)}$$

Items acquired before 11.45 a.m. on 21 September 1999

Examples M8.1 and M8.2 show how the prime cost method of depreciation applies to items acquired before 11.45 a.m. on 21 September 1999. Note that small businesses could still claim the accelerated depreciation rates until 30 June 2001.

Example M8.1

In early 1999 a company purchased new carpet for \$5500 with an effective life estimated to be 5 years. Use the prime cost method to find:

(a) the amount of annual depreciation
(b) the book value after 3 years

Solution

In this case, the original cost price = \$5500 and effective life = 5 years. From Table M8.1, the prime cost method uses a depreciation rate of 27%.

(a) From Equation M8.2:

$$\begin{aligned}\text{Annual depreciation} &= \frac{\text{original cost price}}{\text{effective life}}\\ &= \frac{\$5500}{5}\\ &= \$1100\end{aligned}$$

(b) From Equation M8.4:

$$\begin{aligned}\text{Book value after 3 years} &= \text{original cost price} - (3 \times \text{annual depreciation})\\ &= \$5500 - (3 \times \$1100)\\ &= \$2200\end{aligned}$$

Example M8.2

Using the information in Example M8.1, construct a depreciation schedule for the full 5 years of the effective life of the carpet.

Solution

From Example M8.1(a), the accumulated depreciation *increases* by \$1100 each year and the book value *decreases* by \$1100 each year. Note that in the schedule in Table M8.4 the first row corresponds to 'end of year 0', which represents the time of the original purchase.

Using the prime cost method, note that the book value at the end of the effective life is always *zero*.

Table M8.4 Depreciation schedule for Example M8.2			
End of year	Annual depreciation ($)	Accumulated depreciation ($)	Book value ($)
0	0	0	5500
1	1100	1100	4400
2	1100	2200	3300
3	1100	3300	2200
4	1100	4400	1100
5	1100	5500	0

Items acquired after 11.45 a.m. on 21 September 1999

The formula for calculating the prime cost depreciation rate for an item acquired after 11.45 a.m. on 21 September 1999 can be calculated by reference to the effective life of the item. (See ITAA97 sec 42–165.) Equation M8.5 is equivalent to Equation M8.3 and can be derived by substituting the formula for annual depreciation in Equation M8.2 into Equation M8.3:

$$\text{Annual depreciation rate} = \frac{100\%}{\text{effective life of the item in years}} \qquad \textbf{(M8.5)}$$

Example M8.3

A small business purchases a tea and coffee dispenser for $360 with an ATO effective life of $6\frac{2}{3}$ years. Calculate:

(a) the annual depreciation rate
(b) the book value after 3 years

Solution

(a) Use Equation M8.5 with an effective life of $6\frac{2}{3}$ years. This yields:

$$\begin{aligned}\text{Annual depreciation rate} &= \frac{100\%}{\text{effective life of the item in years}}\\ &= \frac{100\%}{6\frac{2}{3}}\\ &= 15\%\end{aligned}$$

(b) The annual depreciation = 0.15 × $360 = $54.

From Equation M8.4:

Book value at end of year k = original cost − (k × annual depreciation) and so, using $k = 3$:

$$\text{Book value at end of year 3} = \text{original cost} - (3 \times \text{annual depreciation})$$
$$= \$360 - (3 \times \$54)$$
$$= \$198$$

M8.3 DIMINISHING VALUE METHOD (REDUCING BALANCE METHOD)

While the *prime cost method* of calculating depreciation assumes the same amount of depreciation each year, the *diminishing value method* assumes that the depreciation of an asset is highest in the earliest years of its life and calculates it as a fixed percentage of the preceding book value. This is done because many assets (e.g. machines) are more productive when new and become less efficient as they get older. That is:

$$\begin{array}{c}\text{Depreciation}\\\text{during year } k\end{array} = \text{book value at end of year } (k-1) \times \text{depreciation rate} \qquad \textbf{(M8.6)}$$

where $k = 1, 2, 3, \ldots$.

Note that the book value at the end of year 0 is defined to be the original cost. The depreciation rate represents the *fraction* of the book value that is depreciated each year.

The calculation of depreciation using the diminishing value method must be done on a year-by-year basis. The largest amount of depreciation will be during the first year, and depreciation will become progressively smaller in successive years. The Australian Taxation Office calculates the depreciation rate by taking approximately 1.5 times the prime cost depreciation rate. For example, if the prime cost depreciation rate is 15%, the equivalent diminishing value rate is $1.5 \times 15\% = 22.5\%$.

The book value at the end of year k is given by:

$$\begin{array}{c}\text{Book value at}\\\text{end of year } k\end{array} = \text{book value at end of year } (k-1) - \begin{array}{c}\text{depreciation}\\\text{during year } k\end{array} \qquad \textbf{(M8.7)}$$

which may also be shown to equal:

$$\begin{array}{c}\text{Book value at}\\\text{end of year } k\end{array} = (\text{original cost})(1 - \text{depreciation rate})^k \qquad \textbf{(M8.8)}$$

The accumulated depreciation at the end of year k may then be found from:

$$\begin{array}{c}\text{Accumulated depreciation}\\\text{at end of year } k\end{array} = \text{original cost} - \text{book value at end of year } k \qquad \textbf{(M8.9)}$$

Items acquired before 11.45 a.m. on 21 September 1999

Examples M8.4, M8.5 and M8.6 illustrate how the diminishing value method can be used on items acquired before 11.45 a.m. on 21 September 1999.

Example M8.4

According to the ATO, a car acquired after 26 February 1992, but before 11.45 a.m. on 21 September 1999, and having an effective life of between $6\frac{2}{3}$ years and less than 10 years, can be depreciated using the diminishing value method at a rate of 22.5% each year. Suppose that a company purchases such a car costing \$12 500.

(a) Construct a depreciation schedule for the first 3 years. (Give answers to the nearest cent.)
(b) Find the book value at the end of 5 years.

Solution

(a) The depreciation rate expressed as a fraction is 0.225. From Equation M8.8, the following book values can be found:

$$\begin{aligned}\text{Book value at the end of year } k &= (\text{original cost})(1 - \text{depreciation rate})^k \\ &= (\$12\,500)(1 - 0.225)^k \\ &= (\$12\,500)(0.775)^k\end{aligned}$$

From Equation M8.6:

$$\text{Depreciation during year } k = \text{book value at the end of year } (k - 1) \times 0.225$$

Hence, we have:

$$\begin{aligned}\text{Book value at end of year 0} &= \$12\,500 \\ \text{Depreciation during year 1} &= \$12\,500 \times 0.225 \\ &= \$2812.50 \\ \text{Book value at end of year 1} &= (\$12\,500)(0.775) \\ &= \$9687.50 \\ \text{Depreciation during year 2} &= \$9687.50 \times 0.225 \\ &= \$2179.69 \\ \text{Book value at end of year 2} &= (\$12\,500)(0.775)^2 \\ &= \$7507.81 \\ \text{Depreciation during year 3} &= \$7507.81 \times 0.225 \\ &= \$1689.26 \\ \text{Book value at end of year 3} &= (\$12\,500)(0.775)^3 \\ &= \$5818.55\end{aligned}$$

Note that Equation M8.7 could also have been used to obtain the book values at the end of each year. For example:

$$\begin{aligned}\text{Book value at end of year 2} &= \text{book value at end of year 1} - \text{depreciation during year 2} \\ &= \$9687.50 - \$2179.69 \\ &= \$7507.81 \text{ (as before)}\end{aligned}$$

The depreciation schedule for the first 3 years can now be constructed as shown in Table M8.5.

Table M8.5 Depreciation schedule for Example M8.4 using the diminishing value method

End of year	Annual depreciation ($)	Accumulated depreciation ($)	Book value ($)
0	0	0	12 500.00
1	2812.50	2812.50	9 687.50
2	2179.69	4992.19	7 507.81
3	1689.26	6681.45	5 818.55

(b) From Equation M8.8:

$$\begin{aligned}\text{Book value at end of 5 years} &= (\text{original cost})(1 - \text{depreciation rate})^k \\ &= (\$12\ 500)(0.775)^5 \\ &= \$3494.77\end{aligned}$$

If we can estimate the effective life of an asset, we can calculate the rate of depreciation when the diminishing value method is used. In this case, Equation M8.8 may be used, but the unknown quantity is now the depreciation *rate*.

Example M8.5

In August 1999 a company purchased an asset for $12 000, which it decided to depreciate using the diminishing value method. The book value at the end of 3 years was $5062.50. Using the information in Table M8.1:

(a) calculate the rate of depreciation
(b) determine the possible values of the effective life of the asset
(c) construct a depreciation schedule for 5 years

Solution

(a) In this case, the original cost was $12 000. From Equation M8.8:

$$\begin{array}{l}\text{Book value at}\\ \text{end of year } k\end{array} = (\text{original cost})(1 - \text{depreciation rate})^k$$

In this, case $k = 3$ with:

$$\begin{aligned}\text{Book value at end of 3 years} &= \$5062.50 \\ \text{Original cost} &= \$12\ 000\end{aligned}$$

Hence:

$$\begin{aligned}\$5062.50 &= (\$12\ 000)(1 - \text{depreciation rate})^3 \\ (1 - \text{depreciation rate})^3 &= \frac{\$5062.50}{\$12\ 000} \\ &= 0.421\ 875 \\ 1 - \text{depreciation rate} &= \sqrt[3]{0.421\ 875} \\ &= 0.75 \\ \text{Depreciation rate} &= 1 - 0.75 \\ &= 0.25\end{aligned}$$

Therefore, expressed as a percentage, the rate of depreciation based on the *diminishing value method* is 25%.

(b) From Table M8.1, a diminishing value rate of 25% indicates that the effective life of the asset was between 10 years and less than 13 years.

(c) With the rate of depreciation found in (a) and using Equations M8.6 to M8.9, the depreciation schedule can be constructed as in Table M8.6.

Table M8.6 Depreciation schedule for Example M8.5 using the diminishing value method

End of year	Annual depreciation ($)	Accumulated depreciation ($)	Book value ($)
0	0	0	12 000.00
1	3000.00	3000.00	9 000.00
2	2250.00	5250.00	6 750.00
3	1687.50	6937.50	5 062.50
4	1265.62	8203.12	3 796.88
5	949.22	9152.34	2 847.66

Example M8.6

Ruling 2685 of the ATO on the effective life and depreciation rates of items states that a cash register purchased before 11.45 a.m. on 21 September 1999 has an effective life of 7 years and a diminishing value annual depreciation rate of 30% (see Table M8.1). Suppose that a small business purchased such a cash register and that at the end of the third year it had a book value of $1500.

(a) Find the original cost of the cash register.
(b) Construct the depreciation schedule for the first 5 years.
(c) Find the book value at the end of the fifth year.

Solution

(a) We can use Equation M8.8 with $k = 3$, book value at end of 3 years $=$ \$1500 and depreciation rate $=$ 0.30. The unknown is the original cost. Therefore:

$$\begin{aligned} \$1500 &= (\text{original cost})(1 - 0.30)^3 \\ &= (\text{original cost})(0.70)^3 \\ &= (\text{original cost})(0.343) \\ \text{Original cost} &= \frac{\$1500}{0.343} \\ &= \$4373.18 \quad (\text{to the nearest cent}) \end{aligned}$$

Hence, the business paid $4373.18 for the cash register.

(b) With an annual rate of depreciation of 0.30 (or 30%), using Equations M8.6 and M8.8, the depreciation schedule can be constructed as in Table M8.7. In particular, the book value at the end of each year is 70% of the book value at the end of the previous year.

End of year	Annual depreciation ($)	Accumulated depreciation ($)	Book value ($)
0	0	0	4373.18
1	1311.95	1311.65	3061.23
2	918.37	2230.32	2142.86
3	642.86	2873.18	1500.00
4	450.00	3323.18	1050.00
5	315.00	3638.18	735.00

Table M8.7 Depreciation schedule for Example M8.6 using the diminishing value method rate of 30%

(c) From Table M8.7, the book value of the cash register at the end of the fifth year is $735.

Items acquired after 11.45 a.m. on 21 September 1999

The formula for calculating the diminishing value depreciation rate for an item acquired after 11.45 a.m. on 21 September 1999 can be calculated by reference to the effective life of the item (see ITAA97 sec 42–160). The equation is:

$$\text{Annual depreciation rate} = \frac{150\%}{\text{effective life of the item in years}} \qquad \textbf{(M8.10)}$$

Note that this rate is precisely 50% more than the annual depreciation rate using the prime cost method in Equation M8.5.

Example M8.7

Suppose the company in Example M8.5 had purchased its cash register for $4373.18 *after* 11.45 a.m. on 21 September 1999. In particular, suppose it purchased the cash register after 1 January 2001 when the effective life, according to the ATO, was reduced from 7 years to $6\frac{2}{3}$ years. Use Equation M8.10 to determine the annual depreciation rate. Comment.

Solution

Since the effective life is now $6\frac{2}{3}$ years for cash registers, using Equation M8.10 the annual depreciation rate is:

$$\begin{aligned}\text{Annual depreciation rate} &= \frac{150\%}{\text{effective life of the item in years}} \\ &= \frac{150}{6\frac{2}{3}} \\ &= 22.5\%\end{aligned}$$

This means that the annual depreciation rate reduces from 30% to 22.5%.

Note that, as from 1 July 2001, small business taxpayers have been able to claim immediate write-off for depreciable assets that cost less than $1000 and can enter a pooling arrangement for other depreciable assets with an effective life of less than 25 years. The diminishing value rate used in these cases is 30%.

M8.4 UNITS-OF-PRODUCTION METHOD

In the *prime cost* and *diminishing value methods* of depreciation, the value of an asset is reduced after a given period. The *units-of-production method* deals with items that deteriorate more with use rather than age. In this case, their useful life is determined by their total output rather than their age.

Although rarely used in Australia, examples of items for which this method may well be appropriate include tyres on a vehicle, machine parts, cutting blades, photocopiers, printers or any item in continuous use. For such items the taxpayer could perhaps base an argument on the effective life of an asset using this technique.

The units-of-production method of depreciation is similar to the *prime cost method*, with the rate being expressed as a usage rate rather than a time rate. In equation form:

$$\text{Rate of depreciation} = \frac{\text{original cost price}}{\text{effective life in terms of output}} \qquad \textbf{(M8.11)}$$

Example M8.8

An office laser printer costing \$2750 has an effective life of 10 000 pages of output.

(a) Using the *units-of-production method*, calculate the depreciation rate per page of output.
(b) Calculate the book value after the printer has printed 3600 pages of output.

Solution

(a) In this case, original cost price = \$2750 and effective life = 10 000 pages of output. From Equation M8.11:

$$\begin{aligned}\text{Rate of depreciation} &= \frac{\$2750}{10\,000}\\ &= 27.5 \text{ cents per page}\end{aligned}$$

(b) After 3600 pages of output:

$$\begin{aligned}\text{Total depreciation} &= 3600 \times 27.5 \text{ cents}\\ &= \$990.00\end{aligned}$$

Therefore:

$$\begin{aligned}\text{Book value} &= \text{original cost} - \text{accumulated depreciation}\\ &= \$2750 - \$990\\ &= \$1760\end{aligned}$$

Example M8.9

A floodlight with an effective life of 8000 hours and costing \$1800 is used to illuminate the outside of a large office building.

(a) If the floodlight is on for 10 hours per day, how long is its effective life in days?
(b) If the units-of-production method is used, find the rate of depreciation per hour of use.
(c) What is the book value after 5000 hours of use?

Solution

(a) $\text{Effective life in days} = \dfrac{\text{effective life in hours}}{\text{hours of use per day}}$

$$= \frac{8000}{10}$$

$$= 800 \text{ days}$$

(b) In this case, the original cost price = \$1800 and the expected life = 8000 hours. From Equation M8.11:

$$\text{Rate of depreciation} = \frac{\$1800}{8000}$$

$$= 22.5 \text{ cents per hour}$$

(c) After 5000 hours of use:

$$\text{Total depreciation} = 5000 \times 22.5 \text{ cents}$$

$$= \$1125.00$$

Therefore:

$$\text{Book value} = \text{original cost} - \text{accumulated depreciation}$$

$$= \$1800 - \$1125$$

$$= \$675$$

M8.5 RECORD KEEPING

If a deduction for depreciation is claimed in a tax return, the following information must be kept:

- the undeducted cost and written-down (book) value of each asset at the start of the year
- the cost and the date you first used, or installed ready to use, each additional asset you acquired during the income year
- where you dispose of an asset during the year, the cost and sale price, dates of acquisition and disposal, and the undeducted cost and written-down value of the asset
- the adjustments made to cost, undeducted cost and written-down value
- details of balancing charge relief including the alternative treatment of assessable adjustments
- the rate and amount of depreciation claimed for each asset
- the undeducted cost and written-down value of each asset at the end of the income year

The **undeducted cost** of an asset is the cost less the sum of your depreciation deductions and any further amounts that you could not deduct because you used the asset other than for producing assessable income (such as private use) or because a deduction is prevented by the income tax law (such as use for leisure facilities).

The ATO provides special worksheets to assist in recording the necessary information. Other information that must be kept includes:

- details of the basis of your estimates of effective life where you have not adopted the life periods determined by the ATO
- original documents such as suppliers' invoices and receipts for expenditure on plant

You must also keep information and records for the entire period over which you depreciate an item and for a further 5 years from the date of your last claim. The 5 years starts on either 31 October or the date you lodge your tax return, whichever is the later. This period is extended if, at the end of 5 years, you are in a dispute with the ATO that relates to the depreciation claim.

M8.6 REMARKS

Calculations involving the formulae in this chapter may also be performed with the aid of a *financial calculator*. These are special calculators that have inbuilt functions including the use of depreciation formulae.

A computer spreadsheet can be used to prepare depreciation schedules quickly. Discussion of the spreadsheet Microsoft Excel for Windows, which may be used for this purpose, is provided in Appendix C at the end of the statistics section on page S.436.

Further details on depreciation can be obtained in the current edition of *Guide to Depreciation* from the ATO. Other aspects include information on assets previously used only for private purposes, owned for only part of a year or purchased second-hand. Other special cases consist of assets that are sold, lost or destroyed, or those used only partially for producing income.

Quick Quiz

1. A hair salon purchased several washbasins in March 2002 for \$4500, which, according to the ATO, had an effective life of 20 years. Use the *prime cost method* to find the total depreciation by the end of 3 years.
2. A newsagent purchased a computerised cash register costing \$1750 and depreciated it at a *diminishing value* rate of 17.5% each year. Find its book value at the end of the first year.
3. Using the data in Question 2, find the book value at the end of the fourth year.
4. In 1998 a local chemist purchased a set of scales for customers at a cost of \$600. At the end of 2 years the book value was \$216. If the *diminishing value method* was used, find the rate of depreciation (expressed as a percentage).
5. Using the data in Question 4, find the book value at the end of the fourth year.
6. A gymnasium purchased some new weightlifting equipment in early 1999 and at the end of the third year it had a book value of \$4257.25. Depreciation was calculated under the *diminishing value method* at an annual rate of 30%. Find the original cost of the equipment.
7. An ice-cream-making machine has been purchased for \$2200 and is estimated to have an effective life serving 8000 ice-creams. At the end of that time its salvage value is estimated to be \$400. If the *units-of-production method* is used to depreciate the machine, what will be the book value of the machine after it has served 2500 ice-creams?
8. A lathe was purchased for \$15 000 and depreciated 13% each year using the *prime cost method.* What was its book value after 3 years?
9. A hot-water service cost \$1800 in 2002 and, according to the ATO, had an effective life of 20 years. What is its book value after 7 years?
10. The effective life of a washing machine purchased in 2001 is deemed to be $6\frac{2}{3}$ years. If the *prime cost method* of depreciation is used, what is the book value after 3 years?

Multiple Choice Questions

1. An office purchases a photo-typesetting printer for \$1745. Its ATO effective life is 5 years. The annual depreciation using the *prime cost method* is:
(a) \$349 (b) \$948 (c) \$1495
(d) \$500 (e) none of the above

2. For the data in Question 1, the book value at the end of the *second* year (using the *prime cost method*) is:
(a) \$1646 (b) \$948 (c) \$1297
(d) \$500 (e) none of the above

3. For the data in Question 1, suppose that the printer is now depreciated using the *diminishing value method.* The book value at the end of the *first* year is:
(a) \$1646 (b) \$399 (c) \$1276.80
(d) \$1596 (e) none of the above

4. For the data in Question 3, the amount of depreciation during the second year is:
(a) \$399 (b) \$349 (c) \$319.20
(d) \$255.36 (e) none of the above

5. For the data in Question 3, the book value at the end of 5 years (to the nearest cent) is:
(a) \$653.72 (b) \$250.00 (c) \$0.00
(d) \$817.15 (e) none of the above

6. The *diminishing value method* assumes that depreciation of an asset is:
(a) uniform throughout the life of the asset
(b) highest in the earliest years in the life of the asset
(c) highest in the later years in the life of the asset
(d) random throughout the life of the asset
(e) none of the above

7. The *prime cost method* assumes that depreciation of an asset is:
(a) uniform throughout the life of the asset
(b) highest in the earliest years in the life of the asset
(c) highest in the later years in the life of the asset
(d) random throughout the life of the asset
(e) none of the above

8. Which of the following cannot be claimed as a deduction for expenses incurred in earning assessable income?
(a) rent
(b) wages
(c) electricity
(d) the cost of acquiring capital assets
(e) none of the above

9. Which of the following is allowed to depreciate an asset?
(a) the lessee
(b) the hirer
(c) the owner
(d) a hire-purchase customer of the asset
(e) all of the above

10. The *written-down value* of an asset is:
(a) that written down as the original cost
(b) the original value minus all depreciation to date
(c) the sum of all depreciation to date
(d) its value at the end of the effective life
(e) none of the above

Exercises

M8.1* A company purchased a laser printer in 1998 for $4500 with an effective life of 5 years. If the *prime cost method* is used, with the aid of Table M8.1 calculate:

(a) the annual rate of depreciation

(b) the book value at the end of 2 years

M8.2 Answer the questions in Exercise M8.1 assuming that the company paid $3200 for the printer.

M8.3 Using the information in Exercise M8.1, construct a complete depreciation schedule for the first 4 years, assuming that the effective life of the printer is only 4 years.

M8.4* Using the information in Exercise M8.1 and Table M8.1, construct a depreciation schedule for the first 4 years only, assuming the printer cost $900 and has an effective life of 7 years.

M8.5 Consider the information in Exercise M8.1 but now assume the *diminishing value method* is used. With the aid of Table M8.1:

(a) Find the rate of depreciation.

(b) Construct the depreciation schedule for the first 3 years only.

M8.6 On 3 February 2002 a consulting company purchased a portable safe with an effective life of 40 years. If the safe cost $2400, what is the book value after 20 years if:

(a) the *prime cost method* is used?

(b) the *diminishing value method* is used?

M8.7* A restaurateur decided to purchase ducted airconditioning for his restaurant in 1996. At the end of the fourth year the airconditioning had a book value of $2500. Assume that the depreciation was calculated under the *diminishing value method* at 30% each year.

(a) Find the original cost of the airconditioning.

(b) Construct the depreciation schedule for the first 5 years.

(c) Find the book value at the end of the fifth year.

M8.8 Answer the questions in Exercise M8.7 assuming that the book value at the end of 3 years was $1800.

M8.9* A small office purchased a burglar alarm in 1998 for $1650 with an effective life of 6 years. Use the information in Table M8.1 and the *prime cost method* to find:

(a) the annual rate of depreciation

(b) the amount of annual depreciation

M8.10 Answer the questions in Exercise M8.9 assuming that the office paid $800 for the alarm.

M8.11* Using the information in Exercise M8.9, construct a depreciation schedule for the first 3 years only.

M8.12 Using the information in Exercise M8.9, what would be the book value after 3 years if the *diminishing value method* had been used?

M8.13* In March 2002 a motel manager purchased a new video recorder at a cost of \$945 to put in the lobby. The ATO effective life of these recorders is $6\frac{2}{3}$ years. Use the *diminishing value method* to:

(a) construct a depreciation schedule for the first 5 years

(b) find the book value at the end of the sixth year

M8.14 Answer the questions in Exercise M8.13 assuming that the video recorder cost \$699.

M8.15* The manager of a promotion company purchased an eight-ball, coin-operated pool table for the use of her employees and clients. At the end of the second year the table had a book value of \$2023. The depreciation was calculated under the *diminishing value method* using the ATO effective life of 10 years for such a table.

(a) Find the original cost of the table.

(b) Construct the depreciation schedule for the first 4 years.

(c) Find the book value at the end of the sixth year.

M8.16 Answer the questions in Exercise M8.15 assuming that the book value at the end of 2 years was \$1950.75.

M8.17 Answer the questions in Exercise M8.15 assuming that the book value at the end of the third year was \$982.60.

M8.18* A manufacturing plant purchases a steam generator for \$64 000 with an ATO effective life of 20 years. Use the *prime cost method* to calculate, for the first 3 years only:

(a) the amount of depreciation each year

(b) the accumulated depreciation each year

(c) the book value at the end of each year

M8.19 A club purchases a poker machine for \$3000 with an ATO effective life of 5 years. Use the *prime cost method* to construct a depreciation schedule for each year of the poker machine's effective life.

M8.20* A carpenter purchased a set of spanners with an ATO effective life of 5 years. Using the *diminishing value method*, at the end of the third year they had a book value of \$686.

(a) Find the original cost of the spanners.

(b) Construct the depreciation schedule for the first 5 years.

(c) Find the book value at the end of the fifth year.

M8.21* A construction company purchased a front-end loader in 1997 for \$24 000, and after 4 years of using the *diminishing value method* it had a book value of \$11 118. Find:

(a) the annual rate of depreciation (as a percentage)

(b) the range of possible values for the effective life of the front-end loader

(c) the book value at the end of each of the first 3 years

M8.22* A printing business has purchased a batch of pallets at a cost of \$12 000 with an ATO effective life of 5 years. The owner wants to calculate depreciation using the *diminishing value method.*

(a) Find the annual rate of depreciation.

(b) Construct a depreciation schedule.

M8.23 Answer Exercise M8.22 assuming that the pallets cost \$30 000.

M8.24* A factory that makes accessories for domestic animals purchased a fax machine at a cost of \$3000 with an ATO effective life of 5 years. Construct a depreciation schedule for 5 years using the *prime cost method.*

M8.25* Answer Exercise M8.24 assuming that the cost of the fax machine was \$2400.

M8.26 (a) Answer Exercise M8.24 using the *diminishing value method.*

(b) Compare your answer in (a) to that in Exercise M8.24. Which method gives the larger annual depreciation in the early years, and which gives the larger depreciation in the later years? Do you think that this will generally be true? Give a reason.

M8.27 A newsagent purchased a newspaper wrapping machine at a cost of \$4500 with an ATO effective life of 10 years. Construct a depreciation schedule for the first 5 years using the *prime cost method.*

M8.28* An accounting firm purchased an espresso coffee making machine for \$360 with an ATO effective life of $13\frac{1}{3}$ years.

(a) What was the constant rate of depreciation per year if the *prime cost method* were used?

(b) What was the book value of the machine after:

(i) 1 year? (ii) 2 years? (iii) 3 years?

M8.29 On 1 January 2002 a construction company purchased \$250 000 worth of scaffolding with an ATO effective life of 10 years. It now estimates the scaffolding's book value to be \$110 926 (to the nearest dollar). If the company has used the *diminishing value method* of depreciation, what date is it now?

M8.30 On 1 July 2001 Harry purchased a microwave oven for the office at a total cost of \$650 (excluding GST). After using a depreciation rate of 22.5% per annum and the *diminishing value method,* Harry estimates that its current book value is \$109.15.

(a) What effective life is Harry assuming for the microwave oven?

(b) What date is it now?

M8.31* (a) At what annual amount will a 'Music while you work' system depreciate in the first year only, using the *prime cost method,* if it was purchased for \$16 000 with an ATO effective life of 10 years?

(b) Answer (a) if the *diminishing value method* is used.

M8.32 On 30 June 2001, after 2 years of depreciation using the *diminishing value method,* a desktop computer with an ATO effective life of 4 years has a book value of \$1718.75.

(a) What was the original cost of the computer?

(b) What is the book value on 30 June 2003?

M8.33 A real estate investor purchases new carpet for one of her rented properties at a cost of \$2500. The effective life of this carpet is 10 years. Construct the depreciation schedule for the first 5 years using the *diminishing value method.*

M8.34* A judge spends \$1200 on a robe that she does *not* intend to use for court dress for ceremonial occasions. The ATO effective life of the robe is $13\frac{1}{3}$ years. Construct the depreciation schedule for the first 5 years using the *prime cost method.*

M8.35 A hotel spends \$2800 on purchasing a new oven with an ATO effective life of 20 years. Construct a depreciation schedule for the first 4 years using the *prime cost method.*

M8.36* A replacement camera for a street photographer costs \$1400 and has an ATO effective life of 4 years. Find the book value after 2 years using the *prime cost method.*

M8.37 The financial controller of a private college purchases an electronic whiteboard for seminar presentations at a cost of \$3000. The ATO effective life of the whiteboard is 6 years. Find the book value after 3 years using:

(a) the *prime cost method*

(b) the *diminishing value method*

M8.38* A company purchases a power tool for \$1500, which it estimates will have an effective life of 6000 hours.

(a) If the power tool is used for 5 hours each day, how long (in days) is its useful life?

(b) If the *units-of-production method* is used, find the rate of depreciation per hour of use.

(c) What is the book value after 2500 hours of use?

M8.39 Answer the questions in Exercise M8.38 assuming the power tool costs \$4800 and its estimated effective life is 15 000 hours.

M8.40 Suppose that an item is to be depreciated over n years using the *prime cost method.* Show that the ratio of the book value at the end of year k to the book value at the end of year $k + 1$ is:

$$\frac{n - k}{n - k - 1}$$

where $k < n - 1$.

M8.41 Using the result in Exercise M8.40, find the value of the ratio and interpret its value if:

(a) $n = 5$ and $k = 3$

(b) $n = 3$ and $k = 1$

(c) $n = 6$ and $k = 4$

M8.42* Explain why the ratio in Exercise M8.40 is not valid for $k \geq n - 1$.

M8.43 An office photocopier costing \$1650 has an effective life of 6000 pages of copies.

(a) Using the *units-of-production method,* calculate the depreciation rate per page of copies.

(b) Calculate the book value after the photocopier has printed 2400 pages of output.

M8.44 An alarm system costing \$6000 monitors a large warehouse and uses an infra-red beam that operates 24-hours a day. The manufacturer claims that the system has an effective life of 40 000 hours.

(a) If the *units-of-production method* is used, find the rate of depreciation per hour of use.

(b) What is the book value of the alarm system after:

(i) 5000 hours of use?

(ii) 25 000 hours of use?

M8.45* An interstate truck driver purchases a set of 10 tyres for his vehicle at a total cost of $5800. This brand of tyre has an effective life of 15 000 km.

(a) If the *units-of-production method* is used, find the rate of depreciation per kilometre of use.

(b) What is the book value of the set of tyres after:

(i) 2000 km of use?

(ii) 12 000 km of use?

M8.46 A butcher purchases a cutting blade at a cost of $750 for use in his store. With the type of use it will undergo, he estimates that it will depreciate at a rate of 60 cents each time it is used to make a cut.

(a) What is the effective life of the blade (in cuts made)?

(b) If the *units-of-production method* is used, what is the book value of the blade after:

(i) 200 cuts? (ii) 500 cuts?

(c) What is the book value of the blade after half of its effective life has expired?

Chapter M9

Graphing

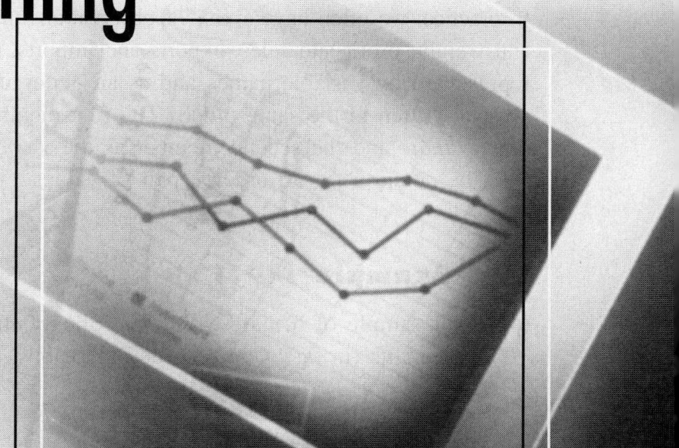

'And some of the bigger bears try to pretend That they came round the corner to look for a friend; And they try to pretend that nobody cares Whether you walk on the lines or squares.'

(A. A. Milne, English writer for children, When We Were Very Young, *1924)*

Learning objectives

In this chapter you will learn to:

- Plot ordered pairs on a graph
- Plot and interpret straight-line graphs
- Solve simple simultaneous equations using graphs
- Use simultaneous equations to solve problems in break-even analysis
- Draw and interpret non-linear graphs (including turning points)

M9.1 INTRODUCTION

One way of illustrating relationships that occur between variables is by means of a **graph**. For example, we may wish to display the manner in which sales or profits have changed over a period of time, or how the number of sales made relates to the price of the items. A graph will often do this very effectively. On other occasions we may be presented with information that is already in graphical form, and we need to interpret the graph. An understanding of the basic ideas concerning graphs is invaluable to the interpretation of such displays.

M9.2 PLOTTING POINTS

In practice, we often have a pair of observations that are matched. Examples are sales and year, height and weight, profit and sales, exports and imports, expenditure and income. These quantities are often represented by variable names and as an **ordered pair** of observations. (That is, the quantities are always written in the same order.) The first member of the ordered pair is usually referred to as the *x-coordinate* and the second member as the *y-coordinate*.

The notation for an ordered pair of values x and y is (x, y).

Example M9.1

A sample of 5 male technical college students was chosen, and their heights (in cm) and weights (in kg) were recorded as below.

Height (cm)	Weight (kg)
175	67
183	81
164	58
172	64
178	75

Represent these observations as 5 ordered pairs of the type (height, weight).

Solution

As ordered pairs the observations are:

(175, 67) (183, 81) (164, 58) (172, 64) (178, 75)

Ordered pairs of observations may be plotted onto a two-dimensional plane; each pair may be represented as a unique point in the plane.

In this plane we draw two perpendicular lines (called *coordinate axes*), with one axis (called the *x-axis*) horizontal and the other (called the *y-axis*) vertical. The point of intersection of these axes is called the *origin*.

On each of the axes there is a *scale*. These scales should be selected such that they are appropriate to the data being presented. *Positive* values of the x-variable are represented in ascending order on the right of the origin, while *negative* values go to the left. *Positive* values of the y-variable are represented proceeding upwards from the origin, while *negative* values proceed downwards.

The coordinate axis system (sometimes called the x–y *plane* if the two variables are x and y) is illustrated in Figure M9.1. Note that each axis is clearly labelled with its scale and variable names.

If both the variables to be represented on the graph are *positive* in value, it is common (and usually desirable) to draw only the upper right-hand section of the system, as in Figure M9.2.

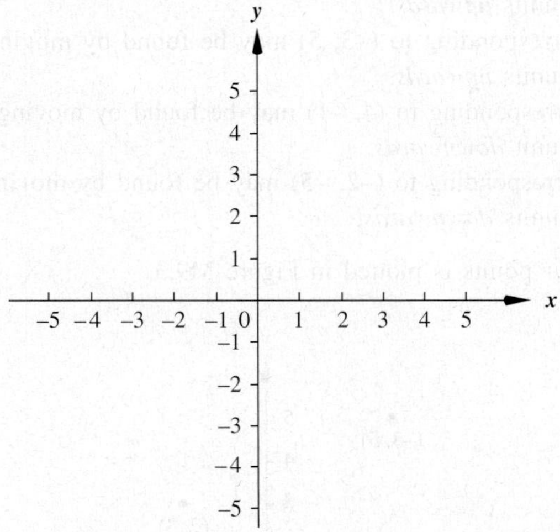

Figure M9.1 A coordinate axes system for two variables, x and y

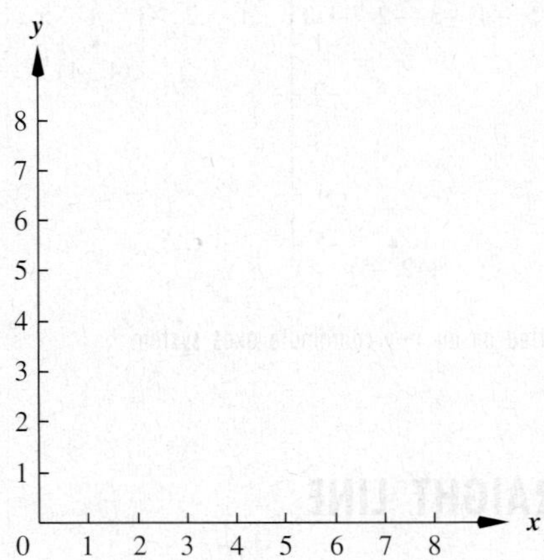

Figure M9.2 The upper right-hand section of an x–y coordinate axes system

Each ordered pair of observations of the type (x, y) corresponds to a point in the x–y plane. In particular, the ordered pair (0, 0) is that point where the axes intersect (the *origin*).

Example M9.2

Describe and plot the points in the x–y plane that correspond to the following ordered pairs. That is, the points that have the following coordinates:

(a) (2, 3) (b) (–3, 5) (c) (4, –1) (d) (–2, –5)

Solution

(a) The point corresponding to (2, 3) may be found by moving 2 units to the *right* of the origin and 3 units *upwards.*

(b) The point corresponding to (–3, 5) may be found by moving 3 units to the *left* of the origin and 5 units *upwards.*

(c) The point corresponding to (4, –1) may be found by moving 4 units to the *right* of the origin and 1 unit *downwards.*

(d) The point corresponding to (–2, –5) may be found by moving 2 units to the *left* of the origin and 5 units *downwards.*

Each of these four points is plotted in Figure M9.3.

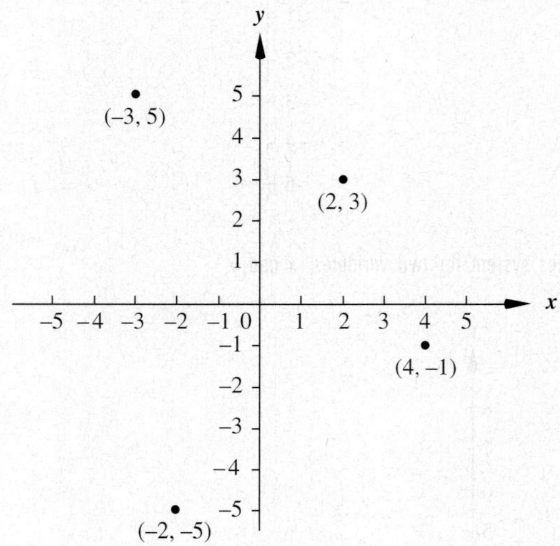

Figure M9.3 Four points plotted on an *x–y* coordinate axes system

M9.3 PLOTTING A STRAIGHT LINE

In Section M3.1 it was stated that a *linear equation* is one in which the largest value of the exponents is 1. A linear equation is also one that may be written in one of the following forms:

$$y = bx + a \quad \text{or} \quad y = a + bx \qquad \textbf{(M9.1)}$$

where a and b are constants. The constant b is called the **slope** or **gradient** of the line, because it represents the rate at which y changes with respect to x. The constant a represents the *y-intercept*, that is the value of y where the line crosses the *y-axis*. (If $a = 0$, the line crosses the y-axis at the point where $y = 0$; that is, the line passes through the **origin**.) It is called a *linear equation* because, when all ordered pairs that satisfy such an equation are plotted on an x–y plane, they form a straight line.

To draw a line we plot a minimum of *two* points that satisfy the equation and draw the straight line that passes through them. The points on that line will then represent all points whose coordinates satisfy the equation of the line. Once the graph is drawn, it is often appropriate to write the equation of the line on the line itself. This is particularly important if there is more than one line drawn on the one diagram.

However, while in theory plotting only two points will suffice, it is strongly recommended that (at least) *three* points be used. If all three points have been plotted correctly, they should lie in a straight line. If they do not lie in a straight line, an error has been made with one (or more) of the points, and all points plotted should be checked.

It does not matter which points on the line are plotted as long as they satisfy the equation. One technique is to select several values of x and find the corresponding values of y by substituting those x-values into the equation. It is advisable not to choose x-values that are too close to each other, because drawing the line accurately may then be difficult.

Example M9.3

Plot on a graph the equation of the line:

$$y = 2x + 3$$

Solution

Suppose we arbitrarily select four values of x and find the corresponding values of y by substituting into the equation. (It is usually convenient to select 0 as one of the x-values.)

If $x = 0$, $y = 2(0) + 3 = 3$
If $x = 2$, $y = 2(2) + 3 = 7$
If $x = -2$, $y = 2(-2) + 3 = -1$
If $x = -4$, $y = 2(-4) + 3 = -5$

We therefore plot the points:

(0, 3) (2, 7) (−2, −1) (−4, −5)

and expect that they will lie on a straight line. This is illustrated in Figure M9.4.

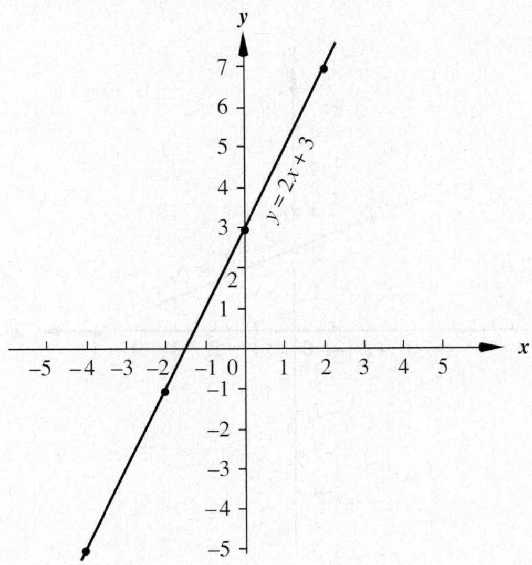

Figure M9.4 Graph of the straight line $y = 2x + 3$

Example M9.4

Write the equation:

$$2x + 5y = 10$$

in the form $y = a + bx$ and draw it on a graph.

Solution

First, put the terms in x on the right-hand side of the equation:

$$5y = 10 - 2x$$

Then divide both sides by 5 in order to make the coefficient of y to be 1:

$$y = 2 - 0.4x$$

(This line has a slope of −0.4 and a y-intercept of 2.)
Suppose we select x-values of −2, 0 and 3 and substitute them into this equation.

If $x = -2$, $y = 2 - 0.4(-2) = 2.8$
If $x = 0$, $y = 2 - 0.4(0) = 2$
If $x = 3$, $y = 2 - 0.4(3) = 0.8$

We therefore plot the points:

(−2, 2.8) (0, 2) (3, 0.8)

This is done in Figure M9.5, and it can be seen that they lie in a straight line.

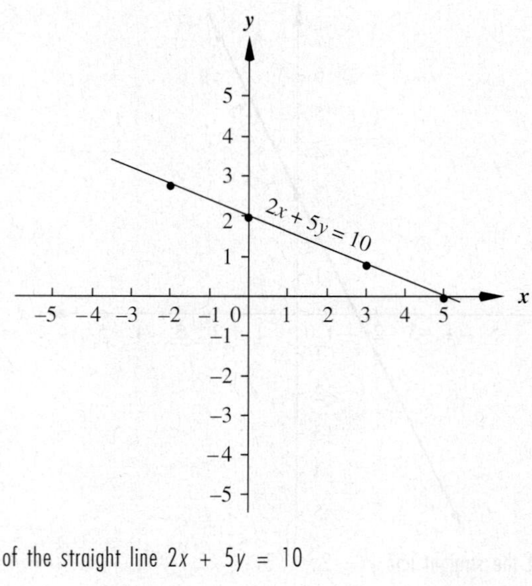

Figure M9.5 Graph of the straight line $2x + 5y = 10$

M9.4 SOLVING SIMULTANEOUS EQUATIONS WITH THE AID OF A GRAPH

Simultaneous equations may be solved by plotting each equation on the same diagram, then finding the coordinates of the point of intersection. The x-coordinate and y-coordinate represent the solution to the equations.

However, when the two lines being plotted have the same *slope*, they are parallel and thus never intersect. In this case, the simultaneous equations have no solution. That is, there are no values of x and y that satisfy both equations.

Example M9.5

Consider the two simultaneous equations in Example M3.11.

$$3x + 4y = 33 \qquad (1)$$
$$2x - 3y = 5 \qquad (2)$$

Solve these equations with the aid of a graph.

Solution

Draw the straight lines representing these graphs on the same diagram. First consider Equation (1). Putting it in the form $y = a + bx$ yields:

$$y = \frac{33}{4} - \frac{3x}{4}$$

Suppose we plot the points where $x = -3$, 0 and 5.

$$\text{If } x = -3,\ y = \frac{33}{4} - \frac{3(-3)}{4} = 10.5$$

$$\text{If } x = 0,\ y = \frac{33}{4} - \frac{3(0)}{4} = 8.25$$

$$\text{If } x = 5,\ y = \frac{33}{4} - \frac{3(5)}{4} = 4.5$$

Therefore, the points to be plotted for this line are:

(–3, 10.5) (0, 8.25) (5, 4.5)

The line passing through these three points is shown in Figure M9.6.

Now consider Equation (2). Putting it in the form $y = bx + a$ yields:

$$y = \frac{2x}{3} - \frac{5}{3}$$

Suppose we plot the points where $x = -5$, 0 and 4.

$$\text{If } x = -5,\ y = \frac{2(-5)}{3} - \frac{5}{3} = -5$$

$$\text{If } x = 0,\ y = \frac{2(0)}{3} - \frac{5}{3} = -\frac{5}{3}$$

$$\text{If } x = 4,\ y = \frac{2(4)}{3} - \frac{5}{3} = 1$$

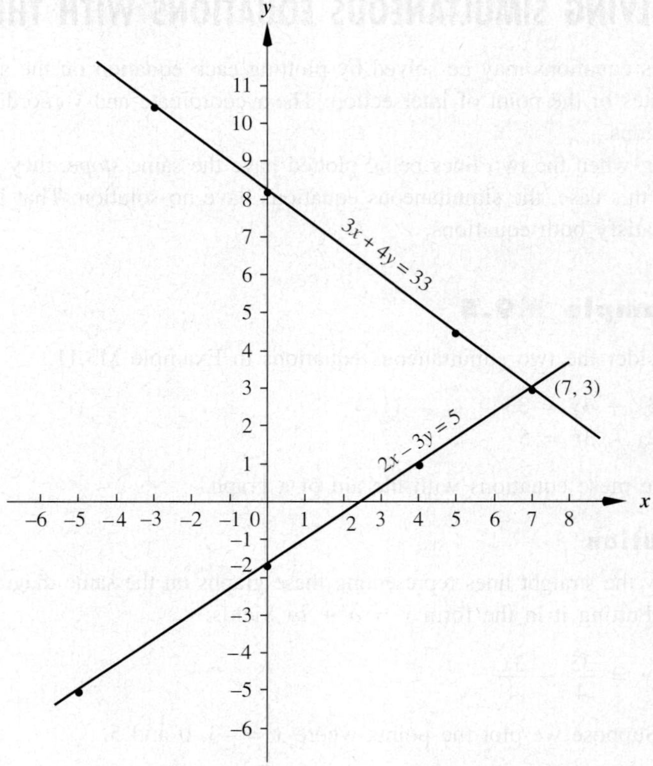

Figure M9.6 Graph of the straight lines $3x + 4y = 33$ and $2x - 3y = 5$

Therefore, the points to be plotted for this line are:

$(-5, -5) \quad (0, -\frac{5}{3}) \quad (4, 1)$

The line passing through these three points is also shown in Figure M9.6.

The point of intersection of these two lines is (7, 3), as shown in Figure M9.6. This means that the solution to the simultaneous equations is $x = 7$ and $y = 3$. As expected, this agrees with the answer found in Example M3.11.

It should be clear that the more neatly and accurately the graph is drawn, the easier it will be to identify the coordinates of the point of intersection of two lines.

M9.5 BREAK-EVEN ANALYSIS

In manufacturing situations it is usually of interest to know how many items must be produced and sold in order to *break even*; that is, to find the number of items where the income gained exactly equals the cost of manufacturing them. This process is known as **break-even analysis** and may be performed either by solving a pair of simultaneous equations or with the aid of a graph.

Costs

Costs can be classified as either *fixed* or *variable*. **Fixed costs** (e.g. rent, maintenance, administration, depreciation, salaries, telephone) are costs that are considered independent of the number of items produced, while **variable costs** (e.g. insurance, labour, materials) are a function of the number produced. The total cost is therefore:

$$\text{Total cost} = \text{variable cost} + \text{fixed cost} \qquad \textbf{(M9.2)}$$

Suppose we let:

x = number of items manufactured
v = variable cost to manufacture each item
f = fixed cost of manufacture
C = total cost

The variable cost is vx, and from Equation M9.2 the total cost C may be expressed as:

$$C = vx + f \qquad \textbf{(M9.3)}$$

This equation is in the same form as Equation M9.1.

Income

Suppose we let:

s = income made from each item
I = total income

The total income is therefore:

$$I = sx \qquad \textbf{(M9.4)}$$

which also is in the same form as Equation M9.1. In this case, however, there is no y-intercept term, so the line will pass through the origin.

The total profit (P) made will be:

$$P = I - C \qquad \textbf{(M9.5)}$$

If the value of P is *negative*, it represents a loss.

Suppose we first consider the graphical solution. This process consists of drawing one line for costs and another line for income on the same diagram and finding their point of intersection. This point represents the break-even point. The technique is demonstrated in Example M9.6. The algebraic method is illustrated in Example M9.7.

Example M9.6

A company manufactures an inexpensive type of scientific calculator. There is a weekly fixed cost of \$500 for producing the calculators and a variable cost of \$8 per calculator. The company receives an income of \$12 for each calculator that it sells.

(a) Find the total cost of manufacturing 80 calculators in a week.
(b) Find the income from selling 80 calculators.

(c) Find the profit (or loss) if the company manufactures and sells 80 calculators in a particular week.
(d) With the aid of a graph, find the point at which total cost is equal to income. That is, find the weekly sales that represent the break-even point. Interpret the value of this point.

Solution

In this case, we have $v = \$8$, $f = \$500$ and $s = \$12$.

(a) From Equation M9.3, with $x = 80$ calculators, the total cost is:

$$\begin{aligned} C &= vx + f \\ &= (\$8)(80) + \$500 \\ &= \$1140 \end{aligned}$$

Hence, the total cost of manufacturing 80 calculators in a week is \$1140.

(b) From Equation M9.4, with $x = 80$ calculators, the total income is:

$$\begin{aligned} I &= sx \\ &= (\$12)(80) \\ &= \$960 \end{aligned}$$

Hence, the income from selling 80 calculators is \$960.

(c) From Equation M9.5, the total profit from manufacturing and selling 80 calculators in a week is:

$$\begin{aligned} P &= I - C \\ &= \$960 - \$1140 \\ &= -\$180 \end{aligned}$$

Since this value of P is *negative*, this represents a loss to the company of \$180.

(d) Suppose x = the number of calculators sold in a week. Then from Equation M9.3, the total cost is:

$$C = 8x + 500$$

From Equation M9.4, the income produced is:

$$I = 12x$$

The *break-even point* occurs at the intersection of these two lines when they are plotted on the same graph. Although their left-hand sides are different (i.e. C and I), they both represent dollar values and so we may think of them as being essentially the same type of variable. The vertical (y) axis may simply be labelled as \$.

The two lines are drawn in Figure M9.7. From the graph in Figure M9.7 we see that the point of intersection is (125, 1500). Therefore, the break-even point of sales is 125 calculators per week, with the total cost and income each equalling \$1500. This means that, if the company sells *less* than 125 calculators in a week, it will make a *loss*; if it sells *more* than 125, it will make a *profit*.

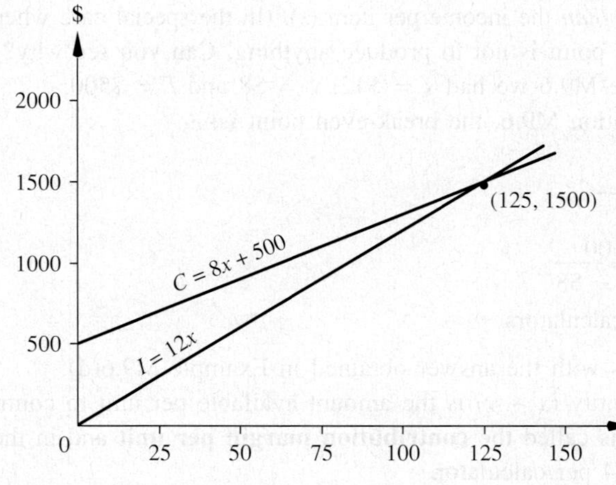

Figure M9.7 Graph of the straight lines $C = 8x + 500$ and $I = 12x$

Example M9.7

Find an algebraic formula for finding the break-even point without the aid of a graph. Verify the solution to Example M9.6 using this formula.

Solution

The break-even point occurs when total cost equals income:

$$I = C$$

Using Equations M9.3 and M9.4, this means:

$$sx = vx + f$$

Solving this equation for x yields:

$$sx - vx = f$$
$$(s - v)x = f$$

which gives:

$$x = \frac{f}{s - v} \qquad \textbf{(M9.6)}$$

That is:

$$\text{Break-even point} = \frac{\text{fixed cost}}{\text{income per item} - \text{variable cost per item}} \qquad \textbf{(M9.7)}$$

We have to be careful in using Equations M9.6 and M9.7. If $s = v$, the income and cost lines are parallel and there is no break-even point. Similarly, if s is *less than* v, there is also no break-even point, because the variable manufacturing cost per item alone exceeds the income per item. Therefore, a break-even point will occur only when the variable cost per item (v) is *less than* the income per item (s). (In the special case where there is *no* fixed cost, the break-even point is not to produce anything. Can you see why?)

In Example M9.6 we had $s = \$12$, $v = \$8$ and $f = \$500$.

From Equation M9.6, the break-even point is:

$$\begin{aligned} x &= \frac{f}{s - v} \\ &= \frac{\$500}{\$12 - \$8} \\ &= 125 \text{ calculators} \end{aligned}$$

This agrees with the answer obtained in Example M9.6(d).

Note: The quantity ($s - v$) is the amount available per unit to contribute to the meeting of fixed costs. It is called the **contribution margin per unit** and in the above example this is $\$12 - \$8 = \$4$ per calculator.

Example M9.8

A small business has to decide whether to lease its own photocopy machine or to continue to have its photocopying done by an outside firm called Ripoff Copies. Ripoff charges the business a constant 15 cents to copy each page (all-inclusive). The leasing cost of a machine is \$75 per week, plus it will cost the business a constant 3 cents per page for paper, toner, etc.

How many pages per week would the business have to photocopy before it would be more profitable to lease a machine?

Solution

For this problem we can treat v (the variable cost of manufacture) as the variable cost per page of leasing, f (the fixed cost of manufacture) as the fixed cost per week of leasing, and s (the income from each item) as the cost per page of Ripoff doing the copying.

Therefore, $v = \$0.03$, $f = \$75$ and $s = 0.15$. Using Equation M9.6, the break-even point x occurs when:

$$\begin{aligned} x &= \frac{f}{s - v} \\ &= \frac{\$75}{\$0.15 - \$0.03} \\ &= 625 \text{ copies} \end{aligned}$$

Hence, the break-even point is 625 copies per week. That is, if the business has *less* than 625 copies to do in a week, it is better off using the services of Ripoff. If it has *more* than 625 copies in a week, it would be more profitable to lease a machine.

M9.6 NON-LINEAR GRAPHS AND TURNING POINTS

On some occasions we may be interested in graphs that are *not* a straight line. Such graphs are called *non-linear* and involve equations that have powers of the x-variable other than 1.

Example M9.9

Some examples of non-linear graphs are:

(a) $y = x^2$ (b) $y = 2x^2 - 4x + 6$

(c) $y = 6 - x^3$ (d) $y = \sqrt{x}$

To plot non-linear graphs we could simply plot as many points as necessary until we obtained the general shape of the curve.

Example M9.10

Draw the graph that represents the equation:

$$y = 8 + 2x - x^2$$

Solution

We will restrict ourselves to positive values of x. Suppose we plot values of x from 0 to 4 at intervals of 0.50. The corresponding values of y are shown below:

x-value	0	0.5	1	1.5	2	2.5	3	3.5	4
y-value	8	8.75	9	8.75	8	6.75	5	2.75	0

The resulting graph is shown in Figure M9.8. Note that the points have been connected using a *smooth curve* since the graph is non-linear.

Figure M9.8 Graph of the equation $y = 8 + 2x - x^2$ for Example M9.10

On some occasions we may have some business data of which we would like to draw a graph. Quite often such graphs are non-linear, and to find the exact equation of the graph would usually be difficult and moreover unnecessary. Instead we could draw the graph and discuss some of its features.

Example M9.11

An airconditioning company has recorded the number of sales of its deluxe units over the past 10 years. The results were:

Year no.	1	2	3	4	5	6	7	8	9	10
Sales (units)	21	34	45	18	29	55	62	40	35	32

Plot sales versus year and comment.

Solution

The resulting graph is shown in Figure M9.9, where the points have been joined by straight lines. (Note that we have not used a smooth curve to join the points in this case since we are not plotting a specific non-linear equation. It is also quite common to join the points on a graph where the data vary over time.)

Note that the graph in Figure M9.9 has points where it rises to reach a *peak* and then falls to a low point (or a *trough*). In fact, these points represent (local) maximum and minimum points on the graph and correspond to positions where the graph changes direction. Such points are called *turning points* and in this case occur in years 3, 4 and 7 when sales = 45 (high), 18 (low), 62 (high), respectively. In the case of Figure M9.8, the turning point occurs when $x = 1$ and $y = 9$.

Example M9.12

Figure M9.10 shows the changing values of the number of job vacancies in Australia between August 1997 and August 2000. Also displayed is a lightly shaded curve used to indicate the long-term trend of the data (this type of curve is discussed in Chapter S11). Note that there are *two* turning points on the job vacancies curve. The first of these occurs around May 1998 when there was a peak (local maximum) of around 95 000, while the other turning point is a trough (local minimum) that took place in November 1998 when it dropped to around 90 000 before steadily rising to reach 120 000 by August 2000.

For the remainder of the graph there is no sign of another turning point occurring, with the data steadily increasing.

Example M9.13

The manager of a small electrical company has been concerned with the fluctuating variable weekly costs of running the business. To try to determine the reasons for this, he recorded these costs on a weekly basis for 10 weeks. The figures are shown in Table M9.1.

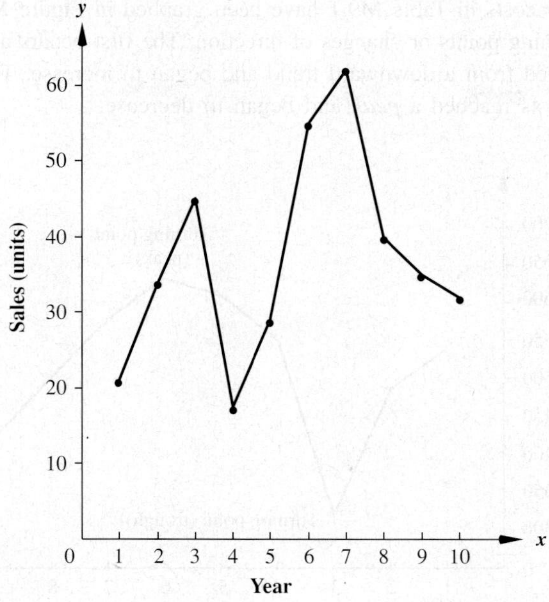

Figure M9.9 Graph of sales data for Example M9.11

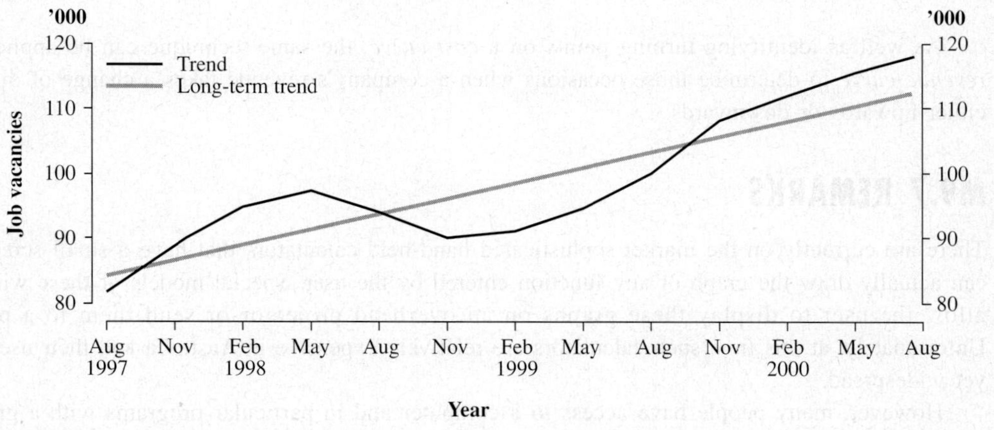

Figure M9.10 The changing value of the number of job vacancies in Australia between August 1997 and August 2000

Source: Australian Bureau of Statistics, *Australian Economic Indicators*, December 2000, Cat. No. 1350.0

Table M9.1 Weekly costs for the electrical company in Example M9.13

Week no.	1	2	3	4	5	6	7	8	9	10
Cost ($)	534	486	320	550	610	630	575	500	425	390

The weekly costs in Table M9.1 have been graphed in Figure M9.11. Note that there are two distinct turning points or changes of direction. The first occurs at week 3 (a *trough*), when the costs changed from a downward trend and began to increase. The second occurs at week 6, when the costs reached a *peak* and began to decrease.

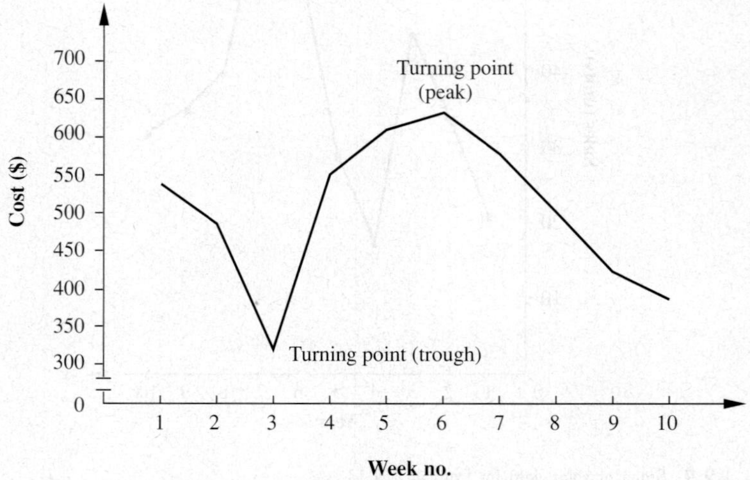

Figure M9.11 Graph of the cost curve for the data in Table M9.1 in Example M9.13

As well as identifying turning points on a *cost curve*, the same technique can be applied to a *revenue curve* to determine those occasions when a company's revenue takes a change of direction either upwards or downwards.

M9.7 REMARKS

There are currently on the market sophisticated hand-held calculators that have a small screen and can actually draw the graph of any function entered by the user. Special models of these will even allow the user to display these graphs on an overhead projector or send them to a printer. Unfortunately, at this time such calculators are relatively expensive in Australia and their use is not yet widespread.

However, many people have access to a computer and in particular programs with a graphics capability. These include spreadsheets such as Microsoft Excel, along with statistical packages such as Minitab and SPSS, which each have their graphics capabilities briefly outlined in Appendix C at the end of the statistics section on page S.436. Another alternative is to use a specialised computer graphics package such as Harvard Graphics®.

Quick Quiz

1. Plot on a graph the equation of the line $2x - 3y = 24$.

2. Plot on a graph the equation of the line $-1.5x + 2.5y = 15$.

3. Solve the following simultaneous equations with the aid of a graph:

$2x + 3y = 18$

$6x - 5y = 14$

4. A company that manufactures bedside lamps has a weekly fixed cost of $1200 for producing the lamps and a variable cost of $45 per lamp. The company receives an income of $75 for each lamp that it sells. Find the total cost of producing 120 lamps per week.

5. For the data in Question 4, find the profit (or loss) if the company makes and sells 150 lamps in a particular week.

6. For the data in Question 4, with the aid of a graph, find the point at which the total cost equals total income. That is, find the weekly sales that represent the break-even point.

7. A sports store pays a supplier $15 for each football and sells them for $25. It also pays the supplier $30 for each hockey stick and sells them for $45. Suppose that the store has a budget of $500 to purchase footballs and hockey sticks. Write two simultaneous equations that can be used to determine how many of each should be purchased so that the total profit made on each type of equipment is the same. (Define your variables.)

8. Draw a graph of the two equations in Question 7. Find the point of intersection and interpret its value.

9. Plot a graph of the equation

$y = x^2 - x$

and make a note of where the curve crosses the *x*-axis.

10. If the fixed cost of producing an item increases by 50%, what is the effect on the break-even point?

Multiple Choice Questions

1. The angle between the positive x-axis and the positive y-axis is:
(a) 60° (b) 90° (c) 180°
(d) 270° (e) none of the above

2. If a straight line passes through the origin, which of the following statements is *always* true?
(a) the slope is zero
(b) the slope is positive
(c) the slope is negative
(d) there is no constant term in the equation
(e) none of the above statements is always true

3. If two lines intersect at the origin, which of the following statements is *always* true?
(a) both lines have positive slopes
(b) both lines have negative slopes
(c) both lines have different slopes
(d) the two lines are parallel
(e) none of the above statements is always true

4. If two lines are parallel, which of the following statements is *always* true?
(a) both lines have positive slopes
(b) at least one line passes through the origin
(c) both lines have the same slope
(d) at least one line has no constant term
(e) none of the above statements is always true

5. A company that manufactures seat belts has a weekly fixed cost of \$2700 and variable costs of \$25 per belt. Each belt produces an income of \$45. The weekly sales that represent the break-even point are:
(a) 1350 belts (b) 600 belts (c) 135 belts
(d) 540 belts (e) none of the above

6. For the data in Question 5, if the company produces 800 belts in 1 week, the result is:
(a) a loss of \$1100
(b) no profit or loss
(c) a profit of \$9300
(d) a profit of \$800
(e) none of the above

7. If the income per item is doubled, the variable cost per item is doubled and the fixed cost is tripled. The effect on the break-even point is:
(a) it stays the same
(b) it increases by a factor of 6
(c) it increases by a factor of 12
(d) it decreases by 50%
(e) none of the above

8. The point of intersection of the two lines:
$2x + 3y = 5$
$20x - 9y = 21$
is:
(a) $x = 1, y = 1$
(b) $x = 1.5, y = 1$
(c) $x = 2.5, y = 0$
(d) $x = 0, y = \frac{5}{3}$
(e) none of the above

9. A small business operating at a market sells watches that it purchases for \$10 each from the supplier and sells for \$16 each. The cost of renting the stall at the market is \$300 for an 8-hour day. The minimum number of watches that the business would have to sell each hour to at least break even is:
(a) 8 (b) 9 (c) 10
(d) 11 (e) none of the above

10. In Question 9, the number of watches that the business would have to sell each day to have a total daily profit (selling price minus purchase price) of *exactly* \$200 is:
(a) 100 (b) 800 (c) 125
(d) 50 (e) none of the above

Exercises

M9.1 The average price of 8 grocery items in Perth is shown in the table below for the September quarters in 1985 and 2000.

Item	Unit	Price in 1985 ($)	Price in 2000 ($)
Milk	1 L	0.74	1.44
Rice	1 kg	0.77	1.77
Dry biscuits	250 g	0.87	1.60
Bananas	1 kg	1.62	2.86
Peaches	825 g can	1.33	2.17
Eggs, 55 g	1 dozen	1.61	2.96
Pet food	400 g	0.71	1.00
Toothpaste	140 g	1.54	2.30

Source: Australian Bureau of Statistics, *Average Retail Prices of Selected Items*, September Quarters 1985 and 2000, Cat. No. 6403.0.

(a) Represent these data as 8 ordered pairs in the form (price in 1985, price in 2000).

(b) Using an appropriate scale on each axis, plot the points in (a) on a graph. Mark the points for each year using a different symbol.

(c) Are there any unusual points on your graph in (b)? If so, identify them and explain why they are unusual.

M9.2* Plot the following points in the x–y plane and label them:

(a) (2, 5) (b) (−2, 4) (c) (0, 5) (d) (3, −3)

M9.3 Plot the following points in the x–y plane and label them:

(a) (−1.5, −3.5) (b) (2.2, 0) (c) (1.2, −2.6) (d) (−0.4, 0.4)

M9.4 Plot on a graph the line represented by the equation $y = x$.

M9.5* Plot on a graph the line represented by the equation $y = x + 3$.

M9.6* Plot on a graph the line represented by the equation $y = 2x - 1$.

M9.7 Plot on a graph the line represented by the equation $y = 4.5 - 1.5x$.

M9.8* Write each of the following equations in one of the forms in Equation M9.1:

(a) $3x + 6y = 24$

(b) $5x - 3y = 13$

(c) $-4x + 7y = 14$

(d) $3x = 2y + 12$

M9.9* Consider the two simultaneous equations below:

$x + y = 6$

$x - y = 2$

(a) Solve these equations with the aid of a graph.

(b) Solve the equations algebraically and check that your answer agrees with that in (a).

M9.10* Consider the two simultaneous equations below:

$2x + y = 11$

$5x - 2y = 5$

(a) Solve these equations with the aid of a graph.

(b) Solve the equations algebraically and check that your answer agrees with that in (a).

M9.11 Consider the two simultaneous equations below:

$4x - 2y = 13$

$7x - y = 19$

(a) Solve these equations with the aid of a graph.

(b) Solve the equations algebraically and check that your answer agrees with that in (a).

M9.12 Consider the two simultaneous equations below.

$2x + y = 4$

$4x + 2y = 8$

Try to solve these equations with the aid of a graph. Explain what happens. (*Hint*: Is there more than one solution? If so, can you write down the general form of *all* solutions?)

M9.13* A small business manufactures greeting cards. There is a weekly fixed cost of \$200 for leasing the machine and a variable cost of \$1.20 per card. The business receives an income of \$2.00 for each card that it sells.

(a) Find the total cost of producing 1000 cards in a week.

(b) Find the profit (or loss) if the business manufactures and sells 1000 cards in a particular week.

(c) With the aid of a graph, find the weekly sales that represent the break-even point.

M9.14 Answer the questions in Exercise M9.13 assuming that the weekly fixed cost of leasing the machine is \$300.

M9.15 Answer the questions in Exercise M9.13 assuming that the business receives an income of \$2.50 for each card that it sells.

M9.16* Use an algebraic formula to find the break-even point for the data in Exercise M9.13.

M9.17 Use an algebraic formula to find the break-even point for the data in Exercise M9.14.

M9.18* Due to student demand, a university union decides to set up a hot-dog stand. The machine for making the hot dogs can be leased for \$140 per week. The union pays the seller a commission of \$0.40 for each hot dog sold, and there is a \$0.75 cost for ingredients. Each hot dog sells for \$1.50.

(a) Find the total cost of selling 330 hot dogs.

(b) Find the profit (or loss) if the stand makes and sells 330 hot dogs in a particular week.

(c) With the aid of a graph, find the weekly sales that represent the break-even point.

M9.19* Find the break-even point in Exercise M9.18(c) by using Equation M9.7. Verify that your answer agrees with that found in Exercise M9.18(c).

M9.20* The leasing company in Exercise M9.18 decides to raise the leasing cost of the hot-dog machine to \$160 per week. Assuming that the variable costs are the same, how many hot dogs would have to be sold in a week to realise a profit of \$50?

M9.21 Consider a relationship between the variables x and y that follows the non-linear equation:

$y = 2x^2 - 6x$

(a) Plot points on the x–y plane for the following values of the x-variable:

$-4,\ -3,\ -2,\ -1,\ 0,\ 1,\ 2,\ 3,\ 4$

(b) Join these points in a smooth curve.

(c) Use the curve in (b) to estimate the value of x for which the value of y is at its minimum. Also find the corresponding value of y. (That is, find the coordinates of the turning point.)

M9.22* Figure M9.12 shows the monthly sales of airconditioners over a 12-month period at a discount store. Write down the values of the turning points of the graph, including the sales and the months in which they occurred.

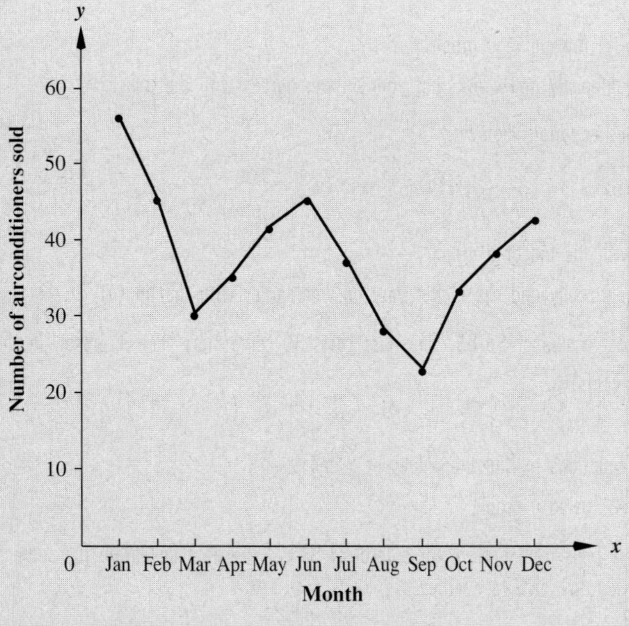

Figure M9.12 Monthly sales graph for Exercise M9.22

M9.23 Plot the following points in the x–y plane and label them:

(a) $(-5, -5)$ (b) $(5, -5)$ (c) $(-5, 5)$ (d) $(5, 5)$

M9.24 Plot on a graph the line represented by the equation $3y = 22 - x$.

M9.25* Plot on a graph the line represented by the equation $2x - y = 4$.

M9.26 Plot on a graph the line represented by the equation $x + 2y = 10$.

M9.27 Plot on a graph the line represented by the equation $-2x + 5y = 20$.

M9.28 Consider the two simultaneous equations below:

$5x - 2y = 1$

$3x + y = 5$

(a) Solve these equations with the aid of a graph.

(b) Solve the equations algebraically and check that your answer agrees with that in (a).

M9.29* Consider the two simultaneous equations below:

$3x - y = 2$

$5x + y = 10$

(a) Solve these equations with the aid of a graph.

(b) Solve the equations algebraically and check that your answer agrees with that in (a).

M9.30 Consider the two simultaneous equations below:

$5x + 2y = 24$

$x + 4y = 21$

(a) Solve these equations with the aid of a graph.

(b) Solve the equations algebraically and check that your answer agrees with that in (a).

M9.31 Consider the two simultaneous equations below:

$2x + y = 9$

$3x - y = 2$

(a) Solve these equations with the aid of a graph.

(b) Solve the equations algebraically and check that your answer agrees with that in (a).

M9.32* Three rulers and 2 erasers cost a sum of \$6.85. One ruler costs 30 cents more than 1 eraser. Define:

x = the cost of 1 ruler (in cents)

y = the cost of 1 eraser (in cents)

(a) Construct simultaneous equations to find the values of x and y.

(b) Solve the equations in (a) using a graph.

M9.33 A cricketer buys 4 bats and 3 balls from a store for a total of \$112. The next day, from the same store, he buys 2 bats and 6 balls for a total of \$74. Define:

x = the cost of 1 bat (in \$)

y = the cost of 1 ball (in \$)

(a) Construct simultaneous equations to find the values of x and y.

(b) Solve the equations in (a) using a graph.

M9.34* A photography teacher buys for her class 2 small cameras and 5 rolls of film from a photographic store for a total of \$107.50. Later that week, from the same store and at the same price per item, she buys 5 small cameras and 8 rolls of film for a total of \$253. Define:

x = the cost of 1 small camera (in \$)

y = the cost of 1 roll of film (in \$)

(a) Construct simultaneous equations to find the values of x and y.

(b) Solve the equations in (a) using a graph.

(c) Use your answer in (b) to find the total cost of 3 small cameras and 2 rolls of film.

M9.35* To raise money at a fete, a school hires a fairy-floss machine at a cost of \$40 per hour. The school must also pay the machine owner a commission of 20 cents for every unit of fairy-floss it sells, and the ingredients of each unit of fairy-floss cost 50 cents. The school estimates that it will sell 50 units per hour. If the desired profit is \$25 per hour, how much should the school charge for each unit of fairy-floss?

M9.36 (a) For the data in Exercise M9.35, use a graph to find the break-even point of the number of units of fairy-floss that must be sold per hour.

(b) Check your answer in (a) algebraically.

M9.37* In Exercise M9.35, calculate the total profit or loss in an 8-hour day if the school decides to charge the following amount for each unit of fairy-floss:

(a) 90 cents (b) \$1.75 (c) \$3.00

M9.38 A company that manufactures wall clocks has a monthly fixed cost of \$1620 for producing the clocks and a variable production cost of \$80 per clock. The company receives \$125 for each clock sold.

(a) Find the annual profit or loss if the company sells 30 clocks per month.

(b) Find the break-even point for monthly sales using a graph.

(c) Find the break-even point for monthly sales using algebra. Check that your answer agrees with that in (b).

M9.39* (a) Draw the graphs of the two equations on the same diagram:

$y = x^2 - 2x$

$y = 2x - 3$

As accurately as you can, find the points of intersection of the two graphs.

(b) Use your answers in (a) to solve for x the equation:

$x^2 - 4x + 3 = 0$

Explain your reasoning.

M9.40 (a) Draw the graphs of the two equations on the same diagram:

$y = x^2 - 2$

$y = 5x - 8$

(b) As accurately as you can, find the points of intersection of the two graphs.

(c) Use your answers in (b) to solve for x the equation:

$x^2 - 5x + 6 = 0$

Explain your reasoning.

M9.41* A stall that sells lemonade pays \$20 per hour to the local council for the licence to operate the stall in the market square. It costs the stall owners 80 cents for the ingredients (including the plastic cup) of one cup of lemonade. If the break-even point is 50 cups per hour, for how much are the stall owners selling each cup?

M9.42* In Exercise M9.41, if the council raises the licence fee to \$30 per hour, and all other data remain the same, find the new break-even point (in cups per hour).

M9.43 In Exercise M9.41, suppose that the council decides to waive all licence fees.

(a) What is the new break-even point if the stall drops its price to $1.00 per cup?

(b) If the price charged per cup is 80 cents, what happens to the break-even point?

M9.44 The weekly takings (in $'000) for a supermarket over a 16-week period are shown in the table below:

Week no.	1	2	3	4	5	6	7	8
Takings ($'000)	25	28	32	35	31	27	22	19

Week no.	9	10	11	12	13	14	15	16
Takings ($'000)	16	20	23	27	31	37	33	25

(a) Plot the data on a graph with week on the horizontal axis and takings on the vertical axis. Join the points with straight lines.

(b) Identify the turning points on the graph in (a) and describe their nature.

Appendixes

Appendix A A TEST OF BASIC MATHEMATICS

This appendix contains a short test of basic mathematical skills that may be required in business. The types of problems covered are similar to those discussed in Chapters M1, M2 and M3 of this text.

1. Write down in words the number represented by the numeral 72 386.

2. Evaluate $12 - 4 + 9 - 5 + 13 - 6$.

3. Evaluate $(3 + 5) \times 4$.

4. Evaluate $3 + (5 \times 4)$.

5. Add $13 + 14$.

6. Divide $\frac{1}{3}$ by $\frac{1}{4}$.

7. Find the lowest common multiple of 9, 16, 24.

8. Multiply $\frac{3}{4}, \frac{14}{5}, \frac{8}{21}$.

9. Divide $\frac{15}{4}$ by $\frac{3}{8}$.

10. Express 0.625 as a fraction.

11. Express $\frac{7}{12}$ as a decimal.

12. Evaluate $0.235 + 0.36 - 0.4$.

13. Evaluate $2^4 \times 2^7 \times 2^{-6}$.

14. Convert $\frac{4}{5}$ to a percentage.

15. Convert 55% to a fraction in its lowest terms.

16. Find 30% of 160.

17. Simplify $5x + 3y - 2x + 4y$.

18. Simplify $(3x - 2)(2x + 1)$.

19. Solve $5x - 2 = 12 - 2x$ for x.

20. Solve these simultaneous equations for x and y:

$$4x + 3y = 22$$
$$6x - 2y = 7$$

Appendix B TRIAL EXAMINATION

Paper A (Chapters M1–M9)

Time allowed: One hour
Number of questions: Five
Show all working
The questions are of equal value

1. A bookstore marks its textbooks 35% over the cost price from the publisher.

(a) If the bookstore pays the publisher $48 for a textbook, for what price will the book retail (including GST)?

(b) If the store sells a textbook for $83.16 (including GST), what is the cost price from the publisher?

(c) If the bookstore sells a book for $38.50, how much is the GST component?

2. A retail store purchased 25 units of a VCR player, these being a mixture of the long-playing (LP) and short-playing (SP) models. The LP models sold for $475 each and the SP models for $350 each. When all the VCR players were sold, the total amount that the store received from the sales was $10 500.

(a) How many each of the LP and SP models were sold?

(b) What would be the total cost of 5 LP models and 3 SP models?

(c) If the store now gave a 10% discount off the LP models and a 15% discount off the SP models, what would be the total cost of the order in (b)?

3. Three directors, Annie, Roula and Nick, agree to share their company's profits under the following arrangement:

- Annie receives 30% of the profit or $25 000, whichever is the smaller amount.
- After Annie takes her share, Roula receives 35% of the remaining profit.
- Nick receives what is left after Annie and Roula take their share.

Calculate how much each director will receive if the company's profit is:

(a) $50 000 (b) $100 000

4. A mechanic purchased a set of screwdrivers with an ATO effective life of 5 years. Using the *diminishing value* method, at the end of the second year they had a book value of $125.44.

(a) Find the original cost of the screwdrivers.

(b) Construct the depreciation schedule for the first 5 years.

(c) Find the book value at the end of the tenth year.

5. A baker invests $15 000 at a simple interest rate of 3.75% per annum. Approximately how long will it take (to the nearest 6 months) for her to double her money?

Tables

- **TABLE 1** AMOUNT AT COMPOUND INTEREST TABLES $(1 + i)^n$
- **TABLE 2** PRESENT VALUE AT COMPOUND INTEREST TABLES $(1 + i)^{-n}$
- **TABLE 3** FUTURE VALUE OF AN ANNUITY TABLES $s_{\overline{n|}\, i} = \dfrac{(1 + i)^n - 1}{i}$
- **TABLE 4** PRESENT VALUE OF AN ANNUITY TABLES $a_{\overline{n|}\, i} = \dfrac{1 - (1 + i)^n}{i}$
- **TABLE 5** TABLE OF COMMON LOGARITHMS $\mathrm{LOG}_{10} N$

Table 1 AMOUNT AT COMPOUND INTEREST TABLES $(1 + i)^n$

$n \backslash i$	$\frac{5}{12}\%$	$\frac{1}{2}\%$	$\frac{2}{3}\%$	$\frac{3}{4}\%$	1%	$1\frac{1}{4}\%$	i / n
1	1.004 166 67	1.005 000 00	1.006 666 67	1.007 500 00	1.010 000 00	1.012 500 00	**1**
2	1.008 350 69	1.010 025 00	1.013 377 78	1.015 056 25	1.020 100 00	1.025 156 25	2
3	1.012 552 16	1.015 075 13	1.020 133 63	1.022 669 17	1.030 301 00	1.037 970 70	3
4	1.016 771 12	1.020 150 50	1.026 934 52	1.030 339 19	1.040 604 01	1.050 945 34	4
5	1.021 007 67	1.025 251 25	1.033 780 75	1.038 066 73	1.051 010 05	1.064 082 15	5
6	1.025 261 87	1.030 377 51	1.040 672 62	1.045 852 24	1.061 520 15	1.077 383 18	6
7	1.029 533 79	1.035 529 40	1.047 610 44	1.053 696 13	1.072 135 35	1.090 850 47	7
8	1.033 823 52	1.040 707 04	1.054 594 51	1.061 598 85	1.082 856 71	1.104 486 10	8
9	1.038 131 11	1.045 910 58	1.061 625 14	1.069 560 84	1.093 685 27	1.118 292 18	9
10	1.042 456 66	1.051 140 13	1.068 702 64	1.077 582 55	1.104 622 13	1.132 270 83	**10**
11	1.046 800 23	1.056 395 83	1.075 827 32	1.085 664 41	1.115 668 35	1.146 424 22	11
12	1.051 161 90	1.061 677 81	1.082 999 51	1.093 806 90	1.126 825 03	1.160 754 52	12
13	1.055 541 74	1.066 986 20	1.090 219 50	1.102 010 45	1.138 093 28	1.175 263 95	13
14	1.059 939 83	1.072 321 13	1.097 487 63	1.110 275 53	1.149 474 21	1.189 954 75	14
15	1.064 356 25	1.077 682 74	1.104 804 22	1.118 602 59	1.160 968 96	1.204 829 18	15
16	1.068 791 06	1.083 071 15	1.112 169 58	1.126 992 11	1.172 578 64	1.219 889 55	16
17	1.073 244 36	1.088 486 51	1.119 584 04	1.135 444 55	1.184 304 43	1.235 138 17	17
18	1.077 716 21	1.093 928 94	1.127 047 94	1.143 960 39	1.196 147 48	1.250 577 39	18
19	1.082 206 70	1.099 398 58	1.134 561 59	1.152 540 09	1.208 108 95	1.266 209 61	19
20	1.086 715 89	1.104 895 58	1.142 125 33	1.161 184 14	1.220 190 04	1.282 037 23	**20**
21	1.091 243 87	1.110 420 06	1.149 739 50	1.169 893 02	1.232 391 94	1.298 062 70	21
22	1.095 790 72	1.115 972 16	1.157 404 43	1.178 667 22	1.244 715 86	1.314 288 48	22
23	1.100 356 52	1.121 552 02	1.165 120 46	1.187 507 23	1.257 163 02	1.330 717 09	23
24	1.104 941 34	1.127 159 78	1.172 887 93	1.196 413 53	1.269 734 65	1.347 351 05	24
25	1.109 545 26	1.132 795 58	1.180 707 18	1.205 386 63	1.282 432 00	1.364 192 94	25
26	1.114 168 36	1.138 459 55	1.188 578 57	1.214 427 03	1.295 256 31	1.381 245 35	26
27	1.118 810 73	1.144 151 85	1.196 502 42	1.223 535 23	1.308 208 88	1.398 510 92	27
28	1.123 472 44	1.149 872 61	1.204 479 11	1.232 711 75	1.321 290 97	1.415 992 30	28
29	1.128 153 58	1.155 621 97	1.212 508 97	1.241 957 09	1.334 503 88	1.433 692 21	29
30	1.132 854 22	1.161 400 08	1.220 592 36	1.251 271 76	1.347 848 92	1.451 613 36	**30**
31	1.137 574 44	1.167 207 08	1.228 729 64	1.260 656 30	1.361 327 40	1.469 758 53	31
32	1.142 314 34	1.173 043 12	1.236 921 17	1.270 111 22	1.374 940 68	1.488 130 51	32
33	1.147 073 98	1.178 908 33	1.245 167 31	1.279 637 06	1.388 690 09	1.506 732 14	33
34	1.151 853 46	1.184 802 88	1.253 468 43	1.289 234 34	1.402 576 99	1.525 566 29	34
35	1.156 652 84	1.190 726 89	1.261 824 89	1.298 903 59	1.416 602 76	1.544 635 87	35
36	1.161 472 23	1.196 680 52	1.270 237 05	1.308 645 37	1.430 768 78	1.563 943 82	36
37	1.166 311 70	1.202 663 93	1.278 705 30	1.318 460 21	1.445 076 47	1.583 493 12	37
38	1.171 171 33	1.208 677 25	1.287 230 00	1.328 348 66	1.459 527 24	1.603 286 78	38
39	1.176 051 21	1.214 720 63	1.295 811 53	1.338 311 28	1.474 122 51	1.623 327 87	39
40	1.180 951 42	1.220 794 24	1.304 450 28	1.348 348 61	1.488 863 73	1.643 619 46	**40**
41	1.185 872 06	1.226 898 21	1.313 146 61	1.358 461 23	1.503 752 37	1.664 164 71	41
42	1.190 813 19	1.233 032 70	1.321 900 92	1.368 649 69	1.518 789 89	1.684 966 77	42
43	1.195 774 91	1.239 197 86	1.330 713 60	1.378 914 56	1.533 977 79	1.706 028 85	43
44	1.200 757 31	1.245 393 85	1.339 585 02	1.389 256 42	1.549 317 57	1.727 354 21	44
45	1.205 760 46	1.251 620 82	1.348 515 59	1.399 675 84	1.564 810 75	1.748 946 14	45
46	1.210 784 46	1.257 878 92	1.357 505 69	1.410 173 41	1.580 458 85	1.770 807 97	46
47	1.215 829 40	1.264 168 32	1.366 555 73	1.420 749 71	1.596 263 44	1.792 943 06	47
48	1.220 895 36	1.270 489 16	1.375 666 10	1.431 405 33	1.612 226 08	1.815 354 85	48
49	1.225 982 42	1.276 841 61	1.384 837 21	1.442 140 87	1.628 348 34	1.838 046 79	49
50	1.231 090 68	1.283 225 81	1.394 069 46	1.452 956 93	1.644 631 82	1.861 022 37	**50**
51	1.236 220 22	1.289 641 94	1.403 363 25	1.463 854 11	1.661 078 14	1.884 285 15	51
52	1.241 371 14	1.296 090 15	1.412 719 01	1.474 833 01	1.677 688 92	1.907 838 72	52
53	1.246 543 52	1.302 570 60	1.422 137 13	1.485 894 26	1.694 465 81	1.931 686 70	53
54	1.251 737 45	1.309 083 46	1.431 618 05	1.497 038 47	1.711 410 47	1.955 832 79	54
55	1.256 953 02	1.315 628 87	1.441 162 17	1.508 266 26	1.728 524 57	1.980 280 70	55
56	1.262 190 33	1.322 207 02	1.450 769 92	1.519 578 25	1.745 809 82	2.005 034 20	56
57	1.267 449 46	1.328 818 05	1.460 441 72	1.530 975 09	1.763 267 92	2.030 097 13	57
58	1.272 730 50	1.335 462 14	1.470 177 99	1.542 457 40	1.780 900 60	2.055 473 35	58
59	1.278 033 54	1.342 139 46	1.479 979 18	1.554 025 83	1.798 709 60	2.081 166 76	59
60	1.283 358 68	1.348 850 15	1.489 845 71	1.565 681 03	1.816 696 70	2.107 181 35	**60**

(continued)

Table 1 AMOUNT AT COMPOUND INTEREST TABLES $(1 + i)^n$ *(continued)*

n \ i	$1\frac{1}{2}\%$	$1\frac{3}{4}\%$	2%	$2\frac{1}{4}\%$	$2\frac{1}{2}\%$	$2\frac{3}{4}\%$	i / n
1	1.015 000 00	1.017 500 00	1.020 000 00	1.022 500 00	1.025 000 00	1.027 500 00	**1**
2	1.030 225 00	1.035 306 25	1.040 400 00	1.045 506 25	1.050 625 00	1.055 756 25	2
3	1.045 678 38	1.053 424 11	1.061 208 00	1.069 030 14	1.076 890 62	1.084 789 55	3
4	1.061 363 55	1.071 859 03	1.082 432 16	1.093 083 32	1.103 812 89	1.114 621 26	4
5	1.077 284 00	1.090 616 56	1.104 080 80	1.117 677 69	1.131 408 21	1.145 273 34	5
6	1.093 443 26	1.109 702 35	1.126 162 42	1.142 825 44	1.159 693 42	1.176 768 36	6
7	1.109 844 91	1.129 122 15	1.148 685 67	1.168 539 01	1.188 685 75	1.209 129 49	7
8	1.126 492 59	1.148 881 78	1.171 659 38	1.194 831 14	1.218 402 90	1.242 380 55	8
9	1.143 389 98	1.168 987 21	1.195 092 57	1.221 714 84	1.248 862 97	1.276 546 02	9
10	1.160 540 83	1.189 444 49	1.218 994 42	1.249 203 43	1.280 084 54	1.311 651 03	**10**
11	1.177 948 94	1.210 259 77	1.243 374 31	1.277 310 50	1.312 086 66	1.347 721 44	11
12	1.195 618 17	1.231 439 31	1.268 241 79	1.306 049 99	1.344 888 82	1.384 783 78	12
13	1.213 552 44	1.252 989 50	1.293 606 63	1.335 436 11	1.378 511 04	1.422 865 33	13
14	1.231 755 73	1.274 916 82	1.319 478 76	1.365 483 43	1.412 973 82	1.461 994 13	14
15	1.250 232 07	1.297 227 86	1.345 868 34	1.396 206 80	1.448 298 17	1.502 198 96	15
16	1.268 985 55	1.319 929 35	1.372 785 71	1.427 621 46	1.484 505 62	1.543 509 44	16
17	1.288 020 33	1.343 028 11	1.400 241 42	1.459 742 94	1.521 618 26	1.585 955 95	17
18	1.307 340 64	1.366 531 11	1.428 246 25	1.492 587 16	1.559 658 72	1.629 569 73	18
19	1.326 950 75	1.390 445 40	1.456 811 17	1.526 170 37	1.598 650 19	1.674 382 90	19
20	1.346 855 01	1.414 778 20	1.485 947 40	1.560 509 20	1.638 616 44	1.720 428 43	**20**
21	1.367 057 83	1.439 536 81	1.515 666 34	1.595 620 66	1.679 581 85	1.767 740 21	21
22	1.387 563 70	1.464 728 71	1.545 979 67	1.631 522 12	1.721 571 40	1.816 353 07	22
23	1.408 377 15	1.490 361 46	1.576 899 26	1.668 231 37	1.764 610 68	1.866 302 78	23
24	1.429 502 81	1.516 442 79	1.608 437 25	1.705 766 58	1.808 725 95	1.917 626 10	24
25	1.450 945 35	1.542 980 54	1.640 605 99	1.744 146 32	1.853 944 10	1.970 360 82	25
26	1.472 709 53	1.569 982 69	1.673 418 11	1.783 389 62	1.900 292 70	2.024 545 75	26
27	1.494 800 18	1.597 457 39	1.706 886 48	1.823 515 88	1.947 800 02	2.080 220 75	27
28	1.517 222 18	1.625 412 90	1.741 024 21	1.864 544 99	1.996 495 02	2.137 426 82	28
29	1.539 980 51	1.653 857 62	1.775 844 69	1.906 497 25	2.046 407 39	2.196 206 06	29
30	1.563 080 22	1.682 800 13	1.811 361 58	1.949 393 44	2.097 567 58	2.256 601 73	**30**
31	1.586 526 42	1.712 249 13	1.847 588 82	1.993 254 79	2.150 006 77	2.318 658 28	31
32	1.610 324 32	1.742 213 49	1.884 540 59	2.038 103 03	2.203 756 94	2.382 421 38	32
33	1.634 479 18	1.772 702 23	1.922 231 40	2.083 960 34	2.258 850 86	2.447 937 97	33
34	1.658 996 37	1.803 724 52	1.960 676 03	2.130 849 45	2.315 322 13	2.515 256 26	34
35	1.683 881 32	1.835 289 70	1.999 889 55	2.178 793 56	2.373 205 19	2.584 425 81	35
36	1.709 139 54	1.867 407 27	2.039 887 34	2.227 816 42	2.432 535 32	2.655 497 52	36
37	1.734 776 63	1.900 086 89	2.080 685 09	2.277 942 29	2.493 348 70	2.728 523 70	37
38	1.760 798 28	1.933 338 41	2.122 298 79	2.329 195 99	2.555 682 42	2.803 558 10	38
39	1.787 210 25	1.967 171 84	2.164 744 77	2.381 602 90	2.619 574 48	2.880 655 95	39
40	1.814 018 41	2.001 597 34	2.208 039 66	2.435 188 97	2.685 063 84	2.959 873 99	**40**
41	1.841 228 68	2.036 625 30	2.252 200 46	2.489 980 72	2.752 190 43	3.041 270 52	41
42	1.868 847 12	2.072 266 24	2.297 244 47	2.546 005 28	2.820 995 20	3.124 905 46	42
43	1.896 879 82	2.108 530 90	2.343 189 36	2.603 290 40	2.891 520 08	3.210 840 36	43
44	1.925 333 02	2.145 430 19	2.390 053 14	2.661 864 44	2.963 808 08	3.299 138 47	44
45	1.954 213 01	2.182 975 22	2.437 854 21	2.721 756 39	3.037 903 28	3.389 864 78	45
46	1.983 526 21	2.221 177 28	2.486 611 29	2.782 995 90	3.113 850 86	3.483 086 06	46
47	2.013 279 10	2.260 047 89	2.536 343 52	2.845 613 31	3.191 697 13	3.578 870 93	47
48	2.043 478 29	2.299 598 72	2.587 070 39	2.909 639 61	3.271 489 56	3.677 289 88	48
49	2.074 130 46	2.339 841 70	2.638 811 79	2.975 106 50	3.353 276 80	3.778 415 35	49
50	2.105 242 42	2.380 788 93	2.691 588 03	3.042 046 40	3.437 108 72	3.882 321 77	**50**
51	2.136 821 06	2.422 452 74	2.745 419 79	3.110 492 44	3.523 036 44	3.989 085 62	51
52	2.168 873 37	2.464 845 66	2.800 328 19	3.180 478 52	3.611 112 35	4.098 785 47	52
53	2.201 406 47	2.507 980 46	2.856 334 75	3.252 039 29	3.701 390 16	4.211 502 08	53
54	2.234 427 57	2.551 870 12	2.913 461 44	3.325 210 17	3.793 924 91	4.327 318 38	54
55	2.267 943 98	2.596 527 85	2.971 730 67	3.400 027 40	3.888 773 03	4.446 319 64	55
56	2.301 963 14	2.641 967 08	3.031 165 29	3.476 528 02	3.985 992 36	4.568 593 43	56
57	2.336 492 59	2.688 201 51	3.091 788 59	3.554 749 90	4.085 642 17	4.694 229 75	57
58	2.371 539 98	2.735 245 03	3.153 624 36	3.634 731 77	4.187 783 22	4.823 321 07	58
59	2.407 113 08	2.783 111 82	3.216 696 85	3.716 513 24	4.292 477 80	4.955 962 39	59
60	2.443 219 78	2.831 816 28	3.281 030 79	3.800 134 79	4.399 789 75	5.092 251 36	**60**

(continued)

Table 1 AMOUNT AT COMPOUND INTEREST TABLES $(1 + i)^n$ *(continued)*

n \ i	3%	$3\frac{1}{2}$%	4%	$4\frac{1}{2}$%	5%	6%	i / n
1	1.030 000 00	1.035 000 00	1.040 000 00	1.045 000 00	1.050 000 00	1.060 000 00	**1**
2	1.060 900 00	1.071 225 00	1.081 600 00	1.092 025 00	1.102 500 00	1.123 600 00	2
3	1.092 727 00	1.108 787 87	1.124 864 00	1.141 166 13	1.157 625 00	1.191 016 00	3
4	1.125 508 81	1.147 523 00	1.169 858 56	1.192 518 60	1.215 506 25	1.262 476 96	4
5	1.159 274 07	1.187 686 31	1.216 652 90	1.246 181 94	1.276 281 56	1.338 225 58	5
6	1.194 052 30	1.229 255 33	1.265 319 02	1.302 260 12	1.340 095 64	1.418 519 11	6
7	1.229 873 87	1.272 279 26	1.315 931 78	1.360 861 83	1.407 100 42	1.503 630 26	7
8	1.266 770 08	1.316 809 04	1.368 569 05	1.422 100 61	1.477 455 44	1.593 848 07	8
9	1.304 773 18	1.362 897 35	1.423 311 81	1.486 095 14	1.551 328 22	1.689 478 96	9
10	1.343 916 38	1.410 598 76	1.480 244 28	1.552 969 42	1.628 894 63	1.790 847 70	**10**
11	1.384 233 87	1.459 969 72	1.539 454 06	1.622 853 05	1.710 339 36	1.898 298 56	11
12	1.425 760 89	1.511 068 66	1.601 032 22	1.695 881 43	1.795 856 33	2.012 196 47	12
13	1.468 533 71	1.563 956 06	1.665 073 51	1.772 196 10	1.885 649 14	2.132 928 26	13
14	1.512 589 72	1.618 694 52	1.731 676 45	1.851 944 92	1.979 931 60	2.260 903 96	14
15	1.557 967 42	1.675 348 83	1.800 943 51	1.935 282 44	2.078 928 18	2.396 558 19	15
16	1.604 706 44	1.733 986 04	1.872 981 25	2.022 370 15	2.182 874 59	2.540 351 68	16
17	1.652 847 63	1.794 675 55	1.947 900 50	2.113 376 81	2.292 018 32	2.692 772 79	17
18	1.702 433 06	1.857 489 20	2.025 816 52	2.208 478 77	2.406 619 23	2.854 339 15	18
19	1.753 506 05	1.922 501 32	2.106 849 18	2.307 860 31	2.526 950 20	3.025 599 50	19
20	1.806 111 23	1.989 788 86	2.191 123 14	2.411 714 02	2.653 297 71	3.207 135 47	**20**
21	1.860 294 57	2.059 431 47	2.278 768 07	2.520 241 16	2.785 962 59	3.399 563 60	21
22	1.916 103 41	2.131 511 58	2.369 918 79	2.633 652 01	2.925 260 72	3.603 537 42	22
23	1.973 586 51	2.206 114 48	2.464 715 54	2.752 166 35	3.071 523 76	3.819 749 66	23
24	2.032 794 11	2.283 328 49	2.563 304 16	2.876 013 83	3.225 099 94	4.048 934 64	24
25	2.093 777 93	2.363 244 98	2.665 836 33	3.005 434 46	3.386 354 94	4.291 870 72	25
26	2.156 591 27	2.445 958 56	2.772 469 78	3.140 679 01	3.555 672 69	4.549 382 96	26
27	2.221 289 01	2.531 567 11	2.883 368 58	3.282 009 56	3.733 456 32	4.822 345 94	27
28	2.287 927 68	2.620 171 96	2.998 703 32	3.429 699 99	3.920 129 14	5.111 686 70	28
29	2.356 565 51	2.711 877 98	3.118 651 45	3.584 036 49	4.116 135 60	5.418 387 90	29
30	2.427 262 47	2.806 793 70	3.243 397 51	3.745 318 13	4.321 942 38	5.743 491 17	**30**
31	2.500 080 35	2.905 031 48	3.373 133 41	3.913 857 45	4.538 039 49	6.088 100 64	31
32	2.575 082 76	3.006 707 59	3.508 058 75	4.089 981 04	4.764 941 47	6.453 386 68	32
33	2.652 335 24	3.111 942 35	3.648 381 10	4.274 030 18	5.003 188 54	6.840 589 88	33
34	2.731 905 30	3.220 860 33	3.794 316 34	4.466 361 54	5.253 347 97	7.251 025 28	34
35	2.813 862 45	3.333 590 45	3.946 088 99	4.667 347 81	5.516 016 37	7.686 086 79	35
36	2.898 278 33	3.450 266 11	4.103 932 55	4.877 378 46	5.791 816 14	8.147 252 00	36
37	2.985 226 68	3.571 025 43	4.268 089 86	5.096 860 49	6.081 406 94	8.636 087 12	37
38	3.074 783 48	3.696 011 32	4.438 813 45	5.326 219 21	6.385 477 29	9.154 252 35	38
39	3.167 026 98	3.825 371 71	4.616 365 99	5.565 899 08	6.704 751 15	9.703 507 49	39
40	3.262 037 79	3.959 259 72	4.801 020 63	5.816 364 54	7.039 988 71	10.285 717 94	**40**
41	3.359 898 93	4.097 833 81	4.993 061 45	6.078 100 94	7.391 988 15	10.902 861 01	41
42	3.460 695 89	4.241 257 99	5.192 783 91	6.351 615 48	7.761 587 56	11.557 032 67	42
43	3.564 516 77	4.389 702 02	5.400 495 27	6.637 438 18	8.149 666 93	12.250 454 63	43
44	3.671 452 27	4.543 341 60	5.616 515 08	6.936 122 90	8.557 150 28	12.985 481 91	44
45	3.781 595 84	4.702 358 55	5.841 175 68	7.248 248 43	8.985 007 79	13.764 610 83	45
46	3.895 043 72	4.866 941 10	6.074 822 71	7.574 419 61	9.434 258 18	14.590 487 48	46
47	4.011 895 03	5.037 284 04	6.317 815 62	7.915 268 49	9.905 971 09	15.465 916 73	47
48	4.132 251 88	5.213 588 98	6.570 528 24	8.271 455 57	10.401 269 65	16.393 871 73	48
49	4.256 219 44	5.396 064 59	6.833 349 37	8.643 671 07	10.921 333 13	17.377 504 03	49
50	4.383 906 02	5.584 926 86	7.106 683 35	9.032 636 27	11.467 399 79	18.420 154 27	**50**
51	4.515 423 20	5.780 399 30	7.390 950 68	9.439 104 90	12.040 769 78	19.525 363 53	51
52	4.650 885 90	5.982 713 27	7.686 588 71	9.863 864 63	12.642 808 26	20.696 885 34	52
53	4.790 412 47	6.192 108 24	7.994 052 26	10.307 738 53	13.274 948 68	21.938 698 46	53
54	4.934 124 85	6.408 832 02	8.313 814 35	10.771 586 77	13.938 696 11	23.255 020 37	54
55	5.082 148 59	6.633 141 14	8.646 366 92	11.256 308 17	14.635 630 92	24.650 321 59	55
56	5.234 613 05	6.865 301 08	8.992 221 60	11.762 842 04	15.367 412 46	26.129 340 89	56
57	5.391 651 44	7.105 586 62	9.351 910 46	12.292 169 93	16.135 783 09	27.697 101 34	57
58	5.553 400 98	7.354 282 15	9.725 986 88	12.845 317 58	16.942 572 24	29.358 927 42	58
59	5.720 003 01	7.611 682 03	10.115 026 35	13.423 356 87	17.789 700 85	31.120 463 07	59
60	5.891 603 10	7.878 090 90	10.519 627 41	14.027 407 93	18.679 185 89	32.987 690 85	**60**

(continued)

Table 1 AMOUNT AT COMPOUND INTEREST TABLES $(1 + i)^n$ *(continued)*

n \ i	7%	8%	9%	10%	11%	12%	i / n
1	1.070 000 00	1.080 000 00	1.090 000 00	1.100 000 00	1.110 000 00	1.120 000 00	**1**
2	1.144 900 00	1.166 400 00	1.188 100 00	1.210 000 00	1.232 100 00	1.254 400 00	2
3	1.225 043 00	1.259 712 00	1.295 029 00	1.331 000 00	1.367 631 00	1.404 928 00	3
4	1.310 796 01	1.360 488 96	1.411 581 61	1.464 100 00	1.518 070 41	1.573 519 36	4
5	1.402 551 73	1.469 328 08	1.538 623 95	1.610 510 00	1.685 058 16	1.762 341 68	5
6	1.500 730 35	1.586 874 32	1.677 100 11	1.771 561 00	1.870 414 55	1.973 822 69	6
7	1.605 781 48	1.713 824 27	1.828 039 12	1.948 717 10	2.076 160 15	2.210 681 41	7
8	1.718 186 18	1.850 930 21	1.992 562 64	2.143 588 81	2.304 537 77	2.475 963 18	8
9	1.838 459 21	1.999 004 63	2.171 893 28	2.357 947 69	2.558 036 92	2.773 078 76	9
10	1.967 151 36	2.158 925 00	2.367 363 67	2.593 742 46	2.839 420 99	3.105 848 21	**10**
11	2.104 851 95	2.331 639 00	2.580 426 41	2.853 116 71	3.151 757 29	3.478 549 99	11
12	2.252 191 59	2.518 170 12	2.812 664 78	3.138 428 38	3.498 450 60	3.895 975 99	12
13	2.409 845 00	2.719 623 73	3.065 804 61	3.452 271 21	3.883 280 16	4.363 493 11	13
14	2.578 534 15	2.937 193 62	3.341 727 03	3.797 498 34	4.310 440 98	4.887 112 29	14
15	2.759 031 54	3.172 169 11	3.642 482 46	4.177 248 17	4.784 589 49	5.473 565 76	15
16	2.952 163 75	3.425 942 64	3.970 305 88	4.594 972 99	5.310 894 33	6.130 393 65	16
17	3.158 815 21	3.700 018 05	4.327 633 41	5.054 470 28	5.895 092 71	6.866 040 89	17
18	3.379 932 28	3.996 019 50	4.717 120 42	5.559 917 31	6.543 552 91	7.689 965 80	18
19	3.616 527 54	4.315 701 06	5.141 661 25	6.115 909 04	7.263 343 73	8.612 761 69	19
20	3.869 684 46	4.660 957 14	5.604 410 77	6.727 499 95	8.062 311 54	9.646 293 09	**20**
21	4.140 562 37	5.033 833 72	6.108 807 74	7.400 249 94	8.949 165 81	10.803 848 26	21
22	4.430 401 74	5.436 540 41	6.658 600 43	8.140 274 94	9.933 574 04	12.100 310 06	22
23	4.740 529 86	5.871 463 65	7.257 874 47	8.954 302 43	11.026 267 19	13.552 347 26	23
24	5.072 366 95	6.341 180 74	7.911 083 17	9.849 732 68	12.239 156 58	15.178 628 93	24
25	5.427 432 64	6.848 475 20	8.623 080 66	10.834 705 94	13.585 463 80	17.000 064 41	25
26	5.807 352 92	7.396 353 21	9.399 157 92	11.918 176 54	15.079 864 82	19.040 072 14	26
27	6.213 867 63	7.988 061 47	10.245 082 13	13.109 994 19	16.738 649 95	21.324 880 79	27
28	6.648 838 36	8.627 106 39	11.167 139 52	14.420 993 61	18.579 901 45	23.883 866 49	28
29	7.114 257 05	9.317 274 90	12.172 182 08	15.863 092 97	20.623 690 61	26.749 930 47	29
30	7.612 255 04	10.062 656 89	13.267 678 47	17.449 402 27	22.892 296 57	29.959 922 12	**30**
31	8.145 112 90	10.867 669 44	14.461 769 53	19.194 342 50	25.410 449 19	33.555 112 78	31
32	8.715 270 80	11.737 083 00	15.763 328 79	21.113 776 75	28.205 598 61	37.581 726 31	32
33	9.325 339 75	12.676 049 64	17.182 028 38	23.225 154 42	31.308 214 45	42.091 533 47	33
34	9.978 113 54	13.690 133 61	18.728 410 93	25.547 669 86	34.752 118 04	47.142 517 48	34
35	10.676 581 48	14.785 344 29	20.413 967 92	28.102 436 85	38.574 851 03	52.799 619 58	35
36	11.423 942 19	15.968 171 84	22.251 225 03	30.912 680 53	42.818 084 64	59.135 573 93	36
37	12.223 618 14	17.245 625 58	24.253 835 28	34.003 948 59	47.528 073 95	66.231 842 80	37
38	13.079 271 41	18.625 275 63	26.436 680 46	37.404 343 44	52.756 162 09	74.179 663 94	38
39	13.994 820 41	20.115 297 68	28.815 981 70	41.144 777 79	58.559 339 91	83.081 223 61	39
40	14.974 457 84	21.724 521 50	31.409 420 05	45.259 255 57	65.000 867 31	93.050 970 44	**40**
41	16.022 669 89	23.462 483 22	34.236 267 86	49.785 181 12	72.150 962 71	104.217 086 89	41
42	17.144 256 78	25.339 481 87	37.317 531 97	54.763 699 24	80.087 568 61	116.723 137 32	42
43	18.344 354 75	27.366 640 42	40.676 109 84	60.240 069 16	88.897 201 15	130.729 913 80	43
44	19.628 459 59	29.555 971 66	44.336 959 73	66.264 076 08	98.675 893 28	146.417 503 46	44
45	21.002 451 76	31.920 449 39	48.327 286 10	72.890 483 69	109.530 241 54	163.987 603 87	45
46	22.472 623 38	34.474 085 34	52.676 741 85	80.179 532 05	121.578 568 11	183.666 116 34	46
47	24.045 707 02	37.232 012 17	57.417 648 62	88.197 485 26	134.952 210 60	205.706 050 30	47
48	25.728 906 51	40.210 573 14	62.585 237 00	97.017 233 78	149.796 953 77	230.390 776 33	48
49	27.529 929 97	43.427 418 99	68.217 908 33	106.718 957 16	166.274 618 68	258.037 669 49	49
50	29.457 025 06	46.901 612 51	74.357 520 08	117.390 852 88	184.564 826 74	289.002 189 83	**50**
51	31.519 016 82	50.653 741 51	81.049 696 88	129.129 938 17	204.866 957 68	323.682 452 61	51
52	33.725 347 99	54.706 040 84	88.344 169 60	142.042 931 98	227.402 323 03	362.524 346 92	52
53	36.086 122 35	59.082 524 10	96.295 144 87	156.247 225 18	252.416 578 56	406.027 268 55	53
54	38.612 150 92	63.809 126 03	104.961 707 90	171.871 947 70	280.182 402 20	454.750 540 78	54
55	41.315 001 48	68.913 856 11	114.408 261 62	189.059 142 47	311.002 466 44	509.320 605 67	55
56	44.207 051 59	74.426 964 60	124.705 005 16	207.965 056 72	345.212 737 75	570.439 078 35	56
57	47.301 545 20	80.381 121 77	135.928 455 63	228.761 562 39	383.186 138 90	638.891 767 76	57
58	50.612 653 36	86.811 611 51	148.162 016 63	251.637 718 63	425.336 614 18	715.558 779 89	58
59	54.155 539 10	93.756 540 43	161.496 598 13	276.801 490 49	472.123 641 74	801.425 833 47	59
60	57.946 426 83	101.257 063 67	176.031 291 96	304.481 639 54	524.057 242 34	897.596 933 49	**60**

Table 2 PRESENT VALUE AT COMPOUND INTEREST TABLES $(1 + i)^{-n}$

n \ i	$\frac{5}{12}\%$	$\frac{1}{2}\%$	$\frac{2}{3}\%$	$\frac{3}{4}\%$	1%	$1\frac{1}{4}\%$	i / n
1	.995 850 62	.995 024 88	.993 377 48	.992 555 83	.990 099 01	.987 654 32	**1**
2	.991 718 46	.990 074 50	.986 798 82	.985 167 08	.980 296 05	.975 461 06	2
3	.987 603 45	.985 148 76	.980 263 73	.977 833 33	.970 590 15	.963 418 33	3
4	.983 505 51	.980 247 52	.973 771 92	.970 554 17	.960 980 34	.951 524 28	4
5	.979 424 57	.975 370 67	.967 323 10	.963 329 20	.951 465 69	.939 777 06	5
6	.975 360 57	.970 518 08	.960 916 99	.956 158 02	.942 045 24	.928 174 88	6
7	.971 313 43	.965 689 63	.954 553 30	.949 040 22	.932 718 05	.916 715 93	7
8	.967 283 08	.960 885 20	.948 231 75	.941 975 40	.923 483 22	.905 398 45	8
9	.963 269 46	.956 104 68	.941 952 07	.934 963 18	.914 339 82	.894 220 69	9
10	.959 272 49	.951 347 94	.935 713 98	.928 003 15	.905 286 95	.883 180 93	**10**
11	.955 292 11	.946 614 87	.929 517 20	.921 094 94	.896 323 72	.872 277 46	11
12	.951 328 24	.941 905 34	.923 361 45	.914 238 15	.887 449 23	.861 508 60	12
13	.947 380 82	.937 219 24	.917 246 48	.907 432 41	.878 662 60	.850 872 69	13
14	.943 449 78	.932 556 46	.911 172 00	.900 677 33	.869 962 97	.840 368 09	14
15	.939 535 05	.927 916 88	.905 137 75	.893 972 54	.861 349 47	.829 993 18	15
16	.935 636 57	.923 300 37	.899 143 46	.887 317 66	.852 821 26	.819 746 35	16
17	.931 754 26	.918 706 84	.893 188 86	.880 712 31	.844 377 49	.809 626 02	17
18	.927 888 06	.914 136 16	.887 273 71	.874 156 14	.836 017 31	.799 630 64	18
19	.924 037 90	.909 588 22	.881 397 72	.867 648 78	.827 739 92	.789 758 66	19
20	.920 203 72	.905 062 90	.875 560 65	.861 189 85	.819 544 47	.780 008 55	**20**
21	.916 385 44	.900 560 10	.869 762 24	.854 779 01	.811 430 17	.770 378 81	21
22	.912 583 01	.896 079 71	.864 002 22	.848 415 89	.803 396 21	.760 867 96	22
23	.908 796 36	.891 621 60	.858 280 35	.842 100 14	.795 441 79	.751 474 53	23
24	.905 025 42	.887 185 67	.852 596 38	.835 831 40	.787 566 13	.742 197 07	24
25	.901 270 13	.882 771 81	.846 950 04	.829 609 33	.779 768 44	.733 034 14	25
26	.897 530 42	.878 379 91	.841 341 10	.823 433 58	.772 047 96	.723 984 34	26
27	.893 806 23	.874 009 86	.835 769 31	.817 303 80	.764 403 92	.715 046 26	27
28	.890 097 49	.869 661 55	.830 234 41	.811 219 66	.756 835 57	.706 218 53	28
29	.886 404 14	.865 334 88	.824 736 17	.805 180 80	.749 342 15	.697 499 78	29
30	.882 726 11	.861 029 73	.819 274 34	.799 186 90	.741 922 92	.688 888 67	**30**
31	.879 063 35	.856 746 00	.813 848 68	.793 237 62	.734 577 15	.680 383 87	31
32	.875 415 78	.852 483 58	.808 458 96	.787 332 62	.727 304 11	.671 984 07	32
33	.871 783 35	.848 242 37	.803 104 92	.781 471 58	.720 103 07	.663 687 97	33
34	.868 165 99	.844 022 26	.797 786 35	.775 654 18	.712 973 34	.655 494 29	34
35	.864 563 65	.839 823 14	.792 502 99	.769 880 08	.705 914 20	.647 401 77	35
36	.860 976 24	.835 644 92	.787 254 63	.764 148 96	.698 924 95	.639 409 16	36
37	.857 403 73	.831 487 48	.782 041 02	.758 460 51	.692 004 90	.631 515 22	37
38	.853 846 04	.827 350 73	.776 861 94	.752 814 40	.685 153 37	.623 718 73	38
39	.850 303 11	.823 234 55	.771 717 16	.747 210 32	.678 369 67	.616 018 50	39
40	.846 774 88	.819 138 86	.766 606 45	.741 647 96	.671 653 14	.608 413 34	**40**
41	.843 261 29	.815 063 54	.761 529 59	.736 127 01	.665 003 11	.600 902 06	41
42	.839 762 28	.811 008 50	.756 486 35	.730 647 16	.658 418 92	.593 483 52	42
43	.836 277 79	.806 973 63	.751 476 50	.725 208 09	.651 899 92	.586 156 56	43
44	.832 807 76	.802 958 84	.746 499 84	.719 809 52	.645 445 46	.578 920 06	44
45	.829 352 12	.798 964 02	.741 556 13	.714 451 14	.639 054 92	.571 772 90	45
46	.825 910 83	.794 989 07	.736 645 16	.709 132 64	.632 727 64	.564 713 97	46
47	.822 483 81	.791 033 90	.731 766 72	.703 853 74	.626 463 01	.557 742 19	47
48	.819 071 02	.787 098 41	.726 920 58	.698 614 14	.620 260 41	.550 856 49	48
49	.815 672 38	.783 182 50	.722 106 54	.693 413 53	.614 119 21	.544 055 79	49
50	.812 287 85	.779 286 07	.717 324 37	.688 251 65	.608 038 82	.537 339 05	**50**
51	.808 917 36	.775 409 02	.712 573 88	.683 128 19	.602 018 64	.530 705 24	51
52	.805 560 86	.771 551 27	.707 854 85	.678 042 86	.596 058 06	.524 153 32	52
53	.802 218 28	.767 712 70	.703 167 07	.672 995 40	.590 156 49	.517 682 29	53
54	.798 889 57	.763 893 24	.698 510 33	.667 985 51	.584 313 36	.511 291 15	54
55	.795 574 68	.760 092 77	.693 884 44	.663 012 91	.578 528 08	.504 978 92	55
56	.792 273 54	.756 311 22	.689 289 18	.658 077 33	.572 800 08	.498 744 61	56
57	.788 986 10	.752 548 47	.684 724 35	.653 178 49	.567 128 79	.492 587 27	57
58	.785 712 30	.748 804 45	.680 189 75	.648 316 12	.561 513 65	.486 505 94	58
59	.782 452 08	.745 079 06	.675 685 18	.643 489 95	.555 954 11	.480 499 70	59
60	.779 205 39	.741 372 20	.671 210 44	.638 699 70	.550 449 62	.474 567 60	**60**

(continued)

Table 2 PRESENT VALUE AT COMPOUND INTEREST TABLES $(1 + i)^{-n}$ *(continued)*

n \ i	$1\frac{1}{2}\%$	$1\frac{3}{4}\%$	2%	$2\frac{1}{4}\%$	$2\frac{1}{2}\%$	$2\frac{3}{4}\%$	i / n
1	.985 221 67	.982 800 98	.980 392 16	.977 995 11	.975 609 76	.973 236 01	**1**
2	.970 661 75	.965 897 77	.961 168 78	.956 474 44	.951 814 40	.947 188 33	2
3	.956 316 99	.949 285 28	.942 322 33	.935 427 32	.928 599 41	.921 837 79	3
4	.942 184 23	.932 958 51	.923 845 43	.914 843 35	.905 950 64	.897 165 73	4
5	.928 260 33	.916 912 54	.905 730 81	.894 712 32	.883 854 29	.873 154 00	5
6	.914 542 19	.901 142 54	.887 971 38	.875 024 27	.862 296 87	.849 784 91	6
7	.901 026 79	.885 643 78	.870 560 18	.855 769 46	.841 265 24	.827 041 28	7
8	.887 711 12	.870 411 57	.853 490 37	.836 938 35	.820 746 57	.804 906 35	8
9	.874 592 24	.855 441 35	.836 755 27	.818 521 61	.800 728 36	.783 363 85	9
10	.861 667 23	.840 728 60	.820 348 30	.800 510 13	.781 198 40	.762 397 91	**10**
11	.848 933 23	.826 268 89	.804 263 04	.782 894 99	.762 144 78	.741 993 10	11
12	.836 387 42	.812 057 88	.788 493 18	.765 667 48	.743 555 89	.722 134 40	12
13	.824 027 02	.798 091 28	.773 032 53	.748 819 05	.725 420 38	.702 807 20	13
14	.811 849 28	.784 364 90	.757 875 02	.732 341 37	.707 727 20	.683 997 28	14
15	.799 851 50	.770 874 59	.743 014 73	.716 226 28	.690 465 56	.665 690 78	15
16	.788 031 04	.757 616 31	.728 445 81	.700 465 80	.673 624 93	.647 874 24	16
17	.776 385 26	.744 586 05	.714 162 56	.685 052 12	.657 195 06	.630 534 54	17
18	.764 911 59	.731 779 90	.700 159 37	.669 977 63	.641 165 91	.613 658 92	18
19	.753 607 47	.719 194 01	.686 430 76	.655 234 84	.625 527 72	.597 234 96	19
20	.742 470 42	.706 824 58	.672 971 33	.640 816 47	.610 270 94	.581 250 57	**20**
21	.731 497 95	.694 667 89	.659 775 82	.626 715 38	.595 386 29	.565 693 98	21
22	.720 687 63	.682 720 28	.646 839 04	.612 924 57	.580 864 67	.550 553 75	22
23	.710 037 08	.670 978 17	.634 155 92	.599 437 24	.566 697 24	.535 818 74	23
24	.699 543 92	.659 438 00	.621 721 49	.586 246 68	.552 875 35	.521 478 09	24
25	.689 205 83	.648 096 32	.609 530 87	.573 346 39	.539 390 59	.507 521 26	25
26	.679 020 52	.636 949 70	.597 579 28	.560 729 97	.526 234 72	.493 937 96	26
27	.668 985 74	.625 994 79	.585 862 04	.548 391 17	.513 399 73	.480 718 21	27
28	.659 099 25	.615 228 29	.574 374 55	.536 323 88	.500 877 78	.467 852 27	28
29	.649 358 87	.604 646 97	.563 112 31	.524 522 13	.488 661 25	.455 330 68	29
30	.639 762 43	.594 247 64	.552 070 89	.512 980 08	.476 742 69	.443 144 21	**30**
31	.630 307 81	.584 027 16	.541 245 97	.501 692 01	.465 114 81	.431 283 91	31
32	.620 992 92	.573 982 47	.530 633 30	.490 652 33	.453 770 55	.419 741 03	32
33	.611 815 68	.564 110 53	.520 228 73	.479 855 58	.442 702 98	.408 507 08	33
34	.602 774 07	.554 408 39	.510 028 17	.469 296 41	.431 905 34	.397 573 80	34
35	.593 866 08	.544 873 11	.500 027 61	.458 969 60	.421 371 07	.386 933 14	35
36	.585 089 74	.535 501 83	.490 223 15	.448 870 02	.411 093 72	.376 577 27	36
37	.576 443 09	.526 291 72	.480 610 93	.438 992 68	.401 067 05	.366 498 56	37
38	.567 924 23	.517 240 02	.471 187 19	.429 332 70	.391 284 92	.356 689 59	38
39	.559 531 26	.508 344 00	.461 948 22	.419 885 28	.381 741 39	.347 143 16	39
40	.551 262 32	.499 600 98	.452 890 42	.410 645 75	.372 430 62	.337 852 22	**40**
41	.543 115 59	.491 008 34	.444 010 21	.401 609 54	.363 346 95	.328 809 95	41
42	.535 089 25	.482 563 48	.435 304 13	.392 772 16	.354 484 83	.320 009 68	42
43	.527 181 53	.474 263 86	.426 768 75	.384 129 25	.345 838 86	.311 444 95	43
44	.519 390 67	.466 106 99	.418 400 74	.375 676 53	.337 403 76	.303 109 44	44
45	.511 714 94	.458 090 40	.410 196 80	.367 409 81	.329 174 40	.294 997 02	45
46	.504 152 65	.450 211 70	.402 153 73	.359 325 00	.321 145 76	.287 101 72	46
47	.496 702 12	.442 468 50	.394 268 36	.351 418 09	.313 312 94	.279 417 73	47
48	.489 361 70	.434 858 48	.386 537 61	.343 685 18	.305 671 16	.271 939 40	48
49	.482 129 75	.427 379 34	.378 958 44	.336 122 42	.298 215 76	.264 661 22	49
50	.475 004 68	.420 028 83	.371 527 88	.328 726 08	.290 942 21	.257 577 83	**50**
51	.467 984 91	.412 804 75	.364 243 02	.321 492 50	.283 846 06	.250 684 02	51
52	.461 068 87	.405 704 92	.357 101 00	.314 418 10	.276 922 98	.243 974 71	52
53	.454 255 05	.398 727 19	.350 099 02	.307 499 36	.270 168 76	.237 444 97	53
54	.447 541 92	.391 869 47	.343 234 33	.300 732 87	.263 579 28	.231 090 00	54
55	.440 928 00	.385 129 70	.336 504 25	.294 115 28	.257 150 52	.224 905 11	55
56	.434 411 82	.378 505 85	.329 906 13	.287 643 30	.250 878 55	.218 885 75	56
57	.427 991 94	.371 995 92	.323 437 38	.281 313 74	.244 759 56	.213 027 49	57
58	.421 666 94	.365 597 96	.317 095 47	.275 123 47	.238 789 82	.207 326 03	58
59	.415 435 41	.359 310 03	.310 877 91	.269 069 40	.232 965 68	.201 777 16	59
60	.409 295 97	.353 130 25	.304 782 27	.263 148 56	.227 283 59	.196 376 79	**60**

(continued)

Table 2 PRESENT VALUE AT COMPOUND INTEREST TABLES $(1 + i)^{-n}$ (continued)

n \ i	3%	$3\frac{1}{2}$%	4%	$4\frac{1}{2}$%	5%	6%	i / n
1	.970 873 79	.966 183 57	.961 538 46	.956 937 80	.952 380 95	.943 396 23	**1**
2	.942 595 91	.933 510 70	.924 556 21	.915 729 95	.907 029 48	.889 996 44	2
3	.915 141 66	.901 942 71	.888 996 36	.876 296 60	.863 837 60	.839 619 28	3
4	.888 487 05	.871 442 23	.854 804 19	.838 561 34	.822 702 47	.792 093 66	4
5	.862 608 78	.841 973 17	.821 927 11	.802 451 05	.783 526 17	.747 258 17	5
6	.837 484 26	.813 500 64	.790 314 53	.767 895 74	.746 215 40	.704 960 54	6
7	.813 091 51	.785 990 96	.759 917 81	.734 828 46	.710 681 33	.665 057 11	7
8	.789 409 23	.759 411 56	.730 690 21	.703 185 13	.676 839 36	.627 412 37	8
9	.766 416 73	.733 730 97	.702 586 74	.672 904 43	.644 608 92	.591 898 46	9
10	.744 093 91	.708 918 81	.675 564 17	.643 927 68	.613 913 25	.558 394 78	**10**
11	.722 421 28	.684 945 71	.649 580 93	.616 198 74	.584 679 29	.526 787 53	11
12	.701 379 88	.661 783 30	.624 597 05	.589 663 86	.556 837 42	.496 969 36	12
13	.680 951 34	.639 404 15	.600 574 09	.564 271 64	.530 321 35	.468 839 02	13
14	.661 117 81	.617 781 79	.577 475 08	.539 972 86	.505 067 95	.442 300 96	14
15	.641 861 95	.596 890 62	.555 264 50	.516 720 44	.481 017 10	.417 265 06	15
16	.623 166 94	.576 705 91	.533 908 18	.494 469 32	.458 111 52	.393 646 28	16
17	.605 016 45	.557 203 78	.513 373 25	.473 176 39	.436 296 69	.371 364 42	17
18	.587 394 61	.538 361 14	.493 628 12	.452 800 37	.415 520 65	.350 343 79	18
19	.570 286 03	.520 155 69	.474 642 42	.433 301 79	.395 733 96	.330 513 01	19
20	.553 675 75	.502 565 88	.456 386 95	.414 642 86	.376 889 48	.311 804 73	**20**
21	.537 549 28	.485 570 90	.438 833 60	.396 787 43	.358 942 36	.294 155 40	21
22	.521 892 50	.469 150 63	.421 955 39	.379 700 89	.341 849 87	.277 505 10	22
23	.506 691 75	.453 285 63	.405 726 33	.363 350 13	.325 571 31	.261 797 26	23
24	.491 933 74	.437 957 13	.390 121 47	.347 703 47	.310 067 91	.246 978 55	24
25	.477 605 57	.423 146 99	.375 116 80	.332 730 60	.295 302 77	.232 998 63	25
26	.463 694 73	.408 837 67	.360 689 23	.318 402 48	.281 240 73	.219 810 03	26
27	.450 189 06	.395 012 24	.346 816 57	.304 691 37	.267 848 32	.207 367 95	27
28	.437 076 75	.381 654 34	.333 477 47	.291 570 69	.255 093 64	.195 630 14	28
29	.424 346 36	.368 748 15	.320 651 41	.279 015 02	.242 946 32	.184 556 74	29
30	.411 986 76	.356 278 41	.308 318 67	.267 000 02	.231 377 45	.174 110 13	**30**
31	.399 987 15	.344 230 35	.296 460 26	.255 502 41	.220 359 47	.164 254 84	31
32	.388 337 03	.332 589 71	.285 057 94	.244 499 91	.209 866 17	.154 957 40	32
33	.377 026 25	.321 342 71	.274 094 17	.233 971 21	.199 872 54	.146 186 22	33
34	.366 044 90	.310 476 05	.263 552 09	.223 895 89	.190 354 80	.137 911 53	34
35	.355 383 40	.299 976 86	.253 415 47	.214 254 44	.181 290 29	.130 105 22	35
36	.345 032 43	.289 832 72	.243 668 72	.205 028 17	.172 657 41	.122 740 77	36
37	.334 982 94	.280 031 61	.234 296 85	.196 199 21	.164 435 63	.115 793 18	37
38	.325 226 15	.270 561 94	.225 285 43	.187 750 44	.156 605 36	.109 238 85	38
39	.315 753 55	.261 412 50	.216 620 61	.179 665 49	.149 147 97	.103 055 52	39
40	.306 556 84	.252 572 47	.208 289 04	.171 928 70	.142 045 68	.097 222 19	**40**
41	.297 628 00	.244 031 37	.200 277 93	.164 525 07	.135 281 60	.091 719 05	41
42	.288 959 22	.235 779 10	.192 574 93	.157 440 26	.128 839 62	.086 527 40	42
43	.280 542 94	.227 805 90	.185 168 20	.150 660 54	.122 704 40	.081 629 62	43
44	.272 371 78	.220 102 31	.178 046 35	.144 172 76	.116 861 33	.077 009 08	44
45	.264 438 62	.212 659 24	.171 198 41	.137 964 37	.111 296 51	.072 650 07	45
46	.256 736 53	.205 467 87	.164 613 86	.132 023 32	.105 996 68	.068 537 81	46
47	.249 258 76	.198 519 68	.158 282 56	.126 338 10	.100 949 21	.064 658 31	47
48	.241 998 80	.191 806 45	.152 194 76	.120 897 71	.096 142 11	.060 998 40	48
49	.234 950 29	.185 320 24	.146 341 12	.115 691 58	.091 563 91	.057 545 66	49
50	.228 107 08	.179 053 37	.140 712 62	.110 709 65	.087 203 73	.054 288 36	**50**
51	.221 463 18	.172 998 43	.135 300 59	.105 942 25	.083 051 17	.051 215 44	51
52	.215 012 80	.167 148 24	.130 096 72	.101 380 14	.079 096 35	.048 316 45	52
53	.208 750 29	.161 495 89	.125 093 00	.097 014 49	.075 329 86	.045 581 56	53
54	.202 670 19	.156 034 67	.120 281 73	.092 836 83	.071 742 72	.043 001 47	54
55	.196 767 17	.150 758 14	.115 655 51	.088 839 07	.068 326 40	.040 567 42	55
56	.191 036 09	.145 660 04	.111 207 22	.085 013 47	.065 072 76	.038 271 15	56
57	.185 471 93	.140 734 33	.106 930 02	.081 352 60	.061 974 06	.036 104 86	57
58	.180 069 84	.135 975 20	.102 817 33	.077 849 38	.059 022 91	.034 061 19	58
59	.174 825 08	.131 377 01	.098 862 82	.074 497 01	.056 212 30	.032 133 20	59
60	.169 733 09	.126 934 31	.095 060 40	.071 289 01	.053 535 52	.030 314 34	**60**

(continued)

Table 2 PRESENT VALUE AT COMPOUND INTEREST TABLES $(1 + i)^{-n}$ *(continued)*

n \ i	7%	8%	9%	10%	11%	12%	i / n
1	.934 579 44	.925 925 93	.917 431 19	.909 090 91	.900 900 90	.892 857 14	**1**
2	.873 438 73	.857 338 82	.841 679 99	.826 446 28	.811 622 43	.797 193 88	2
3	.816 297 88	.793 832 24	.772 183 48	.751 314 80	.731 191 38	.711 780 25	3
4	.762 895 21	.735 029 85	.708 425 21	.683 013 46	.658 730 97	.635 518 08	4
5	.712 986 18	.680 583 20	.649 931 39	.620 921 32	.593 451 33	.567 426 86	5
6	.666 342 22	.630 169 63	.596 267 33	.564 473 93	.534 640 84	.506 631 12	6
7	.622 749 74	.583 490 40	.547 034 24	.513 158 12	.481 658 41	.452 349 22	7
8	.582 009 10	.540 268 88	.501 866 28	.466 507 38	.433 926 50	.403 883 23	8
9	.543 933 74	.500 248 97	.460 427 78	.424 097 62	.390 924 77	.360 610 02	9
10	.508 349 29	.463 193 49	.422 410 81	.385 543 29	.352 184 48	.321 973 24	**10**
11	.475 092 80	.428 882 86	.387 532 85	.350 493 90	.317 283 31	.287 476 10	11
12	.444 011 96	.397 113 76	.355 534 73	.318 630 82	.285 840 82	.256 675 09	12
13	.414 964 45	.367 697 92	.326 178 65	.289 664 38	.257 514 26	.229 174 19	13
14	.387 817 24	.340 461 04	.299 246 47	.263 331 25	.231 994 82	.204 619 81	14
15	.362 446 02	.315 241 70	.274 538 04	.239 392 05	.209 004 35	.182 696 26	15
16	.338 734 60	.291 890 47	.251 869 76	.217 629 14	.188 292 20	.163 121 66	16
17	.316 574 39	.270 268 95	.231 073 18	.197 844 67	.169 632 62	.145 644 34	17
18	.295 863 92	.250 249 03	.211 993 74	.179 858 79	.152 822 18	.130 039 59	18
19	.276 508 33	.231 712 06	.194 489 67	.163 507 99	.137 677 64	.116 106 78	19
20	.258 419 00	.214 548 21	.178 430 89	.148 643 63	.124 033 91	.103 666 77	**20**
21	.241 513 09	.198 655 75	.163 698 06	.135 130 57	.111 742 26	.092 559 61	21
22	.225 713 17	.183 940 51	.150 181 71	.122 845 97	.100 668 70	.082 642 51	22
23	.210 946 88	.170 315 28	.137 781 39	.111 678 16	.090 692 52	.073 787 96	23
24	.197 146 62	.157 699 34	.126 404 94	.101 525 60	.081 704 98	.065 882 10	24
25	.184 249 18	.146 017 90	.115 967 84	.092 296 00	.073 608 09	.058 823 31	25
26	.172 195 49	.135 201 76	.106 392 51	.083 905 45	.066 313 59	.052 520 81	26
27	.160 930 37	.125 186 82	.097 607 81	.076 277 68	.059 741 97	.046 893 58	27
28	.150 402 21	.115 913 72	.089 548 45	.069 343 35	.053 821 60	.041 869 27	28
29	.140 562 82	.107 327 52	.082 154 54	.063 039 41	.048 487 93	.037 383 27	29
30	.131 367 12	.099 377 33	.075 371 14	.057 308 55	.043 682 82	.033 377 92	**30**
31	.122 773 01	.092 016 05	.069 147 83	.052 098 68	.039 353 89	.029 801 72	31
32	.114 741 13	.085 200 05	.063 438 38	.047 362 44	.035 453 95	.026 608 68	32
33	.107 234 70	.078 888 93	.058 200 35	.043 056 76	.031 940 50	.023 757 75	33
34	.100 219 34	.073 045 31	.053 394 81	.039 142 51	.028 775 22	.021 212 27	34
35	.093 662 94	.067 634 54	.048 986 07	.035 584 10	.025 923 63	.018 939 53	35
36	.087 535 46	.062 624 58	.044 941 35	.032 349 18	.023 354 62	.016 910 29	36
37	.081 808 84	.057 985 72	.041 230 59	.029 408 35	.021 040 20	.015 098 48	37
38	.076 456 86	.053 690 48	.037 826 23	.026 734 86	.018 955 13	.013 480 78	38
39	.071 455 01	.049 713 41	.034 702 96	.024 304 42	.017 076 70	.012 036 41	39
40	.066 780 38	.046 030 93	.031 837 58	.022 094 93	.015 384 41	.010 746 80	**40**
41	.062 411 57	.042 621 23	.029 208 79	.020 086 30	.013 859 83	.009 595 36	41
42	.058 328 57	.039 464 11	.026 797 06	.018 260 27	.012 486 33	.008 567 28	42
43	.054 512 68	.036 540 84	.024 584 46	.016 600 25	.011 248 95	.007 649 36	43
44	.050 946 43	.033 834 11	.022 554 55	.015 091 13	.010 134 19	.006 829 78	44
45	.047 613 49	.031 327 88	.020 692 24	.013 719 21	.009 129 90	.006 098 02	45
46	.044 498 59	.029 007 30	.018 983 71	.012 472 01	.008 225 13	.005 444 66	46
47	.041 587 47	.026 858 61	.017 416 25	.011 338 19	.007 410 03	.004 861 31	47
48	.038 866 79	.024 869 08	.015 978 21	.010 307 45	.006 675 70	.004 340 45	48
49	.036 324 10	.023 026 93	.014 658 91	.009 370 41	.006 014 15	.003 875 40	49
50	.033 947 76	.021 321 23	.013 448 54	.008 518 55	.005 418 15	.003 460 18	**50**
51	.031 726 88	.019 741 88	.012 338 11	.007 744 14	.004 881 22	.003 089 45	51
52	.029 651 29	.018 279 52	.011 319 37	.007 040 13	.004 397 49	.002 758 44	52
53	.027 711 48	.016 925 48	.010 384 74	.006 400 11	.003 961 70	.002 462 89	53
54	.025 898 58	.015 671 74	.009 527 28	.005 818 29	.003 569 10	.002 199 01	54
55	.024 204 28	.014 510 87	.008 740 63	.005 289 35	.003 215 41	.001 963 40	55
56	.022 620 83	.013 435 99	.008 018 92	.004 808 50	.002 896 76	.001 753 04	56
57	.021 140 96	.012 440 73	.007 356 81	.004 371 36	.002 609 70	.001 565 21	57
58	.019 757 91	.011 519 20	.006 749 37	.003 973 97	.002 351 08	.001 397 51	58
59	.018 465 33	.010 665 92	.006 192 08	.003 612 70	.002 118 09	.001 247 78	59
60	.017 257 32	.009 875 85	.005 680 81	.003 284 27	.001 908 19	.001 114 09	**60**

Table 3 FUTURE VALUE OF AN ANNUITY TABLES $s_{\overline{n}|i} = \frac{(1+i)^n - 1}{i}$

n \ i	$\frac{5}{12}\%$	$\frac{1}{2}\%$	$\frac{2}{3}\%$	$\frac{3}{4}\%$	1%	$1\frac{1}{4}\%$	i / n
1	1.000 000 00	1.000 000 00	1.000 000 00	1.000 000 00	1.000 000 00	1.000 000 00	**1**
2	2.004 166 67	2.005 000 00	2.006 666 67	2.007 500 00	2.010 000 00	2.012 500 00	2
3	3.012 517 36	3.015 025 00	3.020 044 44	3.022 556 25	3.030 100 00	3.037 656 25	3
4	4.025 069 52	4.030 100 13	4.040 178 07	4.045 225 42	4.060 401 00	4.075 626 95	4
5	5.041 840 64	5.050 250 63	5.067 112 59	5.075 564 61	5.101 005 01	5.126 572 29	5
6	6.062 848 31	6.075 501 88	6.100 893 35	6.113 631 35	6.152 015 06	6.190 654 44	6
7	7.088 110 18	7.105 879 39	7.141 565 97	7.159 483 58	7.213 535 21	7.268 037 62	7
8	8.117 643 97	8.141 408 79	8.189 176 41	8.213 179 71	8.285 670 56	8.358 888 09	8
9	9.151 467 49	9.182 115 83	9.243 770 92	9.274 778 56	9.368 527 27	9.463 374 20	9
10	10.189 598 60	10.228 026 41	10.305 396 06	10.344 339 40	10.462 212 54	10.581 666 37	**10**
11	11.232 055 26	11.279 166 54	11.374 098 70	11.421 921 94	11.566 834 67	11.713 937 20	11
12	12.278 855 49	12.335 562 37	12.449 926 02	12.507 586 36	12.682 503 01	12.860 361 42	12
13	13.230 017 39	13.397 240 18	13.532 925 53	13.601 393 25	13.809 328 04	14.021 115 94	13
14	14.385 559 13	14.464 226 39	14.623 145 03	14.703 403 70	14.947 421 32	15.196 379 88	14
15	15.445 498 96	15.536 547 52	15.720 632 66	15.813 679 23	16.096 895 54	16.386 334 63	15
16	15.509 855 20	16.614 230 26	16.825 436 88	16.932 281 83	17.257 864 49	17.591 163 82	16
17	16.578 646 27	17.697 301 41	17.937 606 46	18.059 273 94	18.430 443 14	18.811 053 36	17
18	18.651 890 63	18.785 787 91	19.057 190 51	19.194 718 49	19.614 747 57	19.046 191 53	18
19	19.729 606 84	19.879 716 85	20.184 238 44	20.338 678 88	20.810 895 04	20.296 768 93	19
20	20.811 813 53	20.979 115 44	21.318 800 03	21.491 218 97	22.019 003 99	22.562 978 54	**20**
21	21.898 529 42	22.084 011 01	22.460 925 36	22.652 403 12	23.239 194 03	23.845 015 77	21
22	22.989 773 30	23.194 431 07	23.610 664 87	23.822 296 14	24.471 585 98	25.143 078 47	22
23	24.085 564 02	24.310 403 22	24.768 069 30	25.000 963 36	25.716 301 83	26.457 366 95	23
24	25.185 920 53	25.431 955 24	25.933 189 76	26.188 470 59	26.973 464 85	27.788 084 03	24
25	26.290 861 87	26.559 115 02	27.106 077 69	27.384 884 12	28.243 199 50	29.135 435 08	25
26	27.400 407 13	27.691 910 59	28.286 784 88	28.590 270 75	29.525 631 50	30.499 628 02	26
27	28.514 575 49	28.830 370 15	29.475 363 44	29.804 697 78	30.820 887 81	31.880 873 37	27
28	29.633 386 22	29.974 522 00	30.671 865 87	31.028 233 01	32.129 096 69	33.279 384 29	28
29	30.756 858 66	31.124 394 61	31.876 344 97	32.260 944 76	33.450 387 66	34.695 376 59	29
30	31.885 012 24	32.280 016 58	33.088 853 94	33.502 901 84	34.784 891 53	36.129 068 80	**30**
31	33.017 866 46	33.441 416 66	34.309 446 30	34.754 173 61	36.132 740 45	37.580 682 16	31
32	34.155 440 90	34.608 623 75	35.538 175 94	36.014 829 91	37.494 067 85	39.050 440 69	32
33	35.297 755 24	35.781 666 86	36.775 097 11	37.284 941 13	38.869 008 53	40.538 571 20	33
34	36.444 829 22	36.960 575 20	38.020 264 43	38.564 578 19	40.257 698 62	42.045 303 34	34
35	37.596 682 68	38.145 378 07	39.273 732 86	39.853 812 53	41.660 275 60	43.570 869 63	35
36	38.753 335 52	39.336 104 96	40.535 557 74	41.152 716 12	43.076 878 36	45.115 505 50	36
37	39.914 807 75	40.532 785 49	41.805 794 79	42.461 361 49	44.507 647 14	46.679 449 32	37
38	41.081 119 45	41.735 449 42	43.084 500 09	43.779 821 70	45.952 723 61	48.262 942 43	38
39	42.252 290 78	42.944 126 66	44.371 730 09	45.108 170 37	47.412 250 85	49.866 229 21	39
40	43.428 341 99	44.158 847 30	45.667 541 63	46.446 481 64	48.886 373 36	51.489 557 08	**40**
41	44.609 293 42	45.379 641 53	46.971 991 91	47.794 830 26	50.375 237 09	53.133 176 54	41
42	45.795 165 47	46.606 539 74	48.285 138 52	49.153 291 48	51.878 989 46	54.797 341 25	42
43	46.985 978 66	47.839 572 44	49.607 039 44	50.521 941 17	53.397 779 36	56.482 308 01	43
44	48.181 753 57	49.078 770 30	50.937 753 04	51.900 855 73	54.931 757 15	58.188 336 87	44
45	49.382 510 88	50.324 164 15	52.277 338 06	53.290 112 15	56.481 074 72	59.915 691 08	45
46	50.588 271 34	51.575 784 97	53.625 853 65	54.689 787 99	58.045 885 47	61.664 637 21	46
47	51.799 055 81	52.833 663 90	54.983 359 34	56.099 961 40	59.626 344 32	63.435 445 18	47
48	53.014 885 21	54.097 832 22	56.349 915 07	57.520 711 11	61.222 607 77	64.228 388 24	48
49	54.235 780 56	55.368 321 38	57.725 581 17	58.952 116 44	62.834 833 85	67.043 743 10	49
50	55.461 762 98	56.645 162 99	59.110 418 37	60.394 257 32	64.463 182 18	68.881 789 89	**50**
51	56.692 853 66	57.928 388 80	60.504 487 83	61.847 214 24	66.107 814 01	70.742 812 26	51
52	57.929 073 88	59.218 030 75	61.907 851 08	63.311 068 35	67.768 892 15	72.627 097 41	52
53	59.170 445 02	60.514 120 90	63.320 570 09	64.785 901 36	69.446 581 07	74.534 936 13	53
54	60.416 988 54	61.816 691 50	64.742 707 22	66.271 795 62	71.141 046 88	76.466 622 83	54
55	61.668 726 00	63.125 774 96	66.174 325 27	67.768 834 09	72.852 457 35	78.422 455 62	55
56	62.925 679 02	64.441 403 84	67.615 487 44	69.277 100 35	74.580 981 92	80.402 736 31	56
57	64.187 869 35	65.763 610 86	69.066 257 36	70.796 678 60	76.326 791 74	82.407 770 52	57
58	65.455 318 81	67.092 428 91	70.526 699 07	72.327 653 69	78.090 059 66	84.437 867 65	58
59	66.728 049 30	68.427 891 05	71.996 877 06	73.870 111 09	79.870 960 25	86.493 340 99	59

(continued)

Table 3 FUTURE VALUE OF AN ANNUITY TABLES $s_{\overline{n}|i} = \dfrac{(1+i)^n - 1}{i}$

(continued)

n \ i	$\frac{5}{12}\%$	$\frac{1}{2}\%$	$\frac{2}{3}\%$	$\frac{3}{4}\%$	1%	$1\frac{1}{4}\%$	i / n
60	68.006 082 84	69.770 030 51	73.476 856 25	75.424 136 93	81.669 669 86	88.574 507 76	**60**
61	69.289 441 52	71.118 880 66	74.966 701 95	76.989 817 95	83.486 366 55	90.681 689 10	61
62	70.578 147 53	72.474 475 07	76.466 479 97	78.567 241 59	85.321 230 22	92.815 210 22	62
63	71.872 223 14	73.836 847 44	77.976 256 50	80.156 495 90	87.174 442 52	94.975 400 34	63
64	73.171 690 74	75.206 031 68	79.496 098 21	81.757 669 62	89.046 186 95	97.162 592 85	64
65	74.476 572 78	76.582 061 84	81.026 072 20	83.370 852 14	90.936 648 82	99.377 125 26	65
66	75.786 891 83	77.964 972 15	82.566 246 01	84.996 133 53	92.846 015 31	101.619 339 33	66
67	77.102 670 55	79.354 797 01	84.116 687 65	86.633 604 53	94.774 475 46	103.889 581 07	67
68	78.423 931 68	80.751 570 99	85.677 465 57	88.283 356 57	96.722 220 21	106.188 200 83	68
69	79.750 698 06	82.155 328 85	87.248 648 67	89.945 481 74	98.689 442 42	108.515 553 34	69
70	81.082 992 64	83.566 105 49	88.830 306 33	91.620 072 85	100.676 336 84	110.871 997 76	**70**
71	82.420 838 44	84.983 936 02	90.422 508 37	93.307 223 40	102.683 100 21	113.257 897 73	71
72	83.764 258 60	86.408 855 70	92.025 325 10	95.007 027 58	104.709 931 21	115.673 621 45	72
73	85.113 276 34	87.840 899 98	93.638 827 26	96.719 580 28	106.757 030 52	118.119 541 72	73
74	86.467 914 99	89.280 104 48	95.263 086 11	98.444 977 14	108.824 600 83	120.596 035 99	74
75	87.828 197 97	90.726 505 00	96.898 173 35	100.183 314 46	110.912 846 84	123.103 486 44	75
76	89.194 148 80	92.180 137 52	98.544 161 18	101.934 689 32	113.021 975 30	125.642 280 02	76
77	90.565 791 08	93.641 038 21	100.201 122 25	103.699 199 49	115.152 195 06	128.212 808 52	77
78	91.943 148 55	95.109 243 40	101.869 129 73	105.476 943 49	117.303 717 01	130.815 468 63	78
79	93.326 245 00	96.584 789 62	103.548 257 26	107.268 020 56	119.476 754 18	133.450 661 99	79
80	94.715 104 35	98.067 713 57	105.238 578 98	109.072 530 72	121.671 521 72	136.118 795 26	**80**
81	96.109 750 62	99.558 052 14	106.940 169 50	110.890 574 70	123.888 236 94	138.820 280 20	81
82	97.510 207 92	101.055 842 40	108.653 103 97	112.722 254 01	126.127 119 31	141.555 533 70	82
83	98.916 500 45	102.561 121 61	110.377 457 99	114.567 670 91	128.388 390 50	144.324 977 87	83
84	100.328 652 53	104.073 927 22	112.113 307 71	116.426 928 45	130.672 274 40	147.129 040 10	84
85	101.746 688 59	105.594 296 85	113.860 729 77	118.300 130 41	132.978 997 15	149.968 153 10	85
86	103.170 633 12	107.122 268 34	115.619 801 30	120.187 381 39	135.308 787 12	152.842 755 01	86
87	104.600 510 76	108.657 879 68	117.390 599 97	122.088 786 75	137.661 874 99	155.753 289 45	87
88	106.036 346 22	110.201 169 08	119.173 203 97	124.004 452 65	140.038 493 74	158.700 205 57	88
89	107.478 164 33	111.752 174 92	120.967 692 00	125.934 486 04	142.438 878 68	161.683 958 14	89
90	108.925 990 02	113.310 935 80	122.774 143 28	127.878 994 69	144.863 267 46	164.705 007 62	**90**
91	110.379 848 31	114.877 490 48	124.592 637 57	129.838 087 15	147.311 900 14	167.763 820 21	91
92	111.839 764 34	116.451 877 93	126.423 255 15	131.811 872 80	149.785 019 14	170.860 867 96	92
93	113.305 763 36	118.034 137 32	128.266 076 85	133.800 461 85	152.282 869 33	173.996 628 81	93
94	114.777 870 71	119.624 308 00	130.121 184 03	135.803 965 31	154.805 698 03	177.171 586 67	94
95	116.256 111 84	121.222 429 54	131.988 658 59	137.822 495 05	157.353 755 01	180.386 231 51	95
96	117.740 512 30	122.828 541 69	133.868 582 98	139.856 163 77	159.927 292 56	183.641 059 40	96
97	119.231 097 77	124.442 684 40	135.761 040 20	141.905 084 99	162.526 565 48	186.936 572 64	97
98	120.727 894 01	126.064 897 82	137.666 113 80	143.969 373 13	165.151 831 14	190.273 279 80	98
99	122.230 926 90	127.695 222 31	139.583 887 90	146.049 143 43	167.803 349 45	193.651 695 80	99
100	123.740 222 43	129.333 698 42	141.514 447 15	148.144 512 01	170.481 382 94	197.072 342 00	**100**
101	125.255 806 69	130.980 366 92	143.457 876 80	150.255 595 85	173.186 196 77	200.535 746 27	101
102	126.777 705 89	132.635 268 75	145.414 262 64	152.382 512 81	175.918 058 74	204.042 443 10	102
103	128.305 946 33	134.298 445 09	147.383 691 06	154.525 381 66	178.677 239 33	207.592 973 64	103
104	129.840 554 44	135.969 937 32	149.366 249 00	156.684 322 02	181.464 011 72	211.187 885 81	104
105	131.381 556 75	137.649 787 01	151.362 023 99	158.859 454 44	184.278 651 84	214.827 734 38	105
106	132.928 979 90	139.338 035 94	153.371 104 15	161.050 900 35	187.121 438 36	218.513 081 06	106
107	134.482 850 65	141.034 726 12	155.393 578 18	163.258 782 10	189.992 652 74	222.244 494 57	107
108	136.043 195 86	142.739 899 75	159.429 535 37	165.483 222 96	192.892 579 27	226.022 550 76	108
109	137.610 042 51	144.453 599 25	159.479 065 60	167.724 347 14	195.821 505 06	229.847 832 64	109
110	139.183 417 69	146.175 867 25	161.542 259 37	169.982 279 74	198.779 720 11	233.720 930 55	**110**
111	140.763 348 59	147.906 746 58	163.619 207 77	172.257 146 84	201.767 517 31	237.642 442 18	111
112	142.349 862 55	149.646 280 32	165.710 002 49	174.549 075 44	204.785 192 48	241.612 972 71	112
113	143.942 986 97	151.394 511 72	167.814 735 84	176.858 193 51	207.833 044 41	245.633 134 87	113
114	145.542 749 42	153.151 484 28	169.933 500 74	179.184 629 96	210.911 374 85	249.703 549 05	114
115	147.149 177 54	154.917 241 70	172.066 390 75	181.528 514 68	214.020 488 60	253.824 843 42	115
116	148.762 299 11	156.691 827 91	174.213 500 02	183.889 978 54	217.160 693 49	257.997 653 96	116
117	150.382 142 03	158.475 287 04	176.374 923 35	186.269 153 38	220.332 300 42	262.222 624 63	117
118	152.008 734 29	160.267 663 48	178.550 756 18	188.666 172 03	223.535 623 43	266.500 407 44	118
119	153.642 104 01	162.069 001 80	180.741 094 55	191.081 168 32	226.770 979 66	270.831 662 53	119
120	155.282 279 45	163.879 346 81	182.946 035 18	193.514 277 08	230.038 689 46	275.217 058 32	**120**

(continued)

Table 3 FUTURE VALUE OF AN ANNUITY TABLES $s_{\overline{n}|i} = \dfrac{(1+i)^n - 1}{i}$ *(continued)*

n \ i	$1\frac{1}{2}\%$	$1\frac{3}{4}\%$	2%	$2\frac{1}{4}\%$	$2\frac{1}{2}\%$	$2\frac{3}{4}\%$	i / n
1	1.000 000 00	1.000 000 00	1.000 000 00	1.000 000 00	1.000 000 00	1.000 000 00	**1**
2	2.015 000 00	2.017 500 00	2.020 000 00	2.022 500 00	2.025 000 00	2.027 500 00	2
3	3.045 225 00	3.052 806 25	3.060 400 00	3.068 006 25	3.075 625 00	3.083 256 25	3
4	4.090 903 37	4.106 230 36	4.121 608 00	4.137 036 39	4.152 515 62	4.168 045 80	4
5	5.152 266 93	5.178 089 39	5.204 040 16	5.230 119 71	5.256 328 52	5.282 667 06	5
6	6.229 550 93	6.268 705 96	6.308 120 96	6.347 797 40	6.387 736 73	6.427 940 40	6
7	7.322 994 19	7.378 408 31	7.434 283 38	7.490 622 84	7.547 430 15	7.604 708 76	7
8	8.432 839 11	8.507 530 45	8.582 969 05	8.659 161 86	8.736 115 90	8.813 838 25	8
9	9.559 331 69	9.656 412 24	9.754 628 43	9.853 993 00	9.954 518 80	10.056 218 80	9
10	10.702 721 67	10.825 399 45	10.949 721 00	11.075 707 84	11.203 381 77	11.332 764 82	**10**
11	11.863 262 49	12.014 843 94	12.168 715 42	12.324 911 27	12.483 466 31	12.644 415 85	11
12	13.041 211 43	13.225 103 71	13.412 089 73	13.602 221 77	13.795 522 97	13.992 137 29	12
13	14.236 829 60	14.456 543 03	14.680 331 52	14.908 271 76	15.140 441 79	15.376 921 07	13
14	15.450 382 05	15.709 532 53	15.973 938 15	16.243 707 88	16.518 952 84	16.799 786 39	14
15	16.682 137 78	16.984 449 35	17.293 416 92	17.609 191 30	17.931 926 66	18.261 780 52	15
16	17.932 369 84	18.281 677 21	18.639 285 25	19.005 398 11	19.380 224 83	19.763 979 48	16
17	19.201 355 39	19.601 606 56	21.012 070 96	21.433 019 57	21.864 730 45	21.307 488 92	17
18	20.489 375 72	21.944 634 68	21.412 312 38	21.892 762 51	22.386 348 71	22.893 444 87	18
19	21.796 716 36	22.311 165 78	22.840 558 63	23.385 349 66	23.946 007 43	24.523 014 60	19
20	23.123 667 10	23.701 611 19	24.297 369 80	24.911 520 03	25.544 657 61	26.197 397 50	**20**
21	24.470 522 11	25.116 389 38	25.783 317 19	26.472 029 23	27.183 274 05	27.917 825 93	21
22	25.837 579 94	26.555 926 20	27.298 983 54	28.067 649 89	28.862 855 90	29.685 566 15	22
23	27.225 143 64	28.020 654 90	28.844 963 21	29.699 172 01	30.584 427 30	31.501 919 21	23
24	28.633 520 80	29.511 016 37	30.421 186 47	31.367 403 38	32.349 037 98	33.368 221 99	24
25	30.063 023 61	31.027 459 15	32.030 299 72	33.073 169 96	34.157 763 93	35.285 848 10	25
26	31.513 968 96	32.570 439 69	33.670 905 72	34.817 316 28	36.011 708 03	37.256 208 92	26
27	32.986 678 50	34.140 422 38	35.344 323 83	36.600 705 90	37.912 000 73	39.280 754 67	27
28	34.481 478 67	35.737 879 77	37.051 210 31	38.424 221 78	39.859 800 75	41.360 975 42	28
29	35.998 700 85	37.363 292 67	38.792 234 51	40.288 766 77	41.856 295 77	43.498 402 24	29
30	37.538 681 37	39.017 150 29	40.568 079 21	42.195 264 02	43.902 703 16	45.694 608 31	**30**
31	39.101 761 59	40.699 950 42	42.379 440 79	44.144 657 46	46.000 270 74	47.951 210 03	31
32	40.688 288 01	42.412 199 55	44.227 029 61	46.137 912 26	48.150 277 51	50.269 868 31	32
33	42.298 612 33	44.154 413 05	46.111 570 20	48.176 015 28	50.354 034 45	52.652 289 69	33
34	43.933 091 52	45.927 115 27	48.033 801 60	50.259 975 63	52.612 885 31	55.100 227 65	34
35	45.592 087 89	47.730 839 79	49.994 477 63	52.390 825 08	54.928 207 44	57.615 483 91	35
36	47.275 969 21	49.566 129 49	51.994 367 19	54.569 618 64	57.301 412 63	60.199 909 72	36
37	48.985 108 74	51.433 536 75	54.034 254 53	56.797 435 06	59.733 947 94	62.855 407 24	37
38	50.719 885 38	53.333 623 65	56.114 939 62	59.075 377 35	62.227 296 64	65.583 930 94	38
39	52.480 683 66	55.266 962 06	58.237 238 41	61.404 573 34	64.782 979 06	68.387 489 04	39
40	54.267 893 91	57.234 133 90	60.401 983 18	63.786 176 24	67.402 553 54	71.268 144 99	**40**
41	56.081 912 32	59.235 731 24	62.610 022 84	66.221 365 21	70.087 617 37	74.228 018 98	41
42	57.923 141 00	61.272 356 54	64.862 223 30	68.711 345 92	72.839 807 81	77.269 289 50	42
43	59.791 988 12	63.344 622 78	67.159 467 77	71.257 351 21	75.660 803 00	80.394 194 96	43
44	61.688 867 94	65.453 153 67	69.502 657 12	73.860 641 61	78.552 323 08	83.605 035 32	44
45	63.614 200 96	67.598 583 86	71.892 710 27	76.522 506 05	81.516 131 16	86.904 173 79	45
46	65.568 413 98	69.781 559 08	74.330 564 47	79.244 262 43	84.554 034 43	90.294 038 57	46
47	67.551 940 18	72.002 736 36	76.817 175 76	82.027 258 34	87.667 885 30	93.777 124 63	47
48	69.565 219 29	74.262 784 25	79.353 519 27	84.872 871 65	90.859 582 43	97.355 995 56	48
49	71.608 697 58	76.562 382 98	81.940 589 66	87.782 511 26	94.131 071 99	101.033 285 44	49
50	73.682 828 04	78.902 224 68	84.579 401 45	90.757 617 76	97.484 348 79	104.811 700 79	**50**
51	75.788 070 46	81.283 013 61	87.270 989 48	93.799 664 16	100.921 457 51	108.694 022 56	51
52	77.924 891 52	83.705 466 35	90.016 409 27	96.910 156 61	104.444 493 95	112.683 108 18	52
53	80.093 764 89	86.170 312 01	92.816 737 46	100.090 635 13	108.055 606 29	116.781 893 65	53
54	82.295 171 36	88.678 292 47	95.673 072 21	103.342 674 42	111.756 996 45	120.993 395 73	54
55	84.529 598 93	91.230 162 59	98.586 533 65	106.667 884 60	115.550 921 36	125.320 714 11	55
56	86.797 542 92	93.826 690 43	101.558 264 32	110.067 912 00	119.439 694 40	129.767 033 75	56
57	89.099 506 06	96.468 657 52	104.589 429 61	113.544 440 02	123.425 686 76	134.335 627 18	57
58	91.435 998 65	99.156 859 02	107.681 218 20	117.099 189 92	127.511 328 93	139.029 856 92	58
59	93.807 538 63	101.892 104 05	110.834 842 57	120.733 921 69	131.699 112 15	143.853 177 99	59

(continued)

Table 3 FUTURE VALUE OF AN ANNUITY TABLES $s_{\overline{n}|i} = \dfrac{(1+i)^n - 1}{i}$ *(continued)*

n \ i	$1\frac{1}{2}\%$	$1\frac{3}{4}\%$	2%	$2\frac{1}{4}\%$	$2\frac{1}{2}\%$	$2\frac{3}{4}\%$	i / n
60	96.214 651 71	104.675 215 88	114.051 539 42	124.450 434 93	135.991 589 95	148.809 140 38	**60**
61	98.657 871 49	107.507 032 15	117.332 570 21	128.250 569 72	140.391 379 70	153.901 391 74	61
62	101.137 739 56	110.388 405 22	120.679 221 61	132.136 207 54	144.901 164 19	159.133 680 02	62
63	103.654 805 65	113.320 202 31	124.092 806 04	136.109 272 21	149.523 693 30	164.509 856 22	63
64	106.209 627 74	116.303 305 85	127.574 662 16	140.171 730 83	154.261 785 63	170.033 877 26	64
65	108.802 772 15	119.338 613 70	131.126 155 41	144.325 594 77	159.118 330 27	175.709 808 89	65
66	111.434 813 74	122.427 039 44	134.748 678 52	148.572 920 66	164.096 288 53	181.541 828 63	66
67	114.106 335 94	125.569 512 63	138.443 652 09	152.915 811 37	169.198 695 74	187.534 228 92	67
68	116.817 930 98	128.766 979 10	142.212 525 13	157.356 417 13	174.428 663 14	193.691 420 22	68
69	119.570 199 95	132.020 401 24	146.056 775 63	161.896 936 51	179.789 379 71	200.017 934 27	69
70	122.363 752 95	135.330 758 26	149.977 911 14	166.539 617 58	185.284 114 21	206.518 427 46	**70**
71	125.199 209 24	138.699 046 53	153.977 469 37	171.286 758 98	190.916 217 06	213.197 684 22	71
72	128.077 197 38	142.126 279 84	158.057 018 75	176.140 711 06	196.689 122 49	220.060 620 54	72
73	130.998 355 34	145.613 489 74	162.218 159 13	181.103 877 05	202.606 350 55	227.112 287 60	73
74	133.963 330 67	149.161 725 81	166.462 522 31	186.178 714 29	208.671 509 31	234.357 875 51	74
75	136.972 780 63	152.772 056 01	170.791 772 76	191.367 735 36	214.888 297 05	241.802 717 09	75
76	140.027 372 34	156.445 566 99	175.207 608 21	196.673 509 41	221.260 504 47	249.452 291 81	76
77	143.127 782 92	160.183 364 41	179.711 760 38	202.098 663 37	227.792 017 09	257.312 229 83	77
78	146.274 699 67	163.986 573 29	184.304 995 58	207.645 883 29	234.486 817 51	265.388 316 15	78
79	149.468 820 16	167.856 338 32	188.992 115 49	213.317 915 67	241.348 987 95	273.686 494 85	79
80	152.710 852 47	171.793 824 24	193.771 957 80	219.117 568 77	248.382 712 65	282.212 873 45	**80**
81	156.001 515 25	175.800 216 17	198.647 396 96	225.047 714 07	255.592 280 47	290.973 727 47	81
82	159.341 537 98	179.876 719 95	203.620 344 90	231.111 287 63	262.982 087 48	299.975 504 98	82
83	162.731 661 05	184.024 562 55	208.692 751 80	237.311 291 60	270.556 639 66	309.224 831 37	83
84	166.172 635 97	188.244 992 39	213.866 606 83	243.650 795 67	278.320 555 66	318.728 514 23	84
85	169.665 225 51	192.539 279 76	219.143 938 97	250.132 938 57	286.278 569 55	328.493 548 37	85
86	173.210 203 89	196.908 717 16	224.526 817 75	256.760 929 69	294.435 533 79	338.527 120 95	86
87	176.808 356 95	201.354 619 71	230.017 354 11	263.538 050 60	302.796 422 13	348.836 616 78	87
88	180.460 482 30	205.878 325 55	235.617 701 19	270.467 656 74	311.366 332 68	359.429 623 74	88
89	184.167 389 54	210.481 196 25	241.330 055 21	277.553 179 02	320.150 491 00	370.313 938 39	89
90	187.929 900 38	215.164 617 18	247.156 656 32	284.798 125 55	329.154 253 28	381.497 571 70	**90**
91	191.748 848 89	219.929 997 98	253.099 789 44	292.206 083 37	338.383 109 61	391.988 754 92	91
92	195.625 081 62	224.778 772 95	259.161 785 23	299.780 720 25	347.842 687 35	404.795 945 68	92
93	199.559 457 84	229.712 401 48	265.345 020 94	307.525 786 45	357.538 754 53	416.927 834 18	93
94	203.552 849 71	234.732 368 50	271.651 921 35	315.445 116 65	367.477 223 39	429.393 349 62	94
95	207.606 142 46	239.840 184 95	278.084 959 78	323.542 631 77	377.664 153 98	442.201 666 74	95
96	211.720 234 59	245.037 388 19	284.646 658 98	331.822 340 99	388.105 757 83	455.362 212 57	96
97	215.896 038 11	250.325 542 48	291.339 592 16	340.288 343 66	398.808 401 77	468.884 673 42	97
98	220.134 478 68	255.706 239 47	298.166 384 00	348.944 831 39	409.778 611 82	482.779 001 94	98
99	224.436 495 86	261.181 098 66	305.129 711 68	357.796 090 10	421.023 077 11	497.055 424 49	99
100	228.803 043 30	266.751 767 89	312.232 305 91	366.846 502 13	432.548 654 04	511.724 448 67	**100**
101	233.235 088 95	272.419 923 83	319.476 952 03	376.100 548 42	444.362 370 39	526.796 871 00	101
102	237.733 615 29	278.187 272 50	326.866 491 07	385.562 810 76	456.471 429 65	542.283 784 96	102
103	242.299 619 51	284.055 549 76	334.403 820 89	395.237 974 01	468.883 215 39	558.196 589 04	103
104	246.934 113 81	290.026 521 88	342.091 897 31	405.130 828 42	481.605 295 78	574.546 995 24	104
105	251.638 125 51	296.101 986 02	349.933 735 26	415.246 272 06	494.645 428 17	591.347 037 61	105
106	256.412 697 40	302.283 770 77	357.932 409 96	425.589 313 88	508.011 563 88	608.609 081 14	106
107	261.258 887 86	308.573 736 76	366.091 058 16	436.165 072 73	521.711 852 98	626.345 830 88	107
108	266.177 771 18	314.973 777 16	374.412 879 32	446.978 786 87	535.754 649 30	644.570 341 23	108
109	271.170 437 74	321.485 818 26	382.901 136 91	458.035 809 57	550.148 515 53	663.296 025 61	109
110	276.237 994 31	328.111 820 07	391.559 159 65	469.341 615 29	564.902 228 42	682.536 666 31	**110**
111	281.381 564 22	334.853 776 93	400.390 342 84	480.901 801 63	580.024 784 13	702.306 424 64	111
112	286.602 287 69	341.713 718 02	409.398 149 70	492.722 092 17	595.525 403 73	722.619 851 31	112
113	291.901 322 00	348.693 708 09	418.586 112 69	504.808 339 24	611.413 538 83	743.491 897 23	113
114	297.279 841 83	355.795 847 98	427.957 834 95	517.166 526 87	627.698 877 30	764.937 924 40	114
115	302.739 039 46	363.022 275 32	437.516 991 65	529.802 773 73	644.391 349 23	786.973 717 32	115
116	308.280 125 05	370.375 165 14	447.267 331 48	542.723 336 14	661.501 132 96	809.615 494 55	116
117	313.904 326 93	377.856 730 53	457.212 678 11	555.934 611 20	679.038 661 29	832.879 920 65	117
118	319.612 891 83	385.469 223 31	467.356 931 67	569.443 139 95	697.014 627 82	856.784 118 46	118
119	325.407 085 21	393.214 934 72	477.704 070 30	583.255 610 60	715.439 993 51	881.345 681 72	119
120	331.288 191 49	401.096 196 08	488.258 151 71	597.378 861 84	734.325 993 35	906.582 687 97	**120**

(continued)

Table 3 FUTURE VALUE OF AN ANNUITY TABLES $s_{\overline{n}|i} = \dfrac{(1+i)^n - 1}{i}$ *(continued)*

n \ i	3%	$3\frac{1}{2}$%	4%	$4\frac{1}{2}$%	5%	6%	i / n
1	1.000 000 00	1.000 000 00	1.000 000 00	1.000 000 00	1.000 000 00	1.000 000 00	**1**
2	2.030 000 00	2.035 000 00	2.040 000 00	2.045 000 00	2.050 000 00	2.060 000 00	2
3	3.090 900 00	3.106 225 00	3.121 600 00	3.137 025 00	3.152 500 00	3.183 600 00	3
4	4.183 627 00	4.214 942 87	4.246 464 00	4.278 191 13	4.310 125 00	4.374 616 00	4
5	5.309 135 81	5.362 465 88	5.416 322 56	5.470 709 73	5.525 631 25	5.637 092 96	5
6	6.468 409 88	6.550 152 18	6.632 975 46	6.716 891 66	6.801 912 81	6.975 318 54	6
7	7.662 462 18	7.779 407 51	7.898 294 48	8.019 151 79	8.142 008 45	8.393 837 65	7
8	8.892 336 05	9.051 686 77	9.214 226 26	9.380 013 62	9.549 108 88	9.897 467 91	8
9	10.159 106 13	10.368 495 81	10.582 795 31	10.802 114 23	11.026 564 32	11.491 315 98	9
10	11.463 879 31	11.731 393 16	12.006 107 12	12.288 209 37	12.577 892 54	13.180 794 94	**10**
11	12.807 795 69	13.141 991 92	13.486 351 41	13.841 178 79	14.206 787 16	14.971 642 64	11
12	14.192 029 56	14.601 961 64	15.025 805 46	15.464 031 84	15.917 126 52	16.869 941 20	12
13	15.617 790 45	16.113 030 30	16.626 837 68	17.159 913 27	17.712 982 85	18.882 137 67	13
14	17.086 324 16	17.676 986 36	18.291 911 19	18.932 109 37	19.598 631 99	21.015 065 93	14
15	18.598 913 89	19.295 680 88	21.023 587 64	21.784 054 29	21.578 563 59	23.275 969 88	15
16	21.156 881 30	21.971 029 71	21.824 531 14	22.719 336 73	23.657 491 77	25.672 528 08	16
17	21.761 587 74	22.705 015 75	23.697 512 39	24.741 706 89	25.840 366 36	28.212 879 76	17
18	23.414 435 37	24.499 691 30	25.645 412 88	26.855 983 70	28.132 384 67	30.905 652 55	18
19	25.116 868 44	26.357 180 50	27.671 229 40	29.063 562 46	30.539 003 91	33.759 991 70	19
20	26.870 374 49	28.279 681 81	29.778 078 58	31.371 422 77	33.065 954 10	36.785 591 20	**20**
21	28.676 485 72	30.269 470 68	31.969 201 72	33.783 136 80	35.719 251 81	39.992 726 68	21
22	30.536 780 30	32.328 902 15	34.247 969 79	36.303 377 95	38.505 214 40	43.392 290 28	22
23	32.452 883 70	34.460 413 73	36.617 888 58	38.937 029 96	41.430 475 12	46.995 827 69	23
24	34.426 470 22	36.666 528 21	39.082 604 12	41.689 196 31	44.501 998 87	50.815 577 35	24
25	36.459 264 32	38.949 856 69	41.645 908 29	44.565 210 15	47.727 098 82	54.864 512 00	25
26	38.553 042 25	41.313 101 68	44.311 744 62	47.570 644 60	51.113 453 76	59.156 382 72	26
27	40.709 633 52	43.759 060 24	47.084 214 40	50.711 323 61	54.669 126 45	63.705 765 68	27
28	42.930 922 52	46.290 627 34	49.967 582 98	53.993 333 17	58.402 582 77	68.528 111 62	28
29	45.218 850 20	48.910 799 30	52.966 286 30	57.423 033 16	62.322 711 91	73.639 798 32	29
30	47.575 415 71	51.622 677 28	56.084 937 75	61.007 069 66	66.438 847 50	79.058 186 22	**30**
31	50.002 678 18	54.429 470 98	59.328 335 26	64.752 387 79	70.760 789 88	84.801 677 39	31
32	52.502 758 52	57.334 502 47	62.701 468 67	68.666 245 24	75.298 829 37	90.889 778 03	32
33	55.077 841 28	60.341 210 05	66.209 527 42	72.756 226 28	80.063 770 84	97.343 164 71	33
34	57.730 176 52	63.453 152 40	69.857 908 51	77.030 256 46	85.066 959 38	104.183 754 60	34
35	60.462 081 81	66.674 012 74	73.652 224 86	81.496 618 00	90.320 307 35	111.434 779 87	35
36	63.275 944 27	70.007 603 18	77.598 313 85	86.163 965 81	95.836 322 72	119.120 866 66	36
37	66.174 222 59	73.457 869 30	81.702 246 40	91.041 344 27	101.628 138 86	127.268 118 66	37
38	69.159 449 27	77.028 894 72	85.970 336 26	96.138 204 76	107.709 545 80	135.904 205 78	38
39	72.234 232 75	80.724 906 04	90.409 149 71	101.464 423 98	114.095 023 09	145.058 458 13	39
40	75.401 259 73	84.550 277 75	95.025 515 70	107.030 323 06	120.799 774 24	154.761 965 62	**40**
41	78.663 297 53	88.509 537 47	99.826 536 33	112.846 687 60	127.839 762 95	165.047 683 56	41
42	82.023 196 45	92.607 371 28	104.819 597 78	118.924 788 54	135.231 751 10	175.950 544 57	42
43	85.483 892 34	96.848 629 28	110.012 381 69	125.276 404 02	142.993 338 66	187.507 577 24	43
44	89.048 409 11	101.238 331 30	115.412 876 96	131.913 842 20	151.143 005 59	199.758 031 88	44
45	92.719 861 39	105.781 672 90	121.029 392 04	138.849 965 10	159.700 155 87	212.743 513 79	45
46	96.501 457 23	110.484 031 45	126.870 567 72	146.098 213 53	168.685 163 66	226.508 124 62	46
47	100.396 500 95	115.350 972 55	132.945 390 43	153.672 633 14	178.119 421 85	241.098 612 10	47
48	104.408 395 98	120.388 256 59	139.263 206 04	161.587 901 63	188.025 392 94	256.564 528 82	48
49	108.540 647 85	125.601 845 57	145.833 734 29	169.859 357 20	198.426 662 59	272.958 400 55	49
50	112.796 867 29	130.997 910 16	152.667 083 66	178.503 028 28	209.347 995 72	290.335 904 58	**50**
51	117.180 773 31	136.582 837 02	159.773 767 00	187.535 664 55	220.815 395 50	308.756 058 86	51
52	121.696 196 51	142.363 236 31	167.164 717 68	196.974 769 46	232.856 165 28	328.281 422 39	52
53	125.347 082 40	148.345 949 58	174.851 306 39	206.838 634 08	245.498 973 54	348.978 307 73	53
54	131.137 494 88	154.538 057 82	182.845 358 65	217.146 372 62	258.773 922 22	370.917 006 20	54
55	136.071 619 72	160.946 889 84	191.159 172 99	227.917 959 38	272.712 618 33	394.172 026 57	55
56	141.153 768 31	167.580 030 99	199.805 539 91	239.174 267 56	287.348 249 24	418.822 348 16	56
57	146.388 381 36	174.445 332 07	208.797 761 51	250.937 109 60	302.715 661 71	444.951 689 05	57
58	151.780 032 80	181.550 918 69	218.149 671 97	263.229 279 53	318.851 444 79	472.648 790 40	58
59	157.333 433 79	188.905 200 85	227.875 658 85	276.074 597 11	335.794 017 03	502.007 717 82	59

(continued)

Table 3 FUTURE VALUE OF AN ANNUITY TABLES $s_{\overline{n}|i} = \dfrac{(1+i)^n - 1}{i}$ *(continued)*

n \ i	3%	$3\frac{1}{2}$%	4%	$4\frac{1}{2}$%	5%	6%	i / n
60	163.053 436 80	196.516 882 88	237.990 685 20	289.497 953 98	353.583 717 88	533.128 180 89	**60**
61	168.945 039 91	204.394 973 78	248.510 312 61	303.525 361 90	372.262 903 78	566.115 871 74	61
62	175.013 391 10	212.548 797 86	259.450 725 11	318.184 003 19	391.876 048 97	601.082 824 05	62
63	181.263 792 84	220.988 005 79	270.828 754 12	333.502 283 33	412.469 851 41	638.147 793 49	63
64	187.701 706 62	229.722 585 99	282.661 904 28	349.509 886 08	434.093 343 98	677.436 661 10	64
65	194.332 757 82	238.762 876 50	294.968 380 45	366.237 830 96	456.798 011 18	719.082 860 77	65
66	201.162 740 55	248.119 577 18	307.767 115 67	383.718 533 35	480.637 911 74	763.227 832 41	66
67	208.197 622 77	257.803 762 38	321.077 800 30	401.985 867 35	505.669 807 33	810.021 502 36	67
68	215.443 551 45	267.826 894 06	334.920 912 31	421.075 231 38	531.953 297 70	859.622 792 50	68
69	222.906 858 00	278.200 835 35	349.317 748 80	441.023 616 79	559.550 962 58	912.200 160 05	69
70	230.594 063 74	288.937 864 59	364.290 458 76	461.869 679 55	588.528 510 71	967.932 169 65	**70**
71	238.511 885 65	300.050 689 85	379.862 077 11	483.653 815 13	618.954 936 25	1027.008 099 83	71
72	246.667 242 22	311.552 464 00	396.056 560 19	506.418 236 81	650.902 683 06	1089.628 585 82	72
73	255.067 259 49	323.456 800 24	412.898 822 60	530.207 057 47	684.447 817 21	1156.006 300 97	73
74	263.719 277 27	335.777 788 24	430.414 775 50	555.066 375 05	719.670 208 07	1226.366 679 03	74
75	272.630 855 59	348.530 010 83	448.631 366 52	581.044 361 93	756.653 718 48	1300.948 679 77	75
76	281.809 781 26	361.728 561 21	467.576 621 18	608.191 358 22	795.486 404 40	1380.005 600 55	76
77	291.264 074 69	375.389 060 85	487.279 686 03	636.559 969 34	836.260 724 62	1463.805 936 59	77
78	301.001 996 93	389.527 677 98	507.770 873 47	666.205 167 96	879.073 760 85	1552.634 292 78	78
79	311.032 056 84	404.161 146 71	529.081 708 41	697.184 400 52	924.027 448 89	1646.792 350 35	79
80	321.363 018 55	419.306 786 85	551.244 976 75	729.557 698 54	971.228 821 34	1746.599 891 37	**80**
81	332.003 909 10	434.982 524 39	574.294 775 82	763.387 794 97	1020.790 262 40	1852.395 884 85	81
82	342.964 026 38	451.206 912 74	598.266 566 85	798.740 245 75	1072.829 775 52	1964.539 637 94	82
83	354.252 947 17	467.999 154 69	623.197 229 52	835.683 556 81	1127.471 264 30	2083.412 016 22	83
84	365.880 535 58	485.379 125 10	649.125 118 70	874.289 316 86	1184.844 827 52	2209.416 737 19	84
85	377.856 951 65	503.367 394 48	676.090 123 45	914.632 336 12	1245.087 068 89	2342.981 741 42	85
86	390.192 660 20	521.985 253 29	704.133 728 39	956.790 791 25	1308.341 422 34	2484.560 645 91	86
87	402.898 440 01	541.254 737 15	733.299 077 53	1000.846 376 85	1374.758 493 45	2634.634 284 66	87
88	415.985 393 21	561.198 652 95	763.631 040 63	1046.884 463 81	1444.496 418 13	2793.712 341 74	88
89	429.464 955 00	581.840 605 81	795.176 282 25	1094.994 264 68	1517.721 239 03	2962.335 082 25	89
90	443.348 903 65	603.205 027 01	827.983 333 54	1145.269 006 59	1594.607 300 98	3141.075 187 18	**90**
91	457.649 370 76	625.317 202 96	862.102 666 88	1197.806 111 89	1675.337 666 03	3330.539 698 41	91
92	472.378 851 89	648.203 305 06	897.586 773 56	1252.707 386 92	1760.104 549 33	3531.372 080 32	92
93	487.550 217 44	671.890 420 74	934.490 244 50	1310.079 219 34	1849.109 776 80	3744.254 405 14	93
94	503.176 723 97	696.406 585 46	972.869 854 28	1370.032 784 21	1942.565 265 64	3969.909 669 45	94
95	519.272 025 68	721.780 815 95	1012.784 648 45	1432.684 259 49	2040.693 528 92	4209.104 249 61	95
96	535.850 186 46	748.043 144 51	1054.296 034 39	1498.155 051 17	2143.728 205 37	4462.650 504 59	96
97	552.925 692 05	775.224 654 57	1097.467 875 77	1566.572 028 47	2251.914 615 64	4731.409 534 86	97
98	570.513 462 81	803.357 517 48	1142.366 590 80	1638.067 769 76	2365.510 346 42	5016.294 106 96	98
99	588.628 866 69	832.475 030 59	1189.061 254 43	1712.780 819 39	2484.785 863 74	5318.271 753 37	99
100	607.287 732 70	862.611 656 66	1237.623 704 61	1790.855 956 27	2610.025 156 93	5638.368 058 58	**100**
101	626.506 364 68	893.803 064 64	1288.128 652 79	1872.444 474 30	2741.526 414 77	5977.670 142 09	101
102	646.301 555 62	926.086 171 91	1340.653 798 90	1957.704 475 64	2879.602 735 51	6337.330 350 62	102
103	666.690 602 29	959.499 187 92	1395.279 950 86	2046.801 177 05	3024.582 872 29	6718.570 171 65	103
104	687.691 320 35	994.081 659 50	1452.091 148 89	2139.907 230 01	3176.812 015 90	7122.684 381 95	104
105	709.322 059 96	1029.874 517 58	1511.174 794 85	2237.203 055 36	3336.652 616 70	7551.045 444 87	105
106	731.601 721 76	1066.920 125 70	1572.621 786 64	2338.877 192 86	3504.485 247 53	8005.108 171 56	106
107	754.549 773 42	1105.262 330 10	1636.526 658 11	2445.126 666 53	3680.709 509 91	8486.414 661 85	107
108	778.186 266 62	1144.946 511 65	1702.987 724 43	2556.157 366 53	3865.744 985 40	8996.599 541 56	108
109	802.531 854 62	1186.019 639 56	1772.107 233 41	2672.184 448 02	4060.032 234 67	9537.395 514 06	109
110	827.607 810 26	1228.530 326 94	1843.991 522 75	2793.432 748 18	4264.033 846 41	10110.639 244 90	**110**
111	853.436 044 56	1272.528 888 39	1918.751 183 66	2920.137 221 85	4478.235 538 73	10718.277 599 60	111
112	880.039 125 90	1318.067 399 48	1996.501 231 00	3052.543 396 83	4703.147 315 66	11362.374 255 57	112
113	907.440 299 68	1365.199 758 46	2077.361 280 24	3190.907 849 69	4939.304 681 45	12045.116 710 91	113
114	935.663 508 67	1413.981 750 01	2161.455 731 45	3335.498 702 93	5187.269 915 52	12768.823 713 56	114
115	964.733 413 93	1464.471 111 26	2248.913 960 71	3486.596 144 56	5447.633 411 29	13535.953 136 37	115
116	994.675 416 35	1516.727 600 15	2339.870 519 14	3644.492 971 07	5721.015 081 86	14349.110 324 56	116
117	1025.515 678 84	1570.813 066 16	2434.465 339 91	3809.495 154 76	6008.065 835 95	15211.056 944 03	117
118	1057.281 149 20	1626.791 523 47	2532.843 953 50	3981.922 436 73	6309.469 127 75	16124.720 360 67	118
119	1089.999 583 68	1684.729 226 79	2635.157 711 64	4162.108 946 38	6625.942 584 14	17093.203 582 31	119
120	1123.699 571 19	1744.694 749 73	2741.564 020 11	4350.403 848 97	6958.239 713 34	18119.795 797 25	**120**

(continued)

Table 3 FUTURE VALUE OF AN ANNUITY TABLES $s_{\overline{n}|i} = \dfrac{(1+i)^n - 1}{i}$ *(continued)*

n \ i	7%	8%	9%	10%	11%	12%	i / n
1	1.000 000 00	1.000 000 00	1.000 000 00	1.000 000 00	1.000 000 00	1.000 000 00	**1**
2	2.070 000 00	2.080 000 00	2.090 000 00	2.100 000 00	2.110 000 00	2.120 000 00	2
3	3.214 900 00	3.246 400 00	3.278 100 00	3.310 000 00	3.342 100 00	3.374 400 00	3
4	4.439 943 00	4.506 112 00	4.573 129 00	4.641 000 00	4.709 731 00	4.779 328 00	4
5	5.750 739 01	5.866 600 96	5.984 710 61	6.105 100 00	6.227 801 41	6.352 847 36	5
6	7.153 290 74	7.335 929 04	7.523 334 56	7.715 610 00	7.912 859 57	8.115 189 04	6
7	8.654 021 09	8.922 803 36	9.200 434 68	9.487 171 00	9.783 274 12	10.089 011 73	7
8	10.259 802 57	10.636 627 63	11.028 473 80	11.435 888 10	11.859 434 27	12.299 693 14	8
9	11.977 988 75	12.487 557 84	13.021 036 44	13.579 476 91	14.163 972 04	14.775 656 31	9
10	13.816 447 96	14.486 562 47	15.192 929 72	15.937 424 60	16.722 008 96	17.548 735 07	**10**
11	15.783 599 32	16.645 487 46	17.560 293 39	18.531 167 06	19.561 429 95	21.654 583 28	11
12	17.888 451 27	18.977 126 46	21.140 719 80	21.384 283 77	22.713 187 24	24.133 133 27	12
13	21.140 642 86	21.495 296 58	22.953 384 58	24.522 712 14	26.211 637 84	28.029 109 26	13
14	22.550 487 86	24.214 920 30	26.019 189 19	27.974 983 36	30.094 918 00	32.392 602 38	14
15	25.129 022 01	27.152 113 93	29.360 916 22	31.772 481 69	34.405 358 98	37.279 714 66	15
16	27.888 053 55	30.324 283 04	33.003 398 68	35.949 729 86	39.189 948 47	42.753 280 42	16
17	30.840 217 30	33.750 225 69	36.973 704 56	40.544 702 85	44.500 842 81	48.883 674 07	17
18	33.999 032 51	37.450 243 74	41.301 337 97	45.599 173 13	50.395 935 51	55.749 714 96	18
19	37.378 964 79	41.446 263 24	46.018 458 39	51.159 090 45	56.939 488 42	63.439 680 75	19
20	40.995 492 32	45.761 964 30	51.160 119 64	57.274 999 49	64.202 832 15	72.052 442 44	**20**
21	44.865 176 78	50.422 921 44	56.764 530 41	64.002 499 44	72.265 143 68	81.698 735 54	21
22	49.005 739 16	55.456 755 16	62.873 338 15	71.402 749 39	81.214 309 49	92.502 583 80	22
23	53.436 140 90	60.893 295 57	69.531 938 58	79.543 024 33	91.147 883 53	104.602 893 86	23
24	58.176 670 76	66.764 759 22	76.789 813 05	88.497 326 76	102.174 150 72	118.155 241 12	24
25	63.249 037 72	73.105 939 95	84.700 896 23	98.347 059 43	114.413 307 30	133.333 870 06	25
26	68.676 470 36	79.954 415 15	93.323 976 89	109.181 765 38	127.998 771 10	150.333 934 46	26
27	74.483 823 28	87.350 768 36	102.723 134 81	121.099 941 92	143.078 635 92	169.374 006 60	27
28	80.697 690 91	95.338 829 83	112.968 216 94	134.209 936 11	159.817 285 87	190.698 887 39	28
29	87.346 529 27	103.965 936 22	124.135 356 46	148.630 929 72	178.397 187 32	214.582 753 88	29
30	94.460 786 32	113.283 211 11	136.307 538 55	164.494 022 69	199.020 877 93	241.332 684 34	**30**
31	102 073 041 37	123.345 868 00	149.575 217 02	181.943 424 96	221.913 174 50	271.292 606 46	31
32	110.218 154 26	134.213 537 44	164.036 986 55	201.137 767 45	247.323 623 69	304.847 719 24	32
33	118.933 425 06	145.950 620 44	179.800 315 34	222.251 544 20	275.529 222 30	342.429 445 55	33
34	128.258 764 81	158.626 670 07	196.982 343 72	245.476 698 62	306.837 436 75	384.520 979 01	34
35	138.236 878 35	172.316 803 68	215.710 754 65	271.024 368 48	341.589 554 80	431.663 496 49	35
36	148.913 459 84	187.102 147 97	236.124 722 57	299.126 805 33	380.164 405 82	484.463 116 07	36
37	160.337 402 02	203.070 319 81	258.375 947 60	330.039 485 86	422.982 490 46	543.598 690 00	37
38	172.561 020 17	220.315 945 40	282.629 782 88	364.043 434 45	470.510 564 41	609.830 532 80	38
39	185.640 291 58	238.941 221 03	309.066 463 34	401.447 777 89	523.266 726 50	684.010 196 74	39
40	199.635 111 99	259.056 518 71	337.882 445 04	442.592 555 68	581.826 066 41	767.091 420 34	**40**
41	214.609 569 83	280.781 040 21	369.291 865 10	487.851 811 25	646.826 933 72	860.142 390 79	41
42	230.632 239 72	304.243 523 42	403.528 132 96	537.636 992 37	718.977 896 43	964.359 477 68	42
43	247.776 496 50	329.583 005 30	440.845 664 92	592.400 691 61	799.065 465 04	1081.082 615 00	43
44	266.120 851 25	356.949 645 72	481.521 774 77	652.640 760 77	887.962 666 19	1211.812 528 80	44
45	285.749 310 84	386.505 617 38	525.858 734 50	718.904 836 85	986.638 559 47	1358.230 032 26	45
46	306.751 762 60	418.426 066 77	574.186 020 60	791.795 320 54	1096.168 801 01	1522.217 636 13	46
47	329.224 385 98	452.900 152 11	626.862 762 45	871.974 852 59	1217.747 369 12	1705.883 752 46	47
48	353.270 093 00	490.132 164 28	684.280 411 07	960.172 337 85	1352.699 579 73	1911.589 802 76	48
49	378.998 999 51	530.342 737 42	746.865 648 07	1057.189 571 63	1502.496 533 50	2141.980 579 09	49
50	406.528 294 47	573.770 156 42	815.083 556 40	1163.908 528 80	1668.771 152 18	2400.018 248 58	**50**
51	435.985 954 54	620.671 768 93	889.441 076 47	1281.299 381 68	1853.335 978 92	2689.020 438 41	51
52	467.504 971 35	671.325 510 44	970.490 773 36	1410.429 319 84	2058.202 936 61	3012.702 891 02	52
53	501.230 319 35	726.031 551 28	1058.834 942 96	1552.472 251 83	2285.605 259 63	3375.227 237 95	53
54	537.316 441 70	785.114 075 38	1155.130 087 83	1708.719 477 01	2538.021 838 19	3781.254 506 50	54
55	575.928 592 62	848.923 201 41	1260.091 795 73	1880.591 424 71	2818.204 240 39	4236.005 047 28	55
56	617.243 594 10	917.837 057 52	1374.500 057 34	2069.650 567 18	3129.206 706 84	4745.325 652 95	56
57	661.450 645 69	992.264 022 13	1499.205 062 51	2277.615 623 90	3474.419 444 59	5315.764 731 31	57
58	708.752 190 89	1072.645 143 90	1635.133 518 13	2506.377 186 29	3857.605 583 49	5954.656 499 06	58
59	759.364 844 25	1159.456 755 41	1783.295 534 76	2758.014 904 92	4282.942 197 68	6670.215 278 95	59
60	813.520 383 35	1253.213 295 84	1944.792 132 89	3034.816 395 41	4755.065 839 42	7471.641 112 42	**60**

Table 4 PRESENT VALUE OF AN ANNUITY TABLES $a_{\overline{n}|i} = \dfrac{1-(1+i)^{n}}{i}$

i / n	$\frac{5}{12}\%$	$\frac{1}{2}\%$	$\frac{2}{3}\%$	$\frac{3}{4}\%$	1%	$1\frac{1}{4}\%$	i / n
1	0.995 850 62	0.995 024 88	0.993 377 48	0.992 555 83	0.990 099 01	0.987 654 32	**1**
2	1.987 569 08	1.985 099 38	1.980 176 31	1.977 722 91	1.970 395 06	1.963 115 38	2
3	2.975 172 53	2.970 248 14	2.960 440 04	2.955 556 24	2.940 985 21	2.926 533 71	3
4	3.958 678 04	3.950 495 66	3.934 211 96	3.926 110 41	3.901 965 55	3.878 057 98	4
5	4.938 102 61	4.925 866 33	4.901 535 06	4.889 439 61	4.853 431 24	4.817 835 04	5
6	5.913 463 18	5.896 384 41	5.862 452 05	5.845 597 63	5.795 476 47	5.746 009 92	6
7	6.884 776 61	6.862 074 04	6.817 005 35	6.794 637 85	6.728 194 53	6.662 725 85	7
8	7.852 059 70	7.822 959 24	7.765 237 10	7.736 613 25	7.651 677 75	7.568 124 29	8
9	8.815 329 16	8.779 063 92	8.707 189 17	8.671 576 42	8.566 017 58	8.462 344 98	9
10	9.774 601 65	9.730 411 86	9.642 903 15	9.599 579 58	9.471 304 53	9.345 525 91	**10**
11	10.729 893 76	10.677 026 73	10.572 420 35	10.520 674 52	10.367 628 25	10.217 803 37	11
12	11.681 222 00	11.618 932 07	11.495 781 80	11.434 912 67	11.255 077 47	11.079 311 97	12
13	12.628 602 83	12.556 151 31	12.413 028 28	12.342 345 08	12.133 740 07	11.930 184 66	13
14	13.572 056 61	13.488 707 77	13.324 200 28	13.243 022 42	13.003 703 04	12.770 552 75	14
15	14.511 587 66	14.416 624 65	14.229 338 02	14.136 994 95	13.865 052 52	13.600 545 92	15
16	15.447 224 22	15.339 925 02	15.128 481 48	15.024 312 61	14.717 873 78	14.420 292 27	16
17	16.378 978 48	16.258 631 86	16.021 670 35	15.905 024 92	15.562 251 27	15.229 918 29	17
18	17.306 866 54	17.172 768 02	16.908 944 05	16.779 181 07	16.398 268 58	16.029 548 93	18
19	18.230 904 43	18.082 356 24	17.790 341 77	17.646 829 84	17.226 008 50	16.819 307 59	19
20	19.151 108 15	18.987 419 15	18.665 902 42	18.508 019 69	18.045 552 97	17.599 316 13	**20**
21	20.067 493 59	19.887 979 25	19.535 664 66	19.362 798 70	18.856 983 13	18.369 694 95	21
22	20.980 076 61	20.784 058 96	20.399 666 88	20.211 214 59	19.660 379 34	19.130 562 91	22
23	21.888 872 97	21.675 680 55	21.257 947 23	21.053 314 73	20.455 821 13	19.882 037 44	23
24	22.793 898 39	22.562 866 22	22.110 543 61	21.889 146 14	21.243 387 26	20.624 234 51	24
25	23.695 168 53	23.445 638 03	22.957 493 65	22.718 755 47	22.023 155 70	21.357 268 65	25
26	24.592 698 95	24.324 017 94	23.798 834 75	23.542 189 05	22.795 203 66	22.081 252 99	26
27	25.486 505 17	25.198 027 80	24.634 604 06	24.359 492 86	23.559 607 59	22.796 299 25	27
28	26.376 602 66	26.067 689 36	25.464 838 47	25.170 712 51	24.316 443 16	23.502 517 78	28
29	27.263 006 80	26.933 024 23	26.289 574 64	25.975 893 31	25.065 785 30	24.200 017 56	29
30	28.145 732 91	27.794 053 97	27.108 848 98	26.775 080 21	25.807 708 22	24.888 906 23	**30**
31	29.024 796 26	28.650 799 97	27.922 697 66	27.568 317 83	26.542 285 37	25.569 290 10	31
32	29.900 212 05	29.503 283 55	28.731 156 62	28.355 650 45	27.269 589 47	26.241 274 18	32
33	30.771 995 40	30.351 525 92	29.534 261 54	29.137 122 03	27.989 692 55	26.904 962 15	33
34	31.640 161 39	31.195 548 18	30.332 047 89	29.912 776 21	28.702 665 89	27.560 456 44	34
35	32.504 725 04	32.035 371 32	31.124 550 88	30.682 656 29	29.408 580 09	28.207 858 22	35
36	33.365 701 28	32.871 016 24	31.911 805 51	31.446 805 25	30.107 505 04	28.847 267 37	36
37	34.223 105 01	33.702 503 72	32.693 846 53	32.205 265 76	30.799 509 94	29.478 782 59	37
38	35.076 951 05	34.529 854 45	33.470 708 48	32.958 080 16	31.484 663 30	30.102 501 33	38
39	35.927 254 16	35.353 089 00	34.242 425 64	33.705 290 48	32.163 032 98	30.718 519 83	39
40	36.774 029 04	36.172 227 86	35.009 032 09	34.446 938 44	32.834 686 11	31.326 933 16	**40**
41	37.617 290 33	36.987 291 41	35.770 561 68	35.183 065 45	33.499 689 22	31.927 835 22	41
42	38.457 052 61	37.798 299 91	36.527 048 03	35.913 712 60	34.158 108 14	32.521 318 74	42
43	39.293 330 40	38.605 273 54	37.278 524 53	36.638 920 70	34.810 008 06	33.107 475 30	43
44	40.126 138 16	39.408 232 38	38.025 024 37	37.358 730 22	35.455 453 52	33.686 395 36	44
45	40.955 490 28	40.207 196 40	38.766 580 50	38.073 181 36	36.094 508 44	34.258 168 25	45
46	41.781 401 11	41.002 185 47	39.503 225 66	38.782 314 01	36.727 236 08	34.822 882 22	46
47	42.603 884 92	41.793 219 37	40.234 992 38	39.486 167 75	37.353 699 09	35.380 624 42	47
48	43.422 955 94	42.580 317 78	40.961 912 96	40.184 781 89	37.973 959 49	35.931 480 91	48
49	44.238 628 32	43.363 500 28	41.684 019 49	40.878 195 42	38.588 078 71	36.475 536 70	49
50	45.050 916 17	44.142 786 35	42.401 343 87	41.566 447 07	39.196 117 53	37.012 875 75	**50**
51	45.859 833 53	44.918 195 37	43.113 917 75	42.249 575 25	39.798 136 17	37.543 580 99	51
52	46.665 394 39	45.689 746 64	43.821 772 60	42.927 618 12	40.394 194 23	38.067 734 31	52
53	47.467 612 67	46.457 459 34	44.524 939 67	43.600 613 51	40.984 350 72	38.585 416 60	53
54	48.266 502 24	47.221 352 58	45.223 450 00	44.268 599 02	41.568 664 08	39.096 707 76	54
55	49.062 076 92	47.981 445 35	45.917 334 44	44.931 611 93	42.147 192 16	39.601 686 67	55
56	49.854 350 46	48.737 756 57	46.606 623 61	45.589 689 26	42.719 992 24	40.100 431 28	56
57	50.643 336 56	49.490 305 05	47.291 347 96	46.242 867 76	43.287 121 02	40.593 018 55	57
58	51.429 048 85	50.239 109 50	47.971 537 71	46.891 183 88	43.848 634 68	41.079 524 49	58
59	52.211 500 93	50.984 188 55	48.647 222 89	47.534 673 82	44.404 588 79	41.560 024 19	59

(continued)

Table 4 PRESENT VALUE OF AN ANNUITY TABLES $a_{\overline{n}|i} = \frac{1-(1+i)^n}{i}$ *(continued)*

n \ i	$\frac{5}{12}\%$	$\frac{1}{2}\%$	$\frac{2}{3}\%$	$\frac{3}{4}\%$	1%	$1\frac{1}{4}\%$	i / n
60	52.990 706 32	51.725 560 75	49.318 433 34	48.173 373 52	44.955 038 41	42.034 591 79	**60**
61	53.766 678 50	52.463 244 53	49.985 198 68	48.807 318 63	45.500 038 03	42.503 300 54	61
62	54.539 430 87	53.197 258 24	50.647 548 36	49.436 544 55	46.039 641 61	42.966 222 75	62
63	55.308 976 80	53.927 620 14	51.305 511 61	50.061 086 40	46.573 902 58	43.423 429 88	63
64	56.075 329 59	54.654 348 39	51.959 117 49	50.680 979 06	47.102 873 85	43.874 992 47	64
65	56.838 502 50	55.377 461 09	52.608 394 86	51.296 257 13	47.626 607 77	44.320 980 22	65
66	57.598 508 71	56.096 976 21	53.253 372 38	51.906 954 97	48.145 156 21	44.761 461 95	66
67	58.355 361 37	56.812 911 65	53.894 078 52	52.513 106 67	48.658 570 50	45.196 505 63	67
68	59.109 073 57	57.525 285 22	54.530 541 58	53.114 746 07	49.166 901 49	45.626 178 40	68
69	59.859 658 32	58.234 114 65	55.162 789 65	53.711 906 77	49.670 199 49	46.050 546 56	69
70	60.607 128 62	58.939 417 56	55.790 850 64	54.304 622 10	50.168 514 35	46.469 675 62	**70**
71	61.351 497 38	59.641 211 51	56.414 752 30	54.892 925 16	50.661 895 39	46.883 630 24	71
72	62.092 777 48	60.339 513 94	57.034 522 15	55.476 848 80	51.150 391 48	47.292 474 31	72
73	62.830 981 72	61.034 342 22	57.650 187 56	56.056 425 61	51.634 050 97	47.696 270 93	73
74	63.566 122 87	61.725 713 66	58.261 775 73	56.631 687 95	52.112 921 75	48.095 082 40	74
75	64.298 213 65	62.413 645 43	58.869 313 63	57.202 667 94	52.587 051 24	48.488 970 27	75
76	65.027 266 70	63.098 154 66	59.472 828 11	57.769 397 46	53.056 486 38	48.877 995 33	76
77	65.753 294 64	63.779 258 36	60.072 345 81	58.331 908 15	53.521 273 64	49.262 217 61	77
78	66.476 310 02	64.456 973 50	60.667 893 19	58.890 231 41	53.981 459 05	49.641 696 40	78
79	67.196 325 33	65.131 316 91	61.259 496 54	59.444 398 42	54.437 088 17	50.016 490 27	79
80	67.913 353 03	65.802 305 38	61.847 182 00	59.994 440 12	54.888 206 11	50.386 657 06	**80**
81	68.627 405 50	66.469 955 61	62.430 975 49	60.540 387 22	55.334 857 53	50.752 253 89	81
82	69.338 495 11	67.134 284 19	63.010 902 81	61.082 270 19	55.777 086 66	51.113 337 17	82
83	70.046 634 13	67.795 307 65	63.586 989 54	61.620 119 30	56.214 937 29	51.469 962 64	83
84	70.751 834 82	68.453 042 44	64.159 261 14	62.153 964 56	56.648 452 76	51.822 185 32	84
85	71.454 109 36	69.107 504 91	64.727 742 85	62.683 835 79	57.077 676 00	52.170 059 58	85
86	72.153 469 91	69.758 711 35	65.292 459 79	63.209 762 57	57.502 649 51	52.513 639 09	86
87	72.849 928 54	70.406 677 96	65.853 436 87	63.731 774 27	57.923 415 35	52.852 976 88	87
88	73.543 497 30	71.051 420 86	66.410 698 88	64.249 900 02	58.340 015 20	53.188 125 31	88
89	74.234 188 18	71.692 956 08	66.964 270 41	64.764 168 75	58.752 490 30	53.519 136 11	89
90	74.922 013 13	72.331 299 58	67.514 175 91	65.274 609 18	59.160 881 48	53.846 060 36	**90**
91	75.606 984 03	72.966 467 25	68.060 439 64	65.781 249 81	59.565 229 19	54.168 948 50	91
92	76.289 112 72	73.598 474 87	68.603 085 74	66.284 118 92	59.965 573 46	54.487 850 37	92
93	76.968 411 01	74.227 338 18	69.142 138 15	66.783 244 58	60.361 953 92	54.802 815 18	93
94	77.644 890 63	74.853 072 82	69.677 620 68	67.278 654 67	60.754 409 82	55.113 891 54	94
95	78.318 563 29	75.475 694 34	70.209 556 96	67.770 376 85	61.142 980 02	55.421 127 44	95
96	78.989 440 62	76.095 218 25	70.737 970 49	68.258 438 56	61.527 702 99	55.724 570 31	96
97	79.657 534 22	76.711 659 95	71.262 884 60	68.742 867 05	61.908 616 82	56.024 266 98	97
98	80.322 855 66	77.325 034 78	71.784 322 45	69.223 689 38	62.285 759 23	56.320 263 68	98
99	80.985 416 42	77.935 357 99	72.302 307 07	69.700 932 39	62.659 167 55	56.612 606 10	99
100	81.645 227 97	78.542 644 77	72.816 861 33	70.174 622 72	63.028 878 77	56.901 339 36	**100**
101	82.302 301 72	79.146 910 21	73.328 007 94	70.644 786 82	63.394 929 47	57.186 508 01	101
102	82.956 649 01	79.748 169 37	73.835 769 48	71.111 450 94	63.757 355 91	57.468 156 06	102
103	83.608 281 17	80.346 437 18	74.340 168 35	71.574 641 13	64.116 193 97	57.746 326 97	103
104	84.257 209 47	80.941 728 54	74.841 226 84	72.034 383 25	64.471 479 18	58.021 063 68	104
105	84.903 445 11	81.534 058 25	75.338 967 06	72.490 702 98	64.823 246 71	58.292 408 57	105
106	85.546 999 28	82.123 441 04	75.833 410 99	72.943 625 79	65.171 531 40	58.560 403 53	106
107	86.187 883 10	82.709 891 58	76.324 580 45	73.393 176 96	65.516 367 72	58.825 089 90	107
108	86.826 107 65	83.293 424 46	76.812 497 14	73.839 381 60	65.857 789 83	59.086 508 55	108
109	87.461 683 97	83.874 054 19	77.297 182 59	74.282 264 61	66.195 831 51	59.344 699 80	109
110	88.094 623 04	84.451 795 22	77.778 658 20	74.721 850 73	66.530 526 25	59.599 703 50	**110**
111	88.724 935 81	85.026 661 91	78.256 945 23	75.158 164 50	66.861 907 18	59.851 559 02	111
112	89.352 633 17	85.598 668 56	78.732 064 80	75.591 230 27	67.190 007 10	60.100 305 20	112
113	89.977 725 98	86.167 829 42	79.204 037 88	76.021 072 23	67.514 858 52	60.345 980 45	113
114	90.600 225 04	86.734 158 62	79.672 885 31	76.447 714 37	67.836 493 58	60.588 622 66	114
115	91.220 141 12	87.297 670 27	80.138 627 79	76.871 180 52	68.154 944 14	60.828 269 30	115
116	91.837 484 93	87.858 378 38	80.601 285 89	77.291 494 31	68.470 241 72	61.064 957 33	116
117	92.452 267 15	88.416 296 90	81.060 880 02	77.708 679 22	68.782 417 55	61.298 723 29	117
118	93.064 498 41	88.971 439 70	81.517 430 48	78.122 758 53	69.091 502 52	61.529 603 25	118
119	93.674 189 29	89.523 820 59	81.970 957 43	78.533 755 36	69.397 527 25	61.757 632 84	119
120	94.281 350 33	90.073 453 33	82.421 480 89	78.941 692 67	69.700 522 03	61.982 847 25	**120**

(continued)

Table 4 PRESENT VALUE OF AN ANNUITY TABLES $a_{\overline{n}|i} = \frac{1-(1+i)^{-n}}{i}$ *(continued)*

n \ i	$1\frac{1}{2}\%$	$1\frac{3}{4}\%$	2%	$2\frac{1}{4}\%$	$2\frac{1}{2}\%$	$2\frac{3}{4}\%$	i / n
1	0.985 221 67	0.982 800 98	0.980 392 16	0.977 995 11	0.975 609 76	0.973 236 01	**1**
2	1.955 883 42	1.948 698 75	1.941 560 94	1.934 469 55	1.927 424 15	1.920 424 34	2
3	2.912 200 42	2.897 984 03	2.883 883 27	2.869 896 87	2.856 023 56	2.842 262 13	3
4	3.854 384 65	3.830 942 54	3.807 728 70	3.784 740 21	3.761 974 21	3.739 427 87	4
5	4.782 644 97	4.747 855 08	4.713 459 51	4.679 452 53	4.645 828 50	4.612 581 86	5
6	5.697 187 17	5.648 997 62	5.601 430 89	5.554 476 80	5.508 125 36	5.462 366 78	6
7	6.598 213 96	6.534 641 39	6.471 991 07	6.410 246 26	6.349 390 60	6.289 408 06	7
8	7.485 925 08	7.405 052 97	7.325 481 44	7.247 184 61	7.170 137 17	7.094 314 41	8
9	8.360 517 32	8.260 494 32	8.162 236 71	8.065 706 22	7.970 865 53	7.877 678 26	9
10	9.222 184 55	9.101 222 91	8.982 585 01	8.866 216 35	8.752 063 93	8.640 076 16	**10**
11	10.071 117 79	9.927 491 81	9.786 848 05	9.649 111 34	9.514 208 71	9.382 069 26	11
12	10.907 505 21	10.739 549 69	10.575 341 22	10.414 778 82	10.257 764 60	10.104 203 66	12
13	11.731 532 22	11.537 640 97	11.348 373 75	11.163 597 87	10.983 184 97	10.807 010 86	13
14	12.543 381 50	12.322 005 87	12.106 248 77	11.895 939 24	11.690 912 17	11.491 008 14	14
15	13.343 233 01	13.092 880 46	12.849 263 50	12.612 165 51	12.381 377 73	12.156 698 92	15
16	14.131 264 05	13.850 496 77	13.577 709 31	13.312 631 31	13.055 002 66	12.804 573 15	16
17	14.907 649 31	14.595 082 82	14.291 871 88	13.997 683 43	13.712 197 72	13.435 107 69	17
18	15.672 560 89	15.326 862 72	14.992 031 25	14.667 661 06	14.353 363 63	14.048 766 61	18
19	16.426 168 37	16.046 056 73	15.678 462 01	15.322 895 90	14.978 891 34	14.646 001 57	19
20	17.168 638 79	16.752 881 30	16.351 433 34	15.963 712 37	15.589 162 29	15.227 252 13	**20**
21	17.900 136 73	17.447 549 19	17.011 209 16	16.590 427 75	16.184 548 57	15.792 946 12	21
22	18.620 824 37	18.130 269 48	17.658 048 20	17.203 352 32	16.765 413 24	16.343 499 87	22
23	19.330 861 45	18.801 247 64	18.292 204 12	17.802 789 55	17.332 110 48	16.879 318 61	23
24	20.030 405 37	19.460 685 65	18.913 925 60	18.389 036 24	17.884 985 83	17.400 796 70	24
25	20.719 611 20	20.108 781 96	19.523 456 47	18.962 382 63	18.424 376 42	17.908 317 95	25
26	21.398 631 72	20.745 731 66	20.121 035 76	19.523 112 60	18.950 611 14	18.402 255 92	26
27	22.067 617 46	21.371 726 44	20.706 897 80	20.071 503 76	19.464 010 87	18.882 974 13	27
28	22.726 716 71	21.986 954 74	21.281 272 36	20.607 827 64	19.964 888 66	19.350 826 40	28
29	23.376 075 58	22.591 601 71	21.844 384 66	21.132 349 77	20.453 549 91	19.806 157 08	29
30	24.015 838 01	23.185 849 34	22.396 455 55	21.645 329 85	20.930 292 59	20.249 301 30	**30**
31	24.646 145 82	23.769 876 50	22.937 701 52	22.147 021 86	21.395 407 41	20.680 585 20	31
32	25.267 138 74	24.343 858 97	23.468 334 82	22.637 674 19	21.849 177 96	21.100 326 23	32
33	25.878 954 42	24.907 969 51	23.988 563 55	23.117 529 77	22.291 880 94	21.508 833 32	33
34	26.481 728 49	25.462 377 89	24.498 591 72	23.586 826 18	22.723 786 28	21.906 407 12	34
35	27.075 594 58	26.007 251 00	24.998 619 33	24.045 795 77	23.145 157 34	22.293 340 26	35
36	27.660 684 31	26.542 752 83	25.488 842 48	24.494 665 79	23.556 251 07	22.669 917 53	36
37	28.237 127 40	27.069 044 55	25.969 453 41	24.933 658 48	23.957 318 12	23.036 416 09	37
38	28.805 051 63	27.586 284 57	26.440 640 60	25.362 991 18	24.348 603 04	23.393 105 68	38
39	29.364 582 88	28.094 628 57	26.902 588 83	25.782 876 46	24.730 344 43	23.740 248 84	39
40	29.915 845 20	28.594 229 55	27.355 479 24	26.193 522 21	25.102 775 05	24.078 101 06	**40**
41	30.458 960 79	29.085 237 89	27.799 489 45	26.595 131 74	25.466 122 00	24.406 911 01	41
42	30.994 050 04	29.567 801 36	28.234 793 58	26.987 903 90	25.820 606 83	24.726 920 69	42
43	31.521 231 57	30.042 065 22	28.661 562 33	27.372 033 16	26.166 445 69	25.038 365 63	43
44	32.040 622 23	30.508 172 21	29.079 963 07	27.747 709 69	26.503 849 45	25.341 475 07	44
45	32.552 337 18	30.966 262 61	29.490 159 87	28.115 119 50	26.833 023 86	25.636 472 09	45
46	33.056 489 83	31.416 474 31	29.892 313 60	28.474 444 50	27.154 169 62	25.923 573 81	46
47	33.553 191 95	31.858 942 81	30.286 581 96	28.825 862 59	27.467 482 55	26.202 991 54	47
48	34.042 553 65	32.293 801 29	30.673 119 57	29.169 547 77	27.773 153 71	26.474 930 94	48
49	34.524 683 39	32.721 180 63	31.052 078 01	29.505 670 19	28.071 369 47	26.739 592 15	49
50	34.999 688 07	33.141 209 46	31.423 605 89	29.834 396 27	28.362 311 68	26.997 169 98	**50**
51	35.467 672 98	33.554 014 21	31.787 848 92	30.155 888 77	28.646 157 74	27.247 854 00	51
52	35.928 741 85	33.959 719 13	32.144 949 92	30.470 306 87	28.923 080 72	27.491 828 71	52
53	36.382 996 90	34.358 446 32	32.495 048 94	30.777 806 23	29.193 249 48	27.729 273 68	53
54	36.830 538 82	34.750 315 79	32.838 283 27	31.078 539 10	29.456 828 76	27.960 363 68	54
55	37.271 466 81	35.135 445 50	33.174 787 52	31.372 654 38	29.713 979 28	28.185 268 79	55
56	37.705 878 63	35.513 951 35	33.504 693 65	31.660 297 68	29.964 857 84	28.404 154 54	56
57	38.133 870 58	35.885 947 27	33.828 131 03	31.941 611 42	30.209 617 40	28.617 182 03	57
58	38.555 537 51	36.251 545 23	34.145 226 50	32.216 734 89	30.448 407 22	28.824 508 06	58
59	38.970 972 92	36.610 855 26	34.456 104 41	32.485 804 29	30.681 372 90	29.026 285 22	59

(continued)

Table 4 PRESENT VALUE OF AN ANNUITY TABLES $a_{\overline{n}|i} = \frac{1-(1+i)^{-n}}{i}$ *(continued)*

n \ i	$1\frac{1}{2}\%$	$1\frac{3}{4}\%$	2%	$2\frac{1}{4}\%$	$2\frac{1}{2}\%$	$2\frac{3}{4}\%$	i / n
60	39.380 268 89	36.963 985 52	34.760 886 68	32.748 952 85	30.908 656 49	29.222 662 01	**60**
61	39.783 516 14	37.311 042 28	35.059 692 82	33.006 310 86	31.130 396 57	29.413 782 98	61
62	40.180 804 08	37.652 130 00	35.352 640 02	33.258 005 73	31.346 728 36	29.599 788 79	62
63	40.572 220 77	37.987 351 35	35.639 843 16	33.504 162 08	31.557 783 77	29.780 816 34	63
64	40.957 852 98	38.316 807 23	35.921 414 86	33.744 901 79	31.763 691 48	29.956 998 87	64
65	41.337 786 18	38.640 596 78	36.197 465 55	33.980 344 05	31.964 577 05	30.128 466 05	65
66	41.712 104 61	38.958 817 48	36.468 103 48	34.210 605 43	32.160 562 98	30.295 344 09	66
67	42.080 891 25	39.271 565 09	36.733 434 78	34.435 799 93	32.351 768 76	30.457 755 81	67
68	42.444 227 83	39.578 933 75	36.993 563 51	34.656 039 05	32.538 310 99	30.615 820 74	68
69	42.802 194 90	39.881 015 97	37.248 591 68	34.871 431 83	32.720 303 40	30.769 655 22	69
70	43.154 871 83	40.177 902 67	37.498 619 29	35.082 084 92	32.897 856 98	30.919 372 47	**70**
71	43.502 336 78	40.469 683 21	37.743 744 41	35.288 102 61	33.071 079 98	31.065 082 70	71
72	43.844 666 77	40.756 445 42	37.984 063 14	35.489 586 91	33.240 078 03	31.206 893 14	72
73	44.181 937 71	41.038 275 60	38.219 669 75	35.686 637 56	33.404 954 17	31.344 908 16	73
74	44.514 224 34	41.315 258 57	38.450 656 62	35.879 352 14	33.565 808 95	31.479 229 36	74
75	44.841 600 34	41.587 477 71	38.677 114 33	36.067 826 05	33.722 740 44	31.609 955 58	75
76	45.164 138 26	41.855 014 95	38.899 131 70	36.252 152 62	33.875 844 33	31.737 183 04	76
77	45.481 909 62	42.117 950 81	39.116 795 78	36.432 423 10	34.025 213 98	31.861 005 40	77
78	45.794 984 85	42.376 364 43	39.330 191 94	36.608 726 75	34.170 940 47	31.981 513 77	78
79	46.103 433 35	42.630 333 59	39.539 403 86	36.781 150 85	34.313 112 65	32.098 796 85	79
80	46.407 323 49	42.879 934 74	39.744 513 59	36.949 780 79	34.451 817 22	32.212 940 98	**80**
81	46.706 722 65	43.125 242 98	39.945 601 56	37.114 700 04	34.587 138 75	32.324 030 15	81
82	47.001 697 20	43.366 332 17	40.142 746 63	37.275 990 26	34.719 159 76	32.432 146 13	82
83	47.292 312 51	43.603 274 86	40.336 026 11	37.433 731 30	34.847 960 74	32.537 368 50	83
84	47.578 633 01	43.836 142 37	40.525 515 79	37.588 001 27	34.973 620 23	32.639 774 69	84
85	47.860 722 18	44.065 004 79	40.711 289 99	37.738 876 55	35.096 214 86	32.739 440 09	85
86	48.138 642 54	44.289 930 99	40.893 421 56	37.886 431 83	35.215 819 38	32.836 438 04	86
87	48.412 455 71	44.510 988 69	41.071 981 92	38.030 740 18	35.332 506 71	32.930 839 94	87
88	48.682 222 37	44.728 244 41	41.247 041 10	38.171 873 04	35.446 348 01	33.022 715 27	88
89	48.948 002 34	44.941 763 55	41.418 667 74	38.309 900 28	35.557 412 69	33.112 131 65	89
90	49.209 854 52	45.151 610 37	41.586 929 16	38.444 890 25	35.665 768 48	33.199 154 89	**90**
91	49.467 836 96	45.357 848 03	41.751 891 33	38.576 909 78	35.771 481 44	33.283 849 05	91
92	49.722 006 86	45.560 538 60	41.913 618 95	38.706 024 23	35.874 616 04	33.366 276 44	92
93	49.972 420 55	45.759 743 10	42.072 175 45	38.832 297 54	35.975 235 16	33.446 497 76	93
94	50.219 133 55	45.955 521 47	42.227 622 99	38.955 792 21	36.073 400 16	33.524 572 02	94
95	50.462 200 54	46.147 932 65	42.380 022 54	39.076 569 40	36.169 170 89	33.600 556 71	95
96	50.701 675 41	46.337 034 55	42.529 433 86	39.194 688 90	36.262 605 74	33.674 507 75	96
97	50.937 611 24	46.522 884 08	42.675 915 55	39.310 209 20	36.353 761 70	33.746 479 56	97
98	51.170 060 34	46.705 537 18	42.819 525 05	39.423 187 48	36.442 694 34	33.816 525 12	98
99	51.390 074 22	46.885 048 82	42.960 318 67	39.533 679 68	36.529 457 90	33.884 695 98	99
100	51.624 703 67	47.061 473 04	43.098 351 64	39.641 740 52	36.614 105 26	33.951 042 32	**100**
101	51.846 998 69	47.234 862 94	43.233 678 08	39.747 423 49	36.696 688 06	34.015 612 96	101
102	52.006 008 56	47.405 270 71	43.366 351 06	39.850 780 92	36.777 256 65	34.078 455 44	102
103	52.281 781 83	47.572 747 62	43.496 422 61	39.951 863 98	36.855 860 14	34.139 616 00	103
104	52.494 366 34	47.737 344 10	43.623 943 73	40.050 722 72	36.932 546 48	34.199 139 66	104
105	52.703 809 20	47.899 109 68	43.748 964 44	40.147 406 09	37.007 362 42	34.257 070 23	105
106	52.910 156 85	48.058 093 05	43.871 533 77	40.241 961 94	37.080 353 58	34.313 450 34	106
107	53.113 455 02	48.214 342 07	43.991 699 77	40.334 437 11	37.151 564 47	34.368 321 50	107
108	53.313 748 79	48.367 903 75	44.109 509 58	40.424 877 37	37.221 038 51	34.421 724 09	108
109	53.511 082 55	48.518 824 32	44.225 009 39	40.513 327 50	37.288 818 06	34.473 697 41	109
110	53.705 500 05	48.667 149 21	44.338 244 50	40.599 831 29	37.354 944 44	34.524 279 72	**110**
111	53.897 044 38	48.812 923 06	44.449 259 32	40.684 431 58	37.419 457 99	34.573 508 24	111
112	54.085 758 01	48.956 189 74	44.558 097 37	40.767 170 25	37.482 398 04	34.621 419 21	112
113	54.271 682 77	49.096 992 37	44.664 801 34	40.848 088 27	37.543 802 97	34.668 047 90	113
114	54.454 859 87	49.235 373 34	44.769 413 08	40.927 225 69	37.603 710 21	34.713 428 61	114
115	54.635 329 93	49.371 374 29	44.871 973 61	41.004 621 70	37.662 156 31	34.757 594 75	115
116	54.813 132 93	49.505 036 16	44.972 523 14	41.080 314 62	37.719 176 88	34.800 578 84	116
117	54.988 308 31	49.636 399 17	45.071 101 12	41.154 341 93	37.774 806 72	34.842 412 49	117
118	55.160 894 88	49.765 502 87	45.167 746 20	41.226 740 27	37.829 079 72	34.883 126 51	118
119	55.330 930 92	49.892 386 11	45.262 496 27	41.297 545 50	37.882 029 00	34.922 750 86	119
120	55.498 454 11	50.017 087 09	45.355 388 50	41.366 792 66	37.933 686 83	34.961 314 71	**120**

(continued)

Table 4 PRESENT VALUE OF AN ANNUITY TABLES $a_{\overline{n}|i} = \frac{1-(1+i)^{-n}}{i}$

(continued)

n \ i	3%	$3\frac{1}{2}$%	4%	$4\frac{1}{2}$%	5%	6%	i / n
1	0.970 873 79	0.966 183 57	0.961 538 46	0.956 937 80	0.952 380 95	0.943 396 23	**1**
2	1.913 469 70	1.899 694 28	1.886 094 67	1.872 667 75	1.859 410 43	1.833 392 67	2
3	2.828 611 35	2.801 636 98	2.775 091 03	2.748 964 35	2.723 248 03	2.673 011 95	3
4	3.717 098 40	3.673 079 21	3.629 895 22	3.587 525 70	3.545 950 50	3.465 105 61	4
5	4.579 707 19	4.515 052 38	4.451 822 33	4.389 976 74	4.329 476 67	4.212 363 79	5
6	5.417 191 44	5.328 553 02	5.242 136 86	5.157 872 48	5.075 692 07	4.917 324 33	6
7	6.230 282 96	6.114 543 98	6.002 054 67	5.892 700 94	5.786 373 40	5.582 381 44	7
8	7.019 692 19	6.873 955 54	6.732 744 87	6.595 886 07	6.463 212 76	6.209 793 81	8
9	7.786 108 92	7.607 686 51	7.435 331 61	7.268 790 50	7.107 821 68	6.801 692 27	9
10	8.530 202 84	8.316 605 32	8.110 895 78	7.912 718 18	7.721 734 93	7.360 087 05	**10**
11	9.252 624 11	9.001 551 04	8.760 476 71	8.528 916 92	8.306 414 22	7.886 874 58	11
12	9.954 003 99	9.663 334 33	9.385 073 76	9.118 580 78	8.863 251 64	8.383 843 94	12
13	10.634 955 33	10.302 738 49	9.985 647 85	9.682 852 42	9.393 572 99	8.852 682 96	13
14	11.296 073 14	10.920 520 28	10.563 122 93	10.222 825 28	9.898 640 94	9.294 983 93	14
15	11.937 935 09	11.517 410 90	11.118 387 43	10.739 545 73	10.379 658 04	9.712 248 99	15
16	12.561 102 03	12.094 116 81	11.652 295 61	11.234 015 05	10.837 769 56	10.105 895 27	16
17	13.166 118 47	12.651 320 59	12.165 668 85	11.707 191 43	11.274 066 25	10.477 259 69	17
18	13.753 513 08	13.189 681 73	12.659 296 97	12.159 991 80	11.689 586 90	10.827 603 48	18
19	14.323 799 11	13.709 837 42	13.133 939 40	12.593 293 59	12.085 320 86	11.158 116 49	19
20	14.877 474 86	14.212 403 30	13.590 326 34	13.007 936 45	12.462 210 34	11.469 921 22	**20**
21	15.415 024 14	14.697 974 20	14.029 159 95	13.404 723 88	12.821 152 71	11.764 076 62	21
22	15.936 916 64	15.167 124 84	14.451 115 33	13.784 424 76	13.163 002 58	12.041 581 72	22
23	16.443 608 39	15.620 410 47	14.856 841 67	14.147 774 89	13.488 573 88	12.303 378 98	23
24	16.935 542 12	16.058 367 60	15.246 963 14	14.495 478 37	13.798 641 79	12.550 357 53	24
25	17.413 147 69	16.481 514 59	15.622 079 94	14.828 208 96	14.093 944 57	12.783 356 16	25
26	17.876 842 42	16.890 352 26	15.982 769 18	15.146 611 45	14.375 185 30	13.003 166 19	26
27	18.327 031 47	17.285 364 51	16.329 585 75	15.451 302 82	14.643 033 62	13.210 534 14	27
28	18.764 108 23	17.667 018 85	16.663 063 22	15.742 873 51	14.898 127 26	13.406 164 28	28
29	19.188 454 59	18.035 767 00	16.983 714 63	16.021 888 53	15.141 073 58	13.590 721 02	29
30	19.600 441 35	18.392 045 41	17.292 033 30	16.288 888 54	15.372 451 03	13.764 831 15	**30**
31	20.000 428 49	18.736 275 76	17.588 493 56	16.544 390 95	15.592 810 50	13.929 085 99	31
32	20.388 765 53	19.068 865 47	17.873 551 50	16.788 890 86	15.802 676 67	14.084 043 39	32
33	20.765 791 78	19.390 208 18	18.147 645 67	17.022 86 207	16.002 549 21	14.230 229 61	33
34	21.131 836 68	19.700 684 23	18.411 197 76	17.246 757 96	16.192 904 01	14.368 141 14	34
35	21.487 220 07	20.000 661 10	18.664 613 23	17.461 012 40	16.374 194 29	14.498 246 36	35
36	21.832 252 50	20.290 493 81	18.908 281 95	17.666 040 58	16.546 851 71	14.620 987 13	36
37	22.167 235 44	20.570 525 42	19.142 578 80	17.862 239 79	16.711 287 34	14.736 780 31	37
38	22.492 461 59	20.841 087 36	19.367 864 23	18.049 990 23	16.867 892 71	14.846 019 16	38
39	22.808 215 13	21.102 499 87	19.584 484 84	18.229 655 72	17.017 040 67	14.949 074 68	39
40	23.114 771 97	21.355 072 34	19.792 773 88	18.401 584 42	17.159 086 35	15.046 296 87	**40**
41	23.412 399 97	21.599 103 71	19.993 051 81	18.566 109 49	17.294 367 96	15.138 015 92	41
42	23.701 359 20	21.834 882 81	20.185 626 74	18.723 549 75	17.423 207 58	15.224 543 32	42
43	23.981 902 13	22.062 688 70	20.370 794 94	18.874 210 29	17.545 911 98	15.306 172 94	43
44	24.254 273 92	22.282 791 02	20.548 841 29	19.018 383 05	17.662 773 31	15.383 182 02	44
45	24.518 712 54	22.495 450 26	20.720 039 70	19.156 347 42	17.774 069 82	15.455 832 09	45
46	24.775 449 07	22.700 918 13	20.884 653 56	19.288 370 74	17.880 066 50	15.524 369 90	46
47	25.024 707 83	22.899 437 80	21.042 936 12	19.414 708 84	17.981 015 71	15.589 028 21	47
48	25.266 706 64	23.091 244 25	21.195 130 88	19.535 606 54	18.077 157 82	15.650 026 61	48
49	25.501 656 93	23.276 564 50	21.341 472 00	19.651 298 13	18.168 721 73	15.707 572 27	49
50	25.729 764 01	23.455 617 87	21.482 184 62	19.762 007 78	18.255 925 46	15.761 860 64	**50**
51	25.951 227 19	23.628 616 30	21.617 485 21	19.867 950 03	18.338 976 63	15.813 076 07	51
52	26.166 239 99	23.795 764 54	21.747 581 93	19.969 330 17	18.418 072 98	15.861 392 52	52
53	26.374 990 28	23.957 260 43	21.872 674 93	20.066 344 66	18.493 402 84	15.906 974 08	53
54	26.577 660 47	24.113 295 10	21.992 956 67	20.159 181 49	18.565 145 56	15.949 975 54	54
55	26.774 427 64	24.264 053 23	22.108 612 18	20.248 020 57	18.633 471 96	15.990 542 97	55
56	26.965 463 73	24.409 713 27	22.219 819 40	20.333 034 04	18.698 544 73	16.028 814 12	56
57	27.150 935 66	24.550 447 60	22.326 749 43	20.414 386 64	18.760 518 79	16.064 918 98	57
58	27.331 005 49	24.686 422 81	22.429 566 76	20.492 236 02	18.819 541 70	16.098 980 17	58
59	27.505 830 58	24.817 799 81	22.528 429 57	20.566 733 03	18.875 754 00	16.131 113 37	59

(continued)

Table 4 PRESENT VALUE OF AN ANNUITY TABLES $a_{\overline{n}|i} = \dfrac{1-(1+i)^{n}}{i}$

(continued)

n \ i	3%	$3\frac{1}{2}$%	4%	$4\frac{1}{2}$%	5%	6%	i / n
60	27.675 563 67	24.944 734 12	22.623 489 97	20.638 022 04	18.929 289 53	16.161 427 71	**60**
61	27.840 353 07	25.067 375 96	22.714 894 21	20.706 241 19	18.980 275 74	16.190 026 14	61
62	28.000 342 79	25.185 870 49	22.802 782 89	20.771 522 67	19.028 834 04	16.217 005 79	62
63	28.155 672 61	25.300 357 96	22.887 291 24	20.833 992 98	19.075 080 04	16.242 458 29	63
64	28.306 478 26	25.410 973 88	22.968 549 27	20.893 773 19	19.119 123 84	16.266 470 09	64
65	28.452 891 52	25.517 849 16	23.046 681 99	20.950 979 13	19.161 070 33	16.289 122 72	65
66	28.595 040 31	25.621 110 30	23.121 809 61	21.005 721 65	19.201 019 36	16.310 493 14	66
67	28.733 048 84	25.720 879 51	23.194 047 70	21.058 106 84	19.239 066 06	16.330 653 90	67
68	28.867 037 71	25.817 274 89	23.263 507 40	21.108 236 22	19.275 301 06	16.349 673 49	68
69	28.997 123 99	25.910 410 52	23.330 295 58	21.156 206 90	19.309 810 48	16.367 616 50	69
70	29.123 421 35	26.000 396 64	23.394 514 98	21.202 111 87	19.342 676 65	16.384 543 87	**70**
71	29.246 040 15	26.087 339 75	23.456 264 40	21.246 040 07	19.373 977 76	16.400 513 08	71
72	29.365 087 52	26.171 342 75	23.515 638 85	21.288 076 62	19.403 788 34	16.415 578 38	72
73	29.480 667 50	26.252 505 08	23.572 729 66	21.328 302 99	19.432 179 38	16.429 790 93	73
74	29.592 881 07	26.330 922 78	23.627 624 68	21.366 797 12	19.459 218 45	16.443 198 99	74
75	29.701 826 28	26.406 688 68	23.680 408 34	21.403 633 60	19.484 969 96	16.455 848 10	75
76	29.807 598 33	26.479 892 44	23.731 161 87	21.438 883 83	19.509 495 20	16.467 781 23	76
77	29.910 289 64	26.550 620 72	23.779 963 33	21.472 616 11	19.532 852 57	16.479 038 89	77
78	30.009 989 94	26.618 957 21	23.826 887 82	21.504 895 80	19.555 097 68	16.489 659 33	78
79	30.106 786 35	26.684 982 81	23.872 007 52	21.535 785 45	19.576 283 51	16.499 678 62	79
80	30.200 763 45	26.748 775 67	23.915 391 85	21.565 344 93	19.596 460 48	16.509 130 77	**80**
81	30.292 003 35	26.810 411 27	23.957 107 55	21.593 631 51	19.615 676 65	16.518 047 90	81
82	30.380 585 77	26.869 962 58	23.997 218 79	21.620 700 01	19.633 977 76	16.526 460 28	82
83	30.466 588 13	26.927 500 08	24.035 787 30	21.646 602 88	19.651 407 39	16.534 396 49	83
84	30.550 085 56	26.983 091 86	24.072 872 41	21.671 390 32	19.668 007 04	16.541 883 48	84
85	30.631 151 03	27.036 803 73	24.108 531 16	21.695 110 35	19.683 816 23	16.548 946 68	85
86	30.709 855 37	27.088 699 26	24.142 818 42	21.717 808 95	19.698 872 60	16.555 610 08	86
87	30.786 267 35	27.138 839 86	24.175 786 94	21.739 530 09	19.713 212 00	16.561 896 30	87
88	30.860 453 74	27.187 284 89	24.207 487 45	21.760 315 88	19.726 868 57	16.567 826 70	88
89	30.932 479 36	27.234 091 68	24.237 968 70	21.780 206 58	19.739 874 83	16.573 421 41	89
90	31.002 407 14	27.279 315 64	24.267 277 60	21.799 240 75	19.752 261 74	16.578 699 45	**90**
91	31.070 298 20	27.323 010 28	24.295 459 23	21.817 455 26	19.764 058 80	16.583 678 72	91
92	31.136 211 84	27.365 227 32	24.322 556 95	21.834 885 42	19.775 294 10	16.588 376 15	92
93	31.200 205 67	27.406 016 73	24.348 612 45	21.851 564 99	19.785 994 38	16.592 807 69	93
94	31.262 335 60	27.445 426 80	24.373 665 82	21.867 526 31	19.796 185 12	16.596 988 39	94
95	31.322 655 93	27.483 504 15	24.397 755 59	21.882 800 30	19.805 890 59	16.600 932 44	95
96	31.381 219 35	27.520 293 87	24.420 918 84	21.897 416 55	19.815 133 90	16.604 653 25	96
97	31.438 077 03	27.555 839 48	24.443 191 19	21.911 403 40	19.823 937 05	16.608 163 44	97
98	31.493 278 67	27.590 183 08	24.464 606 92	21.924 787 94	19.832 321 00	16.611 474 94	98
99	31.546 872 50	27.623 365 29	24.485 198 96	21.937 596 12	19.840 305 71	16.614 599 00	99
100	31.598 905 34	27.655 425 40	24.504 999 00	21.949 852 74	19.847 910 20	16.617 546 23	**100**
101	31.649 422 66	27.686 401 36	24.524 037 50	21.961 581 57	19.855 152 57	16.620 326 63	101
102	31.698 468 60	27.716 329 81	24.542 343 75	21.972 805 33	19.862 050 07	16.622 949 65	102
103	31.746 086 02	27.745 246 19	24.559 945 91	21.983 545 77	19.868 619 11	16.625 424 20	103
104	31.792 316 52	27.773 184 73	24.576 871 07	21.993 823 71	19.874 875 35	16.627 758 68	104
105	31.837 200 51	27.800 178 48	24.593 145 26	22.003 659 05	19.880 833 66	16.629 961 02	105
106	31.880 777 19	27.826 259 40	24.608 793 52	22.013 070 86	19.886 508 25	16.632 038 70	106
107	31.923 084 65	27.851 458 36	24.623 839 92	22.022 077 38	19.891 912 62	16.633 998 77	107
108	31.964 159 86	27.875 805 18	24.638 307 62	22.030 696 06	19.897 059 64	16.635 847 90	108
109	32.004 038 70	27.899 328 68	24.652 218 86	22.038 943 59	19.901 961 56	16.637 592 36	109
110	32.042 756 02	27.922 056 69	24.665 595 06	22.046 835 97	19.906 630 06	16.639 238 07	**110**
111	32.080 345 65	27.944 016 13	24.678 456 79	22.054 388 49	19.911 076 24	16.640 790 63	111
112	32.116 840 43	27.965 232 97	24.690 823 83	22.061 615 78	19.915 310 71	16.642 255 31	112
113	32.152 272 27	27.985 732 34	24.702 715 23	22.068 531 85	19.919 343 53	16.643 637 09	113
114	32.186 672 10	28.005 538 49	24.714 149 26	22.075 150 10	19.923 184 32	16.644 940 65	114
115	32.220 070 00	28.024 674 87	24.725 143 51	22.081 483 34	19.926 842 21	16.646 170 43	115
116	32.252 495 15	28.043 164 13	24.735 714 92	22.087 543 87	19.930 325 91	16.647 330 59	116
117	32.283 975 87	28.061 028 14	24.745 879 73	22.093 343 42	19.933 643 72	16.648 425 08	117
118	32.314 539 68	28.078 288 06	24.755 653 59	22.098 893 22	19.936 803 55	16.649 457 63	118
119	32.344 213 28	28.094 964 31	24.765 051 52	22.104 204 04	19.939 812 90	16.650 431 72	119
120	32.373 022 61	28.111 076 63	24.774 088 00	22.109 286 16	19.942 678 95	16.651 350 68	**120**

(continued)

Table 4 PRESENT VALUE OF AN ANNUITY TABLES $a_{\overline{n}|i} = \frac{1-(1+i)^{n}}{i}$

(continued)

n \ i	7%	8%	9%	10%	11%	12%	i / n
1	0.934 579 44	0.925 925 93	0.917 431 19	0.909 090 91	0.900 900 90	0.892 857 14	**1**
2	1.808 018 17	1.783 264 75	1.759 111 19	1.735 537 19	1.712 523 33	1.690 051 02	2
3	2.624 316 04	2.577 096 99	2.531 294 67	2.486 851 99	2.443 714 72	2.401 831 27	3
4	3.387 211 26	3.312 126 84	3.239 719 88	3.169 865 45	3.102 445 69	3.037 349 35	4
5	4.100 197 44	3.992 710 04	3.889 651 26	3.790 786 77	3.695 897 02	3.604 776 20	5
6	4.766 539 66	4.622 879 66	4.485 918 59	4.355 260 70	4.230 537 85	4.111 407 32	6
7	5.389 289 40	5.206 370 06	5.032 952 84	4.868 418 82	4.712 196 26	4.563 756 54	7
8	5.971 298 51	5.746 638 94	5.534 819 11	5.334 926 20	5.146 122 76	4.967 639 77	8
9	6.515 232 25	6.246 887 91	5.995 246 89	5.759 023 82	5.537 047 53	5.328 249 79	9
10	7.023 581 54	6.710 081 40	6.417 657 70	6.144 567 11	5.889 232 01	5.650 223 03	**10**
11	7.498 674 34	7.138 964 26	6.805 190 55	6.495 061 01	6.206 515 33	5.937 699 13	11
12	7.942 686 30	7.536 078 02	7.160 725 28	6.813 691 82	6.492 356 15	6.194 374 23	12
13	8.357 650 74	7.903 775 94	7.486 903 92	7.103 356 20	6.749 870 40	6.423 548 42	13
14	8.745 467 99	8.244 236 98	7.786 150 39	7.366 687 46	6.981 865 23	6.628 168 23	14
15	9.107 914 01	8.559 478 69	8.060 688 43	7.606 079 51	7.190 869 58	6.810 864 49	15
16	9.446 648 60	8.851 369 16	8.312 558 19	7.823 708 64	7.379 161 78	6.973 986 15	16
17	9.763 222 99	9.121 638 11	8.543 631 37	8.021 553 31	7.548 794 40	7.119 630 49	17
18	10.059 086 91	9.371 887 14	8.755 625 11	8.201 412 10	7.701 616 57	7.249 670 08	18
19	10.335 595 24	9.603 599 20	8.950 114 78	8.364 920 09	7.839 294 21	7.365 776 86	19
20	10.594 014 25	9.818 147 41	9.128 545 67	8.513 563 72	7.963 328 12	7.469 443 62	**20**
21	10.835 527 33	10.016 803 16	9.292 243 73	8.648 694 29	8.075 070 38	7.562 003 24	21
22	11.061 240 50	10.200 743 66	9.442 425 44	8.771 540 26	8.175 739 08	7.644 645 75	22
23	11.272 187 38	10.371 058 95	9.580 206 83	8.883 218 42	8.266 431 60	7.718 433 70	23
24	11.469 334 00	10.528 758 28	9.706 611 77	8.984 744 02	8.348 136 58	7.784 315 81	24
25	11.653 583 18	10.674 776 19	9.822 579 60	9.077 040 02	8.421 744 66	7.843 139 11	25
26	11.825 778 67	10.809 977 95	9.928 972 11	9.160 945 47	8.488 058 26	7.895 659 92	26
27	11.986 709 04	10.935 164 77	10.026 579 92	9.237 223 16	8.547 800 23	7.942 553 50	27
28	12.137 111 25	11.051 078 49	10.116 128 37	9.306 566 51	8.601 621 83	7.984 422 77	28
29	12.277 674 07	11.158 406 01	10.198 282 91	9.369 605 91	8.650 109 76	8.021 806 04	29
30	12.409 041 18	11.257 783 34	10.273 654 04	9.426 914 47	8.693 792 57	8.055 183 97	**30**
31	12.531 814 19	11.349 799 39	10.342 801 87	9.479 013 15	8.733 146 46	8.084 985 69	31
32	12.646 555 32	11.434 999 44	10.406 240 25	9.526 375 59	8.768 600 42	8.111 594 36	32
33	12.753 790 02	11.513 888 37	10.464 440 60	9.569 432 36	8.800 540 92	8.135 352 11	33
34	12.854 009 36	11.586 933 67	10.517 835 41	9.608 574 87	8.829 316 14	8.156 564 38	34
35	12.947 672 30	11.654 568 22	10.566 821 48	9.644 158 97	8.855 239 77	8.175 503 91	35
36	13.035 207 76	11.717 192 79	10.611 762 82	9.676 508 16	8.878 594 38	8.192 414 21	36
37	13.117 016 60	11.775 178 51	10.652 993 42	9.705 916 51	8.899 634 58	8.207 512 69	37
38	13.193 473 45	11.828 868 99	10.690 819 65	9.732 651 37	8.918 589 71	8.220 993 47	38
39	13.264 928 46	11.878 582 40	10.725 522 61	9.756 955 79	8.935 666 41	8.233 029 88	39
40	13.331 708 84	11.924 613 33	10.757 360 20	9.779 050 72	8.951 050 82	8.243 776 68	**40**
41	13.394 120 41	11.967 234 57	10.786 568 99	9.799 137 02	8.964 910 65	8.253 372 04	41
42	13.452 448 98	12.006 698 67	10.813 366 04	9.817 397 29	8.977 396 98	8.261 939 32	42
43	13.506 961 67	12.043 239 51	10.837 950 50	9.833 997 53	8.988 645 93	8.269 588 68	43
44	13.557 908 10	12.077 073 62	10.860 505 04	9.849 088 67	8.998 780 11	8.276 418 46	44
45	13.605 521 59	12.108 401 50	10.881 197 29	9.862 807 88	9.007 910 01	8.282 516 48	45
46	13.650 020 18	12.137 408 80	10.900 181 00	9.875 279 89	9.016 135 15	8.287 961 15	46
47	13.691 607 64	12.164 267 41	10.917 597 25	9.886 618 08	9.023 545 18	8.292 822 45	47
48	13.730 474 43	12.189 136 49	10.933 575 46	9.896 925 53	9.030 220 88	8.297 162 90	48
49	13.766 798 53	12.212 163 41	10.948 234 36	9.906 295 94	9.036 235 03	8.301 038 31	49
50	13.800 746 29	12.233 484 64	10.961 682 90	9.914 814 49	9.041 653 18	8.304 498 49	**50**
51	13.832 473 17	12.253 226 52	10.974 021 01	9.922 558 62	9.046 534 39	8.307 587 94	51
52	13.862 124 46	12.271 506 04	10.985 340 38	9.929 598 75	9.050 931 89	8.310 346 37	52
53	13.889 835 94	12.288 431 52	10.995 725 12	9.935 998 86	9.054 893 59	8.312 809 26	53
54	13.915 734 53	12.304 103 26	11.005 252 40	9.941 817 15	9.058 462 70	8.315 008 27	54
55	13.939 938 81	12.318 614 13	11.013 993 03	9.947 106 50	9.061 678 10	8.316 971 67	55
56	13.962 559 64	12.332 050 12	11.022 011 95	9.951 915 00	9.064 574 87	8.318 724 70	56
57	13.983 700 59	12.344 490 85	11.029 368 76	9.956 286 36	9.067 184 57	8.320 289 91	57
58	14.003 458 50	12.356 010 05	11.036 118 13	9.960 260 33	9.069 535 65	8.321 687 42	58
59	14.021 923 83	12.366 675 97	11.042 310 21	9.963 873 03	9.071 653 73	8.322 935 20	59
60	14.039 181 15	12.376 551 82	11.047 991 02	9.967 157 30	9.073 561 92	8.324 049 28	**60**

Table 5 TABLE OF COMMON LOGARITHMS $\log_{10} N$

N	0	1	2	3	4	5	6	7	8	9
1.0	.0000	.0043	.0086	.0128	.0170	.0212	.0253	.0294	.0334	.0374
1.1	.0414	.0453	.0492	.0531	.0569	.0607	.0645	.0682	.0719	.0755
1.2	.0792	.0828	.0864	.0899	.0934	.0969	.1004	.1038	.1072	.1106
1.3	.1139	.1173	.1206	.1239	.1271	.1303	.1335	.1367	.1399	.1430
1.4	.1461	.1492	.1523	.1553	.1584	.1614	.1644	.1673	.1703	.1732
1.5	.1761	.1790	.1818	.1847	.1875	.1903	.1931	.1959	.1987	.2014
1.6	.2041	.2068	.2095	.2122	.2148	.2175	.2201	.2227	.2253	.2279
1.7	.2304	.2330	.2355	.2380	.2405	.2430	.2455	.2480	.2504	.2529
1.8	.2553	.2577	.2601	.2625	.2648	.2672	.2695	.2718	.2742	.2765
1.9	.2788	.2810	.2833	.2856	.2878	.2900	.2923	.2945	.2967	.2989
2.0	.3010	.3032	.3054	.3075	.3096	.3118	.3139	.3160	.3181	.3201
2.1	.3222	.3243	.3263	.3284	.3304	.3324	.3345	.3365	.3385	.3404
2.2	.3424	.3444	.3464	.3483	.3502	.3522	.3541	.3560	.3579	.3598
2.3	.3617	.3636	.3655	.3674	.3692	.3711	.3729	.3747	.3766	.3784
2.4	.3802	.3820	.3838	.3856	.3874	.3892	.3909	.3927	.3945	.3962
2.5	.3979	.3997	.4014	.4031	.4048	.4065	.4082	.4099	.4116	.4133
2.6	.4150	.4166	.4183	.4200	.4216	.4232	.4249	.4265	.4281	.4298
2.7	.4314	.4330	.4346	.4362	.4378	.4393	.4409	.4425	.4440	.4456
2.8	.4472	.4487	.4502	.4518	.4533	.4548	.4564	.4579	.4594	.4609
2.9	.4624	.4639	.4654	.4669	.4683	.4698	.4713	.4728	.4742	.4757
3.0	.4771	.4786	.4800	.4814	.4829	.4843	.4857	.4871	.4886	.4900
3.1	.4914	.4928	.4942	.4955	.4969	.4983	.4997	.5011	.5024	.5038
3.2	.5051	.5065	.5079	.5092	.5105	.5119	.5132	.5145	.5159	.5172
3.3	.5185	.5198	.5211	.5224	.5237	.5250	.5263	.5276	.5289	.5302
3.4	.5315	.5328	.5340	.5353	.5366	.5378	.5391	.5403	.5416	.5428
3.5	.5441	.5453	.5465	.5478	.5490	.5502	.5514	.5527	.5539	.5551
3.6	.5563	.5575	.5587	.5599	.5611	.5623	.5635	.5647	.5658	.5670
3.7	.5682	.5694	.5705	.5717	.5729	.5740	.5752	.5763	.5775	.5786
3.8	.5798	.5809	.5821	.5832	.5843	.5855	.5866	.5877	.5888	.5899
3.9	.5911	.5922	.5933	.5944	.5955	.5966	.5977	.5988	.5999	.6010
4.0	.6021	.6031	.6042	.6053	.6064	.6075	.6085	.6096	.6107	.6117
4.1	.6128	.6138	.6149	.6160	.6170	.6180	.6191	.6201	.6212	.6222
4.2	.6232	.6243	.6253	.6263	.6274	.6284	.6294	.6304	.6314	.6325
4.3	.6335	.6345	.6355	.6365	.6375	.6385	.6395	.6405	.6415	.6425
4.4	.6435	.6444	.6454	.6464	.6474	.6484	.6493	.6503	.6513	.6522
4.5	.6532	.6542	.6551	.6561	.6571	.6580	.6590	.6599	.6609	.6618
4.6	.6628	.6637	.6646	.6656	.6665	.6675	.6684	.6693	.6702	.6712
4.7	.6721	.6730	.6739	.6749	.6758	.6767	.6776	.6785	.6794	.6803
4.8	.6812	.6821	.6830	.6839	.6848	.6857	.6866	.6875	.6884	.6893
4.9	.6902	.6911	.6920	.6928	.6937	.6946	.6955	.6964	.6972	.6981
5.0	.6990	.6998	.7007	.7016	.7024	.7033	.7042	.7050	.7059	.7067
5.1	.7076	.7084	.7093	.7101	.7110	.7118	.7126	.7135	.7143	.7152
5.2	.7160	.7168	.7177	.7185	.7193	.7202	.7210	.7218	.7226	.7235
5.3	.7243	.7251	.7259	.7267	.7275	.7284	.7292	.7300	.7308	.7316
5.4	.7324	.7332	.7340	.7348	.7356	.7364	.7372	.7380	.7388	.7396
N	0	1	2	3	4	5	6	7	8	9

(continued)

Table 5 TABLE OF COMMON LOGARITHMS $LOG_{10}N$ *(continued)*

N	0	1	2	3	4	5	6	7	8	9
5.5	.7404	.7412	.7419	.7427	.7435	.7443	.7451	.7459	.7466	.7474
5.6	.7482	.7490	.7497	.7505	.7513	.7520	.7528	.7536	.7543	.7551
5.7	.7559	.7566	.7574	.7582	.7589	.7597	.7604	.7612	.7619	.7627
5.8	.7634	.7642	.7649	.7657	.7664	.7672	.7679	.7686	.7694	.7701
5.9	.7709	.7716	.7723	.7731	.7738	.7745	.7752	.7760	.7767	.7774
6.0	.7782	.7789	.7796	.7803	.7810	.7818	.7825	.7832	.7839	.7846
6.1	.7853	.7860	.7868	.7875	.7882	.7889	.7896	.7903	.7910	.7917
6.2	.7924	.7931	.7938	.7945	.7952	.7959	.7966	.7973	.7980	.7987
6.3	.7993	.8000	.8007	.8014	.8021	.8028	.8035	.8041	.8048	.8055
6.4	.8062	.8069	.8075	.8082	.8089	.8096	.8102	.8109	.8116	.8122
6.5	.8129	.8136	.8142	.8149	.8156	.8162	.8169	.8176	.8182	.8189
6.6	.8195	.8202	.8209	.8215	.8222	.8228	.8235	.8241	.8248	.8254
6.7	.8261	.8267	.8274	.8280	.8287	.8293	.8299	.8306	.8312	.8319
6.8	.8325	.8331	.8338	.8344	.8351	.8357	.8363	.8370	.8376	.8382
6.9	.8388	.8395	.8401	.8407	.8414	.8420	.8426	.8432	.8439	.8445
7.0	.8451	.8457	.8463	.8470	.8476	.8482	.8488	.8494	.8500	.8506
7.1	.8513	.8519	.8525	.8531	.8537	.8543	.8549	.8555	.8561	.8567
7.2	.8573	.8579	.8585	.8591	.8597	.8603	.8609	.8615	.8621	.8627
7.3	.8633	.8639	.8645	.8651	.8657	.8663	.8669	.8675	.8681	.8686
7.4	.8692	.8698	.8704	.8710	.8716	.8722	.8727	.8733	.8739	.8745
7.5	.8751	.8756	.8762	.8768	.8774	.8779	.8785	.8791	.8797	.8802
7.6	.8808	.8814	.8820	.8825	.8831	.8837	.8842	.8848	.8854	.8859
7.7	.8865	.8871	.8876	.8882	.8887	.8893	.8899	.8904	.8910	.8915
7.8	.8921	.8927	.8932	.8938	.8943	.8949	.8954	.8960	.8965	.8971
7.9	.8976	.8982	.8987	.8993	.8998	.9004	.9009	.9015	.9020	.9025
8.0	.9031	.9036	.9042	.9047	.9053	.9058	.9063	.9069	.9074	.9079
8.1	.9085	.9090	.9096	.9101	.9106	.9112	.9117	.9122	.9128	.9133
8.2	.9138	.9143	.9149	.9154	.9159	.9165	.9170	.9175	.9180	.9186
8.3	.9191	.9196	.9201	.9206	.9212	.9217	.9222	.9227	.9232	.9238
8.4	.9243	.9248	.9253	.9258	.9263	.9269	.9274	.9279	.9284	.9289
8.5	.9294	.9299	.9304	.9309	.9315	.9320	.9325	.9330	.9335	.9340
8.6	.9345	.9350	.9355	.9360	.9365	.9370	.9375	.9380	.9385	.9390
8.7	.9395	.9400	.9405	.9410	.9415	.9420	.9425	.9430	.9435	.9440
8.8	.9445	.9450	.9455	.9460	.9465	.9469	.9474	.9479	.9484	.9489
8.9	.9494	.9499	.9504	.9509	.9513	.9518	.9523	.9528	.9533	.9538
9.0	.9542	.9547	.9552	.9557	.9562	.9566	.9571	.9576	.9581	.9586
9.1	.9590	.9595	.9600	.9605	.9609	.9614	.9619	.9624	.9628	.9633
9.2	.9638	.9643	.9647	.9652	.9657	.9661	.9666	.9671	.9675	.9680
9.3	.9685	.9689	.9694	.9699	.9703	.9708	.9713	.9717	.9722	.9727
9.4	.9731	.9736	.9741	.9745	.9750	.9754	.9759	.9763	.9768	.9773
9.5	.9777	.9782	.9786	.9791	.9795	.9800	.9805	.9809	.9814	.9818
9.6	.9823	.9827	.9832	.9836	.9841	.9845	.9850	.9854	.9859	.9863
9.7	.9868	.9872	.9877	.9881	.9886	.9890	.9894	.9899	.9903	.9908
9.8	.9912	.9917	.9921	.9926	.9930	.9934	.9939	.9943	.9948	.9952
9.9	.9956	.9961	.9965	.9969	.9974	.9978	.9983	.9987	.9991	.9996
N	0	1	2	3	4	5	6	7	8	9

Summary of useful mathematical formulae

Some of the more commonly used mathematical formulae are listed below, together with their corresponding equation number in the text.

1. A *fraction* is an expression that appears in the form:

$$\frac{a}{b} \qquad \textbf{(M1.1)}$$

2. $a^m \times a^n = a^{m+n}$ **(M1.2)**

3. $a^m \div a^n = a^{m-n}$ **(M1.3)**

4. $(a^m)^n = a^{mn}$ **(M1.4)**

5. $a^{-n} = \dfrac{1}{a^n}$ **(M1.5)**

6. $a^{\frac{1}{k}} = \sqrt[k]{a}$ **(M1.6)**

7. $a^{\frac{m}{n}} = (\sqrt[n]{a})^m$ **(M1.7)**

8. $a^0 = 1$ **(M1.8)**

9. $\log (A \times B) = \log A + \log B$ **(M1.9)**

10. $\log (A \div B) = \log A - \log B$ **(M1.10)**

11. $\log A^n = n \times \log A$ **(M1.11)**

12. The *commission* (C) earned is:

$$C = F + (S \times R) \qquad \textbf{(M2.1)}$$

13. If there is no fixed amount paid, the *commission* earned is:

$$C = S \times R \qquad \textbf{(M2.2)}$$

14. The *rate of discount (R)* is:

$$R = \frac{D}{L} \qquad \textbf{(M2.3)}$$

15. The *actual discount (D)* is:

$$D = R \times L \qquad \textbf{(M2.4)}$$

16. The *discount price* (or *net price*) is:

$$DP = L - D \qquad \textbf{(M2.5)}$$

17. The *actual profit* is:

$$P = SP - CP \quad \textbf{(M2.6)}$$

18. The *actual loss* is:

$$L = CP - SP \quad \textbf{(M2.7)}$$

19. The *profit rate* (as a fraction of selling price) is:

$$P_S = \frac{P}{SP} \quad \textbf{(M2.8)}$$

20. The *profit rate* (as a fraction of cost price) is:

$$P_C = \frac{P}{CP} \quad \textbf{(M2.9)}$$

21. The *loss rate* (as a fraction of selling price) is:

$$L_S = \frac{L}{SP} \quad \textbf{(M2.10)}$$

22. The *loss rate* (as a fraction of cost price) is:

$$L_C = \frac{L}{CP} \quad \textbf{(M2.11)}$$

23. $$\text{Total asset turnover} = \frac{\text{sales}}{\text{total assets}} \quad \textbf{(M4.1)}$$

24. $$\text{Return on investment} = \frac{\text{net profit}}{\text{total assets}} \quad \textbf{(M4.2)}$$

25. $$\text{Stock turnover} = \frac{\text{cost of goods sold}}{\text{average stock}} \quad \textbf{(M4.3)}$$

26. $$\text{Average stock} = \frac{\text{opening stock} + \text{closing stock}}{2} \quad \textbf{(M4.4)}$$

27. $$\text{Debtor turnover} = \frac{\text{gross credit sales}}{\text{average gross accounts receivable}} \quad \textbf{(M4.5)}$$

28. $$\text{Average collection period} = \frac{365}{\text{debtor turnover}} \quad \textbf{(M4.6)}$$

29. $$\text{Current-asset ratio} = \frac{\text{current assets}}{\text{current liabilities}} \quad \textbf{(M4.7)}$$

30. $$\text{Acid-test ratio} = \frac{\text{current assets} - \text{inventory}}{\text{current liabilities} - \text{bank overdraft}} \quad \textbf{(M4.8)}$$

31. *Simple interest* (I) is:

$$I = P \times R \times T \quad \textbf{(M5.1)}$$

32. The *maturity value* (S) of the principal is:

$$S = P + I \quad \textbf{(M5.2)}$$

$$S = P(1 + RT) \quad \textbf{(M5.3)}$$

33. The amount of *principal* (*P*) required is:

$$P = \frac{S}{1 + RT} \quad \textbf{(M5.4)}$$

34. The *effective rate of interest* (*E*) is:

$$E = \frac{2N \times R}{N + 1} \quad \textbf{(M5.5)}$$

35. The *flat rate of interest* (*R*) is:

$$R = \frac{(N + 1) \times E}{2N} \quad \textbf{(M5.6)}$$

36. The *compound interest formula* for the accumulated value (*S*) is:

$$S = P(1 + i)^n \quad \textbf{(M6.1)}$$

37. Accumulation factor $= (1 + i)^n$ **(M6.2)**

38. Amount of compound interest $= P \times$ (accumulation factor $- 1$) **(M6.3)**

39. The *present value* (*P*) at compound interest of an amount *S* is:

$$P = S(1 + i)^{-n} \quad \textbf{(M6.4)}$$

40. Present value factor $= (1 + i)^{-n}$ **(M6.5)**

41. The *compound interest rate* (*i*) per period is:

$$i = \left(\frac{S}{P}\right)^{\frac{1}{n}} - 1 \quad \textbf{(M6.6)}$$

42. The *number of periods* (*n*) required for an amount of *P* to accumulate to an amount of *S*, when the compound interest rate is *i* per period, is:

$$n = \frac{\log\left(\frac{S}{P}\right)}{\log(1 + i)} \quad \textbf{(M6.7)}$$

43. The *future value* (*S*) of an annuity is:

$$S = R \times \frac{(1 + i)^n - 1}{i} \quad \textbf{(M7.1)}$$

44. The *future value* at an interest rate if *i*% of $1 paid at the end of each period for *n* periods is:

$$s_{\overline{n}|i} = \frac{(1 + i)^n - 1}{i} \quad \textbf{(M7.2)}$$

45. The *future value* (*S*) of an annuity is:

$$S = R \times s_{\overline{n}|i} \quad \textbf{(M7.3)}$$

46. The *present value* (*A*) of an annuity is:

$$A = S(1 + i)^{-n} \quad \textbf{(M7.4)}$$

47. Alternatively, the *present value* (A) of an annuity is:

$$A = R \times \left[\frac{1 - (1 + i)^{-n}}{i}\right] \qquad \textbf{(M7.5)}$$

48. The *discounted value* at an interest rate of i% per period of \$1 paid at the end of each period for n periods is:

$$a_{\overline{n}|i} = \frac{1 - (1 + i)^{-n}}{i} \qquad \textbf{(M7.6)}$$

49. An alternative formula for A is:

$$A = R \times a_{\overline{n}|i} \qquad \textbf{(M7.7)}$$

50. The amount of the *annuity payment* (R) made per period is:

$$R = \frac{S}{\dfrac{(1 + i)^n - 1}{i}} \qquad \textbf{(M7.8)}$$

51. An alternative formula for R is:

$$R = \frac{S}{s_{\overline{n}|i}} \qquad \textbf{(M7.9)}$$

52. Book value = original cost of asset − depreciation to date **(M8.1)**

53. $$\text{Annual depreciation} = \frac{\text{original cost price}}{\text{expected life}} \qquad \textbf{(M8.2)}$$

54. $$\text{Annual rate of depreciation} = \frac{\text{annual depreciation}}{\text{original cost price}} \times 100\% \qquad \textbf{(M8.3)}$$

55. If the *prime cost method* is used, at the end of k years:

Book value = original cost − (k × annual depreciation) **(M8.4)**

56. If the *prime cost method* is used:

$$\text{Annual rate of depreciation} = \frac{100\%}{\text{effective life of the item in years}} \qquad \textbf{(M8.5)}$$

57. If the *diminishing value method* is used:

$$\text{Depreciation during year } k = (\text{book value at end of year } (k - 1) \times \text{depreciation rate} \qquad \textbf{(M8.6)}$$

58. If the *diminishing value method* is used:

$$\text{Book value at end of year } k = (\text{book value at end of year } (k - 1) - \text{depreciation during year } k \qquad \textbf{(M8.7)}$$

59. If the *diminishing value method* is used:

$$\text{Book value at end of year } k = \text{original cost } (1 - \text{depreciation rate})^k \qquad \textbf{(M8.8)}$$

60. $$\text{Accumulated depreciation at end of year } k = \text{original cost} - \text{book value at end of year } k \qquad \textbf{(M8.9)}$$

61. If the *diminishing value method* is used:

$$\text{Annual depreciation} = \frac{150\%}{\text{effective life of the item in years}} \qquad \textbf{(M8.10)}$$

62. If the *units-of-production* method is used:

$$\text{Rate of depreciation} = \frac{\text{original cost price}}{\text{expected life in terms of output}} \qquad \textbf{(M8.11)}$$

63. A *linear equation* may be written in one of the following forms:

$$y = bx + a \quad \text{or} \quad y = a + bx \qquad \textbf{(M9.1)}$$

where a and b are constants.

64. In *break-even analysis*,

$$\text{Total cost} = \text{variable cost} + \text{fixed cost} \qquad \textbf{(M9.2)}$$

65. *Total cost* (C) is:

$$C = vx + f \qquad \textbf{(M9.3)}$$

66. *Total income* (I) is:

$$I = sx \qquad \textbf{(M9.4)}$$

67. *Total profit* (P) is:

$$P = I - C \qquad \textbf{(M9.5)}$$

68. *Break-even point* (x) is:

$$x = \frac{f}{s - v} \qquad \textbf{(M9.6)}$$

69. $$\text{Break-even point} = \frac{\text{fixed cost}}{\text{income per item} - \text{variable cost per item}} \qquad \textbf{(M9.7)}$$

Solutions to selected exercises

Chapter M1

M1.1 (a) twenty-three
(b) fifty-nine
(c) three hundred and forty-six
(d) seven hundred and eighty-two

M1.4 (a) $2 \times 7 = 14$ (b) $7 \times 3 = 21$
(c) $8 + 3 = 11$ (d) $2 + 15 = 17$

M1.6 (a) $-\frac{1}{4}$ (b) $-\frac{1}{4}$ (c) $\frac{1}{4}$

M1.8 (a) 60 (b) 300 (c) 90

M1.10 (a) 1 (b) $\frac{7}{20}$ (c) $\frac{3}{20}$

M1.11 (a) $\frac{3}{8}$ (b) $\frac{3}{4}$ (c) 0

M1.13 (a) $\frac{1}{6}$ (b) $\frac{1}{2}$ (c) $\frac{1}{4}$

M1.15 (a) $\frac{1}{5}$ (b) 8 (c) 1

M1.17 (a) $\frac{1}{2}$ (b) $\frac{3}{40}$ (c) $\frac{1}{8}$

M1.19 (a) 0.625 (b) $0.58\dot{3}$ (c) 0.45

M1.21 (a) 0.546 (b) 2.601 (c) 6.9265

M1.22 (a) 3.2 (b) 0.32 (c) 0.125

M1.24 (a) 4 (b) 5 (c) 1020

M1.26 (a) 512 (b) 9 (c) $\frac{1}{5}$

M1.28 (a) 16 (b) 3 (c) $\frac{1}{25}$

M1.31 1

M1.33 (a) −45 (b) 116

M1.35 (a) $1\frac{1}{4}$ cups of oats and $3\frac{1}{4}$ cups of water
(b) 20 cups

M1.36 (a) 2.387×10 (b) 3.59×10^{-1}
(c) 4.6×10^{-4}

M1.38 (a) 0.7520 (b) 1.6830
(c) 0.5514 − 1

M1.40 (a) 19.4816 (b) 13.9919
(c) 1.513

M1.42 (a) 32 (b) −19 (c) −333

M1.46 (a) 9360 (b) 126 (c) 1344

M1.49 (a) 4.56×10^{-2}
(b) 6.7×10^{-5}
(c) 3.4589×10^{2}

M1.53 (a) 5 bottles (b) 12 bottles (c) 17 bottles

M1.55 (a) 120 m^3 (b) 280 m^3 (c) 560 m^3

M1.58 (a) $225 (b) $280 (c) $225

M1.61 (a) 7 hrs 17 min
(b) $210.83

Chapter M2

M2.1 (a) 25% (b) $33\frac{1}{3}$ (c) 12.5%

M2.3 (a) 72% (b) 50.5% (c) 82.73%

M2.4 (a) $\frac{9}{40}$ (b) $\frac{2}{3}$ (c) $\frac{7}{8}$

M2.6 (a) $625 (b) $3150 (c) $7250

M2.7 (a) $650 (b) $2450 (c) $3875

M2.9 (a) 1% (b) 0.75% (c) 3.25%

M2.13 (a) $\$200\,000 + \frac{\$13\,375}{0.02} = \$868\,750$

(b) $\frac{\$868\,750}{0.90} = \$965\,278$

M2.15 List price $= \frac{\$1.70}{0.66667} = \$2.55$

M2.18 GST = $159.50/11
= $14.50
Price excluding GST = $159.50 − $14.50
= $145.00

M2.20 (a) Retail price (excluding GST)

= \$65 × 1.40 = \$91.00

Retail price (including GST)

= \$91.00 × 1.10 = \$100.10

(b) GST = \$107.80/11 = \$9.80

Retail price excluding GST

= \$107.80 − \$9.80 = \$98

Cost price = \$98/1.40 = \$70

M2.22 80% of the list price = \$350. Therefore:

$$\text{List price} = \frac{\$350}{0.80} = \$437.50$$

M2.27 (a) \$21 + (99 × \$3.00) = \$318.00

(b) \$21 + (330 × \$3.00) = \$1011.00

(c) \$21 + (154 × \$3.00) = \$483.00

M2.29 Total discount = \$1000 × 0.35 = \$350

Selling price of store's deal

= \$1000 − \$350 = \$650

From Exercise M2.28, the customer's own deal means a selling price of \$675. Hence, he is \$25 better off with the store's deal.

M2.32 (a) 0 + (0.17 × \$6543) = \$1112.31

(b) \$2380 + (0.30 × \$17 520) = \$7636.00

(c) \$15 580 + (0.47 × \$27 540)

= \$28 523.80

M2.34 (a) \$16 250 × 0.30 = \$4875.00

(b) \$75 320 × 0.30 = \$22 596.00

(c) \$534 860 × 0.30 = \$160 458.00

M2.36 Between \$17 901 and \$18 000

M2.42 (a) \$142 443 (b) \$943 130

M2.43 (a) \$22 500 × 0.95 = \$21 375

(b) \$22 500 × 0.97 = \$21 825

(c) \$22 500

M2.46 $$\text{Original retail price} = \frac{\$1128.96}{0.96 \times 0.98} = \$1200$$

M2.53 \$3 010 000/\$20 170 000 × 100 = 14.92%

M2.55 (a) \$90 + (\$25 000 × 0.025) = \$715

(b) \$90 + (\$56 000 × 0.025) = \$1490

(c) \$90 + (\$125 000 × 0.025) = \$3215

Chapter M3

M3.1 (a) $6x$ (b) $16x - 4y$ (c) $40 - 8x$

M3.3 (a) $x = 2$ (b) $x = 2.5$ (c) $x = 3$

M3.4 (a) $y = 2$ (b) $y = 5$ (c) $y = 3.5$

M3.6 $x = 1, y = 2$

M3.7 $x = 5, y = -3$

M3.8 $x = 25.25, y = 17$

M3.10 Let x = price of each diary (in \$).

Then $3x + 4(1.20) = 11.70$ which yields $x = 2.30$.

Therefore, each diary costs \$2.30.

M3.12 Let x = entrance price of one child.

Then $3x + 2(x + 5) = 24.5$ which yields $x = 2.90$.

Hence, price of one child = \$2.90

and price of one adult = \$2.90 + \$5.00

= \$7.90.

M3.14 Let x = cost of 1 apple (in cents).

Let y = cost of 1 orange (in cents).

Then $x - y = 5$ and $6x + 4y = 100$ which yields $x = 12$ and $y = 7$.

Therefore, each apple costs 12 cents and each orange costs 7 cents.

M3.15 Let x = length and y = breadth.

Then $2x + 2y = 70$ and $2x - 5y = 0$ which yields $x = 25$ and $y = 10$.

(a) Length of field = 25 m

Breadth of field = 10 m

(b) Area of field = 25 × 10

= 250 m^2

M3.17 Let x = number of men and y = number of women.

Then $2x = y - 6$ and $x + 11 = y - 11$

These may be written as:

$$2x - y = -6 \text{ and } x - y = -22.$$

This yields $x = 16$ and $y = 38$.

Therefore, 16 enquiries were from men and 38 were from women.

M3.18 Let x = amount (in \$) paid by Jenny.

Then $x + 2x + 3x = 36$ which yields $x = 6$.

Therefore, Jenny paid \$6.00, Pamela paid \$12.00 and Wilma paid \$18.00.

M3.20 Let x = price of the bucket (in cents) and y = price of the spade (in cents).

Then $x + y = 275$ and $x - y = 35$ which yields $x = 155$ and $y = 120$.

Therefore, 1 bucket costs \$1.55 and 1 spade costs \$1.20.

Hence, 3 buckets and 2 spades cost (3 × \$1.55) + (2 × \$1.20) = \$7.05.

M3.21 Let x = cost of 1 apple (in cents) and y = cost of 1 orange (in cents)

Then $3x + 4y = 78$ and $5x + 2y = 74$

which yields $x = 10$ and $y = 12$.

Therefore, 1 apple costs 10 cents and 1 orange costs 12 cents.

Hence, 10 apples and 8 oranges cost

(10 × 10) + (8 × 12) = 196 cents or \$1.96.

M3.25 (a) $x = \frac{6}{7}$ (b) $x = 1$ (c) $x = \frac{12}{11}$

M3.27 $x = 7, y = 3$

M3.29 $x = 20, y = 12$

M3.33 6

M3.35 (a) \$4.20 (b) \$1.60 (c) \$23.80

M3.38 8 hours

M3.42 Kay \$50 000, Dale \$60 000, Carmel \$40 000

Chapter M4

M4.1 (a) Percentage of 1 L cartons sold

$= 100 \times \frac{17}{40}\% = 42.5\%$

Percentage of 600 mL cartons sold

$= 100 \times \frac{23}{40}\% = 57.5\%$

(b) Number of 1 L cartons sold

$= \frac{17}{40} \times 6600 = 2805$

Number of 600 mL cartons sold

$= \frac{23}{40} \times 6600 = 3795$

M4.3 (a) 4 : 3 (b) 1 : 6 (c) 2 : 5

M4.5 Petrol usage $= \frac{1650}{125}$ km/L

= 13.20 km/L = 1320 km/100 L

M4.7 (a)

January	118.2
February	100.0
March	61.1
April	25.1

(b)

January	18.2% increase
March	38.9% decrease
April	74.9% decrease

M4.9 (a) $\frac{568}{22} \doteqdot 25.82$ pairs/min

(b) $\frac{568}{22} \times 60 \doteqdot 1549$ pairs/hr

(c) $\frac{22 \times 60}{568} \doteqdot 1$ pair every 2.32 seconds

M4.10 Sandra \$13 750, Don \$22 000, Annette \$19 250

M4.12 Total asset turnover $= \frac{\$385\ 000}{\$152\ 000} \doteqdot 2.533$

M4.13 Current-asset ratio $= \frac{\$132\ 000}{\$120\ 000} = 1.10$

This indicates that the company has \$1.10 in assets for every \$1 in current liabilities owed.

M4.14 Current-asset ratio $= \frac{\$52\ 000}{\$65\ 000} = 0.80$

This indicates that the bookstore has 80 cents in current assets for every \$1 in current liabilities owed.

M4.16 Acid-test ratio $= \frac{\$52\ 000 - \$14\ 000}{\$65\ 000 - \$21\ 000} = 0.86$

M4.19 (a) Debtor turnover $= \frac{\$1\ 550\ 000}{\$176\ 000} = 8.807$

(b) Average collection period

$= \frac{365}{8.807}$ days = 41.44 days

M4.20 Carmel \$600, Jack \$1000, Mick \$650, Tanya \$250.

M4.23 (a) Total asset turnover = 237.5%

(b) For every \$1 of total assets, the company can generate sales of \$2.375.

M4.25 (a) Average stock = \$730 000

$$\text{Stock turnover ratio} = \frac{\$1\ 320\ 000}{\$730\ 000} \doteqdot 1.808$$

(b) Stock turns over about every 209 days.

M4.29 (a) Donna \$31 666.67, Michael \$9333.33, Annie \$34 000

(b) Donna \$40 000, Michael \$16 000, Annie \$44 000

(c) Donna \$60 000, Michael \$32 000, Annie \$68 000

M4.34 Goals 20, behinds 45

M4.37 Browsing = 100 customers

Buying = 60 customers

Chapter M5

M5.1 (a) Simple interest

$= \$4000 \times 0.08 \times 4 = \$1280$

(b) Simple interest

$= \$12\,500 \times 0.06 \times 5 = \$3750$

(c) Simple interest

$= \$25\,000 \times 0.07 \times 3 = \$5250$

M5.2 (a) Rate of simple interest

$$= \frac{\$100}{\$1000 \times 5} = 0.02 \text{ or } 2\%$$

(b) Rate of simple interest

$$= \frac{\$1600}{\$5000 \times 3} \doteqdot 0.1066 \text{ or } 10\tfrac{2}{3}\%$$

(c) Rate of simple interest

$$= \frac{\$312.50}{\$2500 \times 5} = 0.025 \text{ or } 2.5\%$$

M5.4 (a) $\text{Period} = \dfrac{\$300}{\$3000 \times 0.05} = 2 \text{ years}$

(b) $\text{Period} = \dfrac{\$1023.75}{\$6500 \times 0.045} = 3.5 \text{ years}$

or 3 years 6 months

(c) $\text{Period} = \dfrac{\$12\ 397.5}{\$18\ 000 \times 0.095} = 7.25 \text{ years}$

or 7 years 3 months

M5.6 (a) Total simple interest $= \$600 \times 0.075 \times 1.25 = \$56.25$

(b) Maturity value of principal

$= \$600 + \$56.25 = \$656.25$

M5.8 Amount she should invest now

$$= \frac{\$3000}{1 + 0.1325 \times 5} \doteqdot \$1804.51$$

M5.12 (a) Total simple interest $= \$3500 \times 0.0625 \times 3 = \$656.25$

(b) Equivalent effective rate of interest

$$= \frac{2 \times 36 \times 0.0625}{36 + 1} \doteqdot 0.1216$$

or 12.16%

M5.14 Equivalent flat rate of interest

$$= \frac{181 \times 0.185}{360} = 0.093 \text{ or } 9.30\%$$

M5.16 (a) Total cost of sewing machine

$= (\$38.50 \times 24) + (0.16 \times \$1000)$

$= \$924 + \$160 = \$1084$

(b) Total interest paid $= \$1084 - \$1000 = \$84$

(c) $\text{Flat rate of interest} = \dfrac{\$84}{\$840 \times 2} = 0.050 \text{ or } 5\%$

(d) Effective rate of interest

$$= \frac{2 \times 24 \times 0.050}{24 + 1} = 0.096$$

or 9.6%

M5.18 $\text{Minimum annual rate of interest} = \dfrac{\$1250}{\$3500 \times 4} = 0.0893 \text{ or } 8.93\%$

M5.19 $\text{Time} = \dfrac{\$15\ 000}{\$25\ 000 \times 0.095} = 6.32 \text{ years}$

M5.20 (a) Equivalent flat rate of interest

$$= \frac{33 \times 0.155}{64} = 0.08 \text{ or } 8.0\%$$

(b) Total simple interest $= \$20\,000 \times 0.08 \times 8 = \$12\,800$

M5.22 Time = $\frac{1}{0.075}$ = 13.33 years
or 13 years 4 months

M5.25 Annual interest rate = 73%

M5.27 (a) 14 September 2002
(b) Total interest = \$149.44
(c) Repayment = \$91.61 per week

M5.29 (a) Amount required = \$8268.73
(b) Repayment = \$1378.12 per month

M5.32 Equivalent flat rate = 7.04%

M5.34 (a) Total interest = \$108 000
(b) Equivalent effective rate = 13.09%

M5.36 (a) Flat rate interest = \$14 250
Effective rate interest = \$13 837.50
(b) Take the effective rate since the total interest payable is lower.

M5.38 (a)

Month	Minimum balance (\$)	Interest (\$)
Jan	0	0
Feb	150	0.41
Mar	150	0.41
Apr	150	0.41
May	650	1.76
Jun	325	0.88
Jul	325	0.88
Aug	325	0.88
Sep	325	0.88
Oct	355	0.96

(b) Total interest = \$7.47

M5.40 No. The total interest will be \$2700, so she will have only a total of \$11 700, still \$300 short of her goal.

Chapter M6

M6.1 P = \$5000 and r = 0.04.
(a) In this case, n = 2.
Accumulation factor = 1.0816
Accumulated value = \$5000 × 1.0816 = \$5408
Total compound interest
= \$5000 × 0.0816 = \$408
(b) In this case, n = 5.
Accumulation factor = 1.216 652 90
Accumulated value = \$5000 × 1.216 652 90
= \$6083.26 (to the nearest cent)
Total compound interest = \$5000 × 0.216 652 90
= \$1083.26 (to the nearest cent)

M6.3 P = \$16 000 and $r = \frac{0.05}{2} = 0.025$.
Total number of interest periods is
$n = 2 \times 5 = 10$.
(a) Accumulation factor = $(1.025)^{10}$ = 1.280 084 54
(b) Accumulated value after 5 years
= \$16 000 × 1.280 084 54
= \$20 481.35 (to the nearest cent)
(c) Total compound interest
= \$16 000 × 0.280 084 54
= \$4481.35 (to the nearest cent)

M6.5 (a) Accumulation factor = $(1.004\,167)^{60}$
= 1.283 359
(b) Accumulated value after 5 years
= \$16 000 × 1.283 359
= \$20 533.74
(c) Total compound interest
= \$16 000 × 0.283 359
= \$4533.74

M6.7 Amount required = $\frac{\$8000}{(1.0075)^{72}}$ = \$4671.39

M6.9 1st \$5000 investment:
Accumulation factor = $(1 + 0.015)^{20}$
= 1.346 855 007
Compound interest = \$5000 × 0.346 855 007
= \$1734.28
2nd \$5000 investment:
Accumulation factor = $(1 + 0.015)^{12}$
= 1.195 618 171

Compound interest = \$6000 × 0.195 618 171

= \$1173.71

Total compound interest = \$1734.28 + \$1173.71

= \$2907.99

M6.11 $n = 20 \times 2 = 40$ and $r = \frac{0.06}{2} = 0.03$

Amount to invest now = \$100 000 ÷ $(1.03)^{40}$

= \$100 000 ÷ 3.262 037 792

= \$30 655.68

M6.12

$$n = \frac{\log\left(\frac{5000}{3000}\right)}{\log\ (1.01)}$$

= 51.34 quarters or about 12.83 years

M6.14 Annual rate of interest = 2.0%

M6.17 Annual rate of interest = 16.65%

M6.19 Present value = \$25 000 × 0.613 913 25

= \$15 347.83

M6.21 Amount Jennifer will receive

= \$3000 × 2.575 082 76

= \$7725.25

M6.24 (a) Plan A: \$25 397.64,

Plan B: \$25 422.94

(b) Choose Plan B.

Extra interest = \$25.30

M6.27 Period of loan = 30 hours

(a) Amount to repay Harry

$$= \$400\left(1 + \frac{0.25}{24}\right)^{30}$$

= \$545.85

(b) No trouble but Rupert will only have \$4.15 left over.

M6.29 (a) 5.15%

(b) 6.82%

(c) 3.71%

M6.31 Annual rate of interest = 5.32%

M6.33 $P = \$8000$, $S = \$10\,000$, $i = \frac{0.04}{4} = 0.01$

Using Equation M6.7:

$$n = \frac{\log\left(\frac{\$10\ 000}{\$8000}\right)}{\log\ (1.01)}$$

$$= \frac{\log 1.25}{\log 1.01}$$

= 22.425 quarters

≑ 5 years 221 days

Hence, Shannon will almost reach her target on about 1 January 2004.

M6.39 Amount to pay back = \$5000$(1.002\,857)^{84}$

= \$6354.07

M6.41 Accumulated value = \$3060.98

Chapter M7

M7.2 Amount in account

= \$100 × 69.770 031 51 = \$6977.00

M7.4 Amount in account

= \$500 × 26.973 464 85 = \$13 486.73

M7.7 Sophie should deposit

\$250 × 22.793 898 39 = \$5698.47

M7.9 Amount after 5 years

= \$500 × 25.544 657 61 = \$12 772.33

M7.11 Present value

= \$7500 × 4.515 052 38 = \$33 862.89

M7.13 $R = \$2000$.

Interest rate per monthly period

$= \frac{0.06}{12} = 0.005$ (or $\frac{1}{2}$% per month).

Total number of interest periods is

$n = 12 \times 5 = 60$.

From Table 4,

$a_{\overline{60}|\,0.005} = 51.725\,560\,75$.

Present value of annuity is

$A = \$240 \times 51.725\,560\,75$

= \$12 414.13 (to the nearest cent).

Equivalent cash price of the car
= \$2000 + \$12 414.13
= \$14 414.13 (to the nearest cent).

M7.15 Quarterly payments
= \$40 000 ÷ 39.380 268 89
= \$1015.74

M7.17 Amount outstanding at end of 10th year
= \$1015.74 × 17.168 638 79
= \$17 438.87

M7.20 Total interest saved
= 60 × (\$2902.60 − \$2580.93)
= \$19 300.20

M7.23 (a) Amount required each month
= \$1500 ÷ 40.535 557 74
= \$37.00

(b) Total interest
= \$1500 − (\$37 × 36) = \$168.00

M7.25 Amount in Karen's account
= \$500 × 6.468 409 88
= \$3434.20

M7.27 Deposit required now by Phyllis
= \$250 000 × 0.228 107 08
= \$57 026.77

M7.28 Present value of 1st annuity
= \$500 × $a_{\overline{3}|0.08}$
= \$500 × 2.577 096 99
= \$1288.55

Value of 2nd annuity 1 year before 1st payment
= \$1000 × $a_{\overline{2}|0.08}$
= \$1000 × 1.783 264 75
= \$1783.26

Present value of 2nd annuity
= $1783.26\ (1.08)^{-3}$
= \$1415.61

Total value of annuity
= \$1288.55 + \$1415.61
= \$2704.16

M7.30 (a) Amount required each month from Jack
= \$9000 ÷ 56.349 915 07
= \$159.72

(b) Total interest = \$9000 − (\$159.72 × 48)
= \$1333.44

M7.32 (a) Amount required each month
= \$40 000 ÷ 95.007 027 58
= \$421.02

(b) Total interest = \$40 000 ÷ (\$421.02 × 72)
= \$9686.56

M7.35 Quarterly payments required from Jeannie
= $a_{\overline{48}|0.01125}$ = 36.932 637
Quarterly payments = \$5415.26

M7.39 Present value = \$1805.03

M7.41 1st year: \$500, 2nd year: \$1020, 3rd year: \$1560.80, 4th year: \$2123.23, 5th year: \$2708.16

M7.43 (a) 6-month payments required from the cricket team

$$= \frac{\$120\,000}{15.589\,162\,29}$$

= \$7697.66

(b) Total interest = \$33 953.11

Chapter M8

M8.1 (a) From Table M8.1, the annual rate of depreciation using the prime cost method is 27%.

(b) Annual depreciation = 0.27 × \$4500 = \$1215

Book value at the end of 2 years
= \$4500 − (2 × \$1215)
= \$2070

M8.4 From Table M8.1, the annual rate of depreciation using the prime cost method is 20%.

End of year	Annual depreciation ($)	Accumulated depreciation ($)	Book value ($)
0	0	0	900
1	180	180	720
2	180	360	540
3	180	540	360
4	180	720	180

M8.7 (a) $2500 = (original cost)$(1 - 0.30)^4$

Therefore, original cost = $10 416.67

(b)

End of year	Annual depreciation ($)	Accumulated depreciation ($)	Book value ($)
0	0	0	10 416.67
1	3125.00	3125.00	7291.67
2	2187.50	5312.50	5104.17
3	1531.25	6843.75	3572.92
4	1071.88	7915.63	2501.05
5	750.31	8665.94	1750.73

(c) Book value at end of 5th year = $1750.73

M8.9 (a) From Table M8.1, the annual rate of depreciation using the prime cost method is 27%.

(b) Annual depreciation = 0.27 × $3300 = $891

M8.11

End of year	Annual depreciation ($)	Accumulated depreciation ($)	Book value ($)
0	0	0	3300
1	891	891	2409
2	891	1782	1518
3	891	2673	627

M8.13 (a) Annual rate of depreciation $= \dfrac{150}{6\frac{2}{3}} = 22.5\%$

End of year	Annual depreciation ($)	Accumulated depreciation ($)	Book value ($)
0	0	0	945.00
1	212.63	212.63	732.37
2	164.78	377.41	567.59
3	127.70	505.11	439.89
4	98.98	604.09	340.91
5	76.70	680.79	264.21

(b) Book value at end of 6th year $= \$945 \times (0.775)^6 = \$204.76$

M8.15 Annual rate of depreciation $= \dfrac{150}{10} = 15\%$

(a) Original cost $= \dfrac{\$2023}{(0.85)^2} = \$2800$

(b)

End of year	Annual depreciation ($)	Accumulated depreciation ($)	Book value ($)
0	0	0	2800
1	420	420	2380
2	357	777	2023
3	303.45	1080.45	1719.55
4	257.93	1338.38	1461.62

(c) Book value at end of 6th year $= \$2800 \times (0.85)^6 = \$1056.02$

M8.18 (a) Annual depreciation $= \dfrac{\$64\,000}{20} = \$3200$

(b) and (c)

End of year	Annual depreciation ($)	Accumulated depreciation ($)	Book value ($)
0	0	0	64 000
1	3200	3200	60 800
2	3200	6400	57 600
3	3200	9600	54 400

M8.20 Annual depreciation rate $= \frac{150}{5} = 30\%$

(a) $686 = (original cost) $(1 - 0.30)^3$

Therefore, original cost = $2000

(b)

End of year	Annual depreciation ($)	Accumulated depreciation ($)	Book value ($)
0	0	0	2000.00
1	600.00	600.00	1400.00
2	420.00	1020.00	980.00
3	294.00	1314.00	686.00
4	205.80	1519.80	480.20
5	144.06	1663.86	336.14

(c) Book value at the end of 5th year = $336.14

M8.21 (a) $11 118 = ($24 000) $(1 - r)^4$

Hence, $r = 0.175$ or annual rate of depreciation = 17.5%

(b) From Table M8.1, the effective life of the front-end loader is between 10 years and less than 13 years.

(c) At end of year 1, book value = ($24 000) $(0.825)^1$ = $19 800

At end of year 2, book value = ($24 000) $(0.825)^2$ = $16 335

At end of year 3, book value = ($24 000) $(0.825)^3$ = $13 476.38

M8.22 Annual rate of depreciation $= \frac{150}{5} = 30\%$

End of year	Annual depreciation ($)	Accumulated depreciation ($)	Book value ($)
0	0	0	12 000
1	3600	3600	8 400
2	2520	6120	5 880
3	1764	7884	4 116

M8.24 Annual depreciation $= \frac{\$3000}{5} = \$600$

End of year	Annual depreciation ($)	Accumulated depreciation ($)	Book value ($)
0	0	0	3000
1	600	600	2400
2	600	1200	1800
3	600	1800	1200
4	600	2400	600
5	600	3000	0

M8.25 Annual depreciation $= \frac{\$2400}{5} = \$480$

End of year	Annual depreciation ($)	Accumulated depreciation ($)	Book value ($)
0	0	0	2400
1	480	480	1920
2	480	960	1440
3	480	1440	960
4	480	1920	480
5	480	2400	0

M8.28 (a) Annual rate of depreciation $= \frac{100}{13\frac{1}{3}} = 7.5\%$

(b) (i) Book value after 1 year $= \$360 \times 0.925 = \$333.00$

(ii) Book value after 2 years $= \$360 \times (0.925)^2 = \$308.03$

(iii) Book value after 3 years $= \$360 \times (0.925)^3 = \$284.92$

M8.31 (a) Annual depreciation using prime cost method$= \$16\,000 \times 0.10 = \$1600$

(b) Annual depreciation using diminishing value method $= \$16\,000 \times 0.15 = \$2400$

M8.34 Annual depreciation $= \frac{\$1200}{13\frac{1}{3}} = \$90$

End of year	Annual depreciation ($)	Accumulated depreciation ($)	Book value ($)
0	0	0	1200
1	90	90	1110
2	90	180	1020
3	90	270	930
4	90	360	840
5	90	450	750

M8.36 Book value after 2 years $= \$1400 - (2 \times \$350)$
$= \$700$

M8.38 (a) Effective life $= 6000/5 = 1200$ days
(b) Depreciation $= \$1500/6000$
$= 25$ cents per hour of use
(c) Book value after 2500 hours of use
$= \$1500 - (0.25 \times 2500) = \$875$

M8.42 If $k \geq n - 1$ then k must have the value n or greater. The ratio refers to the book value in the year $(k + 1)$, which would therefore be at least the year $n + 1$. This would not make sense, since the life of the item is only n years.

M8.45 (a) Rate of depreciation $= \$5800/15\,000$
$= 38.6\dot{6}$ cents/km
(b) (i) Book value
$= \$5800 - (2000 \times 0.386\dot{6})$
$= \$5026.67$
(ii) Book value
$= \$5800 - (12\,000 \times 0.386\dot{6})$
$= \$1160$

Chapter M9

M9.2

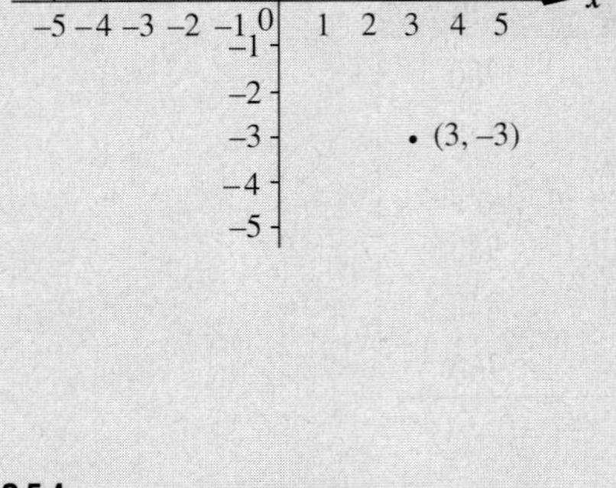

M9.5

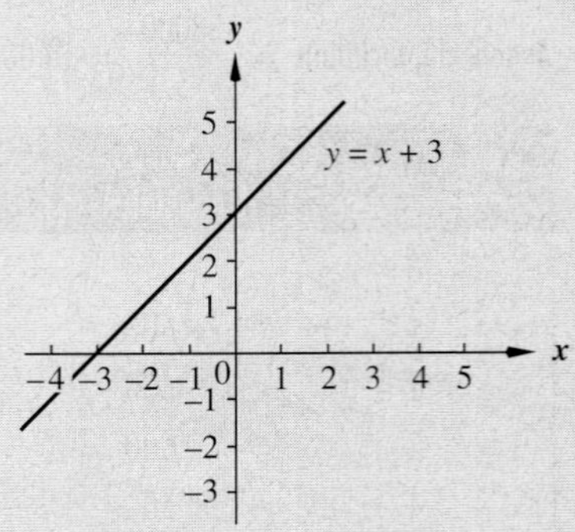

M9.6

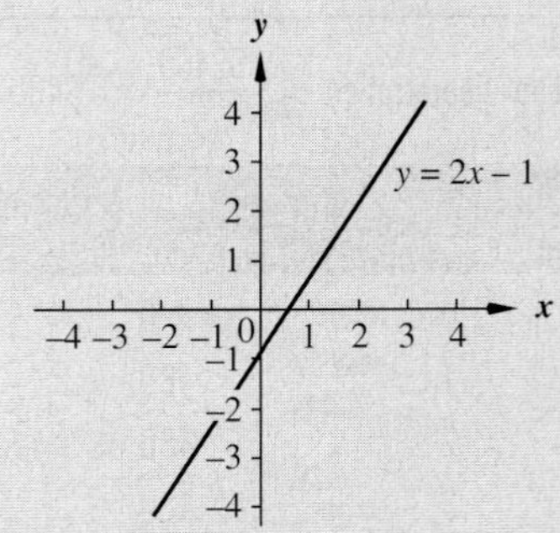

M9.8 (a) $y = 4 - \frac{1}{2}x$
(b) $y = \frac{5}{3}x - 6$
(c) $y = \frac{4}{7}x + 2$
(d) $y = \frac{3}{2}x - 6$

M9.9 (a)

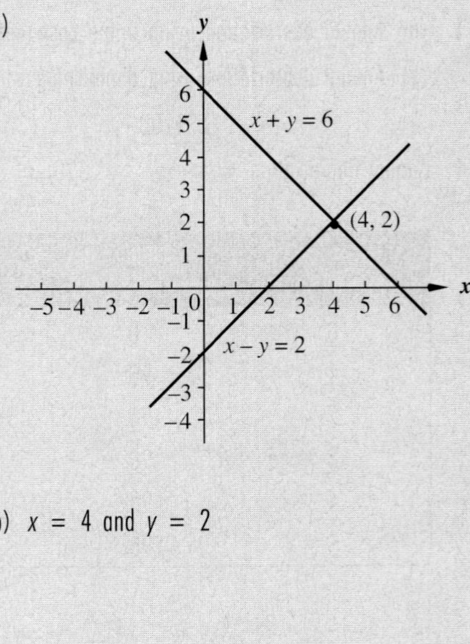

(b) $x = 4$ and $y = 2$

M9.10 (a)

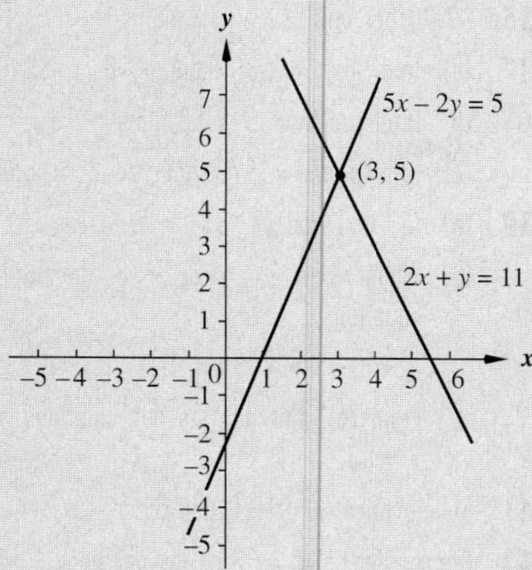

(b) $x = 3$ and $y = 5$

M9.13 (a) Total cost $C = (\$1.20)(1000) + \$200$

$= \$1400$

(b) Income $I = \$2 \times 1000$

$= \$2000$

Profit $= \$2000 - \$1400 = \$600$

(c)

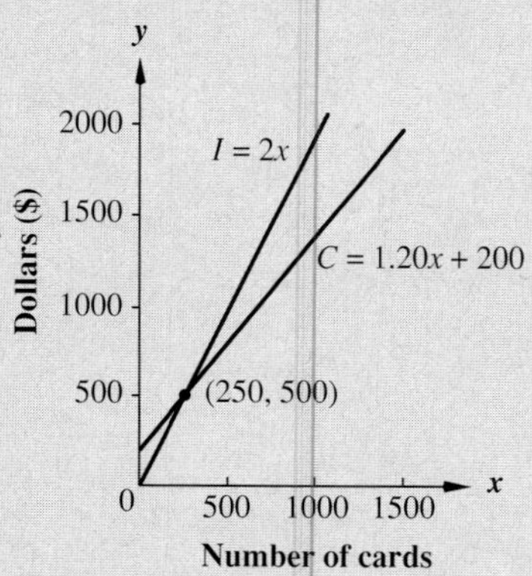

Break-even point is weekly sales of 250 cards.

M9.16 Break-even point $= \dfrac{\$200}{\$2.00 - \$1.20} = \dfrac{\$200}{\$0.80}$

$= 250$ cards per week

M9.18 (a) Total cost $= (\$1.15)(330) + \$140 = \$519.50$

(b) Income $= (\$1.50)(330) = \$495$

Loss $= \$495 - \$519.50 = \$24.50$

(c)

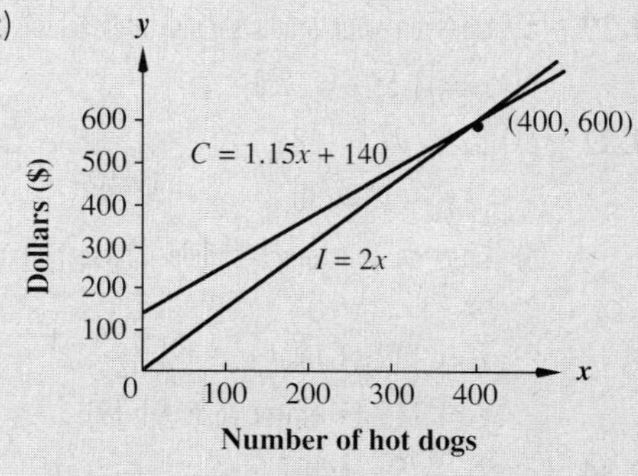

Break-even point is weekly sales of 400 hot dogs.

M9.19 Using Equation M9.7, break-even point is:

$$\frac{f}{s - v} = \frac{140}{1.50 - 1.15} = 400 \text{ hot dogs}$$

which agrees with Exercise M9.18(c).

M9.20 Let x = number of hot dogs sold.

Leasing cost = \$160 and variable cost = \$1.15x.

So total cost = \$160 + \$1.15x.

Since income = \$1.50x,

profit $= \$1.50x - (\$160 + \$1.15x)$

$= \$50.$

This yields $x = 600$, which is the number of hot dogs that would have to be sold per week.

M9.22 Turning points are (March, 30), (June, 45), (Sept, 23).

M9.25

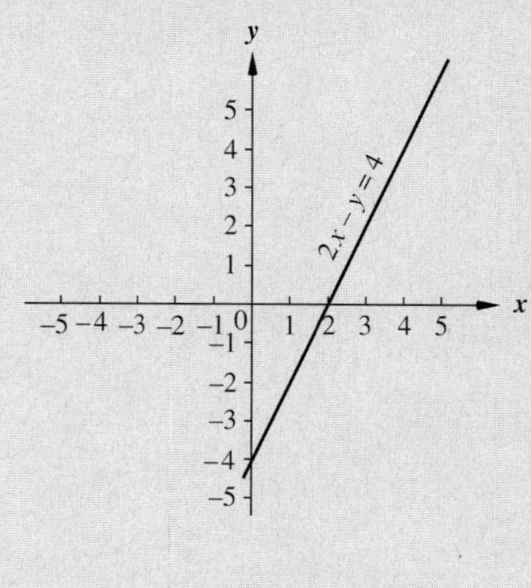

M9.29 (a) Lines on your graph should intersect at (1.5, 2.5).

(b) $x = 1.5, y = 2.5$

M9.32 (a) $3x + 2y = 685$

$x - y = 30$

(b) Lines on your graph should intersect at (149, 119), i.e.

$x = 149$ (1 ruler costs \$1.49)

$y = 119$ (1 eraser costs \$1.19)

M9.34 (a) $2x + 5y = 107.5$

$5x + 8y = 253.0$

(b) Lines on your graph should intersect at (45, 3.50), i.e.

$x = 45, y = 3.50$

(c) Total cost = (3 × \$45) + (2 × \$3.50)

= \$142.00

M9.35 \$2.00 per unit

M9.37 (a) Total *loss* = \$30/hour × 8 = \$240

(b) Total *profit* = \$12.50/hour × 8 = \$100

(c) Total *profit* = \$75/hour × 8 = \$600

M9.39 (a) (1, −1) and (3, 3)

(b) When $x^2 - 2x = 2x - 3$, the result is the equation

$x^2 - 4x + 3 = 0.$

From (a), this has the two solutions, $x = 1$ and $x = 3$.

M9.41 Selling price = \$1.20 per cup

M9.42 Break-even point = 75 cups per hour

Statistics

Chapter S1

Introduction to statistics

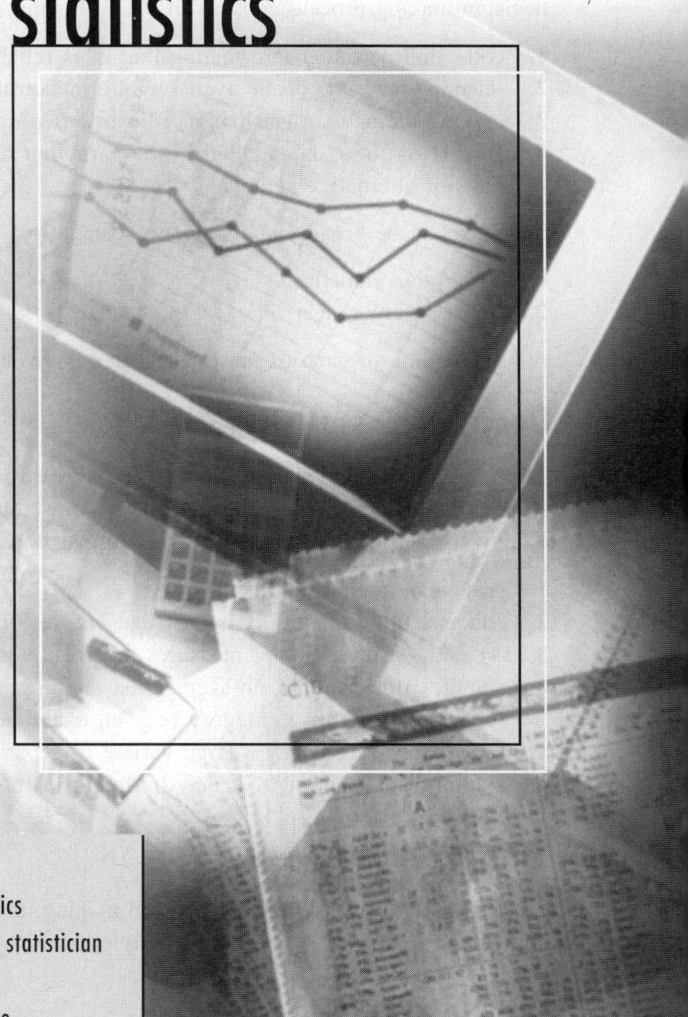

'There are three kinds of lies: lies, damned lies and statistics.'

(Benjamin Disraeli, British Prime Minister, 1804–1881)

Learning objectives

In this chapter you will learn to:

- Identify and understand various types of statistics
- Understand the types of work undertaken by a statistician
- Understand how to apply statistics
- Understand statistics in economics and commerce
- Become aware of publications in statistics
- Understand the role of the Australian Bureau of Statistics
- Become familiar with various types of data

S1.1 THE MEANING OF STATISTICS

Statistics plays an important role in many facets of human endeavour and arises from numerous problems in science and technology. The word 'statistics' occurs remarkably frequently in our everyday lives and yet is one of the most misunderstood terms. It is often thought to be simply a collection of data presented in the form of tables, graphs and diagrams. However, statistics has a far wider involvement in both the physical and social sciences in situations where decisions must be made even though uncertainty exists.

Statistics, in its broad role as the science of decision making in the face of uncertainty, is the scientific method that enables us to make decisions as responsibly as possible. In general, the decision-making process consists of the following steps:

1. collecting pertinent information that is as reliable as possible
2. selecting the parts of the available information that are most helpful to making rational decisions
3. making the actual decisions as sensibly as possible on the basis of the available evidence
4. perceiving the risks entailed in the particular decision made, and evaluating the corresponding risks of alternative actions

Statistics can be broadly split into two categories:

1. descriptive statistics
2. statistical inference

A brief outline of each of these categories is given below.

Descriptive statistics

The collection, condensation and display of information are usually referred to as **descriptive statistics**. Essentially, it is the meaningful presentation of data such that its characteristics can be effectively observed. In particular, descriptive statistics encompasses the following:

1. graphical or pictorial display
2. condensation of large masses of data into a form that is more readily understood (e.g. tables)
3. preparation of summary measures to give a concise description of information that is far more complex; for example, an average may be used to indicate a typical number from among a large set of numbers, or a category (e.g. an examination grade) may be assigned to describe all the individuals covered by a large variety of separate characteristics
4. exhibition of patterns that may be found in sets of information

Statistical inference

Statistical inference relates to decision making and is the subject that leads to future action rather than an inspection of the past. In principle it refers to all types of decisions, but it especially relates to the following:

1. determining whether any apparent characteristics of a situation are really unusual or whether they could have happened by chance
2. estimating the values of unknown numerical quantities and determining the reliability of those estimates
3. using past occurrences to attempt to predict the future

Examples of the kind of problems that a statistician may investigate using statistical inference techniques include:

1. Is there any relationship between certain illnesses or diseases and any particular factors?
2. Is there any difference between males and females in their mathematical ability?
3. Does attending classes really make a difference to final examination marks?
4. How does advertising expenditure relate to sales?
5. Does temperature or altitude affect sporting performance?
6. Are first-born children more intelligent than second-born children?
7. Are the claims made by a manufacturer about its product likely to be true?
8. Are opinion polls and surveys really an accurate reflection of opinion?
9. Is one technique for making scientific measurements better than another?
10. Is there any link between smoking and lung cancer?
11. Does a certain drug really cure a particular psychological affliction?
12. Is there evidence that babies born to mothers who take drugs during their pregnancy are different in some way from other babies?
13. Does using unleaded petrol really make a significant contribution to less air pollution?
14. Is placing more police back 'on the beat' likely to reduce the crime rate?
15. Is there any relationship between the wearing of seat belts, or the introduction of random breath testing for drivers, and the number of people killed on the roads?

Questions such as these can be answered using appropriate statistical testing techniques, of which statisticians have a vast range from which to select. In some cases it is difficult to decide exactly which statistical test is the most appropriate one to use. In Chapter S12 we look at some of the statistical tests most commonly used when dealing with a single sample.

Data collection

The worth of both descriptive and inferential statistics in any given situation depends on the worth of the available data. Collection of reliable data is a fundamental requirement of statistical endeavour, and is often included under the heading of 'experimental design'. The subject of data collection is discussed in Chapter S5.

Variability

To have a full appreciation of statistics it is necessary to understand *variability*. People vary, plants vary, animals vary, weather conditions vary, stock market shares vary, the demand for goods varies, profits vary. Virtually everything varies. This variation occurs both among individuals at any one time and within any one individual as time passes.

It is this variation that affects the reliability of any information obtained from any set of individuals. Conclusions reached from any specific set of individuals may or may not carry over to someone else. In addition, any conclusions made today may not have the same validity in the future.

On the other hand, if we want information about a large group, it may be possible to have some faith in the results of data obtained from some 'representative' portion of that group. In doing so it may be possible to take account of, and possibly control, some of the drastic effects of variability in the larger group.

It is the essence of statistical inference that it must often be decided how much variation is 'acceptable'. For example, if a supposedly unbiased two-sided (head and tail) coin is repeatedly tossed 100 times, we might *expect* to get about 50 heads and 50 tails. However, what if the result were 42 heads and 58 tails? Would this cast doubt on the coin being unbiased? Not necessarily, since obviously some variation from what might be expected may well be quite acceptable. It is the *amount* of variation from what is expected that must be considered, and this question is best answered by means of an appropriate statistical test.

S1.2 WHAT IS A STATISTICIAN?

A **statistician** can broadly be defined as a person who can collect data, present them, analyse them and draw inferences from them. There is a use for statisticians in practically every part of the 'real world' where data collection and analysis are involved.

There is a growing realisation that statistical analysis is necessary in many fields of study, even in those areas (e.g. law, medicine) where it has been relatively slow to gain an involvement. In addition, the advent of fast and efficient computers enabled the mechanical parts of statistical operations to be carried out rapidly and efficiently.

The breadth of statistics can be appreciated by examining just a few of the areas that appear in tertiary education. Examples of some of the major areas of statistics include:

Accounting: auditing; operations research
Anthropology: quantitative investigations of racial identity
Biology: design and analysis of experiments; taxonomy (i.e. classification problems)
Business administration: systems analysis; market research; decision theory
Chemistry: chemical kinetics and thermodynamics
Economics: trend analysis; econometrics; game theory; survey design
Geography: analysis of spatial patterns
Geology: spectral analysis; analysis of spatial patterns
History: pattern identification in data masses
Law: decision models; analysis of forensic data
Linguistics: quantitative analysis of sounds, words and languages
Literature: identification of different styles of authorship
Management: quality management; risk management
Physics: statistical mechanics; astronomical measurement and pattern recognition
Politics: behavioural analysis of voters; opinion polls
Psychology: design and analysis of experiments; measurement of behavioural characteristics ('psychometrics')
Sociology: survey design; categorisation of sociological entities

Statisticians may serve in a consulting capacity for individuals or organisations, or they may lecture, undertake research (e.g. at the CSIRO) or supervise others who are working on projects (e.g. students working on research theses). A statistician may be involved in the application of statistical method (i.e. applied statistics), the development of mathematics required for the theory (i.e. theoretical statistics), or the design of statistical methodology for best effect in particular fields such as economics or biology.

S1.3 STATISTICS IN ECONOMICS AND COMMERCE

There are few areas in which there is more uncertainty than commerce. Apart from the general aspects discussed previously, there are many additional problems not normally encountered. These include:

1. the sheer complexity of the interactions in economic systems, which are extremely difficult to comprehend and even more difficult to describe in mathematical models
2. the extent to which systems change over time owing to demographic factors, technological and social changes, and the political system
3. the incidence of single events that can cause sudden drastic changes; for instance, the election of a new government or a devaluation of currency
4. the difficulty of defining economic quantities in terms precise enough to enable them to be measured, and the difficulties of obtaining the measurements themselves

It follows that statistical activity in the economic/commercial context requires an extremely sound knowledge of the general principles and techniques of decision making in the face of uncertainty. It also requires knowledge of specific techniques that have been developed largely to meet the particular difficulties of the economic environment. In particular it requires:

1. techniques that enable the detection of trends and cycles
2. that genuine structural changes be distinguished from the results of random fluctuation
3. methods to enable the best extrapolation from the past and present to the future, together with an assessment of the possible errors that can result from such extrapolation

S1.4 PUBLICATIONS IN STATISTICS

There are many journals throughout the world that publish statistical research. Some publish mainly general statistical work, while others are more concerned with the application of statistics to specific fields. While the journals are too numerous to list in full, some examples of each type of publication are listed below.

General statistical journals

Applied Statistics
Chance
Computational Statistics
International Statistical Review
Journal of the American Statistical Association
Journal of Applied Probability
Journal of Forecasting
Journal of the Royal Statistical Society (Series A and B)
Journal of Statistical Planning and Inference
Journal of Statistics Education
Journal of Statistical Research
Risk Analysis
Sankhya (Series A and B)
Scandinavian Journal of Statistics

SIAM Review
Teaching Statistics
Technometrics
The American Statistician
The Annals of Statistics
The Canadian Journal of Statistics
The Statistician

Journals in specific areas

American Journal of Epidemiology
Biometrika
Biometrics
Econometrica
Epidemiology
Interfaces
Journal of Applied Econometrics
Journal of Educational Statistics
Journal of Social and Economic Measurement
Journal of Statistical Computation and Simulation
Journal of Statistical Physics
Journal of the Indian Society of Agricultural Statistics
Journal of the Operations Society of America
Psychometrika
Research Quarterly for Exercise and Sport
Statistics in Medicine
Transportation Research

Australian journals

The Australian and New Zealand Journal of Statistics
Bulletin of the Operations Research Society of Australia
Journal of the Australian Mathematical Society (Series A)

S1.5 AUSTRALIAN BUREAU OF STATISTICS (ABS)

The main source of 'official' Australian statistical data is the Australian Bureau of Statistics (ABS). It issues about 1600 monthly, quarterly, annual and irregular publications each year covering almost every field of economic and social activity. These include descriptive matter dealing with Australia's government, international relations, defence, climate, physiography, culture and environment.

Each year the ABS publishes *Year Book Australia* and *Pocket Year Book Australia*, which contain information that relates to the year ended in June or December of the previous year. In addition, individual year books are published annually for each state. To obtain up-to-date information it is preferable to consult relevant individual ABS publications as they are released.

The ABS has been delivering services online for some years and these are continually being enhanced. It has essentially completed the implementation of the government standards relating to online service provision. The ABS has developed a framework for assessing the appropriateness of online service provision and will continue to review all services using this framework. The ABS Web site can be found at the URL www.abs.gov.au.

The ABS is headed by the Australian Statistician, this being a statutory office. There are two Deputy Australian Statisticians, responsible for the Economic Statistics program and the Population & Social Statistics program, respectively. Each Deputy Australian Statistician is supported by a division head, one for the Economic Accounts division and one for the Social and Labour division. Other division heads are responsible for the Corporate Services, Technology Services, Methodology and Information Management divisions.

The ABS has a central office in Canberra and eight regional offices, one in each state and territory capital city. The regional offices have operational responsibility for the national operations of particular statistical collections, but overall program responsibility remains in the central office. Regional offices also allow the ABS to maintain close contact with state clients, including state government agencies, and ABS data providers.

The mission of the ABS is to 'assist and encourage informed decision making, research and discussion within governments and the community, by providing a high quality, objective and responsive national statistical service'.

Coding of ABS publications

Each ABS publication has a five-digit code according to the ABS Catalogue Numbering (*Cat. No.*) system. In this code:

1. The *first* digit defines the broad subject group.
2. The *second* digit defines a subgroup of the broad subject group.
3. The *third* and *fourth* digits are serial numbers that are permanently allocated to publications within categories formed by the first two digits.
4. The *fifth* digit, after the decimal point, represents the state or territory covered by the statistics of the publication. These are 0 (Australia), 1 (NSW), 2 (Victoria), 3 (Queensland), 4 (South Australia), 5 (Western Australia), 6 (Tasmania), 7 (Northern Territory), 8 (ACT) and 9 (External Territories).

The list of first and second digit codes that indicate the subject groups are shown below.

10. Annual reports
11. Directories, catalogues and guides
12. Classifications and work manuals
13. Year books and other multi-subject products
14. Online electronic services
20. Census statistical products and services
29. Census reference products and services
31. Demography—general
32. Population trends and estimates
33. Vital statistics
34. Migration
41. Social statistics—general
42. Education
43. Health

44. Welfare and social services
45. Crime and justice
46. Environment
47. Indigenous statistics
48. Health
49. Social statistics—general
52. National accounts
53. Balance of payments and international investment
54. International trade
55. Public sector accounts
56. Finance
61. Labour statistics—general
62. Labour force
63. Earnings, hours and employment conditions
64. Prices
65. Consumer income and expenditure
71. Agriculture statistics—general
72. Livestock and livestock products
73. Crops and pastures
74. Agricultural land use
75. Agricultural financial statistics and value of products
81. Industry-wide statistics
82. Manufacturing and energy—general
83. Manufacturing commodity production
84. Mining
85–86. Service industries
87. Building and construction
89. Other agencies
92. Transport services
93. Motor vehicle registrations

Example S1.1

Consider the publication ABS Cat. No. 3311.2. The breakdown of the digits is:

First digit: 3 = demography
Second digit: 3 = vital statistics
Third and fourth digits: 11 = serial numbers
Fifth digit: 2 = Victoria

The title of this publication is *Demography, Victoria*.

For further information on the catalogue number of publications, consult the ABS book, *Catalogue of Publications and Products* (current year), Cat. No. 1101.0.

S1.6 TYPES OF DATA

Statistical data can be obtained in a number of ways. These include *measurements* using an instrument, such as a device for measuring blood pressure or a simple ruler, and *counts*, for example where the number of employees absent each day or the number of traffic accidents occurring each year is recorded.

Rank data arise in situations where, for example, individuals or objects are ranked according to some criterion, e.g. where customers are ranked according to risk in requesting a loan or athletes are ranked according to physical fitness. **Categorical data**, or *classification data*, involve placing observations into categories, such as gender, eye colour or country of birth.

Data can also be classified according to whether they are primary or secondary. **Primary data** are information collected by the person or organisation that will be using the information. This includes the case of a researcher who wants to undertake statistical analysis involving information that has not yet been collected. Before commencing on the project, the aim of the study (or hypothesis to be tested) must be identified, along with a suitable method for collecting the data. Techniques for data collection are fully discussed in Chapter S5, including personal interviews, mail surveys and telephone interviews.

Secondary data are information *already* collected by someone else. In this case the researcher has little or no control over the design or method of data collection, which may turn out to be unsatisfactory since it may be inappropriate. If everything is in order, however, much time and cost can be saved by using such material. A good source of secondary data is the Australian Bureau of Statistics, as discussed above.

S1.7 REMARKS

Statistics is a huge field with wide ramifications and many unsolved problems. In the following chapters the main emphasis will be on developing statistical thought from which many statistical techniques can be built.

Problems regarding sampling and statistical description will be developed first, followed by specific topics that have special relevance to business and commerce. It is not possible to become a professional statistician from reading these chapters: that would require many more volumes and much practical experience. The chapters can, however, make you more aware about what statistics entails.

Quick Quiz

1. Name four types of media via which the public obtain statistical information.
2. In what year will the next full Australian census be held?
3. A company claims that 'more people than ever before are using our product'. Explain how this could be misleading.
4. List some of the areas of medical research in which you feel that statistics may be applicable.
5. Describe briefly how you would determine what proportion of voters in your state would vote for the Liberal Party if a state election were to be held next week.
6. If a newspaper claims that the popularity of the Prime Minister has 'risen three points in the past two weeks', what does the newspaper actually mean?
7. What is the difference between *descriptive statistics* and *inferential statistics*?
8. Suppose that you were asked to estimate (off the top of your head) the average age of 40 000 people attending the grand final of a football match and the average age of 66 000 students sitting for the Higher School Certificate (HSC) in Year 12 in New South Wales. Which estimate do you consider might be the more accurate, and why?
9. List three types of company located in Australia that you think might employ a statistics graduate.
10. Would you say that statistics is an 'exact science'? Explain.

Multiple Choice Questions

1. Decision making is sometimes referred to as:
(a) condensation of data
(b) descriptive statistics
(c) statistical inference
(d) graph drawing
(e) none of the above

2. Which of the following best portrays descriptive statistics?
(a) collection of data
(b) condensation of data
(c) display of data
(d) summary of data
(e) none of the above

3. In which of the following fields might a statistician work?
(a) medicine
(b) education
(c) biology
(d) physics
(e) all of the above

4. Which of the following journals is published in the United Kingdom?
(a) *The Annals of Statistics*
(b) *Technometrics*
(c) *Teaching Statistics*
(d) *Sankhya*
(e) none of the above

5. Which of the following is *not* primarily a statistical computer package?
(a) Minitab
(b) SPSS
(c) Lotus 1-2-3
(d) SAS
(e) SYSTAT

6. The main offices of the Australian Bureau of Statistics are located in:
(a) Sydney
(b) Melbourne
(c) Canberra
(d) Brisbane
(e) none of the above

7. *Primary data* are collected by:
(a) primary school children
(b) people doing it for the first time
(c) the actual people who will be using it
(d) mainly elderly people
(e) none of the above

8. *Experimental design* involves:
(a) doing scientific experiments
(b) studying architecture
(c) designing cheap houses
(d) undertaking risky research
(e) none of the above

9. A fundamental requirement of *statistical data* is that:
(a) it always consists of whole numbers
(b) it must be reliable
(c) it is only collected during business hours
(d) it is able to be sold at a profit
(e) none of the above

10. *Demography* is the study of:
(a) people and populations
(b) geography
(c) statistical demonstrations
(d) the statistics of graphs
(e) none of the above

Exercises

S1.1 Scan magazines and newspapers for articles (including advertisements) that use statistics to emphasise a 'fact' or back up a point, but where you have serious doubts as to their validity. List the reservations that you have in each case.

S1.2 Review three decisions you have recently had to make. How many of these were based on complete evidence? Were there any where you had to find the most reasonable decision from evidence that suggested several possible decisions? What criteria did you use to make your decision? Were those criteria rational, and could they have been improved? If so, how?

S1.3 Describe various situations that require:

(a) descriptive statistics (b) inferential statistics (c) some combination of these

S1.4 What sorts of risk would be encountered from bad experimental design in a study intended to see whether the public is likely to buy substantial quantities of a new flavour of toothpaste? Discuss how you would conduct such an investigation.

S1.5 Go to your library and write down the names of all the statistics journals to which it subscribes. Also list the names of three such journals to which it does not subscribe.

S1.6* The field of *operations research* is closely related to that of statistics. Explain (in two or three sentences) to a prospective employer what operations research entails.

S1.7* Suppose that you were asked to determine whether there is a difference between the proportion of males and females at your institution who smoke. Outline briefly how you might go about doing this.

S1.8 Compare the variabilities of the following pairs. Can you say which of a pair is more variable, or can you say that there are *differences* in variability without being able to say that one is more variable than the other?

(a) weights of male adults versus weights of female adults

(b) weights of adults versus weights of 6-year-old children

(c) ages of politicians versus ages of television stars

(d) colours of dogs versus colours of horses

(e) sales of bread from one day to another versus sales of ice-cream

(f) prices of bananas (per kilogram) versus prices of butter (per kilogram)

(g) bank deposits by individuals versus bank deposits by companies

(h) numbers of people using a municipal swimming pool versus numbers using a municipal library

(i) honesty of politicians versus honesty of journalists

S1.9 Suppose that an opinion poll shows that the support for a Prime Minister has dropped from 65% one week to 60% the next week. How should we interpret this result? Is it possible that the support has not really dropped and that these results are merely random fluctuations?

S1.10* Which of the following variables may be influenced by economic and social changes? How do such influences affect the numerical values of the variables?

(a) the average height of the adult male population

(b) the average age of the adult male population

(c) the average income of the adult male population

(d) the average number of hours of television watched per week

(e) the distribution of savings over banks, building societies and other forms of investment

(f) the proportion of incomes spent on electrical appliances

S1.11 Suppose that you have collected data about the price of milk over the last 10 years.

(a) How would you use this information to predict the price of milk in 2 years time?

(b) In what ways would your prediction process in (a) be inadequate?

(c) List any factors that you feel could cause your prediction to be quite incorrect.

S1.12 In some situations absolute data (e.g. actual sales, actual prices, actual numbers of people) are used to illustrate a point or give an impression that a particular trend has occurred. However, if relative data are used instead of absolute data, quite a different picture might emerge. Give some examples of situations where this could occur.

S1.13 Suppose that 3000 drivers in New South Wales were random breath-tested on 11 April and 116 were above the limit of 0.05 blood-alcohol level. On 5 May, 4000 drivers were tested and 98 were above this level.

(a) What additional information would you require before trying to draw conclusions from these data?

(b) What factors, other than a real change in driver behaviour, could cause such a drop in the proportion of above-the-limit drivers?

S1.14 News and financial reports often refer to *seasonally adjusted figures.* Describe how various quantities can be expected to vary from one time of year to another, and suggest ways in which adjustments can be made to the data so that the varying figures can be compared on a common basis.

S1.15 (a) Describe how you would design an experiment to see whether shoppers in different suburbs have different preference patterns for butter versus margarine.

(b) Apart from genuine taste preferences, are there any other factors that may cause shoppers to prefer one product over the other? List some of these.

S1.16* Most modern statisticians do their long and tedious calculations with the aid of a computer. In the late 1970s they used electronic calculators for these tasks. What did they use in, say, the 1940s and 1950s?

S1.17 A new product for removing stains from clothes has just been released onto the market. Describe some of the methods you would use to measure its effectiveness.

S1.18 Find copies of recent newspapers and gather three examples of articles that give a visual display of data (e.g. a graph, diagram, chart). Comment on the way in which the data have been presented, with particular reference to whether it is incorrect, ambiguous or misleading.

S1.19 Name five large companies in Australia that you feel would employ at least one full-time statistician.

S1.20* A failed statistics student was once asked to comment on the results of a survey that showed that 95% of convicted bank robbers like chocolate ice-cream. The student concluded that, if you like chocolate ice-cream, there is a 95% chance that you will be convicted of bank robbery. Comment on the student's conclusion. What conclusion, if any, could *you* draw from the survey?

S1.21* Discuss some of the topics that you would expect to find if you undertook a course in *demography*.

S1.22 Describe some of the activities that might be undertaken by a business involved with *market research*.

S1.23 What is *random variation*?

S1.24 Do you think that it is possible to use statistics to predict the price of shares on the stock exchange? Give reasons.

S1.25 Do you think that it is possible to use statistics to predict the numbers of the balls that will be drawn out on next week's Lotto game in your state? Give reasons.

S1.26 Do you think that it is possible to use statistics to predict the scores in a football game? If so, what information would you require to make such a prediction? If not, state why.

S1.27* What is the difference between *quantitative analysis* and *qualitative analysis*?

S1.28* Statistics plays a prominent part in the field of medicine, particularly in the analysis of clinical trials. What is a *clinical trial*?

S1.29 Suppose that the Bureau of Meteorology announces that the *average* temperature on a certain day was 24.2 degrees. What method do you think they used to work that out?

S1.30 How often is the Australian census undertaken? In what year was the last one taken? How are the census forms distributed and collected?

S1.31 Many publications of the Australian Bureau of Statistics are referenced by a Catalogue Number (Cat. No.). Find the names of the publications that have the following Catalogue Numbers:

(a) 3303.0 (b) 3401.0 (c) 6302.0 (d) 5510.0

(e) 7507.0

S1.32* Suppose that a television commercial asserts 'University tests prove that our product is the safest on the market'. What is this message supposed to convey to viewers? Does it mean anything? Explain.

S1.33 A radio station claims that 85% of callers to an opinion hotline telephone number set up by the station support a certain issue. What can you conclude from that?

S1.34 If you had been asked to forecast the weather in Sydney on the day (15 September) of the opening ceremony of the Olympic Games in the year 2000, what information would you have needed to do this?

S1.35* A branch of statistics is known as *actuarial science*. List some of the activities an *actuary* might undertake. What type of company might employ an actuary?

S1.36 (a) Find a sample of 50 cars (either parked or moving) and record their colour as either (i) red, (ii) blue (iii), silver/grey, (iv) white or (v) other.

(b) Calculate the percentage of cars in each colour category. Which is the most popular colour category?

Chapter S2

Visual presentation of data

'"What's the use of a book", thought Alice, "without pictures or conversations?"'

(Lewis Carroll, Alice's Adventures in Wonderland, *1865)*

Learning objectives

In this chapter you will learn to:

- Understand sources of information
- Construct tables
- Illustrate data using a graph
- Illustrate data using a pie chart
- Illustrate data using a bar chart
- Illustrate data using a pictogram
- Condense raw data using a frequency distribution
- Condense raw data using a grouped frequency distribution
- Construct a histogram and frequency polygon
- Construct and apply ogive curves
- Understand how statistics are misused

S2.1 INTRODUCTION

A problem that arises in many situations is that of condensing large sets of data. In particular, one of three processes may be involved:

1. to present a simple 'neutral' description of the facts of a certain situation
2. to draw conclusions from a set of facts (i.e. make an inference)
3. to refute an inference drawn by someone else, which may be done either by presenting new information or by demonstrating that existing information has been incorrectly interpreted

Some examples of the sources of available information are:

1. *'Official' statistics* These include publications by the Australian Bureau of Statistics (ABS), the Treasury and other government departments, the Government Actuary, the United Nations and its organisations, and so on.
2. *'Semi-official' statistics* These include private publications by firms, such as bank bulletins, stockbroker reports and company annual reports.
3. *Survey reports* These include the results of fact-finding surveys and opinion polls undertaken by market research teams. They are often found in journals and newspapers.
4. *Results* These include data obtained from direct experimentation.
5. *Journals* These include both scientific and other journals.

There are basic rules that should be followed once a set of data has been collected and is ready for presentation. The person who is reading the figures should be convinced that the data presented are both factual and relevant. In particular, it should be possible for the reader to check:

1. the source of the data
2. that the data have been transcribed accurately
3. that the figures are relevant to the problem being considered

On those occasions when there is a case to argue, success is often largely dependent on the way in which the numerical data are presented. In this case it is vital that a correct, comprehensible presentation be given. In this chapter various techniques for the effective presentation of such data are demonstrated, as well as examples of distortions by false representations.

S2.2 TABLES

When constructing a table it is important to determine which relationships should be emphasised. A table that is too difficult to follow or too trivial to be heeded is usually one that presents a confused argument or demonstrates an obvious fact.

On some occasions all the data may be relevant; but if they are all presented at the one time the reader may be unable to see all the important relationships. In some cases, if the table is too large, it can be presented in an appendix. If a table is included in the main text, it should be designed to make only a few points. It may be wise to have more than one table relating to the same set of data.

Some of the rules that should be followed when constructing a table are:

1. Every table should have a clear and unambiguous number. It is common practice to use a decimal number system such as 3.1, 3.2, 3.3, …, 3.9 for tables that are in either the third chapter or section of the work, or relate to the same topic.

2. Each table should have a title that describes the types of information given and includes any relevant dates.
3. Row and column tables should be precise and unambiguous.
4. Categories should not overlap. That is, the same item should not appear in more than one category in any one table.
5. The units of measurement must be clearly stated.
6. Any unimportant figures should be combined or omitted.
7. Any subheadings should be labelled clearly.
8. Any relevant totals, subtotals, percentages, and so on should be shown.
9. The correctness of any calculations should be verified. For example, check that the sum of the column totals equals the sum of the row totals.

Example S2.1

(a) Information concerning participation in swimming, soccer and golf by Australia's adult population under 45 years of age is published by the ABS. If the data were presented in narrative form, they would read something like this:

Australia has a total of 2 351 300 adults under 45 years of age who participate in swimming, soccer or golf, comprising 1 426 300 swimming, 175 900 soccer and 749 100 golf. Of these, there are 590 800 aged 18–24 (350 900 swimming, 89 100 soccer and 150 800 golf), 895 900 aged 25–34 (555 100 swimming, 48 000 soccer and 292 800 golf) and 864 600 aged 35–44 (520 300 swimming, 38 800 soccer and 305 500 golf).

It is very difficult to gain any meaningful information or distinguish any patterns from this narrative form. However, this is not the case if the data are presented in table form, as shown in Table S2.1.

Table S2.1 The number of adults under 45 years of age in Australia who participate in swimming, soccer or golf

Age	Swimming	Soccer	Golf	Total
18–24	350 900	89 100	150 800	590 800
25–34	555 100	48 000	292 800	895 900
35–44	520 300	38 800	305 500	864 600
Total	1 426 300	175 900	749 100	2 351 300

Source: Australian Bureau of Statistics, *Participation in Sport and Physical Education*, Cat. No. 4177.0

(b) Table S2.2 shows data involving the number of registered marriages and the crude marriage rate in Australia. Also provided is the median age at marriage of both bridegrooms and brides. Note how these median ages have been steadily increasing for both groups since the *Family Law Act* was introduced in 1975.

Table S2.2 Selected summary measures of marriages in Australia

Year ended 31 December	Registered marriages		Median age at marriage	
	No.	Crude marriage rate[a]	Bridegroom (years)	Bride (years)
1947	76 457	10.1	26.0	23.0
1954	71 229	7.9	25.6	22.6
1961	76 686	7.3	24.9	21.8
1966	96 061	8.3	24.2	21.5
1971	117 637	9.2	23.8	21.4
1976	109 973	7.9	24.9	22.2
1981	113 905	7.6	25.9	23.3
1986	114 913	7.2	27.3	24.9
1990	116 959	6.9	28.2	25.9
1991	113 869	6.6	28.4	26.0
1992	114 752	6.6	28.7	26.3
1993	113 255	6.4	28.8	26.4
1994	111 174	6.2	29.0	26.6
1995	109 386	6.1	29.2	26.8
1996	106 103	5.8	29.6	27.2
1997	106 735	5.8	29.7	27.5
1998	110 598	5.9	29.8	27.7
1999	114 300	6.0	30.1	27.9

(a) Per 1000 population.

Source: Australian Bureau of Statistics, *Year Book Australia 2001*

S2.3 GRAPHS

Graphs (or *line charts*) do not generally present as many data in detail as do tables, although it is often much easier to see specific relationships with a graph. It is also possible to depict more than one set of data on the same graph.

For economic and business data, the vertical axis often indicates *quantities* (e.g. production, consumption, dollars) or percentages. The horizontal axis often represents units of *time*.

When illustrating data by means of a graph, follow these basic rules:

1. Give the graph a clear title.
2. Label the axes clearly.
3. Do not put too many curves on the same graph, as this will confuse the reader.
4. Quote all sources of data.
5. If possible, accompany the graph with the table of data that are represented by the graph.
6. Use a zero on the scale of the vertical axis if possible. If this is not feasible or convenient, you may break the vertical axis to show the zero point.

Example S2.2

A graph illustrating radio and television consumption in Australia on a working day is shown in Figure S2.1. The horizontal axis represents the time of day between 6.00 a.m. and midnight, while the vertical axis represents the percentage of people tuned to either medium at that time.

Note that the graph for radio is enhanced by the drawing of vertical bars. Interesting research reveals that radio listening peaks at 70% between 8.00 a.m. and 9.00 a.m., while television peaks at 41% between 7.00 p.m. and 9.00 p.m., with radio dominating television as the medium of choice until 6.30 p.m., when television takes over.

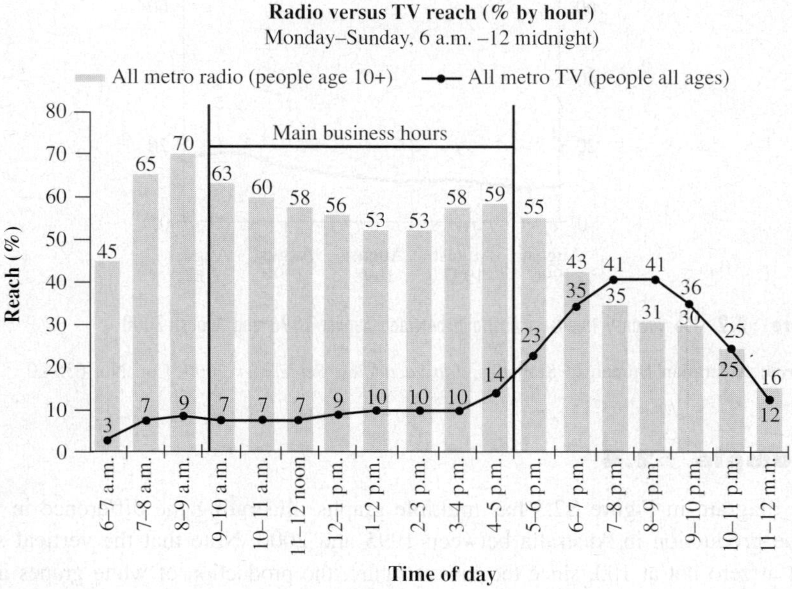

Figure S2.1 Graph illustrating the comparison between the percentage of Australians who tune to radio and television during the period 6.00 a.m. to midnight each day

Source: Radio Marketing Bureau online; ACNielsen

If more than one set of data is illustrated on the same diagram, a *key* that identifies each graph is usually given. The key enables readers to distinguish between various graphs. The alternative to a key is to write on each graph itself, but this can get crowded if a number of graphs appear together. A key is used in Figure S2.1.

Example S2.3

In Figure S2.2 each of the three curves is distinguished by being drawn in a different style to enable the reader to tell them apart. A common technique is to use a *key*, usually written on a blank space on the graph. The key lists the name of each curve with an example of the style written alongside.

This particular graph shows the trend in job vacancies in both the private and public sector, along with the total job vacancies, between August 1996 and August 2000. Note how the vertical scale is given on both the left-hand and right-hand sides for easier reading of the figures on both sides of the diagram.

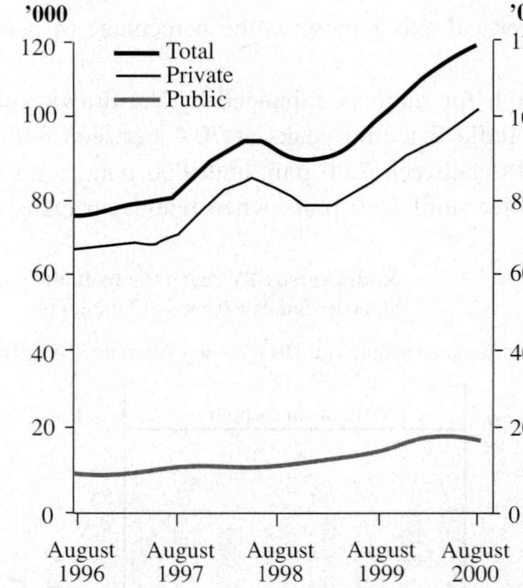

Figure S2.2 Job vacancy trends in Australia between August 1996 and August 2000

Source: Australian Bureau of Statistics, *Job Vacancies, September 2000*, Cat. No. 6354.0

Example S2.4

The diagram in Figure S2.3 has multiple graphs illustrating the difference in white and red grape production in Australia between 1995 and 2000. Note that the vertical scale does not start at zero but at 100, since the lowest figure, the production of white grapes in 1995, is just

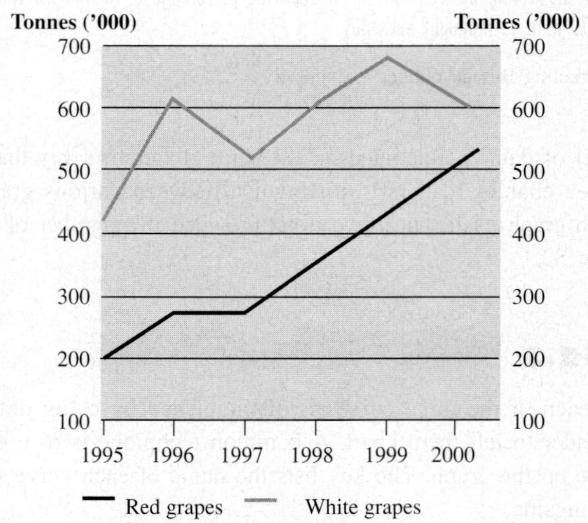

Figure S2.3 Australia's wine grape production between 1995 and 2000

Source: Australian Bureau of Statistics, *Sales of Australian Wine and Brandy by Winemakers*, Cat. No. 8504.0; Australian Wine and Brandy Corporation

under 200 000 tonnes. The curves show how production of both white and red grapes is increasing, but that red grape production has nearly caught up to white.

Example S2.5

According to a study by the Productivity Commission, the number of people in Australia aged 65 or over will jump from 12% of the population to 22% over the next 30 years. However, the Commission predicts that long-term aged care costs will remain at approximately 1% of gross domestic product (GDP). A graph illustrating this percentage since 1997 and projected values until 2031 is represented in Figure S2.4, with the scale on the right-hand vertical axis. Note that as well as the curve there are bars to represent the actual dollar value, and the scale for this is on the left-hand vertical axis. To make the diagram more eye-catching, in the background there is a picture of hands holding a coin.

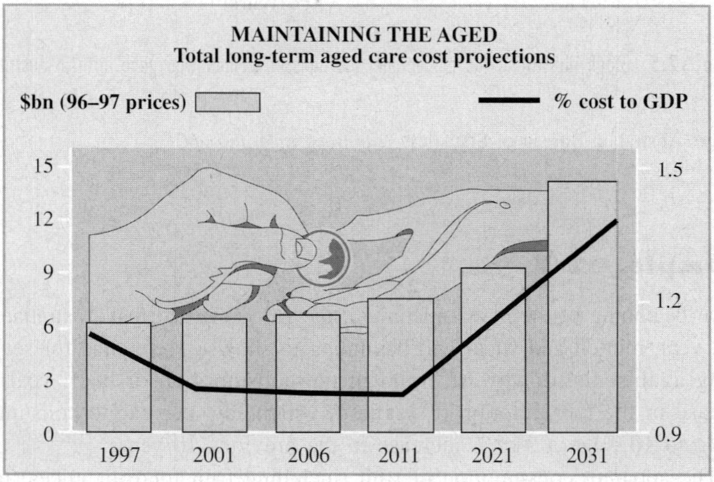

Figure S2.4 Graph illustrating long-term aged care cost projections in Australia until 2031

Source: *Sydney Morning Herald*, 31 October 2000

Example S2.6

The graphs in Figure S2.5 demonstrate the contrast in both the percentage of women and the ages of women who gave birth during 1964 and 1998. Once again, a key is used to identify each graph. Although the graphs have a roughly similar shape, the one for 1964 peaks at around age 22 years, at which age 24% of women gave birth. In 1998 the peak had jumped to 29 years, although only 7% of women gave birth at this age.

In the last 25 years there has been a fall in the proportion of teenage mothers, from 9% in 1977 to just under 5% in 2001. Similarly, the proportion of women having children at 40 and over has increased from under 1% to over 2%, as women have continued to have children later in life.

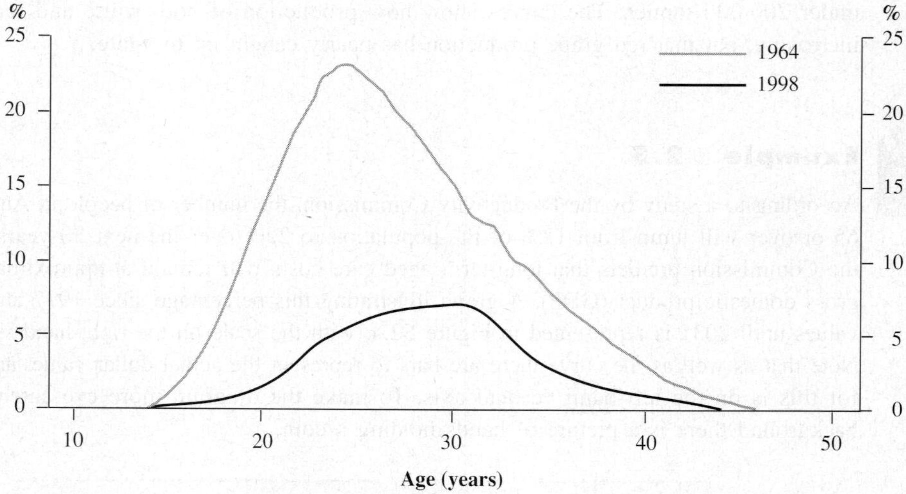

Figure S2.5 Graph illustrating the percentage and ages of women who gave birth in Australia during 1964 and 1998

Source: Australian Bureau of Statistics, *Year Book Australia 2001*

Example S2.7

At the beginning of the new millennium, the ABS revealed that Australians are consuming less beef, veal, mutton and sugar and beginning to show a preference for seafood and vegetables. Going against the downward trend in the consumption of beef, veal and mutton was an increase in the consumption of seafood, which saw per capita consumption increase from 8.3 kg to 10.9 kg, a 31.3% increase in the previous 10 years.

The apparent consumption of fruit (including fruit for fruit juices) increased by 8.3% to 135.0 kg per capita, following an increase of 2.6% to 124.7 kg in the previous year. The recorded per capita consumption has increased about 56.1% since the late 1960s and 71.5% since the late 1930s. The total apparent per capita consumption of vegetables was 162.0 kg, up by 45.0 kg compared with the late 1950s when the recorded consumption was 117.1 kg per capita.

Total apparent sugar consumption decreased in 1998–99 by 8.6% to 43.4 kg per capita, which contrasts with an increased intake of sugar of 5.8% (to 47.5 kg per capita) in 1997–98. This drop is attributable to the drop in the consumption of total cane sugar, which fell by 10.5% from 42.0 kg per capita in 1997–98 to 37.6 kg per capita in 1998–99. Honey consumption also declined by 19.8% to 0.5 kg per capita.

Over the same period, the apparent per capita consumption of low alcohol beer showed a fall to 24.7 litres per capita. Overall per capita beer consumption is now at a level similar to that of the late 1950s, after peaking at over 140 litres per capita in the mid-1970s. This reduction reflects a number of factors including changes in consumer tastes towards wine, the increased availability of quality, competitively priced table wine, and an overall reduction in alcohol intake.

The information regarding these long-term trends in the consumption of selected foodstuffs is shown in Figure S2.6. Note that each of the four graphs spans over 70 years, with a break between 1988–89 and 1996–97. This break is due to the fact that in the earlier years the horizontal axis is spaced at every 10 years while in the later years the data are shown as single years.

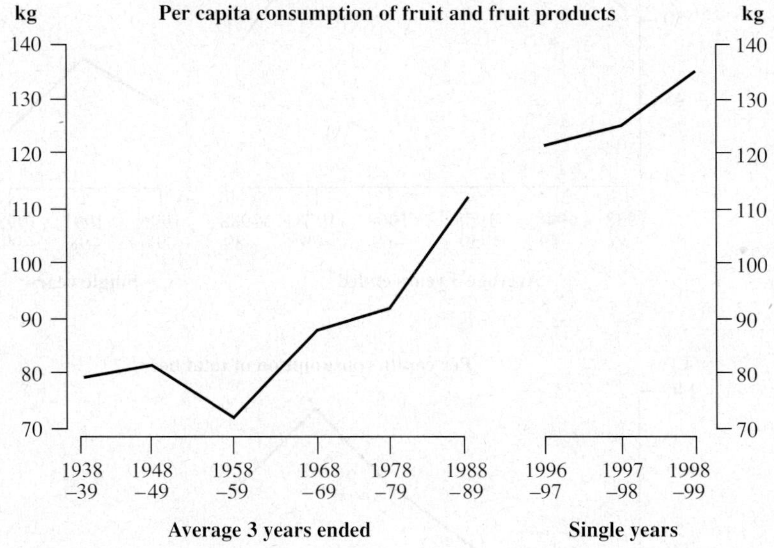

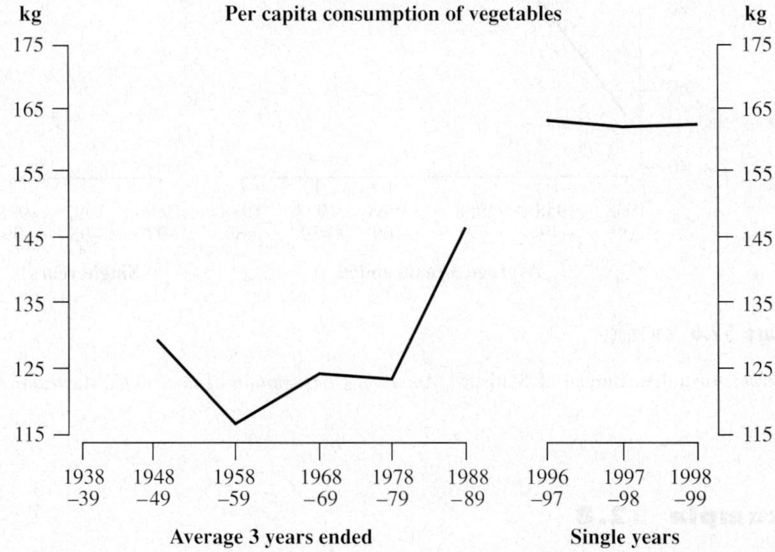

Figure S2.6 Four graphs illustrating the consumption trends of four selected foodstuffs in Australia between 1938–39 and 1998–99

(continued)

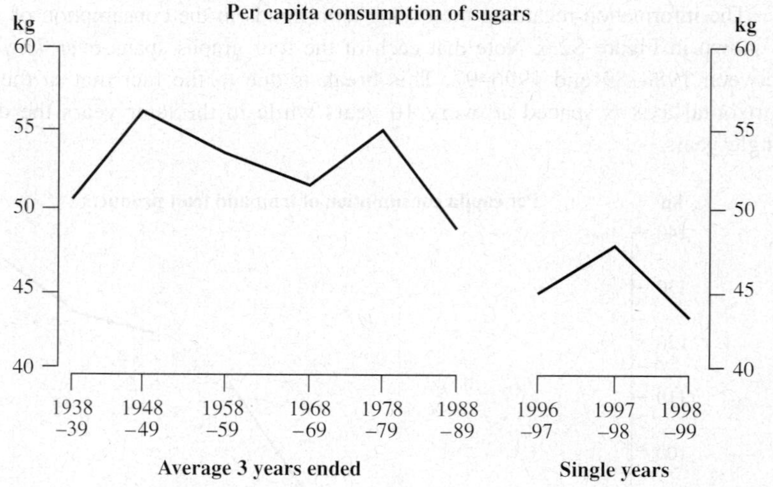

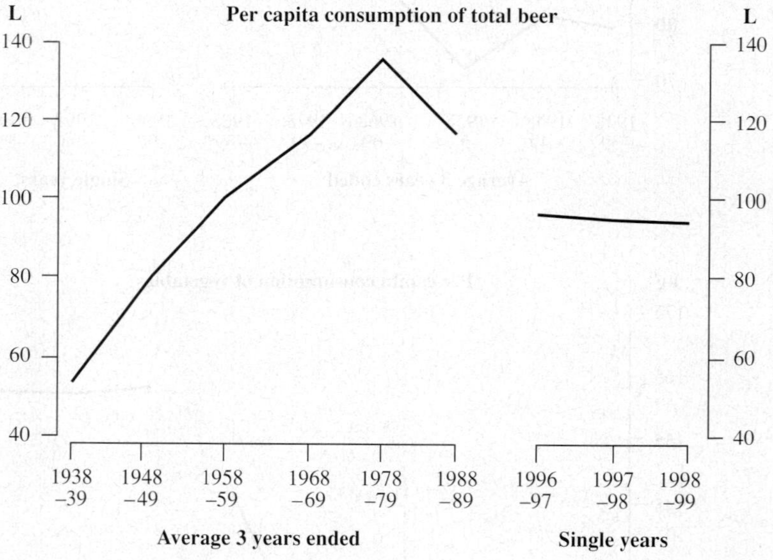

Figure S2.6 continued

Source: Australian Bureau of Statistics, *Apparent Consumption of Foodstuffs, Australia*, Cat. No. 4306.0

Example S2.8

From late 1999, rising oil prices, a plummeting Australian dollar and interest rate hikes played havoc with equities markets. The All Ordinaries Index lost about 6% from a series of record highs in August 1999, while building and transport indicators were down about 14% from their peaks, and the packaging and paper sector shed 16% of its value.

The *Bulletin* magazine drew a diagram of the gloomy situation with a heading 'The big market moves', reproduced exactly as it was shown in Figure S2.7. This essentially showed three graphs that tracked the movement of oil prices, the All Ordinaries Index and the exchange rate of the Australian dollar against the US dollar for the period from 27 September 1999 until 30 August 2000. There are several noteworthy features of this diagram:

- Zero is suppressed on both vertical axes of the diagram.
- The scale on the right-hand vertical axis is not labelled. What does it mean? Is it some type of index and, if so, what kind?
- The horizontal axis is not labelled at all, although presumably it should have read 'Date'. Even so, it is not clear as to why those particular days were chosen (they certainly did not all fall on the same day of the week).

Perhaps you could make your own list of any other features that you find unsatisfactory about the diagram or that you feel may be confusing to the reader.

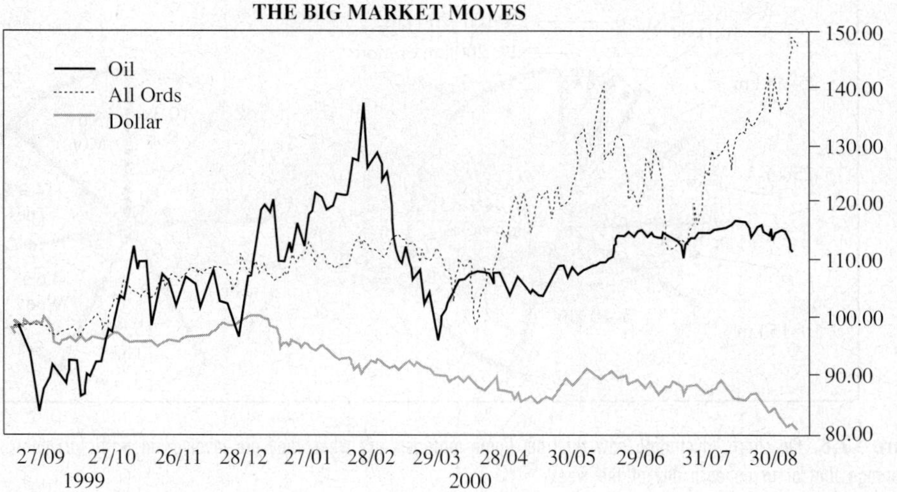

Figure S2.7 Three graphs illustrating the fluctuating trend in Australia of oil prices, the All Ordinaries Index and the exchange rate of the Australian dollar against the US dollar for the period from 27 September 1999 until 30 August 2000

Source: *Bulletin*, 3 October 2000, p. 60

S2.4 PIE CHARTS

A **pie chart** is often used to give a visual presentation of data to indicate the proportions that make up a given total. It is one of a number of so-called *area diagrams*, which consist of geometrical figures (e.g. square, rectangle, ellipse). The pie chart is a circle that is divided by radial lines into sectors in such a way that the area of each sector is proportional to the size of the quantity represented by that sector.

It is not uncommon for the sectors to run clockwise (from 12 o'clock) with the largest sector first, although there is no hard and fast rule about this. Figure S2.8 shows two pie charts, which are used to illustrate how far from home motorists are when they are involved in traffic accidents, along with the percentage that occur on each day of the week. It can be seen that Friday is the most risky day, and that nearly half of all accidents occur within 5 km of home. Other interesting data (not on the chart) are that the worst time for crashes is between 3.00 p.m. and 6.00 p.m. on weekdays, that men are twice as likely to cause an accident as women, that men aged 24 years and younger are responsible for 20% of all accidents, and that the safest drivers are males aged between 35 and 54 years.

In a case where the sectors are large enough to do so, the percentage is often written inside the sector. Otherwise it is written outside the chart. Note here that the artist has separated the sectors to provide a better contrast. When multiple pie charts are drawn either for comparison or in the same diagram, they should be of the same size.

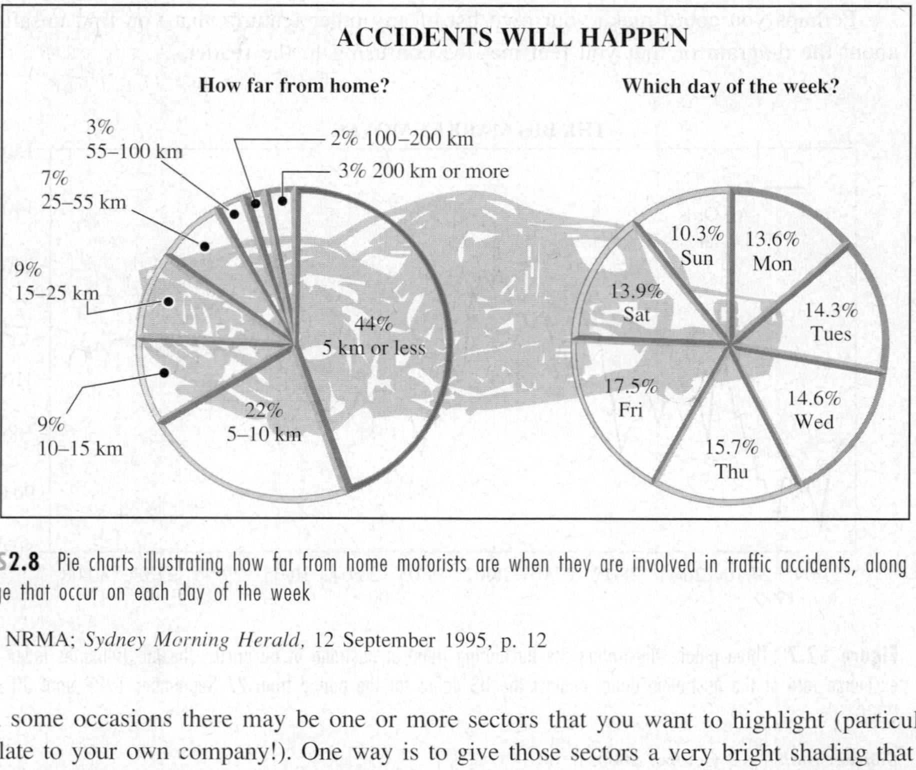

Figure S2.8 Pie charts illustrating how far from home motorists are when they are involved in traffic accidents, along with the percentage that occur on each day of the week

***Source*:** NRMA; *Sydney Morning Herald*, 12 September 1995, p. 12

On some occasions there may be one or more sectors that you want to highlight (particularly if they relate to your own company!). One way is to give those sectors a very bright shading that stands out. Another method is to use an *exploded pie chart* in which those sectors to be highlighted are taken slightly out of (or broken away from) the rest of the pie.

Example S2.9

The two pie charts in Figure S2.9 are both exploded to make the relative size of each sector easier to distinguish. In this case, no piece of either pie chart is highlighted in particular, and the effect is made more interesting by using a $1 Australian coin for the background of each chart. The idea behind the visual display is to explain where the Australian government gets its money from and how it spends it.

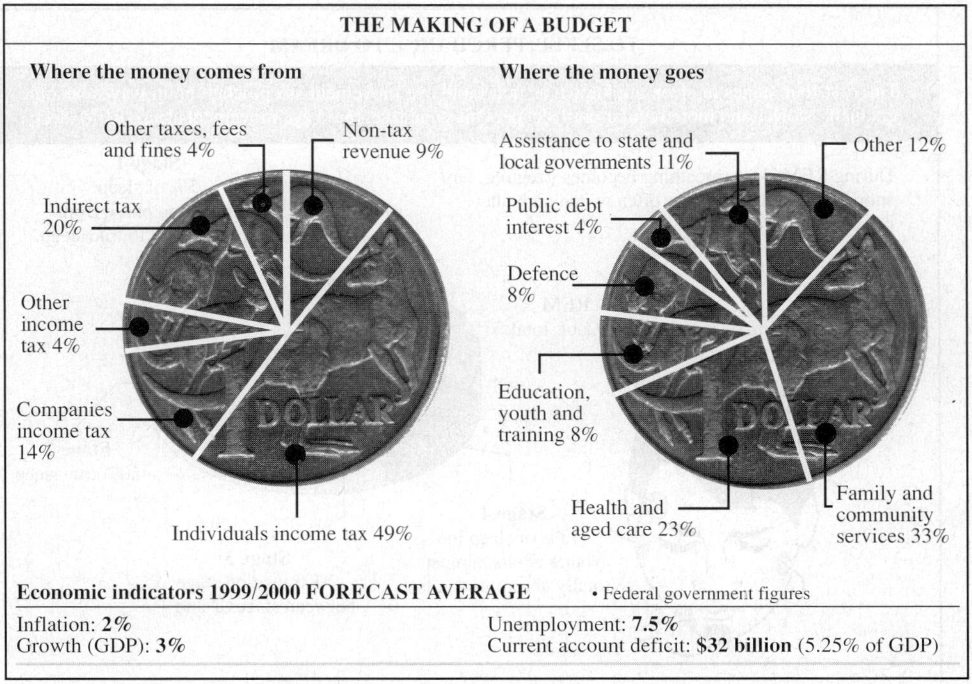

Figure S2.9 Two pie charts that provide an interesting display detailing how the Australian government obtains and spends its money

Source: *Daily Telegraph*

Example S2.10

An interesting pie chart is shown in Figure S2.10 on page S.30, which displays a three-dimensional diagram for a dramatic effect. In some cases a pie chart can be made even more effective by the addition of a visual display. This pie chart shows the picture of a face, since the data relate to the breakdown of the five stages of sleep, including REM (rapid eye movement).

S2.5 BAR CHARTS

A **bar chart** consists of a series of rectangular bars where the lengths of the bars represent the magnitudes of the respective quantities. There are a number of different types of bar charts, which may be used according to the type of data to be displayed.

Follow these guidelines when preparing a bar chart, if possible:

1. Because it is the *lengths* of the bars that are being compared, make all bars the same *width*.
2. If the bar chart is such that an ordering is feasible or desirable, use one of the most common orders, i.e. one in which the bar lengths either increase or decrease as the eye moves down the chart.
3. Clearly label the axes.

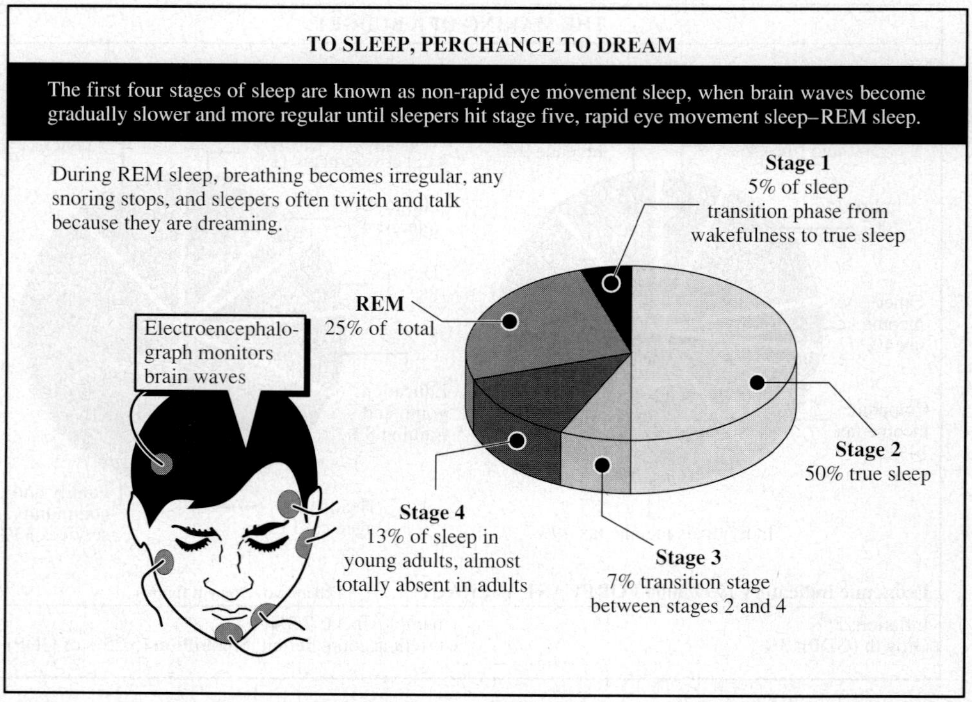

Figure S2.10 A pie chart illustrating the breakdown of the five stages of sleep including REM (rapid eye movement)

Source: *Sydney Morning Herald*

4. Include keys to assist in the interpretation if necessary, e.g. indications of how colours or particular shading patterns refer to the data.
5. Include footnotes, tables and sources of data.

Simple bar charts

A *simple bar chart* is one in which the bars represent *one* quantity or variable only. The length of the bar indicates the number of people or items in that category, and sometimes other information is added to make the bar chart more eye-catching.

Figure S2.11 shows the 10 movies that have had the highest takings in their first week of release in Australian cinemas. Note that instead of a scaled axis, the artist has chosen to write the dollar figure at the end of the bar. Also shown is the number of separate prints made of the movie. To attract the reader, a picture of the Austin Powers character is used with the data.

For movie fans, *The Matrix* just missed out being included in the Top 10, with an opening week of $6.1 million. For trivia fans, the original *Austin Powers* movie took only $3.5 million in its entire cinema run, but its popularity on video built a cult following that enabled its sequel to grab the number three position.

Figure S2.11 A simple bar chart illustrating the 10 movies that have had the highest takings in their first week of release in Australian cinemas

Source: *Sydney Morning Herald*. Photograph courtesy of New Line Cinema and Roadshow Distributions Pty Ltd

Example S2.11

An interesting collection of three simple bar charts is shown in Figure S2.12. The diagram is attempting to illustrate the increase in the cost of fires, vandalism and theft in schools in New South Wales between 1997–1998 and 1999–2000. Once again there is no scale, but the height of each bar is written at the top. Different shading has been used to separate the bars from each other. To add spice to the diagram a picture of a fireman, presumably turning a hose on a burning school, is added on the right-hand-side.

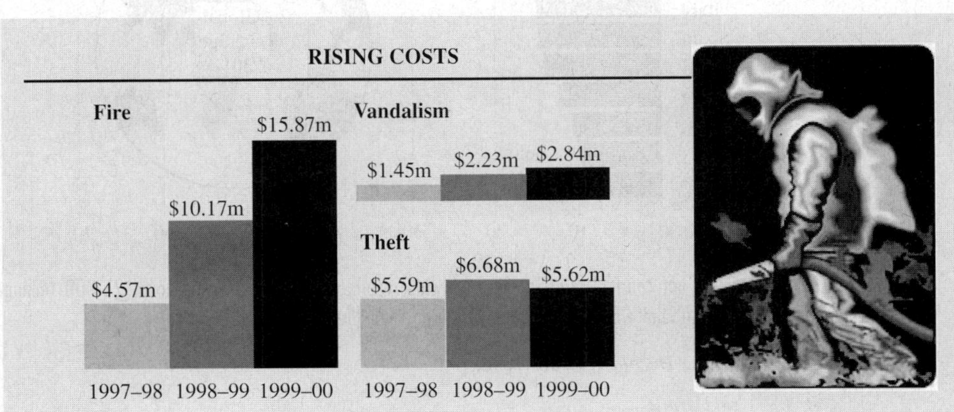

Figure S2.12 Three simple bar charts illustrating the increase the rise in the cost of fires, vandalism and theft in NSW schools between 1997–1998 and 1999–2000

Source: *Sydney Morning Herald*

Example S2.12

Figure S2.13 displays three different sets of simple bar charts, which all use the same scale. The data are from a New York consulting company called the BrainWaves Group, which undertook a survey of more than 25 000 teenagers (including 420 Australians), aged 15 to 18, at schools in 41 countries. Although a number of questions were asked, the bar chart illustrates the percentage of 'yes' responses to the question 'Do you think that the world will improve during your lifetime?' Overall, only 30% said that they thought it would.

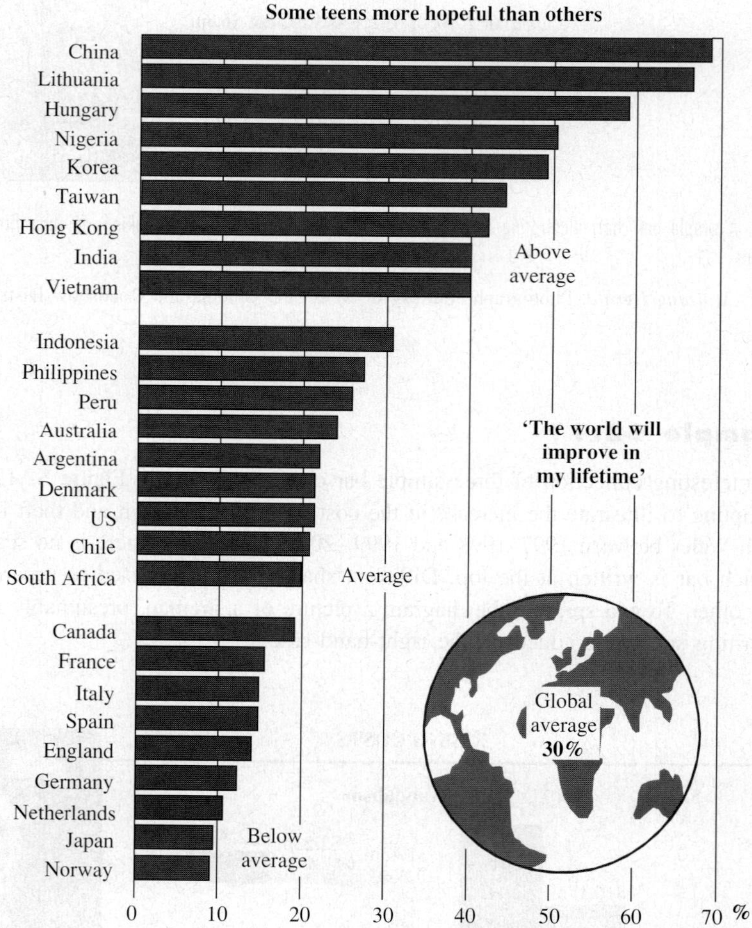

Figure S2.13 A simple bar chart illustrating the worldwide results of a survey in which over 25 000 teenagers were asked whether they felt that the world would improve during their lifetime

***Source*:** *Sydney Morning Herald,* 13 July 1996, p. 3

Multiple bar charts

A *multiple bar chart* is one in which the bars are displayed side by side, often in pairs or triples, in order to emphasise comparisons. This is usually preferable to drawing two separate bar charts of the quantities where immediate comparisons would be made difficult. There is usually no space between the bars for data in the same category.

Example S2.13

In the year 2000 Australia was one of the top five nations in the world online, with 46% of its adult population being connected to the Internet. However, there was a big difference between those residing in capital cities, where about one in three had home Internet access, and those living elsewhere, where less than one in five had home Internet access. This state of affairs is highlighted using a type of multiple bar chart shown in Figure S2.14. Note that the artist could have placed the bars for the same year next to each other and used different shadings for the category of capital city or otherwise. Do you think this would have provided a more effective and informative presentation?

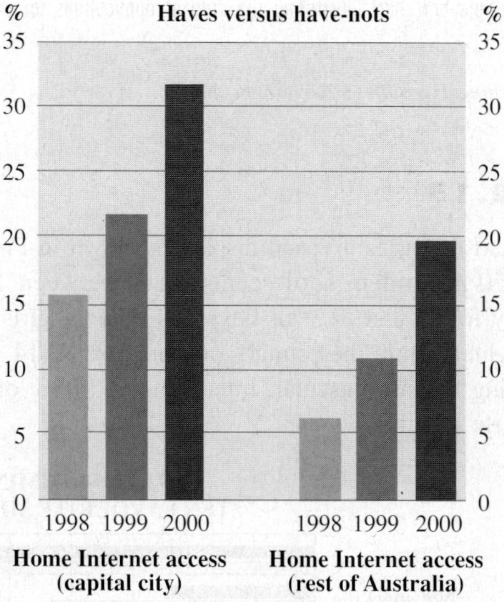

Figure S2.14 A type of multiple bar chart illustrating the trend in home Internet access in Australia and highlighting the difference between those who live in capital cities and those who live elsewhere

Source: *Bulletin*, 5 September 2000

Example S2.14

Figures compiled by the University Admissions Centre in New South Wales revealed a marked decrease in the number of people applying for admission to university between 1997 and 2001. The figures coincided with the introduction of higher fees and a lower repayment threshold. The multiple bar chart in Figure S2.15 has three bars for each of the five years in question. In this case, not only are there bars separating the Year 12 school leavers from non-year 12 applicants, but the total of both groups is shown as well.

Perhaps the figures could have been better presented as a *compound bar chart* (see next section).

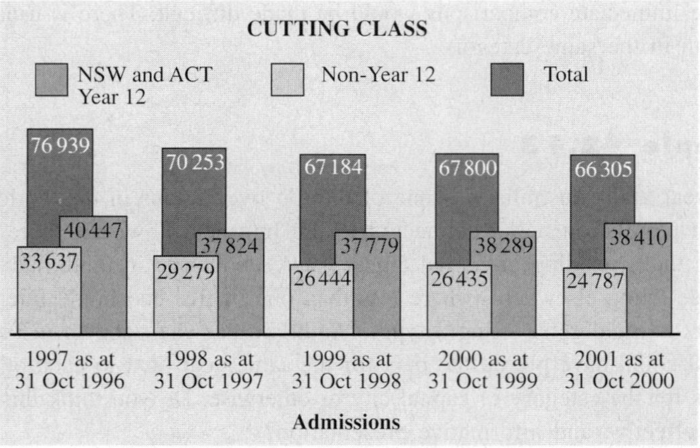

Figure S2.15 A multiple bar chart illustrating the drop of applications for university places in NSW and ACT between 1997 and 2001

Source: *Sydney Morning Herald*, 13 November 2000

Example S2.15

A multiple bar chart that uses a visual display is shown in Figure S2.16 to enhance the results of a survey of 470 Australian adolescents (aged between 11 and 14 years) on their body image. The survey found that 31% of boys and 53% of girls wanted their body weight to be a little or a lot lighter, while the primary concern for 11–14 year old boys was not being fat, followed by looking fit and muscular. Interestingly, 30.5% of boys want to be heavier, while only 14.6% of girls felt this way.

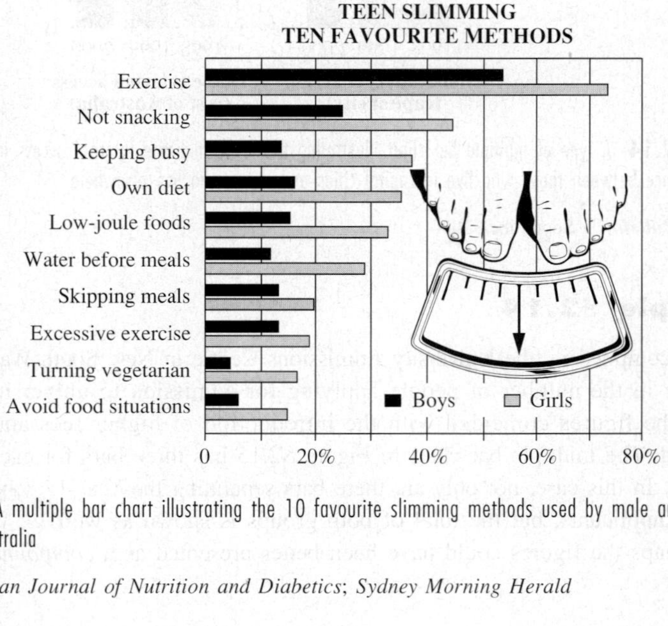

Figure S2.16 A multiple bar chart illustrating the 10 favourite slimming methods used by male and female adolescents in Australia

Source: *Australian Journal of Nutrition and Diabetics*; *Sydney Morning Herald*

The bar chart itself displays the 10 favourite slimming methods used by the males and females in the survey, accompanied by an appropriate drawing of a pair of feet on scales.

Compound bar charts

A *compound bar chart* is one in which some of the bars are stacked on top of each other. This type of bar chart is often used when various totals are also to be compared.

Example S2.16

A *Bulletin-Morgan* poll taken three months after the introduction of the Goods and Services Tax (GST) in Australia asked 917 electors whether they thought Australian families were better off as a result. This survey, conducted face-to-face, gave respondents four choices, as follows:

- better off
- worse off
- neither better off nor worse off
- cannot say

The results of this type of survey are often expressed in terms of percentages that will naturally add up to 100% and a compound bar chart is a useful way of highlighting the proportion of respondents who answer in each category. Figure S2.17 shows the result of this survey, highlighting the difference between Labor and Liberal voters as well as the differences between the states. The first bar shows the breakdown of responses overall.

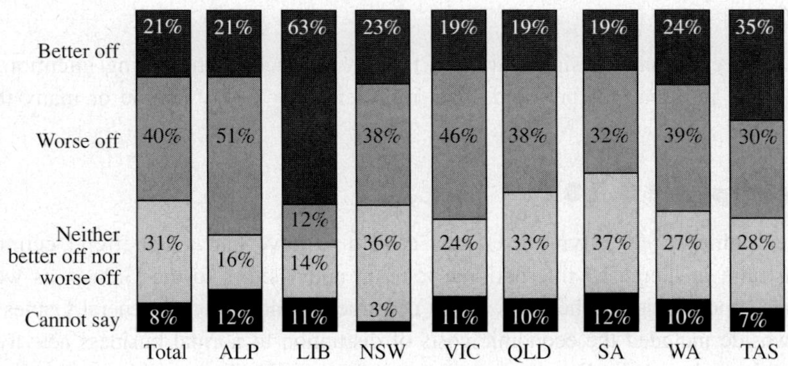

Figure S2.17 The results of a survey in which Australian electors were asked whether they felt families are better or worse off with the GST

Source: *Bulletin*, 10 October 2000

Example S2.17

Another *Bulletin-Morgan* poll taken during 1999 and 2000 asked a large sample of 23 175 voters the question 'Are threats to the environment exaggerated?'. According to the key at the top of the diagram, the responses are displayed in the compound bar chart in Figure S2.18. Note that the bars represent the voting intentions of the respondents, so that it can easily be seen, for example, that there are marked differences between the responses of National Party voters and Greens voters.

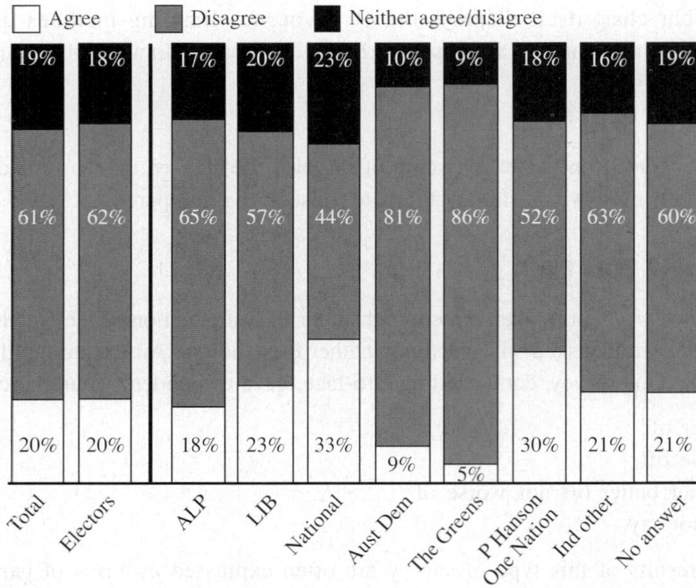

Figure S2.18 The results of a survey in which Australian electors were asked whether they felt that threats to the environment are exaggerated

Source: *Bulletin*, 13 June 2000

However, one missing feature is the *size* of each of the voting intention groups, so that we have no idea whether some bars represent only a few people or many thousands.

Example S2.18

The Sydney 2000 Olympic Games is said to have had a beneficial economic effect for Australia in increased international tourism and visitors to the Games, as well as providing work opportunities in the areas of the construction industry and general Games operations. The downside included the economic costs of disruption to normal business activity in Sydney and elsewhere during the Games, including retailers outside Sydney who suffered significant losses of customers. The compound bar chart in Figure S2.19, based on information from the NSW Treasury, shows the expected economic benefit from staging the Games.

Example S2.19

On only 11 occasions in over 600 cricket Tests until the end of the 2000/01 season has an Australian batsman 'carried his bat' through an entire innings; that is, opened the batting and still been not out at the end of the innings while all other batsmen were either dismissed, absent or retired hurt. (Two players, W. Lawry and W. Woodfull, managed to accomplish this twice each.)

The percentage of the total team runs scored by each such batsman is displayed in the compound bar chart in Figure S2.20, where the height of each bar is always 100%. The bars are arranged in order according to the percentage of runs scored by the opening batsman. At the top of each bar is the total runs scored by the team, while at the bottom is the name of the batsman and the year he performed the feat.

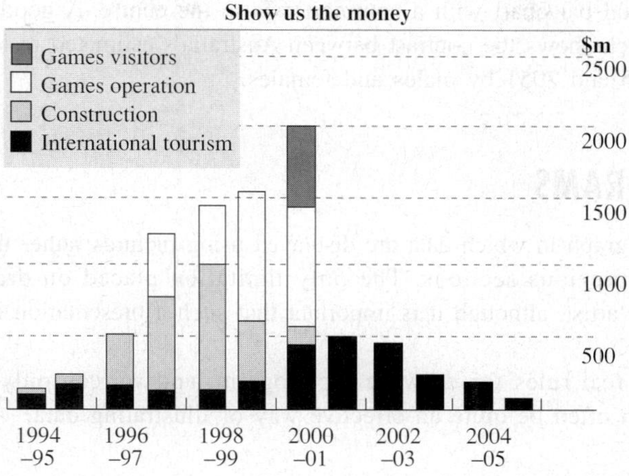

Figure S2.19 A compound bar chart illustrating the amount of money the Sydney 2000 Olympic Games were expected to bring into Australia between 1994–95 and 2004–05

Source: *Sydney Morning Herald*, 26 August 2000

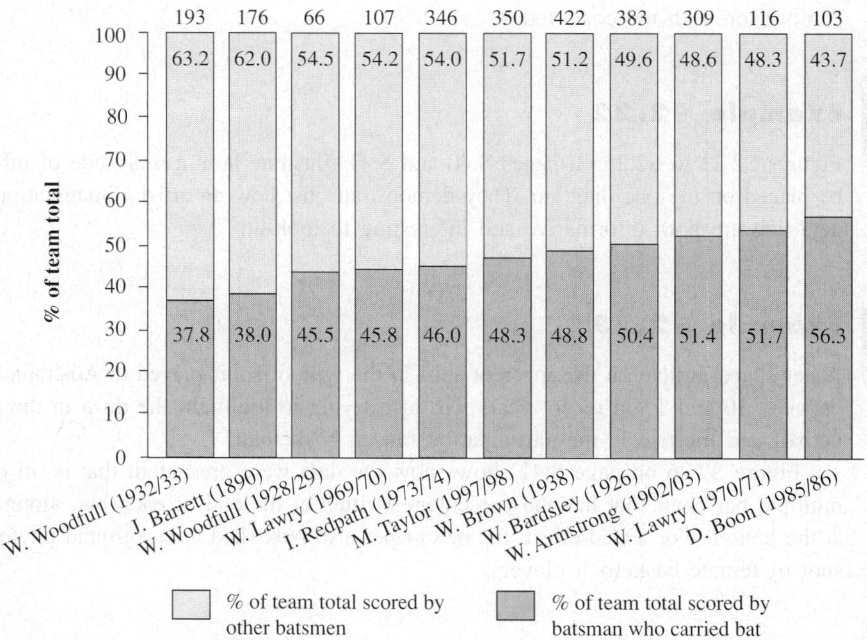

Figure S2.20 A compound bar chart illustrating the percentage of team total runs scored by the ten Australian batsmen who have carried their bat for Australia in Test cricket. The team total is displayed on the top of each bar

Source: *Test Cricket Lists*, compiled by Graham Dawson with statistics by Charlie Watt, 3rd Edition, The Five Mile Press Pty Ltd, Victoria

Example S2.20

A special type of compound bar chart used for comparison purposes is the *population pyramid*, in which each of a set of bar charts share the same axis. In this sense it is a kind of horizontal bar chart with a common axis in the centre. A good example of this is Figure S2.21, which shows the contrast between Australia's estimated resident population at 30 June 1911, 1995 and 2051 by males and females.

S2.6 PICTOGRAMS

A **pictogram** is a graph in which data are displayed using pictures rather than the traditional methods discussed in the previous sections. The only limitation placed on drawing a pictogram is the imagination of the artist, although it is important that such a presentation is not given in such a way as to be deceptive.

There are no real rules for drawing a pictogram, and we can only demonstrate a few here. However, they can often be quite an effective way of illustrating data.

Example S2.21

Figure S2.22 on page S.40 uses a sketch of a vehicle, mechanics and a petrol pump to illustrate the cost of running a new car in the late 1980s. It is interesting to make a comparison with the costs today!

Example S2.22

Figures S2.23 to S2.25 on pages S.40 and S.41 illustrate how a multitude of information can be placed on the one diagram. They demonstrate just how an artist's imagination can present data that are both informative and interesting to look at.

Example S2.23

A newspaper article on the apparent shift in the type of sport played in Australia by girls aged between 10 and 17 in recent years was also trying to highlight the drop in the popularity of netball and the rise in the participation rate in basketball.

Figure S2.26 on page S.42 shows how the data were presented; that is, in the form of a multiple bar chart that has the percentage written at the end of each bar, along with a scale at the bottom. For added effect, the newspaper also published a background photo of an action shot of female basketball players.

Example S2.24

Figure S2.27 on page S.43 uses pictures of pills to show the annual amount spent on the top 10 prescription drugs in Australia. Note that there is no scale, but the total amount spent is shown on the capsule itself, while the average price is shown in a separate column at the right.

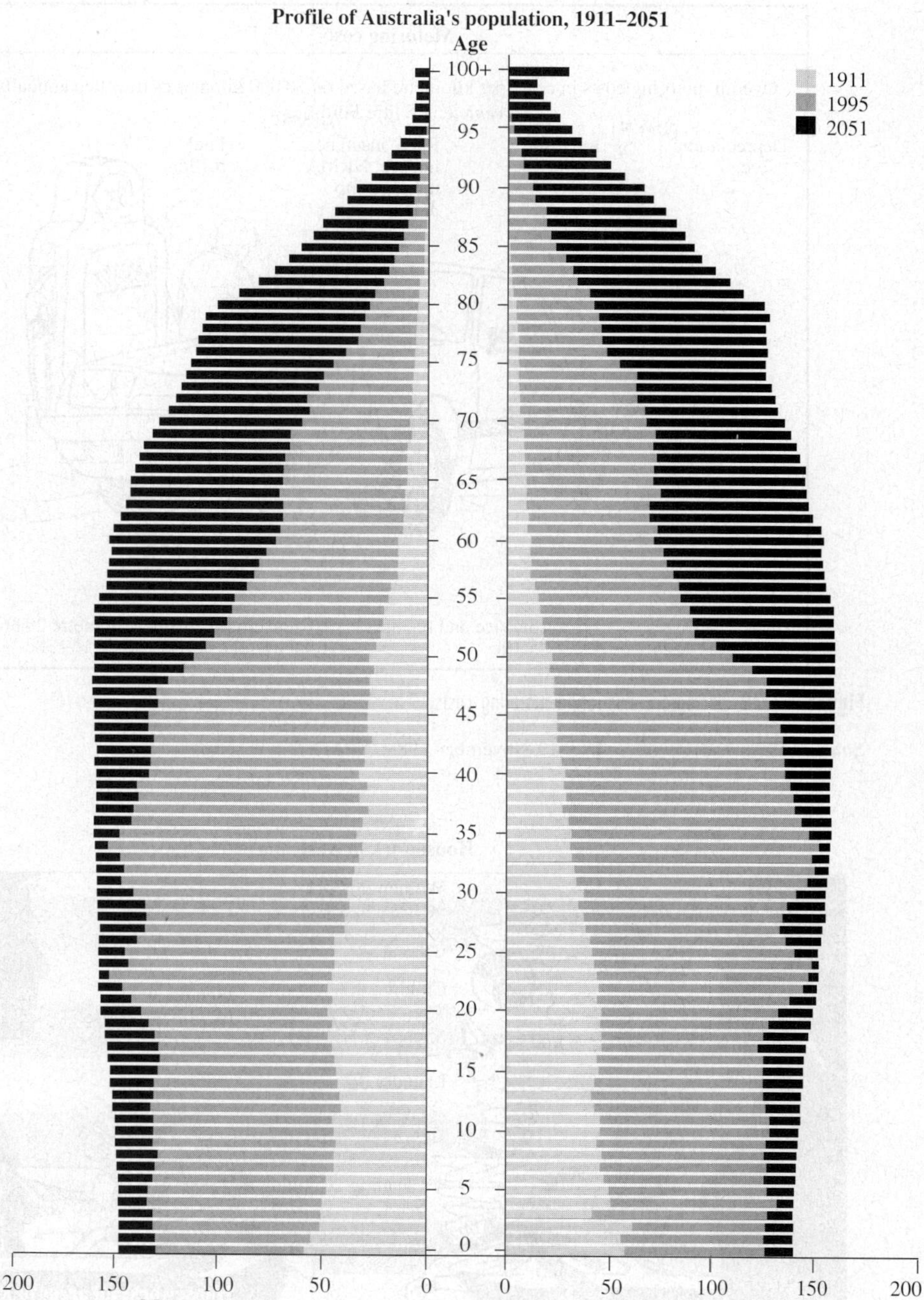

Figure S2.21 Population pyramid bar chart illustrating the estimated resident population of Australia, 30 June 1911, 1995 and 2051

Source: Australian Bureau of Statistics, *Profile of Australia's Population, 1911–2051 by Sex and Age*, Cat. No. 3201.0

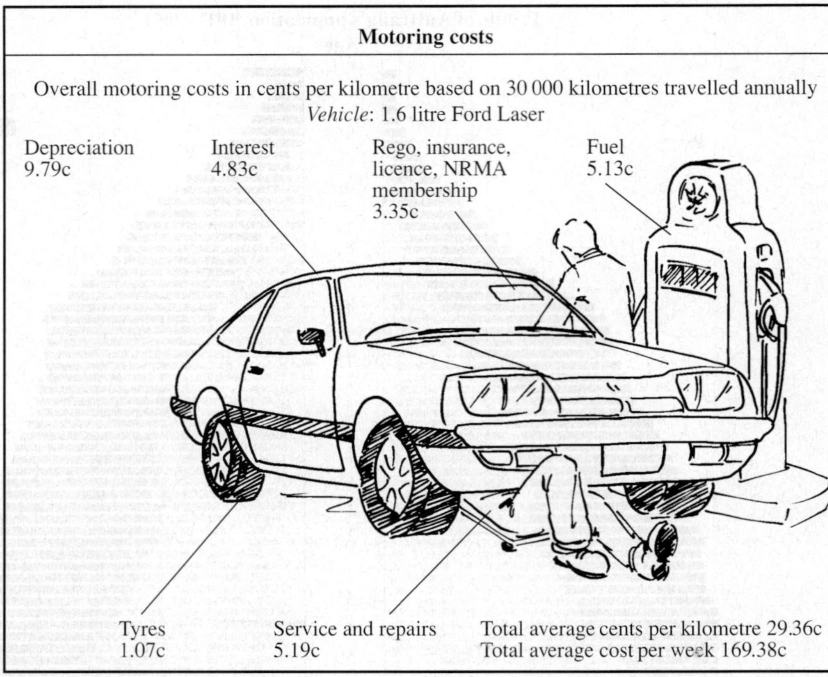

Figure S2.22 Pictogram illustrating motoring costs

Source: *Sydney Morning Herald,* 19 November 1988; NRMA

Housework in Australia

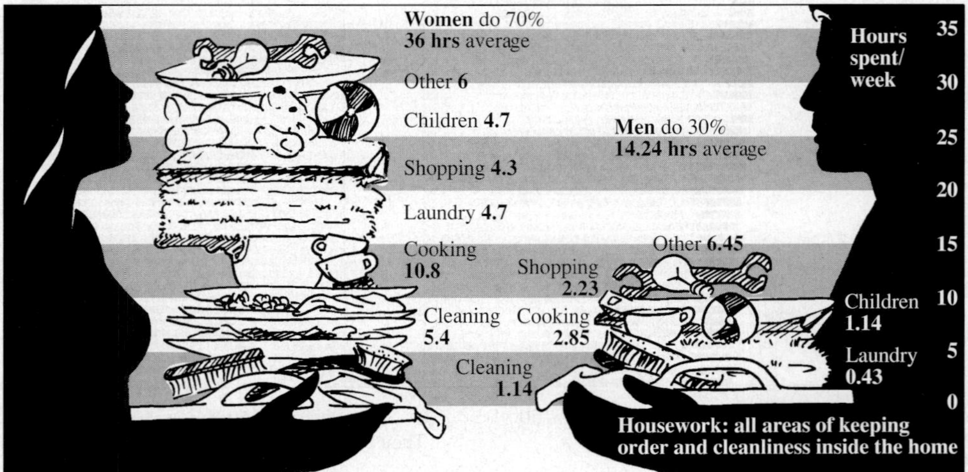

Figure S2.23 Pictogram showing how much information can be placed on one diagram

Source: *Sydney Morning Herald,* 11 May 1992, p. 24; *Juggling Time: How Australian Families Use Time* (Bittman)

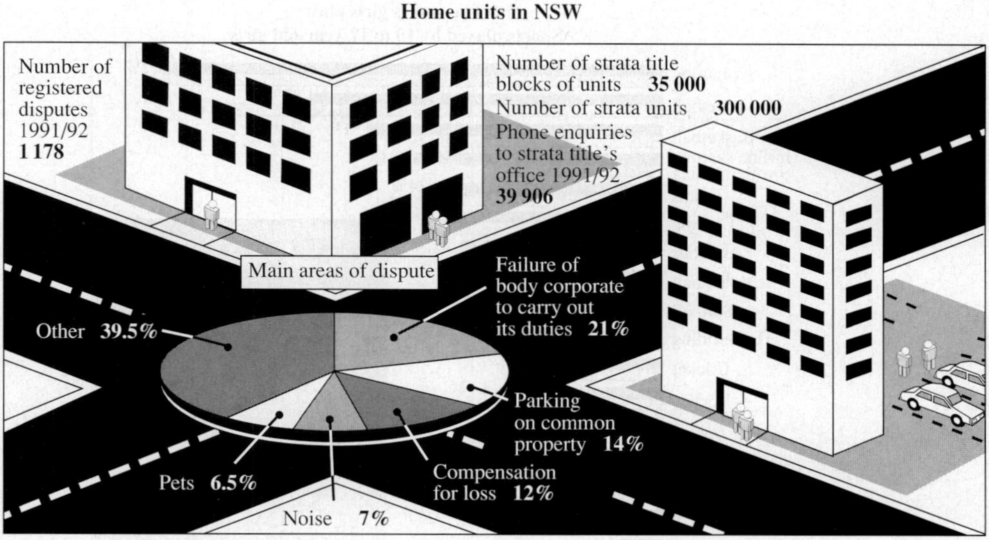

Figure S2.24 Another pictogram, giving detailed information about home units in New South Wales

Source: *Sydney Morning Herald*, 17 August 1992, p. 28; Strata and Tenancy Commissioner's Office

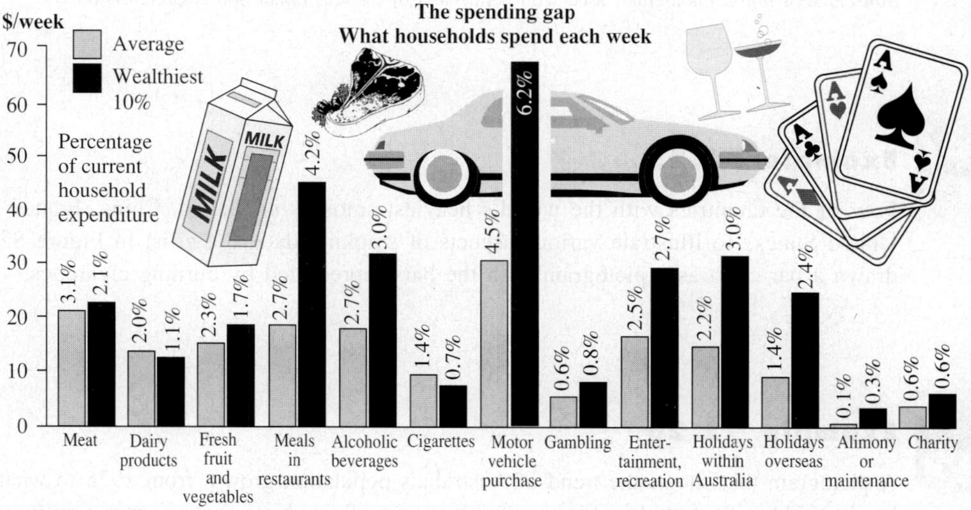

Figure S2.25 A multiple bar chart with drawings that highlight the differences in spending patterns of average Australians and the wealthiest 10%

Source: National Centre for Social and Economic Modelling, University of Canberra; *Australian*

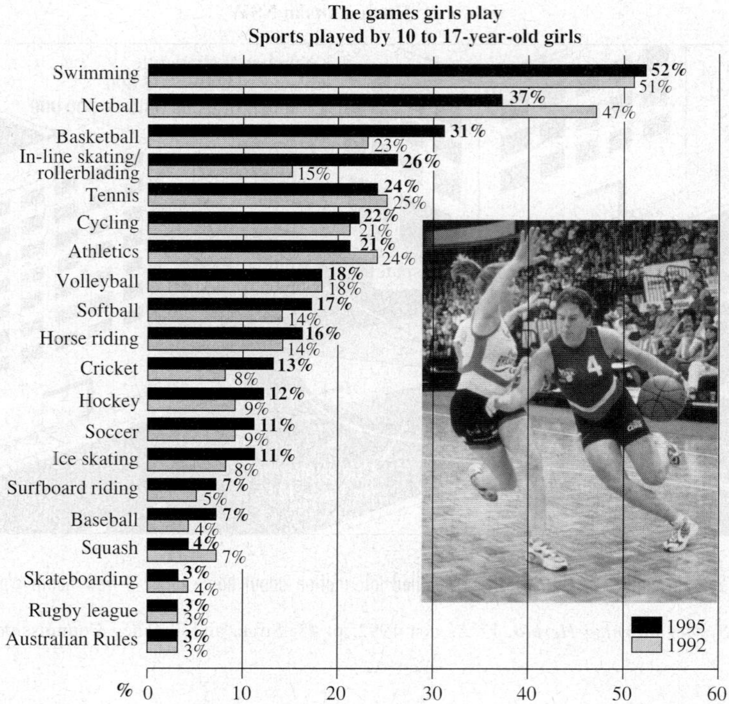

Figure S2.26 A multiple bar chart with a background picture of basketball players, which highlights the changing preferences of sport played in Australia by girls aged between 10 and 17

Source: *Australian*. Photograph used with permission of the Australian Sports Commission

Example S2.25

Four of the countries with the world's heaviest smokers are Japan, China, France and the United States. To illustrate various aspects of smoking data, the artist in Figure S2.28 has drawn a bar chart as a pictogram with the bars represented by burning cigarettes.

Example S2.26

A pictogram that shows the trend of Australia's population growth from 1975 to what it may be in 2051 is illustrated in Figure S2.29 on page S.44. Note that the curve splits into two parts after the year 2000: one based on a low migration intake and one based on a high migration intake. For added effect, the artist has drawn the graph on a map of Australia, which is on an angle facing away from the reader. Consequently, the scale lines showing the percentages are not horizontal but slope upwards.

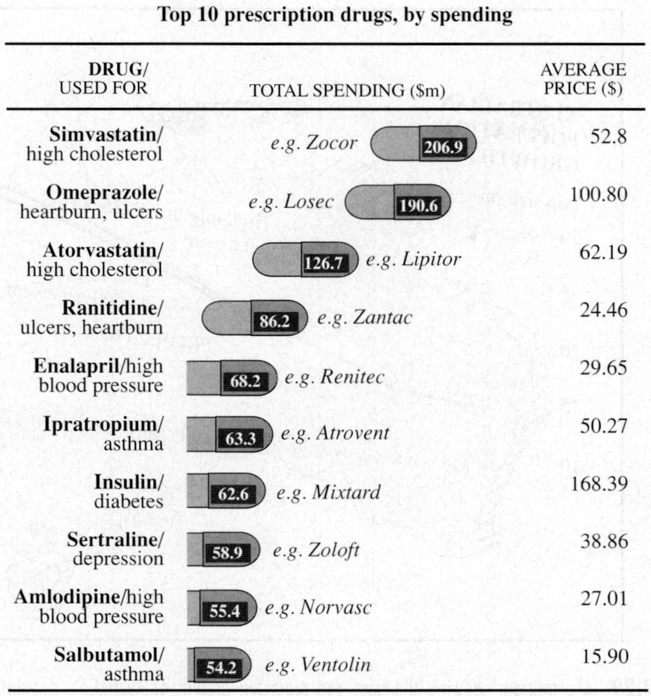

Figure S2.27 A pictogram illustrating the annual amount spent on the top 10 prescription drugs in Australia along with their average price

Source: *Sydney Morning Herald*, 7 September 1999

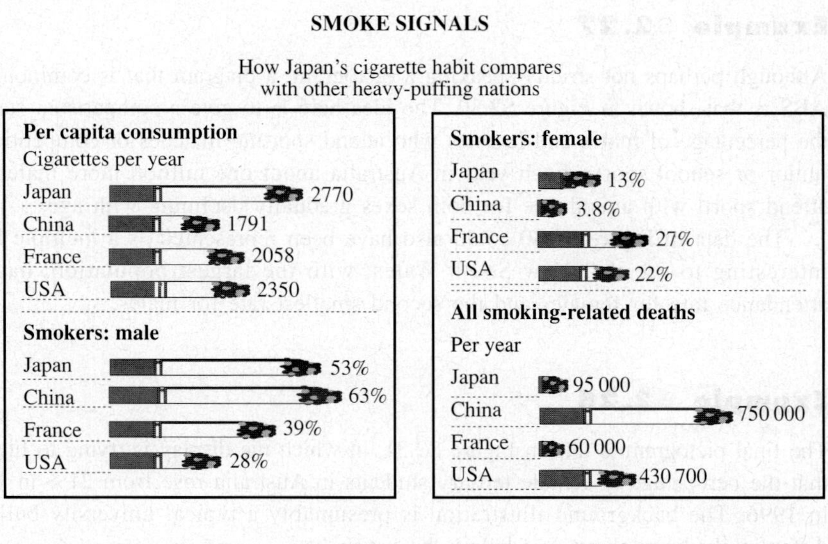

Figure S2.28 A pictogram comparing the smoking habits of people in Japan, China, France and the USA

Source: American Cancer Society; *British Medical Journal*; *Bulletin*

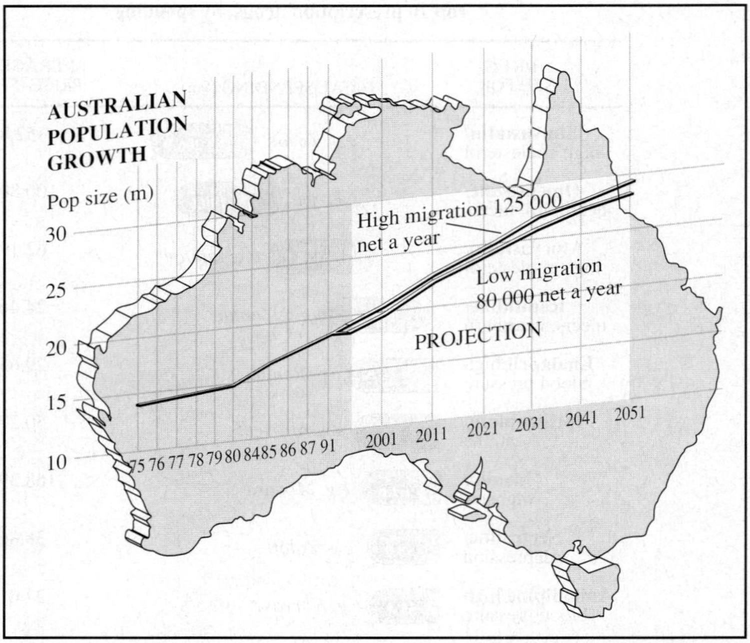

Figure S2.29 A pictogram showing the past and predicted growth of Australia's population based on two different levels of migration intakes

Source: Australian Bureau of Statistics, *Australian Demographic Statistics*, Cat. No. 3101.0; projections by EPAC (Economic Planning Advisory Commission); *Sun-Herald*

Example S2.27

Although perhaps not strictly speaking a pictogram, a diagram that is commonly used by the ABS is that shown in Figure S2.30. The idea here is to give a comparison, state by state, of the percentage of males and females who attend sporting matches or competitions (excluding junior or school sport). Each year in Australia about one million more males than females attend sport, with attendance for both sexes gradually declining with age.

The data in Figure S2.30 could also have been represented as a multiple bar chart. It is interesting to see that New South Wales, with the largest population, has the smallest attendance rate for females and the second smallest rate for males.

Example S2.28

The final pictogram is that in Figure S2.31, in which the display is trying to highlight the fact that the percentage of female tertiary students in Australia rose from 21% in 1949 to 54.3% in 1996. The background illustration is presumably a typical university building, and has different shadings above and below the graph.

This diagram has several poor features, including the fact that there is no real horizontal axis, only the words '1949–1996'. This would make it impossible, for example, to determine

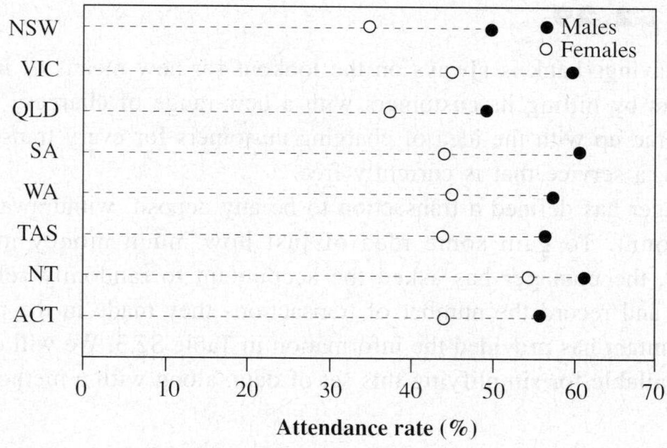

Figure S2.30 A diagram denoting sport attendance in Australia by state or territory of usual residence and sex

***Source*:** Australian Bureau of Statistics, *Sports Attendance, Australia*, Cat. No. 4174.0

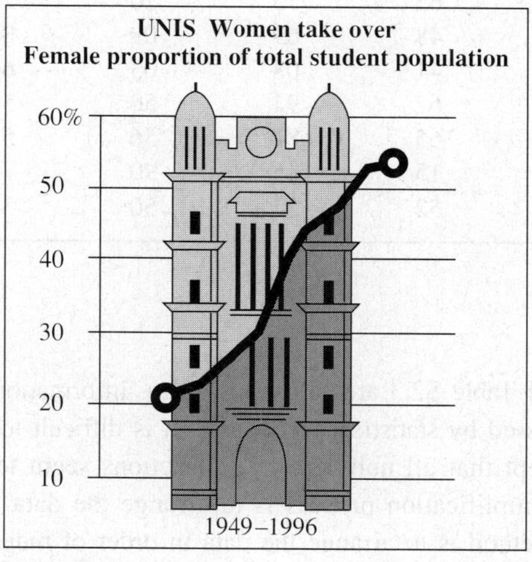

Figure S2.31 A pictogram illustrating the fact that the percentage of female tertiary students in Australia rose from 21% in 1949 to 54.3% in 1996

***Source*:** *Sydney Morning Herald,* 10 December 1996, p. 5

what the percentage was in, say, 1970. Also, note that zero does not appear on the vertical axis. If it did, it would be right down the bottom below the level where the years are written.

S2.7 FREQUENCY DISTRIBUTIONS

Often in practice, data are collected and are initially presented with the *observations* (i.e. sample values) in some random order. In this section, using the data in Example S2.29, various techniques are provided for condensing data into a comprehensible form.

Example S2.29

A particular savings bank is always on the lookout for new means of increasing the profits to shareholders by hitting its customers with a new range of charges. The manager of one branch has come up with the idea of charging customers for every transaction they make on their accounts, a service that is currently free.

The manager has defined a transaction to be any deposit, withdrawal or account enquiry on their account. To gain some idea of just how much money might be gained by this initiative, the manager has asked the accountant to randomly select the accounts of 80 customers and record the number of transactions they made in the previous 60 days.

The accountant has provided the information in Table S2.3. We will consider some of the techniques available for simplifying this set of data, along with a method of determining its shape.

Table S2.3 Number of transactions over a 60-day period for 80 accounts at a savings bank

05	38	20	62	65	60	84	05
01	46	40	92	41	68	25	18
12	58	43	20	62	84	81	35
32	43	65	75	20	26	32	50
50	20	48	69	09	03	37	55
56	47	41	08	03	60	50	43
05	35	62	92	50	50	12	40
52	10	65	81	56	58	48	20
29	40	15	25	90	70	70	38
70	75	52	34	50	35	84	80

Array

Observations in the form of Table S2.3 are called *raw data*. Information before it is analysed is called 'raw' because it is unprocessed by statistical techniques. It is difficult to obtain much information from the data in this form, except that all numbers of transactions seem to be under 100.

The first step in the simplification process is to arrange the data as an array, i.e. to arrange the data in some order. One method is to arrange the data in order of magnitude from the smallest to the largest. This is done in Table S2.4.

Table S2.4 Observations in Table S2.3 arranged in an array

01	12	26	38	47	52	62	75
03	15	29	40	48	55	65	80
03	18	32	40	48	56	65	81
05	20	32	40	50	56	65	81
05	20	34	41	50	58	68	84
05	20	35	41	50	58	69	84
08	20	35	43	50	60	70	84
09	20	35	43	50	60	70	90
10	25	37	43	50	62	70	92
12	25	38	46	52	62	75	92

Frequency distribution

The frequency of an observation is the number of times that observation occurs. Series of statistical observations are sometimes said to have a certain *distribution*. A *frequency distribution* gives a listing of different observations (in order of magnitude), each with the corresponding frequency alongside it. This is done in Table S2.5.

Table S2.5 Observations in Table S2.4 arranged in a frequency distribution

No. of transactions	01	03	05	08	09	10	12	15	18	20	25
Frequency	1	2	3	1	1	1	2	1	1	5	2
No. of transactions	26	29	32	34	35	37	38	40	41	43	46
Frequency	1	1	2	1	3	1	2	3	2	3	1
No. of transactions	47	48	50	52	55	56	58	60	62	65	68
Frequency	1	2	6	2	1	2	2	2	3	3	1
No. of transactions	69	70	75	80	81	84	90	92			
Frequency	1	3	2	1	2	3	1	2		*Total* 80	

When constructing a frequency table, always check that the total frequency equals the number of observations in the raw data, e.g. 80 in the case above.

Grouped frequency distribution

The data may be further reduced by grouping the observations into **class intervals** to form a **grouped frequency distribution**. A class interval is a range of values in which some of the observations may lie. The grouping of data is often desirable because:

1. it reduces the apparent complexity of the data by reducing the number of separate pieces of information
2. it helps to smooth out irregularities in the distribution

A disadvantage of grouping data is that some information will be lost. For example, if we know that there are six observations in an interval labelled 15–20, we cannot say whether they are all at one end of the interval or spread throughout it.

In constructing a grouped frequency distribution, there are several rules that should be followed:

1. The class intervals should be mutually exclusive (i.e. non-overlapping). They should be unambiguously defined so that one observation cannot belong to more than one interval. (In particular, the boundary points between intervals should be precisely specified.)
2. In most circumstances, the intervals should be of the same width. In some cases, such as when there are outlying observations, it may be convenient to have the first and/or the last class interval open-ended, e.g. of the type 'below 250' or 'above 500'. If possible, however, open-ended intervals should be avoided, as further analysis of the data is made more difficult.
3. If there are no observations in a particular interval, it should still be included to avoid a misleading impression of the data.

In practice, it is not unusual to construct a grouped frequency distribution directly from the raw data. It should always be checked that Σf (*sum of the frequencies*) is equal to the number of observations in the raw data.

The numbers of transactions for the 80 accounts in Table S2.3 have been made into grouped frequency distributions in Table S2.6 (class intervals of length 20 transactions) and Table S2.7 (class intervals of length 40 transactions).

In Table S2.8, the data in Table S2.3 have been grouped such that both the first and last intervals are open-ended.

Table S2.6 Observations in Table S2.3 arranged in a grouped frequency distribution with class intervals of length 20 transactions

Class interval	Frequency
0–under 20	13
20–under 40	18
40–under 60	25
60–under 80	15
80–under 100	9
	Σf = 80

Table S2.7 Observations in Table S2.3 arranged in a grouped frequency distribution with class intervals of length 40 transactions

Class interval	Frequency
0–under 40	31
40–under 80	40
80–under 120	9
	Σf = 80

Table S2.8 Observations in Table S2.3 arranged in a grouped frequency distribution with open-ended class intervals

Class interval	Frequency
Under 40	31
40–under 50	12
50–under 60	13
60–under 70	10
70–under 80	5
80 and over	9
	Σf = 80

Histogram

One method of presenting a graphical picture of a grouped frequency distribution is to construct a **histogram**. A histogram is basically a simple (vertical) bar chart in which the frequency of observations within a class interval is represented by the corresponding bar. In general, the class intervals, and hence the width of the bars, will be of equal size, and therefore either the height or the area of a bar above a class interval is proportional to the frequency in that class. To allow for the case of unequal class lengths, however, it is sometimes desirable to use the *area* of a bar as the indicator of the corresponding frequency.

A histogram for the grouped frequency distribution in Table S2.6 is shown in Figure S2.32.

The *relative frequency* of a class interval is the proportion of the total frequency that lies in that class interval. In a relative frequency histogram, the relative frequency (rather than the frequency itself) is shown on the vertical axis. Figure S2.33 is the relative frequency histogram for the data in Table S2.7.

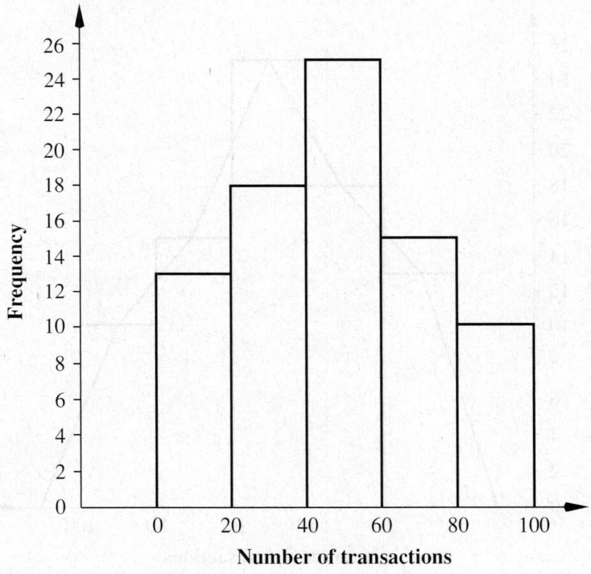

Figure S2.32 Histogram for the data in Table S2.6

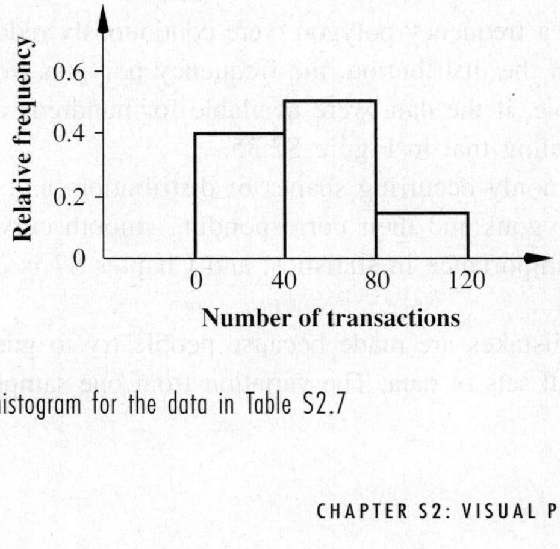

Figure S2.33 Relative frequency histogram for the data in Table S2.7

Frequency polygon

A **frequency polygon** is a line graph of a grouped frequency distribution. It is constructed by marking the point on the top of each histogram bar at the midpoint of the class interval, then joining these points by straight lines. It is also usual to close the polygon at both ends by placing points on the horizontal axis half a class interval to the left of the first class and half a class interval to the right of the last class, and constructing lines from these points to the points on top of the bars that are at the midpoints of the first and last classes.

Example S2.30

An example of a frequency polygon is shown in Figure S2.34. This is the frequency polygon for the histogram in Figure S2.32.

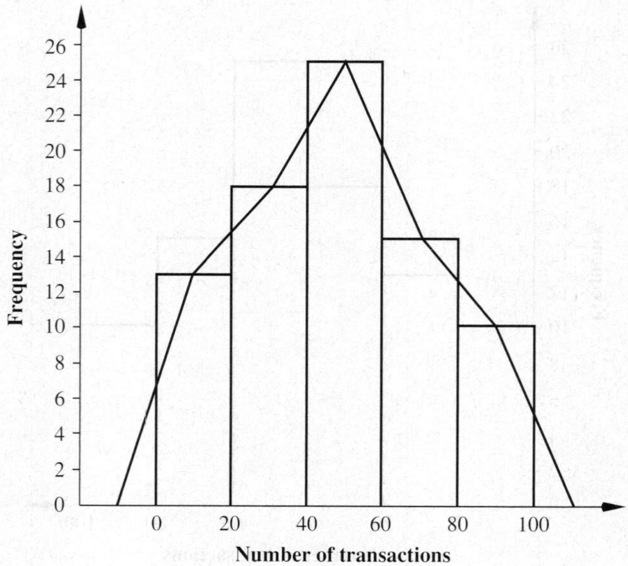

Figure S2.34 Frequency polygon for the histogram in Figure S2.32

If the class intervals of a frequency polygon were continuously reduced in size, and if there were a larger number of items in the distribution, the frequency polygon would more and more resemble a *smooth curve*. For example, if the data were available for hundreds of accounts, the smooth curve might have a shape resembling that in Figure S2.35.

There are several commonly occurring shapes of distribution, and Figures S2.36 to S2.40 show examples of frequency polygons and their corresponding smooth curves. The *normal curve* shown in Figure S2.36 has great importance in statistics, and Chapter S7 is devoted entirely to discussion of it.

Unfortunately, many mistakes are made because people try to guess the shape of distributions merely from observing small sets of data. The variation from one sample to another can be sufficient to cause quite dramatic differences in apparent shapes, and even different choices of class intervals can produce substantially varying impressions. Samples of even 100 observations can still be unreliable in this regard.

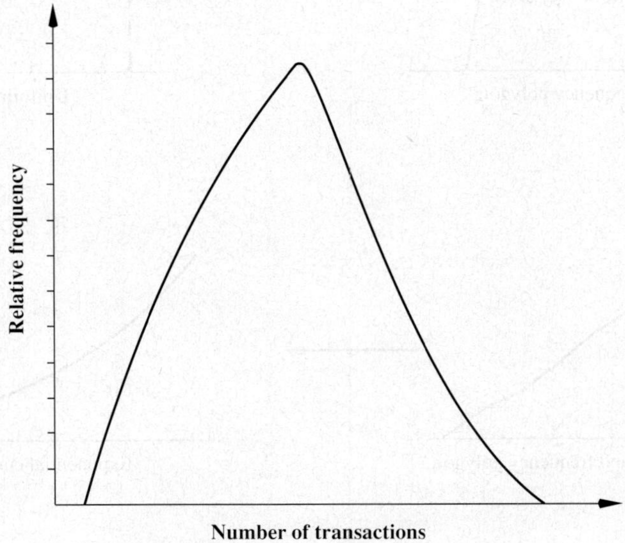

Figure S2.35 Smooth curve corresponding to the frequency polygon in Figure S2.34

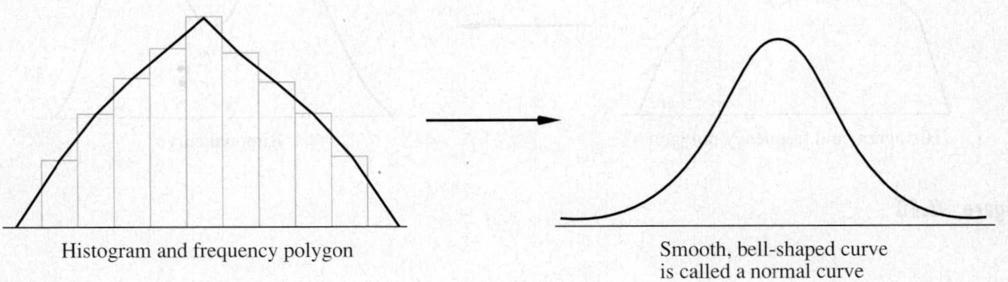

Figure S2.36

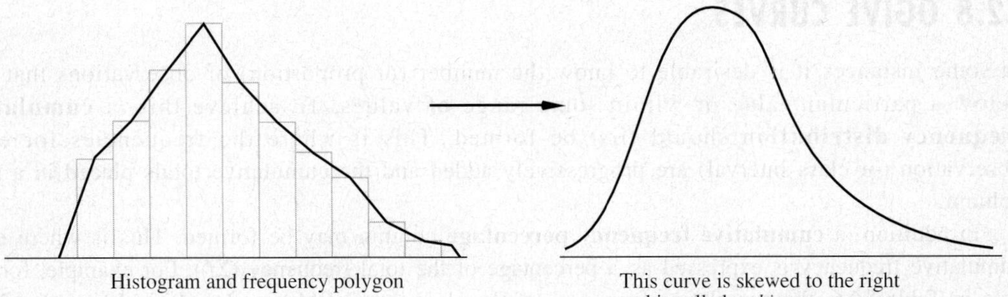

Figure S2.37

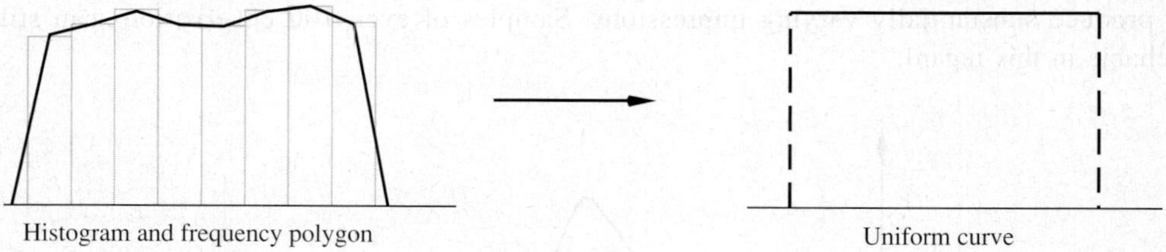

Figure S2.38

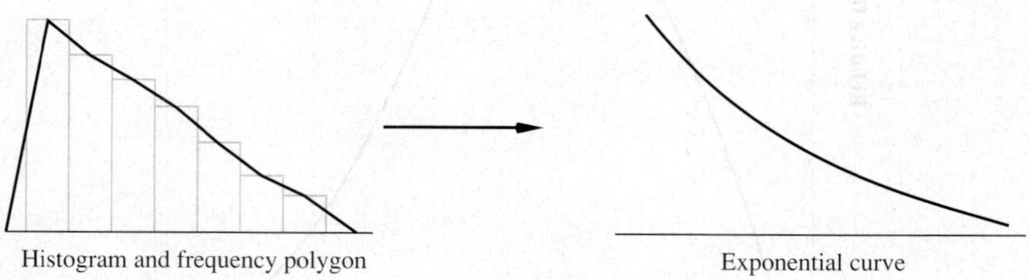

Figure S2.39

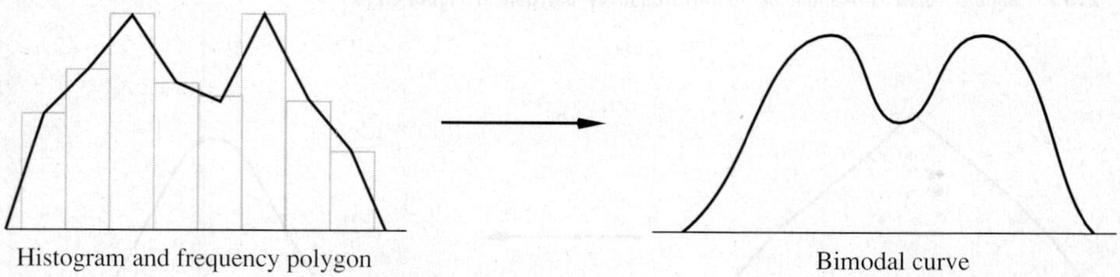

Figure S2.40

S2.8 OGIVE CURVES

In some instances it is desirable to know the number (or proportion) of observations that fall below a particular value or within some range of values. To achieve this, a **cumulative frequency distribution** should first be formed. This is where the frequencies for each observation (or class interval) are progressively added and the cumulative totals placed in a new column.

In addition, a **cumulative frequency percentage** column may be formed. This is where each cumulative frequency is expressed as a percentage of the total frequency (Σf). For example, for the data in Table S2.6, the cumulative frequency of the class interval '40–under 60' is 13 + 18 + 25 = 56. The cumulative frequency percentage is $\frac{56}{80} \times 100 = 70\%$.

Example S2.31

Examples of a cumulative frequency distribution and cumulative frequency percentages are shown in Table S2.9. These are for the data in Table S2.6.

Table S2.9 Cumulative frequency distribution for the data in Table S2.6

Class interval	Frequency	Cumulative frequency	Cumulative frequency percentage
0–under 20	13	13	16.3
20–under 40	18	31	38.8
40–under 60	25	56	70.0
60–under 80	15	71	88.8
80–under 100	9	80	100.0
	$\Sigma f = 80$		

An **ogive curve** is the graphical presentation of a cumulative frequency distribution. In this case, each point on the ogive curve represents the number of accounts that have a number of transactions less than or equal to the corresponding value on the horizontal axis. The cumulative frequency is usually shown on the left vertical axis and the corresponding percentage on the right vertical axis. The ogive curve for the data in Table S2.9 is shown in Figure S2.41.

We might like to know the 'halfway' point in the distribution; that is, such that 50% of the accounts have a number of transactions below that point and the other 50% are above. To estimate this, construct a line from the 50% point on the percentage scale over to where it intersects the ogive curve, then project vertically down to the horizontal axis. This intersection gives the required value. From the dotted lines drawn in Figure S2.41, it appears that the number of transactions required is approximately 46.

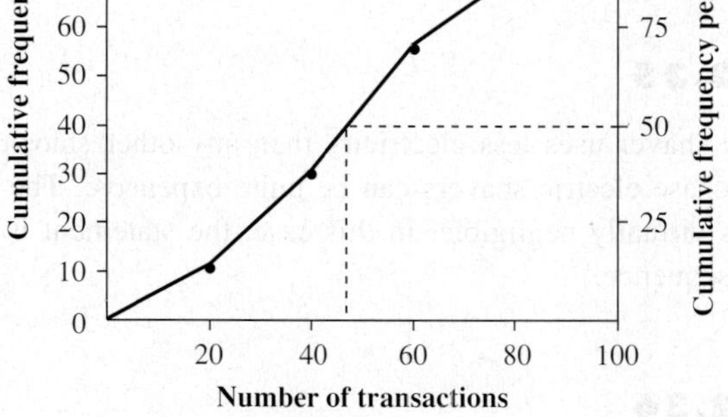

Figure S2.41 Ogive curve for the data in Table S2.9

S2.9 ABUSE OF STATISTICS

There are many instances in practice where statistical information has been presented in such a manner as to purposefully (or otherwise) mislead the reader. This may take the form of a visual display that distorts the true picture, or the quoting of meaningless statistics.

There are two excellent sources of such cases, namely, *How to Lie with Statistics* by Darrell Huff and Irving Geis (Norton C, New York, 1954) and *Use and Abuse of Statistics* by W. J. Reichmann (Methuen & Co. Ltd, London, 1962). Although they were written some years ago, these books give many examples of the misuse of statistics. Here we will demonstrate only a few examples of different types.

Meaningless statements

Consider the following examples.

Example S2.32

'Brand X washing detergent will wash your clothes 50% cleaner and brighter.'

Comment What does '50%' mean? Cleaner and brighter than what? Nothing at all, another product or ....? What is the basis for comparison?

Example S2.33

'Tooth decay can be a very expensive proposition when you visit your dentist. Use Brand Y toothpaste today.'

Comment In this case, a true statement is followed by a statement regarding the product. Although a claim is not explicitly stated, the unwary reader may assume that Brand Y will reduce tooth decay.

Example S2.34

'Coughs and colds clear up 10% faster with Brand Z cough mixture.'

Comment As in Example S2.32, '10%' faster than what? In any case, it is not always clear when a cough or cold 'clears up'. It could simply mean that Brand Z clears them up in 9 days instead of 10.

Example S2.35

'Brand X electric shaver uses less electricity than any other shaver on the market today.'

Comment To purchase electric shavers can be quite expensive. The cost of electricity to run them, however, is virtually negligible. In this case, the statement is highlighting an attribute that is of no consequence.

Example S2.36

'60% of housewives say that they use Brand Y toothpaste.'

Comment Housewives do not 'say' anything of the sort. Was this the result of a properly conducted survey? Why select only housewives? Where did the sample come from, and was it of an appropriate size? It could be that 3 out of a sample of only 5 housewives happened to respond that they used Brand Y.

Example S2.37

'University tests prove that our product is the safest.'

Comment Was the test done at a university or by a university? Who conducted the test? Was it students, academics, gate attendants or unqualified people with no real connection with a university? What does 'prove' mean?

Misleading graphs

One of the most common ways to distort data is by means of a graph.

Example S2.38

Figure S2.42 is a graph explaining how well a pain reliever (Brand X) can assist in ridding one of a headache. However, the graph has no scale, includes meaningless quantities and is quite useless.

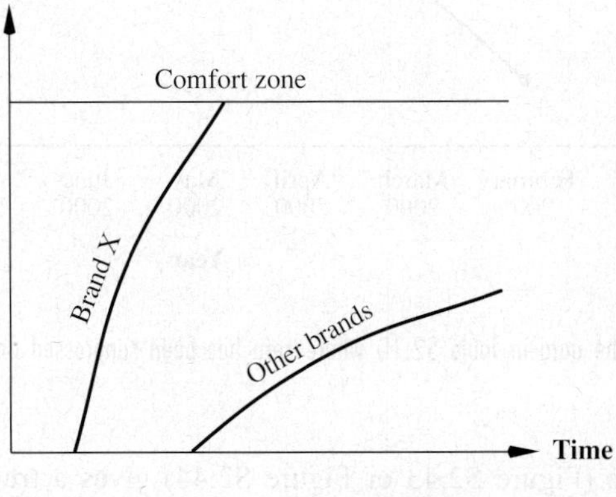

Figure S2.42 Graph illustrating how Brand X pain reliever can assist in relieving a headache

Example S2.39

The total retail turnover (in $million) in the Northern Territory for each of the six months from February 2000 to August 2000 is shown in Table S2.10. Illustrate these data by means of a graph.

Table S2.10 Monthly retail turnover in the Northern Territory between February 2000 and August 2000	
Month	Retail sales ($ million)
February	115
March	125
April	127
May	135
June	142
July	143
August	149

Source: Australian Bureau of Statistics, *Retail Trade Australia*, September 2000, Cat No. 8501.0

Solution

Two graphs that represent the same data in Table S2.10 are shown in Figures S2.43 and S2.44.

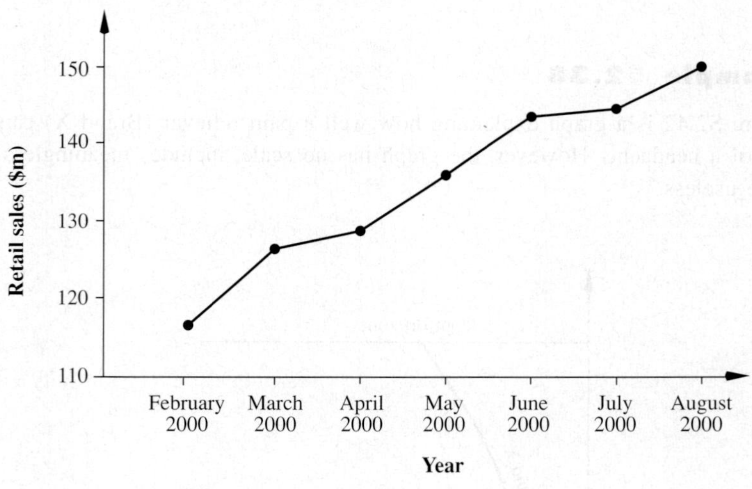

Figure S2.43 Graph of the data in Table S2.10 where zero has been suppressed on the vertical axis

Which of the two graphs (Figure S2.43 or Figure S2.44) gives a truer picture of the monthly retail turnover in the Northern Territory between February and August 2000?

If the intention is to highlight the increase in turnover, then Figure S2.43 shows this in quite a dramatic way. This effect is achieved by not starting the vertical (turnover) axis at zero but at 110, so that the vertical axis is 'shrunk' to include a smaller range. At the very least, the artist should have broken the axis to show that this was done.

On the other hand, if the intention is to show that the increase in turnover is only slight, then the full scale staring from zero can be used, as in Figure S2.44. Although technically correct, there is a lot of wasted space on the graph under the curve.

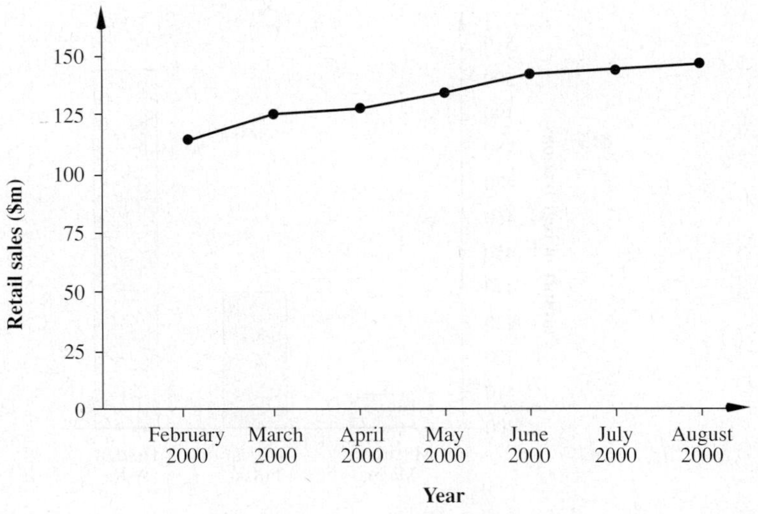

Figure S2.44 Graph of the data in Table S2.10 where zero has not been suppressed on the vertical axis

As to which graph is the more appropriate, the answer is not clear cut. Perhaps the solution lies in a graph that provides a compromise vertical scale, somewhere between these two graphs!

Misleading bar charts

There are a number of ways that data can be presented in a misleading way, limited only by the creativity of the artist. Bar charts are also a common way in which a visual display can convey a meaning that is not in line with reality.

Example S2.40

A motor dealer is sent information on the registrations of three models of new vehicles in Victoria during September 2000 and decides to draw an appropriate diagram for advertising purposes. The actual data are shown in Table S2.11.

Table S2.11 The number of registrations of three models of new vehicles in Victoria during September 2000

Type of vehicle	Number of registrations
Mitsubishi Magna	409
Nissan Pulsar	439
Holden Astra	503

***Source*:** Australian Bureau of Statistics, *New Motor Vehicle Registrations*, Cat. No. 9301.0

The dealer decides to compare the sales in Table S2.11 by using a simple bar chart, as shown in Figure S2.45.

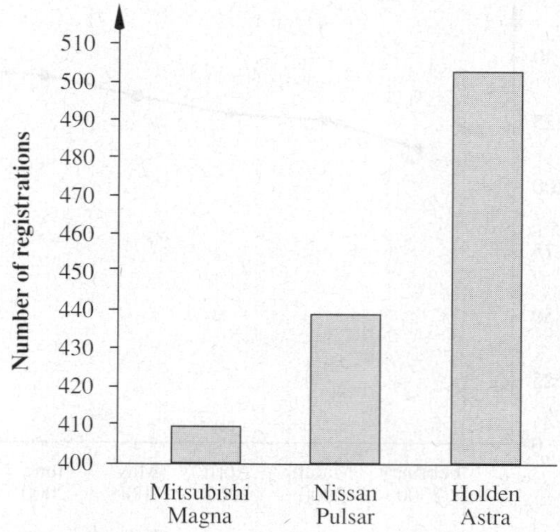

Figure S2.45 Simple bar chart illustrating the data in Table S2.11

Do you think that either Mitsubishi or Nissan would be happy with this bar chart? Do you think they would have a right to complain about the data being misrepresented? It certainly makes an interesting point of discussion.

S2.10 REMARKS

This chapter has presented some of the ways in which otherwise uninteresting data can be displayed in an eye-catching manner. This is particularly important in the media, where graphic and other artists are assigned the task of creating a product that will attract the attention of the reader. A similar situation arises when presenting company reports or even specific business information in which particular figures need to be highlighted.

As a final illustration of just how complex and unusual techniques for presenting statistical information can become, the diagram in the following example appeared in colour in a metropolitan newspaper and occupied about 60 per cent of an entire page.

Example S2.41

In late 2000 the gaming industry in New South Wales (and other states) was under increasing pressure to deal with the issue of problem gamblers. It had been recently announced that poker machines in the state's hotels and clubs were turning over $36.9 billion annually. To highlight this state of affairs, and to emphasise that it was those players from the poorer areas of Sydney that seem to gamble the most, an imaginative artist constructed the graphic in Figure S2.46.

Note that this diagram includes 12 separate simple bar charts that are linked to their corresponding region in Sydney by lines, where the regions are marked both with their name and an icon of a poker machine. These bar charts contain two distinct bars. The left-hand bars

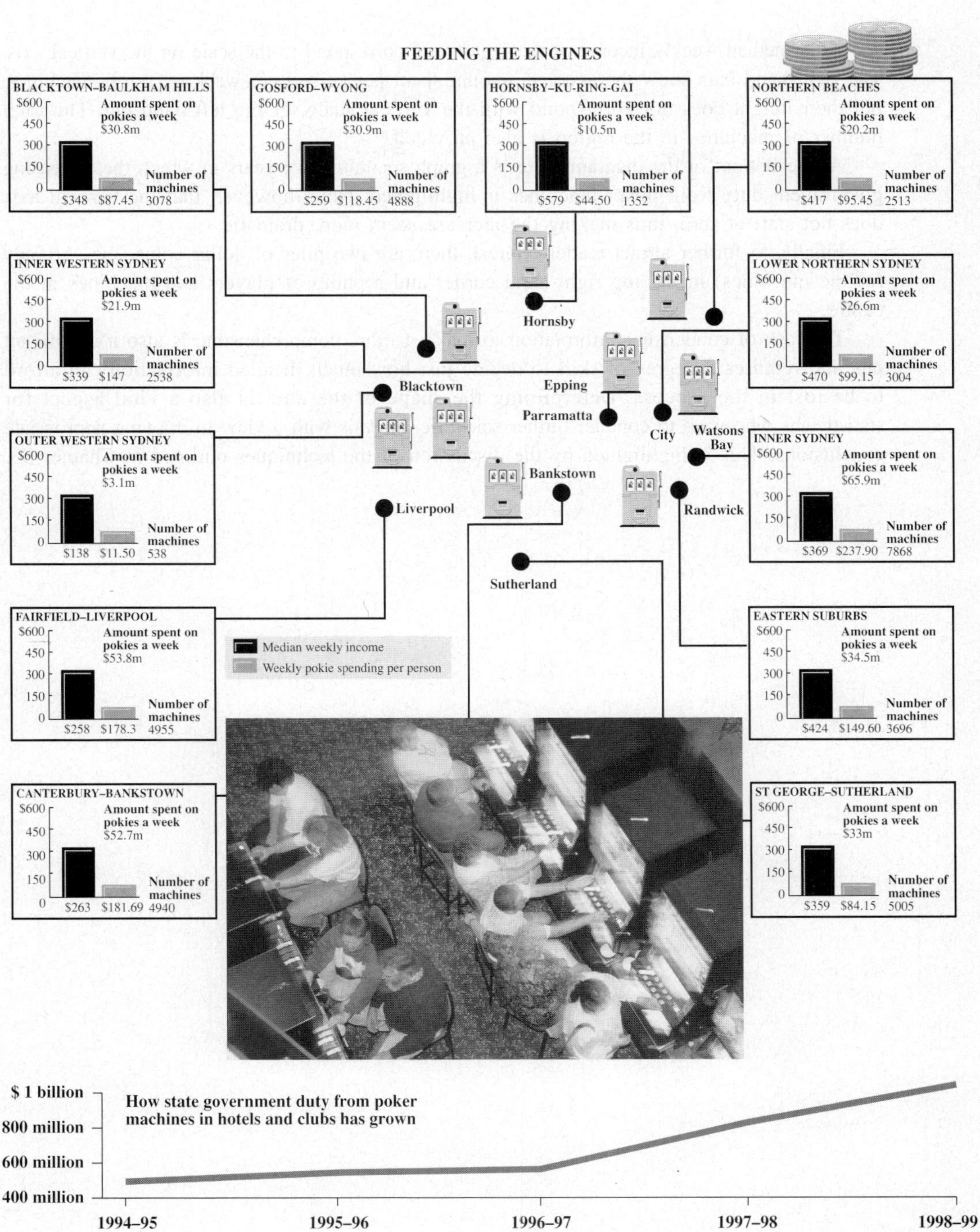

Figure S2.46 Statistics showing the state of poker machines for various regions of Sydney

Source: Australian Bureau of Statistics, NSW Department of Gaming and Racing; *Daily Telegraph*, 20 November 2000, p. 4. ABS data used with permission from the Australian Bureau of Statistics, www.abs.gov.au. Photograph by Peter Solness, Australian Picture Library

show the median weekly income of the region and correspond to the scale on the vertical axis. The right-hand bars show the average amount spent per person per week on poker machines, but their height does not correspond with the vertical scale, or the left-hand bar. The total number of machines in the region is also provided.

At the bottom of the diagram there is a graph spanning five years in which the increasing government duty from poker machines is highlighted. Note, however, that the vertical axis does not start at zero, thus making the increase seem more dramatic.

Finally, to further attract reader interest, there are two piles of dollar coins (the currency for the machines) in the top right-hand corner and a photo of players and machines in the centre.

The task of condensing information to make it more comprehensible is also a significant one and requires a degree of skill to decide just how much detailed information can afford to be lost in the process. Determining the shape of the data is also a vital aspect for statisticians who wish to conduct further scientific analysis with a view to drawing appropriate conclusions. This is highlighted by the discussion of the techniques outlined in Chapter S3.

Quick Quiz

1. Sketch the histogram for a set of data that may be considered bimodal and skewed to the right.

2. Name all the visual displays that may be appropriate to illustrate categorical data.

3. An accounting firm has 60 employees comprising 25 females and 35 males in the following positions:

Position	Number of females	Number of males
Auditor	8	12
Tax consultant	5	17
Management	12	6
Total	25	35

Use a multiple bar chart to illustrate these data.

4. Use two pie charts of the same size to illustrate the data in Question 3.

5. Use a pictogram to illustrate the data in Question 3.

6. An investment analyst records the weekly profits (to the nearest \$1000) made by 40 branches of a chain of grocery stores:

5	7	3	5	1	3	10	7	12	6	8	9	5	6	8	3	2	11	12	10
6	9	6	4	6	3	3	1	7	9	4	12	13	12	7	9	1	10	14	13

Write the data in the form of a frequency distribution.

7. The sales of a competitor's product for each of 4 years are shown below:

Year	2001	2002	2003	2004
Sales	10 500	11 200	12 600	13 000

How might an unscrupulous manager draw a graph to illustrate that the competitor's sales have remained more or less constant over these years? Illustrate your answer by drawing such a graph.

8. Your own company's sales of a product for each of 4 years are shown below.

Year	2000	2001	2002	2003
Sales	10 600	10 650	10 750	10 900

How might you draw a graph to illustrate that your sales have increased dramatically over these years? Illustrate your answer by drawing such a graph.

9. Draw a graph that would indicate that the sales in Question 8 have *not* increased much over the years.

10. A statistician goes to a bank and records for 200 customers the length of time it takes to serve each of them. She places the data in the form of a grouped frequency distribution, draws the corresponding frequency polygon and smooths out the curve. What shape do you think this curve might have?

Multiple Choice Questions

1. An *exploded* pie chart is one in which:
(a) all sectors are different colours
(b) all sectors are of different size
(c) at least one sector is separated from the rest
(d) all sectors are placed in a straight line
(e) none of the above

2. A *population pyramid* consists of:
(a) data on people from ancient Egypt
(b) a pair of bar charts sharing a common axis
(c) a three-dimensional five-sided figure
(d) pie charts stacked on top of one another
(e) none of the above

3. An *ogive curve* is a visual display of:
(a) categorical data
(b) rounded data
(c) data in a frequency distribution
(d) data in a grouped frequency distribution
(e) none of the above

4. A company advertising slogan 'We work harder' probably means:
(a) harder than any of our competitors
(b) harder than we did last year
(c) harder than you would expect we would
(d) harder only when the boss is watching
(e) any of the above

5. If 1000 telephone subscribers were chosen at random and the second-last digit of their telephone number recorded, the *shape* of the distribution of these digits might be expected to be:
(a) normal
(b) uniform
(c) skewed to the right
(d) exponential
(e) none of the above

6. In a grouped frequency distribution the class intervals should be mutually exclusive. This means that they should be:
(a) of the same length
(b) open-ended
(c) not open-ended
(d) not overlapping
(e) none of the above

7. A *frequency polygon* consists of:
(a) a table containing frequencies
(b) a smooth curve passing through a series of frequencies
(c) a series of continuous straight lines
(d) a normal distribution
(e) none of the above

8. A *multiple bar chart* is useful for:
(a) multiplying bars together
(b) making comparisons using side-by-side bars
(c) fitting as many bars as possible on the one chart
(d) enabling a smooth curve to be drawn through all the bars
(e) none of the above

9. *Raw data* are:
(a) difficult to collect
(b) difficult to analyse
(c) data in the form they are collected
(d) data that may not be accurate
(e) none of the above

10. One property of *class intervals* is:
(a) they are always of equal length
(b) the first one must always start from zero
(c) they can never be open-ended
(d) they always have the same frequency
(e) none of the above

Exercises

S2.1* The manager of a department store has decided to erect a video display for a new brand of tiling bond. The display lasts for 25 seconds and is continually repeated. To obtain an indication of how long customers view the video, a random sample of 50 people is selected. A record is kept of the length of time for which each person sustains an interest in the video. These times are shown below (in seconds):

19	18	21	19	23	18	20	33	28	31	25	24	21
23	25	26	19	30	27	32	23	28	19	21	21	23
32	29	27	22	19	35	23	19	18	19	20	28	25
32	24	22	26	24	19	20	30	26	22	22		

(a) Sort the data in order of magnitude.

(b) Construct a frequency distribution.

(c) Construct a cumulative frequency distribution.

S2.2* Use the data in Exercise S2.1 to complete the following:

(a) With class intervals of length 3 seconds, form a grouped frequency distribution.

(b) Construct a histogram from the grouped frequency distribution in (a).

(c) Construct a frequency polygon.

(d) Which, if any, of the smooth curves in Figures S2.36 to S2.40 does the frequency polygon most resemble?

(e) Draw the ogive curve of the grouped frequency distribution in (a).

(f) From the ogive curve, estimate the time below which are 50% of the sample values.

S2.3 Summarise the data in Exercise S2.1 in a manner that would be informative to the manager.

S2.4 The number of divorces granted in Australia from 1993 to 1999 is shown below.

(a) Draw a graph and join the points to illustrate the trend of the data. The scale on the vertical axis is to begin at 48 000 and end at 53 000.

(b) Do you think that the graph in (a) represents the true trend of the number of divorces? Explain.

Year	Number of divorces
1993	48 363
1994	48 312
1995	49 712
1996	52 466
1997	51 288
1998	51 370
1999	52 600

Source: Australian Bureau of Statistics, *Marriages and Divorces 1999*, Cat. No. 3310.0

S2.5 A Time/CNN telephone poll of 1278 adult Americans was taken in 2000 in which respondents were asked 'When parents get divorced, are children harmed?'. The percentage responses are shown in the table below.

Response	Percentage
Almost always harmed	42
Frequently harmed	22
Sometimes harmed	27
Seldom harmed	9

Source: *Time* magazine. © Time Inc., reprinted by permission.

Illustrate the data in the above table using a pie chart.

S2.6* A personnel manager is interested in how long the factory employees spend travelling from home to work each day. A random sample of 100 employees is chosen, and their travel times (in minutes) are shown below in the form of a grouped frequency distribution.

Time taken (minutes)	Frequency
0 – under 10	9
10 – under 20	16
20 – under 30	26
30 – under 40	23
40 – under 50	14
50 – under 60	12
	$\Sigma f = 100$

(a) Construct a histogram of the data.

(b) Construct the frequency polygon for the histogram in (a).

(c) Which, if any, of the smooth curve shapes in Figures S2.36 to S2.40 does the frequency polygon most closely resemble?

S2.7 In a Sydney-based firm, each employee can be classified as living in one of four parts of the metropolitan area. The percentages of employees residing in each area are:

Area	Percentage
North	31.7
South	22.4
East	17.3
West	28.6
Total	100.0

(a) Illustrate the data using a pie chart.

(b) There were 2000 employees in all. Convert the data into frequency form.

(c) The managing director believes that employees are equally likely to live in any of the four areas. Is there any solid evidence to refute her belief? Explain.

S2.8* An advertising agency classified goods into five categories of quality: A (highest), B, C, D and E (lowest). The percentages of heavily advertised brands and moderately advertised brands within each quality grade are shown in the following table.

Quality grade	Percentage of heavily advertised brands	Percentage of moderately advertised brands
A	22.3	18.3
B	19.7	21.1
C	41.2	39.4
D	10.3	12.4
E	6.5	8.8
Total	100.0	100.0

(a) Illustrate the data using two pie charts of equal size.

(b) Illustrate the data using a multiple bar chart.

S2.9 In September 2000 a *Bulletin-Morgan* face-to-face poll of 1055 Australians aged 14 and over was conducted in which respondents were asked 'Do you approve or disapprove of legalised heroin injecting rooms?'. The results for the five age groups are shown in the table below.

Age group	Approve	Disapprove	Neither approve/disapprove
14–17	45	41	14
18–24	34	55	11
25–34	45	41	14
35–49	40	51	9
50 and over	30	60	10

Source: *Bulletin*, 17 October 2000, p. 22

Illustrate the data in the above table using a compound bar chart.

S2.10* The ages at death (in years) of the first 36 presidents of the United States are given below.

Washington	67	Tyler	71	Hayes	70	Harding	57
Adams	90	Polk	53	Garfield	49	Coolidge	60
Jefferson	83	Taylor	65	Arthur	56	Hoover	90
Madison	85	Fillmore	74	Cleveland	71	Roosevelt	63
Monroe	73	Pierce	64	Harrison	67	Truman	88
Adams	80	Buchanan	77	McKinley	58	Eisenhower	78
Jackson	78	Lincoln	56	Roosevelt	60	Kennedy	46
Van Buren	79	Johnson	66	Taft	72	Johnson	64
Harrison	68	Grant	63	Wilson	67	Nixon	81

(a) Construct a grouped frequency distribution for the data. (Use class intervals of length 10 years.)

(b) Draw a histogram from the grouped frequency distribution in (a).

(c) Draw the frequency polygon from the histogram in (b).

(d) Which of the smooth curve shapes does the frequency polygon resemble?

(e) Draw an ogive curve for the data.

(f) Use the curve in (e) to determine the age below which 50% of the presidents died.

S2.11* The number of students enrolled in the first year of a business course at a technical college is shown below for each of 7 years.

Year	1995	1996	1997	1998	1999	2000	2001
Enrolment	160	192	221	301	280	298	316

(a) Draw a graph of the data and join successive points.

(b) Draw a bar chart of the data.

(c) Draw a pictogram of the data.

S2.12 A report by the IUCN (World Conservation Union) gave the percentage of various animal groups that were endangered species. The percentages of each group that are threatened with extinction are shown in the table below.

Group	Percentage threatened with extinction
Fish	34
Primates	33
Amphibians	25
Mammals	23
Reptiles	20
Birds	11

Illustrate the data using a pictogram.

S2.13* The amounts of money (to the nearest dollar) spent by each of the first 40 customers at a small supermarket one Saturday morning were as follows:

15	7	33	25	42	15	6	19	7	18
8	48	51	6	52	48	56	36	52	29
21	50	12	54	18	9	8	4	52	14
18	48	25	47	20	3	48	16	42	23

Use these data to construct:

(a) a frequency distribution

(b) a cumulative frequency distribution

(c) a grouped frequency distribution with class intervals of length $10

(d) a histogram

(e) a frequency polygon

S2.14 Consider the data in Exercise S2.13. Use the answers obtained in that exercise to answer the following:

(a) Describe the *shape* of the distribution.

(b) Draw an ogive curve of the grouped frequency distribution.

(c) From the ogive curve, find the amount of money spent by less than 50% of the sample.

S2.15 The number of gold medals won by the top five countries at the Sydney 2000 Olympic Games is shown in the table below.

Country	Number of gold medals
United States	39
Russia	32
China	28
Australia	16
Germany	14

Illustrate the data in the above table using a simple bar chart.

S2.16 The diagram in Figure S2.47 appeared in the press and was apparently designed to illustrate how the juvenile crime rate had been increasing. List some of the features of this display. In particular, make note of any that you find attractive and any that you find unsatisfactory.

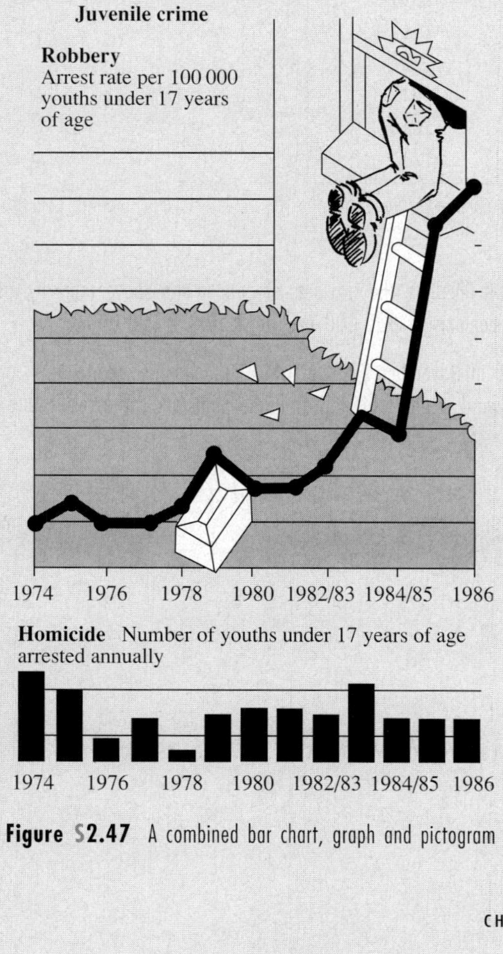

Figure S2.47 A combined bar chart, graph and pictogram for Exercise S2.16

S2.17 The annual production in Australia of men's and boy's shorts is shown in the table below for the years 1994–1995 to 1999–2000.

Year	Production ('000)
1994–95	3757
1995–96	2425
1996–97	1933
1997–98	1646
1998–99	1517
1999–00	1103

Source: The Australian Bureau of Statistics, *Annual Commodity Production*, September Quarter 2000, Cat. No. 8301.0

Illustrate the data in the above table using a graph, joining successive points.

S2.18 The annual production in Australia of women's and girl's long trousers and women's and girl's swimwear is shown in the table below for each of the years 1994–1995 to 1999–2000.

Year ('000)	Women's and girl's long trousers	Women's and girl's swimwear ('000)
1994–95	1858	2559
1995–96	1272	2462
1996–97	1570	2223
1997–98	1044	1619
1998–99	991	1618
1999–00	859	1711

Source: The Australian Bureau of Statistics, *Annual Commodity Production*, September Quarter 2000, Cat. No. 8301.0

Using an appropriate key, on the same diagram illustrate the two sets of data in the above table by means of separate graphs. (Begin the scale on the vertical axis at 800 000 and finish it at 2 600 000.)

S2.19* An opinion poll of 1142 voters was conducted in November 2000 in which respondents were asked 'Are you satisfied or dissatisfied with the way John Howard is doing his job as Prime Minister?'. The results are shown in the table below.

Response	Percentage
Satisfied	38
Dissatisfied	50
Uncommitted	12
Total	100

Source: *Australian*, 7 November 2000, p. 2

Illustrate the data in the above table using an appropriate visual display.

S2.20* A study by the Automotive Research and Marketing Service made a comparison between the number of vehicle faults in Australian cars and imported cars between 1988 and 1992. An illustration of the data is presented in Figure S2.48.

List some of the features of this display. In particular, make note of any features that you find attractive and any that you find unsatisfactory.

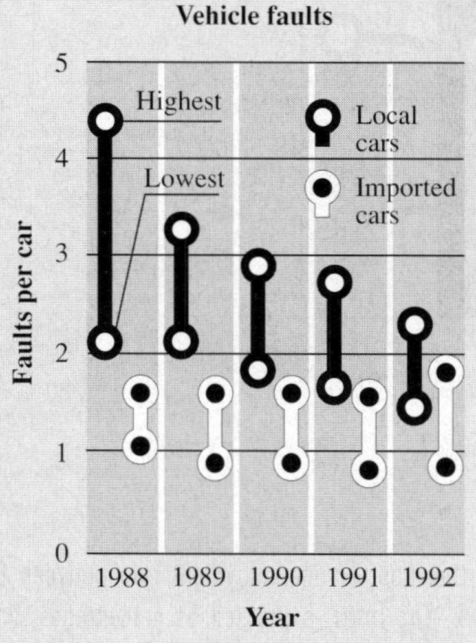

Figure S2.48 A pictogram for Exercise S2.20

Source: Sydney Morning Herald, 2 September 1992, p. 7

S2.21 The number of Australians aged 15 years and over attending a sporting event or competition increased from 6.1 million in 1995 to 6.4 million in 1999. The table below shows the six sports that had the largest change in attendance during that period along with their attendance rates.

Sport	Attendance rate (1995)	Attendance rate (1999)
Australian Rules	13.3	16.8
Cricket	8.3	6.3
Harness racing	4.3	3.6
Basketball	4.9	3.5
Rugby Union	2.5	3.0
Netball	2.2	1.7

Source: Australian Bureau of Statistics, *Sports Attendance*, Cat. No. 4174.0

Illustrate the data in the above table using a multiple bar chart.

S2.22 A report in the *Sydney Morning Herald* on 11 January 1993 said that the Australian share market would be asked to support more than $13 billion in new equity raisings in 1993. The accompanying illustration to this report is shown in Figure S2.49.

List some of the features of this display. In particular, make note of any features that you find attractive and any that you find unsatisfactory. Suggest possible improvements.

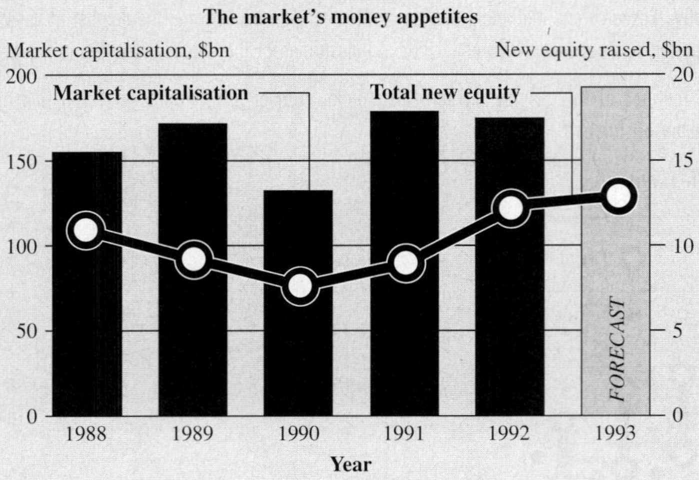

Figure S2.49 A combined bar chart and graph for Exercise S2.22

Source: *Sydney Morning Herald*, 11 January 1993, p. 19

S2.23* The Australian Bureau of Statistics issued the trend series of the monthly unemployment rate between October 2000 and March 2001 for women who were looking for work. The rates, expressed as percentages, are shown in the table below.

Month	Unemployment rate (%)	Month	Unemployment rate (%)
October 2000	6.1	January 2001	6.3
November 2000	6.2	February 2001	6.4
December 2000	6.3	March 2001	6.5

Source: Australian Bureau of Statistics, *Labour Force*, Cat. No. 6203.0

Draw a graph of these data in which the points are joined and use a scale on the vertical axis of between 6.0 and 6.5. Comment as to whether you feel that this graph fairly represents the rise in the rate.

S2.24 According to the Motor Traders Association, in 2000 the five most economical cars in Australia with regard to fuel consumption are those listed in the table below. The table shows the number of litres each car takes to travel 100 km (the lower the litres, the more economical the car).

Model	Fuel consumption (L/100 km)
Daihatsu Cuore	6.0
Toyota Echo	6.1
Daewoo Matiz	6.1
Mazda 121 Metro	7.0
Ford KA	7.2

Although all the cars in the table are very economical, the task is to draw a visual display that would indicate that the Daihatsu Cuore is clearly much more economical than the others. Comment as to whether you feel your diagram is a 'fair' one.

S2.25 In the United States alone, it is estimated that the practice of tipping is worth over AUS$30 billion each year, with tipping customary in 35 professions worldwide. The number of such professions is shown in the table below for a selection of six countries.

Country	Number of professions in which it is customary to tip
United States	32
United Kingdom	23
Italy	20
Australia	13
Norway	9
Japan	4

Source: Michael Lynn, *Personality and Individual Differences*, February 2000

Illustrate the data in the above table using an appropriate visual display.

S2.26 In a study reported in the Australian Journal of Nutrition and Dietetics, 173 boys and 297 girls aged between 11 and 14 were asked whether they wanted their body weight to either increase or decrease.

Also outlined were the favourite slimming methods used by each sex, along with the percentages who had tried them. The figures for the top six methods are shown in the following table.

Slimming method	Percentage of males	Percentage of females
Exercise	52	72
Not snacking	22	42
Keeping busy	12	38
Own diet	14	37
Low joule foods	14	34
Water before meals	11	32

A multiple bar chart of the data is given in Figure S2.16. Now illustrate the data using a pictogram.

S2.27 In a report based on 984 shoppers around the country, the Radio Marketing Bureau found that around 54% of Australians prefer to shop alone, while females and people aged under 45 are more likely to impulse-buy than the rest of the community.

The reasons given in the survey why shoppers purchase on impulse, along with the percentage in each category, are shown in the table below.

Reason	Percentage
Item on special	35
Noticed item	20
New product	13
No reason	9
Item on display	8
Other	15
Total	100

Illustrate the data using a pie chart.

S2.28 The percentages of foreign ownership or control of five Australian products are shown in the table below.

Product	Percentage of foreign ownership or control
Cars	98
Biscuits	90
Confectionery	80
Ice-cream	75
Beer	50
Meat pies	40

Source: *Sun Herald*, 3 September 1995, p. 29

Illustrate the data using a pictogram.

S2.29* The number of burglaries reported in a major city each day over a period of 20 days is shown below:

17	16	26	36	19	8	41	32	26	27
8	38	41	19	26	17	54	16	36	26

(a) Sort the data in an array.

(b) Write the data in the form of a frequency distribution.

S2.30* Construct a grouped frequency distribution for the data in Exercise S2.29 using class intervals of length 10.

S2.31 (a) Use the grouped frequency distribution in Exercise S2.29 to construct a histogram of the data.

(b) Draw a frequency polygon on the histogram in (a). What shape is it?

S2.32 (a) Draw an ogive curve of the grouped frequency distribution in Exercise S2.30.

(b) From the ogive curve, find the number of burglaries that were committed on less than 50% of the days.

S2.33 According to figures released by the Cancer Council in New South Wales, the risk of dying from cancer has fallen to its lowest level in 25 years for men and 17 years for women. The change in risk of contracting fatal cancer for both men and women between the years 1985 and 1995 is shown in Figure S2.50, which appeared in the press.

List some of the features of the display. In particular, make note of any features that you find attractive and any that you find unsatisfactory. Suggest any possible improvements.

S2.34 The 6 countries with the largest divorce rates in 2000 are shown in the following table, along with the rates expressed as a percentage of all marriages. (Australia is seventh with a rate of 46%.)

Country	Divorce rate (as a percentage of all marriages)
Russia	65
Sweden	64
Finland	56
United Kingdom	53
United States	49
New Zealand	47

Source: U.S. National Center for Health Statistics; U.S. Census Bureau; the United Nations

Illustrate the data in the above table using a simple bar chart.

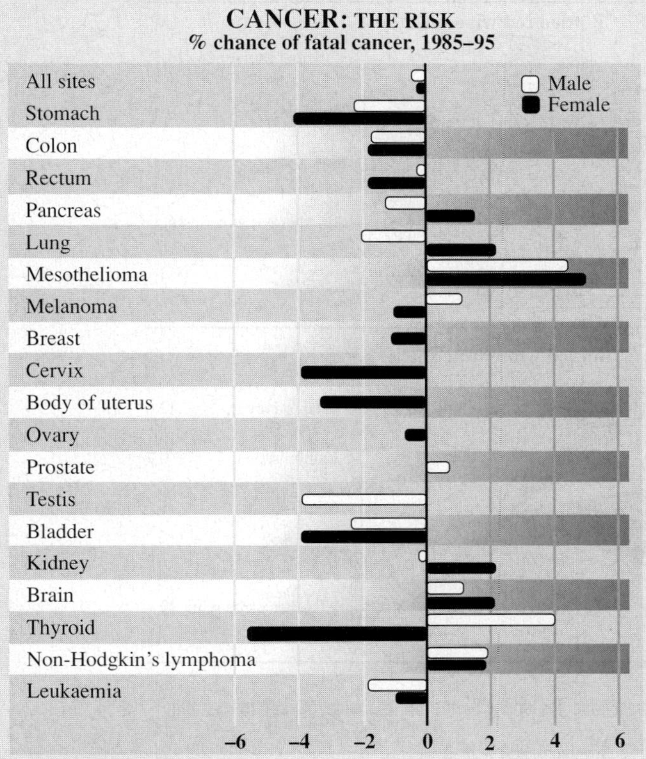

Figure S2.50 The risk of dying from various types of cancer

Source: *Sydney Morning Herald,* 5 June 1998, p. 7

S2.35 The share of the bottled wine market in 2000 in Australia was illustrated by an artist in the two pie charts shown in Figure S2.51 on page S.74.

List some of the features of the display. In particular, make note of any that you find attractive and any that you find unsatisfactory. Suggest any possible improvements.

S2.36 According to the NRMA, in New South Wales its road service division rendered assistance in the year 1999–2000 for a total of 3 374 967 calls. A summary of some of the problems dealt with is shown in the table below.

Problem	Number of calls ('000)
Flat batteries	518
Keys locked in the car	230
Electrical system breakdown	183
Fuel/petrol problems	113
Ignition problems	97
Wheel/tyre problems	87

Source: NRMA Limited, *Concise Annual Report 2000,* p. 11

Illustrate the data in the above table using an appropriate visual display.

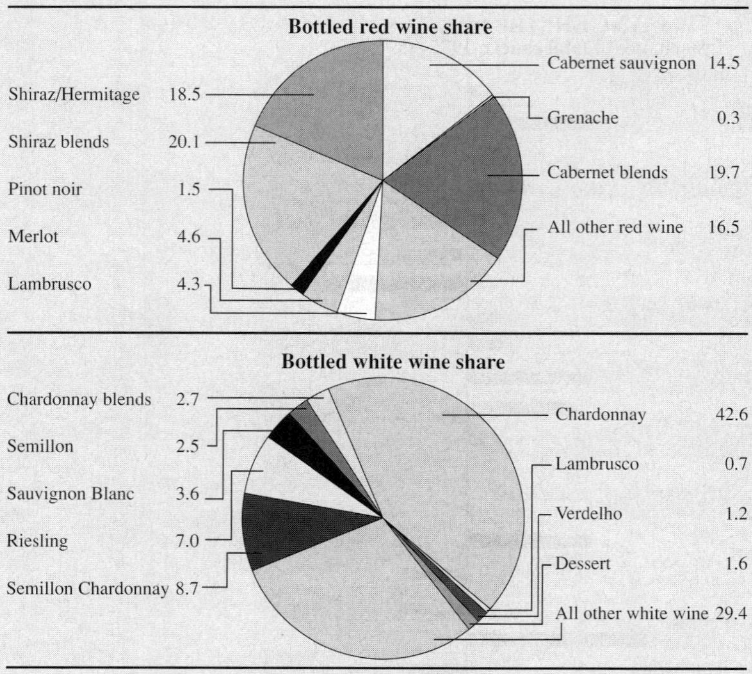

Figure S2.51 Pie charts representing the share of the bottled wine market in Australia

Source: *Sydney Morning Herald*, 17 November 2000, p. 3

S2.37 A visual display of the road toll in New South Wales between 1970 and 2000 is shown in Figure S2.52. One of the aims is to demonstrate how the road toll has been affected by a number of state government initiatives.

(a) List some of the features of the display. In particular, make note of any features that you find attractive and any that you find unsatisfactory. Suggest any possible improvements.

(b) From the display, which of the initiatives do you think had the most dramatic effect on the road toll? Explain your reasoning.

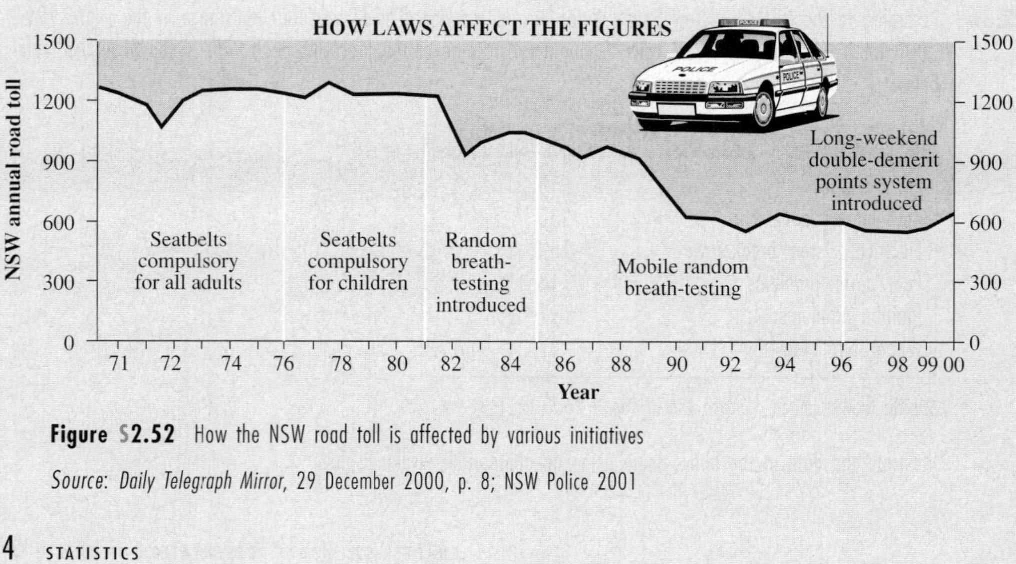

Figure S2.52 How the NSW road toll is affected by various initiatives

Source: *Daily Telegraph Mirror*, 29 December 2000, p. 8; NSW Police 2001

Chapter S3

Measures of central tendency

'In small proportions
we just beauty see
And in short measures
life may perfect be.'

(Ben Johnson,
English playright and poet, 1640)

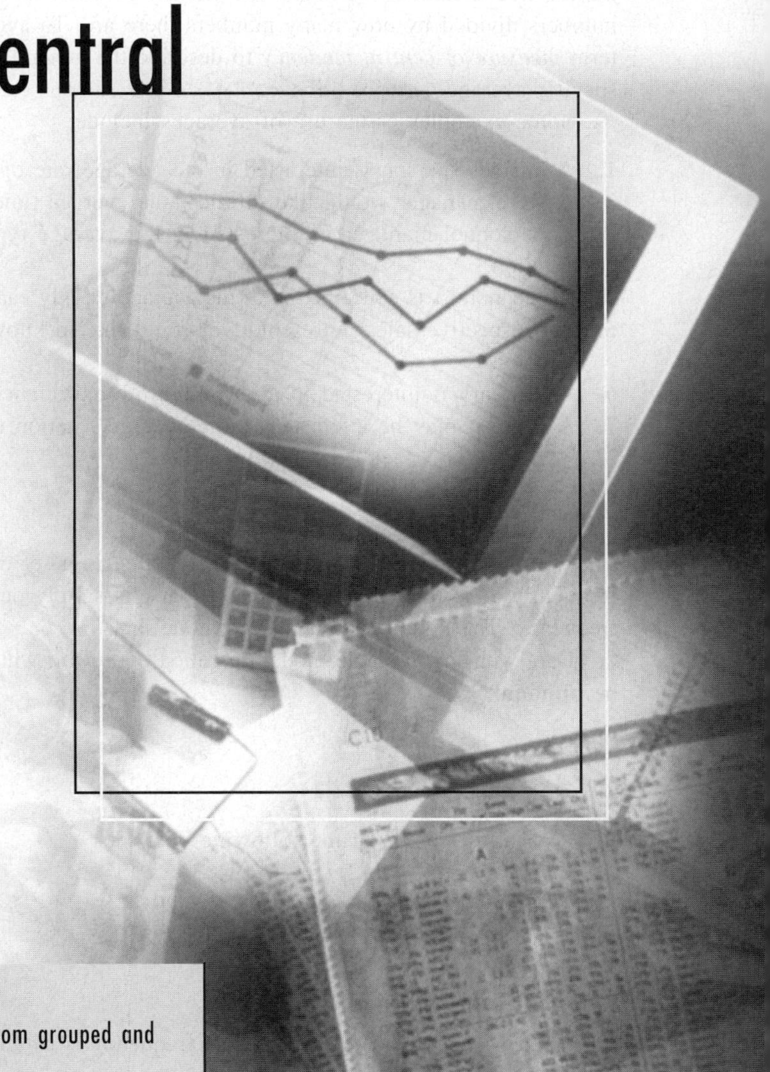

Learning objectives

In this chapter you will learn to:

- Calculate the mode, median and mean from grouped and ungrouped data
- Calculate quartiles, deciles, percentiles and fractiles
- Calculate and interpret the geometric mean
- Determine the significance of the skewness of a distribution

S3.1 INTRODUCTION

Many people have difficulty understanding the significance of large sets of numbers, and for this reason it is common to describe such sets in pictorial or tabular form as discussed in Chapter S2. However, on many occasions it is more convenient to describe a set of numbers by using a *single number*. Calculating a single number is one of the most frequently encountered methods of condensing data, such as when we work out a single 'average' to represent a whole mass of figures.

In its broadest sense, an *average* is simply any single figure that is representative of many numbers. The term is also used specifically to mean an average calculated as the sum of a set of numbers divided by how many numbers there are. To avoid confusion, statisticians simply use the term *measure of central tendency* to describe the general idea of a typical value, and **mean** for the specific average described above.

Some examples of the use of averages include:

1. A business person is interested in the average rate of return on certain types of investments.
2. A store manager would like to know the average time customers must wait for service.
3. A personnel manager is interested in the average typing speed of applicants for a secretarial position.
4. An economist is concerned with the average weekly earnings of employees in a certain profession.
5. A prospective sales representative would like to know the average income to expect from the position.
6. An actuary is interested in the average life expectancies for groups of individuals.
7. A chemist may be interested in the average reaction time of a particular chemical.

S3.2 THE MODE

The simplest and most obvious 'typical' value in a set of numbers is that which occurs the most frequently. It is referred to as the **mode** and is simply defined as that observation with the largest frequency. The notation used for the mode is *Mo*.

Data with just a single mode are called *unimodal*, while if there are two modes they are said to be **bimodal**.

Example S3.1

Find the mode of the following data sets:

(a) 3 6 4 12 5 7 9 3 5 1 5
(b) 23 25 29 33 19 25 29 18
(c) 9 8 5 7 8 25 4 7 25 25 5

Solution

(a) The value with the highest frequency is 5 (which occurs three times). Hence the mode is Mo = 5.
(b) In this case there are two modes, namely 25 and 29. These data are bimodal.
(c) The mode of this data set is 25 (which occurs three times). However, this is not a typical value of these data and illustrates one of the problems with the mode. It is possible that the mode may well be an extreme measure and not at all representative of the overall data.

In practice, the mode is often unreliable as a central measure. However, there are cases where the most frequent result is important regardless of whether it is central or not. For example, a store manager may wish to know the soft drink that has the greatest sales, even though it may not necessarily be described as the 'typical' drink that is purchased.

Calculation of the mode from a frequency distribution

In a frequency distribution the mode may be found from an examination of the frequency column. The observation with the largest frequency is the mode.

Example S3.2

A group of 20 real estate agents were asked how many houses they had sold in the past month. Their responses are summarised in Table S3.1. Find the number of houses sold that represents the mode.

Table S3.1 Sales data for Example S3.2

Number of houses sold	f
0	2
1	5
2	6
3	4
4	2
5	1
Total	20

Solution

The observation with the largest frequency (6) is 2. Hence the mode of these data is 2.

Calculation of the mode from a grouped frequency distribution

It is not possible to calculate the *exact* value of the mode of the original data in a grouped frequency distribution, since information is lost when the data are grouped. However, it is possible to make an *estimate* of the mode. The class interval with the largest frequency is called the *modal class*. This estimate of the mode itself is given in Equation S3.1.

$$\text{Mo} = L + \frac{d_1}{d_1 + d_2}(i) \qquad \textbf{(S3.1)}$$

where:

L = the real lower limit of the modal class
d_1 = the frequency of the modal class minus the frequency of the previous class
d_2 = the frequency of the modal class minus the frequency of the next class above the modal class
i = the length of the class interval of the modal class

Note that this estimate assumes that the data are evenly distributed throughout each class interval. The more this is true, the better the estimate will be.

Example S3.3

The intended length of stay of short-term travellers visiting friends or relatives in Australia is shown in Table S3.2. Estimate the modal length of the stay of such people.

Table S3.2 Intended length of stay of short-term travellers visiting friends or relatives in Australia

Intended length of stay (days)	f
0–6	5 800
7–13	103 900
14–20	102 100
21–30	78 600
31–60	149 900
61–90	47 000
91–182	49 000
183–365	23 700
Total	605 000

Source: Australian Bureau of Statistics, *Year Book Australia 1996*, p. 368

Solution

From Table S3.2 it can be seen that the modal class is 31–60, since it has the largest frequency of 149 900. The following values can now be found:

$$L = 30.5 \text{ (i.e. halfway between 30 and 31)}$$
$$d_1 = 149\,900 - 78\,600 = 71\,300$$
$$d_2 = 149\,900 - 47\,000 = 102\,900$$
$$i = 60.5 - 30.5 = 30$$

From Equation S3.1:

$$\text{Mo} = 30.5 + \frac{71\,300}{71\,300 + 102\,900}(30)$$
$$= 42.8$$

Therefore, the estimate of the modal length of intended stay of short-term travellers visiting friends or relatives in Australia is about 43 days.

S3.3 THE MEDIAN

The **median** is another measure of central tendency. It may be described as the 'middle' observation in a set. In fact, the median is defined as the centre value when the data are arranged in order of magnitude. That is, 50% of the data have a value *less than* the median, and 50% of the data have a value *more than* the median.

Calculation of the median from raw data

The method of calculating the median from a set of raw data (once it has been arranged in order of magnitude) is as follows.

Let n = the number of observations.

1. If n is *odd*, the median is the value of the $\frac{n+1}{2}$th observation.
2. If n is *even*, the median is the mean of the $\frac{n}{2}$th observation and the $\left(\frac{n}{2}+1\right)$th observation. That is, you add these two observations and divide their total by 2.

The symbol for the median of a sample is $\tilde{x}$ (pronounced '*x-tilde*').

Example S3.4

The manager of a hairdressing salon was interested to know how long her customers had to wait before being attended to by staff. On a certain day she recorded the waiting times (in minutes) of 15 randomly chosen customers. They were as follows:

14 28 36 15 29 16 9 40 16 21 36 17 4 15 22

Find the median waiting time.

Solution

First, arrange the data in order of magnitude:

4 9 14 15 15 16 16 17 21 22 28 29 36 36 40

In this case, $n = 15$ (which is an odd number).
Therefore, the median waiting time is:

$$\begin{aligned}\tilde{x} &= \frac{15+1}{2}\text{th observation}\\ &= 8\text{th observation}\\ &= 17\text{ minutes}\end{aligned}$$

Example S3.5

The manager of a supermarket recorded the number of customers who were served at the store between 6 p.m. and 9 p.m. for 12 successive Thursdays. The results were as follows:

285 292 243 286 301 258 286 208 215 263 252 224

Find the median number of customers served on those Thursday nights.

Solution

Arrange the data in order of magnitude:

208 215 224 243 252 258 263 285 286 286 292 301

In this case, $n = 12$ (which is an even number).
Therefore, the median number of customers is:

$$\tilde{x} = \text{mean of 6th and 7th observations}$$
$$= \frac{258 + 263}{2}$$
$$= 260.5$$

Calculation of the median from a frequency distribution

If the data are in the form of a frequency distribution, the same basic formulae for calculating the median apply. However, it is convenient in this case to construct the **cumulative frequency distribution**. This involves constructing an extra column in which the frequencies are cumulated.

Example S3.6

A random sample of 50 local councils was selected, and they were asked how many pieces of heavy-duty equipment they had purchased during the previous 12 months. The results were as follows:

Number of pieces	0	1	2	3	4	5
Frequency	5	10	12	15	6	2

Find the median number of pieces of equipment purchased.

Solution

Construct a cumulative frequency distribution as in Table S3.3.

Table S3.3 Cumulative frequency distribution of equipment pieces purchased for Example S3.6

Number of pieces purchased	Frequency	Cumulative frequency
0	5	5
1	10	15
2	12	27
3	15	42
4	6	48
5	2	50
	$\Sigma f = 50$	

There are $n = \Sigma f = 50$ observations, which are already sorted into numerical order. Since n is even, the median is the average of the 25th and 26th observations.

From the cumulative frequency distribution it can easily be seen that both the 25th and 26th observations are 2 pieces, and hence the median value also is 2 pieces.

Calculation of the median from a grouped frequency distribution

It is not possible to calculate the *exact* value of the median of the original data in a grouped frequency distribution, since information is lost when the data are grouped. However, it is possible to make an *estimate* of the median. The class interval that contains the median is called the *median class*.

The estimate of the median itself is given in Equation S3.2.

$$\tilde{x} = L + \frac{\left(\frac{n}{2} - C\right)}{f}(i) \qquad \text{(S3.2)}$$

where:

$\tilde{x}$ = the median
L = the real lower limit of the median class
$n = \Sigma f$ = the total number of observations in the entire data set
C = the cumulative frequency in the class immediately *before* the median class
f = the frequency of the median class
i = the length of the real class interval of the median class

Example S3.7

One of the most popular recreational pastimes in Australia is a visit to the zoo. Over 3 million people aged 15 years and over visit the zoo each year. The breakdown of their ages is shown in Table S3.4.

Table S3.4 Ages of people 15 years and over who visited zoos in Australia in a 12-month period

Age (years)	f ('000)	Cumulative f ('000)
15–24	719.5	719.5
25–34	864.1	1583.6
35–44	706.6	2290.2
45–54	396.1	2686.3
55–64	237.2	2923.5
65 and over	200.4	3123.9
Total	3123.9	

Source: Australian Bureau of Statistics, *Year Book Australia*

Estimate the median age of these people.

Solution

Since there are 3 123 900 observations, the median age is the average of the 1 561 950th observation and the 1 561 951st observation. An inspection of Table S3.4 shows that both of these observations lie in the class interval 25–34, which is therefore the median class.

The following values can now be found:

L = 24.5
n = 3 123 900
C = 719 500

$$f = 864\ 100$$
$$i = 34.5 - 24.5 = 10$$

Using Equation S3.2:

$$\tilde{x} = 24.5 + \frac{\left(\frac{3\ 123\ 900}{2} - 719\ 500\right)}{864\ 100}(10)$$
$$= 24.5 + 9.7$$
$$= 34.2$$

Hence, we estimate that the median age of people aged over 15 years who visit zoos in Australia is 34.2 years.

S3.4 THE ARITHMETIC MEAN

Perhaps the most widely used average, the **arithmetic mean** is defined as the sum of the observations divided by the number of observations. The formula for computing the arithmetic mean is:

$$\bar{x} = \frac{\Sigma x}{n} \qquad \textbf{(S3.3)}$$

where:

$\bar{x}$ = the symbol for the arithmetic mean calculated from a *sample* (pronounced '*x-bar*')
x = the *symbol* for the variable
Σx = the *sum* of the observations
n = the *number* of observations in the sample

The symbol for the arithmetic mean calculated from a *population* is the Greek letter μ (pronounced 'mew').

Although there are a number of different kinds of means in statistics (such as the **harmonic mean** and the **geometric mean**), the arithmetic mean is by far the most commonly used. For this reason the word 'arithmetic' is rarely used in practice and we simply refer to the 'mean'.

Example S3.8

The number of work days lost due to illness in a manufacturing business for each week over a period of 10 weeks is shown below:

36 28 33 29 28 32 33 33 34 32

Find the mean number of work days lost per week during that period.

Solution

In this problem:

$$n = 10$$
$$\Sigma x = 36 + 28 + 33 + \ldots + 34 + 32 = 318$$

From Equation S3.3, the mean time lost per week is:

$$\bar{x} = \frac{318}{10}$$
$$= 31.8 \text{ work days}$$

Use of an arbitrary origin

In some problems, calculations can be saved by first removing numbers that have no bearing on a calculation, then restoring them at the end. For example, the mean of 1002, 1004 and 1009 is clearly the mean of 2, 4 and 9 with 1000 added (i.e. 5 + 1000 = 1005).

This process is referred to as *using an arbitrary origin*, and the steps involved are as follows:

1. Take a 'guess' at the mean of the data, and call this guess the arbitrary origin. (It is often convenient to choose a number that is a multiple of 10 or 100 or 1000, etc., depending on the data.)
2. Subtract the arbitrary origin from each observation and record the differences.
3. Find the mean of the differences in Step 2, and add this mean to the arbitrary origin to obtain the mean of the original set of data.

Thus, if the differences from the arbitrary origin are recorded as d, then:

$$\bar{x} = \text{arbitrary origin} + \frac{\Sigma d}{n} \qquad \textbf{(S3.4)}$$

Example S3.9

Consider the data in Example S3.5. With the aid of an arbitrary origin, determine the mean number of customers served on Thursday evenings.

Solution

Suppose that an arbitrary origin of, say, 250 is chosen. The differences (d) are:

35 42 −7 36 51 8 36 −42 −35 13 2 −26

Then:

$$\Sigma d = 35 + 42 - 7 + ... + 2 - 26 = 113$$

From Equation S3.4:

$$\bar{x} = 250 + \frac{113}{12}$$
$$= 259.42 \text{ customers}$$

Calculation of the mean from a frequency distribution

As seen in Chapter S2, tabulating data in frequency form is often very convenient. It is therefore useful to be able to calculate a mean directly from such a table. In this case the calculation of the mean is found from the formula:

$$\bar{x} = \frac{\Sigma fx}{\Sigma f} \qquad \text{(S3.5)}$$

where:

f = frequency
Σf = the sum of the frequencies
Σfx = the sum of each observation multiplied by its frequency

Example S3.10

The number of times per week that a particular photocopy machine breaks down was recorded over a period of 60 weeks. The results are given in the frequency table below.

Number of breakdowns	0	1	2	3	4	5
Number of weeks	15	12	16	10	5	2

Find the mean number of breakdowns per week over the 60-week period.

Solution

The usual method of such a calculation is given in Table S3.5.

Table S3.5 Calculations for Example S3.10

x	f	fx
0	15	0
1	12	12
2	16	32
3	10	30
4	5	20
5	2	10
	$\Sigma f = 60$	$\Sigma fx = 104$

Note that the figures in the third (fx) column have been formed by multiplying the corresponding figures in the first two columns.

From Equation S3.5, the mean number of breakdowns per week is:

$$\bar{x} = \frac{\Sigma fx}{\Sigma f} = \frac{104}{60} = 1.73$$

(Reasonableness check: The data are very roughly balanced around 2, which is also the mode. A mean not too far from 2 is therefore reasonable.)

Calculation of the mean from a grouped frequency distribution

It is not possible to calculate the exact mean of the original data in a grouped frequency distribution, since information is lost when the data are grouped. The mean can only be *estimated*.

In this case we assume that the observations are spread evenly throughout each class interval. This essentially means that the calculations are based on the assumption that *all* observations occur at the *midpoint* (m) of their class interval, so the formula for the calculation of the mean from a frequency distribution may be used. That is, in Equation S3.5 we replace x by m. This yields the formula:

$$\bar{x} \div \frac{\Sigma fm}{\Sigma f} \qquad \textbf{(S3.6)}$$

where:

m = the midpoint of the class interval
f = the frequency of the class interval
Σf = the sum of the frequencies

Example S3.11

In the year ending 30 June 2000, over 99% of industrial disputes in Australia were settled in less than 20 days. A summary of the duration of these disputes is shown in the grouped frequency distribution in Table S3.6. Estimate the duration of the industrial disputes in the table.

Table S3.6 Industrial disputes in Australia, 1999–2000

Duration of dispute (days)	Number of disputes
Less than 1	403
1–under 2	146
2–under 5	115
5–under 10	45
10–under 20	19
Total	728

Source: Australian Bureau of Statistics, *Industrial Disputes Australia*, Cat. No. 6321.0

Solution

The usual method for such a calculation is given in Table S3.7. Note that the midpoint (m) of each class interval (although usually obvious) can be found by calculating the mean of the end points of the class interval.

Table S3.7 Industrial disputes in Australia, 1999–2000

Duration of dispute (days)	f	Midpoint (m)	fm
Less than 1	403	0.5	201.5
1–under 2	146	1.5	219.0
2–under 5	115	3.5	402.5
5–under 10	45	7.5	337.5
10–under 20	19	15.0	285.0
Total	728		1445.5

Using Equation S3.6, the mean duration of industrial disputes in Australia during 1999–2000 can be estimated by:

$$
\begin{aligned}
\bar{x} &\doteq \frac{\Sigma fm}{\Sigma f} \\
&= \frac{1445.5}{728} \\
&= 1.99
\end{aligned}
$$

or about two days.

Weighted means

A modification of the usual arithmetic mean is the **weighted arithmetic mean**, which assigns *weights* (or measures of relative importance) to the observations to be averaged.

In particular, if observation x is assigned weight w, the formula for the weighted mean is:

$$\bar{x} = \frac{\Sigma xw}{\Sigma w} \qquad \textbf{(S3.7)}$$

The weights are usually expressed as percentages or fractions. Note that if all n observations have an *equal* weight (e.g. $w = \frac{1}{n}$ for all observations so that $\Sigma xw = \frac{\Sigma x}{n}$ and $\Sigma w = 1$), Equation S3.7 becomes the same as Equation S3.3.

Example S3.12

Suppose that a particular course in business statistics is graded in the following manner:

Assignments	15%
Project	25%
Mid-term exam	20%
Final exam	40%
	100%

Each student is assigned a mark out of 100 for each of the four components of the assessment. One student obtains marks of 83 for assignments, 72 for the project, 41 for the mid-term exam and 47 for the final exam. Find the student's weighted mean mark for the course.

Solution

From Equation S3.7, the student's weighted mean mark is:

$$
\begin{aligned}
\bar{x} &= \frac{83(15) + 72(25) + 41(20) + 47(40)}{15 + 25 + 20 + 40} \\
&= 57.45
\end{aligned}
$$

(Note that percentage weightings have been used.)

Weighted means are of particular importance in business and economics because of their use in price indexes. This topic is discussed in Chapter S10.

Other types of means

Under certain conditions, other types of means are used. For example, the *geometric mean* is used in economic data to average ratios or rates of change (expressed as positive numbers). This is discussed in Section S3.7. Also, the *harmonic mean* is used to average time rates. The calculation of the harmonic mean is beyond the scope of this book.

S3.5 QUARTILES

An extension of the concept of the median (which divides data into two equal parts) is that of **quartiles**, which divide data into *four* equal parts. Quartiles are often used with scores for aptitude tests, examinations and other testing situations. They are also used in commerce and industry when a large number of observations are involved.

If the data are divided into four equal parts, the points of separation are:

1. *First quartile* (Q_1) There are 25% of the observations *below* Q_1 and 75% of the observations *above* Q_1. Q_1 is sometimes called the *lower quartile*.
2. *Second quartile* (Q_2) There are 50% of the observations *below* and 50% of the observations *above* Q_2. Q_2 is obviously the median, and hence the former name is seldom used.
3. *Third quartile* (Q_3) There are 75% of the observations *below* Q_3 and 25% of the observations *above* Q_3. Q_3 is sometimes called the *upper quartile*.

Different opinions exist about the calculation of the quartiles in small samples, since it may not be possible to comply precisely with these definitions. The following rules are recommended for practical purposes:

Let n = the number of observations.

1. Suppose $\frac{n}{4}$ is *not* a whole number. Let m = the next whole number larger than $\frac{n}{4}$. Then the *lower quartile* is the mth observation of the sorted data counting from the *lower end*. The *upper quartile* is the mth observation of the sorted data counting from the *upper end*.
2. Suppose $\frac{n}{4}$ is a whole number. Let $m = \frac{n}{4}$. Then the *lower quartile* is halfway between the mth observation and the $(m + 1)$th observation of the sorted data counting from the *lower end*. The *upper quartile* is similarly defined, counting from the *upper end*.

Example S3.13

The number of enquiries made by customers at the information booth in a department store was recorded for each day over a 10-day period. They were:

52 31 29 90 25 63 43 39 48 66

Find the lower and upper quartiles for these data.

Solution

The sorted observations are:

25 29 31 39 43 48 52 63 66 90

The number of observations = n = 10.

Since $\frac{n}{4} = \frac{10}{4} = 2.5$, we define $m = 3$.

Therefore, the lower quartile = 3rd observation from the lower end
= 31
and the upper quartile = 3rd observation from the upper end
= 63

Example S3.14

A bank manager is interested in the amount of time it takes his tellers to service customers. On a particular morning he records the service time for 20 customers. The times are given below (to the nearest minute).

5	8	3	4	15	10	8	5	3	8
2	10	9	7	5	8	4	10	7	5

(a) Arrange the data in the form of a frequency distribution.
(b) Find the mean service time.
(c) Find the median service time.
(d) Find the service times that represent the lower and upper quartiles.

Solution

(a) Arranging the data into a frequency distribution, and also forming a cumulative frequency distribution, we obtain Table S3.8.

Table S3.8 Cumulative frequency distribution of service times for Example S3.14

Service time (minutes, x)	Frequency (f)	Cumulative frequency
2	1	1
3	2	3
4	2	5
5	4	9
7	2	11
8	4	15
9	1	16
10	3	19
15	1	20
	$\Sigma f = 20$	

(b) It can easily be shown that $\Sigma fx = 136$.
From Equation S3.5, the mean service time is:

$$\bar{x} = \frac{\Sigma fx}{\Sigma f} = \frac{136}{20} = 6.80 \text{ minutes}$$

(c) The median service time = the mean of the 10th and 11th observations
= 7 minutes

(d) Since $n = 20$, we have $m = \frac{n}{4} = 5$. Therefore, the lower quartile is the mean of the 5th and 6th observations from the lower end, i.e. $\frac{4+5}{2} = 4.5$ minutes. Similarly, the upper quartile is the mean of the 5th and 6th observations from the upper end, that is $\frac{8+9}{2} = 8.5$ minutes. $\frac{4+5}{2} = 4.5$ minutes

Calculation of the quartiles from a grouped frequency distribution

It is not possible to calculate the *exact* value of the quartiles of the original data in a grouped frequency distribution, since information is lost when the data are grouped. However, it is possible to make an *estimate* of their value. The class interval that contains the relevant quartile is called the **quartile class**.

The estimate of the lower quartile is given by Equation S3.8 and the upper quartile by Equation S3.9.

$$Q_1 = L + \frac{\left(\frac{n}{4} - C\right)}{f}(i) \quad \textbf{(S3.8)}$$

$$Q_3 = L + \frac{\left(\frac{3n}{4} - C\right)}{f}(i) \quad \textbf{(S3.9)}$$

where:

L = the real lower limit of the quartile class (containing Q_1 or Q_3)
$n = \Sigma f$ = the total number of observations in the entire data set
C = the cumulative frequency in the class immediately *before* the quartile class
f = the frequency of the relevant quartile class
i = the length of the real class interval of the relevant quartile class

Note

1. If the median class happens also to be the *initial* class interval in the data set, then the value of C is *zero*.
2. Suppose the class intervals are in discrete form such as 1–5, 6 –10, 11–15, and so on. In this case, if the median class were the interval, say, 21–25, then the *real* class interval would be defined as 20.5–25.5. In this case

$L = 20.5$
$i = 25.5 - 20.5 = 5$

Example S3.15

Table S3.9 shows the civilian labour force in Australia by age in September 2000. Estimate both the lower and upper quartiles of the age of the civilian labour force.

Table S3.9 Age and cumulative age distributions of the civilian labour force in Australia in September 2000

Age (years)	Frequency ('000)	Cumulative frequency ('000)
15–under 20	391.0	391.0
20–under 25	531.6	922.6
25–under 35	1034.1	1956.7
35–under 45	1070.5	3027.2
45–under 55	939.0	3966.2
55 and over	355.1	4321.3
Total	4321.3	

Source: Australian Bureau of Statistics, *Labour Force Australia*, September 2000, Cat. No. 6203.0

Solution

Since there are 4 321 300 observations, the lower quartile (Q_1) is the mean of the (4 321 300 ÷ 4 =) 1 080 325th and the 1 080 326th observations, counting from the *lower end* of the data. This makes the quartile class for the *lower quartile* to be 25–under 35.

The following values are used to estimate the lower quartile:

$$L = 25.0$$
$$n = 4321.3$$
$$C = 922.6$$
$$f = 1034.1$$
$$i = 35 - 25 = 10$$

Using Equation S3.8:

$$\begin{aligned} Q_1 &= 25.0 + \frac{\left(\frac{4321.3}{4} - 922.6\right)}{1034.1}(10) \\ &= 25.0 + 1.5 \\ &= 26.5 \end{aligned}$$

Hence, the lower quartile of the ages of the civilian labour force in Australia is about 26.5 years. This means that 25% are aged *under* 26.5 years.

Similarly, the upper quartile (Q_3) is the mean of the 1 080 325th and the 1 080 326th observations, counting from the *upper end* of the data (or the mean of the 3 240 974th and 3 240 975th observations counting from the lower end of the data). This makes the quartile class for the *upper quartile* to be 45–under 55.

The following values are used to estimate the upper quartile:

$$L = 45.0$$
$$n = 4321.3$$
$$C = 3027.2$$
$$f = 939.0$$
$$i = 55 - 45 = 10$$

Using Equation S3.9:

$$Q_1 = 45.0 + \frac{\left(\frac{3 \times 4321.3}{4} - 3027.2\right)}{939.0}(10)$$

$$= 45.0 + 2.3$$

$$= 47.3$$

Hence, the upper quartile of the ages of the civilian labour force in Australia is about 47.3 years. This means that 25% are aged *over* 47.3 years.

S3.6 DECILES, PERCENTILES AND FRACTILES

Further division of a distribution into a number of equal parts is sometimes used, the most common of these being **deciles** and **percentiles**.

Deciles divide the sorted data into 10 sections, each of which contains the same number of observations. For example, decile 7 indicates that value which has 70% of the observation values below it. Decile 5 indicates the median.

Percentiles divide the distribution into 100 sections, each of which contains the same number of observations. In general, the xth percentile (denoted by P_x) is that value which has x% of the observation values below it. For example, P_{20} is the 20th percentile, which separates the bottom 20% of the cases from the top 80%. In particular,

P_{25} = lower quartile
P_{50} = median
P_{75} = upper quartile

For small samples, the concept of quartiles, deciles and percentiles is somewhat difficult to apply. Consequently, these terms are usually applicable only to larger sets of data.

Instead of referring to the *percentage* of population values below a given figure, we sometimes refer to the *fraction* of the population below that figure. Thus, instead of using a percentile we would refer to a *fractile*. For example, the 30th percentile is the 0.30 fractile and the 75th percentile (upper quartile) is the 0.75 fractile.

S3.7 THE GEOMETRIC MEAN

On some occasions, when we are dealing with quantities that change over a period, we would like to know the mean rate of change. Examples would be the mean growth rate of savings over several years and the mean ratios of prices from one year to the next. In these circumstances, the *arithmetic mean* can be misleading and an alternative measure, known as the *geometric mean*, should be used. The geometric mean of n observations is the nth root of their product.

The formal definition for the geometric mean of n observations $x_1, x_2, x_3 \ldots x_n$ is given by:

$$\text{Geometric mean} = \sqrt[n]{x_1 x_2 x_3 \ldots x_n} \quad \textbf{(S3.10)}$$

The value of the geometric mean can easily be found using a calculator. Most calculators have a power button labelled x^y or $x^{\frac{1}{y}}$ or something similar, and this can be used to find the nth root of a number.

Note that the value of the geometric mean is undefined for a data set that contains zero or negative observations. This aspect is discussed in Example S3.17.

Example S3.16

Table S3.10 shows the estimated population of Australia as at 30 June in the five years 1996 to 2000, along with the percentage change from one year to the next. The estimated population on 30 June 1995 was 18 196 100.

Table S3.10 Resident population of Australia as at 30 June in the years 1996 to 2000

Year	Population ('000)	Percentage change
1996	18 423.6	1.250
1997	18 618.3	1.056
1998	18 842.2	1.203
1999	19 051.2	1.109
2000	19 162.6	0.585

***Source*:** Australian Bureau of Statistics, *Australian Demographic Statistics*, Cat. No. 3101.0

Although we could find the geometric mean of the population figures themselves, it is more relevant to find this value for the *percentage change* from one year to the next. In this case, note that the percentage changes are all *positive*. The geometric mean of the percentage changes is:

$$\begin{aligned}\text{Geometric mean} &= \sqrt[5]{(1.250)(1.056)(1.203)(1.109)(0.585)} \\ &= \sqrt[5]{1.032} \\ &= 1.006\end{aligned}$$

Using this measure, the mean annual increase in the resident population of Australia between 1996 and 2000 was 1.006%.

Example S3.17

The Australian Bureau of Statistics records information each month on the number of Australian residents who have been overseas for 12 months or more but are now returning home for at least 12 months. Table S3.11 shows this information for each of the five months from February 2000 to June 2000. Also shown is the percentage change in the data from the previous month. (The number of such residents in January 2000 was 6610.) Find the geometric mean of the percentage change of long-term arrivals during the five-month period.

Solution

The difficulty here is that some of the changes in percentages are *negative*, and we have already said that the geometric mean is undefined for negative changes. To overcome a

Table S3.11 Number of long-term arrivals who are Australian resident returning home for the period February–June 2000

Month	Number of long-term arrivals	Percentage change
February	5820	–11.95
March	5310	–8.76
April	6080	14.50
May	4560	–25.00
June	5520	21.05

Source: Australian Bureau of Statistics, *Overseas Arrivals and Departures*, Cat. No. 3401.0

problem where at least one of the changes is negative, we must relate *all* of the changes to 100%. This means that the *negative* percentages in Table S3.11 must each be *subtracted* from 100%, and the *positive* percentages *added* to 100%. This yields (in percentage terms) for the data in Table S3.11:

88.05 91.24 114.50 75.00 121.05

The interpretation of those figures less than 100 is that there was a *reduction* from the previous month. For example, the number of arrivals in March 2000 was only 91.24% of what it was in February 2000.

Using Equation S3.9 we have:

$$\begin{aligned}\text{Geometric mean} &= \sqrt[5]{(88.05)(91.24)(114.50)(75.00)(121.05)} \\ &= 96.46\end{aligned}$$

Since this figure is *less than* 100 (representing a *decrease*), we must subtract its value from 100 to yield 3.54. (If the figure had been more than 100 we would have subtracted 100 from the figure to obtain the mean percentage *increase*.)

This means that the mean monthly change in long-term Australian residents returning home during the five-month period was a *decrease* of 3.54%.

S3.8 SKEWNESS

The **skewness** of a distribution may be measured by comparing the relative positions of the mean, median and mode. If the distribution is *symmetrical*, the values of these three quantities are equal. However, if a distribution has extreme values at one end, the mean also tends to be pulled in that direction.

Figure S3.1 illustrates examples of symmetrical and skewed unimodal distributions. Note in (b) and (c), where the distribution is skewed, that the mean is the measure furthest out towards the tail, while the median lies between the mean and the mode.

There are various measures (beyond the scope of this book) that can be found to indicate the degree of skewness of a distribution. There is also a measure used to gauge the *peakedness* of a distribution. It is called **kurtosis** and shows the extent to which values are concentrated around the mode. That is, kurtosis gives an indication of whether the curve is peaked or flat.

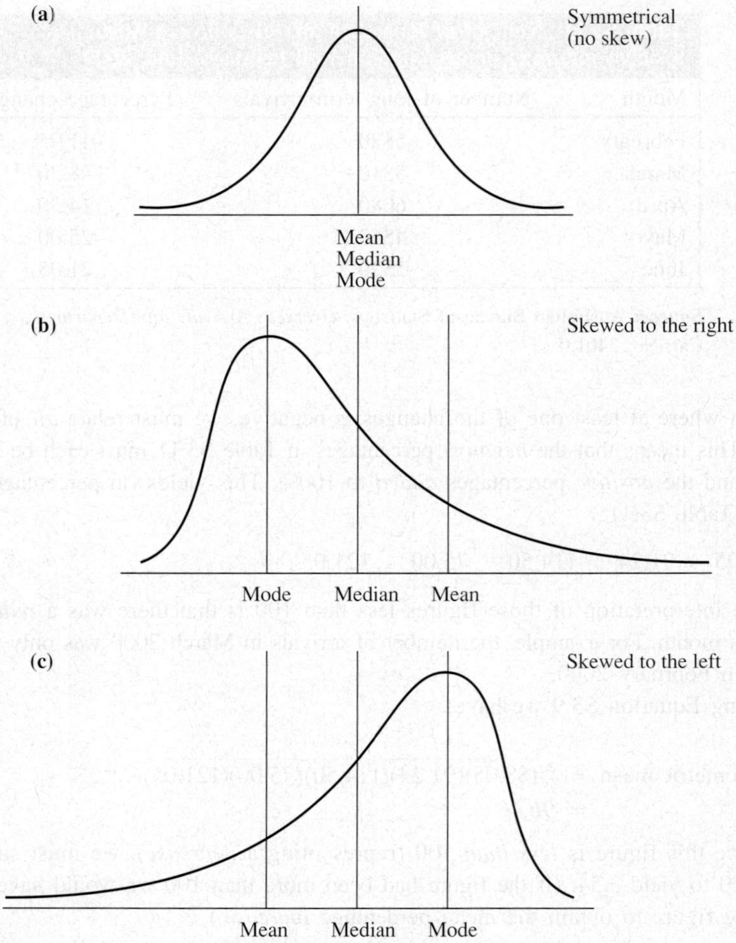

Figure S3.1 Relative positions of the mean, median and mode for various shapes of distributions

S3.9 REMARKS

Unfortunately, there is no general rule that can be used to determine which measure of central tendency is the most appropriate for a given situation. In selecting an average, the characteristics of each measure and the type of data available must be considered. Some characteristics of the various measures are outlined below.

The arithmetic mean

1. It is the most familiar measure.
2. It is the most commonly used measure.
3. It is often used when the data from a sample are used to make inferences about a population.
4. Its value may be distorted by a few outlying observations.
5. Its value cannot be calculated (without further information) for open-ended frequency distributions.

The median

1. It is easy to calculate.
2. It is easy to comprehend.
3. Its value is not unduly affected by outlying observations.
4. It is not used often in statistical inference.
5. Its value can be calculated for open-ended frequency distributions.

The mode

1. It is only rarely used as a measure.
2. It need not be a 'central' value.
3. There may be more than one mode in a set of data.
4. Its value can be calculated for open-ended distributions.
5. Its value is not unduly affected by outlying observations.
6. Its value is strongly influenced by random sampling fluctuations and may vary greatly from one sample to another.

Note that if a distribution is symmetrical, the mean, median and mode all have the same value. If the distribution is skewed, the median usually lies between the mean and the mode.

If a set of data consists of verbal categories rather than numbers, the mode (or modal class) will be the most appropriate measure of location that can generally be determined. It should be remembered that, although the mode represents the value that occurs most frequently, this does not imply that its frequency has a majority out of the total frequency. That is, merely because a value occurs more frequently than any other single value, it does not imply that the value occurs more often than all other values combined.

The median (and quartiles) serves the purpose of dividing the data into equal proportions. This is particularly useful when the data given are open-ended (i.e. when the upper and/or lower limits of the distribution are not known). Consider a researcher who is trying to determine the typical life expectancy of a certain breed of mice in an experiment. If the mean is used, every mouse must die before this figure can be calculated. If the median is used, as soon as half the mice have died the figure can be determined.

The median also is in common usage when referring to the typical price of a house or typical weekly wage. This is because a few very expensive houses or very high wage earners can artificially inflate the mean to make it seem that houses are generally more expensive than they really are or that weekly wages are inflated. Although such figures are often referred to in the media as 'average', they really relate to the median.

Figure S3.2 illustrates how the median price of houses sold at auction in Sydney changed each month between February 1999 and September 2000. Note that the vertical axis starts at $200 000, so that any changes look more dramatic than they actually may be. Note also that there is no bar for January 2000, since so few houses are auctioned in that month that any figure would be misleading. It can be said that each measure of central tendency gives a different conception of the term 'average'. Depending on the situation, one measure may be superior to another, and in some cases all three measures of mean, median and mode can be used in conjunction. In general, however, a single figure to represent an entire collection of observations may not be sufficient. To obtain a picture of how the data relate to the measure chosen, a measure of *dispersion* or spread of the data is usually given also. This aspect is considered in Chapter S4.

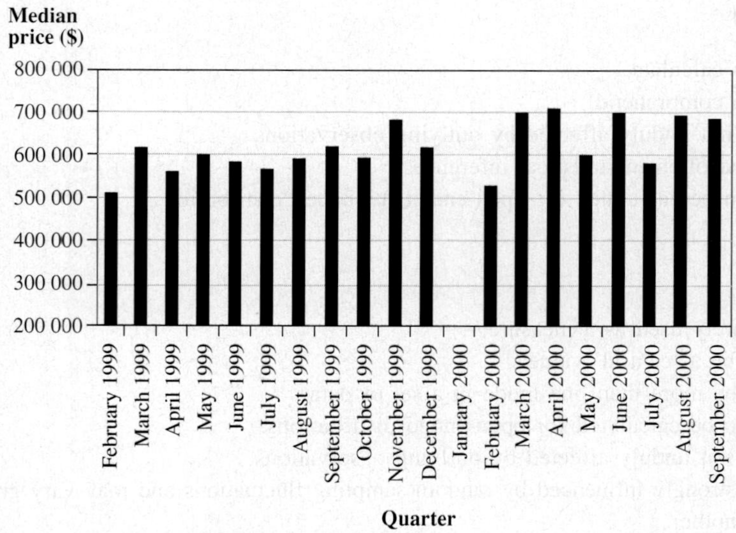

Figure S3.2 The median price of houses sold at auction in Sydney between February 1999 and September 2000

Source: © *Realty Auctioneer*, published by CPM Research (www.cpm.com.au)

Quick Quiz

1. According to a book publisher, the number of enquiries received on 12 successive working days by a large bookshop about a forthcoming novel was:

   28 15 13 9 18 22 17 8 5 15 20 11

   Find the mean number of enquiries received per day.

2. For the data in Question 1, find the median number of enquiries received per day.

3. A neighbourhood watch organiser records the number of burglaries in the district for each of the first 25 weeks of the year. The results are shown in the following frequency distribution:

Number of burglaries	0	1	2	3	4	5	6
Frequency	3	5	6	5	3	2	1

   Find the mean and median number of burglaries per week.

4. Find the lower and upper quartiles for the data in Question 3.

5. A manufacturer of liquid soap sells bottles that are alleged to contain 300 mL. A quality control inspector has recorded the actual contents of a random sample of 30 bottles in the following grouped frequency distribution:

Contents (mL)	*f*
290–under 295	6
295–under 300	8
300–under 305	12
305–under 310	4
Total	30

   Estimate the mean contents of these bottles.

6. Estimate the median of the data in Question 5.

7. A set of six numbers has a mean of 8. If one of the numbers is removed, the mean is now 10. What was the value of the number that was removed?

8. The attributes of honesty, punctuality and neatness are given weightings of 30%, 45% and 25%, respectively, when deciding whether an employee should be given a promotion. (Each employee is given a score out of 10 for each attribute.) Jack obtained scores of 3, 7 and 4 for these attributes, respectively. If Jack requires a *weighted* mean score of *better than* 5.0 to secure a promotion, is he going to reach it? Explain.

9. Using an arbitrary origin of 50, calculate the mean of the following 10 numbers:

   38 68 23 97 50 45 92 58 66 81

10. The number of goals scored by an Australian Rules forward for each of seven seasons was:

    64 29 45 59 62 43 76

    Use the geometric mean to find the mean rate of change in the number of goals kicked for the last six seasons.

Multiple Choice Questions

1. The *mean* of the eight values 5 21 16 13 25 6 35 31 is:
(a) 19 (b) 18.5 (c) 18
(d) 20 (e) none of the above

2. For the data in Question 1 the *median* is:
(a) 19 (b) 18.5 (c) 18
(d) 20 (e) none of the above

3. For the data in Question 1, the *lower quartile* is:
(a) 6 (b) 13 (c) 9.5
(d) 28 (e) none of the above

4. Consider the following frequency distribution of x-values:

x	10	20	30	40
f	4	6	8	2

The *mean* of these data is:
(a) 20 (b) 24 (c) 25
(d) 26 (e) none of the above

5. For the data in Question 4, the *median* is:
(a) 20 (b) 24 (c) 25
(d) 26 (e) none of the above

6. Consider the following grouped frequency distribution:

Class interval	f
0–under 20	3
20–under 40	12
40–under 60	5
Total	20

An estimate of the *mode* (to the nearest whole number) of these data is:
(a) 29 (b) 30 (c) 31
(d) 21 (e) none of the above

7. After playing a poker machine a number of times, a gambler won a mean of \$17.20 for a total win of \$137.60. The number of times the gambler played was:
(a) 4 (b) 8 (c) 16
(d) 120 (e) none of the above

8. The *median* is sometimes used in preference to the *mean* when:
(a) there is no calculator handy
(b) the data have outlying values
(c) the mean may be a very large number
(d) there are only a small number of observations
(e) none of the above

9. It is not possible to calculate the *exact* mean from data in a grouped frequency distribution because:
(a) calculators are not that accurate
(b) the data may contain very large numbers
(c) the median and mean might be different
(d) data are unknown because of the grouping
(e) none of the above

10. Suppose a set of data are *skewed to the left.* Then:
(a) the mean is larger than the median
(b) the mode is less than the median
(c) the median lies between the mean and the mode
(d) the median is greater than the upper quartile
(e) none of the above

Exercises

S3.1* The manager of a small office kept a record of the expenditure from petty cash for each month of 2000. The amounts (in dollars) were:

34.28	52.76	38.26	39.87	42.53	51.63
48.79	50.21	59.88	60.52	57.82	63.45

(a) Determine the mean monthly expenditure by calculating directly from the raw data.

(b) Determine the mean monthly expenditure using an arbitrary origin of $40.

(c) Determine the median monthly expenditure.

S3.2* Consider the following set of data:

8 4 7 2 5 12 10 7

Find:

(a) the mode (b) the median (c) the mean (d) the geometric mean

S3.3* (a) Show that for the following data the median lies between the mode and the mean:

1 2 2 2 3 4 5 6 15

(b) Design a set of data for which the mode lies between the median and the mean.

S3.4 Students' grades in a computer science course are assessed according to their performance in assignments, a mid-course exam and a final exam. The relative weightings of these are:

Assignments	20%
Mid-course exam	25%
Final exam	55%

A particular student obtains marks of 52 out of 60 for assignments, 48 out of 100 for the mid-course exam and 155 out of 200 for the final exam. Based on the above weightings, calculate a weighted mark out of 100 for this student.

S3.5* If you multiply each of a string of numbers by some fixed number, what happens to the median and the mean? For example, what happens if, by multiplying by 10, you change

49 51 55 58

to

490 510 550 580?

S3.6 Consider the following set of data:

16 36 100 4 64

(a) Show that the square root of the mean is not the same as the mean of the square roots.

(b) Show that the square root of the median is the same as the median of the square roots.

(c) Do your answers to (a) and (b) suggest any general statistical relationships? If they do, state what these relationships are.

S3.7* A statistics student was asked to find the median of the following set of data:

1 4 17 5 9

The answer given was 17. What error do you think the student made in arriving at that answer?

S3.8* A small business has 14 part-time employees where the annual salaries received are given below (in dollars):

16 000	15 000	60 000	17 000	16 000	55 000	18 500
17 000	14 500	15 000	17 000	16 000	15 000	15 000

(a) What is the mean salary? (Work directly from the raw data.)

(b) What is the median salary?

(c) What is the modal salary?

(d) Which of the above three measures is the most representative? Justify your answer.

(e) Repeat (a) and (b) on the basis of the data expressed in a frequency distribution format. Verify that you obtained the same answers as when you used the data in raw form.

S3.9 The gross weekly income of families in Australia who earn less than $2000 per week is shown in the table below. (This represents about 97% of all families.)

Gross weekly income ($)	Number of families ('000)
Under 200	1701.0
200–under 400	1999.0
400–under 600	1574.6
600–under 800	1173.5
800–under 1000	823.8
1000–under 1200	538.5
1200–under 1500	565.6
1500–under 2000	453.0
Total	8829.0

Source: Australian Bureau of Statistics, *Income Distribution*, Cat. No. 6523.0, 1997–98

Use the data to answer the following questions.

(a) Find the modal class.

(b) Find the class interval that contains the median.

(c) Estimate the mean gross weekly income.

(d) Estimate the median gross weekly income.

(e) Calculate the percentage of families in the table who have a gross weekly income of:

(i) less than $600 (ii) more than $1000

S3.10* A cricketer's batting average is determined by dividing the total number of runs scored by the total number of (completed) innings played. The batting statistics of 2 cricketers over each of 3 seasons are recorded as follows:

Cricketer	1999–2000 Number of runs	1999–2000 Number of innings	2000–2001 Number of runs	2000–2001 Number of innings	2001–2002 Number of runs	2001–2002 Number of innings
Amos	985	23	864	22	582	16
Taylor	873	31	625	25	643	30

(a) Calculate the batting average of each cricketer for each of the 3 seasons.

(b) Calculate the batting average of the 2 cricketers combined for each of the 3 seasons.

(c) Calculate the batting average of each of the 2 cricketers for the 3 seasons combined.

S3.11* The average speed of a car over a certain distance is calculated by dividing the total distance travelled by the total time taken. Two towns, A and B, are 10 km apart. A motorist travels from A to B at an average speed of 20 km per hour, and then travels from B to A at an average speed of 40 km per hour.

(a) Determine the average speed of the motorist for the two trips combined (i.e. from A to B and then back to A). (*Note*: The answer is *not* 30 km per hour.)

(b) What speed would the motorist have to average on the return trip from B to A to yield an overall speed of 30 km per hour? (Assume the motorist has averaged 20 km per hour from A to B.)

S3.12 In Exercise S3.11, would your answers to (a) and (b) be the same if the towns A and B were *any* distance apart? Explain.

S3.13 The age of the population in Australia (as at 30 June 2000) is shown in the following table in the form of a grouped frequency distribution.

Age (years)	Number ('000)
0–under 5	1263.1
5–under 15	2658.1
15–under 30	4185.0
30–under 50	5706.7
50–under 65	2983.9
65 and over	2360.2
Total	19157.0

Source: Australian Bureau of Statistics, *Australian Demographic Statistics*, June Quarter 2000, Cat. No. 3101.0

(a) Estimate the median age of people in Australia as at 30 June 2000.

(b) Even though there is an open-ended interval, make a reasonable estimate of the mean age of the population of people in Australia as at 30 June 2000.

(c) Estimate the lower quartile of the data.

(d) Estimate the upper quartile of the data.

(e) Estimate the mode of the data.

S3.14 The sales of vacuum cleaners in a department store for the first 10 weeks of the year are:

87 54 65 43 78 32 76 66 54 70

Find the mean weekly sales using an arbitrary origin of:

(a) 50 (b) 30

S3.15* The number of enquiries received at the information counter of a garden nursery were recorded for each 15-minute period during the 8 opening hours on a particular Saturday. The results are shown below in the form of a frequency distribution.

Number of enquiries	Frequency
2	1
5	4
7	6
10	12
12	9
Total	32

For these data find:

(a) the mode (b) the mean (c) the median

(d) Q_1 (e) Q_3

S3.16* The number of stop-work meetings held by the staff of a manufacturing plant were recorded for each week of 2001. The results are shown below in the form of a frequency distribution.

Number of meetings	Frequency
0	7
1	15
2	12
3	9
4	6
5	3
Total	52

For these data calculate:

(a) the mean (b) the median (c) the mode (d) the lower quartile

(e) the upper quartile

S3.17* Rank each of the following values in order of smallest to largest:

(a) the upper quartile (b) decile 4 (c) the 70th percentile

(d) the median (e) decile 6 (f) the lower quartile

(g) the 20th percentile (h) the 0.38 fractile

S3.18 The Go-Slow Construction Company employs 16 labourers at \$800 per week each, 5 architects at \$600 per week each, 1 doctor at \$1500 per week, 2 industrial liaison officers at \$2000 per week each and 20 part-time office workers at \$250 per week each. Find:

(a) the total wage payout

(b) the median wage

(c) the mean wage

(d) the effect on the mean and median if the industrial liaison officers obtain a wage increase of $200 per week each for themselves and $20 per week each for the labourers

S3.19 A personnel consultant has been asked to recommend which of 3 applicants should be appointed to a new position. She believes that the relative importance of various aspects is given by the following weightings:

Professional knowledge	6
General knowledge	6
Personal presentation	4
Quality of tertiary degree	3
Level of ambition	2

The candidates (Brown, Green and Gray) scored as follows on the various aspects, the scores being on a scale from 0 to 20 in each case:

Aspect	Brown	Green	Gray
Professional knowledge	14	12	16
General knowledge	15	14	17
Personal presentation	18	11	16
Quality of tertiary degree	12	12	10
Level of ambition	12	17	11

Calculate an overall weighted mean score for each candidate, and recommend which should be offered the position.

S3.20* Would the same relative positions of the candidates in Exercise S3.19 still apply if:

(a) all 5 aspects received equal weighting?

(b) weightings were in exactly the opposite order (i.e. level of ambition and quality of tertiary degree each receiving weight 6 while professional knowledge has only weight 2)?

S3.21 Give an example of a set of 5 numbers that are *bimodal* and where the value of the mean equals the value of the median.

S3.22* The number of major traffic accidents attended by a rescue squad in each of the first 8 weeks of 2002 were:

14 5 4 19 1 9 15 10

For these data calculate:

(a) the mean (b) the median (c) the mode

(d) the lower quartile (e) the upper quartile

S3.23 The amount of money spent by the first 50 customers on purchases at a sale of ladies' clothes at a large department store is summarised in the following table.

Amount spent ($)	Number of customers
0–under 50	3
50–under 100	6
100–under 150	7
150–under 200	11
200–under 250	15
250–under 300	8
Total	50

(a) Estimate the mean amount spent by the 50 customers.

(b) Estimate the median amount spent by the 50 customers.

(c) Find the modal class.

(d) Estimate the mode.

(e) Estimate the lower quartile.

(f) Estimate the upper quartile.

S3.24 According to the Health Services Association of NSW, the mean length of stay in New South Wales hospitals is 5.34 days. To test this claim, a random sample of 15 patients was chosen, and their lengths of stay (in days) in hospital were recorded as follows:

3 4 5 8 2 1 5 9 4 6 7 9 1 10 6

Are these results consistent with the claim? Why?

S3.25* In a survey of teenage boys, the weightings they attached to various criteria of desirable body attributes were:

Attribute	Percentage weighting
Bigger shoulders	30
Bigger chest	25
Bigger arms	25
Bigger body	15
Smaller waist	5

For each male in the survey a score out of 10 was given for each attribute. A particular male had the following scores:

Attribute	Score
Big shoulders	5
Big chest	7
Big arms	3
Big body	6
Small waist	2

For this male find:

(a) the (unweighted) mean score (b) the weighted mean score

S3.26 The price (in dollars) in June of a 1 kg packet of laundry detergent in Melbourne is shown below for each of 5 years.

Year	1980	1985	1990	1995	2000
Price ($)	1.78	2.96	4.11	4.63	4.65

Source: Australian Bureau of Statistics, *Average Retail Prices of Selected Items*, September Quarters 1980, 1985, 1990, 1995, 2000, Cat. No. 6403.0

(a) For these data find:

(i) the mean price (ii) the median price

(b) Find the percentage increase in the price of laundry detergent between each time period. Use these 4 percentages to calculate the overall geometric mean percentage increase in the price of laundry detergent each year.

S3.27 Construct a set of numbers in which the lower quartile and the mean have the same value, but each is different from the value of the upper quartile.

S3.28* The number of points scored by the top eight Rugby League teams in the NRL for the 2000 season is shown in the table below.

Team	Number of points scored
Broncos	696
Roosters	601
Knights	686
Raiders	506
Panthers	573
Storm	672
Eels	476
Sharks	570

Source: National Rugby League

For these data find:

(a) the mean (b) the median (c) the lower quartile (d) the upper quartile

S3.29 The number of males and females aged 15 years and over looking for full-time work in 6 Australian capital cities in September 2000 is shown in the table below.

City	Number of males	Number of females
Sydney	42 100	20 800
Melbourne	50 300	28 200
Brisbane	28 400	13 200
Adelaide	20 500	12 100
Perth	22 500	7 300
Hobart	5 500	2 000
Total	169 400	83 500

Source: Australian Bureau of Statistics, *Labour Force Australia*, September Quarter 2000, Cat. No. 6203.0

(a) For each of the males and females in the six capital cities, find:

(i) the mean (ii) the median

(iii) the lower quartile (iv) the upper quartile

(b) Construct a total column for each city, and hence find the mean and median of the total number of people looking for full-time work. Check whether either of these is the sum of the two corresponding values calculated in (a).

S3.30 The number of passengers carried on a ferry on each of 12 working days is shown below:

65 78 43 81 67 55 34 68 58 43 59 35

(a) For these data find:

(i) the mean (ii) the median (iii) the mode (iv) the lower quartile
(v) the upper quartile

(b) Recalculate the answers in (a) if each number is multiplied by 5.

(c) Recalculate the answers in (a) if 15 is added to each number.

(d) Explain how your answers in (b) and (c) relate to your answers in (a).

S3.31 Roll a fair six-sided die 10 times and record the outcomes. For your outcomes calculate:

(a) the mean (b) the median (c) the mode (d) the lower quartile
(e) the upper quartile

S3.32* The total number of goals scored in each of the 64 matches of the 1998 Soccer World Cup played in France are shown below in the form of a frequency distribution. (France beat Brazil 3–0 in the final.)

Number of goals	0	1	2	3	4	5	7
Frequency	4	12	12	18	11	6	1

Source: World Cup Soccer; *Sydney Morning Herald*; Hubbies World Cup results page

For these data calculate:

(a) the mean (b) the median (c) the mode (d) the lower quartile
(e) the upper quartile

S3.33 Toss a fair coin 5 times and record the number of heads. Repeat the experiment a further 9 times so that you have 10 values for the number of heads. For your set of 10 numbers, calculate:

(a) the mean (b) the median (c) the mode (d) the lower quartile
(e) the upper quartile

S3.34 Write the 10 values in Exercise S3.33 in the form of a frequency distribution. Recalculate your answers and check that they agree with those obtained previously.

S3.35 The number of litres of petrol sold at a service station was recorded for each of 60 days. The amounts are summarised in the following grouped frequency distribution.

Amount (L)	Frequency
Less than 10 000	6
10 000–under 20 000	14
20 000–under 30 000	21
30 000–under 40 000	13
40 000–under 50 000	6
Total	60

For these data estimate (to the nearest integer):

(a) the mean (b) the median (c) the mode (d) the lower quartile
(e) the upper quartile

S3.36* The heights (in cm) of the 7 players in an international ladies' netball team were:

167 185 190 176 188 175 180

Calculate the mean of these data using an arbitrary origin of:

(a) 100 cm (b) 180 cm

S3.37 State the type of skewness, if any, that you feel may be exhibited by the following data:

(a) the weights of adult females

(b) the heights of adult males

(c) the time it takes to serve customers at a bank

(d) the time it takes people to travel to work each day in their car

(e) the maximum temperature each day

(f) the value of Lotto numbers drawn each week

S3.38* The weekly rent (in dollars) for 3 types of houses in Melbourne with various numbers of bedrooms is shown below for 1999.

Type	Number of bedrooms	Weekly rent ($)
I	1	152
II	2	228
III	3	280

(a) For the 3 types of houses combined, find:

(i) the mean weekly rent (ii) the median weekly rent

(b) For houses of each type, calculate the mean weekly rent per bedroom.

(c) For houses of all types combined, find the mean weekly rent per bedroom. (Assume that there are an equal number of houses of each type.)

S3.39 Suppose in Exercise S3.38 that you are now also given the information in the table below.

Type	Number of bedrooms	Number of houses of this type
I	1	79 800
II	2	386 400
III	3	927 000

Source: Australian Bureau of Statistics, *Australian Housing Survey 1999*, Cat. No. 4182.0

(a) Use this information to find:

(i) the mean and median of the number of bedrooms in these houses combined

(ii) the mean and median weekly rent for houses of all types combined, weighted by the number of each type

(b) Calculate the proportion of houses for which the weekly rent is more than $200.

S3.40 A manager informs 2 employees, Annie and Caroline, that their performance will be rated according to 3 categories with the percentage weightings shown in the table below. The one who obtains the largest weighted mean score will be awarded a salary rise.

Category	Weighting
Initiative	30%
Creativity	50%
Ability to meet deadlines	20%

Annie and Caroline are given a score out of 10 in each category, and their scores are shown below.

Category	Annie's scores	Caroline's scores
Initiative	3.5	6.0
Creativity	8.0	6.5
Ability to meet deadlines	5.5	8.0

(a) Calculate the *unweighted* mean score for each employee.

(b) Calculate the *weighted* mean score for each employee. Who gets the salary rise?

S3.41 Suppose in Exercise S3.40 that Caroline complains that management is incorrect in awarding such a large weighting to creativity. The manager agrees and decides to swap the weightings for initiative and creativity. Calculate the new weighted mean score for each employee. Who gets the salary rise now?

Chapter S4

Measures of dispersion

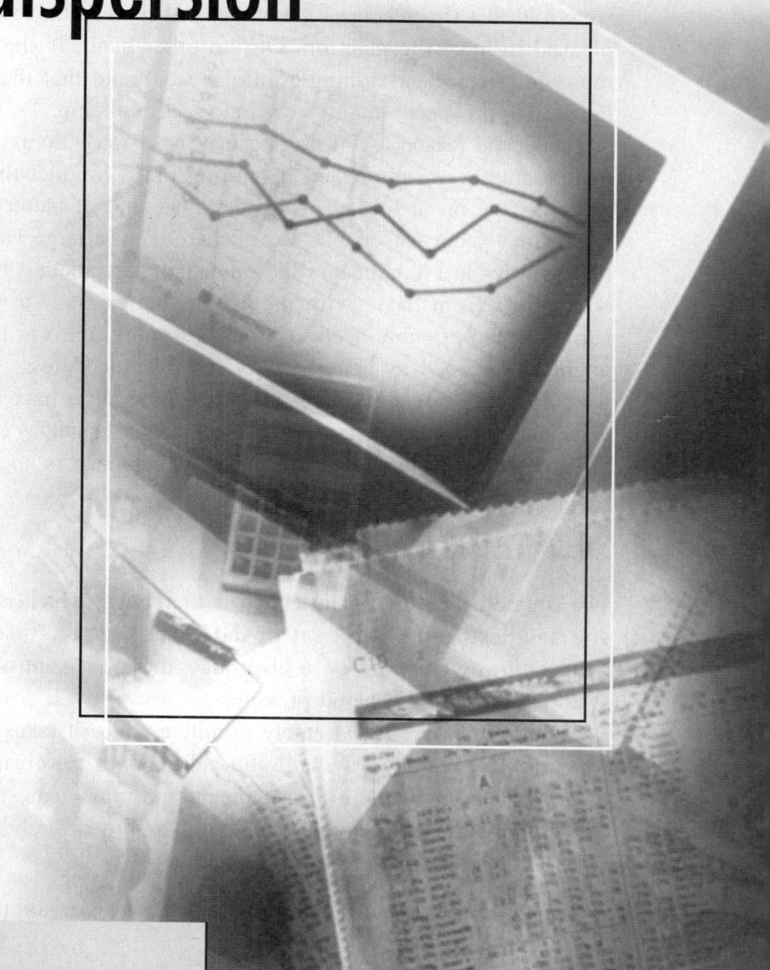

'Variety's the very spice of life that gives it all its flavour.'

(William Cowper, *The Task,* 1785)

Learning objectives

In this chapter you will learn to:

- Calculate common measures of dispersion from grouped and ungrouped data (including the range, interquartile range, mean deviation, and standard deviation)
- Calculate and interpret the coefficient of variation

S4.1 INTRODUCTION

In Chapter S3 various means of summarising a collection of observations with a single representative figure were presented. In general, however, a measure of central tendency in itself is not sufficient to describe a set of data adequately. In addition, a *measure of dispersion* (or spread) of the data is usually required. Such a measure gives an indication of the *internal variation* of the data, that is the extent to which data items vary from one another or from a central point. The term 'measure of dispersion' is used for such a quantity, since it is supposed to represent the extent to which the data are 'dispersed'.

There are several reasons for requiring a measure of dispersion of a set of data. First, it gives an indication of the *reliability* of the average value. For example, if a prospective sales representative were told that the average weekly wage of representatives in her section is \$800, she might be very pleased to have the position. On the other hand, if she were also told that the section has five representatives each earning \$500 per week and that the section head earns \$2300 per week, her immediate prospects might seem much less enticing.

A second reason for measuring the dispersion is to assist in *controlling* unwanted variation. Most manufacturers are interested in turning out a product that will have uniformly high quality. For example, the mean life span of a certain brand of battery used in calculators may be 1000 hours, which may seem reasonable. But if there were a large variability around the mean (say, about half the batteries last 1 hour and the other half last about 1999 hours), there would soon be customer complaints from those with the 1-hour batteries. If a manufacturer measures the extent of the variation, the possible need to produce a more uniform product will be apparent. Such uniformity might be obtained by better quality-control procedures.

In practice, for a given set of data, it is usual to have an indication of two quantities: an average value and a measure of dispersion. This chapter examines some of the more commonly used measures of dispersion and the conditions under which each is appropriate.

S4.2 THE RANGE

The simplest measure of dispersion is the **range**, which is simply the difference between the largest and smallest values in a set of data. The range is used in a number of situations, including temperature fluctuations on a given day, the movement of share prices and the acceptable range of systolic and diastolic blood pressures.

Such a measure is extremely primitive, since it takes account of only the most extreme values, which may not in any way be useful indicators of the great bulk of a population. Extreme values may often result from errors of measurement, and in any case may represent only a very small proportion of the total system. Ease of calculation therefore is rarely sufficient to justify use of the range alone.

The formula for the range is:

$$\text{Range} = \text{largest observation} - \text{smallest observation} \qquad \textbf{(S4.1)}$$

Example S4.1

The estimated readership (people aged 14 years and over) of the 12 metropolitan daily newspapers with the largest circulations in Australia in the year 2000 is shown in Table S4.1.

For these newspapers find:

(a) the mean readership
(b) the range of the readership

Table S4.1 Estimated readership of 12 metropolitan daily newspapers

Newspaper	Readership ('000)
Daily Telegraph (Mon–Fri., NSW)	1168
Weekend Australian (Sat/Sun., national)	934
Daily Telegraph (Sat., NSW)	965
Sunday Telegraph (Sun., NSW)	1893
Saturday Age (Sat., VIC)	1044
Sunday Mail (Sun., QLD)	1429
Sun-Herald (Sun., NSW)	1523
Herald-Sun (Sat., VIC)	1378
Sydney Morning Herald (Mon–Fri., NSW)	850
Sydney Morning Herald (Sat., NSW)	1301
Herald-Sun (Mon–Fri., VIC)	1503
Sunday Herald-Sun (Sun., VIC)	1456

Source: Roy Morgan Research, January 2000–December 2000

Solution

(a) From Equation S3.3, the mean number of readers (in thousands) is:

$$\bar{x} = \frac{1168 + 934 + 965 + \cdots + 1503 + 1456}{12}$$
$$= \frac{15\,444}{12}$$
$$= 1287$$

or 1 287 000 readers

(b) From Table S4.1, we have:

Smallest observation = 850 000 readers
Largest observation = 1 893 000 readers

From Equation S4.1:

$$\text{Range} = 1\,893\,000 - 850\,000$$
$$= 1\,643\,000 \text{ readers}$$

Example S4.2

Table S4.2 shows the amount of energy (in kJ) provided by one serve of each of 5 items on the restaurant menu list. Find the range of these data and comment.

Table S4.2 Energy provided by one serve of each of 5 takeaway menu items

Item	Energy (kJ)
Medium lemonade	712
Large orange juice	723
Medium cola	561
Ice-cream cone	652
Chicken burger	1995

Solution

From Table S4.2, we have:

Smallest observation = 561 kJ
Largest observation = 1995 kJ

From Equation S4.1:

$$\text{Range} = 1995 - 561 = 1434 \text{ kJ}$$

This answer demonstrates the most unfavourable feature of the range as a measure of dispersion. Had it not been for the figure of 1995, the range would have been only 723 – 561 = 162 kJ, giving a much truer indication of the *overall* variation of the bulk of the data. The single value of 1995 for a chicken burger has increased the range to over eight times its value without this item.

Example S4.2 shows that the range is therefore greatly influenced by extreme or outlying values. Such values are often referred to as **outliers** and are defined as values that are inconsistent with the rest of the data. As such, although the range is the quickest and easiest measure of dispersion to calculate, its interpretation should be treated with some caution.

S4.3 THE INTERQUARTILE RANGE (MIDSPREAD)

To overcome the problem of the ranges being affected by outlying observations, a different type of range is sometimes used. Basically, this measures the range of the 'middle' 50% of the values only, rather than the range of all the observations. Such a range is referred to as the **interquartile range** or the **midspread** (since it measures the 'spread' of the middle of the data).

The interquartile range is defined as the difference between the upper and lower quartiles (where the quartiles are as defined in Section S3.5). That is:

$$\text{Interquartile range} = \text{upper quartile} - \text{lower quartile} = Q_3 - Q_1 \qquad \textbf{(S4.2)}$$

The interquartile range is a specific example of an **interfractile range**. An interfractile range is the range of values between two specified fractiles. In the case of the interquartile range it is the range of values between the 0.25 fractile and the 0.75 fractile.

Note that *one-half* of the value of the interquartile range is sometimes referred to as the **quartile deviation** or the **semi-interquartile range**.

Example S4.3

The average retail prices in the September quarter 2000 of 10 grocery items in Melbourne, selected at random, are shown in Table S4.3.

Table 4.3 Average retail prices of 10 selected grocery items in Melbourne during September quarter 2000

Item	Price ($)
Rice (1 kg)	1.59
Butter (500 g)	2.13
Teabags (180 g)	3.36
Pet food (410 g)	0.95
Toilet soap (4 × 125 g)	2.59
Onions (1 kg)	1.10
Dry biscuits (250 g)	1.70
Tomato sauce (600 mL)	1.78
Frozen peas (500 g)	1.33
Leg of pork (1 kg)	5.93

Source: Australian Bureau of Statistics, *Average Retail Prices of Selected Items*, September Quarter 2000, Cat. No. 6403.0

For these data find:

(a) the mean
(b) the range
(c) the interquartile range
(d) the quartile deviation

Solution

(a) From Equation S3.3, the mean of the data is:

$$\bar{x} = \frac{1.59 + 2.13 + ... + 1.33 + 5.93}{10}$$
$$= \$2.25$$

(b) The sorted data are (in dollars):

0.95 1.10 1.33 1.59 1.70 1.78 2.13 2.59 3.36 5.93

From Equation S4.1:

$$\text{Range} = \$5.93 - \$0.95$$
$$= \$4.98$$

(c) Lower quartile = 3rd sorted observation from the *left*
= \$1.33

$$\text{Upper quartile} = \text{3rd sorted observation from the } \textit{right}$$
$$= \$2.59$$

From Equation S4.2:

$$\text{Interquartile range} = \$2.59 - \$1.33$$
$$= \$1.26$$

Note that the interquartile range represents the range of the middle 50% of the data and is considerably less than $4.98, the range of the entire set of data. This is because the range has been distorted by the outlying value of $5.93. If this value had been omitted, the range would be only $2.41.

(d) $$\text{Quartile deviation} = \frac{\text{interquartile range}}{2}$$
$$= \frac{\$1.26}{2}$$
$$= \$0.63 \text{ or } 63 \text{ cents}$$

Example S4.4

Find the approximate interquartile range for the age data given in the grouped frequency distribution in Example S3.15.

Solution

The values of the lower and upper quartiles have already been estimated to be, respectively, $Q_1 = 23.7$ years and $Q_3 = 43.3$ years. Using Equation S4.2, the interquartile range is approximately:

$$\text{Interquartile range} = Q_3 - Q_1$$
$$= 43.3 - 23.7$$
$$= 19.6 \text{ years}$$

The interquartile range may be calculated from grouped frequency distributions that contain open-ended class intervals, but still ignores 50% of the total values of the distribution. It is usually only used with a large number of observations. While it is less affected by extreme values than the range, it is still not among the most commonly used measures of dispersion. Neither the range nor the interquartile range takes into account the value of *every* observation. In the following section we consider some measures that do precisely this.

S4.4 THE MEAN DEVIATION

A measure that does take into account the actual value of each observation is the **mean deviation** (sometimes called the *mean absolute deviation*). In effect, the mean deviation measures the 'average' distance of each observation away from the mean of the data. It gives an equal weight to each observation and is, in general, more sensitive than either the range or interquartile range, since a change in *any* value will affect it.

Suppose that a set of x-values has a mean of $\bar{x}$. Then the *residual* of a particular x-value in the set is defined as:

$$\text{Residual} = x - \bar{x} \tag{S4.3}$$

Example S4.5

For the data in Example S4.3, find the residual of the observations:

(a) \$1.59

(b) \$3.36

Solution

The mean of the data is $\bar{x}$ = \$2.25. From Equation S4.3:

(a) Residual of \$1.59 is \$1.59 − \$2.25 = −\$0.66

(b) Residual of \$2.04 is \$3.36 − \$2.25 = \$1.11

Note that \$1.59 has a *negative* residual since it is *below* the mean, while \$3.36 has a *positive* residual since it is *above* the mean.

The residuals, as defined above, measure the actual deviation (or distance) of each observation from the mean. Suppose we consider the *magnitude* of the residuals, that is the value of the deviations *ignoring the sign*. We would then have the absolute deviation of each observation from the mean, that is $|x - \bar{x}|$.

The *mean deviation* is defined as the mean of these absolute deviations. That is:

$$\text{Mean deviation} = \frac{\Sigma|x - \bar{x}|}{n} \tag{S4.4}$$

The *mean deviation* may be found as follows:

1. Calculate the mean of the data.
2. Subtract the mean from each observation and record the resulting differences.
3. Write down the *absolute value* of each of the differences in Step 2. (That is, record the differences in Step 2 and ignore their signs.)
4. Calculate the mean of the absolute values of the differences in Step 3.

We shall return to the calculation of this quantity later, but it is worth noting now that Steps 1 to 4 can be symbolised as:

1. Find $\bar{x}$.
2. For each x, find $x - \bar{x}$.
3. Now find $|x - \bar{x}|$ for each x.
4. Find the mean of the quantities in Step 3. That is, add them up $(\Sigma|x - \bar{x}|)$ and divide by n.

Example S4.6

The batting scores of two cricketers, Smith and Jones, were recorded over their 10 completed innings to date. Their scores were:

Smith	32	27	38	25	20	32	34	28	40	29
Jones	3	80	64	5	11	87	0	2	53	0

(a) Calculate the batting average (mean score) of each cricketer.
(b) Calculate the mean deviation of the scores of each cricketer.
(c) There is only one batting position left in your team for the next match. Would you select Smith or Jones? Why?

Solution

(a) Smith's batting average is:

$$\bar{x} = \frac{32 + 27 + ... + 29}{10}$$
$$= 30.5 \text{ runs}$$

Jones' batting average is:

$$\bar{x} = \frac{32 + 27 + ... + 29}{10}$$
$$= 30.5 \text{ runs}$$

(b) The usual procedure for calculating the mean deviation is illustrated below. Consider first the scores for Smith as set out in Table S4.4. Note that $\Sigma(x - \bar{x}) = 0$. This will always be true for *any* set of data. From Equation S4.4, the mean deviation of Smith's scores is:

$$\text{Mean deviation} = \frac{\Sigma|x - \bar{x}|}{n}$$
$$= \frac{47.0}{10}$$
$$= 4.7$$

Table S4.4 Calculations for the mean deviation of Smith's scores in Example S4.6

Score (x)	Deviation from mean ($x - \bar{x}$)	Absolute value of deviation $\lvert x - \bar{x}\rvert$
32	+1.5	1.5
27	−3.5	3.5
38	+7.5	7.5
25	−5.5	5.5
20	−10.5	10.5
32	+1.5	1.5
34	+3.5	3.5
28	−2.5	2.5
40	+9.5	9.5
29	−1.5	1.5
	$\Sigma(x - \bar{x}) = 0$	$\Sigma\lvert x - \bar{x}\rvert = 47.0$

The calculation of the mean deviation of Jones' scores is shown in Table S4.5.

Table S4.5 Calculations for the mean deviation of Jones' scores in Example S4.6

Score (x)	Deviation from mean ($x - \bar{x}$)	Absolute value of deviation $\lvert x - \bar{x}\rvert$
3	−27.5	27.5
80	+49.5	49.5
64	+33.5	33.5
5	−25.5	25.5
11	−19.5	19.5
87	+56.5	56.5
0	−30.5	30.5
2	−28.5	28.5
53	+22.5	22.5
0	−30.5	30.5
	$\Sigma(x - \bar{x}) = 0$	$\Sigma\lvert x - \bar{x}\rvert = 324.0$

Note that the value of $\Sigma(x - \bar{x})$ in both calculations is zero, demonstrating again that the mean of the deviations from the mean *must* be zero.

From Equation S4.4, the mean deviation of Jones' scores is:

$$\text{Mean deviation} = \frac{\Sigma\lvert x - \bar{x}\rvert}{n} = \frac{324.0}{10} = 32.4$$

(c) The answer to this question clearly depends on your priorities. If you are looking for a *consistent* batter, the choice will be Smith, since he has by far the smaller mean deviation. While he would probably not make a large score, his past record indicates that he can usually be relied on to make a score fairly close to his average (the mean deviation of his scores is less than 5). If you are looking for a batter who could possibly obtain a large score (and in doing so, considerably help win a match), Jones will be the choice. On the other hand, there also seems a high risk that he would get a very low score.

Calculation of the mean deviation from a frequency distribution

If the data are in the form of a frequency distribution, the mean deviation can be calculated using the following formula:

$$\text{Mean deviation} = \frac{\Sigma f\lvert x - \bar{x}\rvert}{\Sigma f} \qquad \textbf{(S4.5)}$$

where f = the frequency of an observation x, and $n = \Sigma f$ = the sum of the frequencies.

Example S4.7

A quality control inspector has recorded the number of faulty items produced on an assembly line for each of the first 127 working days of the year. The figures are recorded in the following frequency distribution:

Number of faulty items	0	1	2	3	4	5	6	7
Frequency	17	12	19	28	21	19	9	2

Find the mean and mean deviation of the number of faulty items produced each day.

Solution

The method of calculation is shown in Table S4.6.

From Equation S3.5, the mean of the number of faulty items produced each day is $\frac{381}{127} = 3.0$.

From Equation S4.5, the mean deviation of the number of faulty items produced each day is $\frac{188}{127} = 1.48$.

Note that Columns 2 and 3 of Table S4.6 are used to find the mean, and that the mean is then used to formulate Columns 4 and 5.

Table S4.6 Calculations for Example S4.6

x	f	fx	$\lvert x - \bar{x}\rvert$	$f\lvert x - \bar{x}\rvert$
0	17	0	3	51
1	12	12	2	24
2	19	38	1	19
3	28	84	0	0
4	21	84	1	21
5	19	95	2	38
6	9	54	3	27
7	2	14	4	8
	$\Sigma f = 127$	$\Sigma fx = 381$		$\Sigma f\lvert x - \bar{x}\rvert = 188$

To calculate the mean deviation from a *grouped* frequency distribution, assume that the observations in each class interval are concentrated at the midpoint of the interval. Then carry out the above procedure for a frequency distribution.

S4.5 THE STANDARD DEVIATION

Easily the most commonly used measure of dispersion is the **standard deviation**. In a similar manner to the mean deviation, the standard deviation takes into account the value of *every* observation and measures the 'average' deviation of the observations away from their mean.

Although the mean deviation is easier to calculate, it is the absolute value signs that it contains that make it unpleasant for further analysis in more complex statistical analysis. For this reason it is not favoured by statisticians. The standard deviation does not have this problem, although, like the mean deviation, it is still based on the residual values. In general, the values of the mean deviation and standard deviation should be reasonably close, since they are both measuring the same thing; namely, the variation of the observations from their mean.

Population standard deviation

The aim is basically to find an 'average' measure of each observation away from the mean of the set of observations. If the actual residuals are used (i.e. retaining their sign), their mean is *always* zero and therefore quite useless. If we take the mean of the *absolute values* of the residuals, we obtain the mean deviation with the troublesome absolute value signs.

An alternative approach is to work with the *squares* of the residuals. This will eliminate the effect of the signs, since squares of numbers cannot be negative. A first step is therefore to find the sum of the squares of the residuals, and then to find the mean. To take account of the fact that we have squared the residuals, we then take the *square root* of this mean.

The result of this technique is to find the **population standard deviation**, which is denoted by the Greek letter σ (sigma). Its formula is given by:

$$\sigma = \sqrt{\frac{\Sigma(x - \bar{x})^2}{N}} \qquad \textbf{(S4.6)}$$

where N = the size of the population

The *square* of the population standard deviation is also known as the *variance* (σ^2).

Sample standard deviation

In practice, it is rare to calculate the value of σ since populations are usually very large (often thousands of observations). Instead, it is far more likely that the *sample standard deviation* (denoted by s) will be required.

Although it would be tempting to make the formula for calculating s the same as Equation S4.6 with the population size N replaced by the sample size n, there are good theoretical reasons for not doing so. If we did this, and used the value of s to estimate the value of σ, the result would be too small. To correct this error, instead of dividing by n we divide by $(n - 1)$. This results in the following formula for s:

$$s = \sqrt{\frac{\Sigma(x - \bar{x})^2}{n - 1}} \qquad \textbf{(S4.7)}$$

An algebraically equivalent formula for s (which yields exactly the same answer as Equation S4.7) is:

$$s = \sqrt{\frac{\Sigma x^2 - n\bar{x}^2}{n - 1}} \qquad \textbf{(S4.8)}$$

Note on the use of (*n* – 1) in formulae

If the value of n is large, then it will only make a slight difference to your answer whether you divide by n or $(n - 1)$ in Equations S4.7 and S4.8. However, in this book we will use the technically correct formulae for s by always dividing by $(n - 1)$.

You should be aware that some other textbooks do in fact use a formula for s that divides by n, usually on the grounds that it is easier to explain to students! In such cases, any other formulae they use that involve s (such as those in Chapter S12 of this text) will also be slightly different (and slightly incorrect).

The dilemma created by different instructors and textbooks using different formulae for the value of s has not been lost on the manufacturers of calculators. To keep both markets happy, most statistical calculators provide buttons that will supply both answers, and you should be aware of which button is correct for your purpose.

If you want to calculate the value of s from a *sample* using Equation S4.7 or S4.8, the calculator button will usually be indicated by one of σ_{n-1} or $x\sigma_{n-1}$ or s_x or s written either on it or near it. This is the key you will mostly want to use to find the sample standard deviation.

If you want to calculate the value of σ from a *population* using Equation S4.6, the calculator key will usually be indicated by one of σ_n or $x\sigma_n$ or σ_x or σ written either on it or near it.

Important points about the standard deviation

There are several important points to note regarding the value of the standard deviation:

1. The standard deviation cannot be negative.
2. The standard deviation of a set of data is zero if, and only if, the observations are of equal value.
3. As a rough guide to whether a calculated answer is 'reasonable', the standard deviation is often *approximately* 30% of the range.
4. The standard deviation can *never* exceed the range of the data.
5. If a constant k is *added* to each observation in a set of data, the standard deviation of the new set of data has the same value as the standard deviation of the original set of data.
6. If each observation in a set of data is *multiplied* by a constant k, the standard deviation of the new set of data is k times the standard deviation of the original set of data.
7. The more scattered the data, the greater the standard deviation.
8. Because of the squaring operation involved in its calculation, the standard deviation is more influenced by extreme values than is the mean deviation, and thus is usually slightly larger than the mean deviation.
9. The *square* of the standard deviation is called the *variance*.

Calculation of the sample standard deviation

If we use Equation S4.7 to calculate the sample standard deviation of a set of x-values, the steps are as follows:

1. Calculate the mean $\bar{x}$.
2. For each x-value, find the value of the residual. That is, find $x - \bar{x}$.
3. Square the residuals in Step 2. That is, find each $(x - \bar{x})^2$.
4. Calculate the sum of the squares of the residuals in Step 3. That is, find $\Sigma(x - \bar{x})^2$.
5. Divide the sum in Step 4 by $(n - 1)$.
6. Take the square root of the quantity in Step 5. This is the sample standard deviation.

Example S4.8

A market researcher was interested in the discrepancy in the prices charged by supermarkets for a leading brand of pet food. To check this, he selected a random sample of 12 stores and recorded the price displayed for the same brand of a 400 g can. The prices (in cents) were:

89 72 77 78 82 94 76 78 73 80 88 85

Find:

(a) the mean price
(b) the range of the prices
(c) the mean deviation of the prices
(d) the standard deviation of the prices

Solution

(a) From Equation S3.3, the mean price is $\bar{x} = 81$ cents.
(b) The range of the prices is $94 - 72 = 22$ cents.
(c) and (d) Both the mean deviation and the standard deviation can be found using the calculations in Table S4.7.

Table S4.7 Calculations for Example S4.8

Price (cents)	Residual	Absolute value of residual $\lvert x - \bar{x} \rvert$	Square of residual $(x - \bar{x})^2$
89	+8	8	64
72	−9	9	81
77	−4	4	16
78	−3	3	9
82	+1	1	1
94	+13	13	169
76	−5	5	25
78	−3	3	9
73	−8	8	64
80	−1	1	1
88	+7	7	49
85	+4	4	16
	$\Sigma(x - \bar{x}) = 0$	$\Sigma\lvert x - \bar{x} \rvert = 66$	$\Sigma(x - \bar{x})^2 = 504$

Using the results in Table S4.7 and Equation S4.4, the *mean deviation* of the prices is:

$$\text{Mean deviation} = \frac{\Sigma\lvert x - \bar{x} \rvert}{n} = \frac{66}{12} = 5.50 \text{ cents}$$

Using the results in Table S4.7 and Equation S4.7, the *standard deviation* of the prices is:

$$\begin{aligned} s &= \sqrt{\frac{\Sigma(x - \bar{x})^2}{n - 1}} \\ &= \sqrt{\frac{504}{11}} \\ &= \sqrt{45.82} \\ &= 6.77 \text{ cents} \end{aligned}$$

Note

1. The sum of the residuals in Table S4.7 is *zero*. This is true for *any* set of observations. (If the mean is rounded in your calculations, the sum of the values may not be quite zero, due to rounding errors, but should be very close.)
2. To check Point 3 on page S.120, 30% of the range is 0.30 × 22 = 6.6 cents, which is close to our answer of 6.77 cents. Our answer therefore seems *reasonable*.
3. To check Point 4 on page S.120, our answer of 6.77 cents is less than the range of 22 cents.
4. The mean deviation and standard deviation have similar values, with the latter being slightly larger. This is consistent with Point 8 on page S.120.

Example S4.9

Find the standard deviation of the prices for the data in Example S4.8 using the formula in Equation S4.8.

Solution

The calculations are shown in Table S4.8.

Table S4.8 Calculations for Example S4.9

x	x^2
89	7921
72	5184
77	5929
78	6084
82	6724
94	8836
76	5776
78	6084
73	5329
80	6400
88	7744
85	7225
	$\Sigma x^2 = 79\,236$

From Equation S4.8, the standard deviation of the prices is:

$$s = \sqrt{\frac{\Sigma x^2 - n\bar{x}^2}{n-1}}$$
$$= \sqrt{\frac{79\,236 - (12)(81)^2}{11}}$$
$$= \sqrt{45.82}$$
$$= 6.77 \text{ cents}$$

which is the same as the value found using Equation S4.7 in Example S4.8(d).

Calculation of the standard deviation from a frequency distribution

If the data are in the form of a frequency distribution, the standard deviation of a sample can be found using the formula:

$$s = \sqrt{\frac{\Sigma f(x - \bar{x})^2}{\Sigma f - 1}} \qquad \textbf{(S4.9)}$$

Example S4.10

A life assurance policy may be purchased for one, two or three units of insurance cover, each unit being worth \$50 000 on maturity. The numbers of each type of policy purchased by the first 400 customers are shown in Table S4.9.

Table S4.9 Units of cover purchased in Example S4.10

Numbers of units of cover	Frequency (f)
1	85
2	192
3	123
Total	400

(a) Find the mean and standard deviation of units of cover purchased by the customers.
(b) Check that the answer in (a) is reasonable.
(c) Interpret the value of s in (a) in terms of amounts payable on maturity.

Solution

The relevant calculations can be made by constructing Table S4.10.

Table S4.10 Calculations for the mean and standard deviation in Example S4.10

x	f	fx	$(x - \bar{x})$	$(x - \bar{x})^2$	$f(x - \bar{x})^2$
1	85	85	−1.095	1.199	101.92
2	192	384	−0.095	0.009	1.73
3	123	369	0.905	0.819	100.74
	400	838			204.39

(a) From Equation S3.5, we find:

$$\begin{aligned}\bar{x} &= \frac{\Sigma fx}{\Sigma f} \\ &= \frac{838}{400} \\ &= 2.095 \text{ units}\end{aligned}$$

From Equation S4.9:

$$\begin{aligned}s &= \sqrt{\frac{\Sigma f(x-\bar{x})^2}{\Sigma f - 1}} \\ &= \sqrt{\frac{204.39}{399}} \\ &= 0.72 \text{ units}\end{aligned}$$

(b) The range of the data is 3 − 1 = 2 units. We note that 30% of the range is 0.30 × 2 = 0.60 units, which is close to our answer in (a) of 0.72 units. We also verify that our answer is less than the range of 2 units.

(c) Since one unit of insurance cover is worth \$50 000, the typical variation of 0.72 units must correspond to a typical variation in the amounts payable of \$50 000 × 0.72 = \$36 000.

Calculation of the standard deviation from a grouped frequency distribution

As in calculating the mean deviation, when calculating s from a grouped frequency distribution we should assume that the observations in each class interval are concentrated at the *midpoint* of the interval. The procedure in Example S4.10 can then be followed.

The formula for estimating the standard deviation from a grouped frequency distribution is:

$$s \doteq \sqrt{\frac{\Sigma f(m-\bar{x})^2}{\Sigma f - 1}} \qquad \textbf{(S4.10)}$$

where $\bar{x}$ = the estimated mean of the sample
m = the midpoint of the class interval
f = the frequency of the class interval

Example S4.11

One of the ways in which the market value of a used motor vehicle in Australia can be estimated is by using the prices that vehicles of the same type bring in at auction sales. Each vehicle may be classified to be in one of several conditions (e.g. *quite good, good, very good*). Much of the guesswork is taken away from car sale yards and insurance companies, since all they have to do is consult a book (which is usually updated each month) to find the market value for any make, model and year of any vehicle in a particular condition.

Suppose that we had the task of estimating the current market value of a 6-year-old small sedan of which 50 had been sold at auction in the previous month. The data supplied are shown in Table S4.11.

Table S4.11 Car sale prices for Example S4.11

Sale price ($)	Number sold
8 000–under 8 500	4
8 500–under 9 000	10
9 000–under 9 500	21
9 500–under 10 000	12
10 000–under 10 500	3
Total	50

(a) Estimate both the mean and standard deviation of the sale prices.

(b) If a car in very good condition were considered to have a market value of two standard deviations above the mean, what would its value be?

Solution

The required calculations are summarised in Table S4.12. Note that the midpoints of each class interval (m) are found by finding the mean of the two endpoints of the interval.

Table S4.12 Calculations for estimating the mean and standard deviation in Example S4.11

Sale price ($)	f	m	fm	$(m - \bar{x})$	$f(m - \bar{x})^2$
8 000–under 8 500	4	8 250	33 000	−1000	4 000 000
8 500–under 9 000	10	8 750	87 500	−500	2 500 000
9 000–under 9 500	21	9 250	194 250	0	0
9 500–under 10 000	12	9 750	117 000	500	3 000 000
10 000–under 10 500	3	10 250	30 750	1000	3 000 000
Total	50		462 500		12 500 000

(a) From Equation S3.6, the estimate of the mean price is:

$$\bar{x} \doteqdot \frac{\Sigma fm}{\Sigma f} = \frac{462\ 500}{50} = \$9250$$

From Table S4.12 we have:

$$\Sigma f - 1 = 50 - 1 = 49$$
$$\Sigma f(m - \bar{x})^2 = 12\ 500\ 000$$

Using Equation S4.10:

$$s \doteq \sqrt{\frac{12\ 500\ 000}{49}}$$
$$= \sqrt{2\ 551\ 102.0}$$
$$= \$505.08$$

Note that in this question we have only been able to *estimate* the mean and standard deviation of the prices. We cannot calculate their exact values because the data have been grouped and therefore information has been lost.

(b) Using the results in (a), a car of this type in very good condition with a market value of two standard deviations above the mean would be worth:

$$\text{Market value} = \bar{x} + 2s$$
$$= \$9\ 250 + 2(\$505.08)$$
$$= \$10\ 260 \text{ (to the nearest \$10)}$$

S4.6 THE COEFFICIENT OF VARIATION

The **coefficient of variation** is a measure of *relative* variability. It is used to measure the changes that have taken place in a population over time, or to compare the variability of two populations that are expressed in different units of measurement. It is expressed as a percentage rather than in terms of the units of the particular data.

The formula for the coefficient of variation, denoted by V, is

$$V = 100\left(\frac{s}{\bar{x}}\right)\% \qquad \textbf{(S4.11)}$$

where $\bar{x}$ = the mean of the sample
s = the standard deviation of the sample

Example S4.12

Calculate the coefficient of variation for the price of the 400 g cans of pet food in Example S4.8. Interpret the result.

Solution

We have already calculated, for the random sample of 12 cans, that $\bar{x}$ = 81 cents and s = 6.77 cents. Using Equation S4.11, the coefficient of variation is found to be:

$$V = 100\left(\frac{s}{\bar{x}}\right)\%$$
$$= 100\left(\frac{6.77}{81}\right)\%$$
$$= 8.36\%$$

This means that the standard deviation of the price of a 400 g can of pet food is 8.36% of the mean price.

Example S4.13

Two sales representatives, Bob and Nadia, work for a computer sales company. Nadia sells mainframe computers, and over the last 12 months her sales have averaged \$256 000 per month with a standard deviation of \$42 000. Bob sells personal computers, and his sales over the same period have averaged \$36 000 with a standard deviation of \$9500.

Which sales representative is relatively more consistent?

Solution

Using Equation S4.11, the coefficient of variation for Nadia is:

$$\begin{aligned} V &= 100\left(\frac{42\ 000}{256\ 000}\right)\% \\ &= 16.4\% \end{aligned}$$

and for Bob is:

$$\begin{aligned} V &= 100\left(\frac{9500}{36\ 000}\right)\% \\ &= 26.4\% \end{aligned}$$

Thus, the *relative variation* of Bob's sales is *larger* than that of Nadia's, although the *standard deviation* of Bob's sales is smaller than that of Nadia's.

The coefficient of variation indicates that, relative to their mean, the monthly sales of Bob are more variable than those of Nadia.

Note that the coefficient of variation is usually only appropriate for data in which all values are *positive*. Examples of such data include sales, height, weight and volume. If some of the values can be *negative* (e.g. profit, loss, measurements of change), the coefficient of variation should *not* be used.

S4.7 REMARKS

As with measures of central tendency, no ideal measure of dispersion exists. Among the more important characteristics of the standard deviation are:

1. It is the most frequently used measure of dispersion, and because of its mathematical properties it has widespread use in problems involving statistical inference.
2. Without additional information, it cannot be computed from an open-ended distribution. That is, if the mean cannot be calculated, neither can the standard deviation.
3. Its value is affected by the value of every observation in the data. That is, if the value of any observation changes, so will the value of the standard deviation.
4. If the data have a number of extreme values, the value of s may be distorted so as not to be a good 'representative' measure of dispersion.

In contrast, the mean deviation satisfies Points 2 and 3; it is less popular in respect of Point 1, but less dangerous in respect of Point 4. The range is bad in most of these respects, but the interquartile range is much better in respect of Points 2 and 4 than either the mean deviation or the standard deviation.

Whichever measure is chosen, remember that it represents a measure of internal variation of the data. In particular, the mean deviation and the standard deviation indicate a *typical variation* away from the mean. They should be considered in that light when either evaluating a quoted figure or checking to see whether a calculated value is reasonable.

In Appendix B on page S.430 there is a note on how to find the standard deviation of a sample using the pre-programmed functions of a calculator. If a calculator has this option, all that is required is for the data to be entered into the calculator using the data entry key. Most statistical calculators will then, at the press of an appropriate button, yield the values of n, $\bar{x}$, s, Σx and Σx^2.

Quick Quiz

1. A supporter of bicycle lanes around the city recorded the number of bicycles that passed by her house between 7 a.m. and 9 a.m. for each of 10 successive weekdays. The results were:

   15 18 23 22 21 18 14 20 25 12

   Find the mean, median and range of these data.
2. Find the mean deviation of the data in Question 1.
3. Find the standard deviation of the data in Question 1.
4. The number of cars requiring towing was recorded by a breakdown service for each of 60 days. The results were:

Number of cars	3	4	5	6	7	8
Frequency	6	10	15	14	9	6

   Find the mean, range and mean deviation of these data.
5. Find the standard deviation of the data in Question 4.
6. The systolic blood pressure (in mm) was recorded for 40 salespeople at 5 p.m. on a Friday afternoon. The results are shown below.

Blood pressure (mm)	f
110–under 120	5
120–under 130	9
130–under 140	15
140–under 150	11
Total	40

   Estimate the mean and the standard deviation of the blood pressures for these 40 salespeople.
7. A set of data has an interquartile range of 20 and a lower quartile of 6. If the data are symmetrical, what is the value of the median?
8. A set of 6 numbers has a mean deviation of 10. If one of the numbers with a value equal to the mean is removed, what is the new value of the mean deviation? Explain your answer.
9. A set of data has a standard deviation of 1.5 times that of the mean. What is the value of the coefficient of variation?
10. If you double the value of each value in a set of data, what happens to the value of the coefficient of variation? Justify your answer.

Multiple Choice Questions

1. Consider the following set of data:

3 19 27 15 9 22 12 5

The *range* of these data is:

(a) 12 (b) 27 (c) 24
(d) 8 (e) none of the above

2. For the data in Question 1, the *mean deviation* is:

(a) 8.37 (b) 7.83 (c) 6.75
(d) 8.62 (e) none of the above

3. For the data in Question 1, the *standard deviation* (to two decimal places) is:

(a) 8.37 (b) 7.83 (c) 70.00
(d) 6.75 (e) none of the above

4. Consider the following frequency distribution of *x*-values:

x	1	2	3	4	5
f	5	3	3	5	4

The *mean deviation* is:

(a) 1.69 (b) 1.30 (c) 1.48
(d) 1.52 (e) none of the above

5. For the data in Question 4, the *standard deviation* (to two decimal places) is:

(a) 1.20 (b) 2.32 (c) 1.48
(d) 1.52 (e) none of the above

6. If, for a set of data, the standard deviation and the mean are equal, it *automatically* follows that:

(a) all data values are equal
(b) all data values are zero
(c) the shape of the data is bell-shaped
(d) all values in the data set are positive
(e) none of the above

7. To calculate the *quartile deviation* you must:

(a) divide the interquartile range by 2
(b) divide the interquartile range by 4
(c) divide the standard deviation by 4
(d) divide the standard deviation by 2
(e) none of the above

8. The *standard deviation* is a measure of:

(a) the standard of quality of the data
(b) the variation of the data
(c) the deviation of the data around the median
(d) the size of the population
(e) none of the above

9. The *range* of a set of data:

(a) can be distorted by outlying observations
(b) is always difficult to find without a calculator
(c) is the most popular measure of variation
(d) is usually about equal to the value of the standard deviation
(e) none of the above

10. The *mean deviation* of a set of data:

(a) gives a good estimate of the mean
(b) gives different weights to different observations
(c) can usually be found as a function key on calculators
(d) is a popular measure of variation because it uses absolute value signs
(e) none of the above

Exercises

S4.1* The number of complaints received by the manager of a supermarket was recorded for each of the last 10 working days before Christmas. They were:

21 15 18 5 10 17 21 19 25 28

For this set of data calculate:

(a) the mean

(b) the median

(c) the range

(d) the lower and upper quartiles

(e) the interquartile range

(f) the mean deviation

(g) the standard deviation (check that your answer is reasonable)

(h) the coefficient of variation

S4.2* A company secretary has recorded the amount of petty cash used by the records department for each of the past 10 weeks. The amounts were (in dollars):

32.50 56.00 48.00 43.20 78.00

21.60 34.90 66.00 40.00 18.80

For these data calculate:

(a) the range
(b) the mean deviation
(c) the interquartile range
(d) the quartile deviation
(e) the standard deviation
(f) the coefficient of variation

S4.3* A newsagent has recorded how many outstanding accounts there were for each month over a 2-year period. The results are shown in the following table.

Number of outstanding accounts	Frequency
0	1
1	9
2	7
3	3
4	4
Total	24

For these data calculate:

(a) the mean
(b) the median
(c) the mode
(d) the range
(e) the interquartile range
(f) the mean deviation
(g) the standard deviation

S4.4 The following data show the weight gains of 12 guinea pigs that were fed for a week on a diet consisting of milk and chocolate. The gains in grams were:

55 66 43 52 34 28 43 57 66 39 45 48

(a) Sort the data in order of increasing magnitude.

(b) Find the mean value.

(c) Find the median value.

(d) Round the data to the nearest multiple of 10.

(e) Find the mean of the rounded data in (d).

(f) Find the standard deviation of the rounded data in (d).

(g) Find the standard deviation of the original values.

(h) Find the mean deviation of the original values.

S4.5* The number of absentees from a section of an industrial plant was recorded each week over a period of 26 weeks. The numbers were:

3 4 0 2 1 4 5 5 0 1 2 2 3

5 4 3 0 2 1 0 3 5 5 0 1 1

(a) Arrange the data in the form of a frequency distribution.

(b) Calculate the range of the data.

(c) Calculate the mean of the data from the frequency distribution in (a).

(d) Calculate the mean deviation of the data from the frequency distribution in (a).

(e) Calculate the standard deviation of the data from the frequency distribution in (a).

S4.6* The lengths (in minutes) of the speeches of 10 politicians in Parliament on a certain day were:

20 2 15 7 43 10 27 36 9 22

(a) Find the mean length.

(b) Find the median length.

(c) Find the range of the lengths.

(d) Find the mean deviation of the lengths.

(e) Find the standard deviation of the lengths.

(f) Find the coefficient of variation of the lengths.

S4.7 Fifteen typists were each given the same contract to type using the same word processor, and the times taken (to the nearest minute) were recorded. The results were:

29 23 18 22 15 27 18 21

33 19 24 26 17 28 25

(a) Find the slowest time.

(b) Find the fastest time.

(c) Find the range of the times.

(d) Find the mean time.

(e) Find the median time.

(f) Find the times that represent the upper and lower quartiles.

(g) Find the standard deviation of the times.

(h) Check that your answer in (g) is reasonable.

S4.8* The earnings were recorded for each of 50 used-car sales representatives in the last week before Christmas. The results are shown in the following grouped frequency distribution:

Weekly earnings ($)	Frequency
0 – under 350	12
350 – under 400	9
400 – under 450	10
450 – under 500	8
500 – under 550	6
550 – under 600	5
Total	50

(a) Estimate the mean weekly earnings.

(b) Estimate the range of the weekly earnings.

(c) Estimate the interquartile range of the weekly earnings.

(d) Estimate the mean deviation of the weekly earnings.

(e) Estimate the standard deviation of the weekly earnings.

S4.9 In addition to the 50 sales representatives in Exercise S4.8, the weekly earnings of a further 5 representatives were recorded and found to be each over $700.

(a) With only this information, is it now possible to estimate the mean or standard deviation of the weekly earnings of the 55 sales representatives combined? If not, state why. If it is possible, do so.

(b) List all the measures of central tendency and dispersion that can be estimated for these new data.

S4.10* A national business school administered an examination to its 60 final-year MBA students. The results are summarised in the following grouped frequency distribution:

Exam score	Number of students
40 – under 50	6
50 – under 60	12
60 – under 70	14
70 – under 80	19
80 – under 90	6
90 – under 100	3
Total	60

(a) Estimate the mean score.

(b) Estimate the standard deviation of the scores.

(c) Construct a histogram and frequency polygon for the data.

(d) What percentage of the students would you estimate obtained a score within one standard deviation of the mean?

S4.11* A real estate agent kept a record of the number of houses becoming available for sale in her district for each of the first 16 working days after an Australia Day holiday. The numbers were:

3 2 0 4 2 3 5 2 0 1 5 3 4 2 1 6

(a) Put the data in the form of a frequency distribution:

Using the distribution in (a), calculate:

(b) the mean
(c) the median
(d) the mode
(e) the range
(f) the interquartile range
(g) the mean deviation
(h) the standard deviation
(i) the quartile deviation

S4.12 A hospital administrator is interested in the number of outpatients treated in the casualty section of a large hospital. The medical records officer is asked to record the number of outpatients for a period of 100 days. She presents the information to the administrator in the following form:

Number of outpatients	Frequency (days)
Less than 10	6
10 – under 20	12
20 – under 30	20
30 – under 40	31
40 – under 50	23
50 – under 60	8
Total	100

Using this distribution, answer the following:

(a) Estimate the mean.
(b) Estimate the median.
(c) Estimate the mode.
(d) Estimate the standard deviation.
(e) Estimate the interquartile range.
(f) Explain why the open-ended class interval does not cause a particular problem.

S4.13* The time between breakdowns for a photocopy machine was recorded over a period of several months. During this period, 20 breakdowns were observed. The times are shown in the following table:

Time between breakdowns (days)	Frequency
0–under 5	3
5–under 10	5
10–under 15	7
15–under 20	3
20–under 25	2
Total	20

(a) Calculate the mean time between breakdowns.
(b) Calculate the standard deviation of the time between breakdowns.
(c) Check that your answer in (b) is reasonable.

S4.14* Sir Donald Bradman's scores in his first 12 completed innings in Test cricket were:

18 79 112 40 58 123 8 131 254 1 334 14

For these scores find:

(a) the mean, median and mode
(b) the range
(c) the interquartile range
(d) the mean deviation
(e) the standard deviation

S4.15 Construct a set of six numbers (none of which can be zero) such that the standard deviation of the six numbers is zero.

S4.16* Construct a set of at most three numbers for which the standard deviation is at least 70% of the range. Verify your answer.

S4.17 The annual production of canned meat (in tonnes) in Australia is shown in the table below for each of the six years 1994–1995 to 1999–2000.

Year	Canned meat production (tonnes)
1994–1995	16 780
1995–1996	19 822
1996–1997	24 772
1997–1998	18 319
1998–1999	16 140
1999–2000	20 783

Source: Australian Bureau of Statistics, *Manufacturing Production*, September Quarter 2000, Cat. No. 8301.0

For these data find:

(a) the range
(b) the mean
(c) the mean deviation
(d) the standard deviation
(e) the coefficient of variation

S4.18 The percentage of births that were to teenage mothers in 1999 is shown below for each of the 8 regions of Australia.

Region	NSW	VIC	QLD	SA	WA	TAS	NT	ACT
Rate	4.64	2.95	6.12	4.28	5.67	7.71	13.31	3.22

Source: Australian Bureau of Statistics, *Births, Australia*, 1999, Cat. No. 3301.0

(a) For these data find:

(i) the minimum and maximum values
(ii) the range
(iii) the mean
(iv) the lower and upper quartiles
(v) the quartile deviation
(vi) the standard deviation

(b) Remove the figure for the Northern Territory (NT) and recalculate the values in (a). Note any statistic that changed markedly.

S4.19* Show that $\sum(x - \bar{x}) = 0$ for *any* set of x-values.

S4.20 According to *Forbes Magazine* (16 June 2000) there are 482 billionaires (in US$) in the world, with the richest 200 having a combined worth of over AUS$2 000 000 000 000. The table below shows the world's 10 richest people and their worth (in AUS$billion).

Name	Worth (AUS$billion)
Bill Gates	120
Larry Ellison	94
King Fahd of Saudi Arabia	60
Warren Buffett	56
S. Robert Walton	40
Prince Alwaleed Bin Talai Alsud	40
Theo and Karl Albrecht	40
Masayoshi Son	39
Michael Deli	36
Kenneth Thompson	32

(a) With the exception of royalty members of the list, briefly describe the background of the people named in the table.

(b) For these data find:

(i) the mean
(ii) the median
(iii) the range
(iv) the lower and upper quartiles
(v) the interquartile range
(vi) the standard deviation

S4.21 Give an example of three non-zero numbers where the mean, median and range are all equal.

S4.22* Give an example of five non-zero numbers (*not* all the same) where the interquartile range is equal to the mean.

S4.23 Give an example of three non-zero numbers where the mean deviation is zero.

S4.24* The circulation (to the nearest 1000) for the period ended December 2000 of 10 magazines sold in Australia is shown below.

Title	Circulation ('000)
Australian Women's Weekly	704
Cleo	206
Better Homes and Gardens	312
Girlfriend	119
Woman's Day	605
Reader's Digest	463
Who Weekly	199
Ralph	91
Australian Penthouse	63
Picture	119

Source: Audit Bureau of Circulations 2000

(a) Make a list of any of these magazines that you have ever purchased.

(b) For the data in the table find:

(i) the mean | (ii) the median

(iii) the range | (iv) the mean deviation

(v) the standard deviation

S4.25 Calculate how many standard deviations the following magazines (with their circulations shown below in parentheses) are above or below the mean circulation of the 10 magazines in Exercise S4.24.

(a) *Dolly* (163 000) (b) *That's Life* (405 000)

(c) *Cosmopolitan* (206 000) (d) *Bulletin* (79 000)

S4.26* For each of 8 weeks a restaurateur recorded how many customers ordered her special dish of lobster tails. The results were:

12 10 8 16 5 9 13 7

For these data calculate:

(a) the range (b) the mean deviation (c) the standard deviation

S4.27 The number of goals scored by a soccer team in each game of its 12-week competition was:

0 2 2 1 6 3 2 1 3 0 1 3

(a) Put the data in a frequency distribution.

(b) Using the frequency distribution in (a) find:

(i) the range (ii) the interquartile range

(iii) the mean deviation (iv) the standard deviation

S4.28* For the soccer data in Exercise S4.27 calculate the standard deviation from the raw data, and check that your answer agrees with that in (b)(iv).

S4.29 To relieve the boredom of walking to TAFE each day, a business student counts how many people she walks past who wear a hat of some type. She records this number for 5 days of a certain week and calculates the standard deviation to be 2.55. Unfortunately, she loses the value for one of the days. The data for the other 4 days were:

10 7 4 5

From this information, can you find the value for the missing day using the information given? Is so, then do it. If not, explain why.

S4.30* If each value in a data set is increased by 10%, what happens to the value of:

(a) the mean? (b) the range?

(c) the mean deviation? (d) the standard deviation?

S4.31 Two rival pizza parlours (A and B) are fighting to gain the larger share of sales in their district. Daily sales were recorded for the previous month. Pizza Parlour A sold a daily average of 47.6 pizzas with a standard deviation of 4.8 pizzas. Pizza Parlour B sold a daily average of 51.2 pizzas with a standard deviation of 6.6 pizzas.

(a) Find the coefficient of variation for each pizza parlour.

(b) Interpret the values in (a).

S4.32* Consider the pizza parlour problem in Exercise S4.31. On one particular day, each pizza parlour sold two standard deviations below its mean. Which parlour sold the more pizzas on that day? Explain.

S4.33 The management of a department store asked the lift operator to record the number of people who travelled in each ride of the lift starting from the ground floor. The operator did this over a period of 100 rides, with the results shown below.

Number of people	Frequency
1–2	16
3–4	28
5–6	34
7–8	13
9–10	9
Total	100

Use the data in the table to answer the following:

(a) Estimate the mean.
(b) Estimate the standard deviation.

(c) Check that your answer in (b) is reasonable.

S4.34 Using the data and your answers in Exercise S4.33, estimate the percentage of time that the lift carried a number of people within 1.50 standard deviations of the mean.

S4.35* Suppose that for a set of 7 observations you are told that the value of the standard deviation is 12.8. Find the value of $\sum(x - \bar{x})^2$.

S4.36 Roll a fair six-sided die 6 times and record the outcomes.

(a) Write down the 6 outcomes.

(b) For the 6 outcomes in (a) calculate:

(iv) the standard deviation

S4.37 In Exercise S4.36, the theoretical values are 3.50 for the mean and 1.87 for the standard deviation. In each case, find the percentage distance that you were away from these values in your answers to (b)(i) and (b)(iv).

S4.38 Select any commercial television program shown on any weeknight between 7.00 p.m. and 7.30 p.m.

(a) Record how many commercials are shown and their lengths (in seconds).

(b) For the length data in (a) calculate:

(i) the range (ii) the interquartile range (iii) the mean

(iv) the standard deviation

S4.39 Consider the following set of data:

12 4 7 10 9

These data have a mean of 8.40 and a standard deviation of 3.05. Use this information to find the mean and standard deviation of the following data:

40 16 25 34 31

Explain your reasoning.

Chapter S5

Sampling

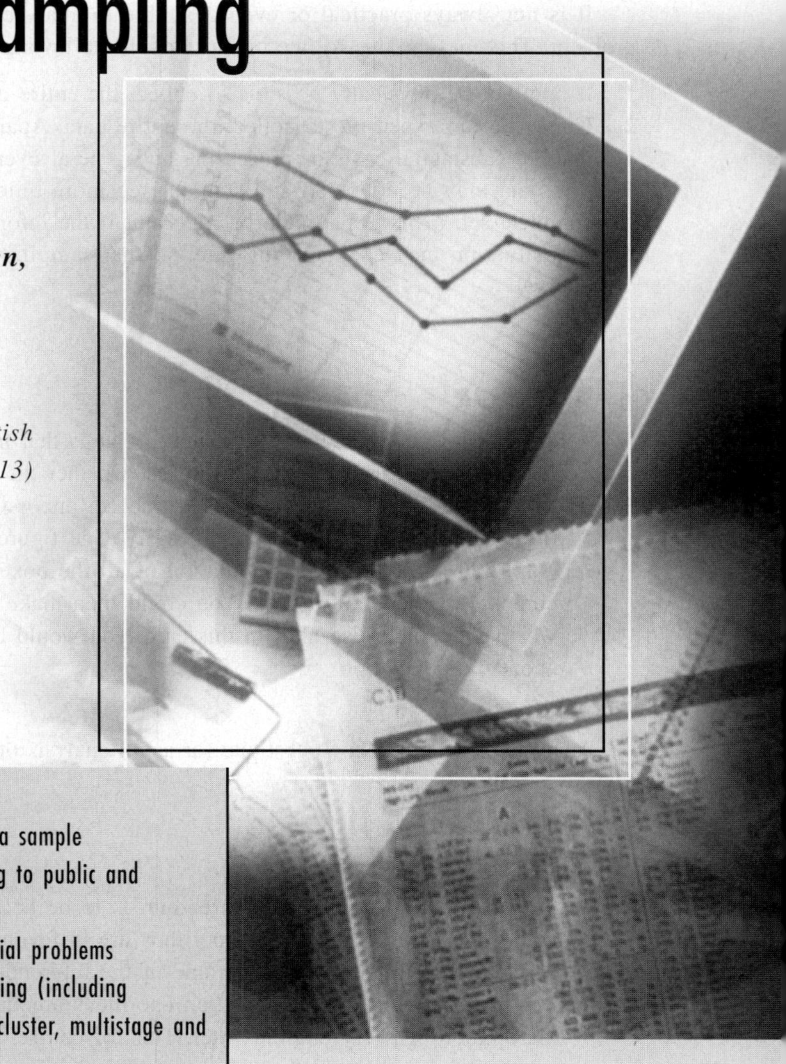

'Many a word, at random spoken,

May soothe a wound or heart

that's broken.'

(Sir Walter Scott, Scottish novelist and poet, 1813)

Learning objectives

In this chapter you will learn to:

- Distinguish between a population and a sample
- Understand the importance of sampling to public and private enterprise
- Apply sampling techniques to commercial problems
- Understand common methods of sampling (including random, systematic, quota, stratified, cluster, multistage and importance)
- Measure and round numbers
- Design questionnaires and surveys

S5.1 INTRODUCTION

It is rare to have access to all the information that we would like to know about a given situation. Usually, we need to examine some *portion* of the total system, then extend our knowledge of that portion to the total system. Such a portion is said to be a *sample*, and the total system is referred to as the *population*. Formal definitions are:

1. A **population** is the complete set that we seek to obtain information about.
2. A **sample** is some part of the population (i.e. a 'subset' of the population) that is actually available as a source of information.

It is not always practical or even desirable to analyse the entire data relating to a particular problem. This may be due to one or more of the following:

1. It may not be physically possible to collect the entire data.
2. It may be too expensive to collect the entire data. Apart from the direct collection costs, there may be substantial costs of processing and general overheads.
3. The data may be collected and summarised more quickly by using a sample rather than the complete set. (This is particularly important if the information is required urgently.)
4. The situation may change with time, so that sampling needs to be confined to a short time interval.

Example S5.1

A company that manufactures matches claims that in each box the 'average' contents is 50 matches. Suppose that you wish to test whether the claim is true.

It would be certainly impractical (indeed, impossible) to take *every* box of matches, count the contents of each and arrive at an average figure. Instead, you might draw a sample of (say) 200 boxes from the population of all the boxes manufactured, and measure the mean contents of these 200 boxes. You could then make a decision based on your results as to whether the claim is valid. In this case, you would be using a sample to make an inference about the population.

Analysis of a sample is a substitute for a full analysis of the population from which it is drawn.

Example S5.2

Suppose that a nationwide referendum is to be held in two weeks time where 10 million voters will indicate whether or not they are in favour of uranium mining. You are given the task of predicting what the outcome of the referendum might be.

To do this, you could spend an enormous amount of time, money and effort to canvass the opinion of every eligible voter. This, of course, would be totally impractical. Instead, a sample of voters could be taken from the population of all eligible voters, and the voting intentions of those in the sample recorded. An estimate of the number of voters in the population who will vote for either choice could be based on the information from the sample.

For most referendums and elections, the counting of votes cannot commence until after the close of voting, which in Australia is generally at 6.00 p.m. local time. This can cause an anomaly in national votes, where the result may well be known for the eastern states while those on the west coast are still voting!

On some occasions, a guide to the voting outcome is gained by undertaking an *exit poll*. With an exit poll, random samples of voters are selected as they come out of the voting booth and asked how they voted. Some statisticians feel that this method gives a more accurate reflection of the true voting pattern, since any intention of the voter *before* they vote can be altered. Of course, even if the exit polls are reasonably accurate, they can often only make a reasonable prediction of the outcome, usually within a matter of hours before it would be known anyway.

Example S5.3

Suppose that you are the purchasing officer for a large chain of supermarkets and have placed an order for 10 000 bags of potatoes, each of which is claimed to weigh approximately 4 kg. The shipment arrives, but before accepting the load you wish to be convinced that the claim is valid.

To do this, a *sample* of, say, 50 bags could be chosen from the *population* of 10 000 bags, and the contents of each of the bags in the sample weighed. Based on the weights of the bags in your sample, a decision could be made on whether to accept the entire load.

Example S5.4

A company that manufactures fireworks wishes to have some kind of quality control for its skyrockets to ensure that only a small percentage are 'duds' (i.e. skyrockets that fail to operate). The company turns out skyrockets in batches of 5000 per week. You are given the task of deciding whether a particular weekly batch is acceptable.

It would, of course, be possible to test the entire population of 5000, but this would only be destructive, as any skyrocket tested cannot be sold. Instead, a *sample* might be chosen from the batch and these skyrockets tested. Based on the results of the sample, a decision could be made regarding the entire batch.

S5.2 BIAS

A sample should be **unbiased**. That is, it should be genuinely representative of the population. If there is any **bias** (i.e. a tendency for certain individuals to have a larger or smaller chance of being selected in the sample), the result will be valueless. Unfortunately, bias does not always arise in obvious ways.

Example S5.5

Consider, for example, the difficulties that could arise in Example S5.2. Where would you go to select an unbiased sample? A house-to-house survey during weekday daylight hours would possibly yield a sample consisting largely of housewives, unemployed people and shift workers. These would not be representative of the population as a whole.

Suppose that the sample were selected at suburban shopping centres on a Saturday morning. There could still be bias. For example, young people may be more likely to shop then than older people, people with certain religious views may not shop on Saturdays, and two-income families may have to shop on Saturdays since opportunities for weekday shopping are restricted.

Bias may also be generated even if the sample selected is sufficiently representative. A common example is in interviewing situations, where the wording of the questions and behaviour or appearance of the interviewer can combine to distort the outcome of sample surveys.

In practice, it is important to be extremely careful to eliminate sources of bias in any sampling process. This is not always an easy task, but it needs to be done in interpreting other people's sampling as well as your own.

Example S5.6

One of the classic examples of biased sampling occurred in 1936 when *Literary Digest* magazine conducted a poll of 10 million people in the United States to determine who would win the upcoming presidential race. Of the 2.4 million responses, 57% were in favour of Alfred Landon and 43% were for Franklin Roosevelt. In fact, Roosevelt won easily, polling 62% of the popular vote and losing only two states. Shortly after its effort, the *Literary Digest* went bankrupt.

What went wrong? For a start, the poll was conducted by mailing questionnaires to people listed in telephone books and to magazine subscribers, generally omitting poorer voters. It has been said that in 1936 political preference was related to economic level and that poorer voters, who were ignored by the sample, tended to vote for Roosevelt. Also, relying on respondents who took the time to return a mailed questionnaire was in itself very dangerous. In practice, only about 25% of people surveyed by mail, even today, bother to return such questionnaires. In any case, those who do respond are usually *not* a representative sample of the population, and any conclusion reached from their responses must be viewed with suspicion.

Example S5.7

From time to time it is quite usual for a radio or television station to conduct a 'phone-in' poll to determine public opinion on some current issue. This often takes the form of the respondents either dialling one telephone number to register a YES vote or another number to register a NO vote.

The results of such polls should be viewed with extreme caution. The respondents are limited to those who listen to or watch the particular station, and this in itself can make them very non-representative of the population. Also, they may be heavily influenced by the opinion of the commentator who posed the question on air and the way in which the question was framed. And, of course, there is nothing to stop respondents from ringing many times to register a vote!

A similar situation exists for Web sites that ask users to register their vote on an issue of the day. These results will often be extremely biased, since not only are they restricted to those who log onto that particular site, but there is usually no barrier/check in place to stop one person from voting numerous times. Such surveys can give an extremely false impression of so-called 'public opinion'.

There are several simple rules that can be followed to help eliminate bias in sampling. These include:

1. Do *not* use only people who volunteer to be in the sample.
2. Do *not* choose a sample using a method that omits segments of the population.
3. Do *not* use people in the sample only because they are 'handy' or readily available.
4. The person selecting the sample should not be one who has a vested interest in the results of the sample.

Even following a set of rules, in practice it is very difficult to eliminate bias completely. Quite often, the best we can hope for is to limit it as much as possible.

S5.3 SAMPLING ERROR

Even if a sampling process were completely free of bias, there would still be fluctuations due to naturally occurring random variation. In general, no two samples will be identical (except by a pure fluke), and it is necessary to assess how much variation can be expected to occur from one sample to another. This variation between samples means that information from any one sample will not be an *exact* representation of the population.

This random discrepancy between a measurement from a sample (called a **sample statistic**) and the population quantity being estimated (the **population parameter**) is called **sampling error**.

In fact, this error due to random sampling *can* be assessed. There are mathematical formulae that enable statisticians to calculate *in theory* the amount of error (sometimes called the **standard error**) that should occur when a sample is used to estimate the value of a population parameter (such as a mean, median or proportion). It also helps to construct an interval estimate (called a **confidence interval**) for these parameters, in which a range of possible values for the population parameter is calculated and the probability that the true value of the parameter is one of those values is given. The topics of standard errors and confidence intervals are covered in Section S7.8.

Any estimate of a population parameter should be accompanied by an indication of the probable magnitude of the sampling error. This will give the reader some indication of just how accurate the estimate really is.

S5.4 SELECTION OF A RANDOM SAMPLE

It is often assumed that samples are drawn *randomly*. Choosing a **random sample** from a population means that each member of the population has the same chance of being selected, with no one member being favoured over any other. This is intended to ensure that the sample is unbiased, although of course there will still be sampling error (in the sense of the previous section) resulting from the random fluctuations in any sampling process.

When selecting any sample, you must answer several questions before taking any action:

1. What precisely are the members from which the sample is to be chosen (i.e. the population)?

2. What should be the *size* of the sample? That is, how many members should be included in the sample?
3. *How* should the members be selected for inclusion in the sample? (The method of selection determines how reliable the sample will be.)

The answer to question 1 is not always as easy as we may think, as the following example shows.

Example S5.8

A clothing manufacturer has designed a new style of men's socks and would like to test possible market reaction. It selects a random sample of adult males and finds that a large proportion of those chosen state that they would wear such socks. However, when the product comes onto the market, very few pairs are sold.

The problem here may be one of population identification. In practice, although men will be the ones who wear the socks, they may *not* be the ones who actually *buy* them. (Married men's wives may be the ones who purchase socks. Also, many socks are purchased as gifts.) In this case, and numerous others, the user is *not* the one who makes the purchase.

The problem of sample size depends on the type of problem and the desired accuracy of any conclusion. Although large samples (in general) yield more accurate estimates, they cost more time and money. Also, it can be shown, for example, that doubling the size of a sample does *not* double the accuracy. And, if a sample is biased, increasing the sampling size only increases the bias. The actual methods for determining sample sizes according to specific criteria involve statistical techniques that are beyond the scope of this book.

Deciding which method of sampling will yield a random sample is not always easy. Often, random samples are more expensive than non-random ones, since all members of the population must be listed and interviewers must sometimes travel long distances to ensure that all groups are represented. However, random samples occur in many facets of everyday life.

Example S5.9

In a card game, a random sample of cards is distributed to each player by means of shuffling the deck. However, in practice, studies suggest that cards are often *not* well shuffled between hands. In particular, card players who claim remarkable hands are sometimes shown to have had these as a result of poor shuffling either from a new (ordered) pack of cards or where the cards have been ordered as a result of the previous hand.

Example S5.10

In certain states of Australia, police undertake 'random' breath-testing (RBT) of motorists. This generally means that vehicles are randomly selected and their drivers subjected to a test to determine whether they exceed some specified blood-alcohol limit. (It is a matter of speculation whether the vehicles that are stopped are actually selected 'randomly'.)

Example S5.11

Suppose that one of two people must be selected for a particular task. One method is to toss a coin and select one of the people according to whether it comes down heads or tails. If there were, say, 6 people involved, each could be assigned a number from 1 to 6. A 'fair' (i.e. unbiased) six-sided die could then be rolled, and the person whose number appeared on the uppermost face would be selected.

Lottery method

This method is similar to the method used some years ago when selecting prize winners in a lottery. Each member of the population was assigned a number. These numbers were transferred to 'marbles', which were well mixed; then a sample was chosen randomly from them. Nowadays, the marble technique is no longer used (since 1980 the winning numbers have been computer-generated), but the selection process works on a similar principle. In other situations the 'marbles' may, of course, take other physical forms, such as equally sized pieces of paper.

Another example of this method is that used to draw numbers from various Lotto-type games. For instance, in New South Wales, players must select 6 numbers from a card that contains the numbers 1 through to 44. The draw for the correct 6 numbers is made on live television and is done by a machine designed to select 6 balls from a box containing 44 balls, each numbered with exactly one of the numbers 1 through to 44. In other words, viewers can see the random sample selection procedure in action.

The potential prize money for obtaining the correct six numbers is enormous (on occasions over $15 million), and there are countless schemes available to 'help' players to predict the next draw's numbers. (There are about 7 million possible combinations.) However, the scope for predicting a set of numbers that are random is left to your imagination!

Random numbers

A process similar to the lottery method is selecting the identifying numbers from a *Table of random numbers* as published in books or generated by a computer. In their simplest form, these tables consist of a series of the digits 0 to 9 that have (as far as can be ascertained) an equal chance of occurring. Often, they are presented in 'blocks' that are five digits wide and five digits long. An example of a small set of random numbers is given in Table S5.1.

When selecting random numbers from a random number table it is desirable to adopt a specific pattern. Of course, a different pattern should be used each time, in order to avoid generating one sample repeatedly.

Suppose, for example, that seven random two-digit numbers are required. Using Table S5.1, some patterns might be:

1. Select the first two digits in the upper left-hand corner of each block, starting at the top left-hand corner of the table and working along the rows. (This yields 35, 30, 97, 09, 68, 32, 01.)
2. Starting with column 3, select the last two digits and proceed down the column. (This yields 81, 18, 84, 99, 93, 64, 10.)
3. Starting with the third row of blocks, select the third and fourth digits in the second row of each block and proceed along the rows. (This yields 93, 05, 82, 72, 51, 91, 24.)

Similar techniques may be used to generate a sequence of three-, four- and five-digit random numbers. Note that one-digit numbers will lie between 0 and 9, two-digit between 00 and 99,

Table S5.1 A table of random numbers

35997	30761	97081	09501	68887	32876	01705	34260	95065	45528
88241	30402	12318	52430	40139	96986	84900	72408	42027	31676
54382	73370	26184	14024	57444	57660	52173	30274	93448	63273
77681	74946	02099	69091	19372	66961	14595	58642	75760	52253
53148	26074	52293	65359	63971	04833	86492	01227	54505	19515
89889	46933	13364	33883	83389	36952	52505	67513	40071	31001
03105	87912	29610	75108	37363	28479	43546	89992	19550	54863
82633	19209	21548	35022	21960	57961	11815	95867	00559	26428
69386	57453	70147	73538	49562	46806	64550	36653	25718	68792
31113	07607	48037	71020	22666	65957	11141	39227	07990	19849
65972	74528	40888	55386	95918	92088	91125	53648	66122	00138
79933	71058	34826	97725	69513	22915	18246	52244	91161	40861
40374	13239	56162	04703	95851	22824	41271	28202	62852	84238
46625	20031	08524	20077	65817	21174	29279	57712	22401	67500
30980	74485	26480	21343	30031	61921	35744	57308	71196	01865
49234	62616	54921	29008	83672	85839	96836	45077	80900	66906
63526	93824	71820	11033	20183	85704	04683	63512	39144	56880
64424	95979	17709	94849	31771	05737	84286	16757	46256	24476
73180	59978	08524	78963	95437	86351	33824	32540	18357	02668
99260	21284	81351	70961	10255	06911	47394	72408	23827	59865
96395	30665	43699	03593	29165	23388	26628	92402	16731	86740
29493	09069	78653	90094	42735	33682	95041	89887	92369	57949
81585	50593	14698	04737	72551	57271	59433	00156	33966	58773
59108	49578	18100	59836	73221	21110	01650	11058	47770	66141
84576	58388	40915	94507	32209	17272	65674	95552	25685	05345

three-digit between 000 and 999, four-digit between 0000 and 9999, and five-digit between 00000 and 99999. Any numbers that are repeated in the sample may be ignored.

Example S5.12

Suppose that there are 74 297 items from which a random sample of 50 items is to be chosen for testing.

For convenience, the items should be labelled from 00001 to 74297. Since 74297 is a five-digit number, it is necessary to generate a sequence of 50 five-digit random numbers. The numbers selected will then correspond to the items in the population that should be included in the sample.

If a random number table such as that in Table S5.1 is used, a selection system must first be established. Suppose that, starting with the second row in the first block of numbers, the numbers are chosen going across the rows. This yields the sequence:

88241 30402 12318 52430 40139 96986 84900 72408, etc.

Note that the numbers 96986 and 84900 do *not* correspond to actual items, so will be discarded. The selection procedure is repeated until 50 actual numbers are generated. These will then correspond to the items in the population that should be included in the random sample.

Example S5.13

A suburb contains 786 medical practitioners from whom it is desired to obtain a random sample of 10.

To do this, the practitioners are numbered, for convenience, from 001 to 786. Since 786 is a three-digit number, it is necessary to generate a sequence of 10 three-digit random numbers.

Suppose that the random numbers in Table S5.1 are used. The selection system adopted is to choose the last three digits in the numbers starting at the first block in the second last column. This yields the numbers 065, 027, 448, 760, 505, 071, 550, 559, 718, 990. These numbers correspond to those practitioners in the suburb who have been assigned those numbers. The exception is the number 990, which does not correspond to a medical practitioner and may be ignored. It should be replaced by the next number, which is 122.

As well as random number tables, many calculators have their own inbuilt random number generator, which will generate a sequence of random numbers of desired length. The technique used (as is the case with sequences generated by computers) is a **mixed congruential method**, which actually generates **pseudo random numbers**. That is, numbers that for all intents and purposes are random, but are actually generated by a mathematical formula.

For some calculators the random number generator is indicated by the letters *RAN #*, often near the decimal point key. The generator operates in conjunction with the *SHIFT* key and will display a random number in the form of a 0 followed by a decimal point followed by three digits (e.g. 0.639). Repeated pressings of the *SHIFT* key and the decimal point key will display a random sequence of numbers in this form. Check your calculator instruction manual for further details.

S5.5 SELECTION OF OTHER TYPES OF SAMPLES

It is of course possible for a random sample to be very non-representative of the population, due simply to an extreme case of random fluctuation. Moreover, truly random samples can sometimes be expensive to obtain. It is therefore often argued that sampling methods should be tried that are not truly random but that give good results in practice. The justification of this argument varies from one situation to another.

Systematic sampling

In **systematic sampling**, the members of the population are put in the form of an **array**. That is, they are arranged in some natural order (usually ascending or descending). Then every *n*th item is selected for the sample. The value of *n* should be chosen so that it will yield a sufficiently large sample. The procedure would involve using a random number table to select a number between, say, 1 and 100 for the starting point. Every *n*th item is then selected, starting from that item. For example, if the random number were 63 and we selected every 50th item, we would select items numbered 63, 113, 163, 213, …, etc. This clearly is not truly random. For example, if the ordering is alphabetical and there are two people called Zapatopoulos, they cannot both be selected as they could be in a true random sample. For most purposes, though, such a method gives fairly good results.

The greatest problem seems to arise when the listing of the population is incomplete. Taking every tenth number out of the telephone directory may give a reasonable selection of the people listed there, but it discriminates very forcefully against people who do not have a phone and those who have 'silent' numbers. It may also give biases of a more subtle kind, since a family of six people with one phone has the same status as a person living alone.

More sophisticated methods can improve the representativeness of a non-random sample and lead to a reduction in sampling error. Common techniques include quota sampling, stratified sampling, cluster sampling, multistage sampling and importance sampling. These methods are often applied in market research and opinion polling. Each of these types of sample is discussed below.

Quota sampling

Quota sampling uses selection rules that include certain types of people and at the same time control the mix of people who can be included in the sample. The basic aim is to reduce the cost of data collection.

One difficulty in using a random sample is that, once the sample is selected, interviewers must commit themselves to contacting the selected persons. In some cases this can be an expensive process, and in the end some of them are never actually interviewed. Moreover, those who are not interviewed may represent a particular class of respondents who may be missed out altogether.

An example of quota sampling might include the case where the interviewer is instructed to obtain a sample of 40 people with the following characteristics:

5 males aged under 25
5 males aged 25 or over
10 females who are unemployed
20 females who are employed

The above quotas are based on sex, age and employment status, although other characteristics could be used. When using quota sampling in the above example, an interviewer may call on units in a sample housing block and try first of all to fill the most difficult quotas (say, males aged under 25). If a unit is unattended at the time the interviewer calls, the interviewer will go to the next unit. Using this method, each sample block obtains the set of interviews as specified in the quota.

A number of factors can affect the cost of conducting a survey. These include the length of the interview, the number of interviewers used, telephone charges, the travel expenses of interviewers, the number of call-backs required, and so on. However, studies suggest that the average cost of interviewing a selected random sample may be up to three times as much as that of using a quota sampling method.

Stratified sampling

The **stratified sampling** method divides the population into *strata* (subsets), then takes samples of equal size from each stratum. It is particularly useful when the important strata can be identified and subjects are homogeneous within each stratum. The selection of subjects from each stratum is done randomly.

There are some advantages in using a stratified sample over just using a plain random sample. It may cost more to obtain than the selection of a random sample of similar size, but the precision of results is usually greater. Also, stratified sampling provides valuable information about the characteristics of the strata themselves, as well as overall characteristics of the population. The clients of market research firms often request that stratified samples be taken to ensure the fair representation of all subgroups of the population.

Example S5.14

A town consists of 5000 adults, comprising 2800 females and 2200 males. A sample of size 200 adults is to be chosen and their opinion asked as to whether they approve of having a nuclear reactor built near the town.

To use stratified sampling with respect to the two sexes, the sample would consist of $\left(\frac{2800}{5000} \times 200\right) = 112$ females and $\left(\frac{2200}{5000} \times 200\right) = 88$ males. The 112 females would be chosen randomly from the 2800 females in the town, and the 88 males would be chosen randomly from the 2200 males.

Example S5.15

Suppose that you wish to draw a sample of size 30 from a university course that has an enrolment of 600 students. Assume that this population of 600 students comprises 100 accounting majors, 200 economics majors and 300 mathematics majors.

Using the stratified method, you would choose 5 accounting majors, 10 economics majors and 15 mathematics majors from the respective groups to form the sample.

Cluster sampling

In **cluster sampling** also the population is divided into strata, and then certain strata are selected randomly and the sample is chosen from only those strata.

Example S5.16

If the population falls into natural strata (e.g. suburbs), a sample of suburbs could be chosen randomly and a random sample drawn from only those suburbs chosen.

Cluster sampling is also used when the population in question is concentrated in defined *areas*. For example, to interview car drivers it might make sense to concentrate on service stations and car parks; but if the interview questions relate to the effectiveness of shark alarms, the most likely places to select the sample would be various popular swimming spots.

Multistage sampling

An extension of cluster sampling is **multistage sampling**, in which the population is first broken down into a set of distinct groups, from which a number of groups are selected randomly. The groups selected are again broken down according to another characteristic, and a random selection is again made. This process is repeated until all desired stages have been considered.

The advantage of multistage sampling over, say, quota sampling is that selection bias on the part of the interviewer is eliminated.

Example S5.17

Suppose that an opinion poll is to be conducted on a national scale across Australia, and the intention is to make a separate study of the three geographical regions of the eastern, central and western states.

Within each region, all the cities or towns of similar sizes are grouped together (e.g. all cities with a population between 20 000 and 100 000). From each group a random sample of cities is selected to complete the *first* stage of the process. These are the cities to be included in the sample. No interviews will be conducted in other cities in that group. If the poll is to be conducted on a face-to-face basis, interviewers will be stationed in the selected cities. Often, however, many interviews are conducted using the telephone to cut down on costs.

The chosen cities are then broken down into smaller groups, say, suburbs. At the *second* stage some suburbs are selected at random from each of the chosen cities.

Each of the chosen suburbs is then divided into smaller sections, say, council wards. In the *third* stage, some wards are selected at random from the chosen suburbs.

At the *fourth* stage, some households are selected at random from the chosen wards.

Finally, some members of the chosen households are interviewed as part of the poll, with instructions often being quite specific as to who should answer the questions. Depending on the nature of the poll, the interviewer may be instructed, for example, to speak only with all females between the ages of 13 and 18 or only all males over the age of 45. If such persons are not available, the interviewer will have a fallback set of instructions. For example, the interviewer may be told to survey the oldest male over the age of 18, but if none is available then the oldest female over the age of 18 may be interviewed instead.

Importance sampling

Another technique of great social relevance is *importance sampling*, in which the probability of selecting a particular individual (or larger subset) is dependent on its importance. For example, in election predictions, there is often little purpose in sampling electorates where the result is a 'foregone conclusion'. If the 'swinging' electorates can be identified, the efforts of the pollsters can be concentrated on effective sampling from these.

S5.6 MEASUREMENT AND ROUNDING

Statistical data may be either *discrete* or *continuous*. **Discrete data** are those in which the possible values are distinct and isolated.

Example S5.18

Very commonly, discrete data are obtained where the observations are placed into certain categories. For example, in a survey to test public reaction to a new brand of soft drink, three possible responses may be that you (a) like it, (b) dislike it or (c) are indifferent.

Example S5.19

Another type of discrete data is obtained where the response is to a question such as 'How many …?' For instance, a survey may be taken to determine how many cars each family in a certain district owns. In this case, the response of each family in the sample will be a distinct integer, such as 0, 1, 2, …, and so on.

Continuous data are *not* made up of naturally distinct units. Examples are length, height, weight and time. In this case, it is not possible to record the exact observation with total precision. It does not matter how fine a scale or measuring instrument is used; there can be a finer one still. For instance, heights are commonly measured to the nearest centimetre, but in principle far finer measurement would be possible without ever attaining complete accuracy. Inevitably, when continuous data are *rounded* (i.e. approximated in some way), errors are going to occur as a result.

Approximation to the nearest number above or below

In this method of approximating data, the digits, 1, 2, 3 and 4 are rounded *down*, while the digits 6, 7, 8 and 9 are rounded *up*.

A borderline case that ends in a 5 presents a problem. One method of resolution is to round such cases to the nearest *even* value. For example, if you are rounding numbers to the nearest integer, 38.5 would become 38 and 43.5 would become 44. Such a convention avoids a consistent upward bias as obtained by the popular method of *always* rounding up all 5s. However, this convention tends to create larger errors when dealing with the totals of rounded numbers.

Example S5.20

Consider the following set of data. It is rounded to the nearest multiple of 10 above or below.

Actual	Approximation
155	160
112	110
79	80
63	60
88	90
102	100
65	60
91	90
755	750

Types of error

The *error* made in recording or approximating continuous data is the difference between the computed value and the true value. There are several methods of expressing these errors.

1. The *absolute error* is simply the difference between the approximated value and the actual value. In this case, the sign of the difference is ignored.
2. The **relative error** is the ratio of the absolute error to the actual value.
3. The **relative error percentage** is the absolute error expressed as a percentage of the actual value. (This is also 100 × relative error.)

Example S5.21

Calculate the three types of error made in the total of the approximations in Example S5.20.

Solution

$$\text{Absolute error} = 755 - 750 = 5$$

$$\text{Relative error} = \frac{5}{755} = 0.0066$$

$$\text{Relative error percentage} = 100 \times 0.0066\% = 0.66\%$$

Example S5.22

The first 8 sales at a suburban pharmacy on a Monday morning were:

\$1.65 \$12.64 \$4.75 \$6.76 \$15.35 \$8.43 \$7.50 \$2.85

(a) Round the sales data to the nearest dollar.
(b) Round the sales data to the nearest multiple of 10 cents.
(c) Find the relative error percentage in the sum as a result of the rounding in (a).
(d) Find the relative error percentage in the sum as a result of the rounding in (b).

Solution

(a) The rounded sales data are:

\$2 \$13 \$5 \$7 \$15 \$8 \$8 \$3

(b) The rounded sales data are:

\$1.60 \$12.60 \$4.80 \$6.80 \$15.40 \$8.40 \$7.50 \$2.80

(c)

$$\text{Actual sum} = \$59.93$$

$$\text{Rounded sum} = \$61.00$$

$$\text{Absolute error} = \$59.93 - \$61.00 = \$1.07 \text{ (ignoring the sign)}$$

$$\text{Relative error} = \frac{\$1.07}{\$59.93} = 0.0179$$

$$\text{Relative error percentage} = 100 \times 0.0179\% = 1.79\%$$

(d)

$$\text{Actual sum} = \$59.93$$

$$\text{Rounded sum} = \$59.90$$

$$\text{Absolute error} = \$59.93 - \$59.90 = \$0.03$$

$$\text{Relative error} = \frac{\$0.03}{\$59.93} = \$0.0005$$

$$\text{Relative error percentage} = 100 \times 0.0005\% = 0.05\%$$

S5.7 QUESTIONNAIRE AND SURVEY DESIGN

Most large companies from time to time are interested in the opinion of actual and potential consumers of their product. The reasons for this may be self-evident but may vary depending on what the company actually wants to know. Examples include the likely market share of a new product, the reasons consumers buy the opposition product in preference to the company product, and also why they may not like the opposition product.

Companies determine consumer opinion by considering the results of a random sample of their particular population of interest. While some companies may undertake the sampling procedure themselves, it is quite common for them to employ the services of a *market research firm*, which will assist in designing the survey questions and in collecting the actual data. This firm may employ its own statistician, but in general such firms contract out the statistical analysis work to a consultant statistician.

Much of the market research collection of data is done by *telephone*. Such firms have large banks of telephones with professional operators dialling the members of the previously selected random sample. Considerations such as sample response rate, and the effort made to try repeatedly to contact sample members, are important issues. This process is sometimes known as **cold calling** (i.e. calling a respondent without warning) and is an art in itself, requiring much patience. It may also subject the caller to a certain amount of abuse at times!

Another common technique is to employ a team of professional *interviewers* who will select likely candidates (in a supposedly random manner) at appropriate places such as shopping centres. Often, these interviewers are asked to select only people in a certain age group. For example, it may be known by a company that the large majority of consumers of a particular hair shampoo are women between the ages of 18 and 35. The interviewer is asked to survey only such people, since to question others in the population would be a waste of resources.

Whether the results of the survey are meaningful is a function not only of the randomness of the sample but also of the actual questions asked. Most people will balk at answering a long survey, but may give the interviewer, say, 5 or even 10 minutes of their time if they are told beforehand that this is the maximum time it will take. If the survey takes longer than this, there is a risk of biasing the sample towards respondents who have plenty of time on their hands.

Many surveys take the form of a *questionnaire* in which the respondent must answer a series of questions. These may be asked by the interviewer, or the respondent may be asked to fill in a form. There are two basic types of questions: multiple choice questions and open-ended questions.

Multiple choice questions

A common type of question is the *multiple choice question*, in which a series of choices is put to respondents. They must simply tick a box, and there is usually no room for further comment. An example of such a question would be:

Do you like reading about the royal family in our magazine?

- ☐ Never
- ☐ Rarely
- ☐ Sometimes
- ☐ Often
- ☐ Always

Respondents are asked to tick exactly one box. This type of question should be unambiguous and mutually exclusive. That is, the possible responses should cover all likely possibilities without overlapping. One disadvantage is that the possible responses given sometimes suggest alternatives to the respondents that they would not otherwise have thought of, and that they may not have chosen if no suggestions had been made.

An advantage of multiple choice questionnaires is that they can be easily coded and entered into a computer database. There is no need for interpretation of the response at the data entry stage.

Open-ended questions

Another disadvantage of multiple choice questionnaires is that they restrict any suggestion or comment from the respondent. For example, suppose that a telephone company wanted to know whether its customers were happy with the service provided. In this case, an *open-ended question* could be asked. Examples might be:

'How would you describe the service that our company provides?'
'What are some of the good and bad things about our service?'

When requesting the respondent's own words there will obviously be a multitude of replies, which must be classified into a limited number of categories before any statistical analysis can be undertaken. This classification can be difficult when hundreds or even thousands of replies to open-ended questions are received. For this reason, such questions are often used only in small-scale studies, where their use may lie in determining general opinion and helping to frame the allowable responses to a larger-scale multiple choice questionnaire.

In some cases a questionnaire may combine both multiple choice and open-ended questions. An example might be the use of a category 'Other (please specify)' for some questions, where the respondent may have an answer that is not one of the choices provided. Alternatively, at the end of what is basically a multiple choice questionnaire, there may simply be a space labelled 'Comments', for anything that respondents may like to add in their own words.

Rules for designing questions

There are a number of basic rules that should be borne in mind when framing survey questions. Some of these are listed below.

Order of questions

The first questions should be relatively simple, sometimes serving mainly to put the respondent at ease.

If more than one topic is involved, it is advisable to finish one topic completely before starting on questions on another topic.

Questions that may evoke an emotional response (e.g. that mention specific brand names or that refer to problems the respondent may have had) should be placed near the end of the questionnaire if possible, since they may colour later responses.

Direct questions

Some respondents have difficulty in recalling events or may simply exaggerate their response to an interviewer in order to raise their level of self-esteem. For example, the question 'What high school did you attend?' may be embarrassing to those who did not attend high school. An alternative question might be 'What was the name of the last school you attended?' This question avoids embarrassing admissions; respondents are able simply to provide an answer to the question without further explanation.

Unbiased questions

It is very important that the questions are not *leading* or *loaded*. A leading question is one that leads the respondent in a particular direction. Examples might be: 'Do you agree that high-quality premium unleaded petrol in your car would give better performance than ordinary unleaded petrol?' or 'With so much controversy about the effects of smoking these days, do you think that tobacco sponsorship of all major sporting events should be banned?' Sometimes, leading questions are somewhat more subtle (and purposefully so), but companies may still use the results from these for commercial purposes if it suits their cause.

Unambiquity and clarity of questions

Questions should be framed as simply and clearly as possible so that they have the same meaning to all respondents. Many surveys have come to grief because, after the event, when the results are analysed, the interviewer has realised that the questions meant different things to different people. The results are then useless, since it is impossible to tell just how any particular respondent interpreted the question. An example of such a question might be: 'A judge recently stated that he could not be too hard on drug offenders. Do you agree with the judge's sentiments?' (Think about the ambiguity of the wording!)

Pre-sampling

Before embarking on a lengthy and expensive survey it is advisable to undertake a small-scale sample to test how the questionnaire is going to work out. Most market researchers design their questions, take a pre-sample of between 50 and 200 respondents and carefully consider their responses. By doing so, they are readily able to identify any of the problems listed above and thereby avoid difficulties that may not have been apparent beforehand. After fixing any errors, the researchers can be reasonably confident that the questionnaire is ready to use on their large-scale sample.

S5.8 MODERN PRACTICES

There are three main types of surveys, namely mail surveys, telephone interviews and in-person interviews. While each has its advantages and disadvantages, they are undergoing great change with the increasing use of computers and the advances in telecommunications. A useful insight into these types of surveys is provided by the following summary of the Survey Research Methods section of the American Statistical Association.

Mail surveys are currently the most common example of self-reported data collection. One reason is that they can be relatively low cost, although they are not necessarily easy to carry out. Planning the questionnaires for mail surveys is often more difficult than for surveys that use interviewers. For example, care is needed to anticipate issues that respondents may have and to deal with them ahead of time.

Using the mail system can be particularly effective in business surveys, while mail surveys also work well when they are directed towards specific groups, such as subscribers to a specialised magazine or members of a professional organisation.

The manner in which self-reported data are obtained has begun to move away from the traditional mail-out/mail-back approach. The use of fax machines, and more recently the Internet, is rapidly increasing. Fax numbers and Internet addresses are being added to specialised membership lists and other lists. As a by-product, they can be used along with more conventional items like names and mailing addresses in building potential sampling frames.

As the use of computers and telecommunications becomes more widespread, touch-tone applications or applications involving respondents' computers 'talking' directly to the survey organisations' computers will increase significantly. This increase is already well underway in health surveys, where samples of patient records are often supplied electronically.

Interview surveys, whether face-to-face or by telephone, offer distinct advantages over self-reported data collection. The 'presence' of an interviewer can increase cooperation rates and make it possible for respondents to get immediate clarification.

The main requirement for good interviewers is the ability to approach strangers, in person or on the telephone, and persuade them to participate in a survey. Once a respondent's cooperation is acquired, the interviewer must maintain it, while collecting the necessary data, that is, information that must be obtained in exact accordance with specific instructions.

For high-quality data to be collected, interviewers must be carefully trained through classroom instruction, self-study, or both. In training, good interviewer techniques are stressed, such as how to make initial contacts, how to conduct interviews in a professional manner and how to avoid influencing or biasing responses. Training generally involves practice interviews to familiarise interviewers with the variety of situations they are likely to encounter.

Time must be spent going over survey concepts, definitions and procedures. A question-by-question approach is needed to ensure that interviewers can deal with any misunderstandings that might arise. In most reputable survey organisations, interviewers are also required to take a strict oath of confidentiality before beginning work.

Survey materials must be prepared and issued to interviewers. For traditional paper and pencil in-person interviews, interviewers must receive sample copies of the questionnaire, a reference manual, information about the identification and location of the households, and any cards or pictures to be shown to the respondents.

Before conducting in-person interviews, survey organisations frequently send an advance letter to the sample respondents, explaining the purpose of the survey and that an interviewer will be calling soon.

In many surveys, especially those sponsored by governments, information must be given to the respondents regarding the voluntary or mandatory nature of the survey and how the answers are to be used. Visits to respondents are scheduled with attention to considerations such as the best time of day to call or visit, and allowance is made for repeated attempts (i.e. call-backs) in not-at-home situations.

The use of computers in survey interviewing is becoming quite common. In the United States, for example, most large-scale telephone surveys are now conducted via CATI (**Computer-Assisted**

Telephone Interviews). With CATI, the interviewers use a computer terminal. The questions to be asked appear on the computer screen, and the interviewers use the keyboard to directly enter the respondents' replies as they are given. This technique is also common in Australia.

CATI's important advantages are in quality and speed, not in cost savings. CATI can cost more for small, non-repeated surveys, due to the need to program the questionnaire. CATI's cost per interview decreases as sample size increases, so that in large and/or repeated surveys, it is cost competitive with conventional telephone methods.

The CATI interviewer's computer screen is programmed to show questions in a planned order, so that interviewers cannot inadvertently omit questions or ask them out of sequence. For example, the answers to some questions require 'branching' (i.e. answers to prior questions determine which other questions are asked). CATI can be programmed to do the correct branching automatically. In non-computer-assisted telephone interviewing, incorrect branching has sometimes been an important source of errors, especially omissions.

In the CATI setting, the computer can be programmed to edit replies. The computer can check whether the reply to a particular question is consistent with other information reported. If the editing indicates that a problem might exist, the respondent is asked to confirm or correct earlier answers. CATI can produce statistical results more quickly than traditional methods of data collection. For example, CATI eliminates the need for a separate computer data-entry step. Furthermore, with CATI, some organisations are able to provide summaries of results as each questionnaire is completed or at the end of each day.

In recent years, there has been a trend towards the use of notebook computers for in-person interview surveys. Notebook computers can be taken into the field, and either the interviewer or the respondent directly enter data in response to the questions. Data collection carried out in this way is referred to as CAPI (*Computer-Assisted Personal Interviews*).

CAPI notebooks are not directly connected to a centralised computer. Nonetheless, most of CATI's quality and speed advantages also apply to CAPI. The use of CAPI methods is expected to expand in the next few years.

As electronic technology becomes more sophisticated and more widely used, traditional paper and pencil survey methods may eventually disappear altogether, at least in government surveys.

No matter what type of data collection is used, a number of 'back-end' processes may be required to get the data into a format whereby aggregated totals, averages or other statistics can be computed.

For mail surveys and conventional paper and pencil interviews, this may involve coding after the questionnaires have been completed. Coded paper questionnaires are entered into a computer (e.g. keyed onto a disk) so that a computer file can be created. At this point, most of the remaining back-end steps are common to all surveys, whether or not a computer was used initially for data collection.

Once a computer file has been generated, additional computer editing, separate from clerical editing, can be undertaken to alter inconsistent or impossible entries.

Decisions are usually required about how to handle missing items, such as in cases where the respondent did not know the answer or refused to provide one or where the question simply was not asked. Preferred practice for missing items is to provide special codes indicating why the data are not included. When resources are available, the 'filling in' or imputation of these missing data items should be undertaken to reduce any biases arising from their absence.

When the file is 'clean' the survey data are ready for analysts to begin summarising what has been learned. It is a good idea to use commercially available software packages to carry out this step rather than using your own specially written computer programs.

Often, the best way to start the analysis is with simple counts and related percentages for each question. Next, it is common to produce tables of growing complexity. Eventually, there may be a need for even more sophisticated forms of data presentation to address the concerns outlined when the survey was initially conceived.

The results of surveys are usually printed in publications and presented at staff briefings or in more formal settings. Additional analysis can also be undertaken by making unidentifiable computer data files available to other researchers at a nominal cost.

Conducting a credible survey entails scores of activities, each of which must be carefully planned and controlled. Taking shortcuts can invalidate results and badly mislead the sponsor and other users. Avoid these three short cuts, which crop up often:

- not pre-testing field procedures
- not sufficiently following up on non-respondents
- sloppy field work and inadequate quality controls

A pre-test of the questionnaire and field procedures is the only way of finding out whether everything works properly, especially if a survey employs new techniques or a new set of questions. Because it is rarely possible to foresee all the potential misunderstandings or biasing effects of different questions and procedures, it is vital for a well-designed survey operation to include provision for a pre-test. There should usually be a series of small-scale pilot studies to test the feasibility of the individual techniques (if new) or to perfect the questionnaire concepts and wording.

This should be followed by a full-scale 'dress rehearsal' to find out whether everything connects together as intended.

Failure to follow up non-respondents can ruin an otherwise well-designed survey. It is not uncommon for the initial response rate in many surveys to be under 50%. To deal with this possibility, survey plans should include returning to sample households where no one was home (perhaps at a different time or on a weekend), attempting to persuade persons who are inclined to refuse, and so on. In the case of mail surveys, it is usually necessary to conduct several follow-up mailings spaced, for example, about three weeks apart.

There is some evidence that responses to subsequent mailings may differ from responses to the first mailing. Thus, biases can result without the extra effort of follow-ups. Depending on the circumstances, it may even be necessary to contact a sub-sample of the remaining non-respondents by telephone or personal visit.

A low response rate does more damage in rendering a survey's results questionable than a small sample, because there may be no valid way of scientifically inferring the characteristics of the population represented by the non-respondents.

Sloppy execution of a survey in the field can seriously damage results. Controlling the quality of the field work is done in several ways, most often through observation or by supervisory or senior personnel redoing a small sample of interviews. There should be at least some questionnaire-by-questionnaire checking while the survey is being carried out. This is essential if omissions or other obvious mistakes in the data are to be uncovered before it is too late to fix them.

In other words, to ensure that the proper execution of a survey corresponds to its design, every facet of the survey must be looked at carefully during implementation. This includes re-examining the sample selection and redoing some of the interviews, as well as assessing the editing and coding of the responses.

Without proper checking, errors may go undetected. On the other hand, with good procedures, it might even be possible to prevent them.

Furthermore, insisting on high standards in the recruiting and training of interviewers is crucial to conducting a quality survey.

S5.9 REMARKS

The main aims of this chapter were to provide an understanding of samples and populations, and to examine some of the techniques available for collecting a reasonable sample. If a sample is truly representative of a population, it is usually possible to use the sample data to make inferences regarding the population parameters.

You may have noticed, in various newspaper and other media opinion polls, that details of the manner in which the sample was chosen, along with the size of the sample taken, are often provided as a footnote. On many occasions the questions asked are of a categorical nature, such as 'For which political party would you vote if an election were held today?', with the percentages of the sample choosing each party being recorded.

Although techniques for determining the appropriate size of a selected sample are beyond the scope of this book, it is not uncommon for a random sample of around 1000 to be chosen by the media in such polls. One reason for this is that any percentage calculated from the sample has a 95% chance of being within 1.5% of the true percentage for the population. For example, if 42.5% of a random sample of 1000 voters claimed to support the Prime Minister, we could be 95% sure that between 41.0% and 44.0% of the population felt the same way.

In Chapter S2, various methods of presenting the data obtained from samples and populations were discussed.

Quick Quiz

1. List some of the advantages and disadvantages of using a telephone questionnaire system rather than a mailed questionnaire system to determine voting preferences.
2. A factory has 225 workers on the shop floor. Describe how you would select a random sample of 20 workers to form an advisory committee.
3. Suppose you had to conduct a survey of your statistics class to determine the proportion of students who found the subject enjoyable. How many students would you select in your sample to get a reasonably accurate picture of the true population proportion? Give reasons for your answer.
4. Round the following set of costs to the nearest 10 cents:

   \$1.87 \$2.35 \$6.95 \$3.44 \$1.95 \$4.25 \$0.96
5. Find the relative error percentage made in the total when the rounding is made in Question 4.
6. An insurance company has 120 staff, comprising 16 executive management, 40 middle management and 64 clerical staff. If stratified sampling is to be used to select a random sample of 15 staff, how many of each of the 3 groups should be included in the sample?
7. Explain how you would use a coin to randomly select 1 person from a group of 3 people.
8. If you wanted a random sample of 50 people from the telephone book, what might be wrong with selecting the first 50 people whose last name begins with the letter S?
9. Using a calendar, write down the date of every 25th day of the current year.
10. Using Table S5.1, describe how you would select a random sample of 20 owners of lawn-mowers in a district where there are 1848 such owners.

Multiple Choice Questions

1. The absolute error in the sum when the numbers 27, 35, 48, 75 and 95 are rounded to the nearest multiple of 10 is:
(a) 10 (b) 15 (c) 20
(d) 25 (e) none of the above

2. For the data in Question 1, the relative error percentage (to two decimal places) is:
(a) 6.67% (b) 7.14% (c) 3.57%
(d) 3.33% (e) none of the above

3. Which of the following variables is *not* continuous?
(a) height
(b) weight
(c) brand of camera purchased
(d) area
(e) all of the above

4. In a random sample of 100 teenage males, all of whom either smoked or drank or did both, 20 smoked but did not drink, while 50 drank but did not smoke. Therefore, the number in the sample who *both* smoked and drank is:
(a) 20 (b) 30 (c) 40
(d) 50 (e) none of the above

5. An advantage of multiple choice questions is:
(a) they are easy to make up
(b) the results always follow a normal distribution
(c) they can be easily coded
(d) they prevent the respondent from making a comment
(e) none of the above

6. Suppose you have selected a random sample of 1000 in order to conduct a survey. Of these, only 220 bother to respond. Would you:
(a) consider this response rate high enough?
(b) bribe the 780 non-respondents with a gift to respond?
(c) take another random sample of size 1000?
(d) abandon the survey?
(e) do none of the above?

7. The accuracy of any estimates about a population made from the sample depends on:
(a) the size of the sample
(b) the variation in the population
(c) how the sample was conducted
(d) the ability of the statistician
(e) all of the above

8. The sampling technique whereby members of the population are placed in array and every tenth member is selected is an example of:
(a) random sampling
(b) cluster sampling
(c) systematic sampling
(d) stratified sampling
(e) none of the above

9. The first government lottery conducted in Australia was in New South Wales in:
(a) 1821 (b) 1881 (c) 1901
(d) 1931 (e) none of the above

10. The *mixed congruential method* is used to:
(a) mix balls in a barrel
(b) divide a population into groups
(c) omit undesirable elements of a sample
(d) generate random numbers
(e) none of the above

Exercises

S5.1* Which of the following variables are discrete and which are continuous?

(a) exam scores

(b) temperature

(c) sex

(d) volume

(e) brand of cigarettes preferred

(f) number of children per family

(g) share prices

(h) the value of the Australian dollar in terms of the United States dollar

(i) interest rates

(j) mass

S5.2* Suppose you have access to the Higher School Certificate (HSC) UAI (University Admissions Index) scores of all New South Wales students in the previous year. Construct an example in which this set of data is:

(a) the population (b) the sample

S5.3 You are given complete information regarding the salaries of the representatives of a certain life assurance company. Construct an example in which this set of data is:

(a) the population (b) the sample

S5.4 List some of the advantages and disadvantages of each of the sampling techniques discussed in this chapter.

S5.5* Discuss some methods that may be used to obtain a random sample of 50 students from a college with a total enrolment of 1500.

S5.6 Explain how you would obtain a random sample of 20 bottles of wine from a batch of 1000 bottles.

S5.7* A glass manufacturer wishes to know which factors affect consumer choice of glass jars in a certain test area. The names and addresses of everyone on the electoral roll in that area are available. Discuss whether this roll could be used to choose a random sample of customers for questioning.

S5.8 Consider the following set of data:

48 23 78 99 35 66 78 12 45 54

(a) Round the data to the nearest multiple of 20.

(b) Find the absolute error made in the total when the rounding is made in (a).

(c) Find the relative error made in the total when the rounding is made in (a).

(d) Find the relative error percentage made in the total when the rounding is made in (a).

S5.9* (a) Round the data in Exercise S5.8 to the nearest multiple of 10.

(b) Find the absolute error made in the total when the rounding is made in (a).

(c) Find the relative error made in the total when the rounding is made in (a).

(d) Find the relative error percentage made in the total when the rounding is made in (a).

S5.10 (a) Round the data in Exercise S5.8 to the nearest multiple of 5.

(b) Find the absolute error made in the total when the rounding is made in (a).

(c) Find the relative error made in the total when the rounding is made in (a).

(d) Find the relative error percentage made in the total when the rounding is made in (a).

S5.11* Consider the following set of data, which represent the prices in 10 districts of a 1 kg pack of dried skim milk:

\$2.15 \$2.87 \$2.42 \$1.98 \$1.78 \$2.13 \$2.44 \$2.91 \$1.63 \$1.50

(a) Round the data to the nearest 10 cents.

(b) Determine both the absolute and the relative error and the relative error percentage made in the sum of the prices due to the rounding in (a).

(c) Round the data to the nearest whole dollar.

(d) Determine both the absolute and the relative error and the relative error percentage made in the sum of the prices due to the rounding in (c).

S5.12* Suppose there are 92 people on the electoral roll in a retirement village. Describe how you would use the random numbers in Table S5.1 to select a random sample of 10 people from the village.

S5.13 If you were trying to interview a random sample of 100 families from among all the families in the Sydney metropolitan area, would you be justified in taking a detailed map of the area, selecting 100 locations by randomly sticking 100 pins into the map, and interviewing one family at each location? Justify your answer.

S5.14* Suppose that the 487 workers in a large company are each identifiable by a three-digit identification number from 001 to 487. Use the random numbers in Table S5.1 to select a random sample of 10 workers.

S5.15* A random sample of 2000 people includes 60% females. Of the females, 75% expressed opposition to smoking in public places; of the males, 60% expressed opposition. In the population as a whole, there are 50% males and 50% females. What is your best estimate of the proportion of the complete population (males and females combined) who oppose smoking in public places?

S5.16* Suppose that a city consists of four suburbs (A, B, C and D) and that the populations of these suburbs are:

A: 250 000 B: 100 000 C: 150 000 D: 300 000

Describe how you could use stratified sampling to select a random sample of 200 people from the city.

S5.17 Suppose you want to investigate the likely demand for a new flavour of ice-cream called Lollipop Special. Discuss the possible biases (if any) in the following proposed sampling techniques:

(a) Ask all the children in a certain school whether they would buy an ice-cream called Lollipop Special.

(b) Give all the children in the school a sample of Lollipop Special, and then ask them whether they like it.

(c) Go to suburban shopping centres on a Saturday morning, randomly give out free samples of Lollipop Special, and see whether people then buy the product from the supermarket.

(d) Go to a cricket Test match, sell Lollipop Special 10 cents cheaper than the other ice-creams, and see whether people buy it.

Can you propose any better method than the above suggestions?

S5.18 Ask a random sample of 10 students from your college whether they were watching television at 7 p.m. on the previous Monday night. Use the responses to estimate the proportion of students in the whole college who were watching television at that time. Do you feel that your estimate is likely to be accurate? Why?

S5.19 The manager of a car saleyard claims that 35% of cars driven in Australia are white. To test this claim, take a random sample of 50 cars and find the sample proportion that are white. Based on your results, do you think it likely that the manager's claim is true? Give reasons.

S5.20* The amounts spent by 12 children at a school canteen one lunchtime were recorded as follows:

\$0.40	\$1.52	\$0.95	\$0.77	\$1.36	\$1.55
\$0.75	\$0.35	\$0.66	\$0.89	\$1.74	\$0.15

(a) Round the data to the nearest 10 cents.

(b) Determine the absolute error, the relative error and the relative error percentage made in the sum of the amounts due to the rounding in (a).

S5.21 Describe how you would use the random numbers in Table S5.1 to select a random sample of 15 people from a population of 83 people.

S5.22 A medical researcher claims to have discovered a cure for patients who are diagnosed as HIV-positive. You are engaged to find a random sample of patients for testing the alleged cure. Describe precisely the sampling procedure you would use.

S5.23 The average price in Canberra of 1 kg of oranges was \$1.28 in June 1986, \$1.10 in June 1991 and \$1.04 in June 1996. (*Source*: Australian Bureau of Statistics, *Average Retail Prices of Selected Items*, June Quarters 1986, 1991 and 1996, Cat. No. 6403.0.) An economics student claims that this is clear evidence that the price of fruit in Canberra is on the way down. Do you agree? Explain why or why not. (*Hint*: The price in the March quarter 2001 was \$2.04.)

S5.24 Design a multiple choice questionnaire consisting of 5 questions that you might use to assess public opinion on the introduction of the Goods and Services Tax (GST) into Australia.

S5.25 Design a multiple choice questionnaire consisting of 5 questions that you might use to assess whether the students in your statistics class feel that they are getting value for money in their education.

S5.26 Test your questionnaire from Exercise S5.25 on 10 students in your class. Give a brief summary of the results and any conclusions that you can make from them.

S5.27* The staff of a large company comprises 60 males and 140 females. There are 40 executives, of whom 30 are male and 10 female, while the remaining staff in the company are all clerical.

(a) How many male and female clerical staff are there?

(b) The company wishes to take a representative sample of 40 people from the company and would like the sample to represent both the gender of staff and their types of position in the proportions in which they exist in the company. Describe precisely how you would achieve such a sample.

S5.28 Your company has asked you to mail out a questionnaire (with a return stamped addressed envelope) in order to determine how many consumers use its particular brand of washing powder. After mailing to a selected random sample of 100 consumers, you find that only 15 have bothered to return the questionnaire. Describe what your next plan of action might be in order to increase the response rate to a reasonable level.

S5.29* Discuss the relative merits of using a telephone interview technique as opposed to a personal interview method. Make particular reference to the types of questions and surveys that may be more suitable for each type of sampling.

S5.30 On occasions when there has been a controversial program on television, the station involved reports that 'Our switchboard has been jammed with complaints from viewers'. Often, this leads to the conclusion that the feelings of these viewers reflect the community's view as a whole.

Do you feel that viewers who ring television stations to air their views are representative of the population? Give reasons for your answer.

S5.31 A used-car yard has 84 cars for sale on its lot. A motor transport inspector wants to select a random sample of 6 cars to inspect for possible defects, to give an indication as to the general state of all cars in the yard. Explain how this could be done.

S5.32 (a) Investigate how police in your state select drivers to undertake random breath-testing (RBT) to determine whether the drivers have a blood alcohol reading above the legal limit.

(b) Do you feel that the method used is really random? Explain.

(c) How could you improve on the method currently used by police in your state to select drivers?

S5.33 In a certain town, 20 pregnant women all have miscarriages in the same month. A newspaper reporter claims that the miscarriages must have occurred because, for the first time, the town switched to daylight-saving at the beginning of the month.

What does the reporter's conclusion illustrate?

S5.34* From an office staff of 4 people, the manager has to select 1 person to undertake an unpleasant duty. To be fair, she wants to give everyone an equal chance of being selected. Explain how this could be done using:

(a) a coin (b) a fair six-sided die

S5.35 Answer Exercise S5.34 for a staff of 8 people.

S5.36 A company employs 24 executives, 75 managers, 7 section heads and 194 clerical staff. A consultant investigating work practices requires a sample, stratified on job level, of 25 employees to interview regarding stress leave.

(a) Approximately how many of each level should be in the sample?

(b) Explain how you would select such a sample.

S5.37* The populations of each of the 8 states and territories of Australia (in '000) in June 2000 are shown in the table below.

State	Population ('000)
New South Wales	6445.7
Victoria	4753.7
Queensland	3566.4
South Australia	1497.6
Western Australia	1883.9
Tasmania	470.4
Northern Territory	195.5
Australian Capital Territory	310.8

Source: Australian Bureau of Statistics, *Australian Demographic Statistics*, September Quarter 2000, Cat. No. 3101.0

Round the data in the table (in '000) to:

(a) the nearest whole number

(b) the nearest multiple of 10

(c) the nearest multiple of 100

S5.38 According to the Australian Bureau of Statistics (*Pocket Year Book Australia 2000*, p. 57), 28.9% of people in Australia aged 18–44 years are smokers compared to only 11.3% in the age group 65 years or more.

(a) Explain how you would select a random sample of people from each age group to verify these figures.

(b) Can you think of any reasons why the percentages would be so different for the two age groups?

S5.39* A law student wants to check a media claim that convicted armed bank robbers in Australia spend an average of only 25 months in prison. Explain how the student could estimate this average herself.

S5.40 Explain how you could estimate the mean continuous life of a certain brand of torch battery.

S5.41 A confectionery company claims that it fills its popular fun bags with an average of 25 chocolates. A teenager with a sweet tooth goes to his local store and buys 5 such bags. He counts the contents of each bag and discovers that they contain 25, 24, 26, 22 and 23 chocolates. It does not take him long to calculate that the mean of his bags is only 24 chocolates, so he writes to the company accusing it of false advertising. Is he justified in his complaint? Explain.

S5.42* The El Cheapo and Rock Bottom grocery stores in the same town both claim to have the cheapest prices. Explain how you could determine which store is the overall cheapest.

S5.43 Explain how you would select a random sample to estimate the mean length of commercials played between 7.00 p.m. and 9.00 p.m. on a weeknight on all commercial television stations in your city.

S5.44* Explain how you would estimate the mean of the length of words used by the author in this book.

S5.45 The times it takes (in seconds) for 6 amateur runners to cover a distance of 200 m are shown below.

28.35 31.45 34.27 22.55 26.75 25.46

(a) Sort the data.

(b) Round each original value to the nearest 0.1 second.

(c) Round each value in (b) to the nearest second.

(d) Round each original value to the nearest second.

(e) Are the answers to (c) and (d) identical? Explain.

Chapter S6

Elementary probability

'It's true that hard work never killed anybody, but I figure why take the chance?'

(Ronald Reagan, interview in the Guardian, *31 March 1987)*

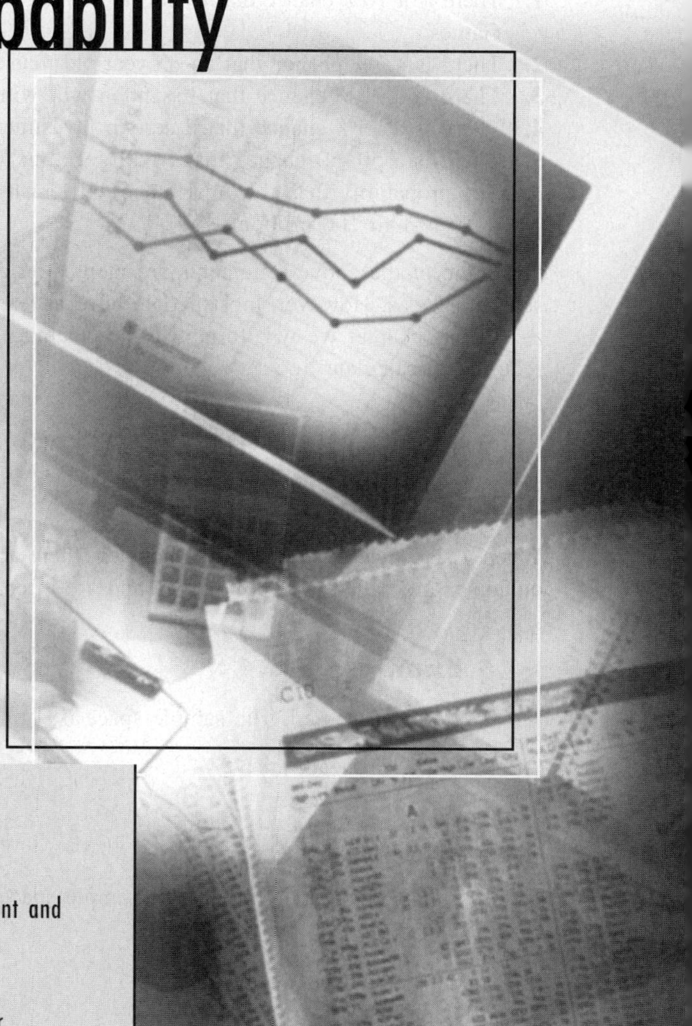

Learning objectives

In this chapter you will learn to:

- Understand elementary probability concepts
- Calculate the probability of events
- Distinguish between mutually exclusive, dependent and independent events
- Calculate conditional probabilities
- Understand and use the general addition law for probabilities
- Understand and apply Venn diagrams
- Understand and apply probability tree diagrams

S6.1 INTRODUCTION

In everyday language we often refer to the *probability* that certain events will happen. These could include the probability that it will rain, the probability that a certain sporting team will win a match and the probability that we will get to an appointment on time. Indeed, we also use the word 'chance' as a substitute for probability on some occasions.

While we all use the word 'probability' in our language, there would be few people who could provide a formal definition of its meaning. Consider the following statements and try to put your own interpretation on their intention:

1. There is a 10% chance that it will rain during the opening ceremony of the Athens 2004 Olympic Games.
2. There is a 7% chance that an 85-year-old man alive today will still be alive in 12 months time.
3. There is a 30% chance that Essendon will win the AFL premiership in the year 2004.
4. There is a 25% chance that a certain investment will yield a profit in the coming year.
5. There is a 50–50 chance that I will get a tax refund next year.
6. The probability that a 767 jet plane will crash into the Sydney Harbour Bridge before the year 2010 is 1 in 100 million.

Some of the above statements are merely reflections of a personal point of view, with no real scientific basis. However, most people have at least a rough feeling for what is meant.

In this chapter we will formalise the meaning of probability and look at some of the formulae required to calculate it.

S6.2 PROBABILITY OF EVENTS

Sample space

When a statistical experiment is conducted, there are a number of possible outcomes. These possible outcomes are called a **sample space** and are often denoted by the letter S.

Example S6.1

A coin is tossed. The sample space is:

$S = \{\text{head, tail}\}$

Example S6.2

A six-sided die is rolled. The sample space is:

$S = \{1, 2, 3, 4, 5, 6\}$

Example S6.3

A person is chosen at random for an experiment and the sex of the person is recorded. The sample space is:

$S = \{\text{male, female}\}$

Example S6.4

A salesperson has 3 table tennis tables left in the store to sell. The total number of tables sold in the next 5 days is recorded. The sample space is:

$S = \{0, 1, 2, 3\}$

Example S6.5

A survey is undertaken in which the gross annual income of accountants in 2002/2003 is recorded. The figure recorded is whether this income exceeded $100 000. The sample space is:

S = {gross income less than or equal to $100 000, gross income more than $100 000}

Example S6.6

A stockbroker makes a listing of all his stocks. For each one he records whether he feels it will rise or fall in value in the next 12 months. The sample space is:

S = {rise in value, fall in value}

In order to deal with the relevant aspects of sample spaces, it is necessary to formalise some of the common terms that are used.

Events

Definition S6.1

An event is some subset of a sample space.

Example S6.7

A coin is tossed. Define an event A to be:

A = outcome is a head

Example S6.8

A six-sided die is rolled. Define an event A to be:

A = outcome is an odd number
= 1 or 3 or 5

Example S6.9

In Example S6.4, let event A be that the salesperson sells *at most* two table tennis tables in the next 5 days. Then:

A = 0 or 1 or 2

More than one event can be defined from a sample space. A pack (or deck) of 52 playing cards is useful in illustrating this. These 52 cards consist of 4 *suits*: hearts, diamonds, clubs and spades. Hearts and diamonds are *red* cards, while clubs and spades are *black* cards. Each suit contains 13 cards, namely 2, 3, 4, 5, 6, 7, 8, 9, 10, jack, queen, king, ace. The jack, queen and king are also known as *face cards* (since there is a face drawn on them).

Example S6.10

Suppose a card is drawn at random from a pack of 52 playing cards. Define events A, B and C:

A = card drawn is an ace
B = card drawn is red
C = card drawn is a face card

Definition S6.2

The *impossible event* (or *empty set*) is one that contains no outcomes. It is often denoted by the Greek letter Φ (phi).

Example S6.11

A hand of 5 cards is dealt from a deck. Let A be the event that the hand contains 5 aces. Since there are only 4 aces in the deck, event A cannot occur. Hence, A is an impossible event.

Probability

Definition S6.3

If A is an event, the probability that it occurs is denoted by $P(A)$.

Definition S6.4

The probability (or *chance*) that an event A occurs is the proportion of possible outcomes in the sample space that yield the event A. That is:

$$P(A) = \frac{\text{Number of outcomes that yield event } A}{\text{Total number of possible outcomes}}$$

Note

1. Definition S6.4 makes sense only if the number of possible outcomes is finite (i.e. we can count them). We will restrict the discussion in this chapter to such cases.
2. If an event can *never* occur, its probability is 0. An event that *always* happens has probability 1.
3. The value of a probability must *always* lie between 0 and 1. It is a serious error to claim that a probability lies outside the range 0 to 1.
4. A probability may be expressed as a decimal or a fraction.
5. All outcomes are considered equally likely.

Example S6.12

In Example S6.8, $P(A) = \frac{3}{6} = \frac{1}{2} = 0.50$

Example S6.13

In Example S6.9, $P(A) = \frac{3}{4} = 0.75$

Example S6.14

In Example S6.10,
$$P(A) = \frac{4}{52} = \frac{1}{13}$$
$$P(B) = \frac{26}{52} = \frac{1}{2}$$
$$P(C) = \frac{12}{52} = \frac{3}{13}$$

Mutually exclusive events

Definition S6.5

Two events A and B are said to be **mutually exclusive** if they cannot occur simultaneously.

Example S6.15

Suppose a coin is tossed. Define events A and B where:

A = outcome is a head
B = outcome is a tail

Since A and B cannot both occur, the events are mutually exclusive.

Example S6.16

A person is selected at random from the population. Define events A and B to be:

A = person is a non-smoker
B = person smokes at least 5 cigarettes per day

Since A and B cannot both occur, the events are mutually exclusive.

If two events A and B are mutually exclusive, then the following relationship holds.

$$P(A \text{ or } B) = P(A) + P(B) \qquad \textbf{(S6.1)}$$

That is, the probability that event A or event B occurs is the sum of the two individual probabilities that each event occurs.

Example S6.17

A card is drawn from a deck of 52 cards. Let events A and B be:

A = card is a face card
B = card has a points value less than 6

The events A and B are clearly mutually exclusive since they cannot both occur at the same time. We have:

$$P(A) = \frac{12}{52} = \frac{3}{13}$$

$$P(B) = \frac{16}{52} = \frac{4}{13}$$

Hence, using Equation S6.1, the probability that the card drawn is a face card *or* has a points value less than 6 is:

$$\begin{aligned} P(A \text{ or } B) &= P(A) + P(B) \\ &= \frac{3}{13} + \frac{4}{13} \\ &= \frac{7}{13} \end{aligned}$$

Example S6.18

An accountant found in a study that receivables fell into one of 4 categories:

A = paid on time
B = paid early
C = paid late
D = uncollectable

Of a sample of 120 receivables, she found that 35 were paid on time, 40 were paid early, 28 were paid late and the remainder were uncollectable.

Find the probability that a receivable was paid early or on time.

Solution

The events A and B are mutually exclusive. We have:

$$P(A) = \frac{35}{120}$$

$$P(B) = \frac{40}{120}$$

Using Equation S6.1:

$$\begin{aligned} P(A \text{ or } B) &= P(A) + P(B) \\ &= \frac{35}{120} + \frac{40}{120} \\ &= \frac{75}{120} \\ &= \frac{5}{8} \end{aligned}$$

Hence, the probability that a receivable was paid early or on time is $\frac{5}{8}$.

The formula in Equation S6.1 can be extended to any number of mutually exclusive events.

Definition S6.6

Suppose that $A_1, A_2, A_3 \ldots A_n$ are n mutually exclusive events. Then:

$$P(A_1 \text{ or } A_2 \text{ or } A_3 \ldots \text{ or } A_n) = P(A_1) + P(A_2) + P(A_3) + \ldots + P(A_n) \qquad \textbf{(S6.2)}$$

Example S6.19

A card is drawn from a deck of 52 cards. Define events:

A = card is 2 or 3 or 4
B = card is 7 or 8
C = card is a face card

Then:

$$P(A) = \frac{12}{52}$$

$$P(B) = \frac{8}{52}$$

$$P(C) = \frac{12}{52}$$

Since the events A, B and C are mutually exclusive, using Equation S6.2:

$$\begin{aligned} P(A \text{ or } B \text{ or } C) &= P(A) + P(B) + P(C) \\ &= \frac{12}{52} + \frac{8}{52} + \frac{12}{52} \\ &= \frac{32}{52} \\ &= \frac{8}{13} \end{aligned}$$

Hence, the probability that the card drawn is a 2 or 3 or 4 or 7 or 8 or a face card is $\frac{8}{13}$.

Independent events

Definition S6.7

Two events A and B are **independent** if the occurrence of one does not alter the likelihood of the other event occurring. Events that are *not* independent are called **dependent**.

Example S6.20

1. The outcomes of successive tosses of a coin are independent since the result of the first toss does not affect the second.
2. If a truly random sample is selected from a population, the eye colours of any two members of the sample are independent.
3. Incomes of people who live in adjacent houses are *not* independent. Even though they may not interact in any way, people who live in a particular district tend to belong to a similar socioeconomic level.

If two events A and B are independent, the following relationship holds:

$$\text{P}(A \text{ and } B) = \text{P}(A) \times \text{P}(B) \qquad \textbf{(S6.3)}$$

That is, the probability that event A and event B *both* occur is the *product* of the two probabilities that each event occurs.

Example S6.21

Two coins are tossed independently. Find the probability that both coins show heads.

Solution

Define events A and B:

A = head on first coin
B = head on second coin

Then $\text{P}(A) = \text{P}(B) = \frac{1}{2}$.

Since the events A and B are independent, using Equation S6.3:

$$\begin{aligned}\text{P}(A \text{ and } B) &= \text{P}(A) \times \text{P}(B)\\ &= \frac{1}{2} \times \frac{1}{2}\\ &= \frac{1}{4}\end{aligned}$$

Hence, there is a probability of $\frac{1}{4}$ that two independent tosses of a coin will both yield heads.

Example S6.22

It is known that 40% of the adult population of a city favours Australia becoming a republic. If 2 adults are selected at random, find the probability that both of them are in favour.

Solution

Define events A and B:

A = first adult is in favour
B = second adult is in favour

Since the events A and B are independent, using Equation S6.3:

$$\begin{aligned} P(A \text{ and } B) &= P(A) \times P(B) \\ &= 0.40 \times 0.40 \\ &= 0.16 \end{aligned}$$

Hence, there is a probability of 0.16 that both adults will be in favour.

The formula in Equation S6.3 can be extended to any number of independent events.

Definition S6.8

Suppose that $A_1, A_2, A_3, \ldots, A_n$ are n independent events. Then:

$$P(A_1 \text{ and } A_2 \text{ and } A_3 \ldots \text{ and } A_n) = P(A_1) \times P(A_2) \times P(A_3) \times \ldots \times P(A_n) \qquad \textbf{(S6.4)}$$

Example S6.23

Suppose a coin is tossed, a six-sided die is rolled and a card is drawn randomly from a deck of 52 cards. Define events A, B and C:

A = coin is a tail
B = die is a 3
C = card is a heart

Then:

$$P(A) = \frac{1}{2}$$

$$P(B) = \frac{1}{6}$$

$$P(C) = \frac{1}{4}$$

To find the probability that *all three* events occur we use Equation S6.4, where:

$$
\begin{aligned}
P(A \text{ and } B \text{ and } C) &= P(A) \times P(B) \times P(C) \\
&= \frac{1}{2} \times \frac{1}{6} \times \frac{1}{4} \\
&= \frac{1}{48}
\end{aligned}
$$

Hence, provided that the three events are independent, there is a probability of $\frac{1}{48}$ that all three events will occur.

Complementary events

Definition S6.9

The *complements* of an event are those outcomes of a sample space for which the event does *not* occur. Two events that are complements of each other are said to be **complementary**.

Example S6.24

Suppose a six-sided die is thrown. Define an event A:

A = outcome is 1 or 2

Then the complement of event A would be event B, where:

B = outcome is 3 or 4 or 5 or 6

If two events A and B are mutually exclusive and are complements of each other, then:

$$P(A) + P(B) = 1 \qquad \textbf{(S6.5)}$$

Example S6.25

Consider Example S6.24. We have:

$$P(A) = \frac{2}{6} \text{ and } P(B) = \frac{4}{6}$$

These two events are mutually exclusive and are complements of each other. Hence:

$$P(A) + P(B) = \frac{2}{6} + \frac{4}{6} = 1$$

Example S6.26

Equation S6.5 is particularly useful when calculating the probability that, for example, at least one person (or object) in a random sample has a particular characteristic. Define the two events:

A = no one has the characteristic
B = at least one person has the characteristic

Then A and B are mutually exclusive and cover all possibilities. Using Equation S6.4:

$$\text{P}(\textit{at least one} \text{ person has the characteristic}) = 1 - \text{P}(\textit{no} \text{ person has the characteristic}) \qquad \textbf{(S6.6)}$$

Example S6.27

A coin is tossed 6 times. Find the probability of obtaining at least 1 head.

Solution

Define events A and B:

A = at least 1 head
B = no heads

Since the 6 tosses are independent, using Equation S6.4:

$$\begin{aligned} \text{P}(B) &= \frac{1}{2} \times \frac{1}{2} \times \frac{1}{2} \times \frac{1}{2} \times \frac{1}{2} \times \frac{1}{2} \\ &= \frac{1}{64} \end{aligned}$$

Since the two events A and B are mutually exclusive and cover all possibilities, using Equation S6.6:

$$\begin{aligned} \text{P}(A) &= 1 - \text{P}(B) \\ &= 1 - \frac{1}{64} \\ &= \frac{63}{64} \end{aligned}$$

Hence, the probability of at least 1 head in 6 independent tosses of a coin is $\frac{63}{64}$.

Conditional probabilities

Definition S6.10

The probability that event A occurs, given that an event B has occurred, is called the conditional probability that A occurs given that B occurs. The notation for this **conditional probability** is $\text{P}(A|B)$.

For any two events, A and B, the following relationship holds:

$$\text{P}(A|B) = \frac{\text{P}(A \text{ and } B)}{\text{P}(B)} \qquad \textbf{(S6.7)}$$

Note: If two events A and B are *independent*, we recall Equation S6.3, which states:

$$\text{P}(A \text{ and } B) = \text{P}(A) \times \text{P}(B)$$

Substituting this result into Equation S6.7 yields:

$$\begin{aligned} P(A|B) &= \frac{P(A) \times P(B)}{P(B)} \\ &= P(A) \end{aligned}$$

That is, the conditional probability that event A occurs, given that event B occurs, is simply the probability that event A occurs. This result is what we would expect.

Example S6.28

Table S6.1 shows the sex and age characteristics of a population of 20 males and 30 females.

Table S6.1 Data for Example S6.28

	Sex		
	Male	Female	*Total*
Age under 30	8	20	28
Age 30 or over	12	10	22
Total	20	30	50

A person is chosen at random from the population. Define events:

A = person is male
B = person is female
C = person is aged under 30
D = person is aged 30 or over

Find:

(a) P(A)	(b) P(B)	(c) P(C)
(d) P(D)	(e) P(A and D)	(f) P(B and C)
(g) P($A \mid D$)	(h) P($C \mid B$)	(i) P(A or C)

Solution

From Definition S6.4:

(a) $P(A) = \frac{20}{50} = \frac{2}{5}$

(b) $P(B) = \frac{30}{50} = \frac{3}{5}$

(c) $P(C) = \frac{28}{50} = \frac{14}{25}$

(d) $P(D) = \frac{22}{50} = \frac{11}{25}$

(e) Note that the events A and D are *not* independent, so we cannot simply multiply the individual probabilities $P(A)$ and $P(D)$. Instead, we must take a careful look at what the question is asking.

In this case, $P(A \text{ and } D)$ means 'the probability that the person is male *and* aged 30 or over'. From Table S6.1, there are 12 people in this category out of the total of 50 people. Using Definition S6.4:

$$P(A \text{ and } D) = \frac{12}{50} = \frac{6}{25}$$

(f) Note that the events B and C are *not* independent, so we cannot simply multiply the individual probabilities $P(B)$ and $P(C)$. Once again, we need to look at what the question is asking.

In this case, $P(B \text{ and } C)$ means 'the probability that the person is female *and* aged under 30'. From Table S6.1, there are 20 people in this category out of the total of 50 people. Using Definition S6.4:

$$P(B \text{ and } C) = \frac{20}{50} = \frac{2}{5}$$

(g) Using Equation S6.7 and the result in (e):

$$\begin{aligned} P(A|D) &= \frac{P(A \text{ and } D)}{P(D)} \\ &= \frac{\frac{12}{50}}{\frac{22}{50}} \\ &= \frac{6}{11} \end{aligned}$$

Hence, the probability that the person is a male, given that they are aged 30 or over, is $\frac{6}{11}$.

(h) Using Equation S6.7 and the result in (f):

$$\begin{aligned} P(C|B) &= \frac{P(C \text{ and } B)}{P(B)} \\ &= \frac{\frac{20}{50}}{\frac{30}{50}} \\ &= \frac{2}{3} \end{aligned}$$

Hence, the probability that the person is aged under 30, given that she is female, is $\frac{2}{3}$.

(i) Note that the events A and C are *not* mutually exclusive, so we cannot simply add the individual probabilities $P(A)$ and $P(C)$. Again, we need to take a careful look at what the question is asking.

In this case, P(A or C) means 'the probability that the person is male *or* aged under 30'. From Table S6.1, there are 8 + 12 + 20 = 40 people in this category out of the total of 50 people. Using Definition S6.4:

$$P(A \text{ or } C) = \frac{40}{50} = \frac{4}{5}$$

General addition law

When two events are *not* mutually exclusive we should use the following **general addition law**:

$$P(A \text{ or } B) = P(A) + P(B) - P(A \text{ and } B) \qquad \textbf{(S6.8)}$$

Note: If the events A and B are mutually exclusive, P(A and B) = 0. In this case, Equation S6.8 becomes Equation S6.1.

Example S6.29

Use Equation S6.8 and the data in Table S6.1 to find:

(a) P(A or D) (b) P(B or C)

In each case, interpret the results.

Solution

(a) From Example S6.28 we found:

$$P(A) = \frac{2}{5} \qquad P(D) = \frac{11}{25} \qquad P(A \text{ and } D) = \frac{6}{25}$$

Using Equation S6.8:

$$\begin{aligned} P(A \text{ or } D) &= P(A) + P(D) - P(A \text{ and } D) \\ &= \frac{2}{5} + \frac{11}{25} - \frac{6}{25} \\ &= \frac{3}{5} \end{aligned}$$

Hence, the probability that the person is male *or* aged 30 or over is $\frac{3}{5}$.

(b) From Example S6.28 we found:

$$P(B) = \frac{3}{5} \qquad P(C) = \frac{14}{25} \qquad P(B \text{ and } C) = \frac{2}{5}$$

Using Equation S6.8:

$$\begin{aligned} P(B \text{ or } C) &= P(B) + P(C) - P(B \text{ and } C) \\ &= \frac{3}{5} + \frac{14}{25} - \frac{2}{5} \\ &= \frac{19}{25} \end{aligned}$$

Hence, the probability that the person is female *or* aged under 30 is $\frac{19}{25}$.

S6.3 VENN DIAGRAMS

Sample spaces and events are often presented in a visual display called a **Venn diagram**. While there are several variations as to how these diagrams are drawn, we will use the following conventions.

1. A *sample space* is represented by a *rectangle*.
2. *Events* are represented by regions within the rectangle. This is usually done using *circles* (or parts of circles).

Definition S6.11

(a) The **union** of two events A and B is the set of all outcomes that are in event A *or* event B. The notation is:

Union of event A and event $B = A \cup B$

Hence, we could write, for example, $P(A \cup B)$ instead of $P(A \text{ or } B)$.

(b) The **intersection** of two events A and B is the set of all outcomes that are in *both* event A and event B. The notation is:

Intersection of event A and event $B = A \cap B$

Hence, we could write, for example, $P(A \cap B)$ instead of $P(A \text{ and } B)$.

Venn diagrams are used to assist in presenting a picture of the union and intersection of events, and in the calculation of probabilities.

Example S6.30

Use a Venn diagram to represent the following events:

(a) event A
(b) the complement of event A
(c) the union of events A and B
(d) the intersection of events A and B
(e) the intersection of events A, B and C

Solution

The solutions are shown in Figure S6.1 on page S.182. Note that the relevant region has been shaded.

Example S6.31

A bank manager classifies customers into various categories, namely:

1. easy to deal with
2. a longstanding customer
3. has over $10 000 in the bank

(Note that these categories are *not* mutually exclusive.) Suppose we define these categories as events A, B and C:

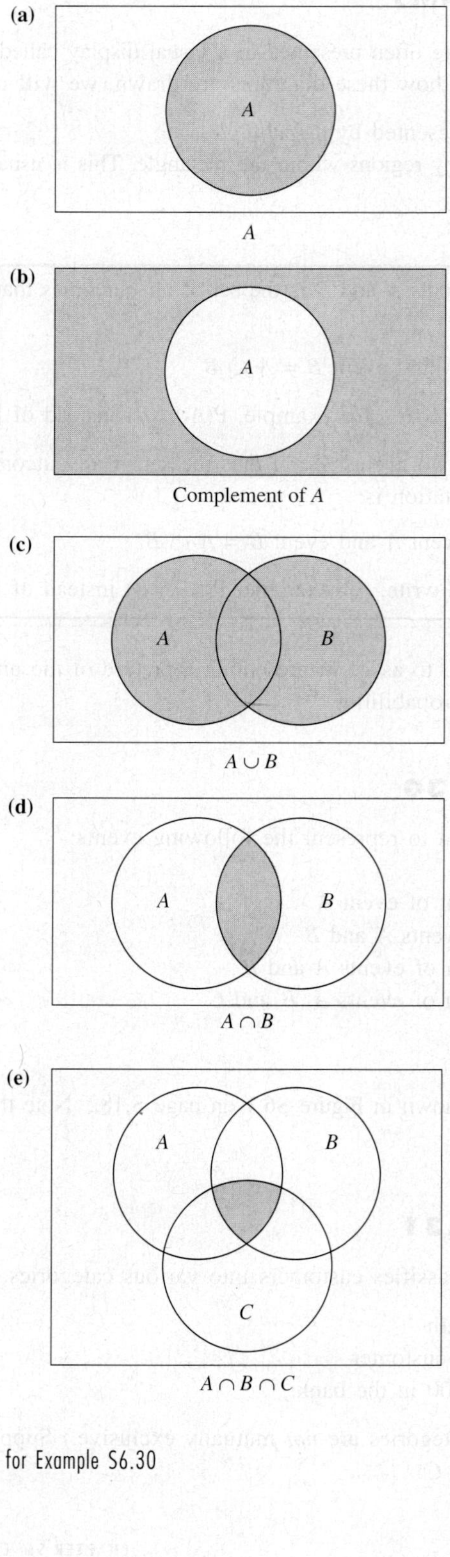

Figure S6.1 Venn diagrams for Example S6.30

A = customer is easy to deal with
B = customer is longstanding
C = customer has over \$10 000 in the bank

The Venn diagram in Figure S6.2 gives a visual display of these data for 55 customers, together with the number of customers in each category (or combination of categories) written in the appropriate part of the diagram.

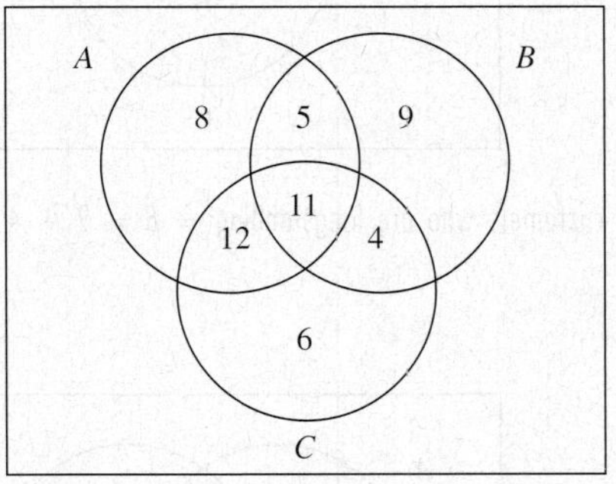

Figure S6.2 Venn diagram for Example S6.31

Use the Venn diagram to find the number of customers who:

(a) are easy to deal with
(b) are longstanding
(c) have over \$10 000 in the bank
(d) are easy to deal with *or* longstanding
(e) are easy to deal with *and* have over \$10 000 in the bank
(f) are easy to deal with *and* have over \$10 000 in the bank *but* are *not* of long standing
(g) are easy to deal with *or* are longstanding *and* who have over \$10 000 in the bank
(h) are easy to deal with *and* are longstanding *and* have over \$10 000 in the bank

In each case, shade the appropriate region.

Solution

Inspection of the Venn diagram in Figure S6.2 gives us the shaded Venn diagrams in Figures S6.3 to S6.10, with the following results:

(a)

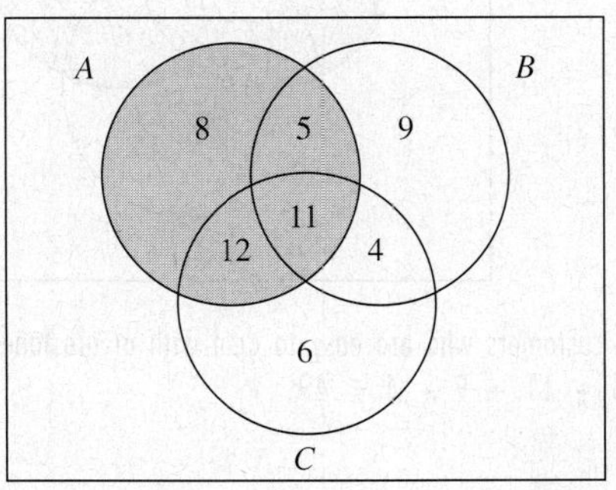

Figure S6.3 Number of customers who are easy to deal with = A = 8 + 12 + 11 + 5 = 36

(b)

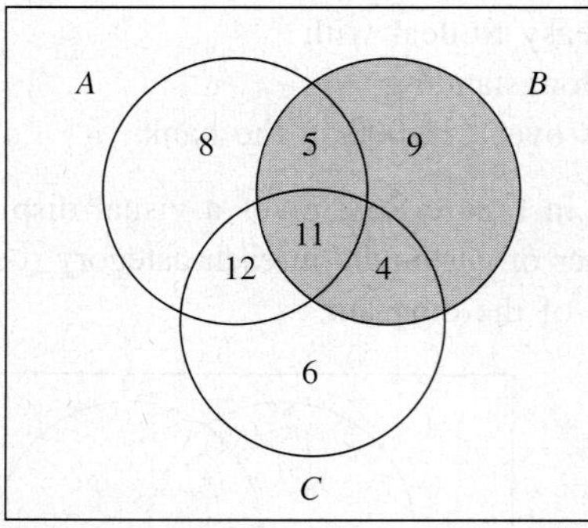

Figure S6.4 Number of customers who are longstanding = $B = 9 + 4 + 11 + 5 = 29$

(c)

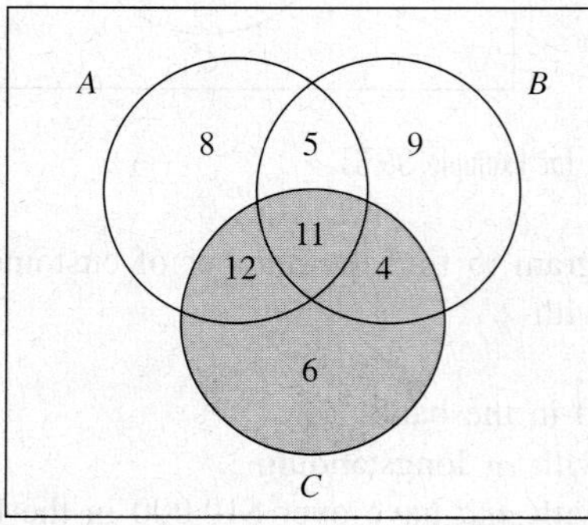

Figure S6.5 Number of customers who have over $10 000 in the bank = $C = 6 + 12 + 11 + 4 = 33$

(d)

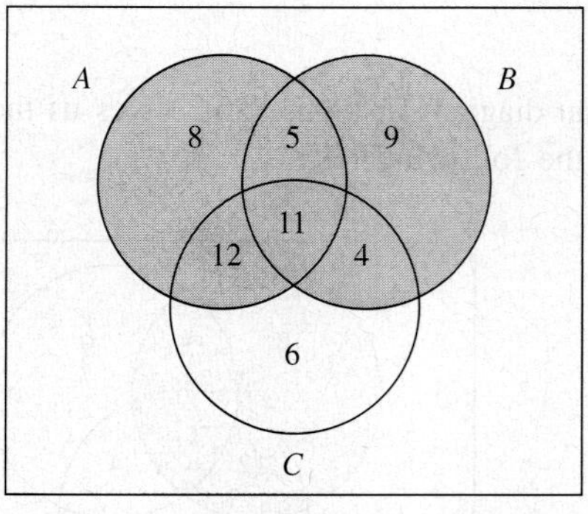

Figure S6.6 Number of customers who are easy to deal with or are longstanding =
$A \cup B = 8 + 12 + 5 + 11 + 9 + 4 = 49$

(e)

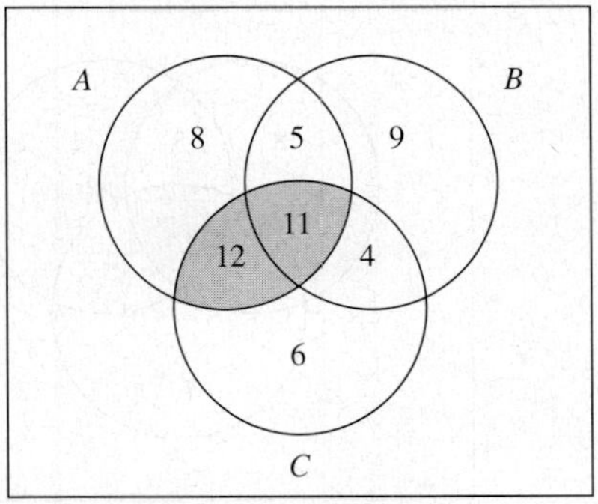

Figure S6.7 Number of customers who are easy to deal with and have over $10 000 in the bank $= A \cap C = 12 + 11 = 23$

(f)

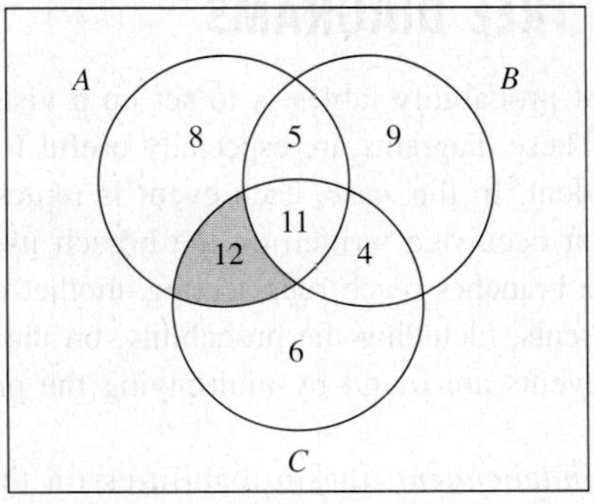

Figure S6.8 Number of customers who are easy to deal with and have over $10 000 in the bank but are not longstanding $= A \cap C \cap (\text{not } B) = 12$

(g)

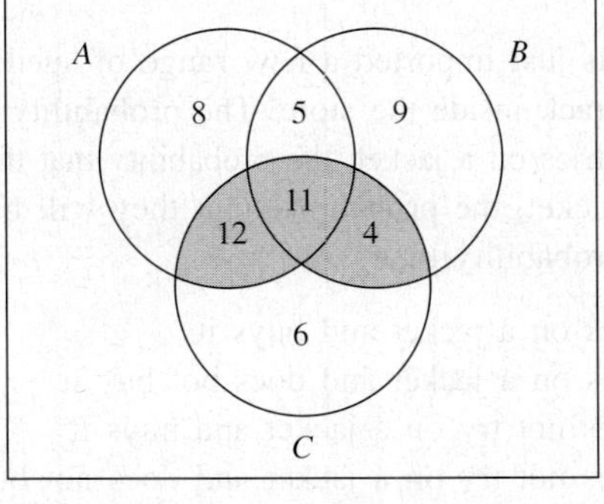

Figure S6.9 Number of customers who are easy to deal with or are longstanding and who have over $10 000 in the bank $= (A \cup B) \cap C = 12 + 11 + 4 = 27$

(h)

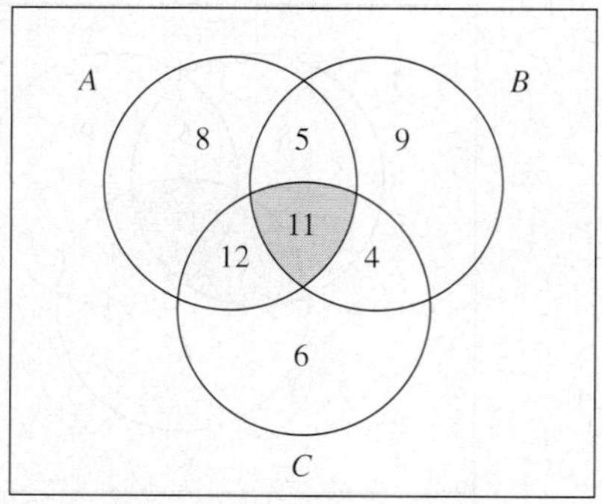

Figure S6.10 Number of customers who are easy to deal with and are longstanding and have over $10 000 in the bank = $A \cap B \cap C = 11$

S6.4 PROBABILITY TREE DIAGRAMS

An alternative to using joint probability tables is to set up a visual display of the probabilities using a *probability tree* diagram. These diagrams are especially useful for determining probabilities involving events that are *not* independent. In this case, each event is represented by a *branch* of the tree, with the probability of each event occurring written on the branch itself. At the end of each branch there is a *fork* where one or more branches (each representing another event) are drawn. It is usual to write which event a branch represents, including the probability, on the branch itself. The joint probabilities for combinations of these events are found by multiplying the probabilities along the branches from the beginning of the tree.

If the events are *not independent*, the probabilities on the second tier of branches will be **conditional probabilities**, since their values will depend on what happened in the first event. We will consider this case only, since the probabilities for independent events may be found using the formulae in Section S6.2.

Example S6.32

A clothing store has just imported a new range of suede jackets that it has advertised at a bargain price on a rack inside the store. The probability that a customer tries on a jacket is 0.40. If a customer tries on a jacket, the probability that they will buy it is 0.70. If a customer does *not* try on a jacket, the probability that they will buy it is 0.15.

Calculate the probability that:

(a) a customer tries on a jacket and buys it
(b) a customer tries on a jacket and does not buy it
(c) a customer does not try on a jacket and buys it
(d) a customer does not try on a jacket and does not buy it

Solution

This problem is one in which a probability tree diagram may be used effectively. Define events:

A = customer tries on a jacket
B = customer buys a jacket

We are given the following probability:

$$P(A) = 0.40$$

It follows that:

$$\begin{aligned} P(\textit{not } A) &= 1 - P(A) \\ &= 1 - 0.40 \\ &= 0.60 \end{aligned}$$

These probabilities are written on the initial branches of the probability tree.

We are also given the following conditional probabilities:

$$\begin{aligned} P(B|A) &= 0.70 \\ P(B|\textit{not } A) &= 0.15 \end{aligned}$$

Since the events $B|A$ and *not* $B|A$ are complementary, it follows that:

$$\begin{aligned} P(\textit{not } B|A) &= 1 - P(B|A) \\ &= 1 - 0.70 \\ &= 0.30 \end{aligned}$$

Since the events $B|$*not* A and *not* $B|$*not* A are complementary, it follows that:

$$\begin{aligned} P(\textit{not } B|\textit{not } A) &= 1 - P(B|\textit{not } A) \\ &= 1 - 0.15 \\ &= 0.85 \end{aligned}$$

These conditional probabilities are written on the second tier of branches.

The probability tree diagram for this problem is shown in Figure S6.11 on page S.188. From Figure S6.11 it can be seen there are four possible combinations of the events. We can use the values above, along with Equation S6.7, to answer the questions asked.

(a) P(customer tries on a jacket *and* buys it)

$$\begin{aligned} &= P(A \text{ and } B) \\ &= P(B|A)P(A) \\ &= 0.70 \times 0.40 \\ &= 0.28 \end{aligned}$$

(b) P(customer tries on a jacket *and* does *not* buy it)

$$\begin{aligned} &= P(A \text{ and } \textit{not } B) \\ &= P(\textit{not } B|A)P(A) \\ &= 0.30 \times 0.40 \\ &= 0.12 \end{aligned}$$

(c) P(customer does *not* try on a jacket *and* buys it)

$$\begin{aligned} &= P(\textit{not } A \text{ and } B) \\ &= P(B|\textit{not } A)P(\textit{not } A) \\ &= 0.15 \times 0.60 \\ &= 0.09 \end{aligned}$$

(d) P(customer does *not* try on a jacket *and* does *not* buy it) $= \text{P}(not\ A \text{ and } not\ B)$
$= \text{P}(not\ B \mid not\ A)\text{P}(not\ A)$
$= 0.85 \times 0.60$
$= 0.51$

Note that these four probabilities add up to 1.

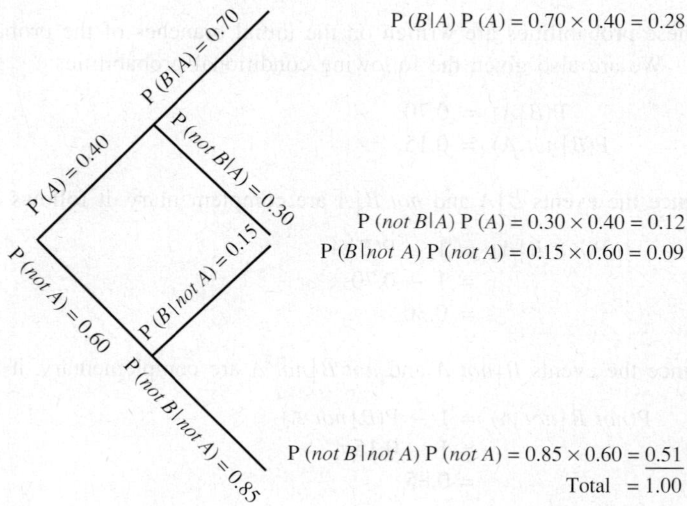

Figure S6.11 Probability tree diagram for Example S6.32

In Example S6.32 we saw how to calculate probabilities involving events that are *not* independent. We can also use the results to calculate the second-tier probabilities that are not conditional.

Example S6.33

Suppose we want simply to calculate the straight probability that a customer buys a jacket, that is to find the value of P(*B*). To do this we must use the following formula:

$$\text{P}(B) = \text{P}(B \mid A)\text{P}(A) + \text{P}(B \mid not\ A)\text{P}(not\ A) \qquad \textbf{(S6.9)}$$

In Example S6.32 we have:

$\text{P}(B \mid A) = 0.70 \quad \text{P}(A) = 0.40 \quad \text{P}(B \mid not\ A) = 0.15 \quad \text{P}(not\ A) = 0.60$

Therefore, using Equation S6.9:

$$\text{P}(B) = (0.70)(0.40) + (0.15)(0.60) = 0.37$$

That is, 37% of customers will buy a jacket.

A probability tree can also be used to answer questions when random samples are drawn *without replacement*. That is, after the first member of the sample is drawn out, it is not replaced before the second member is drawn out. Such problems involve events that are *not* independent.

Example S6.34

Consider the problem of a small business that employs 4 males and 6 females. Due to a current financial crisis, management must reluctantly retrench 2 employees. All employees are considered to be of equal ability and have similar lengths of service to the business. To be fair to all concerned, management decides to write each of the 10 names on equal-sized pieces of paper and place them in a box. These papers will then be mixed and a name drawn out. That person will be retrenched, and their piece of paper will *not* be put back in the box. Then, from the remaining 9 pieces of paper in the box, a second name will be drawn out to determine the second employee who will be retrenched.

Find the probability that:

(a) the second name drawn is male
(b) the second name drawn is female
(c) both names drawn are male
(d) both names drawn are female

Solution

In this case, the outcomes of the draws are *not* independent, since the first name will *not* be replaced in the box. That is, the outcome of the second draw will depend on the outcome of the first draw.

Define the following events:

A_1 = first name drawn is male
A_2 = second name drawn is male
B_1 = first name drawn is female
B_2 = second name drawn is female

In drawing the probability tree in this case, the probabilities on the *second* tier of branches will be *conditional probabilities*, since their value will depend on what happened to the *first* draw.

To draw the probability tree we need first to calculate the appropriate conditional probabilities, as below.

If the first name drawn out is a male, this leaves 3 male names and 6 female names in the box. Therefore:

$$P(A_2|A_1) = \frac{3}{9} = \frac{1}{3}$$

If the first name drawn out is a male, this leaves 3 male names and 6 female names in the box. Therefore:

$$P(B_2|A_1) = \frac{6}{9} = \frac{2}{3}$$

If the first name drawn out is a female, this leaves 4 male names and 5 female names in the box. Therefore:

$$P(A_2|B_1) = \frac{4}{9}$$

If the first name drawn out is a female, this leaves 4 male names and 5 female names in the box. Therefore:

$$P(B_2|B_1) = \frac{5}{9}$$

These four conditional probabilities are drawn on the second-tier branches of the probability tree shown in Figure S6.12.

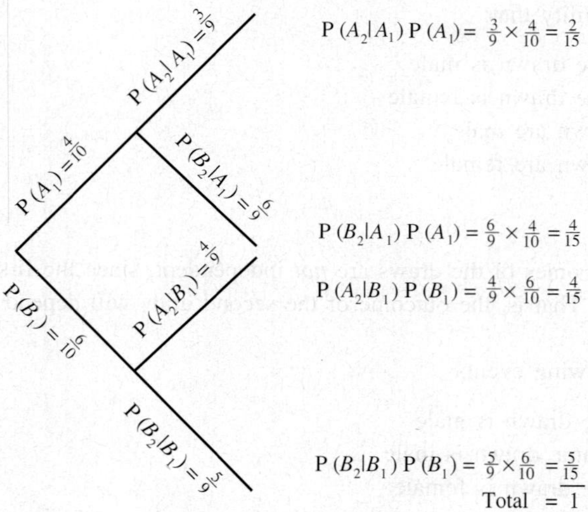

Figure S6.12 Probability tree for Example S6.34

(a) We must find $P(A_2)$. This may be calculated using Equation S6.9:

$$\begin{aligned} P(A_2) &= P(A_2|A_1)P(A_1) + P(A_2|B_1)P(B_1) \\ &= \left(\frac{3}{9}\right)\left(\frac{4}{10}\right) + \left(\frac{4}{9}\right)\left(\frac{6}{10}\right) \\ &= \frac{36}{90} \\ &= \frac{2}{5} \end{aligned}$$

That is, the probability that the second name drawn is male is $\frac{2}{5}$.

(b) We must find $P(B_2)$. This also may be calculated using Equation S6.9:

$$\begin{aligned} P(B_2) &= P(B_2|A_1)P(A_1) + P(B_2|B_1)P(B_1) \\ &= \left(\frac{6}{9}\right)\left(\frac{4}{10}\right) + \left(\frac{5}{9}\right)\left(\frac{6}{10}\right) \\ &= \frac{54}{90} \\ &= \frac{3}{5} \end{aligned}$$

That is, the probability that the second name drawn is female is $\frac{3}{5}$.

Note that $P(A_2) + P(B_2) = 1$ as expected, since A_2 and B_2 cover all possibilities of outcomes on the second draw.

(c) We must find $P(A_1 \text{ and } A_2)$. We can use Equation S6.7 to write:

$$\begin{aligned} P(A_1 \text{ and } A_2) &= P(A_2|A_1)P(A_1) \\ &= \left(\frac{3}{9}\right)\left(\frac{4}{10}\right) \\ &= \frac{2}{15} \end{aligned}$$

That is, the probability that both names drawn are male is $\frac{2}{15}$.

(d) We must find $P(B_1 \text{ and } B_2)$. We can use Equation S6.7 to write:

$$\begin{aligned} P(B_1 \text{ and } B_2) &= P(B_2|B_1)P(B_1) \\ &= \left(\frac{5}{9}\right)\left(\frac{6}{10}\right) \\ &= \frac{1}{3} \end{aligned}$$

That is, the probability that both names drawn are female is $\frac{1}{3}$.

Probability trees can also be used when there are a number of tiers of events, although they can become quite cumbersome when many tiers or probabilities are involved. Nevertheless, in simple cases they are very useful in determining what can sometimes be quite tricky probabilities if such a visual display is not used.

S6.5 REMARKS

The probabilities considered in this chapter have relied on our being able (at least in theory) to write down the sample space for an experiment. This will be true for discrete data, but the definitions do not hold if the data are continuous. This is because the sample space will then have an infinite number of points, and our formulae for probability as we have presented them must be revised.

However, if the data are continuous, there are other ways in which questions regarding probability can be answered. Examples are given in Chapter S7, where the continuous data follow a normal distribution.

You should be aware, however, that not all probabilities can be calculated in a straightforward manner. Consider the following example.

Example S6.35

Suppose you are given a drawing pin. When tossed, it can land in one of two possible ways, as shown in Figure S6.13.

Find the probability that a drawing pin will land in position (a) when tossed.

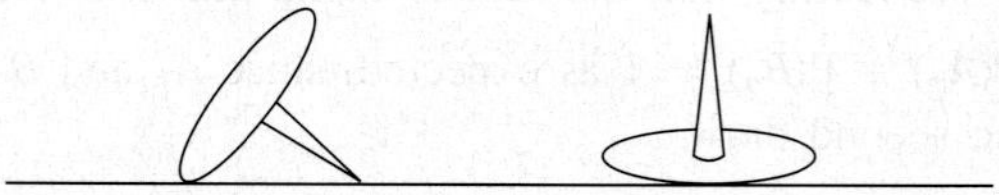

Figure S6.13 Two possible ways a drawing pin can land, for Example S6.35

Solution

In this case, by 'probability' we really mean the *long-run proportion of times that the drawing pin will land in position (a).*

We will never know the *exact* answer to this question. The best we can do is toss the drawing pin a number of times and record the proportion of times it lands in position (a). This will be our best estimate of the true probability. The more times we toss the pin, the closer our estimate will be to the true (unknown) probability.

For example, if we toss the pin 200 times and it lands in position (a) 124 times, we estimate that the *probability* that the pin will land in position (a) is $\frac{124}{200} = 0.62$.

There is no set number of times that the pin should be tossed. It really depends on how accurate you want the estimate to be.

Quick Quiz

1. Give two examples of an experiment in which the outcomes are all equally likely.
2. A six-sided die is rolled 3 times. Find the probability of obtaining at least one 6.
3. A population of 50 cars consists of 23 coloured red, 12 coloured white, 6 coloured blue, 4 coloured green and the remainder coloured brown. A car is selected at random. Find the probability that it is coloured white or brown.
4. In Question 3, a random sample of 3 cars is chosen. Find the probability that none of them is coloured red or green.
5. In Question 3, a car is chosen at random. Find the probability that it is coloured blue, given that it is not coloured white.
6. A single card is drawn from a deck of 52 cards. Find the probability that it is *either* a face card *or* a heart.
7. Two cards are drawn at random from a deck of 52 cards. If the first card is not replaced before the second card is drawn, what is the probability that the second card drawn is black, given that the first card drawn is black?
8. Suppose that the probability that it rains on any given day in July in Perth is 0.27. What is the probability that, on a random sample of 6 such days, it rains at most once? Are your assumptions likely to be true? Explain.
9. Two fair six-sided dice are rolled. Find the probability that the sum of the faces is 7, given that it is greater than 5.
10. A room contains 5 men and 3 women. Two people are selected without replacement. Find the probability that at least one is a woman.

Multiple Choice Questions

1. A coin is tossed 5 times. The probability of obtaining at least 1 head is:

(a) $\frac{7}{8}$ (b) $\frac{15}{16}$ (c) $\frac{31}{32}$

(d) $\frac{63}{64}$ (e) none of the above

2. A company staff consists of 20 accountants, 12 economists and 4 secretaries. If 2 staff members are chosen at random, the probability that they are *both* either an economist or a secretary is:

(a) $\frac{4}{21}$ (b) $\frac{16}{81}$ (c) $\frac{4}{5}$

(d) $\frac{16}{25}$ (e) none of the above

3. An accountant has 40 clients on a 'special' file. They comprise 15 executives (10 male, 5 female), 20 tycoons (8 male, 12 female) and 5 politicians (3 male, 2 female). The accountant selects one of these clients at random to prepare their income tax return. The conditional probability that an executive is chosen, given that the person is female, is:

(a) $\frac{5}{19}$ (b) $\frac{15}{40}$ (c) $\frac{19}{40}$

(d) $\frac{1}{8}$ (e) none of the above

4. For the data in Question 3, the conditional probability that a male is chosen, given that the person is a tycoon, is:

(a) $\frac{21}{40}$ (b) $\frac{1}{2}$ (c) $\frac{2}{5}$

(d) $\frac{1}{5}$ (e) none of the above

5. For the data in Question 3, the probability that a tycoon *or* a male is chosen is:

(a) $\frac{1}{2}$ (b) $\frac{33}{40}$ (c) $\frac{21}{40}$

(d) $\frac{7}{10}$ (e) none of the above

6. For the data in Question 3, the probability that an executive *or* a tycoon, but *not* a female, is chosen is:

(a) $\frac{7}{8}$ (b) $\frac{9}{20}$ (c) $\frac{6}{7}$

(d) $\frac{21}{40}$ (e) none of the above

7. A day of the week is selected at random. The probability that it starts with the letter 't', given that it contains the letter 's', is:

(a) $\frac{1}{2}$ (b) $\frac{1}{4}$ (c) $\frac{2}{7}$

(d) 1 (e) none of the above

8. A fair coin and a die are tossed simultaneously. The probability that the coin lands heads *and* the die has a value of at least 2 is:

(a) $\frac{1}{12}$ (b) $\frac{1}{6}$ (c) $\frac{5}{6}$

(d) $\frac{5}{12}$ (e) none of the above

9. In a certain population, the probability that a person is a smoker, given that they are female, is 0.35. Forty per cent of the population are female. The proportion of female smokers in the population is:

(a) 0.05 (b) 0.0875 (c) 0.175

(d) 0.14 (e) none of the above

10. Suppose that the probability that it rains on at least one of two days is 0.36. The probability that it rains on any one single day is:

(a) 0.60 (b) 0.40 (c) 0.64

(d) 0.80 (e) none of the above

Exercises

S6.1* A coin is tossed 3 times. Find the probability that the outcome is:

(a) all heads (b) all tails (c) 2 heads and 1 tail

S6.2 A six-sided die is rolled 3 times. Find the probability that all 3 outcomes are greater than 4.

S6.3 Suppose we know that 20% of females have (naturally) blond hair and that 40% of females have blue eyes. Can we use this information to calculate the probability of females who have *both* blond hair and blue eyes? If so, then do it. If not, explain why.

S6.4 State, with reasons, whether you feel that the following pairs of events are independent:

(a) age of first child in a family, age of second child in a family

(b) the lengths of two adjacent books on a library shelf

S6.5* A box contains 3 red balls, 5 black balls and 8 blue balls. A ball is chosen at random. Find the probability that the ball is:

(a) red (b) black (c) blue (d) blue or red

S6.6 Go to an appropriate road and record the colours of 50 cars that drive past. Record their colour as either red, white, blue, brown or other.

(a) Use your results to estimate the probability that cars driven down that road are each of the above colours.

(b) Compare your answers with those of other students. Are you in general agreement?

S6.7* Among 20 applicants for a position in a law firm, there are 8 who have tertiary qualifications. If 2 of the applicants are chosen at random for interviews, find the probability that:

(a) neither has tertiary qualifications

(b) both have tertiary qualifications

(c) at least one has tertiary qualifications

S6.8 One hundred clients of a registered tax agent fall into the categories of personal (25 male, 30 female), corporate (10 male, 15 female) and small business (8 male, 12 female). The agent selects one of the clients at random to begin work. Find the probability that the client chosen is:

(a) male (b) female (c) personal (d) corporate

(e) small business

S6.9* For Exercise S6.8, find the probability that the client chosen is:

(a) male but *not* personal

(b) personal but *not* male

(c) *both* female and corporate

(d) *both* personal and small business

(e) *not* small business

(f) *not* male and *not* corporate

S6.10 For Exercise S6.8, find the *conditional probability* that the client chosen is:

(a) male, given that the client is small business

(b) corporate, given that the client is female

(c) personal, given that the client is male

(d) small business, given that the client is *not* corporate

S6.11 People who work for a modelling agency may have one or more of the following three attributes:

A: blue eyes *B*: over 170 cm tall *C*: blond hair

Draw Venn diagrams and shade in the appropriate region for those who:

(a) have blue eyes and blond hair but are *not* over 170 cm tall

(b) have blond hair but *not* blue eyes

(c) are over 170 cm tall and have blue eyes

S6.12* If 15% of the residents of a city own a video camera, find the probability that none of 4 independently chosen residents will own a video camera.

S6.13 Give an example of two events whose intersection is the empty set.

S6.14 Of a random sample of 25 company accounts it is found that 9 are delinquent.

(a) Estimate the probability that a company account is delinquent.

(b) If the company had 350 accounts, how many do you estimate would be delinquent?

S6.15* An integer is selected at random from all the integers from 1 to 100. What is the probability that it will contain the digit 1?

S6.16 An integer is selected at random from all the integers from 1 to 100. What is the probability that it will contain *at least one* of the digits 2 and 3?

S6.17 Suppose there are 22 horses starting in the Melbourne Cup, and you have drawn one of the horses at random in a sweep. What is the probability that your horse will win? What assumption are you making?

S6.18* A box contains 5 red balls, 6 white balls and 9 black balls. Two balls are drawn at random. (The first ball is *not* replaced.) Find the probability that both balls are of the same colour.

S6.19* After the airing of a particular controversial program, a television station received 80 letters of complaint. Writers were classified as follows: very angry (10 male, 6 female), angry (15 male, 23 female) and annoyed (16 male, 10 female).

The station's general manager asks to read a 'typical' letter of complaint, and the producer is told to select a letter at random. Find the probability that the writer of the letter chosen is:

(a) male (b) female (c) very angry (d) angry

(e) annoyed

S6.20* For Exercise S6.19, find the probability that the writer of the letter chosen is:

(a) male but *not* very angry (b) very angry but *not* male

(c) *both* female and annoyed (d) *both* male and very angry

(e) *not* annoyed (f) *not* male and *not* very angry

S6.21 For Exercise S6.19, find the *conditional probability* that the writer of the letter chosen is:

(a) male, given that the writer is angry

(b) very angry, given that the writer is female

(c) annoyed, given that the writer is male

(d) very angry, given that the writer is *not* annoyed

S6.22 Two six-sided dice are rolled. What is the probability that the sum of the digits on their faces is either 7 or 11?

S6.23 How do you rate your probability of passing your course in statistics in the current session? Explain how you arrived at your estimate.

S6.24* A student rates his chances of passing each subject in his degree at 90% for each subject. If he does 12 subjects for his degree, what is the probability that he will pass them all? What assumption are you making?

S6.25 (a) If a weather forecaster says 'There is a 40% chance of rain today', what do you think she means?

(b) How do you think professional weather forecasters arrive at probabilities such as the one in (a)?

S6.26* Find the probability of obtaining *at least* 3 heads in 4 tosses of a fair coin.

S6.27 Consider the drawing pin exercise in Example S6.35. Throw a drawing pin 25 times and record the number of times it falls in each of the two positions. Use these results to find the probability that it falls in each of these positions.

S6.28* A fair *five*-sided die (with sides numbered 1, 2, 3, 4, 5) is rolled 3 times. Find the probability that at least two of the outcomes are odd.

S6.29 Two people are chosen at random, one after the other, from a room containing 10 people of whom 6 are male. If the first person chosen is female, what is the probability that the second person chosen is male?

S6.30* According to the Australian Bureau of Statistics, 34.0% of people aged 18–24 in New South Wales are smokers. (*Source*: *Year Book Australia 1996*, p. 109.) If a random sample of 6 people of this category is selected, find the probability that:

(a) all 6 are smokers (b) none is a smoker (c) at least one is a smoker

S6.31 In Australia during 1999 there were 52 600 divorces granted, of which 16 306 were initiated by men. (*Source*: Australian Bureau of Statistics, *Marriages and Divorces Australia*, 1999, Cat. No. 3310.0.) If 3 divorces are selected at random, find the probability that:

(a) all were initiated by men

(b) at least 1 was initiated by the man

(c) at most 2 were not initiated by the man

S6.32* In a certain city, if an assault takes place, the probability that some person will be arrested for the crime is 0.40. The probability that the person arrested will be convicted is 0.25. What is the probability for an assault that some person will be arrested and convicted? What assumption(s) are you making?

S6.33 A student estimates that the probability that she will pass both her mathematics and history final exams is 0.45. She also estimates that the probability that she will pass the history final exam is 0.60. If the two exams are independent, what is the probability that she will pass the mathematics final exam?

S6.34 In Exercise S6.33, the student discovers that she has passed the mathematics exam. What is the probability that she will pass the history exam?

S6.35* Two cards are drawn at random from a deck of 52 cards. If the first card is *not* replaced before the second card is drawn, what is the probability that *both* cards drawn will be aces?

S6.36* In Exercise S6.35, if the first card *is* replaced before the second card is drawn, what is the probability that both cards drawn will be aces?

S6.37 One card is drawn from a deck of 52 cards. Find the probability that:

(a) the card is red, given that it is a queen

(b) the card is a queen, given that it is red

(c) the card is a red queen

S6.38 In a survey of 30 males and 50 females it was found that 15 males and 20 females exercise regularly. If a person is selected at random, what is the probability that:

(a) they are male

(b) they are female

(c) they are a male who exercises regularly

(d) given that they are female, they exercise regularly

(e) given that they do *not* exercise regularly, they are male

S6.39 A biologist has a cage containing 4 male and 6 female rats. If 2 rats are selected at random, without replacement, find the probability that:

(a) the first rat selected is male

(b) the second rat selected is female, given that the first one selected is male

(c) the second rat selected is male, given that the first one selected is female

(d) the 2 rats selected are of the same sex

(e) the 2 rats selected are of opposite sexes

S6.40* A surgery is operated by doctors A and B. Doctor A is absent on 10% of days, and Doctor B is absent on 5% of days. In a random sample of 200 days, on how many days would you expect that:

(a) both doctors will be absent?

(b) both doctors will be present?

(c) at least one doctor will be absent?

(d) at least one doctor will be present?

S6.41 In the game of Australian Rules football, a kick at goal can either be successful (score a goal), score a behind or miss altogether. Of a random sample of 100 kicks by a full-forward from 50 m out from the posts, 55 are successful, 30 score a behind and the remainder miss altogether. If the player has 2 further kicks from the same position, find the probability that:

(a) both kicks miss altogether

(b) at least one scores a goal

(c) one is a goal and one is a behind

(d) the second is a behind, given that the first is a goal

(e) at most, one goal is kicked

S6.42 Draw a Venn diagram to illustrate the data in Exercise S6.41.

S6.43 In a beachside suburb it rains on 20% of days and is windy on 30% of days. If rain and wind are independent, on any particular day find the probability that:

(a) it rains and is windy

(b) it does *not* rain and is *not* windy

(c) it rains or is windy

(d) it does *not* rain or is *not* windy

S6.44 Draw a Venn diagram to illustrate the data in Exercise S6.43.

S6.45 In an article published in the *Sydney Morning Herald* it was reported from companies that 25% of chief information officers (CIOs) report to the chief financial officer (CFO), 46% report to the chief executive officer (CEO) and the remainder report to someone other than the CFO or CEO.

Suppose that 2 CIOs are selected at random from 2 independent companies. Find the probability that:

(a) both report to the CEO

(b) both report to someone other than the CFO or CEO

(c) 1 reports to the CFO and 1 reports to someone other than the CFO or CEO

(d) at least 1 reports to the CEO

S6.46* In a survey of households in a suburb of Melbourne it was found that 30% had a personal computer, with 23% of those also having a modem. All households with a modem also had a personal computer.

In a random sample of 2000 households, how many households would you expect to have:

(a) a personal computer but no modem?

(b) a personal computer and a modem?

(c) no personal computer and no modem?

S6.47 Illustrate the data in Exercise S6.46 using a Venn diagram.

S6.48 Using the data in Exercise S6.46, a keen statistics student calculates that the answer to (b) is 1060 households. What fundamental error has she made in arriving at this answer?

S6.49* A TAFE science student must enrol in one elective subject and has the choice of either electronics or astronomy. Since he likes both subjects, he decides to roll a fair six-sided die to help him decide. If he rolls a 1 or 2, he will enrol in electronics. If he rolls a 3 or 4 or 5 or 6, he will enrol in astronomy.

If he enrols in electronics, there is a probability of 0.70 that he will pass. If he enrols in astronomy, there is a probability of 0.80 that he will pass.

(a) Draw a probability tree for the problem.

(b) Calculate the probability that:

(i) he enrols in electronics and passes it

(ii) he enrols in astronomy and does *not* pass it

(iii) he passes the elective subject

S6.50 Answer the questions in Exercise S6.49 if the student is *equally likely* to enrol in either electronics or astronomy.

S6.51* Two players, Anna and Sarah, plan to play a tennis match on Sunday. The probability that the weather will be hot on Sunday is 0.75. If the weather is hot, there is a probability of 0.60 that Anna will win. If the weather is *not* hot, there is a probability of 0.80 that Sarah will win.

(a) Draw a probability tree for the problem.

(b) Calculate the probability that:

  (i) the weather is hot and Sarah wins the match

  (ii) the weather is *not* hot and Anna wins the match

  (iii) Anna wins the match

  (iv) Sarah wins the match

S6.52 Answer the questions in Exercise S6.51 if the probability that the weather will be hot on Sunday is 0.20.

S6.53 A crate contains 20 identical cartons, each of which contains a camera. The manufacturer knows that exactly 6 of these cameras are defective and not worth repairing, while the remaining 14 cameras are perfectly all right. The crate is taken to a salesroom where customers are allowed to select one carton from the crate after paying an amount of money.

The first customer is Kate, who selects her carton and leaves the room in order to inspect the camera it contains. The second customer is David, who selects his carton from the remaining 19 in the crate.

(a) Draw a probability tree for the problem.

(b) Calculate the probability that:

  (i) Kate selects a carton containing a defective camera

  (ii) David selects a carton containing a defective camera, given that Kate's camera is *not* defective

  (iii) both Kate and David select a carton containing a defective camera

  (iv) both Kate and David select a carton containing a camera that is *not* defective

Chapter S7

The normal distribution

'Thank God we're normal!'

(John Osborne, English playwright,
The Entertainer, *1957)*

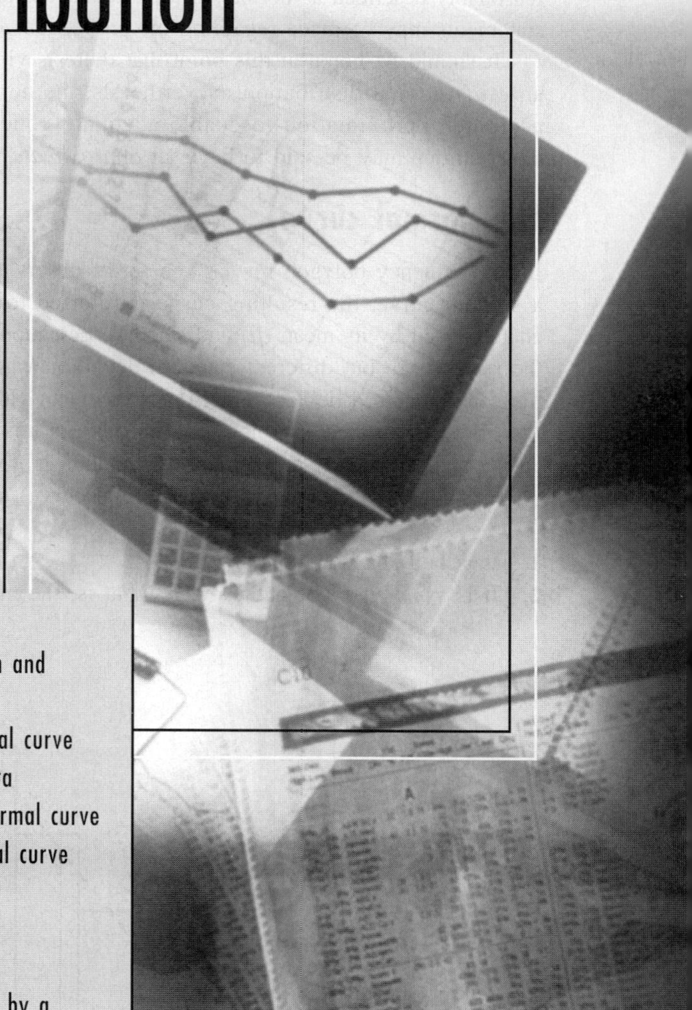

Learning objectives

In this chapter you will learn to:

- Identify the properties of the normal distribution and normal curve
- Identify the characteristics of the standard normal curve
- Understand examples of normally distributed data
- Read *z*-score tables and find areas under the normal curve
- Find the *z*-score given the area under the normal curve
- Compute proportions
- Check whether data follow a normal distribution
- Understand and apply the Central Limit Theorem
- Solve business problems that can be represented by a normal distribution
- Calculate estimates and their standard errors
- Calculate confidence intervals for the population mean

S7.1 INTRODUCTION

When the frequencies of observations for a large population are arranged in the form of a histogram with short interval lengths, often the resulting frequency polygon appears to follow the pattern of a smooth bell-shaped curve (Figure S7.1). Many naturally occurring distributions have this shape, which is referred to as a **normal distribution**.

The term **Gaussian distribution** is also often applied because of the work of the famous mathematician Carl Gauss (1777–1855).

Measurements made on characteristics such as height, blood pressure, test scores, IQ, crop yield per hectare and many others often have (or at least are alleged to have) a distribution that may be reasonably described as normal. If a particular characteristic does follow this distribution, many useful statements may be made regarding the proportions of the population that lie in appropriate intervals.

It should be noted at this point that there is virtually no naturally occurring system that follows an *exact* normal distribution. Nevertheless, the normal distribution can often provide a sufficiently accurate approximation to enable adequate conclusions to be drawn. For this reason, a set of observations may be said to have an *approximate* normal distribution.

The normal curve

If the frequency polygon for a given set of observations that have a normal distribution is made into a smooth curve, the resulting curve is referred to as a **normal curve**. A specific normal curve is characterised by its mean (μ) and standard deviation (σ). Figure S7.2 shows three normal curves with the same mean but different standard deviation values. Figure S7.3 shows three normal curves with the same standard deviation but different mean values.

For our purposes, it is not essential to write a mathematical equation to describe the normal curve, and consequently only a graphical presentation is given in this chapter. The main features of a normal distribution (as represented by the normal curve) are:

1. It is bell-shaped.
2. It is *symmetric* about the mean. That is, the mean and median are equal.

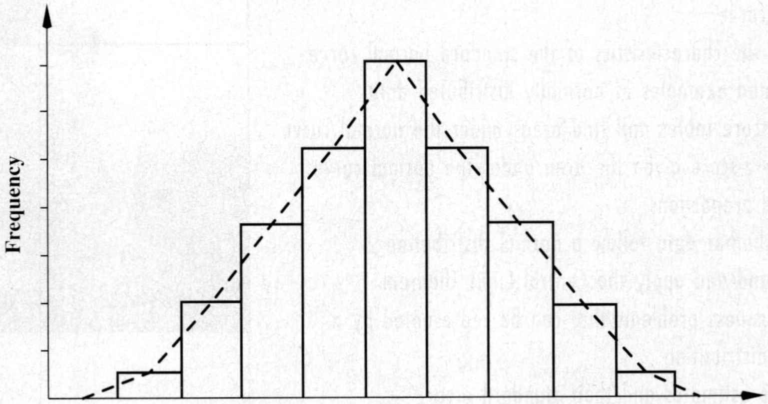

Figure S7.1 An example of hypothetical data with a bell-shaped frequency polygon

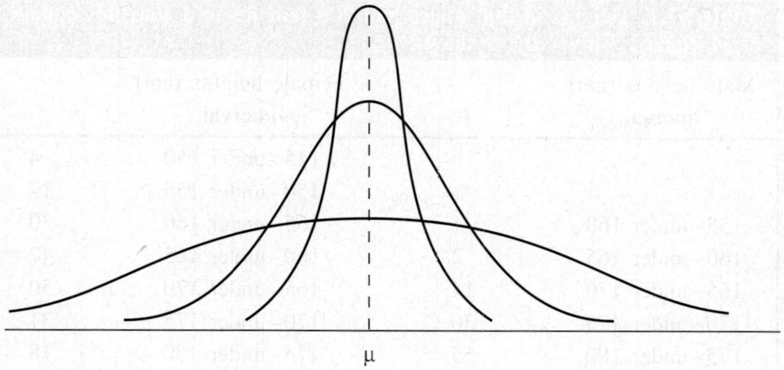

Figure S7.2 Normal curves with the same mean but different standard deviation values

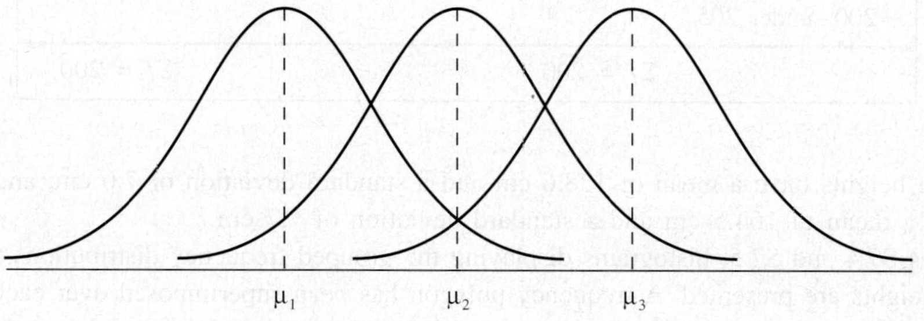

Figure S7.3 Normal curves with the same standard deviation but different mean values

3. It is *asymptotic* to the horizontal axis. In other words, the curve gets closer and closer to (but does not touch) the horizontal axis as it moves outwards.
4. Approximately 68% of the distribution lies within one standard deviation of the mean; about 95% lies within two standard deviations of the mean; and about 99.7% lies within three standard deviations of the mean.
5. The location and dispersion will depend on the values of μ and σ (see Figures S7.2 and S7.3).
6. The total area under any normal curve is 1. Thus, Point 4 above translates to:

The area of the curve within 1 standard deviation of the mean = 0.68 approximately.
The area within 2 standard deviations of the mean = 0.95 approximately.
The area within 3 standard deviations of the mean = 0.997 approximately.

S7.2 AN EXAMPLE OF DATA THAT HAVE AN APPROXIMATE NORMAL DISTRIBUTION

As part of a study by the author, random samples of 200 males and 200 females were selected from the students enrolled in a large introductory statistics course at Macquarie University, Sydney. Each student was asked to record his or her height (to the nearest cm). The results are shown in Table S7.1.

Table S7.1 Heights of 200 males and 200 females in a statistics course at Macquarie University, Sydney			
Male heights (cm) interval	f	Female heights (cm) interval	f
		145–under 150	4
		150–under 155	17
155–under 160	1	155–under 160	30
160–under 165	2	160–under 165	47
165–under 170	14	165–under 170	50
170–under 175	30	170–under 175	31
175–under 180	55	175–under 180	18
180–under 185	57	180–under 185	3
185–under 190	28		
190–under 195	10		
195–under 200	2		
200–under 205	1		
	$\Sigma f = 200$		$\Sigma f = 200$

The male heights have a mean of 178.6 cm and a standard deviation of 7.0 cm, and the female heights have a mean of 164.5 cm and a standard deviation of 7.7 cm.

In Figures S7.4 and S7.5, histograms displaying the grouped frequency distributions for the male and female heights are presented. A frequency polygon has been superimposed over each histogram, and the close approximation of the data to a normal distribution can be seen.

This technique of constructing the histogram and corresponding frequency polygon is the simplest method of determining whether a set of observations may be viewed as following an approximately normal distribution.

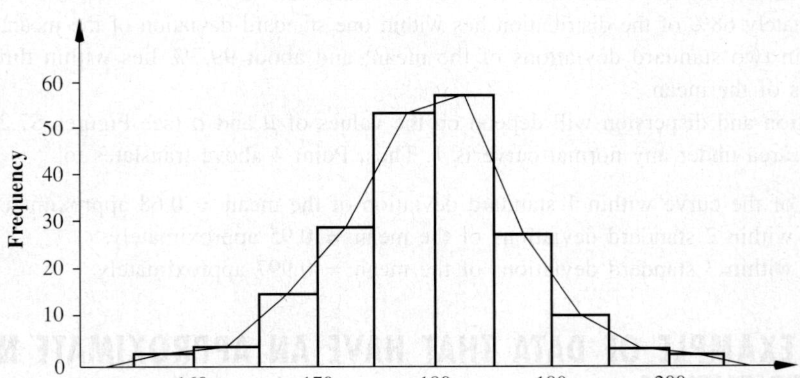

Figure S7.4 Histogram and frequency polygon for the heights of 200 males in a statistics course at Macquarie University, Sydney

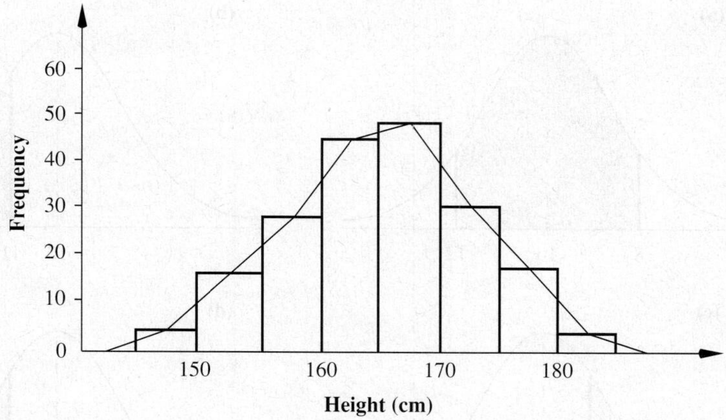

Figure S7.5 Histogram and frequency polygon for the heights of 200 females in a statistics course at Macquarie University, Sydney

S7.3 AREAS UNDER THE NORMAL CURVE

As remarked earlier, the proportions of observations that take on certain values are represented by areas under the distribution curve. That is, the proportion of observations that take on a value between a and b is the area under the curve between two vertical lines erected at a and b.*

Example S7.1

The time taken to repair defective components produced on an assembly line was monitored by the supervisor each day over a 6-month period. It was found that the daily repair time followed an approximate normal distribution, with a mean of 10 hours and a standard deviation of 2 hours.

Sketch normal curves, and shade in the region that represents the proportion of days on which the repair time for defective components was:

(a) between 8 and 12 hours
(b) more than 11 hours
(c) less than 9 hours
(d) between 11 and 14 hours

Solution

The relevant regions for each part are shown in Figure S7.6.

* In Chapter S6 we dealt only with *discrete* probability distributions, that is those where we could write down all possible values and their corresponding probabilities. Since the normal distribution is continuous, we need a new definition of the term *probability*. We cannot find the probability of an *individual* value occurring, since that would be zero.

However, we will discuss the *proportion* of times that events take place, and this may be taken to mean the probability that the event takes place.

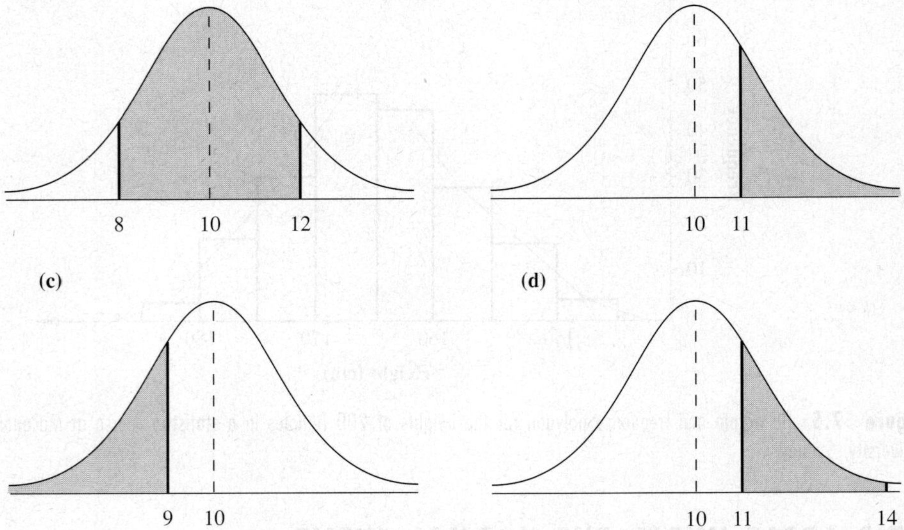

Figure S7.6 Normal curve for repair time (in hours) for Example S7.1

In Example S7.1, *one* curve was used to represent a *single* distribution, and various proportions of the observations were indicated by shading under this curve. We could calculate these areas and thus obtain values for the proportions. In fact, after detailed calculation, we could construct a table to show values for all the different areas under this curve. In practice, we often need values for areas under many normal curves (i.e. with different means and different standard deviations), so this particular table would have very limited use.

It is, of course, impractical to construct such a table for *every* normal curve. Suppose, for example, that the weights of elephants and the weights of mice are both normally distributed. They would clearly have different means and different standard deviations, and therefore different normal curves. Moreover, other species of elephants (or mice) probably have different parameters. It is unrealistic to expect that anyone would construct tables of areas under the curves for all these other species, and even these tables would not cover the weights of other animals. It is even more unrealistic to expect anyone to construct tables that give values for all the areas under all possible normal curves.

Standard score (*z*-score)

The above problem can be overcome by converting all our measurements into **standard score** or ***z*-score** form. The *z*-score of a measurement is defined as the number of standard deviations the measurement is away from the mean. In particular, if the measurement is *above* the mean, the corresponding *z*-score is positive; but if the measurement is *below* the mean, the corresponding *z*-score is *negative*. The formula is:

$$\text{Standard score} = z = \frac{\text{observed value} - \text{mean}}{\text{standard deviation}} \qquad \textbf{(S7.1)}$$

Thus, if a distribution has a mean of μ and a standard deviation of σ, the corresponding z-score of an observation x is:

$$z = \frac{x - \mu}{\sigma} \qquad \text{(S7.2)}$$

It immediately follows from the definition that the mean itself has a z-score of zero and that a value exactly one standard deviation from the mean has $z = +1$ or -1. The larger a positive z-value, the further the given x is *above* the mean, while large negative z-values correspond to extreme values *below* the mean. Since the standard deviation is by its nature a typical deviation from the mean, it is reasonable to assume that not many data items will differ from the mean by many times the standard deviation. It is found in practice with most data sets that the vast majority of observations have a z-score in the range roughly -2 to $+2$.

The standard normal distribution and standard normal curve

If we follow the above technique, it is possible to transform a measurement x from *any* normal distribution with a mean μ and standard deviation σ into a standard score (or z-score) using Equation S7.2. The z-score of an observation expresses its value in *relative units*. Consequently, raw scores from different normal distributions can be compared by converting them to z-scores from their own normal distribution. For example, a student who obtained a mark corresponding to a z-score of 1.60 for a particular exam performed better, *relative to other students who did the same exam*, than a student who obtained a mark corresponding to a z-score of, say, 1.30 in another exam.

The particular normal distribution of z-scores is called the **standard normal distribution**. It is a special normal distribution that has $\mu = 0$ and $\sigma = 1$. The corresponding curve is called the *standard normal curve*. It is the only normal curve for which statistical tables exist, although these tables can have several forms. One of the more common presentations of such a table is given in Table 1 on page S.453, which is discussed in the next section.

Quite elaborate mathematics were required to produce Table 1, although it is fairly easy to use. The values in the table do *not* need to be memorised since, apart from being too lengthy, every practising statistician always has such a table handy. The reason that a table exists for *one* normal distribution only is that it is simply impractical and unnecessary to have any others. A problem involving *any* normal curve can be solved by transforming it into the corresponding problem involving the standard normal curve.

Example S7.2

Suppose that the weights of a certain species of elephant are normally distributed with mean 20 tonnes and standard deviation 2 tonnes, as are the weights of mice with mean 30 g and standard deviation 3 g.

The question, 'What is the probability that an elephant has a weight between 20 tonnes and 22 tonnes?' then has the same answer as 'What is the probability that a mouse has a weight between 30 g and 33 g?', because both questions are equivalent to 'What is the probability that a standard score in a normal distribution is between 0 and 1?' The appropriate regions are shown in Figure S7.7.

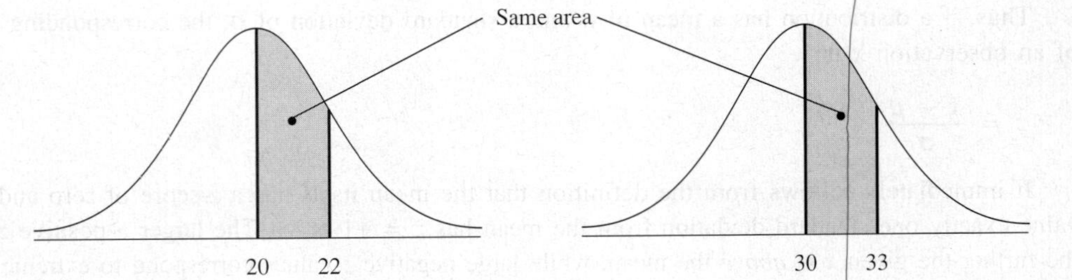

Figure S7.7 Regions for elephants (left, in tonnes) and mice (right, in grams)

S7.4 USING THE TABLE OF THE STANDARD NORMAL CURVE

Table 1 gives the area under the standard normal curve between the mean (zero) and any positive number z. Such a region is illustrated in Figure S7.8.

Techniques for calculating various types of areas are given below and in Section S7.5.

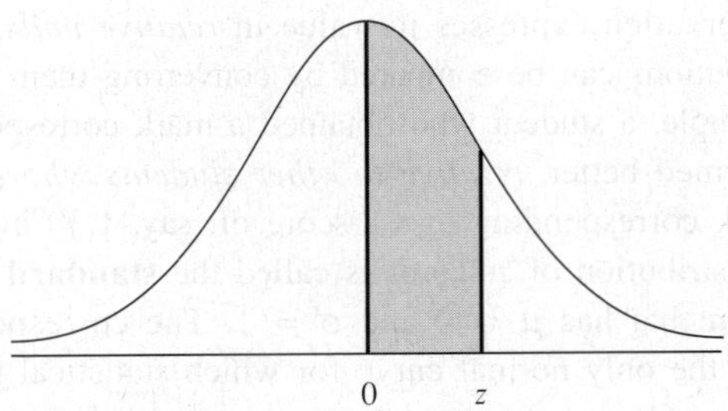

Figure S7.8 The region represented in Table 1 for a value z

Example S7.3

Using Table 1, determine the area under the standard normal curve for the following:

(a) between $z = 0$ and $z = 1.50$
(b) between $z = -2.10$ and $z = 0$
(c) between $z = 0.60$ and $z = 1.80$
(d) between $z = -0.30$ and $z = 2.25$
(e) to the right of $z = 1.95$

Solution

When solving problems of this type it is always advisable to sketch a normal curve and shade in the appropriate region. This is done in each of the solutions below.

(a) From Table 1, the area between $z = 0$ and $z = 1.50$ is 0.4332 (see Figure S7.9).
(b) In order to find the area to the left of $z = 0$, the symmetry of the normal curve can be used. In this case, the area between $z = -2.10$ and $z = 0$ is the same as the area between $z = 0$ and $z = +2.10$. From Table 1, this area is 0.4821 (see Figure S7.10).

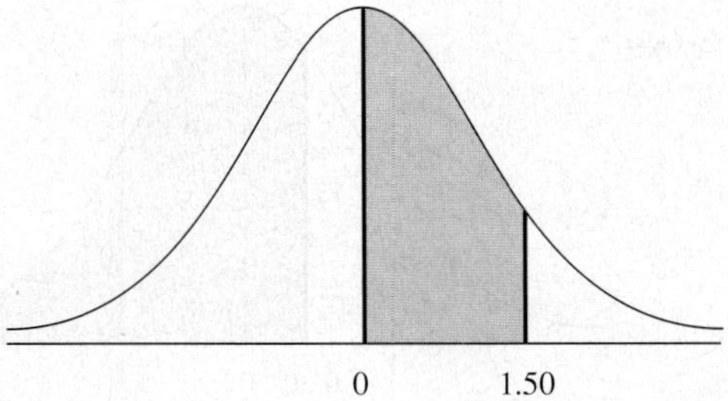

Figure S7.9

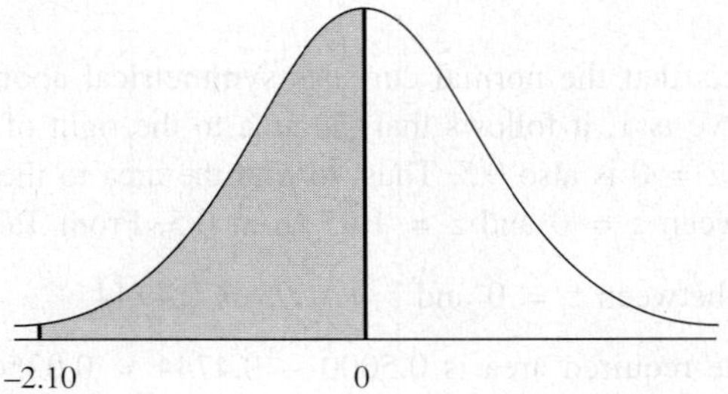

Figure S7.10

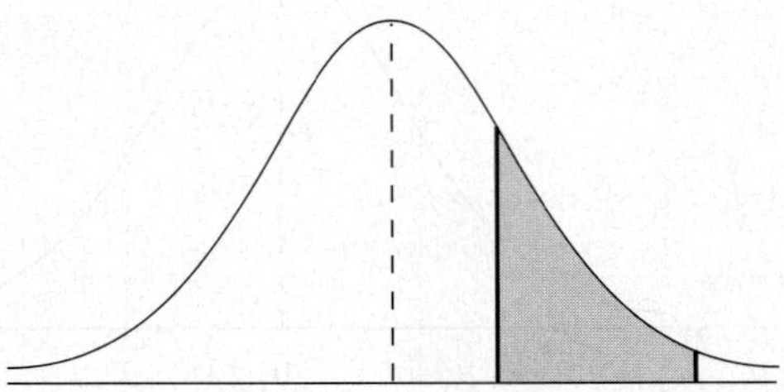

Figure S7.11

(c) The area between z = 0.60 and z = 1.80 may be found by subtracting the area between z = 0 and z = 0.60 from the area between z = 0 and z = 1.80. From Table 1:

- the area between z = 0 and z = 1.80 is 0.4641
- the area between z = 0 and z = 0.60 is 0.2257

Therefore, the required area is 0.4641 – 0.2257 = 0.2384 (see Figure S7.11).

(d) The area between z = –0.30 and z = 2.25 may be found by adding the area between z = –0.30 and z = 0 to the area between z = 0 and z = 2.25. The area between z = –0.30 and z = 0 equals the area between z = 0 and z = +0.30 (by symmetry). From Table 1:

- the area between z = 0 and z = 0.30 is 0.1179
- the area between z = 0 and z = 2.25 is 0.4878

Therefore, the required area is 0.1179 + 0.4878 = 0.6057 (see Figure S7.12).

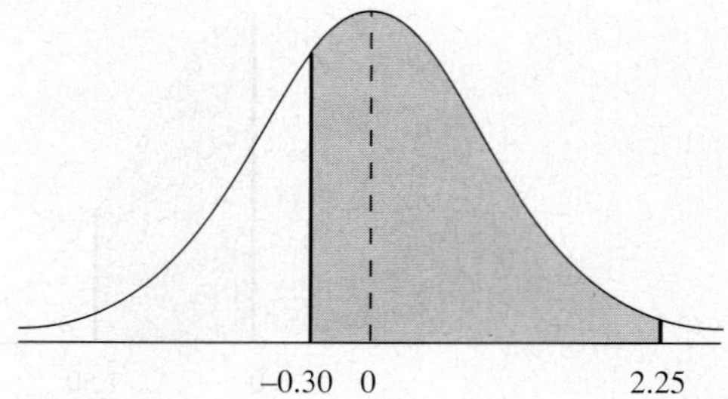

Figure S7.12

(e) From the facts that the normal curve is symmetrical about the mean and the total area under the curve is 1, it follows that the area to the right of $z = 0$ is 0.5 and that the area to the left of $z = 0$ is also 0.5. Thus, to find the area to the right of $z = 1.95$, we subtract the area between $z = 0$ and $z = 1.95$ from 0.5. From Table 1:

- the area between $z = 0$ and $z = 1.95$ is 0.4744

Therefore, the required area is 0.5000 – 0.4744 = 0.0256 (see Figure S7.13).

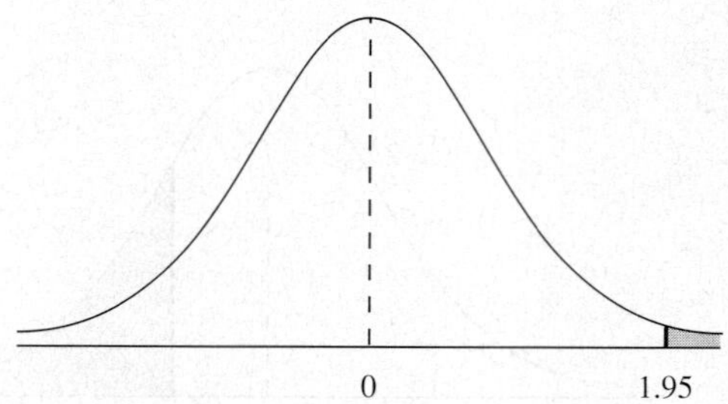

Figure S7.13

Conversion to raw scores

In order to determine appropriate areas under *any* normal curve, the z-score (or standard score) may be calculated as shown in Section S7.3. The z-score for an item of data indicates how far away from the mean it is in 'standard deviation' units. For example, an observation that is 2.5 standard deviations *below* the mean has a z-score of –2.5, and an observation that is 1.8 standard deviations *above* the mean has a z-score of +1.8. The z-scores express the given problem in 'standard form' so that the standard normal curve can be used.

1. To convert a raw score of x (from a distribution with mean μ and standard deviation σ) to a z-score, subtract the mean from x and divide by the standard deviation. This is given in Equation S7.2.
2. To convert a z-score to a raw score x, multiply the z-score by the standard deviation and add this product to the mean. In equation form:

$$x = \mu + z\sigma \qquad \textbf{(S7.3)}$$

Note that Equations S7.2 and S7.3 represent the same equation expressed in different forms.

Example S7.4

The manager of a small movie theatre snack bar has been keeping a record of the amount that each customer spends at the interval. She has found that these amounts have a mean of \$5.00 and a standard deviation of \$0.80.

(a) Find the z-score for each of the following amounts spent:

(i) \$7.20 (ii) \$3.70

(b) Find the amounts spent that correspond to z-scores of:

(i) 1.20 (ii) −0.60 (iii) 0

Solution

For this problem, $\mu = 5$ and $\sigma = 0.8$.

(a) (i) From Equation S7.2:

$$z = \frac{7.2 - 5.0}{0.8} = 2.75$$

(ii) From Equation S7.2:

$$z = \frac{3.7 - 5.0}{0.8} = -1.625$$

(b) (i) From Equation S7.3:

$$x = 5.0 + (1.20)(0.8) = \$5.96$$

(ii) From Equation S7.3:

$$x = 5.0 + (-0.60)(0.8) = \$4.52$$

(iii) From Equation S7.3:

$$x = 5.0 + (0)(0.8) = \$5.00$$

The calculation in this last case is of course unnecessary, because a z-score of zero automatically corresponds to the mean of the population.

S7.5 COMPUTATION OF PROPORTIONS

Proportion within a given interval

If a set of data has a normal distribution, the proportion of observations that lie in a particular *interval* can be found using the following procedure:

1. Determine the z-score for each endpoint of the interval.
2. Find the area (from Table 1) for each z-value. (If the z-value is negative, *ignore the sign.*)
3. If the endpoints of the interval lie on *opposite* sides of the mean, *add* the two areas found in Step 2. If the two endpoints lie on the *same* side of the mean, *subtract* the smaller area from the larger one.

Example S7.5

One of the tasks of the national marketing manager of a pet food company is to keep track of the number of cans of dog food returned each week by dissatisfied customers. Over a 2-year period it has been found that the number of cans returned per week followed an approximate normal distribution, with a mean of 250 cans and a standard deviation of 40 cans.

Find the proportion of weeks in which the number of cans returned is between:

(a) 170 and 350
(b) 260 and 280
(c) 200 and 240

Solution

For this problem, $\mu = 250$ and $\sigma = 40$. The shaded area in each figure below is the one that is required.

(a) A raw score of 170 corresponds to a z-score of:

$$z = \frac{170 - 250}{40}$$
$$= -2.0$$

A raw score of 350 corresponds to a z-score of:

$$z = \frac{350 - 250}{40}$$
$$= 2.5$$

Because the extremities of the interval are on *opposite* sides of the mean, *add* together the areas for $z = 2.5$ and $z = 2.0$. From Table 1:

- the area between $z = -2.0$ and $z = 0$ is 0.4772
- the area between $z = 0$ and $z = 2.5$ is 0.4938

Therefore, the area between $z = -2.0$ and $z = 2.5$ is $0.4938 + 0.4772 = 0.9710$. Hence, the proportion of weeks in which the number of cans returned is between 170 and 350 is 0.9710 or 97.10% (see Figure S7.14).

(b) A raw score of 260 corresponds to a z-score of:

$$z = \frac{260 - 250}{40}$$
$$= 0.25$$

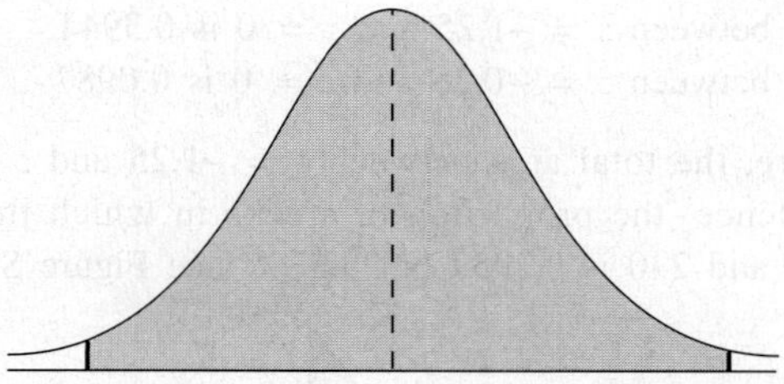

Figure S7.14

A raw score of 280 corresponds to a z-score of:

$$z = \frac{280 - 250}{40}$$
$$= 0.75$$

Because the extremities of the interval are on the *same* side of the mean, *subtract* the area for $z = 0.25$ from the area for $z = 0.75$. From Table 1:

- the area between $z = 0$ and $z = 0.25$ is 0.0987
- the area between $z = 0$ and $z = 0.75$ is 0.2734

Therefore, the total area between $z = 0.25$ and $z = 0.75$ is $0.2734 - 0.0987 = 0.1747$. Hence, the proportion of weeks in which the number of cans returned is between 260 and 280 is 0.1747 or 17.47% (see Figure S7.15).

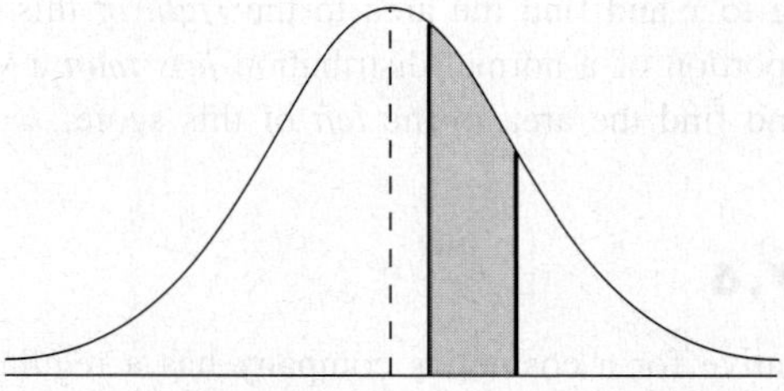

Figure S7.15

(c) A raw score of 200 corresponds to a z-score of:

$$z = \frac{200 - 250}{40}$$
$$= -1.25$$

A raw score of 240 corresponds to a z-score of:

$$z = \frac{240 - 250}{40}$$
$$= -0.25$$

Because the extremities are on the *same* side as the mean, *subtract* the area for $z = 0.25$ from the area for $z = 1.25$. (Note that the area between $z = -1.25$ and $z = 0$ is larger than the area between $z = -0.25$ and $z = 0$.) From Table 1:

- the area between $z = -1.25$ and $z = 0$ is 0.3944
- the area between $z = -0.25$ and $z = 0$ is 0.0987

Therefore, the total area between $z = -1.25$ and $z = -0.25$ is $0.3944 - 0.0987 = 0.2957$. Hence, the proportion of weeks in which the number of cans returned is between 200 and 240 is 0.2957 or 29.57% (see Figure S7.16).

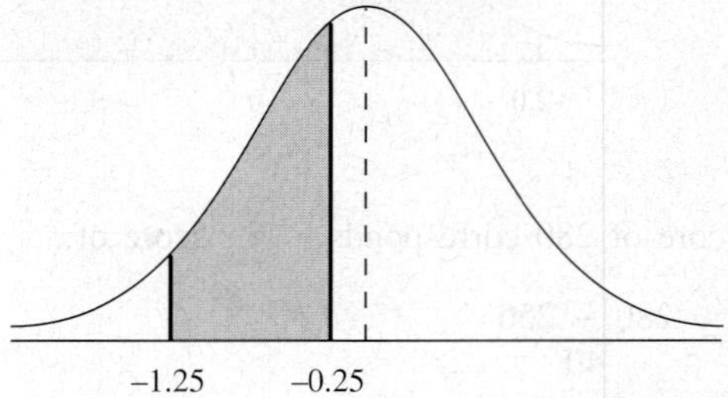

Figure S7.16

Proportion greater or less than a given value

Table 1 can also be used to determine what proportion of a normal population is *greater than* a certain value or *less than* a certain value. For each such problem it is advisable to sketch the region required. As in Example S7.5, it is necessary first to convert the raw score to a z-score.

1. To determine the proportion of a normal distribution *greater than* a value of x, calculate the z-score corresponding to x and find the area to the *right of* this score.
2. To determine the proportion of a normal distribution *less than* a value of x, calculate the z-score corresponding to x and find the area to the *left* of this score.

Example S7.6

A sales representative for a cosmetics company has a regular customer to whom she must make deliveries every Monday morning. To ease the boredom of the drive she has recorded the times that it takes her to travel the distance each week. She finds that these times follow an approximate normal distribution, with a mean of 20 minutes and a standard deviation of 3 minutes.

Find the proportion of mornings in which the time taken for the drive:

(a) exceeds 21 minutes 48 seconds
(b) is less than 16 minutes
(c) exceeds 18 minutes

Solution

For this problem, $\mu = 20$ and $\sigma = 3$. Recall that each of the areas to the right and left of $z = 0$ is 0.5000.

(a) A time of 21 minutes 48 seconds is 21.8 minutes. A raw score of 21.8 corresponds to a z-score of:

$$z = \frac{21.8 - 20}{3}$$
$$= 0.60$$

The required proportion (shaded in Figure S7.17) may be found from Table 1 by subtracting the area between $z = 0$ and $z = 0.60$ from 0.5000:

- the area between $z = 0$ and $z = 0.60$ is 0.2257

Therefore, the area to the right of $z = 0.60$ is $0.5000 - 0.2257 = 0.2743$. Hence, on approximately 27.43% of mornings, the time exceeds 21.8 minutes.

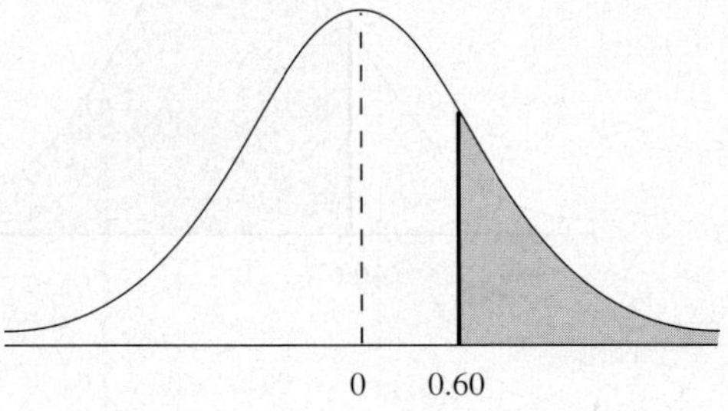

Figure S7.17

(b) A raw score of 16 corresponds to a *z*-score of:

$$z = \frac{16 - 20}{3}$$
$$= -1.33$$

The required proportion (shaded in Figure S7.18) may be found by subtracting the area between $z = -1.33$ and $z = 0$ from 0.5000. From Table 1:

- the area between $z = -1.33$ and $z = 0$ is 0.4082

Therefore, the area to the left of $z = -1.33$ is $0.5000 - 0.4082 = 0.0918$. Hence, on approximately 9.18% of mornings, the time taken is less than 16 minutes.

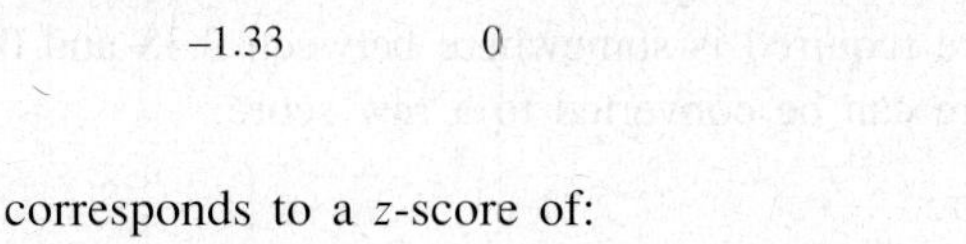

Figure S7.18

(c) A raw score of 18 corresponds to a *z*-score of:

$$z = \frac{18 - 20}{3}$$
$$= -0.67$$

The required proportion (shaded in Figure S7.19) may be found by adding the area between $z = -0.67$ and $z = 0$ to the area to the right of $z = 0$ (i.e. 0.5000). From Table 1:

- the area between $z = -0.67$ and $z = 0$ is 0.2486

Therefore, the area to the right of $z = -0.67$ is $0.2486 + 0.5000 = 0.7486$. Hence, on approximately 75% of mornings, the time taken exceeds 18 minutes.

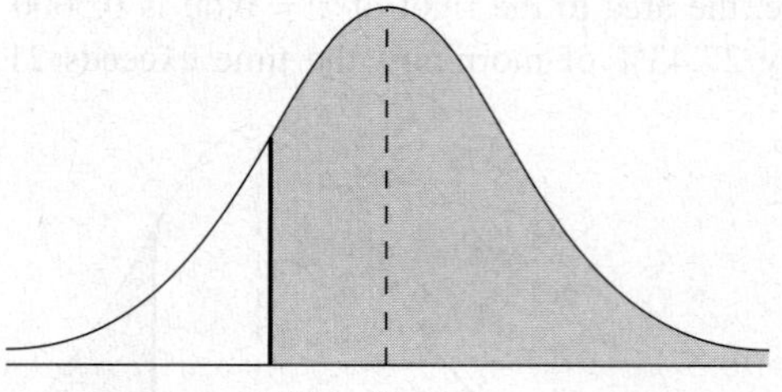

Figure S7.19

Value corresponding to a given proportion

For some problems it is necessary to find the z-score for a given area rather than the other way around. An example of such a problem is given below.

Example S7.7

Suppose that a particular MBA entrance exam is designed so that the marks will be normally distributed with a mean of 300 and a standard deviation of 60. The policy is that the top 35% gain admission to the program. What is the lowest mark that would gain admission?

Solution

In order to calculate the mark, the corresponding z-score must be found, i.e. the point z on the normal curve that has 35% of the area to the right of it. From Figure S7.20 it can be seen that the required value of z is that which corresponds to a tabled area of 0.1500 (recall that the area to the right of $z = 0$ is 0.5000).

From Table 1, $z = 0.38$ gives an area of 0.1480 while $z = 0.39$ gives an area of 0.1517. Thus, the z-score required is somewhere between 0.38 and 0.39, say 0.385. From Equation S7.3, this z-score can be converted to a raw score:

$$\begin{aligned} x &= \mu + \sigma z \\ &= 300 + (60 \times 0.385) \\ &= 323.1 \end{aligned}$$

Hence, the lowest mark that would gain admission is approximately 323.

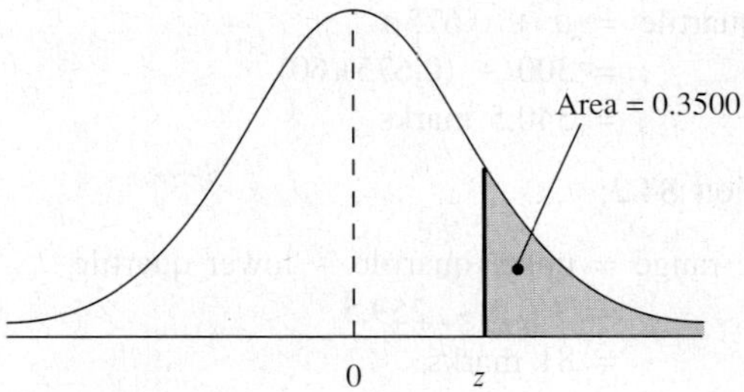

Figure S7.20

Example S7.8

For the data in Example S7.7, find:

(a) the value of the marks that represent the lower and upper quartiles
(b) the interquartile range of the marks

Solution

Recall from Section S3.5 that the lower quartile is the value that cuts off the bottom 25% of marks and the upper quartile is the value that cuts off the top 25% of marks. To find these values, we can use Equation S7.3 with the appropriate values of z.

The value of z for the upper quartile is the value that cuts off the top 25% of the area under the standard normal curve. From Table 1, look up the closest z-value in the table that corresponds to an area of 0.2500. Note that $z = 0.67$ has an area of 0.2486 and $z = 0.68$ corresponds to an area of 0.2517. Hence, we will estimate that $z = 0.675$ will correspond approximately to an area of 0.2500.

In Equation S7.3 we will use a value of $z = 0.675$ for the upper quartile and $z = -0.675$ for the lower quartile. These values hold when finding the values of the quartiles for *any* normal distribution with *any* value of μ and σ. A graphical illustration of these values is shown in Figure S7.21.

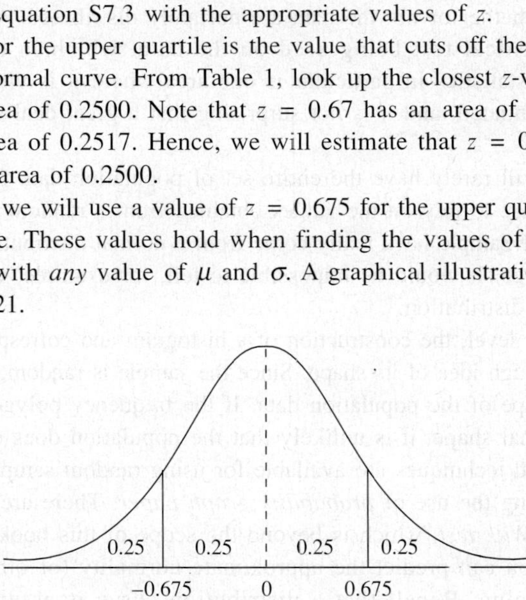

Figure S7.21 A standard normal curve indicating the cut-off values of z for the lower and upper quartiles

(a) Hence, using Equation S7.3:

$$\begin{aligned}\text{Lower quartile} &= \mu - 0.675\sigma \\ &= 300 - (0.675)(60) \\ &= 259.5 \text{ marks}\end{aligned}$$

$$\begin{aligned}\text{Upper quartile} &= \mu + 0.675\sigma \\ &= 300 + (0.675)(60) \\ &= 340.5 \text{ marks}\end{aligned}$$

(b) Using Equation S4.2:

$$\begin{aligned}\text{Interquartile range} &= \text{upper quartile} - \text{lower quartile} \\ &= 340.5 - 259.5 \\ &= 81 \text{ marks}\end{aligned}$$

S7.6 CHECKING WHETHER DATA ARE NORMALLY DISTRIBUTED

Table 1 provides areas calculated to four decimal place accuracy. Even if the distribution is believed to be normal, it is rare for the mean μ and standard deviation s to be known precisely, so the order of accuracy apparent from the table is rarely justified in practice. You should be cautious of claiming in real-life situations that (for instance) 84.13% of a population lies within a certain range. However accurate your manipulations of the numbers may be, the inherent problems of knowledge of μ and σ may imply that you should claim only 'roughly 85%' rather than '84.13%'.

When dealing with everyday problems, statisticians are often concerned whether certain populations follow a normal distribution. A major reason for this is that much statistical testing depends on whether or not the data are normally distributed. A wrong assumption can lead to an incorrect statistical test being used on the data, which in turn can lead to a doubtful conclusion. Medical statistics in particular is one area where it is vitally important that any such assumption be substantiated, and it is not surprising that a great deal of effort is put into checking this aspect.

In practice, we will rarely have the entire set of population data from a particular distribution. It is therefore necessary to rely on the values obtained from a random sample to determine whether data obtained from the sample have really come from a normal distribution. Of course, such a sample must be relatively large. It would be impossible to tell, for example, whether only six values have come from a normal distribution.

On an elementary level, the construction of a histogram and corresponding frequency polygon of the data can give a rough idea of its shape. Since the sample is random, the shape of the sample data should reflect the shape of the population data. If the frequency polygon for a large sample clearly does *not* have a normal shape, it is unlikely that the population does either.

More sophisticated techniques are available for using random sample values to test a population for normality, including the use of *probability graph paper*. There are also formal test procedures, such as the *Shapiro–Wilk test*, which is beyond the scope of this book.

Often, though, you can predict the approximate normality (or otherwise) of a distribution by commonsense reasoning. Recall that a distribution curve is essentially a *relative frequency* graph, indicating how frequently various values occur relative to other values. Among its other characteristics, a normal curve is *symmetric about the mean*. That is, a value a certain distance *above* the mean has the same relative frequency as a value that distance *below* the mean, so the frequencies at $(\mu - A)$ and $(\mu + A)$ are the same. The region between these values is shown shaded in Figure S7.22.

Thus, a simple test for possible normality in a given situation is to see whether this is true for various choices of the distance A.

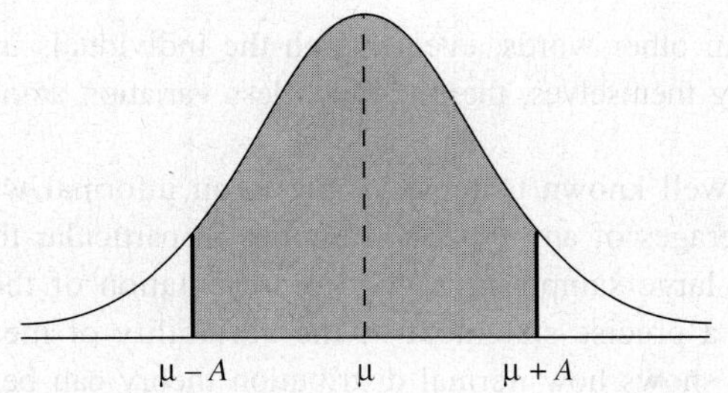

Figure S7.22 The area between μ and $\mu + A$ is the same as the area between μ and $\mu - A$

Example S7.9

Are the weights of adult Australian males normally distributed?

Solution

Suppose we consider the random sample of 200 male students selected in Section S7.2 as being fairly typical of Australian adult males. Each was asked to record his weight (to the nearest kg) as well as his height. The mean weight of these males was 70.3 kg.

If the weights are normally distributed, we would expect a weight of (say) 40 kg *above* the mean to occur as frequently in the population as one 40 kg *below* the mean. That is, a weight of 70.3 + 40 = 110.3 kg should be as common as a weight of 70.3 – 40 = 30.3 kg.

This is clearly untrue. Very few adult males have a weight of 30.3 kg, whereas 110.3 kg is not nearly as rare. This distribution is therefore not even symmetric and hence is not normal. A weight distribution may be fairly symmetric near its centre, and this often causes people to assume that the whole of the distribution is symmetric. In fact, the weights of various populations often have a distribution that is skewed to the right.

Example S7.10

Are personal incomes in Australia normally distributed?

Solution

Again, the distribution is far from symmetric. There are a few very large incomes of size many times the mean income, so the distribution has an extreme right-hand tail.

According to the Australian Bureau of Statistics (Cat. 6301.0), the average annual income of full-time adult workers in Australia in August 2000 was $41 464. However, while a number of these workers would have an income of $50 000 *more* than the mean, it is impossible to have an income of $50 000 *less* than the mean (since it would be negative). This skewed to the right distribution is substantially non-normal.

S7.7 THE CENTRAL LIMIT THEOREM

Many people frequently refer to the *law of averages*, although exactly what is meant by this expression is not always clear. From a statistical point of view, it could refer to the fact that the averages of fairly large samples have a reliability that makes them good indicators of the corresponding population. In other words, even though the individuals in a population may have a considerable variation among themselves, there is much less variation among the averages of samples of many individuals.

This principle is quite well known to most people in an informal way. It is indeed one of the reasons why people use averages of any type, and implies in particular that it is reasonable to have some faith in a mean of a large sample as a good representation of the mean of the population. Statistical theory provides a precise statement of the variability of means relative to that of the component individuals, and shows how normal distribution theory can be applied to solve problems about means.

Specifically, this idea is justified by perhaps the most widely used theorem in statistics: the **Central Limit Theorem**. It is this theorem that explains the importance of the normal distribution.

THE CENTRAL LIMIT THEOREM

If random samples of size n are selected from a population with mean μ and standard deviation σ, the *means* of the samples are *approximately normally distributed* with mean μ and standard deviation $\frac{\sigma}{\sqrt{n}}$, even if the population itself is *not* normally distributed, provided that n is not too small. The approximation becomes more and more accurate as the sample size n is increased.

The broad application of this theorem is seen in the fact that there need *not* be a normal distribution of the individuals in the population, although of course the theorem still holds if the population *does* have a normal distribution. How large the sample size n must be in order to get a good approximation depends on the population from which the sample is drawn and on how the approximation is to be used. If the original population distribution of the individuals is close to normal, n can be quite small (e.g. 10), whereas a very skewed original population may require n to be 25 or even more. If the original population is *not* normally distributed, we will assume that n must be at least 25 for the Central Limit Theorem to hold.

The Central Limit Theorem also says that any unusual shapes of populations are suppressed. That is, even if *individual* values come from a very non-normal population (such as one that is skewed or bimodal), the sample *means* tend to have a normal distribution.

In the following examples the notation 'P(z-score takes on particular values)' will mean 'the proportion of time that a z-score takes on those particular values'. For example, P(z-score is less than 0.57) = the proportion of z-scores that are less than 0.57. This is simply the area under the standard normal curve to the left of z = 0.57.

Conversion of a sample mean $\bar{x}$ to a z-score

When dealing with problems to which the Central Limit Theorem applies, the principles for finding areas under a normal curve are exactly the same as in Section S7.5. However, the formula given in Equation S7.2 applies only when converting a raw score of x to a z-score. The formula for converting a mean of $\bar{x}$ to a z-score is given by:

$$z = \frac{\bar{x} - \mu}{\frac{\sigma}{\sqrt{n}}} \qquad \textbf{(S7.4)}$$

where μ = the population mean
σ = the population standard deviation
n = the size of the sample
$\bar{x}$ = the mean of the sample

Example S7.11

Assume that the weights of adult Australian males are skewed to the right with a mean of 70.3 kg and a standard deviation of 2.74 kg. (These figures are based on the data for the 200 male students in Section S7.2.)

If a random sample of 25 adult males is chosen, what is the probability that their *mean* weight is greater than 71.0 kg?

Solution

For this problem μ = 70.3, σ = 2.74 and n = 25. Since the sample size is *large* (at least 25), the Central Limit Theorem applies even though the population of adult male weights is *not* normally distributed (as we found in Example S7.9). Convert the mean of $\bar{x}$ = 71.0 to a z-score using Equation S7.4:

$$z = \frac{71.0 - 70.3}{\frac{2.74}{\sqrt{25}}} = 1.28$$

Therefore, $P(\bar{x} > 71.0) = P(z > 1.28) = 0.5000 - 0.3997 = 0.1003$.

Thus, the probability that the mean weight of the 25 adult males is greater than 71.0 kg is about 10%.

Example S7.12

The contents of nominal 750 mL bottles of soft drink are subject to some variation in the manufacturing process. The mean (μ) is known to be 755 mL, and the standard deviation is σ = 12 mL. From 36 bottles, what is the chance that the mean contents will be less than the nominal value of 750 mL?

Solution

For this problem μ = 755, σ = 12 and n = 36. Even though the population may *not* be normally distributed, the Central Limit Theorem still applies since the sample size is *large* (at least 25). Convert the mean of $\bar{x}$ = 750 to a z-score using Equation S7.4:

$$z = \frac{750 - 755}{\frac{12}{\sqrt{36}}} = -2.5$$

Therefore, $P(\bar{x} < 750) = P(z < -2.5) = 0.5000 - 0.4938 = 0.0062$.

Thus, the chance that the mean contents of 36 bottles will be less than 750 mL is about 0.6%. That is, around 6 bottles in 1000 will not reach the nominal level of 750 mL.

Example S7.13

Apples grown in a certain orchard are known to have a weight distribution with a mean of 150.0 g and a standard deviation of 10 g, although this distribution is *not* normal. Crates containing 400 apples are packed ready for delivery to the markets. Approximately what proportion of these crates will contain apples with a total weight more than 59.6 kg?

Solution

This problem is not about *means* but about *totals*. However, any problem involving totals can be converted to one involving means and checked as to whether the Central Limit Theorem applies. In this case, although the population is *not* normally distributed, the sample size of $n = 400$ is certainly large enough for the Central Limit Theorem to apply.

We have $\mu = 150.0$, $\sigma = 10$ and $n = 400$. The steps required are as follows.

Proportion of crates with apples that weigh more than 59.6 kg

= proportion of crates with apples that weigh more than 59 600 g

= P(*total* weight of 400 apples exceeds 59 600 g)

= P(*mean* weight of 400 apples exceeds $\frac{59\ 000}{400}$ g)

= P(*mean* weight of 400 apples exceeds 149.0 g)

= $P(\bar{x} > 149.0)$

Convert this mean of $\bar{x} = 149.0$ to a z-score using Equation S7.4:

$$z = \frac{149.0 - 150.0}{\frac{10}{\sqrt{400}}}$$

$$= -2.0$$

Therefore, $P(\bar{x} > 149.0) = P(z > -2.0)$.

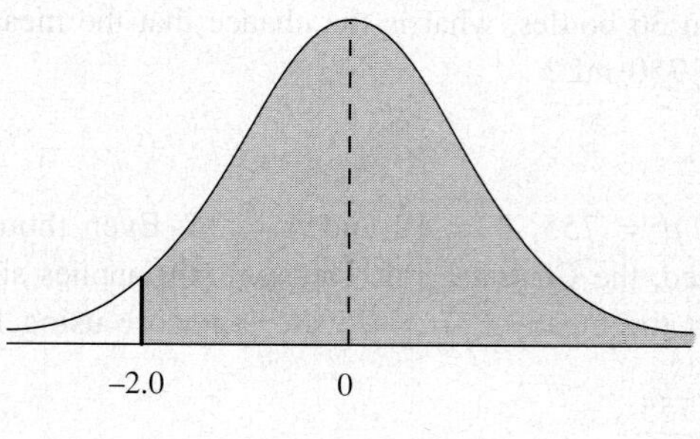

Figure S7.23

Thus, the area required is that to the right of $z = -2.0$ (see Figure S7.23). It may be found by adding the area between $z = -2.0$ and $z = 0$ to the area to the right of $z = 0$ (i.e. 0.5000). From Table 1:

- the area between $z = -2.0$ and $z = 0$ is 0.4772

Therefore, the area to the right of $z = -2.0$ is 0.4772 + 0.5000 = 0.9772. Hence, the proportion of crates that have a total apple weight exceeding 59.6 kg (or a mean apple weight exceeding 149.0 g) is 0.9772 or 97.72%.

Example S7.14

With reference to the data in Example S7.12, *how many* bottles out of a random sample of 36 would you expect to have a contents less than 750 mL?

Solution

Note that this is *not* a problem about means (or totals) of samples and therefore *not* one that involves the Central Limit Theorem. Instead, it simply requires the calculation of an expected frequency. This type of problem is a two-stage process:

1. Calculate the *proportion* of bottles that have a contents less than 750 mL.
2. Multiply this proportion by the sample size $n = 36$.

Convert $x = 750$ to a z-score using Equation S7.2:

$$z = \frac{x - \mu}{\sigma} = \frac{750 - 755}{12} = -0.42$$

Therefore, the proportion of bottles with contents less than 750 mL is:

$$P(z < -0.42) = 0.5000 - 0.1628 = 0.3372$$

Hence, for a sample of 36 bottles we would expect $36 \times 0.3372 = 12.14$ (i.e. about 12) to have a contents less than 750 mL.

You should take special note of the difference between the two types of problem in Examples S7.12 and S7.14.

S7.8 CONFIDENCE INTERVALS FOR AN UNKNOWN POPULATION MEAN

Point estimates

If we want to give a *single* estimate of an unknown population mean (μ), we simply use the value of the mean ($\bar{x}$) obtained from a random sample taken from that population. Naturally, if we take a different random sample, we would usually expect to obtain a different value of $\bar{x}$ and hence a different estimate of μ. A single estimate is sometimes referred to as a **point estimate**.

Example S7.15

The executives of a bus company want to obtain a reliable estimate of the mean number of passengers who travel weekdays on a given route. To do this, they count the number of passengers on a random number of weekdays. They use the mean of this sample to estimate the mean for *all* weekdays.

Example S7.16

An egg farmer wants to estimate the mean weight of all the eggs that her hens lay. To do this, she takes a random sample of eggs and measures the mean weight of these eggs. She uses this mean to estimate the mean weight of *all* eggs laid by her hens.

The question that naturally arises is just how accurate is a point estimate. The answer to this depends on two factors:

1. how variable the data are in the population
2. the size of the random sample

In general, the *more variable* the data in the population, the *less* accurate the point estimate. Also, the *larger* the size of the random sample, the *more* accurate the point estimate.

The standard error of the mean

The precise measure of accuracy of a point estimate of the mean is provided by the **standard error of the mean**, which describes just how these two factors combine. The formal definition is:

$$\text{Standard error of the mean} = \frac{\sigma}{\sqrt{n}} \qquad \text{(S7.5)}$$

where σ = the population standard deviation
n = the size of the random sample

Note that, in practice, the value of σ is usually not known. In this case, if the random sample is *large* (i.e. $n \geq 25$), the value of σ can be replaced by the *sample* standard deviation s. In this case, Equation S7.5 becomes:

$$\text{Standard error of the mean} \div \frac{s}{\sqrt{n}} \qquad \text{(S7.6)}$$

where s = the sample standard deviation
n = the size of the random sample

Example S7.17

A travelling salesman recorded the weekly cost of petrol for his vehicle for each of a random sample of 30 weeks. He found that the mean was $45.50 and that the standard deviation was $9.20.

Estimate his general mean weekly expenditure on petrol, and find the standard error of this estimate.

Solution

In this case, we have $\bar{x}$ = \$45.50, s = \$9.20 and n = 30. Hence, the point estimate of the population mean is $\bar{x}$ = \$45.50.

Using Equation S7.6, the standard error of this estimate is:

$$\text{Standard error} \doteq \frac{\$9.20}{\sqrt{30}} = \$1.68$$

Example S7.18

Do the results in Example S7.17 indicate that the estimate was fairly accurate?

Solution

As a rough rule of thumb, an estimate will be reasonably accurate if the standard error is relatively small compared to the point estimate. In this case, the standard error of the mean was \$1.68, which is only small compared to the point estimate of the mean of \$45.50.

We would therefore conclude that this estimate is fairly accurate.

Meaning of confidence intervals

The rule given in Example S7.18 is somewhat ad hoc, and we now provide a more scientific approach to the problem. Instead of providing a single point estimate for μ, we now consider a *range of values* or an *interval* within which we feel the value of μ may lie. Moreover, we will be able to provide a *probability* or *level of confidence* that this interval does indeed contain the true value of μ.

Such intervals are called **confidence intervals**, and we will see how to construct them below. The probability that the true value of μ *is* in the interval is determined before we construct the interval. Common probabilities to use are 0.90, 0.95 and 0.99, although any probability could be used.

A confidence interval that contains the true value of μ with probability 0.95 is called a *95% confidence interval for* μ. Similar expressions are used where the 95% is replaced by 90% if the probability is 0.90 and 99% if the probability is 0.99.

Confidence intervals for μ are based on the values from random samples and, as before, if we take a different random sample, we will obtain a different confidence interval. For a given random sample, if we want to be more confident that we have the true value of μ in our interval, we can use a larger level of confidence, but we will also get a larger interval in turn.

Example S7.19

What is meant by the statement that 'there is a probability of 0.95 that a 95% confidence interval contains the true value of μ'?

Solution

First, a word of caution is necessary. Once you have obtained your random sample and constructed the 95% confidence interval for μ, either the interval *will* contain the true value of μ or it *won't*. The value of μ is a constant and does not change!

However, what the '95% confidence' means is the following: 'If we were to repeat the experiment and take many random samples from the population and construct many 95% confidence intervals for μ, we would expect that approximately 95% of those intervals would contain the true value of μ and 5% would not.'

In practice, of course, we usually only construct *one* 95% confidence interval based on *one* random sample. It may be one of the 95% of intervals that do contain μ, or it may be one of the 5% that do not. We will usually never know whether it does contain μ, since we will often never know the true value of μ. However, the term does reflect the probability that our particular interval is one of those that *do* contain the true value of μ.

Construction of confidence intervals for μ

The construction of confidence intervals for μ can be complex. We will consider only two cases, although one of these is quite common in practice. Before we decide on how to construct a confidence interval for μ, we must examine the available data to see which case is appropriate.

Case I

Suppose that we have the following information:

1. The population has a normal distribution.
2. The population standard deviation σ is known.
3. The sample size n can be of *any* size.

Then a 95% confidence interval for μ is:

$$\left(\bar{x} - 1.96\frac{\sigma}{\sqrt{n}},\ \bar{x} + 1.96\frac{\sigma}{\sqrt{n}}\right) \qquad \text{(S7.7)}$$

where $\bar{x}$ = the mean of the sample
σ = the population standard deviation
n = the size of the sample

Note

1. For a 90% confidence interval for μ, we replace the 1.96 in Equation S7.7 by 1.645.
2. For a 99% confidence interval for μ, we replace the 1.96 in Equation S7.7 by 2.58.

Example S7.20

Experience has shown that the systolic blood pressures of adult populations follow a normal distribution, with a standard deviation of σ = 8.50 mm, but the mean is known to vary from population to population.

A medical researcher is keen to estimate the mean blood pressure of a particular population. To do this, she takes a random sample of 15 adults and records their systolic blood pressures. She finds that the mean of this sample is 125.00 mm.

(a) Construct a 95% confidence interval for the population mean blood pressure.
(b) Construct a 99% confidence interval for the population mean blood pressure.

Solution

In this case, we have all the assumptions necessary for Case I. The data are:

the population is normal
$\bar{x} = 125.00$ mm
$\sigma = 8.50$ mm
$n = 15$

(a) Using Equation S7.7, a 95% confidence interval for μ is:

$$\left(125.00 - 1.96 \times \frac{8.50}{\sqrt{15}}, 125.00 + 1.96 \times \frac{8.50}{\sqrt{15}}\right)$$
$$= (125.00 - 4.30, 125.00 + 4.30)$$
$$= (120.70, 129.30)$$

Thus, we are 95% confident that the true population mean blood pressure lies between 120.70 mm and 129.30 mm.

(b) Using Equation S7.7, a 99% confidence interval for μ is:

$$\left(125.00 - 2.58 \times \frac{8.50}{\sqrt{15}}, 125.00 + 2.58 \times \frac{8.50}{\sqrt{15}}\right)$$
$$= (125.00 - 5.66, 125.00 + 5.66)$$
$$= (119.34, 130.66)$$

Thus, we are 99% confident that the true population mean blood pressure lies between 119.34 mm and 130.66 mm. Note that this is *wider* than the 95% confidence interval in (a), but the trade-off is that we are *more* confident that the true value of μ lies in this interval.

Case II

Suppose that we have the following information:

1. The population has *any* distribution.
2. The population standard deviation σ is known or unknown. (If the value of σ *is* known, it should be used. If the value of σ is *not* known, it should be replaced by the sample standard deviation s.)
3. The sample size n must be *large* (i.e. $n \geq 25$).

Then a 95% confidence interval for μ is:

$$\left(\bar{x} - 1.96 \frac{s}{\sqrt{n}}, \bar{x} + 1.96 \frac{s}{\sqrt{n}}\right) \qquad \textbf{(S7.8)}$$

where $\bar{x}$ = the mean of the sample
s = the sample standard deviation
n = the size of the sample

Note

1. For a 90% confidence interval for μ, we replace the 1.96 in Equation S7.8 by 1.645.
2. For a 99% confidence interval for μ, we replace the 1.96 in Equation S7.8 by 2.58.
3. The assumption that n is large is important. This is because the formula in Equation S7.8 relies on the Central Limit Theorem, which we assume holds since the population may *not* be normally distributed. Also, because n is large we can approximate an unknown value of σ by s.

Example S7.21

A random sample of weekly cash takings was recorded for each of 30 weeks at a men's clothing store. It was found for this sample that the mean cash takings were $3485.00 with a standard deviation of $490.00.

(a) Construct a 95% confidence interval for the population mean weekly cash takings.
(b) Construct a 90% confidence interval for the population mean weekly cash takings.

Solution

In this case, we have all the assumptions necessary for Case II. The data are:

the population distribution is unknown (i.e. we do not know whether it is normal or not)
$\bar{x}$ = $3485.00
σ is unknown, but s = $490.00
n = 30 (which is large)

(a) Using Equation S7.8, a 95% confidence interval for μ is:

$$\left(3485.00 - 1.96 \times \frac{490.00}{\sqrt{30}}, 3485.00 + 1.96 \times \frac{490.00}{\sqrt{30}}\right)$$

$$= (3485.00 - 175.34, 3485.00 + 175.34)$$

$$= (3309.66, 3660.34)$$

Thus, we are 95% confident that the true population mean weekly cash takings lies between $3309.66 and $3660.34.

(b) Using Equation S7.8, a 90% confidence interval for μ is:

$$\left(3485.00 - 1.645 \times \frac{490.00}{\sqrt{30}}, 3485.00 + 1.645 \times \frac{490.00}{\sqrt{30}}\right)$$

$$= (3485.00 - 147.16, 3485.00 + 147.16)$$

$$= (3337.84, 3632.16)$$

Thus, we are 90% confident that the true population mean weekly cash takings lies between \$3337.84 and \$3632.16. Note that this is *narrower* than the 95% confidence interval in (a), and as a consequence we are *less* confident that the true value of μ lies in this interval.

S7.9 REMARKS

There are other situations for which we could construct confidence intervals for μ, but these are beyond the scope of this book. One example is the case where we have only a small sample from a normal population and the value of σ is unknown. In this case, a confidence interval based on the *t-statistic* would be used. A discussion of the *t-statistic* is given in Chapter S12.

An interesting application of *z*-scores is provided by Melbourne scientist Charles Davis, who used them to try to answer the question as to who is the greatest cricketer of all time. To do this, Davis calculated the *z*-score of each Test cricketer who had played the game up until the 2000/2001 series. The top 10 cricketers (taking into account batsmen, bowlers and all-rounders and ranking them on a single scale) are shown in Table S7.2.

Table S7.2 The 10 Test cricketers with the greatest z-scores

Cricketer	Country	*z*-score
D. G. Bradman	Australia	4.4
G. S. Sobers	West Indies	3.3
I. Khan	Pakistan	3.1
R. J. Hadlee	New Zealand	3.0
S. F. Barnes	England	2.5
I. T. Botham	England	2.4
S. R. Tendulkar	India	2.2
G. A. Headley	West Indies	2.2
T. L. Goddard	South Africa	2.2
C. L. Walcott	West Indies	2.1

Source: *Sydney Morning Herald*, 10 November 2000, p. 42

From Table S7.2 it can be seen that Bradman has a *z*-score of 4.4, off the scale in Table 1. In fact, when the *z*-scores of all Test cricketers were graphed as a bar chart, with the height of the bar representing the *z*-score of each player, the shape was fairly normal, the only exception being Bradman. This chart is shown in Figure S7.24.

The probability of obtaining a *z*-score of 4.4 or more is around 1 in 200 000 so that, on average, 1 in every 200 000 Test cricketers will be like Bradman. Since it will be many, many years before this number of cricketers play at Test level, it seems very unlikely that another player of Bradman's standing will arise in the foreseeable future.

Finally, for comparison, Davis gives a listing of the all-time greats with the highest *z*-scores in their respective sports. These are shown in Table S7.3 and, based on this criterion, it appears that Bradman is relatively the greatest sportsperson of all time.

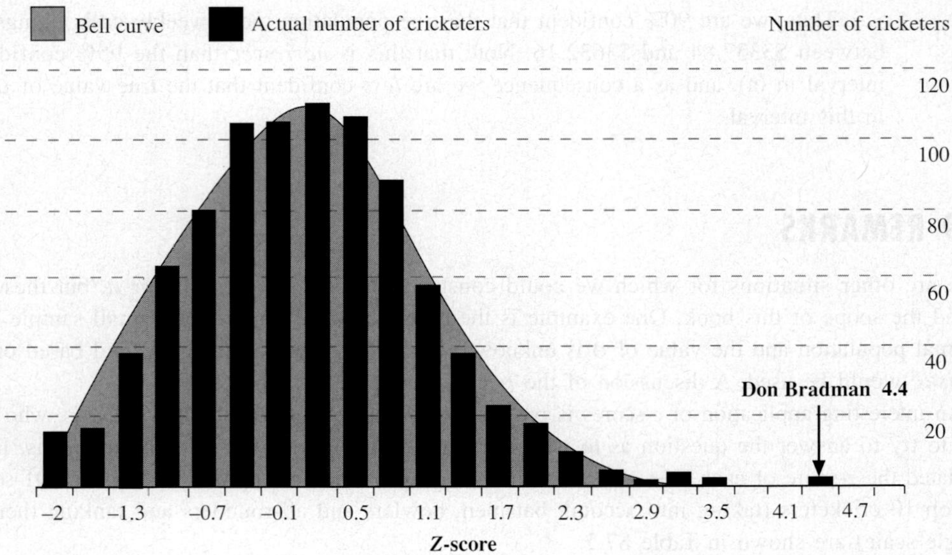

A *z*-score of 2 is exceptional and 3 is near the top of the all-time greats

Figure S7.24 A bar chart of *z*-scores for Test cricketers

Source: *Sydney Morning Herald*, 10 November 2000, p. 42

Table S7.3 The seven sportspeople with the greatest z-scores in their respective sport

Sportsperson	Sport	*z*-score
Don Bradman	Cricket	4.4
Pele	Soccer	3.7
Ty Cobb	Baseball	3.6
Jack Nicklaus	Golf	3.5
Michael Jordan	Basketball	3.4
Bjorn Borg	Tennis	3.2
Joe Montana	American football	3.1

Source: *Sydney Morning Herald*, 10 November 2000, p. 42

Quick Quiz

1. Find the area under the standard normal curve between $z = -2.56$ and $z = 1.58$.
2. If the heights of a certain population of adult females follow a normal distribution with $\mu = 173$ cm and $\sigma = 10$ cm, in a random sample of 500 adult females, how many would you expect to have a height below 150 cm?
3. For the population in Question 2, find the proportion of females who have heights above 160 cm.
4. Marks in a class test in accounting follow a normal distribution with $\mu = 65$ and $\sigma = 6$. Find the mark (to the nearest whole mark) that cuts off the top 5% of students.
5. Find the interquartile range of a normal distribution that has $\mu = 100$ and $\sigma = 20$.
6. In an annual fun-run the finishing times for competitors followed a non-normal distribution with a mean of 65 minutes and a standard deviation of 12 minutes. If a random sample of 36 runners is chosen, what is the probability that their average finishing time is less than 68 minutes?
7. A normal distribution has an interquartile range of 10. If the mean is 50, find the value of the lower quartile.
8. A random sample of 25 is selected from a population. Find the *length* of a 95% confidence interval for the population mean if the sample standard deviation is 8.50.
9. Answer Question 8, except this time using a 99% confidence interval.
10. A normal distribution has a mean of 75 and an upper quartile value of 81.58. What is the value of the standard deviation?

Multiple Choice Questions

1. The area under the standard normal curve between $z = 1.35$ and $z = 2.67$ is:
(a) 1.3200 (b) 0.0847 (c) 0.9077
(d) 0.0923 (e) none of the above

2. The heights of adult males are assumed to follow a normal distribution with $\mu = 170$ cm and $\sigma = 8.0$ cm. The *proportion* (to four decimal places) of adult males with a height between 150 cm and 180 cm is:
(a) 0.9938 (b) 0.8882 (c) 0.0994
(d) 0.8944 (e) none of the above

3. In Question 2, the *expected number* (to the nearest whole number) of adult males with a height greater than 180 cm in a random sample of 500 is:
(a) 53 (b) 11 (c) 197
(d) 0 (e) none of the above

4. The upper quartile of a normal distribution with $\mu = 500$ and $\sigma = 40$ is:
(a) 530 (b) 473 (c) 510
(d) 527 (e) none of the above

5. The length of time customers queue for service at a supermarket checkout follows a normal distribution with $\mu = 8.00$ minutes and $\sigma = 2.50$ minutes. A random sample of 25 customers is chosen. The probability that their *mean* waiting time (to two decimal places) in the queue is *less than* 7.00 minutes is:
(a) 0.50 (b) 0.48 (c) 0.02
(d) 0.34 (e) none of the above

6. For Question 5, the probability (to two decimal places) that their *total* waiting time in the queue is *more than* 168.75 minutes is:
(a) 0.99 (b) 0.01 (c) 0.19
(d) 0.81 (e) none of the above

7. A normal distribution has a mean of 40 and a standard deviation of 8. The *z*-score corresponding to an observation of 23.4 is:
(a) 2.20 (b) 16.60 (c) −2.20
(d) 2.925 (e) none of the above

8. The fraction of the area under a standard normal curve that lies to the *left* of −2.65 is:
(a) 0.9960 (b) 0.4960 (c) 0.3500
(d) −0.4960 (e) none of the above

9. In a study of elite long-jumpers, it was found that the distance that they jump follows a normal distribution with a mean of 7.80 m and a standard deviation of 0.24 m. The proportion of time that they jump at least 8.16 m is:
(a) 0.4332 (b) 0.9332 (c) 0.3600
(d) 0.0668 (e) none of the above

10. The weight of hailstones found in a Brisbane storm was found to follow a normal distribution with a mean of 120 g and a standard deviation of 30 g. In a random group of 200 hailstones, the number (to the nearest hailstone) you would expect to weigh *more* than 192 g is:
(a) 2 (b) 98 (c) 198
(d) 8 (e) none of the above

Exercises

S7.1 An aptitude test for computer programming was administered to 400 office workers in a large company. The test was marked on a scale of 1 to 100. The results are given below in the form of a grouped frequency distribution.

Test mark	Frequency
1–10	2
11–20	18
21–30	40
31–40	57
41–50	84
51–60	83
61–70	59
71–80	38
81–90	16
91–100	3
Total	400

(a) Construct a histogram for the data with class intervals of length 10 marks.

(b) Construct the frequency polygon.

(c) Modify the frequency polygon into a smooth curve. Does it follow the bell-shaped normal distribution?

(d) Estimate the mean mark from a visual inspection of the curve.

(e) Estimate the mean mark from the data and compare this with the estimate in (d).

S7.2* The lengths of time that a plant supervisor has to wait for a lift to take him to the sixth floor of his office building each morning were recorded over a long period. They were found to follow an approximate normal distribution with a mean of 4.5 minutes and standard deviation of 1.0 minutes.

(a) Sketch normal curves and shade in the regions that represent the proportion of times that he has to wait for a lift for:

(i) more than 6.0 minutes

(ii) between 5.2 and 6.8 minutes

(iii) between 3.5 and 5.5 minutes

(iv) less than 4.8 minutes

(b) Use Table 1 to determine the areas of the above regions and hence the corresponding proportions.

S7.3 The morning announcer on a radio station sometimes has to operate a talk-back segment in which various listeners call the station to express their views. To avoid any defamatory statements going to air, these calls are usually placed on a 5-second or 7-second delay. If the caller makes a defamatory statement, the announcer can press a 'panic button' to cut off the call before it goes to air.

The station manager is very concerned about unsatisfactory calls going to air and has devised a test for the announcer to test her reaction time between recognising such a call and pressing the panic button. Over a series of trials her reaction times are found to be normally distributed with a mean of 3.80 seconds and a standard deviation of 1.10 seconds.

(a) Find the z-scores that correspond to reaction times of length:

(i) 4.00 seconds (ii) 2.85 seconds (iii) 1.50 seconds (iv) 5.00 seconds

(v) 7.00 seconds

(b) Find the reaction times that correspond to z-scores of:

(i) 0 (ii) 1.96 (iii) −0.80 (iv) 2.18

(v) −1.55 (vi) 0.63

(c) The station manager estimates that about 150 calls each year should be terminated by the panic button. Of these calls, approximately how many are likely to go to air if:

(i) a 5-second delay is used? (ii) a 7-second delay is used?

S7.4* The board of directors of a national soft-drink manufacturer has been alarmed by the amount of criticism it has been receiving in the press regarding the contents of its Lime Cola bottles. To investigate the situation it collects a very large random sample of bottles that should contain 750 mL. The board finds that the actual amounts in the bottles follow a normal distribution with a mean of 751.5 mL and a standard deviation of 5.0 mL.

Find the proportion of Lime Cola bottles that have contents in the following ranges:

(a) less than 755 mL (b) more than 749 mL

(c) less than 750 mL (d) more than 760 mL

(e) between 740 mL and 765 mL (f) *not* between 746 mL and 754 mL

S7.5* The amount of fuel that an estate agent used in driving prospective buyers around the city to inspect home units was recorded each week for 200 weeks. The amounts used were found to follow an approximate normal distribution with a mean of 75 L and a standard deviation of 12 L.

(a) Estimate the approximate *number* of weeks in which the fuel consumption was in the following ranges:

(i) less than 89 L (ii) more than 70 L

(iii) more than 80 L (iv) less than 61 L

(v) between 70 L and 80 L (vi) between 55 L and 80 L

(b) What fuel consumption was exceeded in only 20 out of the 200 weeks?

S7.6 The birth weights of female babies born at a certain hospital are approximately normally distributed with a mean of 3.40 kg and a standard deviation of 0.48 kg.

(a) Find the proportion of female babies who have birth weights in the following ranges:

(i) more than 3.60 kg (ii) more than 2.96 kg

(iii) less than 2.50 kg (iv) less than 3.00 kg

(v) less than 3.94 kg (vi) between 3.00 kg and 4.00 kg

(vii) between 2.60 kg and 3.00 kg (viii) between 2.50 kg and 3.70 kg

(b) If babies are to be defined as 'underweight' when they are in the lowest 5% of the distribution, what will be the cut-off point for defining underweight?

S7.7* In a university entrance examination, the results in a certain subject are scaled so that the marks will have a normal distribution with a mean of 100 and a standard deviation of 25. The percentages of students who will be awarded the various grades are shown below.

Grade 1: top 10% of marks

Grade 2: next 20% of marks

Grade 3: next 40% of marks

Grade 4: next 20% of marks

Grade 5: bottom 10% of marks

(a) If 2000 students sit for the examination, approximately how many students will be allocated each grade?

(b) What marks will be the cut-off points for the various grades?

S7.8* The owner of a timber yard was concerned about the amount of wastage that was occurring in the building materials section. To investigate the matter she obtained numerous discarded offcuts of standard-width timber and recorded their length. They were found to have an approximate normal distribution with a mean of 0.80 m and a standard deviation of 0.25 m.

Find the lengths that represent:

(a) the median

(b) the upper quartile

(c) the lower quartile

(*Hint*: Read Example S7.8 on page S.217.)

S7.9* The amount of time required for a certain mechanic to replace the clutch of a vehicle is normally distributed with a mean of 90 minutes and a standard deviation of 12 minutes. The mechanic promises a customer that his clutch will be repaired within 2 hours, but does not commence work on it for 8 minutes.

Find the probability that the car will be ready as promised.

S7.10* In order to organise divisions for a bowling league, the coordinator asks each of the 950 prospective entrants to bowl one game and record the score. It is found that the scores follow a distribution with a mean of 126 and a standard deviation of 27, although the distribution is clearly not normal. The 950 entrants are then divided into 38 randomly chosen groups, each of size 25. The mean scores for the players in each of the 38 groups are calculated and the 38 means recorded.

What can be said about the distribution of these means?

S7.11 A college that specialises in training people for top-level management positions administers various types of physical and psychological tests. In one such test, objects must be assembled and pictures arranged in order. Although the maximum possible score for this test is 130, the college has found that results in the past have followed an approximately normal distribution with a mean = 91.7 and a standard deviation = 8.3.

What percentage of people obtain scores in the following ranges?

(a) above 105

(b) below 74

(c) above 83

(d) between 100 and 110

(e) between 80 and 100

(f) above 125

S7.12 A section supervisor of an insurance company records the lengths of time that his secretary spends having morning tea. He finds that the times are approximately normally distributed with a mean of 20 minutes and a standard deviation of 6 minutes.

What proportion of the times taken by the secretary are in the following ranges?

(a) less than 10 minutes

(b) more than 15 minutes

(c) more than 27 minutes

(d) between 10 minutes and 30 minutes

(e) more than 38 minutes

(f) more than 45 minutes

S7.13* It is known that the inorganic sodium content of males who are in the 30 to 39-year age bracket and who weigh approximately 78 kg follows a distribution with a mean of 83.4 g and a standard deviation of 11.2 g. (*Source*: *Geigy Scientific Tables*, 8th edn, vol. 1, p. 143.)

Assuming this distribution to be normal, how many such males in a random sample of 500 would you expect to have a sodium content *larger than* 100 g?

S7.14* A farmer claims that the daily average amount of milk she has obtained from her favourite cow Clarabelle over the years is 3.50 L. She also claims that 30% of the time Clarabelle's daily output of milk exceeds 4.00 L. Assuming that Clarabelle's daily output of milk follows a normal distribution, find an approximate value for the standard deviation of this distribution.

S7.15 Which of the following variables do you believe are approximately normally distributed?

(a) prices of houses

(b) ages of public servants

(c) delays to commuters travelling on suburban train services

(d) prices of books

(e) lengths of time before new washing machines need repair

(f) daily numbers of people visiting a tourist attraction

In any case where you do *not* believe that an approximately normal distribution applies, write down what you feel the shape of the distribution might be.

S7.16 A company claims that the light bulbs it manufactures have an average lifetime of 400 hours and a standard deviation of 25 hours. If the claim is true and randomly chosen batches of 100 light bulbs are tested, in what proportion of the batches will the mean life of the light bulbs be in the following ranges?

(a) greater than 404.5 hours

(b) less than 396.0 hours

(c) greater than 392.8 hours

(d) less than 406.8 hours

(e) between 392.0 and 402.5 hours

(f) between 400.0 and 405.0 hours

S7.17* The times spent by a customer placing bets at a certain TAB agency are known to be approximately normally distributed with a mean of 90 seconds and standard deviation of 25 seconds. A lady joins a queue and sees that there are 9 people to place bets before her.

What is the probability that she will begin placing her bet within 12 minutes?

S7.18 In a report in the Number Crunch column (*Melbourne Age* and *Sydney Morning Herald*, 16 September 2000), the mean time that Australians spend each week on exercise or sports is 70 minutes.

Assuming that this time follows a normal distribution with a standard deviation of 20 minutes, what proportion of Australians will spend:

(a) between 30 minutes and one hour?

(b) between 0 and 20 minutes?

(c) more than two hours?

S7.19* For the data in Exercise S7.18, find the probability that the mean weekly time spent by a group of 50 Australians on exercise or sports will be less than 68 minutes.

S7.20 According to specifications (see the URL *www.toyota.com.au*), the average fuel consumption per 100 km of travel around the city by a 2001 Toyota Camry CSi V6 Wagon is 11.5 L. Assuming that this fuel consumption has a standard deviation of 1.5 L, of a random sample of 30 such cars, how many would you expect to have a fuel consumption more than 12.0 L per 100 km?

S7.21* The blood cholesterol levels of adult males in a certain occupation are known to follow a normal distribution with a mean of 5.2 mmol/L and a standard deviation of 1.2 mmol/L. According to the National Heart Foundation, the desirable blood cholesterol level is less than 5.5 mmol/L.

What percentage of males in this occupation fall into this category?

S7.22* An Environmental Protection Authority survey indicates that residents in the Sydney suburb of Manly recycle an average of 103.3 kg of material per year. If this amount follows a normal distribution with a standard deviation of 20.0 kg, how much material is recycled by the resident whose recycling amount represents the 95th percentile?

S7.23* A financial controller is interested in the number of defective items produced each hour by a machine in the factory. A random sample of 25 hours produced the following number of defective items per hour:

10	6	5	7	8	4	5	5	6	8	4	3	7
8	4	10	5	3	2	0	5	8	9	3	7	

Find:

(a) a point estimate of the population mean number of defective items that the machine produces per hour

(b) the standard error of the estimate in (a)

S7.24* Construct a 95% confidence interval for the mean number of defective items that the machine in Exercise S7.23 produces per hour.

S7.25 In addition to the data given in Exercise S7.23, if you were now told that the number of defective items that the machine produces per hour follows a normal distribution, would it change your answer to Exercise S7.24? If so, find the new answer. If not, explain why.

S7.26 In addition to the data given in Exercise S7.23, if you were told that the population standard deviation of the number of defectives that the machine produces per hour is 1.80, would it change your answer to Exercise S7.24? If so, find the new answer. If not, explain why.

S7.27 Construct a 90% confidence interval for the mean number of defective items that the machine in Exercise S7.23 produces per hour.

S7.28 Construct a 99% confidence interval for the mean number of defective items that the machine in Exercise S7.23 produces per hour.

S7.29* A police department statistician has noted over the years that the weekly number of on-the-spot parking fines issued in the town follows a normal distribution with a standard deviation of 12.0, irrespective of the size of the fine. She notes, however, that the mean number of fines per week varies according to the current size of the fine.

With the government just changing the size of the fines, she wants to estimate the mean of this new distribution. To do this, she records the number of fines for each week over a period of 16 weeks and finds that there was a mean of 84.6 fines per week.

Construct a 95% confidence interval for the mean number of fines issued per week.

S7.30 For the problem in Exercise S7.29, suppose that the statistician has noted that the daily number of traffic fines does *not* follow a normal distribution. Does this change your answer to Exercise S7.29? Explain.

S7.31 Find a 99% confidence interval for the problem in Exercise S7.29.

S7.32* Find a 90% confidence interval for the problem in Exercise S7.29.

S7.33 A suburban post office recorded the number of parcels sent from the office for each of 25 weeks. It found that the sample had a mean of 74 parcels and a standard deviation of 13 parcels.

(a) Find both 95% and 99% confidence intervals for the mean number of parcels sent from the post office each week.

(b) Use your answers in (a) to determine whether you believe the population mean could be 80 parcels. Explain.

(c) What assumption are you making about the sample?

S7.34* A fast-food outlet wishes to determine how long its customers entering the queue in the store have to wait for their order to be taken. A random sample of 60 customers was observed. Their mean waiting time was 7.20 minutes with a standard deviation of 2.50 minutes.

(a) Find both 95% and 99% confidence intervals for the mean waiting time in the queue for all customers.

(b) Use your answers in (a) to determine whether you believe the population mean waiting time could be 6.00 minutes. Explain.

S7.35 (a) Watch commercial television channels for 25 random 15-minute time intervals (perhaps your favourite shows!), and record the number of commercials shown in each period. (This can be done on several occasions to collect all the data.)

(b) Use the data in (a) to find a 95% confidence interval for the mean number of commercials shown in a 15-minute time interval.

S7.36 (a) Go to a fast-food outlet during a busy period (i.e. lunch or dinner) and count the number of customers who come through the door for each of 25 five-minute periods.

(b) Use the data in (a) to find a 95% confidence interval for the mean number of customers who come through the door in a five-minute period.

S7.37 Find the interquartile range of:

(a) the standard normal distribution

(b) a normal distribution with $\mu = 100$ and $\sigma = 15$

(c) a normal distribution with $\sigma = 8$

S7.38* A normal distribution has $\mu = 90$ and an upper quartile of 103.5. What is the value of σ?

S7.39 The number of days it rains in a certain city in February follows a normal distribution with $\mu = 12$ and $\sigma = 3.6$. Find the proportion of Februaries when the number of rainy days is:

(a) more than 13 (b) less than 5 (c) between 3 and 20

S7.40 A certain lift has a sign that reads 'Total weight of passengers not to exceed 816 kg or the lift will not operate'. It is believed that the weights of passengers follow a normal distribution with $\mu = 68$ kg and $\sigma = 6.0$ kg. Find the probability that the lift will *not* operate with:

(a) 11 passengers (b) 12 passengers (c) 13 passengers

S7.41* If you were now told in Exercise S7.40 that the weights of passengers do *not* follow a normal distribution, would you still be able to answer the questions? If so, find the answers. If not, explain why.

S7.42 In Exercise S7.40, find a 95% confidence interval for the *total* weight of:

(a) 11 passengers (b) 12 passengers (c) 13 passengers

S7.43 During 2000 the mean number of hours worked by full-time employed persons in Australia was 41.3 hours per week. (*Source*: Australian Bureau of Statistics, *Labour Statistics in Brief*, Cat. No. 6104.0.) Suppose that the distribution is normally distributed with a standard deviation of 8.0 hours. If a random sample of 200 such workers were selected, how many of them (to the nearest whole number) would you expect to work each week *more than*:

(a) 29.0 hours? (b) 35.0 hours? (c) 50.0 hours? (d) 65.0 hours?

S7.44 For the data in Exercise S7.43, if a random sample of 500 such workers were selected, how many of them (to the nearest whole number) would you expect to work each week *less than*:

(a) 20.0 hours? (b) 30.0 hours? (c) 45.0 hours? (d) 60.0 hours?

S7.45 The weights of suitcases taken by travellers on international journeys are known to follow a distribution that is skewed to the right with μ = 18 kg and σ = 5 kg. A certain flight has a total weight limit on suitcases of 3500 kg. What is the probability that this weight limit will be *exceeded* if the *total number* of suitcases on board is:

(a) 190? (b) 200? (c) 202? (d) 204?

(e) 205?

S7.46* A random sample of 50 elderly diabetes patients revealed systolic blood pressures that had a mean of 160 mm and a standard deviation of 6.8 mm. Find both 95% and 99% confidence intervals for the mean blood pressures of all elderly diabetes patients.

S7.47 In Exercise S7.46, would it change your answer if you were now told that the distribution of blood pressures was:

(a) normal? (b) skewed to the right?

In each case, explain your reasoning.

S7.48 A recent study revealed that the mean amount of television viewed each week by a random sample of 200 teenagers was 26.4 hours. If these times had a standard deviation of 5.4 hours, find 95% and 99% confidence intervals for the mean amount of television watched each week by all teenagers.

S7.49 Using your results in Exercise S7.48, would you believe that teenagers watch an average of 28 hours of television each week? Explain your answer.

S7.50* Based on a random sample of cans of peaches, a 95% confidence interval for the mean weight of all cans has length 8.33 g. If the standard deviation of the weights in the sample is 17.0 g, how many cans were in the sample?

S7.51 Based on a random sample of 96 packages of ground beef, a 99% confidence interval for the mean weight of all packages has length 12.64 g. What is the standard deviation of the weights in the sample?

S7.52 In a nationwide mathematics examination, the marks for males were normally distributed with μ = 65 and σ = 5.3, while the marks for females were normally distributed with μ = 60 and σ = 8.2. A group of 1000 males and a group of 1000 females were selected at random. In which group would you expect to find the most candidates with a mark over 75? Justify your answer.

S7.53* The amount of time that a drug designed to give temporary relief from migraine headaches takes to be effective varies from patient to patient. However, it is known that this time follows a normal distribution with a mean of 35.7 minutes and a standard deviation of 8.0 minutes. For a random sample of 36 patients, find the probability that the mean amount of time taken for the drug to be effective is:

(a) more than 37.0 minutes (b) less than 39.0 minutes

(c) between 32.0 minutes and 38.0 minutes

S7.54 Select a random sample of 25 cars (parked or moving) and record the *last* digit of their number plates. (Ignore plates that do not contain any digits.)

(a) Write down the values in your sample.

(b) Use the values in (a) to find:

(i) a 95% confidence interval for the mean of the last digit of *all* number plates

(ii) a 99% confidence interval for the mean of the last digit of *all* number plates

S7.55 (a) Using the 95% interval for the mean of the last digit of *all* number plates found in Exercise S7.54(b)(i), would you believe that the population mean could be 5.0? Explain.

(b) Do you believe that this is actually the case? Why?

S7.56 Go to your city telephone directory and select a page at random. From that page select 30 telephone numbers at random.

(a) Write down the *last two* digits of the telephone numbers selected.

(b) Use the values in (a) to find:

(i) a 95% confidence interval for the mean of the last two digits of *all* telephone numbers

(ii) a 99% confidence interval for the mean of the last two digits of *all* telephone numbers

S7.57 (a) Using the 95% interval for the mean of the last two digits of *all* telephone numbers found in Exercise S7.56(b)(i), would you believe that the population mean could be 50.0? Explain.

(b) Do you believe that this is actually the case? Why?

Chapter S8

Correlation

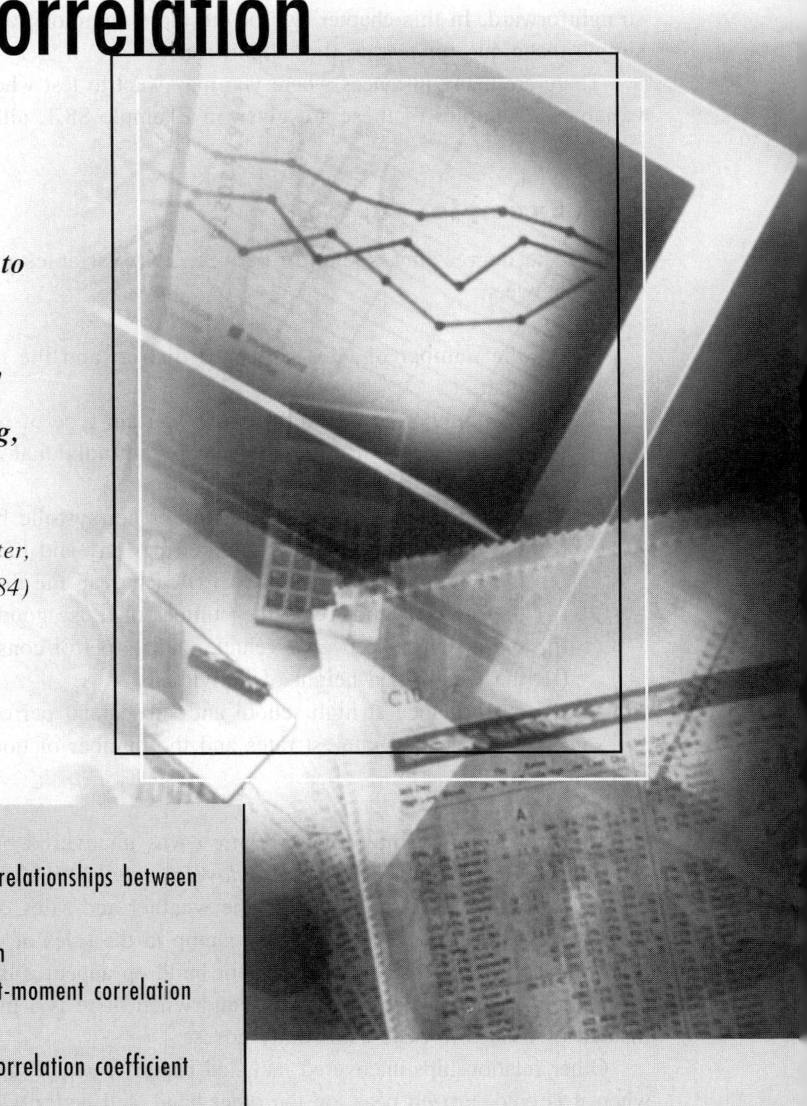

'A very quiet and tasteful way to be famous is to have a famous relation. Then you cannot only be nothing, you can do nothing, too.'

(P. J. O'Rourke, American writer, Modern Manners, *1984)*

Learning objectives

In this chapter you will learn to:

- Understand correlation analysis and relationships between variables
- Draw and interpret a scatter diagram
- Understand and calculate the product-moment correlation coefficient
- Understand and calculate the rank correlation coefficient
- Recognise spurious correlation
- Test a correlation coefficient for significance

S8.1 INTRODUCTION

Up to this point, the only problems considered have been those in which there was only *one* variable. That is, for each member of the sample only a single measurement was taken. In many problems, however, there is more than one variable of interest, and in this chapter we look at those cases where *two* variables are measured. The question to be considered now is whether there is any *relationship* or *association* between the two variables.

The topic of assessing any such relationship is called *correlation analysis*. Statisticians have defined an index called the **correlation coefficient**, which defines the strength of the association between two variables. There are several such coefficients, and their interpretation is not always straightforward. In this chapter two of the more common coefficients are discussed, together with a simple technique for testing their significance.

There are many instances where you may want to test whether there is a relationship between two variables. Examples of these are given in Example S8.1, although the possible list is endless.

Example S8.1

The degree of association between two variables may be of interest in the relationships between:

(a) the number of CDs sold in a district and the number of teenagers who live in that district
(b) the number of breakdowns of a certain type of machine and the age of the machine
(c) the level of education obtained by an individual and his or her income in later working life
(d) the age of an individual and his or her systolic blood pressure
(e) the number of cigarettes smoked per day and life expectancy
(f) the Consumer Price Index (CPI) each year and the average weekly earnings for that year
(g) the price of goods and the number of those goods sold
(h) the driving speed of a vehicle and its petrol consumption
(i) the weight and height of individuals
(j) performance at high school and subsequent performance at tertiary level
(k) housing loan interest rates and the number of houses sold

An interesting example of correlation was uncovered by a team of researchers in the United Kingdom in 1997 (reported in the *Sunday Telegraph*, 16 March 1997, p. 21). They found that there is a significant relationship between the weather and sales of various grocery items. For example, figures suggest that there is a dramatic slump in the sales of cat and dog food when the temperature is low, since pets stay indoors and do not build up appetites by venturing outside. On the other hand, toilet paper sales soar in cold weather and when there is a high pollen count, because people use it instead of tissues to deal with runny noses.

Other relationships uncovered included bananas, which sell well when the weather is fine but not when it is cold. Frozen peas, on the other hand, sell better when it is cold, since people then turn to the traditional meal of meat and vegetables. In fact, one supermarket claimed that it could save $6000 a week by using weather forecasts to help it decide what stock to put on the shelves.

In a more bizarre correlation, other researchers claimed to have found that there is a statistical relationship between the intelligence of adults and whether they sucked dummies as babies. In a test of the IQ of 1000 men and women, they found that those who had sucked dummies as a baby recorded a 7% lower IQ score than those who had not. While the conclusions drawn from such data may be arguable, the researchers suggested that one possible reason for this result was that dummies are used by less intelligent parents who, more than likely, will have less intelligent children!

If there is a relationship between any two variables, it may be possible to predict the value of one of the variables from the value of the other. If, for example, there is a relationship between the price of apples and the number of apples sold per day at a certain fruit stall, the owner may be able to predict how many apples might be sold on a day when the price was set at a particular level. The topic of establishing an equation to enable such a prediction is called *regression analysis* and is discussed fully in Chapter S9.

Dependent and independent variables

In practice, to establish whether there is a relationship between two variables, an appropriate random sample must be taken and, for each individual in the sample, a measurement recorded of each of the two variables. Thus, for each individual we have *two* data points. Such data are said to be *bivariate* since they consist of two variables.

These data may be written as **ordered pairs** where they are expressed in a specific order for each individual (first variable value, second variable value). Although for correlation purposes it does not really matter which of the two variables is called the first variable and which is called the second variable, there is a basic rule that makes things easier when regression analysis is used.

We have already stated that one of the purposes of correlation analysis is to establish whether the value of one variable can perhaps be predicted from the value of the other. In this case, the following definitions are appropriate.

Definition S8.1

The *dependent* variable is the one whose value is to be predicted. It is usually denoted by the letter y.

The *independent variable* is the one whose value is used to make the prediction. It is usually denoted by the letter x.

For example, if a quantity is observed over a period, it may be desired to predict what the value of that quantity will be at some future time. In this case, 'time' will be the independent variable and 'quantity' will be the dependent variable.

In some instances it may not be entirely clear which of the two variables is the one to be predicted, but most of the time it is fairly obvious. Once this issue has been settled, then the ordered pairs take on the form (x-value, y-value).

Example S8.2

A human resource manager is concerned with the number of sick days taken by employees in the company and feels that it may be related to the age of the employee. He selects a random sample of 6 employees and records the number of days absent last year and their age (in years) as at 30 June last year. The results are:

Age (years)	18	26	52	41	60	33
Number of sick days	3	4	9	6	12	4

For this problem it seems reasonable that the human resource manager may want to predict the number of sick days from the age of the employee rather than the other way around. In this case, the dependent variable (y) will be 'number of sick days' and the independent variable (x) will be 'age'.

Hence, the data may be expressed as a set of ordered pairs in the form (age, number of sick days) and written:

(18, 3) (26, 4) (52, 9) (41, 6) (60, 12) (33, 4)

In this form the ordered pairs resemble points on a graph. In fact, that is precisely one of the reasons for writing them that way, since the next step is to produce a visual display of what these data look like. In doing so, it is possible to get a 'feel' for what the actual relationship between the variables may be, even before any calculations are undertaken.

S8.2 SCATTER DIAGRAMS

A **scatter diagram** is a display in which the ordered pairs of measurements are plotted on a coordinate axes system. The independent variable (x) is represented on the horizontal axis and the dependent variable (y) on the vertical axis. The points representing the data are usually plotted either by dots or crosses. Most statistical computer packages will perform a scatter diagram of bivariate data.

Example S8.3

An insurance company manager is concerned about the health of female adults, since the company is prepared to give a reduced premium rate to those who have a certain level of fitness. In particular, he would like to investigate how their height is related to their weight, with a view to possibly using these measurements as a fitness criterion.

To this end, he selects a random sample of 12 adult females and measures both their height (in cm) and weight (in kg). The results are shown in Table S8.1.

Table S8.1 Data for Example S8.3

Female number	1	2	3	4	5	6	7	8	9	10	11	12
Height (cm)	167	168	165	165	160	156	169	166	162	158	168	168
Weight (kg)	71.8	72.0	69.3	70.0	64.2	58.1	74.0	70.0	59.3	59.0	67.1	64.0

Draw a scatter diagram of the data.

Solution

Although it may be open to debate, in this case it seems more likely that a person's weight may be predicted from their height rather than the other way around. (In other words, your weight depends on your height; your height does not depend on your weight.) In that case, weight is the dependent variable and height is the independent variable.

This means that height will be on the horizontal axis and weight on the vertical axis. Figure S8.1 shows the scatter diagram for these data. Note that the height axis is broken to start at 150 cm, and the weight axis is broken to start at 55 kg, as there are no data points with values below these levels.

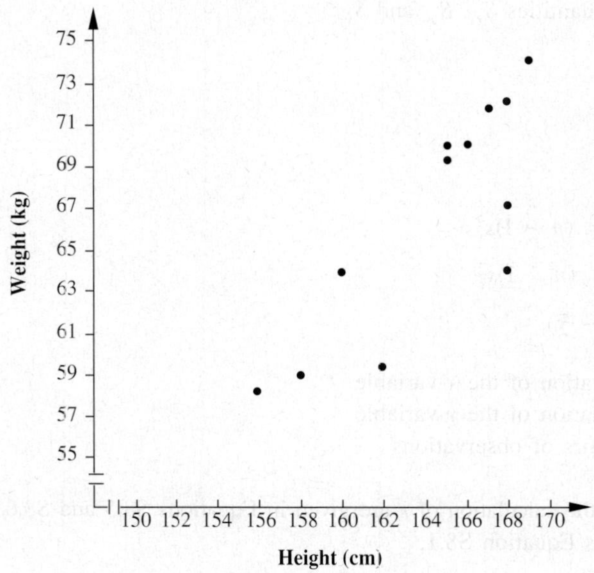

Figure S8.1 Scatter diagram for the data in Table S8.1

S8.3 THE PEARSON PRODUCT-MOMENT CORRELATION COEFFICIENT

The numerical measure of the degree of association between two variables is given by the **product-moment correlation coefficient** and is due to the work of the statistician Karl Pearson in 1896. This index provides a quantitative measure of the extent to which the two variables are associated. There are other types of correlation coefficient as well as the 'product-moment' but, as this is by far the most common one, it is usually referred to simply as the 'correlation coefficient'.

The value of the correlation coefficient is calculated from the bivariate data by means of a formula that involves the values of the data points.

The value of the correlation coefficient calculated from a *sample* is denoted by the letter r. The value of the correlation coefficient calculated from a *population* is denoted by the Greek letter ρ (pronounced 'roe').

As is the case with most population parameters, the value of ρ will rarely be known. Instead, the value of r (from the sample) is calculated from the sample and, based on this value, certain inferences can be made about what the value of ρ may actually be.

Value of r

There are several alternative formulae for finding the value of r, but lamentably all of them require tedious calculations. Fortunately, many relatively inexpensive calculators today have a button that calculates the value of r directly after entering the data points. (Check your calculator handbook for details, or see whether there is a button on your calculator that has the letter 'r' or 'ρ' either on the button or below it.)

One formula for calculating r is given in Equation S8.1, where the calculations are broken down into the three separate quantities S_{xx}, S_{yy} and S_{xy}:

$$r = \frac{S_{xy}}{\sqrt{(S_{xx})(S_{yy})}} \quad \text{(S8.1)}$$

where:

$$S_{xx} = \Sigma(x - \bar{x})^2 = (n - 1)s_x^2 \quad \text{(S8.2)}$$

$$S_{yy} = \Sigma(y - \bar{y})^2 = (n - 1)s_y^2 \quad \text{(S8.3)}$$

$$S_{xy} = \Sigma(x - \bar{x})(y - \bar{y}) \quad \text{(S8.4)}$$

and s_x = standard deviation of the x-variable
s_y = standard deviation of the y-variable
n = number of pairs of observations

Alternative formulae for the calculation of r are given in Equations S8.5 and S8.6, both of which will yield the same answer as Equation S8.1.

$$r = \frac{S_{xy}}{(n - 1)s_x s_y} \quad \text{(S8.5)}$$

$$r = \frac{\Sigma xy - n\bar{x}\bar{y}}{(n - 1)s_x s_y} \quad \text{(S8.6)}$$

Note

1. The value of r must *always* lie between −1 and +1 (both inclusive). If you obtain a value outside this range, you have made a mistake.
2. If r = +1, the two variables have *perfect positive* correlation. On a scatter diagram, this means that the points all lie on a straight line that has a *positive* slope.
3. If r = −1, the two variables have *perfect negative* correlation. On a scatter diagram, this means that the points all lie on a straight line that has a *negative* slope.
4. If the two variables are *positively* correlated, but not perfectly so, the coefficient lies between 0 and 1.
5. If the two variables are *negatively* correlated, but not perfectly so, the coefficient lies between −1 and 0.
6. If the two variables have no overall upward or downward trend whatsoever, the coefficient is 0.

Positive and negative correlation

Figure S8.2 shows examples of scatter diagrams illustrating various degrees of correlation.

It can be seen that, if correlation exists, it can be in one of two directions: positive or negative. In practical terms these have the following meanings:

1. If two variables x and y are *positively* correlated, this means that:
   (a) *large* values of x are associated with *large* values of y, and
   (b) *small* values of x are associated with *small* values of y
2. If two variables x and y are *negatively* correlated, this means that:
   (a) *large* values of x are associated with *small* values of y, and
   (b) *small* values of x are associated with *large* values of y

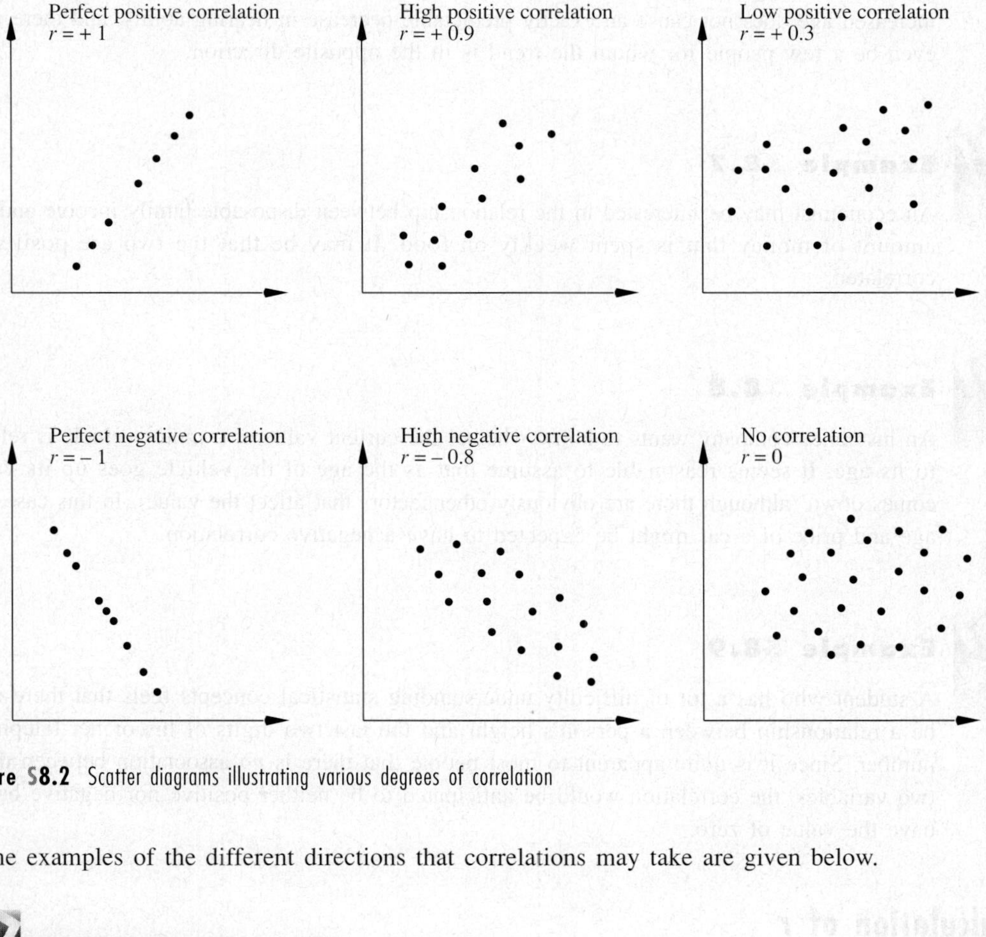

Figure S8.2 Scatter diagrams illustrating various degrees of correlation

Some examples of the different directions that correlations may take are given below.

Example S8.4

If a person starts off with a particular amount of money, the amount spent and the amount remaining are perfectly negatively correlated. As one goes up, the other goes down in precise proportion.

Example S8.5

The height and weight of human beings are positively correlated, but not perfectly so. Large heights tend to be associated with large weights, and small heights with small weights, but there is not a precise relationship. Knowing a person's exact height does not enable automatic prediction of his or her exact weight.

Example S8.6

There is said to be a negative correlation between age and hearing ability, in that as age increases hearing ability decreases. It is not a perfect negative correlation, however, since increased age does not cause an exactly predictable decrease in hearing ability, and there may even be a few people for whom the trend is in the opposite direction.

Example S8.7

An economist may be interested in the relationship between disposable family income and the amount of money that is spent weekly on food. It may be that the two are positively correlated.

Example S8.8

An insurance company wants to know whether the current value of a motor vehicle is related to its age. It seems reasonable to assume that as the age of the vehicle goes up its value comes down (although there are obviously other factors that affect the value). In this case, the age and price of a car might be expected to have a negative correlation.

Example S8.9

A student who has a lot of difficulty understanding statistical concepts feels that there may be a relationship between a person's height and the last two digits of his or her telephone number. Since it is quite apparent to most people that there is *no* association between these two variables, the correlation would be anticipated to be neither positive nor negative but to have the value of zero.

Calculation of *r*

As has been noted, the calculation of the value of r can be quite tedious, and it is very easy to make an arithmetic error, which can lead to an incorrect answer. Whichever of the formulae in Equations S8.1, S8.5 or S8.6 is used to find the value of r, it is usually advisable to set up the calculations in the form of a table.

Example S8.10

Using Equation S8.1, calculate the correlation coefficient for the data in Table S8.1.

Solution

In this case, we denote *height* by x and *weight* by y. The calculations are shown in Table S8.2. Note that the first two columns contain the original data, which must first be used to calculate the values of $\bar{x}$ and $\bar{y}$ using Equation S3.3. We have $n = 12$. Therefore:

$$\bar{x} = \frac{1972}{12} = 164.3$$

$$\bar{y} = \frac{798.8}{12} = 66.6$$

both rounded to one decimal place. Because of this rounding, the total of the columns $(x - \bar{x})$ and $(y - \bar{y})$ are not *exactly* zero (although they should be very close).

Table S8.2 Calculations required for r for the data in Table S8.1

x	y	$(x - \bar{x})$	$(x - \bar{x})^2$	$(y - \bar{y})$	$(y - \bar{y})^2$	$(x - \bar{x})(y - \bar{y})$
167	71.8	2.7	7.29	5.2	27.04	14.04
168	72.0	3.7	13.69	5.4	29.16	19.98
165	69.3	0.7	0.49	2.7	7.29	1.89
165	70.0	0.7	0.49	3.4	11.56	2.38
160	64.2	−4.3	18.49	−2.4	5.76	10.32
156	58.1	−8.3	68.89	−8.5	72.25	70.55
169	74.0	4.7	22.09	7.4	54.76	34.78
166	70.0	1.7	2.89	3.4	11.56	5.78
162	59.3	−2.3	5.29	−7.3	53.29	16.79
158	59.0	−6.3	39.69	−7.6	57.76	47.88
168	67.1	3.7	13.69	0.5	0.25	1.85
168	64.0	3.7	13.69	−2.6	6.76	−9.62
1972	798.8	0.4	206.68	−0.4	337.44	216.62

From Table S8.2 we have:

$$S_{xx} = \Sigma(x - \bar{x})^2 = 206.68$$
$$S_{yy} = \Sigma(y - \bar{y})^2 = 337.44$$
$$S_{xy} = \Sigma(x - \bar{x})(y - \bar{y}) = 216.62$$

Using Equation S8.1, we obtain:

$$r = \frac{216.62}{\sqrt{(206.68)(337.44)}} = 0.82$$

Great care must be exercised in interpreting the meaning of a value of r. This point is discussed in Sections S8.5 and S8.6.

Example S8.11

The chief of police has given the department statistician the task of determining whether air temperature is related to the number of traffic accidents in the city each day. On each of 8 randomly selected days the statistician records the maximum temperature and the number of traffic accidents that took place in the city. The results are:

Maximum temperature (°C)	14	23	16	22	30	34	19	27
Number of traffic accidents	4	6	3	6	8	11	5	7

Use Equation S8.6 to find the correlation coefficient between maximum temperature and the number of traffic accidents.

Solution

It is clear in this case that the dependent variable (y) is 'number of traffic accidents' and the independent variable (x) is 'maximum temperature'.

Equation S8.6 requires the value of the mean and standard deviation of each of the two variables. Suppose this is done either using a calculator with this facility or by using the techniques described in Chapters S3 and S4. In this case, the results are:

$n = 8 \qquad \bar{x} = 23.125 \qquad s_x = 6.896 \qquad \bar{y} = 6.250 \qquad s_y = 2.493$

The only remaining part of Equation S8.6 to find is Σxy, and this is done in Table S8.3.

Table S8.3 Calculations of Σxy required to find the value of r for the data in Example S8.11, using Equation S8.6

x	y	xy
14	4	56
23	6	138
16	3	48
22	6	132
30	8	240
34	11	374
19	5	95
27	7	189
185	50	1272

From Table S8.3 the value of $\Sigma xy = 1272$. Therefore, using Equation S8.6, the value of the correlation coefficient is:

$$r = \frac{1272 - (8)(23.125)(6.250)}{(7)(6.896)(2.493)}$$

$$= 0.962$$

This result suggests that there may be a positive relationship between temperature and the number of traffic accidents, although it must be properly tested for significance, as will be seen in Section S8.6.

S8.4 THE SPEARMAN RANK CORRELATION COEFFICIENT

An alternative measure of the degree of association between two variables is the **rank correlation coefficient**, which is usually attributed to Charles Spearman in 1904. This coefficient does not strictly measure the degree of association between the actual observations, but rather the association between the *ranks* of the observations. However, the information in the ranks is usually very similar to that in the observations themselves because the concept of correlation is largely based on relative magnitudes.

On many occasions data are presented not in the form of actual measurements, but in rank order. For example, entrants in a contest may be ranked by several different judges, a particular group of students may be ranked by examiners in various subjects, and a group of consumers may be asked to rank different brands of a certain item.

The formula for calculating the value of r_s, the rank correlation coefficient is:

$$r_s = 1 - \frac{6\Sigma d^2}{n^3 - n} \qquad \textbf{(S8.7)}$$

where d = the difference between the corresponding pairs of rankings
n = the number of pairs of observations

Note

Being a correlation coefficient, r_s has the following properties:

1. $r_s = +1$ for *perfect positive* correlation of the ranks, that is when the x-rank = the y-rank in each case, and hence $\Sigma d^2 = 0$.
2. $r_s = -1$ for *perfect negative* correlation of the ranks, that is when they run in precisely opposite order to each other, and hence $\Sigma d^2 = \frac{n^3 - n}{3}$.
3. All other values of r_s lie between −1 and +1.

The subscript s in r_s stands for 'Spearman' and is used to distinguish it from the Pearson product-moment correlation coefficient. Although it looks different, in fact it is an ordinary product-moment correlation coefficient. However, the observations are ranks, and it has been changed algebraically for the simplest possible calculation. The method of calculating r_s is illustrated in Example S8.12.

Example S8.12

A bank has two loan officers who must each give approval to customers who desire a loan. The manager currently has 10 such customers and has asked each loan officer independently to rank the applications in order of merit. The rankings are shown below.

Customer	A	B	C	D	E	F	G	H	I	J
Officer X ranking	4	6	2	1	5	10	7	3	9	8
Officer Y ranking	4	8	1	3	7	9	6	2	10	5

Calculate the rank correlation coefficient of the two sets of rankings.

Solution

The first step is to calculate the differences, d, between the pairs of rankings of each of the 10 customers. These 10 differences are:

0 –2 1 –2 –2 1 1 1 –1 3

Note that $\Sigma d = 0$ (as should always be the case) and:

$$\Sigma d^2 = 0 + 4 + 1 + 4 + 4 + 1 + 1 + 1 + 1 + 9 = 26$$

From Equation S8.7:

$$r_s = 1 - \frac{6 \times 26}{990}$$
$$= 0.84$$

Also note that the same value of r_s is obtained if the x-rankings are subtracted from the y-rankings or vice versa.

This result suggests that there is a positive correlation between the rankings of the two loan officers. However, as is the case with values of r, the value of r_s must be tested for significance. This is done in Section S8.6.

It is also possible (and often desirable) to calculate a rank correlation coefficient for data consisting of ordinary numerical values. In this case, the x-values are ranked and (separately) the y-values are ranked, and a rank correlation coefficient is calculated from the two sets of rankings.

If *tied ranks* occur, they should be replaced by the average (arithmetic mean) of the ranks that they replace. For example, if two observations are placed equal fifth, they replace the positions 5 and 6 and should each be given a rank of 5.5. If three observations are placed equal sixth, they replace the positions 6, 7 and 8 and should each be assigned the average rank of 7.

Example S8.13

Calculate the rank correlation coefficient of the data in Table S8.1.

Solution

In this case, the heights (x) and weights (y) must be ranked. These rankings and the value of d and d^2 are shown in Table S8.4.

Note, for example, in Table S8.4 that the weight of 165 kg occupies the ranks of 5 and 6, so it has the average rank of 5.5. Also, the weight of 168 kg occupies the ranks of 9, 10 and 11, so it has the average rank of 10.

From Table S8.4 we have $\Sigma d = 0$. (This is always the case. Although we do not have to calculate this quantity, it does serve as a brief check that a mistake has not been made with the rankings.) We also have $\Sigma d^2 = 75.5$ from the table, and $n = 12$.

Therefore, from, Equation S8.7:

$$r_s = 1 - \frac{6 \times 75.5}{12^3 - 12}$$
$$= 0.74$$

Table S8.4 Calculations required for r_s for the data in Table S8.1

x	y	Rank of x	Rank of y	d	d^2
167	71.8	8	10	−2	4
168	72.0	10	11	−1	1
165	69.3	5.5	7	−1.5	2.25
165	70.0	5.5	8.5	−3	9
160	64.2	3	5	−2	4
156	58.1	1	1	0	0
169	74.0	12	12	0	0
166	70.0	7	8.5	−1.5	2.25
162	59.3	4	3	1	1
158	59.0	2	2	0	0
168	67.1	10	6	4	16
168	64.0	10	4	6	36
				0	75.5

Note that the value of $r_s = 0.74$ is similar to the product-moment correlation coefficient of $r = 0.82$ found in Example S8.10. This will often be the case, although there are some exceptions, particularly when there are unusual observations in the sample.

Again, the value of r_s must be tested for significance. (See Section S8.6.)

S8.5 SPURIOUS CORRELATION

Before beginning a formal procedure for interpreting a correlation coefficient, we should emphasise some important points.

1. If two variables are significantly correlated, this does *not* imply that one must be the *cause* of the other. Association is not sufficient to establish a causal relationship.
2. The degree of association is *not* directly proportional to the magnitude of the correlation coefficient. For example, a coefficient of 0.40 does *not* mean that there is a 40% association between the two variables. Statements such as this are almost meaningless. It is often in fact the *square* of the coefficient that indicates the 'strength' of an association. For example, a coefficient of 0.4 may indicate $(0.4)^2 = 0.16$ or 16% as the subjective 'strength' of the relationship. (See Section S9.3.)
3. The correlation coefficient is subject to variations in sampling. If different samples are drawn from a population, it is quite likely that different values for the correlation coefficient will be found. In particular, any interpretation of a correlation coefficient must take into account the size of the sample.

Consider the following three examples:

Example S8.14

A group of male school children aged between 6 and 13 was chosen, and their heights and reading comprehension abilities were recorded. A high positive correlation was found between the two variables.

Example S8.15

The price of wine casks and the annual sales of wine casks in Australia were recorded over a period of 15 years. A high positive correlation was found between the two variables.

Example S8.16

A group of 50 adults was chosen at random and the lengths of their left arms and right legs were recorded. A high positive correlation was found to exist between the two variables.

In each of these three examples, the high degree of correlation does not establish a causal relationship. In Example S8.14, height does not 'cause' reading comprehension nor vice versa. The correlation is due merely to the fact that both physical and mental attributes develop as a child gets older. In Example S8.15, it is not plausible that a price rise in wine casks would cause sales to rise, although the reverse is possible. Presumably both wine cask prices and sales are increasing with the passage of time and economic conditions. In Example S8.16, it makes no sense to suggest that the length of the left arm is a 'cause' of the length of the right leg.

Correlation between two variables that is really induced by other external variables is referred to as **spurious correlation** or *false correlation*. Rational assessment of causality cannot simply be based on significantly high values of the correlation coefficient, and considerable experience is essential in the experimental environment if correct conclusions are to be made. In particular, it is very important that attention be given to other factors that may affect both the variables being investigated.

S8.6 INTERPRETATION OF THE CORRELATION COEFFICIENT

In this section is a method for determining whether an obtained correlation coefficient is *significant* (i.e. whether the two variables *are* correlated). In particular, the correlation coefficient of a sample will be tested to find whether it provides significant evidence that the correlation coefficient of the corresponding population is different from zero.

You should note that the test of significance on the Pearson *product-moment* correlation coefficient (r) is valid only if the samples are drawn from a population that has a bivariate *normal* distribution. If the population is *not* normal, the observations should be ranked and the *rank* correlation coefficient (r_s) calculated. The test of significance for r_s is valid for *any* distribution of the population from which the samples are drawn, with the one proviso that there should not be too many tied observations.

Testing a value of r

The mathematical theory is omitted in explaining this procedure. Only the formula for testing the significance of a value of r is presented. The steps are:

1. Assume that the two variables are *uncorrelated*.
2. Calculate the correlation coefficient (r).
3. Calculate the value of the expression:

$$z = r\sqrt{n - 1} \qquad \textbf{(S8.8)}$$

where n = the number of pairs of observations

4. If $|z| > 1.96$, there is strong evidence to suggest that the assumption in Step 1 is *incorrect*, and that in fact a significant degree of correlation *does* exist between the two variables. The assumption is therefore *rejected*.
5. If $|z| < 1.96$, there is no strong reason to reject the assumption that the two variables are uncorrelated.

Example S8.17

Consider the data in Example S8.10 where the value of r was found to be 0.82. Assuming that the normality assumptions hold, test the value of r for significance.

Solution

Assume that there is *no* correlation between the height and weight of adult females. In this case, $n = 12$. From Equation S8.8:

$$z = 0.82\sqrt{11} = 2.72$$

Since $|z| = |2.72| = 2.72 > 1.96$, we conclude that the value of r *is* significant.

Therefore, based on this sample, there does appear to be a *positive* correlation between height and weight for adult females. That is, taller women tend to be heavier, and shorter women tend to be lighter.

Example S8.18

A financial analyst is interested in determining whether there is any relationship between sales revenue and the number of employees for companies in the computer-hardware manufacturing industry. To test this, a random sample of 25 such companies was chosen, and a correlation coefficient (r) of –0.53 was found. Is this significant?

Solution

Assume that there is *no* correlation between sales revenue and the number of employees. In this case, $n = 25$. From Equation S8.8:

$$z = -0.53\sqrt{24} = 2.60$$

Since $|2.60| > 1.96$, we *reject* our assumption.

Therefore, based on this sample, there does appear to be a *negative* relationship between sales revenue and the number of employees. That is, those companies with fewer employees tend to have the larger sales revenues, and those companies with the greater number of employees tend to have the smaller sales revenues.

Example S8.19

Test the correlation coefficient found in Example S8.11 for significance.

Solution

In this case, $r = 0.962$, calculated from $n = 8$ pairs of observations. From Equation S8.8:

$$z = 0.962\sqrt{7} = 2.55$$

Since $|z| = 2.55 > 1.96$, we conclude that the value of r *is* significant.

Therefore, based on this sample, there does appear to be a *positive* relationship between the number of traffic accidents in the city and the temperature. That is, the hotter the day, the more accidents there tend to be, and the cooler the day, the fewer accidents there tend to be.

Testing a value of r_s

The steps for testing a value of r_s for significance are:

1. Assume that the two variables are *uncorrelated*.
2. Find the *critical value* of r_s for the given value of n, using Table 3 on page S.455.
3. If $|r_s| >$ critical value, the assumption in Step 1 is *rejected* and a significant relationship *does* exist between the two sets of rankings.
4. If $|r_s| <$ critical value, the assumption in Step 1 is *accepted* and a significant relationship does *not* exist between the two sets of rankings.

In Table 3 the critical values are shown for values of n up to 30. If the value of n exceeds 30, the method for testing the significance of r (as outlined earlier in this section) may be used.

Example S8.20

The number of umbrellas sold and the number of hours of sunshine each day were recorded over a 20-day period. The rank correlation coefficient was found to be $r_s = -0.49$. Is this significant?

Solution

Assume that the number of umbrellas sold and hours of sunshine are uncorrelated. From Table 3, the critical value (for $n = 20$) is 0.4451. Since:

$$|r_s| = |-0.49| = 0.49 > 0.4451$$

we *reject* the assumption. We therefore conclude that a significant correlation exists between the sale of umbrellas and the number of hours of sunshine. Since this correlation is *negative*, a large number of hours of sunshine in a day is associated with small sales of umbrellas, and a small number of hours of sunshine is associated with large sales of umbrellas.

Example S8.21

Test the value of the rank correlation coefficient in Example S8.12 for significance.

Solution

Assume that there is no relationship between the two sets of rankings of the loan officers. From Table 3, the critical value (for $n = 10$) is 0.6364. Since:

$$| r_s | = 0.84 > 0.6364$$

we *reject* the assumption. We therefore conclude that a significant correlation exists between the way in which the two loan officers have ranked the customers. Since the correlation is *positive*, the two officers are in substantial *agreement*.

S8.7 REMARKS

Correlation is a statistical technique that is often misused and misinterpreted. The correlation coefficient merely gives an indication of the extent to which large values of one variable are associated with large or small values of the other, and cannot assess whether there is a causal relationship.

While the correlation coefficient may tell us whether two variables are significantly related, it does not reveal *how* they are related. This is of great importance when the objective is to predict the value of one variable for a given value of the other. Such problems come under the category of *regression analysis*, discussed in Chapter S9.

It is also important to remember that *independent* variables are always uncorrelated, but the reverse does not apply. The Pearson product-moment correlation coefficient is essentially a measure of *linear relationship*, and the Spearman rank correlation coefficient measures the extent to which the variables have the same *ordering*. There are other possible forms of dependence that will not be detected by such coefficients. These generally occur when the two variables have a relationship that is not linear.

Quick Quiz

1. The age in years and current market value of a certain type of car are shown below.

Age (years)	1	2	3	4	5	6	7
Market value ($'000)	20	16	14	14	12	13	9

Draw a scatter diagram of the data.

2. Calculate the product-moment correlation coefficient for the data in Question 1.
3. Test whether the value of r in Question 2 is significant. State your conclusion.
4. In a talent quest two judges, X and Y, are asked to rank 8 contestants in order of talent. The rankings are shown below.

Contestant	1	2	3	4	5	6	7	8
Judge X	3	5	1	2	4	8	6	7
Judge Y	8	3	4	7	1	2	3	6

Calculate the value of r_s for these data.

5. Test the value of r_s found in Question 4 for significance.
6. What conclusions can you draw from your test in Question 5?
7. Consider two sets of 6 rankings for which $\Sigma d^2 = 30$. Calculate the value of r_s.
8. A keen statistics student found that there was a significant correlation coefficient between the number of house burglaries and the type of weather over a period of 30 weeks. Do you think that this correlation is spurious? Discuss.
9. The product moment correlation coefficient between the lean body mass (kg) and resting metabolic rate for 100 women was found to be 0.28. Test this for significance and state your conclusion.
10. An economist found that when she looked at the relationship between the interest rate and the unemployment rate for each of the last 25 years the value of r was 0.56. Interpret the meaning of this finding.

Multiple Choice Questions

1. The product-moment correlation coefficient between the number of packets of cigarettes sold and the number of cans of beer sold for each day over a period of 26 days is 0.46. The value (to two decimal places) of the corresponding *z*-test statistic for significance is:
(a) 11.96 (b) 2.35 (c) 2.30
(d) 17.63 (e) none of the above

2. In Question 1 the conclusion would be:
(a) the variables are positively correlated
(b) the variables are negatively correlated
(c) the variables are uncorrelated
(d) the more cigarettes you smoke, the less beer you drink
(e) none of the above

3. Two business executives, Eva and Harry, ranked 6 investments in terms of value for money. The results were:

Investment	1	2	3	4	5	6
Eva's rank	5	1	3	2	4	6
Harry's rank	4	5	3	6	2	1

The *rank* correlation coefficient (to two decimal places) is:
(a) 0.44 (b) −0.56 (c) 0.23
(d) −0.77 (e) none of the above

4. For Question 3, the critical value required to test r_s for significance is:
(a) 0.9000 (b) 0.8286 (c) 0.7450
(d) 0.3620 (e) none of the above

5. For Question 3, the conclusion from this test statistic is:
(a) the two executives agree
(b) the two executives disagree
(c) the two executives neither agree nor disagree
(d) the two executives both agree and disagree
(e) none of the above

6. If a value of r is *exactly* zero between two variables X and Y, then this means that:
(a) X and Y have the same mean
(b) X and Y have the same standard deviation
(c) X and Y are correlated
(d) the points on the scatter plot form a circle
(e) none of the above

7. For two variables the value of r^2 is 0.64. Then the value of r:
(a) must be 0.80
(b) could be −0.80
(c) could be 0.4096
(d) could be 0.36
(e) none of the above

8. The value of r indicates:
(a) whether the data follow a normal distribution
(b) whether the data are variable
(c) whether any correlation might be spurious
(d) whether further calculations might be difficult
(e) none of the above

9. If you find that a value of r is 1.28, then:
(a) you have discovered a rare occurrence
(b) you have definitely made a mistake
(c) the correlation is very highly significant
(d) the scatter plot of the data would probably be curved
(e) none of the above

10. Suppose the scatter plot of two variables, X and Y, lies on a horizontal straight line. This means that:
(a) X and Y have a perfect positive correlation
(b) X and Y have a perfect negative correlation
(c) X and Y are related
(d) X and Y are not related
(e) none of the above

Exercises

S8.1* An experiment was undertaken to determine whether there was an association in male adults between consumption of beer and number of cigarettes smoked per day. Ten male adults were chosen at random, and data were obtained on the average number of cans of beer consumed per week and the average number of cigarettes smoked per day. The results are shown in the table below.

Person	Average weekly consumption of cans of beer (x)	Average number of cigarettes smoked per day (y)
1	18	0
2	0	0
3	3	2
4	9	15
5	2	6
6	0	3
7	4	9
8	2	0
9	15	25
10	9	16

(a) Draw a scatter diagram of the data.

(b) Calculate the value of the correlation coefficient r.

(c) Test the value of r found in (b) for significance. What conclusion can you draw?

S8.2 (a) Rank each of the variables in Exercise S8.1.

(b) Calculate the rank correlation coefficient r_s from the rankings in (a).

(c) Test the value of r_s found in (b) for significance. What conclusions can you draw?

S8.3* The product-moment correlation coefficient between weight (in kg) and time (in seconds) taken to run 100 m was calculated for 22 females, and found to be -0.42.

(a) Is the value of r significant?

(b) What conclusions can be drawn?

S8.4* In a certain factory, the rank correlation coefficient between the number of industrial accidents in each of 12 sections last year and the number of employees in that section was found to be $r_s = 0.63$.

Is this value of r_s significant? What conclusions can be drawn?

S8.5* Two horseracing tipsters (X and Y) were asked to nominate the order in which 8 horses would finish in the main race. The rankings they gave are shown below.

Horse	A	B	C	D	E	F	G	H
Tipster X	2	7	8	3	5	1	6	4
Tipster Y	5	4	8	1	7	3	6	2

Were the two tipsters in general agreement?

S8.6 In the problem in Exercise S8.5, the actual order in which the horses finished was:

Horse	A	B	C	D	E	F	G	H
Place	1	8	6	4	3	2	7	5

Test whether you feel that either (or both) of the tipsters was accurate in their ranking.

S8.7 Show that the expressions for r in Equations S8.1 and S8.5 are equivalent.

S8.8 The population of a certain outer suburb of Brisbane and the number of stray dogs captured in a certain municipality in Melbourne were recorded for each of 8 years. The results are shown in the table below.

Year	Population of suburb	Number of stray dogs
1995	863	21
1996	1051	25
1997	1280	26
1998	1562	28
1999	1897	34
2000	2051	36
2001	2463	39
2002	2937	42

(a) Draw a scatter diagram of the data.

(b) Calculate the value of the correlation coefficient r.

(c) Test the value of r found in (b) for significance. What conclusions can you draw?

S8.9* The sales volume of petrol and the maximum daily temperature were recorded for each of 12 days at a Melbourne service station and found to have a correlation coefficient of $r = 0.52$. The same data were recorded for each of 25 days at an Adelaide service station and found to have a correlation coefficient of $r = 0.42$.

(a) Test these two values of r for significance.

(b) What do your results in (a) indicate about simply using the value of r alone to determine significance?

S8.10 A real estate agent wanted to test the degree of association between the price of a house in a certain suburb and the floor space of the house. She found that for a random sample of 25 houses the correlation coefficient was $r = 0.46$.

Test this value of r for significance and draw an appropriate conclusion.

S8.11* The manager of an insurance company wishes to determine whether there is any association between the length of time an agent has been with the company and the amount of new business (in dollars insured) written. For a random sample of 50 agents he finds that the correlation coefficient between the two variables is -0.38.

What conclusions can he draw?

S8.12 There are two types of lifts in weightlifting, namely the Snatch and the Clean and jerk. Each type of lift classifies the lifter into a weight division, where the lifter's own weight must not exceed the division limit. The table below shows the weight limits (x) and world records for each of the Snatch (y) and the Clean and jerk (z).

Weight limit in kg (x)	Snatch record in kg (y)	Clean and jerk record in kg (z)
56	138.0	168.0
62	152.5	180.5
69	165.0	196.5
77	170.5	210.0
85	181.0	218.0
94	188.0	232.5
105	197.5	242.5

Source: Adapted from the International Weightlifing Federation World Records

In answering these questions you may use the following statistics:

$$\bar{x} = 78.29 \qquad \bar{y} = 170.36 \qquad \bar{z} = 206.86$$

$$s_x = 17.59 \qquad s_y = 20.64 \qquad s_z = 27.00$$

$$S_{xx} = 1855.4 \qquad S_{yy} = 2555.9 \qquad S_{zz} = 4373.9$$

$$S_{xy} = 2136.3 \qquad S_{xz} = 2817.4 \qquad S_{yz} = 3323.5$$

(a) Draw a scatter plot of Snatch record versus weight limit and join successive points. Comment on the shape.

(b) Draw a scatter plot of Clean and jerk record versus weight limit and join successive points. Comment on the shape. Note any unusual observations.

(c) Draw a scatter plot of Clean and jerk record versus Snatch record and join successive points. Comment on the shape.

(d) Find the correlation coefficient between Snatch record and weight limit and test for significance.

(e) Find the correlation coefficient between Clean and jerk record and weight limit and test for significance.

(f) Find the correlation coefficient between Clean and jerk record and Snatch record and test for significance.

S8.13 Checkout operators to be employed by a supermarket chain are given a 1-week training period, then given a speed and accuracy test. Separate scores are recorded for the two aspects. For 10 randomly chosen operators, the results were as follows (recorded on scales of 0 to 100):

Speed score (x)	Accuracy score (y)
81	27
90	38
62	37
80	60
43	65
76	52
58	82
82	47
90	56
36	18

(a) Draw a scatter diagram of the data.

(b) Calculate the value of the correlation coefficient *r*.

(c) Test the value of *r* found in (b) for significance. What conclusions can you draw?

S8.14* (a) Rank each of the variables in Exercise S8.13.

(b) Calculate the rank correlation coefficient r_s from the rankings in (a).

(c) Test the value of r_s found in (b) for significance. What conclusions can you draw?

S8.15 Two consumer organisations, X and Y, were each asked to rank 8 brands of mobile phone for value for money. The results were as follows:

Brand of mobile phone	A	B	C	D	E	F	G	H
Ranking by X	5	2	7	3	4	1	6	8
Ranking by Y	2	4	1	6	3	8	7	5

(a) Calculate the rank correlation coefficient r_s.

(b) Test the value of r_s found in (a) for significance. What conclusions can you draw? Did the organisations agree or disagree?

S8.16* The general manager of a TAB agency wants to investigate whether there is a relationship between the number of people who attend a race meeting and the amount of money that people place on bets at the TAB. This information is recorded for 10 race meetings in the table below.

Attendance ('000)	Amount invested ($'000)
15	250
12	146
20	375
16	200
25	280
9	125
18	310
21	350
14	190
20	304

(a) Draw a scatter diagram of the data.

(b) Calculate the value of the correlation coefficient *r*.

(c) Test the value of *r* found in (b) for significance. What conclusions can you draw?

S8.17 (a) Rank each of the variables in Exercise S8.16.

(b) Calculate the rank correlation coefficient r_s from the rankings in (a).

(c) Test the value of r_s found in (b) for significance. What conclusions can you draw?

S8.18* The table below shows the University Admissions Index (UAI) (based on high school work) and grade point average (GPA) of 7 students after the first semester at college.

UAI	GPA
75	3.12
84	3.92
68	1.45
72	2.01
79	2.58
90	3.60
65	1.50

(a) Draw a scatter diagram of the data.

(b) Calculate the value of the correlation coefficient r.

(c) Test the value of r found in (b) for significance. What conclusions can you draw?

S8.19 For the data in Example S3.18, suppose that you are now told that there was an error in recording the data, and that the student with a UAI of 90 had in fact a GPA of 0.60, *not* 3.60.

(a) Calculate the new value of the correlation coefficient r.

(b) Test the value of r found in (a) for significance. What conclusions can you now draw?

(c) Do you think that it is reasonable to keep this new data point in the sample? Give reasons.

S8.20* Recording data for each year over the 20-year period 1983 to 2002, a researcher noted that the correlation coefficient between the total beer consumption in Victoria and the average weekly wage of schoolteachers in that state was 0.925.

Test this value for significance and state carefully any conclusions that you can draw.

S8.21* The manager of a department store asked the information desk in the store to record the number of enquiries made each day about the 4 L cans of paint that were on sale. She then asked the paint department to record each day the number of cans that were actually sold. The results (noted over the 6 days of the sale) are shown below.

Number of enquiries	Number of cans sold
12	6
3	4
10	5
5	2
13	9
11	4

(a) Draw a scatter diagram of the data.

(b) Calculate the value of the correlation coefficient r.

(c) Test the value of r found in (b) for significance. What conclusions can you draw?

S8.22 The number of job vacancies in Australia in August 1999 and August 2000 for various industries are shown in the table below.

Industry	Job vacancies in August 1999	Job vacancies in August 2000
Manufacturing	14 100	13 100
Construction	6 900	5 000
Wholesale trade	8 700	5 100
Retail trade	9 400	21 200
Education	2 900	4 700
Transport and storage	3 100	3 700
Health and community service	11 100	10 600
Property and business services	25 100	37 900

Source: Australian Bureau of Statistics, *Job Vacancies in Australia, August 2000*, Cat. No. 6350.0

(a) Draw a scatter diagram of the data. (Put the August 2000 data on the vertical axis and the August 1999 figures on the horizontal axis.) Identify any unusual features or observations.

(b) You are now told that the product-moment correlation coefficient between the two variables is $r = 0.896$. Test this value for significance. What conclusions can you draw?

(c) What other factors might be taken into account when considering the relationship between the two variables?

S8.23* The fact that two variables are each strongly correlated with something else may sometimes camouflage more subtle correlations. Invent an example where two variables, x and y, are strongly *positively* correlated with one another because of their mutual association with a third variable, but where x and y would be *negatively* correlated if that third variable were held fixed.

S8.24* The foreman of a manufacturing plant was concerned about the amount of absenteeism at the plant on Monday mornings. He had a suspicion that it was related to the performance of the local Australian Rules football team, which played its games on Sunday. Over a period of 8 weeks he recorded the number of points scored by the team and the number of absentees the next day. The results were:

Number of points scored	Number of absentees
97	5
132	2
156	0
66	8
87	6
112	4
99	4
59	7

(a) Draw a scatter diagram of the data.

(b) Calculate the value of the correlation coefficient r.

(c) Test the value of r found in (b) for significance. What conclusions can you draw?

S8.25 Select a random sample of 9 cars (either parked or moving) and record the *last* digit (x) and the *second-last* digit (y) on their number plates. (Ignore number plates with no digits.)

(a) Write down the 9 ordered pairs of values for x and y.

(b) Draw a scatter diagram of the data. Do you detect a pattern?

(c) Calculate the value of the correlation coefficient r.

(d) Test the value of r found in (c) for significance. What conclusions can you draw?

S8.26 Explain why, in order for a product-moment correlation coefficient r to be significant, it is necessary for the sample size n to satisfy the following inequality:

$$n > 1 + \left(\frac{1.96}{r}\right)^2$$

S8.27* The number of fires in a national park and the total rainfall for that region were recorded over an 8-year period. The results were:

Number of fires	12	19	8	25	23	20	12	33
Rainfall (cm)	20	15	30	16	14	16	25	8

(a) Plot the data for number of fires versus rainfall on a scatter diagram. Estimate the value of the correlation coefficient r by an inspection of your diagram.

(b) Calculate the value of r.

(c) Test the value of r found in (b) for significance. What conclusions can you draw?

S8.28 The median ages (in years) at first marriage for bridegrooms and brides in Australia are shown in the table below for each of the eight years 1992 to 1999.

Year	1992	1993	1994	1995	1996	1997	1998	1999
Age of bridegroom (years)	26.9	27.0	27.2	27.3	27.6	27.8	27.9	28.2
Age of bride (years)	24.7	24.8	25.1	25.3	25.3	25.9	26.2	26.4

Source: Australian Bureau of Statistics, *Marriages and Divorces Australia 1999*, Cat. No. 3310.0

(a) Using a scale between 24.0 and 27.0 for brides on the horizontal axis, and a scale between 26.0 and 29.0 for bridegrooms on the vertical axis, draw a scatter diagram of the data. Join the points.

(b) Calculate the value of the correlation coefficient r.

(c) Test the value of r found in (b) for significance. What conclusions can you draw?

S8.29 The number of balls bowled and the number of wickets taken by the 8 leading wicket-takers for Australia in Test cricket (as at 31 March 2001) are shown in the table below.

Name	Number of balls bowled	Number of wickets taken
Shane Warne	24 415	376
Dennis Lillee	18 467	355
Glenn McGrath	16 784	326
Craig McDermott	16 586	291
Richie Benaud	19 108	248
Graham McKenzie	17 681	246
Ray Lindwall	13 650	228
Clarrie Grimmett	14 513	216

Source: Australian Cricket Board

(a) Find the cricketer who has the smallest ratio of balls bowled to wickets taken. Interpret his result.

(b) Rank each of the two variables in the table.

(c) Calculate the value of the rank correlation coefficient r_S from your rankings in (b).

(d) Test the value of r_S found in (c) for significance. What conclusions can you draw?

S8.30* In a council election at which 7 candidates stood for office, newspapers conducted opinion polls, after which they predicted the order in which the candidates would finish in order of votes received. Both their predicted order and the actual order are shown below.

Candidate	A	B	C	D	E	F	G
Newspaper prediction	6	5	7	4	2	3	1
Actual	1	2	3	4	5	6	7

(a) Calculate the value of the rank correlation r_S.

(b) Test the value of r_S found in (a) for significance. What conclusions can you draw?

S8.31 A random sample of 6 drivers was selected at a Motor Registry, and their age and the last 2 digits of their telephone number were recorded. The results are shown below.

Driver	A	B	C	D	E	F
Age (years)	23	52	19	35	48	67
Last two telephone digits	88	70	99	65	65	04

(a) Draw a scatter diagram of the data.

(b) A trainee statistician believes that he has found a remarkable relationship between the age of drivers and the last 2 digits of their telephone number. Calculate the value of the correlation coefficient r.

(c) Test the value of r found in (b) for significance. Do you believe the trainee statistician? Give reasons.

S8.32* (a) Rank each of the variables in the table in Exercise S8.31.

(b) Calculate the value of the rank correlation coefficient r_S from your rankings in (a).

(c) Test the value of r_S found in (b) for significance. What conclusions can you draw?

S8.33 (a) Construct 8 pairs of numbers that have a *perfect positive* correlation coefficient of $r = 1$. Illustrate your answer by calculating the value of r using an appropriate formula.

(b) Plot your data in (a) on a scatter diagram.

S8.34 (a) Construct 8 pairs of numbers that have a *perfect negative* correlation coefficient of $r = -1$. Illustrate your answer by calculating the value of r using an appropriate formula.

(b) Plot your data in (a) on a scatter diagram.

S8.35* Construct 8 pairs of numbers that have a *perfect positive* rank correlation coefficient of $r_S = 1$. Illustrate your answer by calculating the value of r_S using the appropriate formula.

S8.36 Construct 8 pairs of numbers that have a *perfect negative* rank correlation coefficient of $r_S = -1$. Illustrate your answer by calculating the value of r_S using the appropriate formula.

S8.37 In a series of trials, a car was driven over a fixed distance of 200 km at 5 different constant speeds, and the number of litres of fuel used in each trial was recorded. The results were:

Trial	1	2	3	4	5
Speed (km/h)	30	40	50	60	70
Fuel (L)	16	18	20	18	16

(a) Draw a scatter diagram of the data. Put fuel on the vertical axis.

(b) Show that the product-moment correlation coefficient between speed and fuel consumption is *zero*. Explain why the correlation is zero, despite the fact that from the diagram in (a) a relationship can clearly be seen.

S8.38 Give three different examples of pairs of real-world data for which a *spurious correlation* may exist. (You do *not* have to give actual values.)

S8.39 (a) If a group of 8 men were all *exactly* 5 years older than their partners, what would be the product-moment correlation coefficient between the ages of the men and their partners?

(b) Illustrate your answer in (a) by choosing 8 such pairs of ages and drawing the scatter diagram.

(c) Calculate the value of the correlation coefficient r for your data in (b).

S8.40 The life expectancies (in years) of male and female babies born between 1995 and 2000 in 12 countries are shown in the table below.

Country	Males (x)	Females (y)
Cameroon	53	56
Papua New Guinea	57	59
Israel	62	63
Vietnam	65	70
Thailand	66	72
The Philippines	67	70
Malaysia	70	74
New Zealand	74	80
United Kingdom	75	80
Australia	76	81
Hong Kong	76	81
Japan	77	83

Source: *United Nations Statistics Division*, 2000

(a) Using a scale between 50 and 80 for males on the horizontal axis, and a scale between 55 and 85 for females on the vertical axis, draw a scatter diagram of the data.

(b) For these data:

$$S_{xx} = 693.5 \qquad S_{yy} = 927.0 \qquad S_{xy} = 795.4$$

Calculate the value of the correlation coefficient r.

(c) Test the value of r in (b) for significance. What conclusions can you draw?

S8.41 The median prices of houses sold during the March quarter 2000 for 8 areas in Sydney, along with their postcodes, are shown in the table below.

Suburb	Postcode (x)	Median price ($'000) ($y$)
Campbelltown	2560	167.0
Penrith	2750	187.0
Blacktown	2148	195.0
Manly	2095	700.0
North Sydney	2060	732.5
Hunters Hill	2110	820.0
Woollahra	2025	880.5
Mosman	2088	950.0

Source: *Real Estate Institute of NSW*, Property Market View, *June Quarter 2000*

(a) Using a scale between 2000 and 2800 for postcode on the horizontal axis, and a scale between 160 and 1000 for median price ($'000) on the vertical axis, draw a scatter diagram of the data.

(b) For these data:

S_{xx} = 400 181.7 S_{yy} = 795 690.9 S_{xy} = −283 836.9

Calculate the value of the correlation coefficient r.

(c) Test the value of r in (b) for significance. What conclusions can you draw?

S8.42 A movie fan in the United States is interested in whether there is a relationship between the amount that a new movie takes in at the box office in its first week of release and how much it has taken in after four months. She is aware that some movies start out well but do not live up to their early promise, while for others the reverse is true.

To determine the situation, she selects a random sample of 10 movies that were released during the first half of the year 2000 and ranks them in order of gross takings (US$) for both their first week of release and after four months. These ranks are shown in the table below, with rank 1 indicating the highest takings.

Movie	Rank of the gross takings after the first week	Rank of the gross takings after four months
Mission Impossible 2	1	1
Gladiator	2	2
Scream 3	3	3
U–571	5	4
Mission to Mars	4	5
Fantasia 2000	10	6
Keeping the Faith	8	7
Hanging Up	6	8
Battlefield Earth	7	9
American Psycho	9	10

Source: *The Movie Times* (from the *URL* **www.the-movie-times.com**)

(a) Make a list of the above movies that you have seen and rank them in order of how much you enjoyed them. Do you feel that you basically agree or disagree with either of the sets of ranking in the table?

(b) Calculate the value of the rank correlation coefficient r_s for the data in the table.

(c) Test the value of r_s in (b) for significance. What conclusions can you draw?

(d) Suppose you are now told that, taking the actual gross takings into account, the product-moment correlation coefficient is $r = 0.886$. Test this value for significance. What conclusions can you draw?

(e) Compare your two conclusions in (c) and (d) and comment.

S8.43 The average crowd per match and the position where the team finished at the end of the regular season is shown in the following table for each of the 10 years between 1991 and 2000 for the Australian Rules team the Sydney Swans.

Year	Average crowd per match	Finishing position
1991	11 140	12
1992	9 963	15
1993	9 423	15
1994	9 814	15
1995	15 926	12
1996	26 850	1
1997	35 818	6
1998	32 354	3
1999	30 539	8
2000	25 563	10

Source: *Sunday Telegraph*, 18 March 2001, p. 60

(a) Draw a scatter plot of average crowd per match versus finishing position. Comment.

(b) Find the product moment correlation coefficient between average crowd per match and finishing position.

(c) Test the coefficient in (b) for significance. What conclusions can you draw?

Chapter S9

Regression analysis

'You can only predict things after they have happened.'

(Eugene Ionesco, playwright, Le Rhinoceros, *1959)*

Learning objectives

In this chapter you will learn to:

- Understand linear regression
- Calculate the least-squares regression equation
- Understand and calculate the goodness of fit of an equation
- Use the regression line for prediction
- Understand other forms of regression

S9.1 INTRODUCTION

In Chapter S8 we examined the problem of measuring the closeness of the relationship between two numerical variables. The strength of this relationship was measured by the calculation of a correlation coefficient.

Suppose that there is a significant correlation between two variables. Then, in general, it is possible to predict (with better than chance accuracy) the value of one of the variables from knowledge of the value of the other. If the correlation coefficient is zero, a knowledge of x is no help in predicting y; however, if the correlation coefficient is +1 or –1, the value of y may be predicted with perfect accuracy. The magnitude of the correlation is a guide to the likely *accuracy* of prediction, although the value of the coefficient does not indicate *how* the prediction should be made.

An equation of the form:

y = some algebraic expression involving x

is required to predict y in terms of x. With rare exceptions, the prediction will not be totally accurate, and the best we can hope for is that the prediction will be correct on average. In other words, even though the prediction will sometimes be too high and sometimes too low, we seek the best prediction that can reasonably be attained overall. A predictive equation of this kind is called a *regression equation* or *regression line*.

Linear regression is the case where the relationship between the variables is represented by a straight line. An example is shown in Figure S9.1. The dots are the actual data points graphed as a scatter diagram, and the straight line has been chosen to pass through the points as centrally as practicable.

If we wish to predict the value of y when x is known to take the value x_0, we find the point on the regression line corresponding to $x = x_0$. The corresponding value y_0 then gives the predicted value of y. It can be seen that some actual y-values are higher than they would be predicted to be on the basis of their corresponding x-values, and that others are lower, but on average the regression line seems to give reasonable predictions.

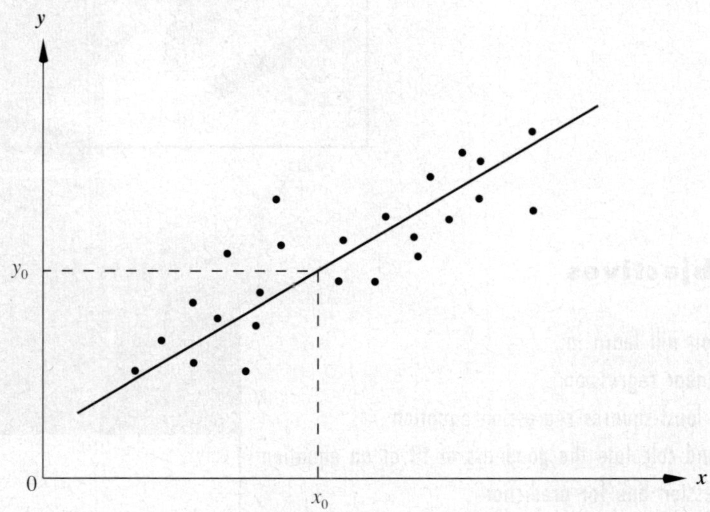

Figure S9.1 Example of a straight line fitted to a scatter diagram

In many cases, the relationship between two variables may well be described as linear and the methods described in this chapter will be appropriate. On some occasions, however, the relationship may *not* be linear, and fitting a straight line to the data in that case would be of little use (and probably dangerous!).

A simple way to check whether a straight line may be appropriate is to draw a scatter diagram of the data. Then some subjective judgement can be made as to whether the points fall roughly in a straight line or not.

Example S9.1

The financial controller of a company has recorded the total amount of stock held (as at 30 June) by the company for each of the years 1994 to 2002. The results are shown in Table S9.1.

Table S9.1 Year and stock data for Example S9.1

Year	Amount of stock held ($'000)
1994	21
1995	25
1996	32
1997	34
1998	30
1999	28
2000	25
2001	23
2002	20

The scatter diagram for these data is shown in Figure S9.2.

The points on the scatter diagram clearly do *not* lie in a straight line. It would therefore be a very poor idea to attempt to fit a straight line to these data, particularly if the aim is to use the line to describe the trend of the data or, even worse, to make a prediction.

There are regression techniques available to deal with situations where the data are non-linear, but these are beyond the scope of this book. They often involve some transformation of the data points, and a computer package is usually used.

The main point of this discussion is that data must be examined first before attempting to fit a straight line simply because that is the easiest type of line to fit. To this end, a visual display of the data using a scatter diagram is an essential beginning.

S9.2 CALCULATION OF THE REGRESSION EQUATION

You will recall that in Chapter S8 we had two variables, which we labelled y (the *dependent* variable) and x (the *independent* variable). The same convention will be used in this chapter.

It will also be assumed that we are trying to find the line that will predict the value of the y-variable from a knowledge of the x-variable. This is known as the **least-squares regression line** of y on x.

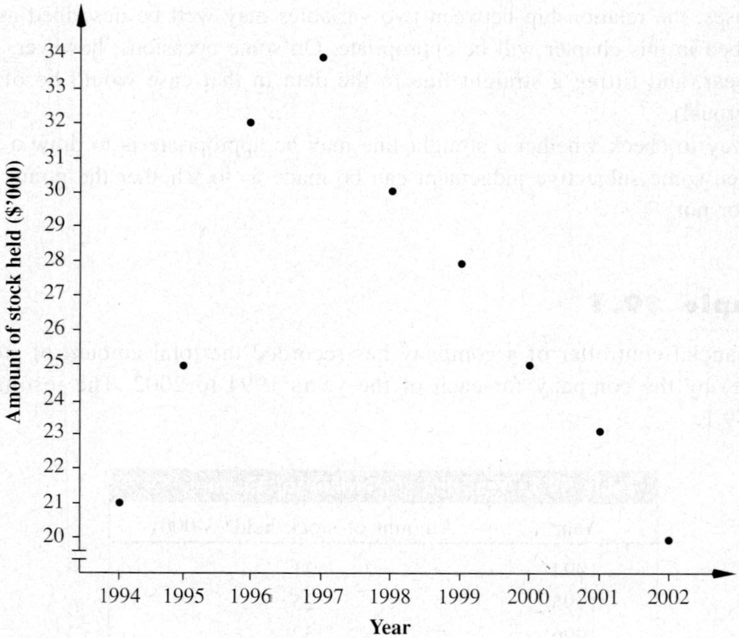

Figure S9.2 Scatter diagram for the data in Example S9.1

However, you should be aware that a totally *different* line should be used to find the line that will predict the value of the x-variable from a knowledge of the y-variable. This is known as the *least-squares regression line* of x on y. In practice, you would not usually find the equation of this line, since it is the dependent variable y that we wish to predict, and the selection of y is made by the user.

To calculate the least-squares regression line in any problem, we can base our calculations only on the data from a random sample of values. In that sense, the equation that we find will not be exactly the same as it might be if we were to use the entire population. Indeed, if we changed the random sample, we would get a different set of data points for x and y and hence a different regression line.

Suppose we *did* have access to *all* data points in the population. Then the least-squares regression line of y on x would have the form:

$$\hat{y} = \alpha + \beta x \tag{S9.1}$$

where α (alpha) and β (beta) are constants and $\hat{y}$ is the predicted value of y.

In practice, we will usually never know the true values of α and β, since we cannot measure the entire population. Instead, we use whatever sample values for x and y we have to make an *estimate* of α and β.

Suppose we denote the estimate of α by 'a' and the estimate of β by 'b', where a and b are constants, and we also use $\hat{y}$ to denote the best prediction of y that we can make for a given value of x. Then the least-squares regression line of y on x for our sample may be written as:

$$\hat{y} = a + bx \tag{S9.2}$$

The geometric interpretation of a and b is shown in Figure S9.3.

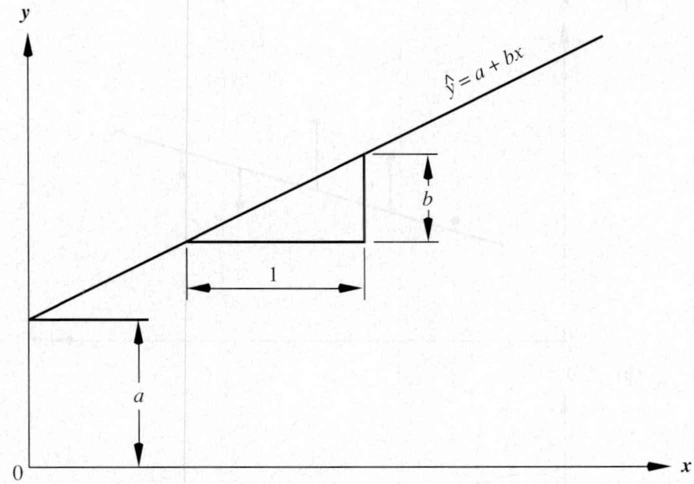

Figure S9.3 Interpretation of the regression coefficients *a* and *b*

It can be seen that a is the value of y where the line intersects the y-axis, that is where $x = 0$. The value of b denotes the *slope* of the line, representing the increase in the value of y when the value of x is increased by 1.

The problem is how to find the 'best' values of a and b from a set of data, that is the values of a and b that best approximate the values of α and β, respectively.

The method we use is the 'least-squares' criterion mentioned earlier in this section. Before becoming involved in formulae, it is useful to consider a visual display of the situation.

In Figure S9.4(a) there are five points on the scatter diagram and a line has been drawn through the centre of them. The *vertical* distance from each point to the line has also been drawn. These distances represent *errors* that the line would make in predicting the value of y for the x-value of each data point. Clearly, the closer the data points are to the line drawn, the smaller the errors will be.

Our aim is to construct a line that will minimise these errors for our particular set of data points. In Figure S9.4(a) the line drawn does not look too bad. However, the line drawn in Figure S9.4(b) appears to be a poor fit of the points, and it is not too hard to visualise another line for these data that would result in smaller errors.

From a mathematical point of view, these errors (or *residuals* as they are sometimes called) are the differences between the *actual* y-values and the *predicted* y-values. These differences may be denoted by the residuals $(y - \hat{y})$, and we need to find an *overall* minimum for their value.

While there are many possible ways of interpreting this statement, by far the most common method is the *least-squares*. This simply states that *the sum of the squares of the errors* should be made as small as possible. That is, we want to find the straight line that will minimise the value of the expression $\Sigma(y - \hat{y})^2$, which is also called the *sum of the squares of the residuals*.

The method for finding this minimum requires calculus (or at least sophisticated algebra) and is beyond the scope of this book. However, it can be shown that the values of a and b that minimise this sum of the squares of the residuals are given by:

$$b = \frac{S_{xy}}{S_{xx}} \qquad \textbf{(S9.3)}$$

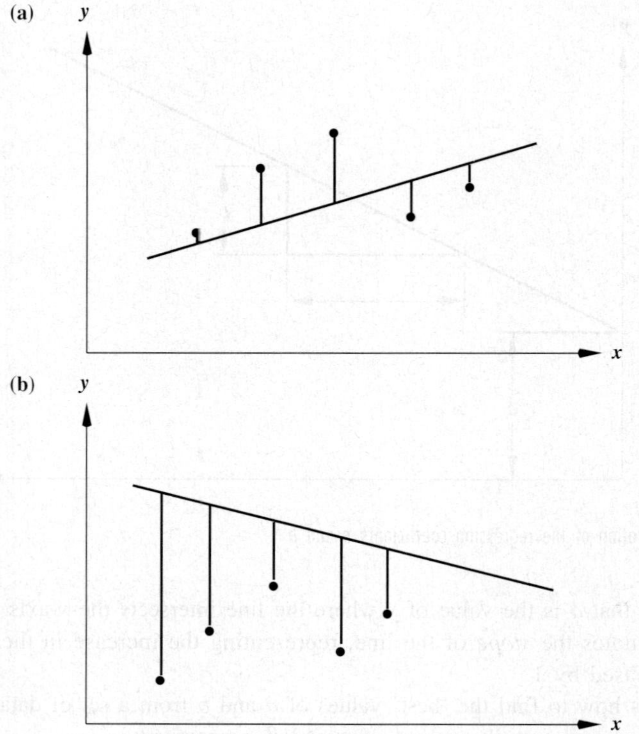

Figure S9.4 (a) The distances of the points from the line are small: the line is a good fit
(b) The distances from the points from the line are large: the line is a very poor fit

where $S_{xy} = \Sigma(x - \bar{x})(y - \bar{y})$

$S_{xx} = \Sigma(x - \bar{x})^2$

and $\bar{x}$ = mean of the x-values

$\bar{y}$ = mean of the y-values

Note that the values of S_{xy} and S_{xx} are the same as those found in Equations S8.4 and S8.2, respectively, when calculating the value of the product-moment correlation coefficient r.

Once the value of b is determined, the value of a can be found from:

$$a = \bar{y} - b\bar{x} \tag{S9.4}$$

Example S9.2

A car manufacturer believes that the quarterly repair bill (spare parts only) for a popular model of car is strongly influenced by the age of the car. To investigate this a random sample of 8 cars of varying ages is chosen and the repair bill for the previous 12 months recorded. The results are shown in Table S9.2.

Table S9.2 Age and repair bill data for Example S9.2	
Age (years)	Quarterly repair bill ($)
1	70
3	140
5	230
8	350
7	300
12	570
8	380
4	200

Fit the least-squares regression line of repair bill on age to the data.

Solution

In this case, the dependent (y) variable is repair bill and the independent variable (x) is age. The relevant calculations are shown in Table S9.3.

Table S9.3 Calculations for Example S9.2

x	y	$(x - \bar{x})$	$(y - \bar{y})$	$(x - \bar{x})^2$	$(x - \bar{x})(y - \bar{y})$
1	70	–5	–210	25	1050
3	140	–3	–140	9	420
5	230	–1	–50	1	50
8	350	2	70	4	140
7	300	1	20	1	20
12	570	6	290	36	1740
8	380	2	100	4	200
4	200	–2	–80	4	160
48	2240	0	0	84	3780

From Table S9.3, with $n = 8$, the following quantities can be found:

$$\bar{x} = \frac{\Sigma x}{n} = \frac{48}{8} = 6$$

$$\bar{y} = \frac{\Sigma y}{n} = \frac{2240}{8} = 280$$

$$S_{xy} = \Sigma(x - \bar{x})(y - \bar{y}) = 3780$$

$$S_{xx} = \Sigma(x - \bar{x})^2 = 84$$

From Equation S9.3:

$$b = \frac{S_{xy}}{S_{xx}} = \frac{3780}{84} = 45$$

From Equation S9.4:

$$\begin{aligned} a &= \bar{y} - b\bar{x} \\ &= 280 - (45)(6) \\ &= 10 \end{aligned}$$

Hence, from Equation S9.2 the least-squares regression line of y on x is:

$$\hat{y} = 10 + 45x$$

This may also be written as:

Quarterly repair bill (in \$) = 10 + 45 (age of car in years)

The scatter diagram and least-squares regression line are illustrated in Figure S9.5. Note that the calculated line is indeed a close fit to the data points.

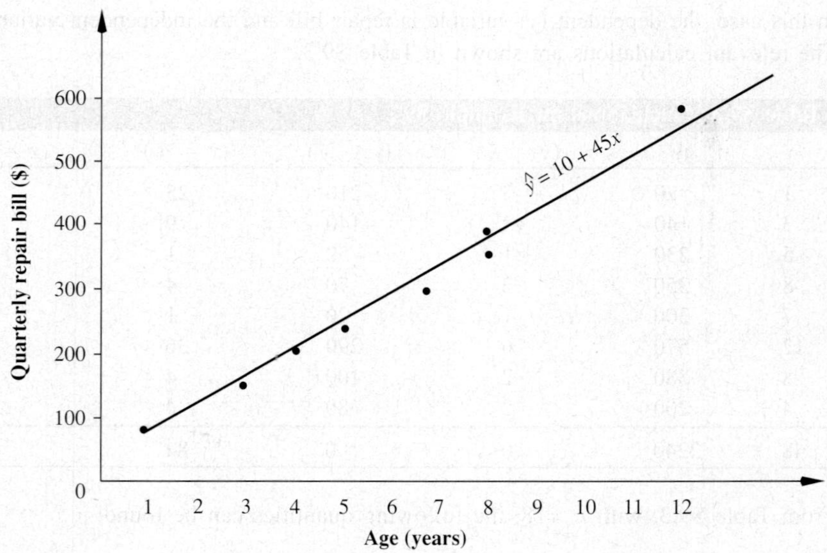

Figure S9.5 Scatter diagram and regression line for the data in Example S9.2

Example S9.3

Interpret the value of the slope of the regression line $b = 45$ in Example S9.2.

Solution

The value of b represents the increase in the value of y when x increases by 1. Since x represents the age of the car (in years) and y represents the value (in dollars), this means that, if the age of the car increases by 1 year, the quarterly repair bill for spare parts will increase by, on average, \$45.

Example S9.4

An accountant wishes to undertake a cost analysis of electricity consumption of office heating for various maximum daily temperatures. He recorded the data over a period of 8 random days, as shown below.

Day	1	2	3	4	5	6	7	8
Maximum temperature (°C)	26	31	25	26	14	20	16	34
Electricity consumption ('000 units)	35	20	37	24	42	41	40	17

Find the least-squares regression line of electricity consumption on maximum temperature.

Solution

It seems from the question that we are trying to predict electricity consumption from the maximum daily temperature. Accordingly, we let:

x = maximum daily temperature (°C)

y = electricity consumption ('000 units)

The relevant calculations are shown in Table S9.4. From the table, with $n = 8$, the following quantities can be found:

$$\bar{x} = \frac{\Sigma x}{n} = \frac{192}{8} = 24$$

$$\bar{y} = \frac{\Sigma y}{n} = \frac{256}{8} = 32$$

$$S_{xy} = \Sigma(x - \bar{x})(y - \bar{y}) = -439$$

$$S_{xx} = \Sigma(x - \bar{x})^2 = 338$$

Table S9.4 Calculations for Example S9.4

x	y	$(x - \bar{x})$	$(y - \bar{y})$	$(x - \bar{x})^2$	$(x - \bar{x})(y - \bar{y})$
26	35	2	3	4	6
31	20	7	–12	49	–84
25	37	1	5	1	5
26	24	2	–8	4	–16
14	42	–10	10	100	–100
20	41	–4	9	16	–36
16	40	–8	8	64	–64
34	17	10	–15	100	–150
192	256	0	0	338	–439

From Equation S9.3:

$$b = \frac{S_{xy}}{S_{xx}} = \frac{-439}{338} = -1.30$$

From Equation S9.4:

$$a = \bar{y} - b\bar{x}$$
$$= 32 - (-1.30)(24)$$
$$= 63.2$$

Hence, from Equation S9.2 the least-squares regression line of y on x is:

$$\hat{y} = 63.2 - 1.30x$$

In other words:

Electricity consumption (in '000 units) = 63.2 − 1.30 (maximum daily temperature in °C)

Example S9.5

Draw a scatter diagram of the data in Example S9.4, and draw the least-squares regression line on it.

Solution

The scatter diagram and least-squares regression line are shown in Figure S9.6. Note that the slope of the line is *negative* and that the line is not too bad a fit.

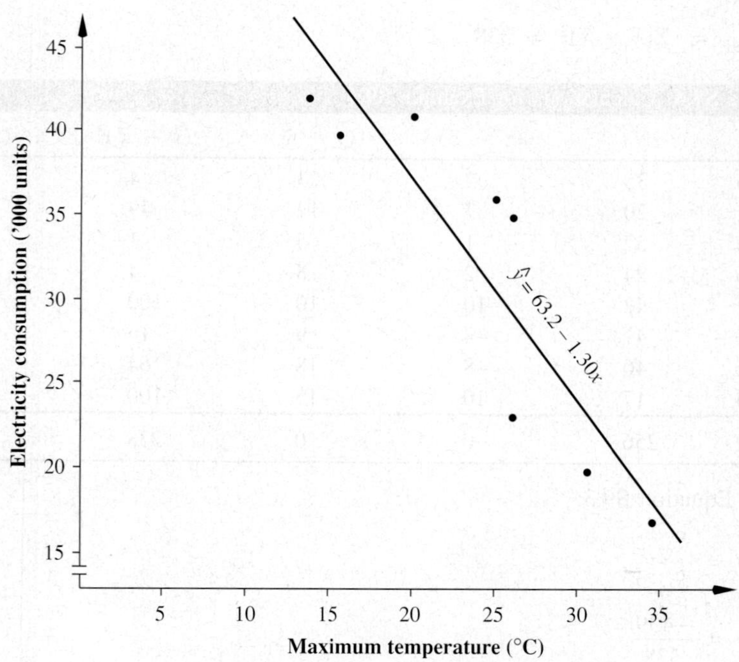

Figure S9.6 Scatter diagram and least-squares regression line for the data in Example S9.4

Example S9.6

Give an interpretation of the slope of the line found in Example S9.4.

Solution

The slope of the line is –1.30, that is a *negative* value. This means that for every *increase* in the daily maximum temperature there will be, on average, a *decrease* in the electricity consumption of office heating of $1.3 \times 1000 = 1300$ units.

Alternative formula for the slope of the least-squares regression line

A formula for the slope of the least-squares regression line b was given in Equation S9.3. However, there is a mathematically equivalent expression that may be more convenient if you happen to know the correlation coefficient between x and y and their respective standard deviations. This alternative expression is:

$$b = r\left(\frac{s_y}{s_x}\right) \qquad \textbf{(S9.5)}$$

where r = correlation coefficient between x and y
s_x = standard deviation of the x-variable
s_y = standard deviation of the y-variable

Example S9.7

Find the least-squares regression line of weight (y) on height (x) for the data in Example S8.3 (Table S8.1).

Solution

For these data we have already calculated, in Example S8.10, that:

$$\begin{aligned} n &= 12 \\ r &= 0.82 \\ \bar{x} &= 164.33 \text{ (to } \textit{two} \text{ decimal places)} \\ \bar{y} &= 66.57 \text{ (to } \textit{two} \text{ decimal places)} \\ S_{xx} &= 206.68 \\ S_{yy} &= 337.44 \end{aligned}$$

From Equation S8.2 we have $S_{xx} = (n - 1)\,s_x^2$. Hence:

$$\begin{aligned} s_x &= \sqrt{\frac{S_{xx}}{n-1}} \\ &= \sqrt{\frac{206.68}{11}} \\ &= 4.33 \end{aligned}$$

From Equation S8.3 we have $S_{yy} = (n - 1)s_y^2$. Hence:

$$s_y = \sqrt{\frac{S_{yy}}{n-1}}$$
$$= \sqrt{\frac{337.44}{11}}$$
$$= 5.54$$

Therefore, from Equation S9.5:

$$b = (0.82)\left(\frac{5.54}{4.33}\right)$$
$$= 1.05$$

From Equation S9.4:

$$a = 66.57 - (1.05)(164.33)$$
$$= -106.0 \text{ (to } one \text{ decimal place)}$$

Hence, from Equation S9.2 the least-squares regression line of weight on height is:

$$\hat{y} = -106.0 + 1.05x$$

or:

$$\text{Weight (in kg)} = -106.0 + 1.05 \text{ (height in cm)}$$

Note that exactly the same regression line would have been found had we used Equation S9.3 instead of Equation S9.5.

S9.3 GOODNESS OF FIT

How can we determine how well a regression line fits the data? Our criterion for determining the least-squares regression line was to choose the line that has the smallest sum of the squares of errors. However, to determine whether that line is a *good fit* we can use the **coefficient of determination**, denoted by r^2. This quantity is defined as:

$$r^2 = \frac{(S_{xy})^2}{S_{xx}S_{yy}} \qquad \textbf{(S9.6)}$$

In practice, from Equation S8.1 it can be seen that the coefficient of determination is simply the square of the correlation coefficient r. (Note that it is not unusual for the coefficient of determination to be written as R^2, that is using an upper case 'R'.) Since the value of r always lies between –1 and 1, the value of r^2 must always lie between 0 and 1.

A brief interpretation of the value of r^2 is as follows:

1. If the value of r^2 is close to 1, a straight line fits the data well.
2. If the value of r^2 is close to 0, a straight line fits the data poorly.

However, a valuable aspect of the coefficient of determination is that it provides information as to how much of the variation in the y-variable (dependent variable) is attributable to the x-variable (independent variable).

In particular, we know that variables will vary in value for a variety of reasons. For example, not all companies make the same profit, not all accounts are paid on time, and not all financial analysts are paid the same salary. Statisticians often want to know the *reason* for any variation, and how much of it is due to particular factors.

To this end we can use the value of r^2. To be precise, the value of $100r^2$ is the *percentage* of variation in the y-variable that may be attributed to the x-variable.

Almost all modern computer packages will automatically provide the value of r^2 whenever a regression analysis is undertaken.

Example S9.8

Find the coefficient of determination for the age and repair bill data in Example S9.2 (Table S9.2) and interpret its value.

Solution

We have already found in Example S9.2 that:

$$S_{xy} = 3780$$
$$S_{xx} = 84$$

and it can be shown that:

$$S_{xy} = 172\,000$$

Using Equation S9.6:

$$\begin{aligned} r^2 &= \frac{(S_{xy})^2}{S_{xx}S_{yy}} \\ &= \frac{(3780)^2}{(84)(172\,000)} \\ &= 0.989 \end{aligned}$$

Since r^2 is close to 1, the line appears to be a very good fit to the data. (This is also verified by an examination of Figure S9.5.)

An interpretation of r^2 is that 98.9% of the variation in the quarterly repair bill of a car is attributable to the age of the car.

Example S9.9

Find the coefficient of variation for the height and weight data in Example S8.3 (Table S8.1) and interpret its value.

Solution

In Example S8.10 we found that $r = 0.82$. Using Equation S9.6, the coefficient of determination is:

$$\begin{aligned} r^2 &= (0.82)^2 \\ &= 0.672 \end{aligned}$$

An interpretation of r^2 is that 67.2% of the variation in the weight of an adult female is due to her height. The other 32.8% of the variation in her weight is due to other factors.

Example S9.10

Find the coefficient of variation for the temperature and traffic accidents data in Example S8.11, and interpret its value.

Solution

In Example S8.11 we found that $r = 0.962$. Using Equation S9.6, the coefficient of determination is:

$$\begin{aligned} r^2 &= (0.962)^2 \\ &= 0.925 \end{aligned}$$

An interpretation of r^2 is that 92.5% of the variation in the number of traffic accidents is due to the maximum temperature. The other 7.5% of the variation in the number of traffic accidents is due to other factors.

S9.4 USING THE REGRESSION LINE FOR PREDICTION

Suppose that we have found the least-squares regression line of y on x and we wish to find the value of $\hat{y}$ (i.e. the predicted value of y) for a specific value of x.

If the regression line is a *poor* fit of the data (as we could see from an inspection of the line drawn on a scatter diagram or from a low value of r^2), the prediction will be of little use.

Even if the regression line is a *good* fit of the data, it is always dangerous to make a prediction of y for an x-value that was outside the limits (i.e. smallest and largest) of the x-values used in finding the equation of the line. (However, this does not stop people from making such predictions!) The reason for this is that we have no guarantee that the trend of points outside these limits will be linear, and indeed the equation may well provide ridiculous answers. (See Example S9.11.)

However, to make a simple 'point' prediction, just substitute the x-value in question into the regression equation.

Example S9.11

Consider the regression line of weight on height found in Example S9.7. Use it to predict the weight of a female who is:

(a) 122 cm tall
(b) 100 cm tall

Solution

The original weight and height data are given in Example S8.3 (Table S8.1). It can be seen that weights (y-values) ranged between 58.1 kg and 74.0 kg, and heights (x-values) between 156 cm and 169 cm.

(a) Because 122 cm is well outside the limits of the height data, we would be suspicious of making a prediction using this x-value. We have already found the regression equation to be:

$$\hat{y} = -106.0 + 1.05x$$

If we substitute $x = 122$ into this equation, we get:

$$\hat{y} = -106.0 + 1.05(122) = 22.1$$

The line therefore predicts that a female who is 122 cm tall would weigh only about 22 kg. Such a low weight would be extremely *unlikely*, even for someone with that low a height.

(b) The x-value of 100 cm is even further outside the limits of the height data. If we substitute $x = 100$ into the regression equation we get:

$$\hat{y} = -106.0 + 1.05(100) = -1.0$$

The line therefore predicts that a female who is 100 cm tall would have a negative weight! This demonstrates that the regression line can even give *impossible* predictions if the x-value is too far outside the limits of the original data.

Example S9.12

For the age and repair bill data in Example S9.2 (Table S9.2), use the least-squares regression line to estimate the annual repair bill on a 10-year-old car.

Solution

The least-squares regression line of y on x was found to be:

$$\hat{y} = 10 + 45x$$

If $x = 10$:

$$\hat{y} = 10 + 45(10)$$
$$= 460$$

That is, the predicted annual repair bill on a 10-year-old car is $460.

Example S9.13

For the temperature and electricity consumption data in Example S9.4, predict the electricity consumption of office heating when the maximum temperature is 30°C.

Solution

The least-squares regression line was:

Electricity consumption (in '000 units) = 63.2 – 1.30 (maximum daily temperature in °C)
If maximum daily temperature = 30°C:
Electricity consumption (in '000 units) = 63.2 – (1.30)(30)
= 24.2

That is, the regression line predicts that the electricity consumption will be 24 200 units.

S9.5 OTHER FORMS OF REGRESSION

It was mentioned in Section S9.1 that the methods used in least-squares regression are appropriate only if a *straight line* is the best fit. For some data there may be other non-linear relationships (e.g. quadratic), and these are more difficult to determine, particularly by hand. The calculations are usually very cumbersome and best left to a computer.

In some cases, it may be desired to find the relationship between a dependent variable y and *several* independent x-variables, say $x_1, x_2, x_3, \ldots x_k$. This process is known as *multiple regression*. This is a very common problem in practice, and it is essential that a computer perform the many tedious calculations required.

S9.6 REMARKS

Regression is one of the most used (and most misused) statistical techniques. While it can be extremely valuable, its difficulties should also be recognised. The following are some of the most severe limitations:

1. This chapter has considered the regression line of y on x. As stated previously, this equation *cannot* be used to predict a value of x from a value of y. For example, to predict the age of a car in Example S9.2 that had a quarterly repair bill of \$400, we could *not* use the equation $\hat{y} = 10 + 45x$. We could calculate another regression line in which repair bill is the x-variable and age is the y-variable, and use the equation of this line.
2. *Extrapolation*, that is, trying to make predictions outside the limits of x-values for which data have been obtained, is dangerous, since there is no guarantee that previous patterns will continue.
3. There is rarely any really substantial reason to believe that the true relationship is a straight line. You should always draw a scatter diagram and examine the coefficient of determination r^2 to measure the goodness of fit.
4. Least-squares criteria create unsatisfactory results when there are *outliers*, that is values that are relatively isolated from the main body of the data. In particular, if there is just one case that is extreme in its deviation from a straight line that closely fits the rest of the data, the least-squares result may be the sort of compromise line that is of little use.

Quick Quiz

1. Consider the following least-squares regression equation of y cn x:

$\hat{y} = 125.80 - 13.05x$

Find the value that the line predicts for y when $x = 6.40$.

2. Suppose that you are given the following information:

(i) the *slope* of the least-squares regression line of y on x is negative

and

(ii) the value of r^2 is 0.5476

Find the value of the correlation coefficient r between x and y.

3. Consider the following pairs of values for two variables x and y:

x	1	2	3	4	5	6
y	7	7	7	7	7	7

Find the least-squares regression line of y on x.

4. A tax agent was trying tc find a relationship between the gross income (in \$'000) of her clients and the fee (in dollars) that she charged them to do their tax returns. She found that when she did a least-squares regression line of gross income on fee charged the slope of the line was 75.0. Give an interpretation of what the figure of 75.0 means.

5. Suppose for a given set of data we have the following information for two variables x and y:

$$n = 20$$
$$\Sigma x = 280$$
$$\Sigma y = 160$$
$$\Sigma(x - \bar{x})(y - \bar{y}) = -225$$
$$\Sigma(x - \bar{x})^2 = 450$$

Find the least-squares regression line of y on x.

6. In Question 5, suppose you are now given the additional information that $\Sigma(y - \bar{y})^2 = 305$. Find the value of r^2 and state, with reasons, whether you feel that the least-squares regression line is a good fit of the data.

7. Consider a least-squares regression line of y on x. If $x = 5$, the line predicts $y = 18.4$; and if $x = 20$, the line predicts $y = 12.1$. What is the equation of the line?

8. Does a non-significant value of the correlation coefficient r between two variables x and y indicate that a linear regression line would provide a poor prediction of y from x? Discuss.

9. Consider two variables x cnd y that have standard deviations of 5.48 and 10.32, respectively. If the correlation coefficient r between the variables is -0.356, find the slope cf the least-squares regression line of y on x.

10. The *attendance* at Australian Rules matches was regressed on *ticket prices* for 16 teams. The value of the coefficient of determination r^2 was found to be 0.1764. Give an interpretation of this value.

Multiple Choice Questions

1. A high value of r^2 between two variables x and y indicates:
(a) x and y are uncorrelated
(b) x and y must have a positive correlation
(c) x and y must have a negative correlation
(d) the variation in y is largely attributable to x
(e) none of the above

2. If the regression line of y on x is horizontal, this means:
(a) x and y are correlated
(b) as x increases then y increases
(c) the line has a positive slope
(d) x is useful in predicting y
(e) none of the above

3. An accountant records the size of the mortgage of customers at a bank and the age of the borrower. The results for 6 borrowers were:

Mortgage ($'000)	110	92	123	100	105	96
Age (years)	36	38	28	41	36	37

The *slope* (to one decimal place) of the least-squares regression line of *mortgage* on *age* is:
(a) 2.2 (b) −2.2 (c) 1.5
(d) −1.6 (e) none of the above

4. In Question 3, the value of the *constant* term (to one decimal place) in the least-squares regression line of mortgage on age is:
(a) 130.6 (b) 155.4 (c) 117.0
(d) 182.1 (e) none of the above

5. In Question 3, suppose a 30-year-old borrower has a mortgage. The least-squares regression line will predict a mortgage (to the nearest $1000) of:
(a) $117 000 (b) $125 000 (c) $127 000
(d) $121 000 (e) none of the above

6. In Question 3, the percentage coefficient of determination (to the nearest whole number) is:
(a) 66% (b) 71% (c) 53%
(d) 84% (e) none of the above

7. In a typing test, the regression line of the *number of typing errors made per page* on the *number of words typed per minute* was found to be:

Number of errors per page = −3.2 + 1.4 (number of words typed per minute)

For every extra 10 words typed each minute, the average increase in errors per page is:
(a) 3.2 (b) −3.2 (c) 1.4
(d) 1.96 (e) none of the above

8. In Question 7, suppose that a typist simply types nothing. The line predicts that the number of errors per page would be:
(a) 3.2 (b) −3.2 (c) 0
(d) 1.4 (e) none of the above

9. Suppose that a regression line of y on x is *exactly* horizontal. This means that:
(a) y can be exactly predicted from x
(b) x can be exactly predicted from y
(c) x and y are perfectly correlated
(d) x and y probably have a non-linear relationship
(e) none of the above

10. The fuel consumption (L) of a certain vehicle and its average speed (km/hour) over 100 kilometres is given by the following equation:
Fuel consumption = 6.2 + 0.025 (average speed)
If the vehicle has an average speed of 150 km/hr, the line predicts that the fuel consumption will be:
(a) 6.2 L (b) 6.575 L (c) 3.0 L
(d) 3.575 L (e) none of the above

Exercises

S9.1* An insurance manager believes that the number of new policies written annually by his agents is related to the number of years of selling experience that these agents have. A random sample of 12 agents revealed the following data:

Number of years of experience	Number of new policies written annually
2	17
5	22
7	34
6	37
12	50
9	41
5	13
20	48
13	39
4	20
10	35
20	63

(a) Find the least-squares regression line of number of new policies on number of years of experience.

(b) Find the coefficient of determination for the line in (a) and comment.

(c) Estimate the number of new policies written annually by an agent with 15 years of experience.

S9.2* What can be said about the slope of a least-squares regression line if the two variables involved have a correlation coefficient of $r = 0$?

S9.3 In Exercise S9.1, can we reliably estimate the number of policies written annually by an agent with 1 year of experience? Explain.

S9.4* Interpret the slope of the least-squares regression line found in Exercise S9.1.

S9.5* Under what condition do you think that the least-squares regression line of y on x would have exactly the same equation as the least-squares regression line of x on y?

S9.6 The expenditure on child-care facilities in the previous year by a random sample of 6 local councils, and the number of children under age 5 living in the electorates, are shown below.

Council	1	2	3	4	5	6
Expenditure ($'000)	125	180	154	90	102	63
Number of children	1723	2510	1856	1525	1624	920

(a) Draw a scatter diagram of the data.

(b) Find the least-squares regression line of expenditure on number of children.

(c) Draw the line found in (b) on the scatter diagram in (a). Comment on whether you feel that the line is a good fit of the data.

(d) Use the line found in (b) to predict the expenditure of a local council that has 1250 children under age 5 living in the electorate.

S9.7* For the data in Exercise S9.6:

(a) Find the value of the coefficient of determination r^2.

(b) Interpret the meaning of this value of r^2 in terms of the problem.

S9.8* The number of freight tonnes (in millions) carried on government railways in Australia between 1965 and 1990 is shown at 5-year intervals in the table below.

Year	1965	1970	1975	1980	1985	1990
Number of freight tonnes (millions)	66	84	104	126	160	186

Source: Australian Bureau of Statistics, *Year Book Australia 1992*, p. 742

(a) Draw a scatter diagram of the data.

(b) Find the least-squares regression line of number of freight tonnes on year. (Use only the last two digits of the year.)

(c) Draw the line in (b) on the scatter diagram in (a). Comment on whether you feel that the line is a good fit of the data.

(d) Use your line found in (b) to predict the number of freight tonnes carried on government railways in Australia in 1995.

(e) You are now told that the *actual* figure for 1995 was 215.0 million tonnes. What percentage error did you make in (d)?

S9.9 In Exercise S9.8, suppose you are now told that the *actual* number of freight tonnes carried on government railways in Australia in 1983 was 124 million.

(a) What does your regression line found in Exercise S9.8(b) predict for the year 1983?

(b) What percentage error did your prediction in (a) make? Comment.

S9.10 The number of short-term arrivals ('000) from overseas to Australia for six months of the year 2000 are shown in the table below.

Month	March	April	May	June	July	August
Number of arrivals	411.9	405.6	319.3	347.9	436.8	368.3

Source: Australian Bureau of Statistics, *Overseas Arrivals and Departures September 2000*, Cat. No. 3401.0

(a) Plot the data on a scatter diagram. Do you see any linear trend?

(b) Find the least-squares regression line of arrivals on month where the months are labelled as 1, 2, 3, 4, 5 and 6. (*Hint*: The *slope* of the line is -2.74.)

(c) Use the line in (b) to estimate the number of short-term arrivals to Australia for September 2000. In view of your diagram in (a), do you feel that this answer will be fairly accurate? Comment.

S9.11 (a) You are now told that the number of short-term arrivals from overseas to Australia for September 2000 was actually 406 500. Find the percentage error that you made using your estimate in Exercise S9.10(c).

(b) From the table in Exercise S9.10, find the mean number of arrivals per month during the six-month period. Is this figure closer to the actual number of 406 500 than the estimate in Exercise S9.10(c)?

(c) What do you feel the problem might be in using linear regression to make predictions on data such as these? Was there any event in September 2000 that may have affected the figures for that month?

S9.12 Find the value of r^2 for the electricity consumption data in Example S9.4 and give an interpretation of this value.

S9.13* (a) Find six different pairs of data such that the value of r^2 is exactly 1.

(b) Plot your points in (a) on a scatter diagram. What characteristic does it have?

S9.14* Answer the following questions based on your results in Exercise 8.1:

(a) Find the least-squares regression line of y on x.

(b) Find the coefficient of determination and interpret it.

(c) If an adult male drinks an average of 12 cans of beer per week, what would you predict for the average number of cigarettes that he smokes per day?

S9.15* An adult male smokes an average of 5 cigarettes per week. Could you use the regression line in Exercise S9.14(a) to predict the average number of cans of beer he drinks per week? If so, make this prediction. If not, state why.

S9.16 A bank manager is interested in the relationship between the number of housing loan applications and the prevailing credit card interest rate. She took a random sample of 8 months from the previous 3 years with the following results:

Number of loan applications	52	40	32	64	21	28	43	24
Interest rate (%)	12.3	13.7	13.5	12.0	14.8	14.0	13.3	14.4

(a) Find the least-squares regression line that allows for prediction of loan applications from interest rate.

(b) Draw a scatter diagram of the data and draw the line found in (a).

(c) What does the slope of the line signify in economic terms?

(d) Is the line a good fit to the data? Why or why not?

S9.17 The bank manager in Exercise S9.16 anticipates that the interest rate on credit cards is going to drop to 11.5%. Is she justified in using the regression line found in Exercise S9.16(a) to predict what the number of housing loan applications might be after the drop? Give reasons.

S9.18* Consider the following set of paired data for the variables x and y:

x	1	2	3	4	5	7	8	9
y	1	4	9	16	25	49	64	81

(a) Plot the data on a scatter diagram.

(b) Find the least-squares regression line of y on x.

(c) Draw the line found in (b) on the diagram in (a).

(d) Use the line found in (b) to predict the value of y for an x-value of 6. Can you make a better prediction *without* using this line? Why or why not?

S9.19 The police department is interested in determining the effect of alcohol on driving ability. Twelve male drivers of similar weight, drinking habits and driving experience were randomly assigned to three groups of equal size. The first group was asked to consume 2 cans of beer within 20 minutes, and the second 4 cans of beer in

20 minutes. The third did not consume any beer. Each of the 12 then undertook a driving test under similar conditions, and their individual scores were recorded (the higher the score, the better). The results were:

Number of cans of beer consumed	Points scored
0	78
0	82
0	75
0	67
2	85
2	42
2	50
2	55
4	27
4	48
4	39
4	42

(a) Draw a scatter diagram of the data.

(b) Find the least-squares regression line of points scored on beer consumption and draw the line on the diagram in (a).

S9.20 Use the least-squares regression line found in Exercise S9.19(b) to predict the points scored for males of similar background if the number of cans of beer consumed is:

(a) 0.5

(b) 2.5

(c) Which of the two predictions found in (a) and (b) would you expect to be the more accurate? Why?

S9.21* A leading garden nursery advertises that it will deliver 'high-grade topsoil to your front door'. The nursery manager has devised the following price list for the amount delivered:

Amount delivered (tonnes)	1	2	3	4	5	6	7	8
Price per tonne ($)	26	22	20	19	18	15	21	27

(a) Plot the data on a scatter diagram.

(b) Find the least-squares regression line of price per tonne on amount delivered.

(c) Draw this line on the scatter diagram in (a).

(d) Find the coefficient of determination for the line found in (b). Comment.

S9.22* Explain why the sign of the product-moment correlation coefficient between two variables x and y must always be the same as the sign of the slope of the least-squares regression line of y on x.

S9.23 Consider the equation $y = 12 - x$, which relates two variables x and y.

(a) Find the corresponding values of y for each value of

$x = 1, 2, 3, 4, 5.$

(b) Plot the 5 ordered pairs in (a) on a scatter diagram.

(c) Find the value of the product-moment correlation coefficient r between x and y.

(d) Find the least-squares regression line of y on x using the formulae in this chapter. What do you notice?

S9.24* Consider the equation $y = 6 - \frac{x}{2}$, which relates two variables x and y.

(a) Find the corresponding values of y for each of $x = 2, 4, 6, 8, 10$.

(b) Plot the 5 ordered pairs in (a) on a scatter diagram.

(c) Find the value of the product-moment correlation coefficient r between x and y.

(d) Find the least-squares regression line of y on x using the ordered pairs in (a) and the formulae in this chapter. What do you notice?

S9.25 What relationship do you notice between the equations in Exercises S9.23 and S9.24?

S9.26 A farmer records the amount of rainfall (in mm) each year on his farm and the average yield of corn (in bushels/ha) for that year over a 6-year period. The results are as follows:

Year	1	2	3	4	5	6
Rain (mm)	350	180	150	400	580	140
Average yield (bushels/ha)	240	100	90	280	60	80

(a) Draw a scatter diagram of the data with rain on the horizontal axis.

(b) Identify any unusual observations from the scatter diagram in (a).

(c) Find the least-squares regression line of average yield on rain.

(d) Draw the line found in (b) on the scatter diagram in (a). Comment on whether you feel that it is a good fit.

S9.27 (a) In Exercise S9.26, remove any clearly unusual observation(s). Recalculate the least-squares regression line of average yield on rain.

(b) Draw the new line found in (a) on the scatter diagram with the unusual point(s) removed. Comment on whether you feel that it is a significantly better fit than previously.

S9.28 Use the least-squares regression line in Exercise S9.27(a) to predict the average yield of corn per hectare if the annual rainfall is:

(a) 300 mm (b) 190 mm (c) 360 mm

S9.29* A rugby union team believes that there is a relationship between the rank of where it finishes on the competition ladder and its average crowd per match. The results for the previous 6 seasons were:

Rank	5	1	2	3	6	4
Average crowd	2100	3000	3700	2800	1500	2500

(a) Draw a scatter diagram of the data. (Put rank on the horizontal axis.)

(b) Find the least-squares regression line of average crowd on rank.

(c) Draw the line found in (b) on the scatter diagram in (a). Do you feel that it is a good fit? Why or why not?

S9.30 (a) Use the least-squares regression line in Exercise S9.29 to predict the average crowd per match if the rugby union team were to finish first on the competition ladder.

(b) Suppose that there were 20 teams in the competition in Exercise S9.29. What does the regression line found in Exercise S9.29(b) predict that the average crowd per match might be if the team were to finish last? What does this result illustrate?

S9.31 The number of new motor vehicle registrations in Australia for the 7 months July 2000 to January 2001 are shown in the table below.

Month	Number of new vehicle registrations
July 2000	78 986
August 2000	76 368
September 2000	63 986
October 2000	66 263
November 2000	71 388
December 2000	72 740
January 2001	55 143

Source: Australian Bureau of Statistics, *New Motor Vehicle Registrations*, Cat. No. 9301.0

(a) Plot the data on a scatter diagram and join the points. Do you see any linear trend?

(b) Find the least-squares regression line of *number of new vehicle registrations* on *month* where the months are labelled as 1, 2, 3, 4, 5, 6, 7. (*Hint*: The *slope* of the line is −2549 and $r^2 = 0.457$.)

(c) Use the line in (b) to estimate the number of new vehicle registrations in Australia for February 2001. In view of your diagram in (a), do you feel that this answer will be fairly accurate? Comment.

S9.32 (a) You are now told that the number of new vehicle registrations in Australia for February 2001 was actually 58 626. Find the percentage error that you made using your estimate in Exercise S9.31(c).

(b) From the table in Exercise S9.31, find the mean number of new registrations per month during the 7-month period. Is this figure closer to the actual number of 60 789 than the estimate in Exercise S9.31(c)? Comment.

S9.33 The exchange rate between the Australian dollar (AUS$) and the Canadian dollar (C$) on the first of the month is shown in the following table for 6 months during the year 2000. The data show how many cents Canadian that AUS$1 would purchase.

Month	June	July	August	September	October	November
Canadian cents	86.0	88.5	87.0	85.0	82.0	80.4

Source: *Australian Financial Review*, June–November 2000

(a) Plot the data on a scatter diagram and join the points. (Put month on the horizontal axis.) Do you see any linear trend?

(b) Find the least-squares regression line of *Canadian cents* on *month* where the months are labelled as 1, 2, 3, 4, 5, 6. (*Hint*: The *slope* of the line is −1.414.)

(c) Draw the line found in (b) on the scatter diagram in (a). Do you think it is a good fit?

S9.34 (a) An office manager is planning a trip to Canada early in the year 2001 and wants to use the line in Exercise S9.33 to predict the number of Canadian cents that AUS$1 will buy at the beginning of:

(i) January 2001 (ii) April 2001

(b) The value of AUS\$1 on 1 January 2001 was, in fact, 83.4 cents Canadian. Find the absolute and percentage errors made by the prediction in (a)(i). Comment.

(c) The value of AUS\$1 on 1 April 2001 was, in fact, 78.5 cents Canadian. Find the absolute and percentage errors made by the prediction in (a)(ii). Comment.

S9.35 Using the regression line in Exercise S9.33(b), at the beginning of which month does it predict that one Australian dollar will be worth zero Canadian cents? Do you trust this prediction? Explain.

S9.36* The number of enrolments in economics at high schools and universities in New South Wales was recorded in each of the years from 1991 to 1996. The results are shown below.

Year	1991	1992	1993	1994	1995	1996
Enrolments	4650	4710	4730	4400	4180	4040

Source: *The Australian*, 2 October 1996, p. 28

(a) Draw a scatter diagram of the data and join the points. (Put year on the horizontal axis.)

(b) Find the least-squares regression line of enrolments on year. (Label the years 1, 2, 3, 4, 5, 6.)

(c) Draw the line found in (b) on the scatter diagram in (a). Do you feel that it is a good fit? Why or why not?

S9.37 The data in Exercise S9.36 are of great concern to an economics professor in New South Wales who has been informed that the university will close down the economics department altogether if the total enrolments drop below 3000. If she is due to retire during 2006, does the regression line found in (b) predict that the department will be closed down before she goes? Do you have any confidence in this prediction? Explain.

S9.38 On 9 August 1994, the residents of Sydney shivered as they created a New South Wales weekday morning record of 9004 Mw of electricity consumption. The hourly temperatures (°C) between 1 a.m. and 10 a.m. are shown below.

Time (a.m.)	1	2	3	4	5	6	7	8	9	10
Temperature (°C)	7.2	6.8	6.2	5.8	5.5	5.3	5.3	6.4	8.3	10.3

Source: Bureau of Meteorology

(a) Draw a scatter diagram of the data. (Put time on the horizontal axis and use a scale for temperature between 5.0°C and 10.0°C on the vertical axis.) Join successive points on the diagram.

(b) How would you describe the shape of the graph drawn in (a)? Do you feel that linear regression would have been of much help in using these data to predict the temperature at 11 a.m.? Explain.

S9.39 Suppose that a least-squares regression line of temperature on time was found for the data in Exercise S9.38. If it had been used to predict the temperature at 2 p.m., do you think that it would have *underestimated* or *overestimated* the actual value? Explain. (You do *not* have to find the equation of the line to answer this question.)

S9.40 (a) Suppose that you were to track the world record for the men's 200-m freestyle swimming as at 1 January each year over a period of 20 years. Do you think that linear regression would be a good method of predicting what the world record in this event might be 20 years into the future? Explain.

(b) What sort of downward trend would you expect the world records in (a) to have? Explain.

Chapter S10

Index numbers

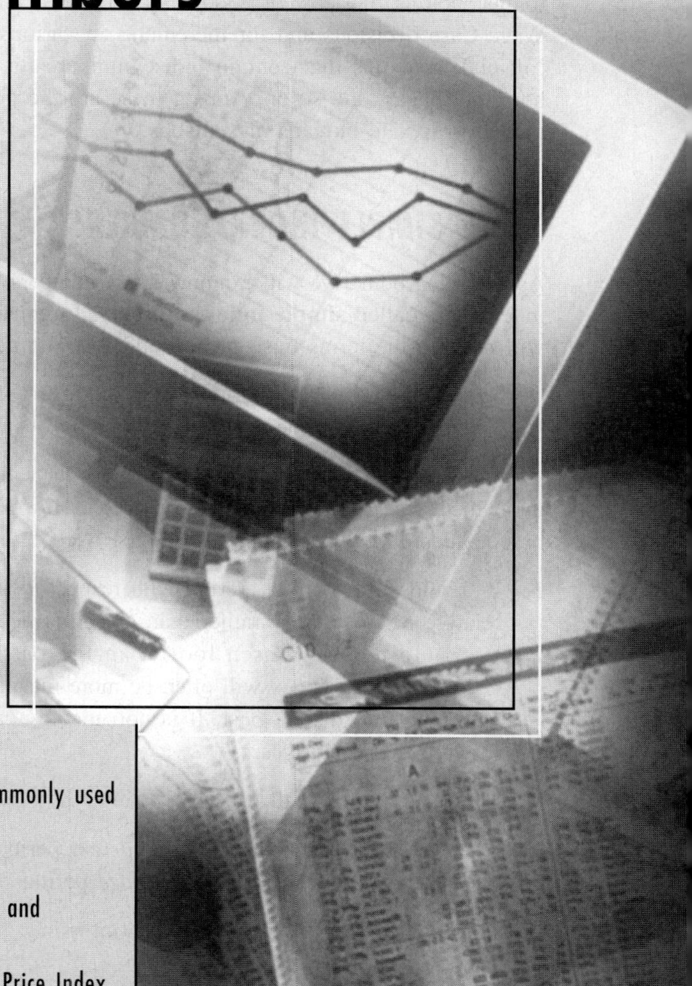

'Every time somebody's autobiography comes out I turn to the index to see if my name occurs. Of course it never does.'

(James Agate, British drama critic and novelist, diary entry, 16 September 1932)

Learning objectives

In this chapter you will learn to:

- Interpret and use a range of index numbers commonly used in the Australian business sector
- Define an index number and explain its use
- Perform calculations involving simple, composite and weighted index numbers
- Understand the basic structure of the Consumer Price Index (CPI) and perform calculations involving its use
- Understand other indexes used in the Australian business sector

S10.1 INTRODUCTION

Essentially, index numbers is the first topic that deals with data analysed over time, the second being time series, which is discussed in Chapter S11. In this chapter, the basic types of index numbers used in business and economic problems are examined in detail.

An **index number** is a statistical value that measures the change in a variable with respect to time. Two variables that are often considered in this analysis are *price* and *quantity*. With the aid of index numbers, the average price of several articles in one year may be compared with the average price of the same quantity of the same articles in a number of different years.

Although index numbers may not be able to provide a precise estimate of how variables change over time, they nevertheless can be utilised (with reservations) as some measure of the change in things such as 'the cost of goods and services'. They have also been used in Australia in National Wage Case decisions and for increments in various types of pensions. There are several sources of 'official' statistics that contain index numbers for quantities such as food prices, clothing prices, housing, wages, and so on. Among these are the publications of the Australian Bureau of Statistics and several state bureaus of statistics.

S10.2 SIMPLE INDEX NUMBERS

As a starting point, we will examine index numbers that are constructed from a *single item only*. Such indexes are called **simple index numbers**. In considering all the index numbers in this chapter we will refer to the terms **base period** and **current period**. Formally they are defined as:

Current period = the period for which you wish to find the index number
Base period = the period with which you wish to compare prices in the current period

The current period does not necessarily have to be the period of time actually occurring at the moment. It can, in fact, be any period at all. However, it is the choice of the *base period* that should be considered very carefully. The choice itself often depends on economic factors.

1. It should be a 'normal' period with respect to the relevant index. That is, it should be a time of relative price stability rather than one in which there were marked price fluctuations.
2. It should not be chosen too far in the past. Because of the rapid changes in economic development, an index will often be more meaningful if it compares current conditions with those in the not too distant past. It is not uncommon for the base period for an index to be moved forward every few years.

The notation we shall use is:

p_n = the price of an item in the *current* period
p_o = the price of an item in the *base* period

You may like to think of the subscript n in p_n as standing for *new* and the subscript o in p_o as standing for *old*.

Price relative

The **price relative** of an item is the ratio of the price of the item in the current period to the price of the same item in the base period. The formal definition is:

$$\text{Price relative} = \frac{p_n}{p_o} \qquad \textbf{(S10.1)}$$

Example S10.1

Table S10.1 shows the average retail price of one dozen 55 g eggs in Hobart for the September quarter at 5-year intervals between 1985 and 2000.

Table S10.1 Average retail price of one dozen 55 g eggs in Hobart

Year	1985	1990	1995	2000
Price ($)	1.74	2.05	2.68	2.64

***Source*:** Australian Bureau of Statistics, *Average Retail Prices of Selected Items*, September Quarters 1985, 1990, 1995 and 2000

(a) Find the price relative of eggs in Hobart in 2000 using 1985 as the base year.
(b) Find the price relative of eggs in Hobart in 2000 using 1990 as the base year.
(c) Find the price relative of eggs in Hobart in 2000 using 1995 as the base year.

Solution

In each case, p_n = price in 2000 = $2.64.

(a) p_o = price in 1985 = $1.74.
From Equation S10.1:

$$\text{Price relative} = \frac{p_n}{p_o} = \frac{\$2.64}{\$1.74} = 1.517$$

Since the price relative is about 1.5, this tells us that the price of eggs in 2000 was about 50% more expensive than the price in 1985.

(b) p_o = price in 1990 = $2.05.
From Equation S10.1:

$$\text{Price relative} = \frac{p_n}{p_o} = \frac{\$2.64}{\$2.05} = 1.288$$

Since the price relative is about 1.3, this tells us that the price of eggs in 2000 was about 30% more expensive than the price in 1990.

(c) p_o = price in 1995 = $2.68.
From Equation S10.1:

$$\text{Price relative} = \frac{p_n}{p_o} = \frac{\$2.64}{\$2.68} = 0.985$$

Since the price relative is 0.015 *under* 1, this tells us that the price of eggs in 2000 was 1.5% *less* expensive than the price in 1995.

Simple price index

The price relative provides a ratio that indicates the change in price of an item from one period to another. A more common method of expressing this change is to use a **simple price index**, which is just the price relative expressed as a percentage. The formal definition is:

$$\text{Simple price index} = \text{price relative} \times 100$$

$$= \frac{p_n}{p_o} \times 100 \qquad \textbf{(S10.2)}$$

The simple price index allows us to find the *percentage change* in the price of an item from one period to another.

1. If the simple price index is *more than* 100, subtract 100 from the simple price index. The result is the percentage *increase* in price from the base period to the current period.
2. If the simple price index is *less than* 100, subtract the simple price index from 100. The result is the percentage by which the item cost *less* in the base period than it does in the current period.

Example S10.2

1. A simple price index of 107.3 means that there was a 7.3% *increase* in price between the base period and the current period.
2. A simple price index of 86.4 means that there was a 13.6% *decrease* in price between the base period and the current period.

Example S10.3

Table S10.2 shows the average retail prices of 500 g of butter, 500 g of frozen peas and a 150 g can of instant coffee in Sydney for the September quarter at 5-year intervals between 1980 and 2000.

Table S10.2 Average retail price of butter, frozen peas and instant coffee in Sydney

		Price of one unit ($)				
Item	Unit	1980	1985	1990	1995	2000
Butter	500 g	1.09	1.58	1.88	1.60	1.89
Frozen peas	500 g	0.55	0.82	1.30	1.26	1.34
Instant coffee	150 g	2.95	4.21	4.38	5.80	6.24

Source: Australian Bureau of Statistics, *Average Retail Prices of Selected Items*, September Quarters 1980, 1985, 1990, 1995 and 2000

(a) Find the simple price index for each item for all years relative to the base year 1980.

(b) Use your answers in (a) to find the simple price index of each item in 2000 relative to a base year of 1990.

Solution

(a) The simple price indexes relative to the base year 1980 can be found using Equation S10.2. For example:

- The simple price index of frozen peas in 1995 is $\frac{1.26}{0.55} \times 100 = 229.1$
- The simple price index of instant coffee in 1990 is $\frac{4.38}{2.95} \times 100 = 148.5$

Note that the simple price index for each item in the base year (1980) will always be 100. The complete set of simple price indexes is shown in Table S10.3.

Table S10.3 Simple price indexes for Example S10.3

		Simple price index				
Item	Unit	1980	1985	1990	1995	2000
Butter	500 g	100.0	145.0	172.5	146.8	173.4
Frozen peas	500 g	100.0	149.1	236.4	229.1	243.6
Instant coffee	125 g	100.0	142.7	148.5	196.6	211.5

From Table S10.3, for example, the simple price index is 173.4 for butter in 2000. This means that butter was just over 1.7 times more expensive in 2000 than it was in 1980. In other words, it has increased in price by 73.4%.

(b) We can use Table S10.3 to find the simple price indexes for 2000 relative to a base year of 1990 by finding:

$$\frac{\text{Simple price index in 2000}}{\text{Simple price index in 1990}} \times 100$$

In this case we have:

$$\text{Butter} = \frac{173.4}{172.5} \times 100 = 100.5$$

$$\text{Frozen peas} = \frac{243.6}{236.4} \times 100 = 103.0$$

$$\text{Instant coffee} = \frac{211.5}{148.5} \times 100 = 142.4$$

Note that these results could also be obtained using Equation S10.2.

Example S10.4

In Example S10.3(b), the simple price index for frozen peas in 2000 relative to a base year of 1990 is 103.0. Interpret its value.

Solution

Since the value of 103.0 is *above* 100, we *subtract* 100 from its value to obtain 3.0. This means that the price of frozen peas was 3.0% *more* expensive in 2000 than it was in 1990.

Example S10.5

Use the figures in Table S10.3 to find the simple price index for frozen peas in 1995 relative to a base year of 1990. Interpret its value.

Solution

From Table S10.3, the simple price index for frozen peas in 1995 relative to a base year of 1990 is:

$$\frac{\text{Simple price index of frozen peas in 1995}}{\text{Simple price index of frozen peas in 1990}} \times 100 = \frac{229.1}{236.4} \times 100$$
$$= 96.9$$

Since the value of 96.9 is *below* 100, we *subtract* its value from 100 to obtain 3.1. This means that the price of frozen peas *dropped* by 3.1% between 1990 and 1995.

S10.3 COMPOSITE INDEX NUMBERS

A **composite index number** is constructed from changes in a *number* of different items. Many such indexes can be made, although some of these have severe drawbacks. In this section we discuss some of the more commonly used composite indexes and their strengths and weaknesses.

Simple aggregate index

The first composite index considered is the **simple aggregate index**. This has appeal because its nature is simplistic and it is easy to find. The formal definition is:

$$\text{Simple aggregate index} = \frac{\text{sum of prices in the current period}}{\text{sum of prices in the base period}} \times 100$$
$$= \frac{\Sigma p_n}{\Sigma p_o} \times 100 \qquad \textbf{(S10.3)}$$

where Σp_n = the sum of the prices in the *current* period
Σp_o = the sum of the prices in the *base* period

Basically, this index is found by totalling the price of a basket of goods in the current year, dividing it by the price of the same basket of goods in the base year, and multiplying the answer by 100. The interpretation of its value is the same as that shown previously for a simple index number.

Even though the simple aggregate index is easy to calculate, it has serious disadvantages:

1. An item with a relatively large price can dominate the index.
2. If prices are quoted for different quantities, the simple aggregate index will yield a different answer.
3. It does not take into account the quantity of each item sold.

Disadvantage 2 is perhaps the worst feature of this index, since it makes it possible, to a certain extent, to manipulate the value of the index. One possibility is to agree on the same units to be used year after year, but the index still does not take into account other factors such as the *quantity* of each item used. Consequently, the simple aggregate index is not a recommended or commonly used index, but represents a useful starting point for more advanced indexes.

Example S10.6

The average retail prices in Melbourne for 5 items in the September quarters of 1990, 1995 and 2000 are shown in Table S10.4.

Table S10.4 Average retail prices in Melbourne for 5 items in 1990, 1995 and 2000

		Price of one unit ($)		
Item	Unit	1990	1995	2000
Frozen chicken	1 kg	2.89	2.92	3.53
Tomato sauce	600 mL	1.69	1.83	1.78
Pink salmon (can)	210 g	2.75	2.26	2.60
Chocolate (milk)	250 g	1.98	2.44	3.06
Pineapple (can, sliced)	450 g	0.83	0.90	1.17

Source: Australian Bureau of Statistics, *Average Retail Prices of Selected Items*, September Quarters 1990, 1995 and 2000

Use Table S10.4 to calculate the simple aggregate index for 2000 using:

(a) 1990 as the base year

(b) 1995 as the base year

Solution

(a) If 1990 is the base year:

$$\Sigma p_o = 2.89 + 1.69 + 2.75 + 1.98 + 0.83 = \$10.14$$
$$\Sigma p_n = 3.53 + 1.78 + 2.60 + 3.06 + 1.17 = \$12.14$$

Hence, the total cost of the 5 items in 1990 is \$10.14 and in 2000 it is \$12.14. From Equation S10.3:

$$\text{Simple aggregate index} = \frac{\$12.14}{\$10.14} \times 100 = 119.7$$

(b) If 1995 is the base year:

$$\Sigma p_o = 2.92 + 1.83 + 2.26 + 2.44 + 0.90 = \$10.35$$
$$\Sigma p_n = \$12.14, \text{ as in (a)}$$

Hence, the total cost of the 5 items in 1995 is \$10.35. From Equation S10.3:

$$\text{Simple aggregate index} = \frac{\$12.14}{\$10.35} \times 100 = 117.3$$

Example S10.7

Interpret the values of the two simple aggregate indexes found in Example S10.6(a) and (b).

Solution

(a) The simple aggregate index was 119.7. Since this is larger than 100, we subtract 100 to obtain 19.7. This means that the price of this group of 5 items increased overall in price by 19.7% between 1990 and 2000.

(b) The simple aggregate index was 117.3. Since this is larger than 100, we subtract 100 to obtain 17.3. This means that the price of this group of 5 items increased overall in price by 17.3% between 1995 and 2000.

Example S10.8

Suppose that we change the units for frozen chicken to 5 kg and for chocolate to 1 kg in Table S10.4. In this case, the price of frozen chicken in Table S10.4 is multiplied by 5 and the price of chocolate is multiplied by 4. The new set of prices is shown in Table S10.5.

Table S10.5 Average retail prices using different units for the data in Table S10.4

Item	Unit	Price of one unit ($)		
		1990	1995	2000
Frozen chicken	5 kg	14.45	14.60	17.65
Tomato sauce	600 mL	1.69	1.83	1.78
Pink salmon (can)	210 g	2.75	2.26	2.60
Chocolate (milk)	1 kg	7.92	9.76	12.24
Pineapple (can, sliced)	450 g	0.83	0.90	1.17

Use Table S10.5 to calculate the simple aggregate index for 2000 using:

(a) 1990 as the base year

(b) 1995 as the base year

Solution

Using the data in Table S10.5, we now have:

(a) If 1990 is the base year:

$$\Sigma p_o = 14.45 + 1.69 + 2.75 + 7.92 + 0.83 = \$27.64$$
$$\Sigma p_n = 17.65 + 1.78 + 2.60 + 12.24 + 1.17 = \$35.44$$

Hence, the total cost of the 5 items in 1990 is $27.64 and in 2000 it is $35.44. From Equation S10.3:

$$\text{Simple aggregate index} = \frac{\$35.44}{\$27.64} \times 100 = 128.2$$

(b) If 1995 is the base year:

$$\Sigma p_o = 14.60 + 1.83 + 2.26 + 9.76 + 0.90 = \$29.35$$
$$\Sigma p_n = \$35.44, \text{ as in (a)}$$

Hence, the total cost of the 5 items in 1995 is \$29.35. From Equation S10.3:

$$\text{Simple aggregate index} = \frac{\$35.44}{\$29.35} \times 100 = 120.7$$

Example S10.9

Interpret the values of the two simple aggregate indexes found in Example S10.8(a) and (b), and compare these with the interpretation in Example S10.7. Comment.

Solution

(a) The simple aggregate index was 128.2. This implies that the 5 items rose overall in price by 28.2% between 1990 and 2000. This differs from the 19.7% increase credited to the same items (using different units) in Example S10.7(a).

(b) The simple aggregate index was 120.7. This implies that the 5 items rose overall in price by 20.7% between 1995 and 2000. This differs from the 17.3% increase credited to the same items in Example S10.7(b).

The different values obtained for the simple aggregate index when the units are changed demonstrate a weakness in its use.

Average of relative prices

One method of overcoming the problem of the change in value of the simple aggregate index when the units change is to use a composite index known as the **average of relative prices** (or *simple relative price index*). This index also does not take into account the quantity of each item sold, but it is still a vast improvement on the simple aggregate index.

Essentially, the average of relative prices is the average of the individual simple price indexes of all items. The formal definition is:

$$\text{Average of relative prices} = \frac{\text{sum of the simple prices indexes}}{k}$$

$$= \frac{\Sigma\left(\frac{p_n}{p_o} \times 100\right)}{k} \qquad \text{(S10.4)}$$

where k = the number of items
p_n = the price of an item in the *current* period
p_o = the price of an item in the *base* period

Example S10.10

Using 1990 as the base year, use the data in Table S10.4 to calculate the average of relative prices for 2000.

Solution

We first calculate the individual simple price indexes using Equation S10.2. These are shown in Table S10.6.

Table S10.6 Simple price indexes for items in Table S10.4

Item	Simple price index
Frozen chicken	$3.53/2.92 \times 100 = 120.9$
Tomato sauce	$1.78/1.83 \times 100 = 97.3$
Pink salmon (can)	$2.60/2.26 \times 100 = 115.0$
Chocolate (milk)	$3.06/2.44 \times 100 = 125.4$
Pineapple (can, sliced)	$1.17/0.90 \times 100 = 130.0$

Using Equation S10.4, with $k = 5$, we have:

$$\text{Average of relative prices} = \frac{120.9 + 97.3 + 115.0 + 125.4 + 130.0}{5}$$
$$= \frac{588.6}{5}$$
$$= 117.7$$

Example S10.11

Give an interpretation of the value found for the average of relative prices found in Example S10.10.

Solution

The average of relative prices in Example S10.10 was 117.7. Since this is larger than 100, we subtract 100 to obtain 17.7. This implies that the price of this group of items had an overall increase in price of 17.7% between 1990 and 2000. This differs from both the 19.7% increase credited to the same items in Example S10.7(a) and the 28.2% in Example S10.9(a).

Even though the average of relative prices overcomes some of the problems of the simple aggregate index, it still gives a distorted value for a real increase in the cost of goods and services. This is because it does not take into account the quantity of each item sold. That is, it unrealistically assumes that each item has an equal number of sales. To take account of this, in many types of index an explicit weighting is given to each item, often to correspond to the quantities of each item consumed.

S10.4 WEIGHTED INDEX NUMBERS

The use of a **weighted index** allows greater importance to be attached to some items. Information other than simply the change in price over time can then be used, and can include such factors as quantity sold or quantity consumed of each item.

One difficulty is determining which factors to use in these weightings. There are a number of possibilities, but we will restrict ourselves to the most common ones.

Laspeyres index

The most commonly used weighted index is the **Laspeyres index** (also known as the **average of weighted relative prices**). In this case, the weights used are the quantities of each item bought in the base period. The formula is:

$$\text{Laspeyres index} = \frac{\Sigma p_n q_o}{\Sigma p_o q_o} \times 100 \qquad \textbf{(S10.5)}$$

where p_n = the price in the *current* period
p_o = the price in the *base* period
q_o = the quantity *bought* (or sold) in the *base* period

This index is in fact the ratio of the sums of the actual prices of the items in the two periods, weighted by the respective numbers of items sold in the base period.

In fact, it follows that if the quantities sold (q_o) in the base period are considered:

$\Sigma p_o q_o$ = the total price of q_o items using *base* period prices
$\Sigma p_n q_o$ = the total price of q_o items using *current* period prices

Thus, the Laspeyres index measures the *relative* change in the cost of purchasing these items in the quantities specified in the base period.

Example S10.12

The average retail prices in Adelaide of a 500 g jar of strawberry jam, 1 kg of sausages and 500 g of processed cheese (sliced) are shown in Table S10.7 for each of the September quarters in the years 1990, 1995 and 2000. Also shown are the (hypothetical) quantities of each item sold in 1990.

Table S10.7 Average retail prices and quantities sold in Adelaide for 3 items in 1990, 1995 and 2000

		Quantity sold	Price ($)		
Item	Unit	in 1990 (q_o) (millions)	1990 (p_o)	1995	2000
Strawberry jam	500 g	30	2.01	2.45	2.39
Sausages	1 kg	120	3.61	3.74	4.54
Cheese (sliced)	500 g	50	3.32	3.27	3.17

Source: Australian Bureau of Statistics, *Average Retail Prices of Selected Items*, September Quarters 1990, 1995 and 2000

Using 1990 as the base year, find the Laspeyres index for:
(a) 1995
(b) 2000

Solution

(a) In this case, the current year is 1995. The relevant calculations are shown in Table S10.8 (in millions).

Table S10.8 Calculations for Example S10.12(a)

Item	$p_o q_o$ ($)	$p_n q_o$ ($)
Strawberry jam	2.01 × 30 = 60.30	2.45 × 30 = 73.50
Sausages	3.61 × 120 = 433.20	3.74 × 120 = 448.80
Cheese (sliced)	3.32 × 50 = 166.00	3.27 × 50 = 163.50
Total	659.50	685.80

From Table S10.8 we have $\Sigma p_o q_o$ = \$659.50 and $\Sigma p_n q_o$ = \$685.80. Using Equation S10.5:

$$\text{Laspeyres index} = \frac{\$685.80}{\$659.50} \times 100$$
$$= 104.0$$

Hence, according to the Laspeyres index, the weighted average retail price of these items in Adelaide rose by 4.0% between 1990 and 1995.

(b) In this case, the current year is 2000. The relevant calculations are shown in Table S10.9 (in millions).

Table S10.9 Calculations for Example S10.12(b)

Item	$p_o q_o$ ($)	$p_n q_o$ ($)
Strawberry jam	2.01 × 30 = 60.30	2.39 × 30 = 71.70
Sausages	3.61 × 120 = 433.20	4.54 × 120 = 544.80
Cheese (sliced)	3.32 × 50 = 166.00	3.17 × 50 = 158.50
Total	659.50	775.00

From Table S10.9 we have $\Sigma p_o q_o$ = \$659.50 and $\Sigma p_n q_o$ = \$775.00. Using Equation S10.5:

$$\text{Laspeyres index} = \frac{\$775.00}{\$659.50} \times 100$$
$$= 117.5$$

Hence, according to the Laspeyres index, the weighted average retail price of these items in Adelaide rose by 17.5% between 1990 and 2000.

One advantage of the Laspeyres index is that it is always possible to make price comparisons between periods other than the current period and the base period.

One of the major drawbacks of the Laspeyres index, however, is that the fixing of quantities in the base period yields information that will eventually become out of date. For this reason, it is sometimes argued that a more realistic index would be obtained if the weights used reflected the quantities sold in the *current period* rather than the base period.

Using current period consumption seems an attractive idea, but it has some practical difficulties. The main problem is that obtaining a detailed appropriate list of weights for every period would be very time-consuming and costly. Another difficulty is that we could no longer measure the price changes from period to period since the weights would be different for each period.

Paasche index

The **Paasche index** does use consumption in the *current* period. It therefore measures the change in the cost of purchasing items, in terms of quantities relating to the current period. As we are using different weights, the Paasche and Laspeyres indexes will generally not yield the same result, although they may not be too dissimilar since they are both basically trying to measure the same thing.

The formal definition of the Paasche index is:

$$\text{Paasche index} = \frac{\Sigma p_n q_n}{\Sigma p_o q_n} \times 100 \qquad \textbf{(S10.6)}$$

where p_n = the price in the *current* period
p_o = the price in the *base* period
q_n = the quantity bought (or sold) in the *current* period

If the quantities sold (q_n) in the current period are considered:

$\sum p_n q_n$ = the total price of q_n items using *current* period prices
$\sum p_o q_n$ = the total price of q_n items using *base* period prices

Example S10.13

Suppose for the data in Example S10.12 that the quantities of strawberry jam, sausages and processed cheese sold in 2000 were 55 million, 150 million and 76 million units, respectively. Find the Paasche index for 2000 using:

(a) 1990 as the base year
(b) 1995 as the base year

Solution

(a) In this case, the current year is 2000 and the base year is 1990. The relevant calculations are shown in Table S10.10 (in millions).

Table S10.10 Calculations for Example S10.13(a)

Item	$p_o q_n$ (\$)	$p_n q_n$ (\$)
Strawberry jam	2.01 × 55 = 110.55	2.39 × 55 = 131.45
Sausages	3.61 × 150 = 541.50	4.54 × 150 = 681.00
Cheese (sliced)	3.32 × 76 = 252.32	3.17 × 76 = 240.92
Total	904.37	1053.37

From Table S10.10 we have $\Sigma p_o q_n$ = \$904.37 and $\Sigma p_n q_n$ = \$1053.37. Using Equation S10.6:

$$\text{Paasche index} = \frac{\$1053.37}{\$904.37} \times 100$$
$$= 116.5$$

Hence, according to the Paasche index, the weighted average retail price of these items in Adelaide rose by 16.5% between 1990 and 2000.

(b) In this case, the current year is 2000 and the base year is 1995. The relevant calculations are shown in Table S10.11 (in millions).

Table S10.11 Calculations for Example S10.13(b)

Item	$p_o q_n$ (\$)	$p_n q_n$ (\$)
Strawberry jam	2.45 × 55 = 134.75	2.39 × 55 = 131.45
Sausages	3.74 × 150 = 561.00	4.54 × 150 = 681.00
Cheese (sliced)	3.27 × 76 = 248.52	3.17 × 76 = 240.92
Total	944.27	1053.37

From Table S10.11 we have $\Sigma p_o q_n$ = \$944.27 and $\Sigma p_n q_n$ = \$1053.37. Using Equation S10.6:

$$\text{Paasche index} = \frac{\$1053.37}{\$944.27} \times 100$$
$$= 111.6$$

Hence, according to the Paasche index, the weighted average retail price of these items in Adelaide rose by 11.6% between 1995 and 2000.

Comparison of the Laspeyres and Paasche indexes

Because the Laspeyres and Paasche indexes usually yield different results, it is useful to make a comparison of them.

1. The Laspeyres index measures the ratio of expenditures on *base year* quantities in the current year to expenditures on those quantities in the base year.
2. The Paasche index measures the ratio of expenditures on *current year* quantities in the current year to expenditures on those quantities in the base year.
3. Since the Laspeyres index uses base period weights, it may *overestimate* the rise in the cost of living (because people may have reduced their consumption of items that have become proportionately dearer than others).
4. Since the Paasche index uses current period weights, it may *underestimate* the rise in the cost of living.
5. The Laspeyres index is usually larger than the Paasche index.
6. With the Paasche index it is difficult to make year-to-year comparisons since every year a new set of weights is used.

7. The Paasche index requires that a new set of weights be obtained each year, and this information can be expensive to obtain.
8. Because of Points 6 and 7, the Laspeyres index is the one most commonly used.

Fisher's ideal index

As a result of Points 3 and 4 above, a compromise index was formulated by the statistician Irving Fisher. He established what has become known as **Fisher's ideal index**, which is the geometric mean of the Laspeyres and Paasche indexes. Although it has little use in practice, it does demonstrate the many different types of index that can be used. The formal definition is:

$$\text{Fisher's index} = \sqrt{(\text{Laspeyres index})\ (\text{Paasche index})}$$

$$= \sqrt{\frac{\Sigma p_n q_o \Sigma p_n q_n}{\Sigma p_o q_o \Sigma p_o q_n}} \times 100 \qquad \textbf{(S10.7)}$$

Example S10.14

Using the results in Examples S10.12(b) and S10.13(a), calculate Fisher's ideal index for 2000 where the base year is 1990.

Solution

We already know from Example S10.12(b) that the Laspeyres index is 117.5 and from Example S10.13(a) that the Paasche index is 116.5. Using Equation S10.7:

$$\text{Fisher's ideal index} = \sqrt{(\text{Laspeyres index})\ (\text{Paasche index})}$$

$$= \sqrt{(117.5)(116.5)}$$

$$= 117.0$$

According to Fisher's ideal index, the weighted average retail price of these items in Adelaide rose by 17.0% between 1990 and 2000.

S10.5 CONSUMER PRICE INDEX (CPI)

The measure most commonly used in Australia as a general indicator of the rate of price change for consumer goods and services is the **Consumer Price Index (CPI)**. Although the CPI is often thought of as a 'cost-of-living' index, this is not really correct. A true cost-of-living index would consider changes in the standard of living and also the substitutions that consumers make in order to maintain their standard of living when market conditions change. In fact, the Australian CPI assumes the purchase of a constant 'basket' of goods and services and measures price changes in that basket alone. A number of publications from the Australian Bureau of Statistics provide details on the CPI. One of these is *Introduction to the 14th Series Australian CPI* (Cat. No. 6456.0), and there is always an explanation in the annual *Year Book Australia* (Cat. No. 1301.0). Extracts from these publications are shown below.

Prices are a key factor in the operation of the economy, and the indexes that measure movements in the various categories of prices are used extensively by a wide range of users who need to be aware of and to analyse price movements.

Consumer Price Index

The Consumer Price Index (CPI) measures quarterly changes in the retail price of a 'basket' of goods and services that account for a high proportion of expenditures by metropolitan wage and salary earner households. Since September 2000 this 'basket' has covered a wide range of goods and services arranged in the following 11 groups: food; alcohol and tobacco; clothing and footwear; housing; household furnishings, supplies and services; health; transportation; communication; recreation; education and miscellaneous. These 11 groups in turn are divided into various subgroups and expenditure classes, as shown in Table S10.14 on page S.314.

Retail prices of food and groceries and average rentals of houses have been collected by the Australian Bureau of Statistics for the years extending back to 1901 and, in some cases, by the statistical offices of various states for earlier years.

Retail prices of a more extensive range of commodities (including clothing) and certain services in common demand have been collected at frequent and regular intervals since 1923. (Comparable information was collected for the month of November in each year from 1914 to 1922.) The range of items for which retail prices data are obtained was considerably expanded in 1948 and in later years. Average retail prices of certain food and grocery items in the relevant periods were published in the annual *Labour Report* (now discontinued). An historical perspective giving more detail of these earlier indexes was published in *Year Book Australia 1988*.

From its inception in 1960, the CPI covered the 6 state capital cities. In 1964 the geographical coverage of the CPI was extended to include Canberra. From June quarter 1982 geographic coverage was further extended to include Darwin.

Index numbers at the Group and All Groups levels are published for each capital city and for the weighted average of eight capital cities. Sub-group index numbers are published for the weighted average of 8 capital cities. Expenditure class index numbers are available for each city plus the weighted average of 8 capital cities.

Table S10.12 shows the CPI from 1989/90 to 1999/00 in all capital cities as well as their weighted average. Table S10.13 shows the CPI for each of the 8 groupings used from 1983/84 to 1999/00. Note that Table S10.13 cannot be continued in this exact form beyond 1999/00 since the number of groupings changed from 8 to 11 in September 2000.

Table S10.12 Consumer Price Index: All Groups index numbers (reference base year 1989/90 = 100.0)

Year	Sydney	Melbourne	Brisbane	Adelaide	Perth	Hobart	Canberra	Darwin	Weighted average of 8 capital cities
1989/90	100.0	100.0	100.0	100.0	100.0	100.0	100.0	100.0	100.0
1990/91	104.9	105.8	104.9	106.2	105.1	104.9	105.1	105.7	105.3
1991/92	106.7	108.1	107.0	108.9	105.9	107.1	107.8	108.0	107.3
1992/93	107.7	108.9	108.5	111.2	106.2	108.5	109.5	109.5	108.4
1993/94	109.2	111.1	110.6	113.4	108.5	111.7	111.4	111.5	110.4
1994/95	113.0	114.1	114.7	116.9	112.3	115.2	115.1	114.7	113.9
1995/96	118.7	118.4	119.1	121.2	116.7	119.6	120.3	119.5	118.7
1996/97	120.4	119.9	121.0	122.3	118.3	121.4	121.2	121.6	120.3
1997/98	120.5	119.8	121.6	121.6	118.0	121.3	120.4	121.3	120.3
1998/99	122.5	120.9	122.9	123.2	120.1	122.5	121.5	122.4	121.8
1999/00	125.4	124.1	125.0	126.3	122.9	124.8	124.2	124.2	124.7

Source: Australian Bureau of Statistics, *Consumer Price Index*, Cat. No. 6401.0; United States Bureau of Labor Statistics

Table S10.13 Consumer Price Index: Group index numbers: weighted average of eight capital cities (base of each index: year 1989/90 = 100.0)

Period	Food	Clothing	Housing	Household equipment and operation	Transportation	Tobacco and alcohol	Health and personal care	Recreation and education	All groups
1983/84	66.2	65.3	57.6	69.0	64.4	61.6	73.4	66.5	65.0
1984/85	69.7	69.3	61.9	72.3	68.6	66.8	61.0	69.1	67.8
1985/86	75.0	75.1	67.2	78.1	74.3	72.9	66.6	75.2	73.5
1986/87	81.1	82.3	72.2	83.9	82.6	80.6	77.3	82.2	80.3
1987/88	85.3	88.7	77.2	89.7	89.1	87.4	86.0	88.7	86.3
1988/89	93.4	95.0	86.9	94.8	92.5	92.3	93.3	94.2	92.6
1989/90	100.0	99.9	100.0	100.0	100.0	100.0	100.0	100.0	100.0
1990/91	103.2	104.6	103.5	105.0	106.8	108.8	109.6	105.0	105.3
1991/92	105.7	106.4	98.9	107.5	108.8	115.0	121.3	106.8	107.3
1992/93	107.4	107.5	94.6	107.3	111.3	124.4	124.0	109.1	108.4
1993/94	109.4	106.7	94.2	107.8	113.8	133.7	129.0	111.9	110.4
1994/95	112.1	106.7	100.0	109.2	117.5	141.0	135.5	114.6	113.9
1995/96	116.0	107.0	105.9	111.7	122.6	156.1	141.7	117.7	118.7
1996/97	119.7	107.3	101.6	113.5	124.3	161.4	149.1	110.7	120.3
1997/98	121.8	107.4	94.5	113.8	123.5	164.6	153.6	123.6	120.3
1998/99	126.5	106.7	95.8	113.7	122.1	168.7	153.7	126.2	121.8
1999/00	129.2	105.5	99.9	113.3	128.9	175.2	151.9	128.2	124.7

Source: Australian Bureau of Statistics, *Consumer Price Index*, Cat. No. 6401.0; United States Bureau of Labor Statistics

Conceptual basis for measuring price changes

The CPI is a quarterly measure of the change in average retail price levels. It provides a method of comparing the average price level for a quarter with the average price level of the reference base year or changes in the average price level from one quarter to the next. In measuring price changes, the CPI aims to measure only pure price changes (i.e. price changes excluding the effects of any changes in the quality or quantity of the goods or services concerned); in other words to measure, each quarter, the change in the cost of purchasing an identical basket of goods and services. The CPI is a measure of changes in transaction prices, the prices actually paid by consumers for the goods and services they buy. It is not concerned with nominal, recommended or list prices (unless they are the prices consumers actually pay).

The CPI is often loosely, and incorrectly, referred to as a cost-of-living index. However, a true cost-of-living index (if such a measure could be compiled) would not be the same thing as a fixed-weight retail price index like the CPI. A cost-of-living index would need to take account of changes in standards of living and the substitutions that consumers make in response to changing market conditions, such as changes in supply, or in response to disparate price movements.

The CPI measures price change over time and does not provide comparisons between relative price levels at a particular date. For example, it does not show whether beef is dearer than lamb, or whether bus fares are dearer than train fares. The fact that the index number for any particular component is higher than that for another component in a particular quarter does not mean that the first component is more expensive than the second. It merely means that, since the reference base year, prices for the first component have risen more than prices for the second component. Similarly, the CPI does not provide any basis for measuring relative price levels between the different cities. For example, it says nothing about whether prices are higher in Sydney or Perth; it shows only whether prices have changed more in Sydney or in Perth.

Index population

Because the spending patterns of various groups in the population differ somewhat, the pattern of one large group, fairly homogeneous in its spending habits, is chosen for the purpose of calculating the CPI. The CPI population group is, in concept, metropolitan employee households. For this purpose employee households are defined as those households that obtain the major part of their household income from wages and salaries; metropolitan means the 8 capital city Statistical Divisions.

Details of the 14th series Consumer Price Index

The 14th series CPI was introduced in the September quarter 2000 with weights updated to be based on the 1998–99 Household Expenditure Survey (HES). The quantities underpinning the 1998–99 expenditures were preserved by revaluing these expenditures to June quarter 2000 prices. The most notable change to the 14th series CPI structure was an increase in the number of food groups from 8 to 11, although the overall item coverage is unchanged from the 13th series.

Each of the 89 expenditure classes in the current CPI has a fixed weight (i.e. the measure of its relative importance). Details of the weighting pattern are shown in Table S10.14.

Table S10.14 Consumer Price Index 14th series: weighting pattern at June Quarter 2000, weighted average of 8 capital cities

Groups and subgroups	Percentage contribution to the All Groups CPI
Food	*17.72*
Dairy and related products	1.51
Bread and cereal products	2.20
Meat and seafood	2.62
Fruit and vegetables	2.30
Non-alcoholic drinks and snack food	2.48
Meals out and takeaway foods	4.93
Other food	1.69
Alcohol and tobacco	*7.41*
Alcoholic drinks	5.14
Tobacco	2.27
Clothing and footwear	*5.19*
Men's clothing	0.98
Women's clothing	1.80
Children and infant's clothing	0.47
Footwear	0.83
Clothing accessories, supplies and services	1.10

(continued)

Table S10.14 (continued)

Groups and subgroups	Percentage contribution to the All Groups CPI
Housing	*19.75*
Rents	5.60
Utilities	3.23
Other housing	10.91
Household furnishings, supplies and services	*8.09*
Furniture and furnishings	3.58
Household appliances, utensils and tools	1.98
Household supplies	1.91
Household services	0.62
Health	*4.69*
Health services	3.55
Pharmaceuticals	1.14
Transportation	*15.25*
Private motoring	14.40
Urban transport fares	0.85
Communication	*2.88*
Postal	0.15
Telecommunication	2.73
Recreation	*12.29*
Audio, visual and computing	2.70
Books, newspapers and magazines	1.08
Sport and other recreation	4.16
Holiday travel and accommodation	4.35
Education	*2.69*
Preschool and primary	0.50
Secondary education	0.94
Tertiary education	1.25
Miscellaneous	*4.04*
Insurance services	1.46
Personal care	2.14
Child-care	0.44
Total All Groups	*100.00*

Source: Australian Bureau of Statistics, *Year Book Australia*, Cat. No. 1301.0

Changes in the weighting pattern have been made at approximately 5-yearly intervals to take account of changes in household spending patterns. The CPI, in fact, comprises 11 separate series that have been linked to form a continuous series. For example, the 11th series was introduced in the March quarter 1987, with a weighting pattern based on estimated household expenditure in 1984. The 12th series was introduced in the September quarter 1992, with a weighting pattern based on estimated household expenditure in 1988/89. The 13th series commenced in the September quarter 1998, while the 14th series commenced in the September quarter 2000.

The CPI 'basket of goods and services' covers items that are considered representative of metropolitan households' spending habits and whose prices can be associated with an identifiable and specific quantity of a commodity or service.

Income taxes and personal savings, on the other hand, do not form part of the CPI because they cannot be clearly associated with the purchase of a specific quantity of any goods or services.

Collecting prices

Since the CPI is designed to measure the impact of changing prices on metropolitan wage and salary earner households, price movements are monitored in the kinds of retail outlets or other establishments where such households normally purchase goods and services. This involves collecting prices from many sources including supermarkets, department stores, footwear stores, restaurants, motor vehicle dealers and service stations, dental surgeries, hotels and clubs, schools, hairdressers, travel agents, airlines, bus operators, house builders, real estate developers, electricians and plumbers. Items such as bus, rail and air fares, electricity and gas charges, telephone charges and local authority rates are collected from the appropriate authorities. Information on rents is obtained from property management companies and government housing authorities. In total, around 100 000 separate price quotations are collected each quarter.

Prices of the goods and services included in the CPI are generally collected quarterly. However, some important items are priced monthly or more frequently (e.g. bread, fresh meat and fish, fresh fruit and vegetables, petrol, alcohol and tobacco) and a small number are collected annually (e.g. seasonal clothing, local government rates and charges).

The prices used in the CPI are those that any member of the public would have to pay on the pricing day to purchase the specified good or service. Any sales or excise taxes that the consumer must pay when purchasing specific items are included in the CPI price. Sale prices, discount prices and 'specials' are reflected in the CPI so long as the items concerned are of normal quality (i.e. not damaged or shopsoiled) and are offered for sale in reasonable quantities. To ensure that the price movements reflect the experience of the bulk of the metropolitan population, the brands and the varieties of the items that are priced are generally those that sell in greatest volume.

Periodic revision of the CPI

The CPI is periodically revised in order to ensure it continues to reflect current conditions. CPI revisions have usually been carried out at approximately 5-yearly intervals, the most recent completed in September 2000. Previous revisions of the CPI were introduced at: December quarter 1963; December quarter 1968; December quarter 1973; September quarter 1974; September quarter 1976; March quarter 1982; March quarter 1987; September quarter 1992; and September quarter 1998. Following each revision, the new series, with its changed composition and weighting pattern, is linked to the previous series to form one continuous series. The process of linking ensures that the continuous series reflects only price variations and not differences in costs of the old and new baskets.

Changes in quality

Since the CPI aims to measure the price change of a constant basket of goods and services over time, it is necessary to ensure that identical or equivalent items are priced in successive time periods. This involves evaluating changes in the quality of goods and services included in the index and removing the effects of such changes so that the index reflects only the price change.

Long-term linked series

To provide an approximate long-term measure of retail price change for the period since the first Australian retail price index was compiled, a single series of index numbers has been constructed by linking together selected retail price index series. The index numbers are expressed on a reference base 1945 = 100, which was the end of a period of relative price stability during World War II. The successive series linked together to produce this long-term series of index numbers are:

- from 1901 to 1914, the A Series Index
- from 1914 to 1946/47, the C Series Index
- from 1946/47 to 1948/49, a combination of the C Series Index (excluding rent) and the housing group of the CPI
- from 1948/49 onwards, the CPI

Table S10.15 shows the value of the Australian retail price index from 1901 to 1999.

Table S10.15 Retail price index numbers,[a] long-term linked series (base year 1945 = 100)

Year	Index number	Year	Index number	Year	Index number
1901	47	1934	73	1967	286
1902	50	1935	74	1968	293
1903	49	1936	75	1969	302
1904	46	1937	78	1970	313
1905	48	1938	80	1971	332
1906	48	1939	82	1972	352
1907	48	1940	85	1973	385
1908	51	1941	89	1974	443
1909	51	1942	97	1975	510
1910	52	1943	101	1976	579
1911	53	1944	100	1977	650
1912	59	1945	100	1978	702
1913	59	1946	102	1979	766
1914	61	1947	106	1980	844
1915	70	1948	117	1981	926
1916	71	1949	128	1982	1028
1917	75	1950	140	1983	1132
1918	80	1951	167	1984	1177
1919	91	1952	196	1985	1257
1920	103	1953	205	1986	1370
1921	90	1954	206	1987	1487
1922	87	1955	211	1988	1594
1923	89	1956	224	1989	1714
1924	88	1957	229	1990	1839
1925	88	1958	233	1991	1898
1926	90	1959	237	1992	1917
1927	89	1960	245	1993	1952
1928	89	1961	252	1994	1989
1929	91	1962	251	1995	2082
1930	87	1963	252	1996	2136
1931	78	1964	258	1997	2141
1932	74	1965	268	1998	2159
1933	71	1966	276	1999	2191

(a) The index numbers relate to the weighted average of 6 state capital cities up to 1980; from 1981 they relate to the weighted average of 8 capital cities.

Source: Australian Bureau of Statistics, *Consumer Price Index*, Cat. No. 6401.0; United States Bureau of Labor Statistics

International Consumer Price Indexes

A comparison of the CPI for a number of countries is shown in Table S10.16 from the base year 1989/90 to 1999/00. During this 11-year period, of the countries listed, Japan had the smallest increase in CPI (11.6%), while Indonesia had the largest (267.1%).

Table S10.16 Consumer Price Index: international comparisons(a)

Year	USA	Japan	Germany(b)	Canada	UK	Hong Kong
1989/90	100.0	100.0	100.0	100.0	100.0	100.0
1990/91	105.5	103.4	102.7	105.6	107.7	110.8
1991/92	108.7	105.9	106.9	109.0	115.0	121.7
1992/93	112.1	106.8	110.5	110.9	118.6	130.8
1993/94	114.8	107.9	113.9	112.0	122.1	140.1
1994/95	118.0	107.8	116.4	113.4	124.8	151.4
1995/96	120.9	107.3	116.3	116.0	128.3	160.8
1996/97	124.3	108.2	118.2	118.8	131.5	168.2
1997/98	125.8	112.4	120.3	120.6	134.6	175.1
1998/99	127.2	112.4	120.7	122.0	137.2	174.0
1999/00	130.9	111.6	121.8	125.0	139.3	165.8

Year	Republic of Korea	Taiwan	Indonesia	Australia	New Zealand(c)
1989/90	100.0	100.0	100.0	100.0	100.0
1990/91	109.1	104.0	108.7	105.6	104.0
1991/92	117.9	107.6	120.0	108.8	106.5
1992/93	123.5	111.4	129.0	111.0	108.7
1993/94	130.4	114.2	137.8	113.5	109.4
1994/95	138.0	119.1	150.3	116.5	110.5
1995/96	144.4	122.5	163.7	121.1	111.9
1996/97	151.3	125.7	174.1	123.9	113.7
1997/98	162.1	127.2	232.7	125.4	114.9
1998/99	169.0	128.2	368.3	126.9	116.9
1999/00	172.1	129.3	367.1	129.4	118.7

(a) All groups excluding housing.
(b) The statistics for Germany refer to Western Germany (Federal Republic of Germany before the unification of Germany).
(c) From March quarter 1994 the statistics for New Zealand refer to 'all groups excluding housing and credit services'.

Source: Australian Bureau of Statistics, *Consumer Price Index*, Cat. No. 6401.0; United States Bureau of Labor Statistics

S10.6 USING THE CONSUMER PRICE INDEX

There are a number of situations where the CPI is used to make adjustments to prices charged and payments made. Examples include changes in rent, pension payments and payments to the Child Support Agency. Such adjustments are often made by the relevant government agency, but if members of the community wish to use the CPI for such purposes, it is their responsibility to ensure that this index is suitable.

Example S10.15

According to a tenancy agreement, the rent on a Melbourne home unit is to be adjusted annually, based on the percentage change in the Melbourne All Groups CPI between the current year figure and the figure in the previous year.

The weekly rent in 1993/94 was \$275. Find the adjusted weekly rent in 1999/00.

Solution

From Table S10.12, the relevant figures for the Melbourne All Groups CPI are:

1993/94 111.1 1999/00 124.1

The adjusted weekly rent in 1999/00 can be found from:

$$\frac{\text{Melbourne CPI in 1999/00}}{\text{Melbourne CPI in 1993/94}} \times \text{weekly rent in 1993/94} = \frac{124.1}{111.1} \times \$275$$

$$= \$307 \text{ (to the nearest \$)}$$

S10.7 INDEX NUMBERS AS A MEASURE OF DEFLATION

One of the uses for price indexes is to measure the changes in the purchasing power of the dollar. This is known as deflation. It allows us to express the value of goods, services, salaries, and so on in terms of real dollars.

Deflation is important since values that are stated in current figures can change significantly with time solely as a result of inflation. In order to eliminate the effect of inflation and obtain a clear picture of the 'real' change, the values must be deflated.

Example S10.16

A faculty manager has been working at a university in Sydney since the end of 1994. At that time, she commenced on what she thought was an attractive salary of \$58 000. Even though she was awarded a raise in salary each year, she feels that the purchasing power of her salary decreased over the next few years.

Table S10.17 shows her actual annual salary and the Sydney All Groups CPI between 1994/95 and 1999/00 from Table S10.12.

Table S10.17 Sydney All Groups CPI and annual salary for the manager in Example S10.16

Year	Sydney All Groups CPI	Actual annual salary (\$)
1994/95	113.0	58 000
1995/96	118.7	58 900
1996/97	120.4	59 300
1997/98	120.5	60 200
1998/99	122.5	61 500
1999/00	125.4	62 300

(a) Using the Sydney All Groups CPI as a measure of purchasing power, calculate her 'real' annual salary each year in terms of 1994/95 dollars. That is, find what her salary was each year in terms of the purchasing power she had in 1994/95.

(b) Also using the Sydney All Groups CPI, calculate what her annual salary would have been each year had it been adjusted in line with this CPI. That is, find what her 'adjusted' salary should have been each year to maintain the purchasing power she had in 1994/95.

Solution

(a) This question requires us to put each of her actual salaries into equivalent salaries in each of the other years. That is, we must deflate the actual salaries for 1994/95 to 1999/00 and express them in terms of *'real' (or equivalent) salaries* of the base year 1994/95.

To find a 'real' salary in a current year we use Equation S10.8:

$$\text{'Real' salary} = \text{actual salary in current year} \times \frac{\text{CPI in base year}}{\text{CPI in current year}} \qquad \textbf{(S10.8)}$$

For example, using Equation S10.8, the real salary in 1999/00 would be:

$$\begin{aligned}\text{'Real' salary in 1999/00} &= \text{actual salary in 1999/00} \times \frac{\text{CPI in 1994/95}}{\text{CPI in 1999/00}} \\ &= \$62\,300 \times \frac{113.0}{125.4} \\ &= \$56\,140\end{aligned}$$

That is, an annual salary of \$62 300 in 1999/00 is equivalent to an annual salary of \$56 140 in 1994/95. This is below the actual starting salary of \$58 000, so for 1999/00 she has less purchasing power than she did in 1994/95.

The complete set of 'real' salaries is given in Table S10.18.

Table S10.18 Calculation of 'real' annual salaries for the manager in Example S10.16

Year	Sydney All Groups CPI	Actual annual salary (\$)	'Real' annual salary (\$)
1994/95	113.0	58 000	58 000
1995/96	118.7	58 900	56 071
1996/97	120.4	59 300	55 655
1997/98	120.5	60 200	56 453
1998/99	122.5	61 500	56 731
1999/00	125.4	62 300	56 140

It can be seen from the last column of the table that the manager's 'real' annual salary (in terms of 1994/95 dollars) has never kept pace with what it should have been according to changes in the Sydney All Groups CPI. Therefore her feelings on her reduced purchasing power are well founded.

(b) To calculate what the manager's 'adjusted' annual salary would have been each year had it kept in line with the Sydney All Groups CPI, we use Equation S10.9.

$$\text{'Adjusted' annual salary} = \text{base year salary} \times \frac{\text{CPI in current year}}{\text{CPI in base year}} \qquad \textbf{(S10.9)}$$

For example, using Equation S10.9 the 'adjusted' annual salary in 1999/00 would be:

$$\text{'Adjusted' annual salary in 1999/00} = \$58\,000 \times \frac{\text{CPI in 1999/00}}{\text{CPI in 1994/95}}$$

$$= \$58\,000 \times \frac{125.4}{113.0}$$

$$= \$64\,365$$

This is somewhat more than the actual annual salary of $62 300 she was actually receiving in 1999/00. The complete set of 'adjusted' salaries is shown in Table S10.19.

Table S10.19 Calculation of 'adjusted' annual salaries for the manager in Example S10.16

Year	Sydney All Groups CPI	Actual annual salary ($)	'Adjusted' annual salary ($)
1994/95	113.0	58 000	58 000
1995/96	118.7	58 900	60 926
1996/97	120.4	59 300	61 798
1997/98	120.5	60 200	61 850
1998/99	122.5	61 500	62 876
1999/00	125.4	62 300	64 365

It can be seen from the last column of Table S10.19 that, after the first year, the manager's actual annual salary is always below what the annual salary would have been had it been adjusted in line with the Sydney All Groups CPI. That is, since the actual salary is always less than the 'adjusted' salary, the manager's salary has not kept up with changes to the CPI.

S10.8 OTHER TYPES OF AUSTRALIAN INDEXES

As mentioned previously, the Consumer Price Index is only one such series used to measure changes in the price of goods. A few examples of other indexes are listed below, together with a brief description of each.

- *Price index of materials used in house building (base year 1989/90)*
  This index measures the changes in prices of selected materials used in the construction of houses in each capital city. The value in 1999/00 was 122.8 (Cat. No. 6408.0).
- *Price indexes of materials used in building other than house building (base year 1989/90)*
  These indexes measure the changes in prices of selected materials used in the construction of buildings other than houses in metropolitan areas. The value in 1999/00 was 116.1 (Cat. No. 6407.0).
- *Price indexes of materials used in manufacturing industries (base year 1988/89)*
  These indexes measure changes in the prices of materials (including fuels) used by establishments classified to the Manufacturing Division of the ASIC, 1983 edition. The values in 1999/00 were 118.8 for all imported materials and 114.5 for all home-produced materials (Cat. No. 6411.0).

- *Price indexes of articles produced by manufacturing industries (base year 1988/89)*
  These indexes measure movements in the prices of articles produced by those establishments classified to the Manufacturing Division of the ASIC, 1983 edition. The value in 1999/00 was 120.6 (Cat. No. 6412.0).
- *Export price index (base year 1989/90)*
  This index measures changes in the price of all exports of merchandise from Australia, including re-exports. Export prices are obtained from major exporters of selected items. The index numbers for each month relate to the prices of exports shipped during that month. The value in 1999/00 was 98.0 (Cat. No. 6405.0).
- *Import price index (base year 1989/90)*
  This index measures changes in the prices of imports of merchandise into Australia on a free-on-board, country-of-origin basis and excludes items that are difficult to value such as jewellery, ships, antiques and works of art. The index numbers for each month relate to the prices of imports landed in Australia during that month. The value in 1999/00 was 120.2 (Cat. No. 6414.0).

All Ordinaries index

The *All Ordinaries index* measures share price movements on the Australian stock exchange. It is derived from the market values of a large number of leading mining and industrial companies that are listed on the Australian exchange.

The Australian Stock Exchange (ASX) has 31 indexes and Index Number 30 is the composite index for the All Ordinaries. It is the index most commonly referred to in Australia when the share price index is discussed and is based on the aggregate market value (AMV) of a wide selection of the companies quoted on the ASX. The market value, or market capitalisation, of any company in the All Ordinaries index portfolio is the number of shares on issue multiplied by the current price per share in that company. The AMV of the All Ordinaries index share portfolio is simply the sum of the market values of the companies included.

The All Ordinaries index was given a base value of 500 at the end of trading on the last trading day in December 1979, at which time the ASX included the operations of state-based exchanges. Its value grew to 2200 in September 1996, 2620 in July 1997, and was 3425 in July 2001.

Trade-weighted index

The *trade-weighted index* (TWI) provides an indication of movements in the average value of the Australian dollar against the currencies of our main trading partners. From September 1974 to 8 December 1984, the exchange rate for the Australian dollar was set in relation to the average value of a basket of currencies weighted in proportion to each country's trade with Australia.

Since the float of the Australian dollar in December 1984, the value of the Australian dollar has been determined by supply and demand. Hence, the trade-weighted index is no longer used as a tool for managing the exchange rate. However, the method of calculating the trade-weighted index has been basically unchanged since 1974. The changes that have been made include the updating at intervals of the weights of different currencies to reflect changing trade patterns.

Figure S10.1 illustrates the trend of the trade-weighted index between January 1989 and January 2001. Since 1998 the value of the TWI has been steadily falling and by July 2001 was 50.5, about half its value in the base year of 1970.

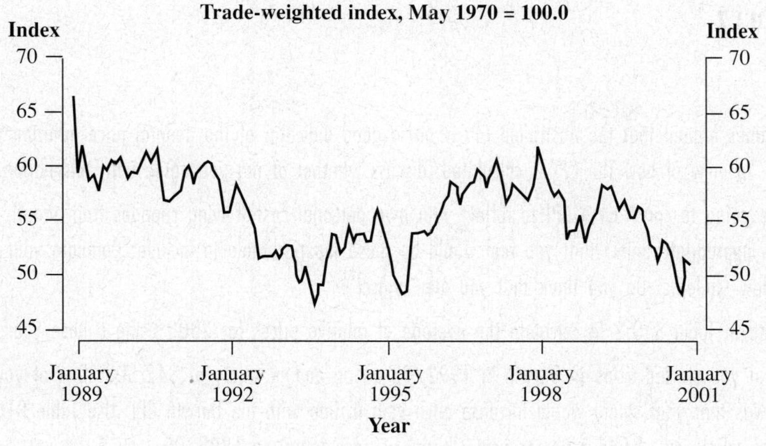

Figure S10.1 The trend of the Australian trade-weighted index between January 1989 and January 2001

Source: Australian Bureau of Statistics, *Australian Economic Indicators*, March 2001, Cat. No. 1350.0

Quick Quiz

1. It is sometimes argued that the Australian CPI is *not* a good indicator of the general price inflation that confronts households. In view of how the CPI is calculated, discuss whether or not you agree with this notion.
2. If you were going to construct a CPI to reflect your own personal cost-of-living changes from month to month, list 10 items (with appropriate units) that you feel would be those most suitable to include. Compare your list with the lists of your fellow students. Do you think that you are 'typical'?
3. Use the data in Table S10.2 to calculate the average of relative prices for 2000 using a base year of 1985.
4. Suppose that you started a job in Darwin in 1992/93 on an annual salary of \$42 500. One of your working conditions was that your salary would increase each year in line with the Darwin CPI. Use Table S10.12 to determine what annual salary you should have received (to the nearest dollar) in 1999/00.
5. Table S10.14 shows the CPI weighting pattern at June quarter 2000, weighted average of 8 capital cities. Consider your own personal average weekly expenditure. Write down the percentage of your disposable income that you spend on the 11 major groups (i.e. food, alcohol and tobacco, housing, etc.). Highlight any group where you feel that your personal percentage differs significantly from that of the population.
6. Give a hypothetical example of 4 items, including unit size and prices, where the simple aggregate index can be distorted by a single item with a relatively large price. Calculate the index with both 1 unit of this item and 2 units of this item, and comment on the difference in value.
7. Suppose that you are told that $\sum p_o q_n = 2.15$ and the Paasche index = 118.6. Find the value of $\sum p_n q_n$.
8. If all the unit sizes used are doubled, what is the effect on the value of the simple aggregate index?
9. What does the value of the NASDAQ index represent?
10. Suppose an item had a retail price of \$15 in 1968. If its price had moved in line with the index given in Table S10.15, what would its price have been in 1998?

Multiple Choice Questions

1. The prices of 1 kg of tomatoes, 1 kg of sausages and 250 g of tea were recorded at the beginning of each of the years 1992 and 2002. They were:

Item	Unit	1992	2002
Tomatoes	1 kg	0.89	1.53
Sausages	1 kg	1.25	2.95
Tea	250 g	1.02	1.68

The index number for sausages (to one decimal place) in 2002 relative to a base year of 1992 is:
(a) 236.0 (b) 171.9 (c) 164.7
(d) 2.4 (e) none of the above

2. For Question 1, the simple aggregate index (to one decimal place) for 2002 relative to a base year of 1992 is:
(a) 51.3 (b) 194.9 (c) 2.0
(d) 151.6 (e) none of the above

3. For Question 1, the average of relative prices (to one decimal place) in 2002 relative to a base year of 1992 is:
(a) 236.0 (b) 171.9 (c) 194.9
(d) 190.9 (e) none of the above

4. For Question 1, suppose that the number of units of tomatoes, sausages and tea sold in 1992 were (in millions) 1.2, 1.5 and 2.3, respectively. The Laspeyres index in 2002 relative to a base year of 1992 is:
(a) 194.9 (b) 191.4 (c) 171.9
(d) not calculable (e) none of the above

5. Using the data in Question 1 and Question 4, the Paasche index in 2002 relative to a base year of 1992 is:
(a) 194.9 (b) 191.4 (c) 171.9
(d) not calculable (e) none of the above

6. The current Australian CPI uses:
(a) the Paasche index
(b) the Laspeyres index
(c) Fisher's ideal index
(d) a combination of the indexes in (a), (b) and (c)
(e) none of the above

7. A *composite index number* is:
(a) always at least 100
(b) only applicable if prices change each quarter
(c) constructed from changes in a number of different items
(d) composed of changes in less than 10 items at a time
(e) none of the above

8. The 14th series of the Australian CPI was introduced in the year:
(a) 1980
(b) 1990
(c) 2000
(d) 2002
(e) none of the above

9. Among the countries in Table S10.16, the country that had the smallest percentage change in its CPI between 1996/97 and 1999/00 was:
(a) USA
(b) Taiwan
(c) New Zealand
(d) Germany
(e) none of the above

10. The *Hang Seng index* is usually associated with:
(a) the USA
(b) Australia
(c) China
(d) the United Kingdom
(e) none of the above

Exercises

S10.1 Shown in the table below are the prices of 1 nip of scotch in a public bar, 1 kg of potatoes and 500 g of butter in Sydney for each of the September quarters 1980, 1990 and 2000.

Item	Unit	Price of one unit ($)		
		1980	1990	2000
Scotch	1 nip	0.81	2.53	3.16
Potatoes	1 kg	0.41	0.93	1.09
Butter	500 g	1.09	1.88	1.89

Source: Australian Bureau of Statistics, *Average Retail Prices of Selected Items, September Quarters 1980, 1990 and 2000*, Cat. No. 6403.0

Using a base year of 1980, find the price relative for each item in:

(a) 1990 (b) 2000

S10.2 (a) Using the data in Exercise S10.1, and a base year of 1980, find the simple price index for each item in:

(i) 1990 (ii) 2000

(b) Use your answers in (a) to determine which item proportionately increased the most in price between:

(i) 1980 and 1990 (ii) 1980 and 2000 (iii) 1990 and 2000

S10.3* Using the data in Exercise S10.1, and a base year of 1980, find the simple aggregate index for:

(a) 1990 (b) 2000

S10.4 Interpret the values found in Exercise S10.3(a) and (b).

S10.5 (a) Using the data in Exercise S10.1, and a base year of 1980, find the simple aggregate index for 2000, except this time change the unit of scotch to 5 nips.

(b) Compare your answer in (a) with that for Exercise S10.3(b). Why should it be different?

(c) What does your answer in (b) demonstrate about the simple aggregate index?

S10.6* (a) Using the data in Exercise S10.1, and a base year of 1980, find the average of relative prices for:

(i) 1990 (ii) 2000

(b) Interpret the values found in (a).

S10.7 Suppose that you are now told that the quantities sold of each item in Exercise S10.1 during 1980 and 2000 were as shown in the following table.

Item	Units (million) sold in 1980	Units (million) sold in 2000
Scotch	36.9	43.5
Potatoes	42.0	60.8
Butter	22.5	27.6

(a) Find the Laspeyres index in 2000 using 1980 as the base year.

(b) Find the Paasche index in 2000 using 1980 as the base year.

(c) Interpret the values found in (a) and (b).

S10.8 Use the values found in Exercise S10.7(a) and (b) to find the value of Fisher's ideal index in 2000 using 1980 as the base year. Interpret its value.

S10.9 Shown in the table below are the prices of 1 L of super grade petrol, 1 kg of T-bone steak and a 150 g jar of instant coffee in Melbourne for each of the September quarters 1990, 1995 and 2000.

		Price of one unit ($)		
Item	Unit	1980	1995	2000
Petrol (super)	1 L	0.63	0.74	0.92
T-bone steak	1 kg	9.89	10.78	11.40
Coffee, instant	150 g	4.32	5.69	6.23

Source: Australian Bureau of Statistics, *Average Retail Prices of Selected Items, September Quarters 1990, 1995 and 2000*, Cat. No. 6403.0

Using a base year of 1990, find the price relative for each item in:

(a) 1995 (b) 2000

S10.10 (a) Using the data in Exercise S10.9, and a base year of 1990, find the simple price index for each item in:

(i) 1995 (ii) 2000

(b) Use your answers in (a) to determine which item proportionately increased the most in price between:

(i) 1990 and 1995 (ii) 1995 and 2000 (iii) 1990 and 2000

(c) Interpret the simple price index for coffee in 2000 using a base year of 1995.

S10.11* Using the data in Exercise S10.9, and a base year of 1990, find the simple aggregate index for:

(a) 1995 (b) 2000

S10.12 Interpret the values found in Exercise S10.11(a) and (b).

S10.13 (a) Using the data in Exercise S10.9, and a base year of 1990, find the simple aggregate index for 2000, except this time change the unit of petrol to 10 L.

(b) Compare your answer in (a) with that for Exercise S10.11(a). Why should it be different?

(c) What does your answer in (b) demonstrate about the simple aggregate index?

S10.14 (a) Using the data in Exercise S10.9, and a base year of 1995, find the simple aggregate index for 2000, except this time change the unit of coffee to 750 g.

(b) Interpret the value found in (a). Do you think that this is a reasonable guide to the changes in prices for these items between 1995 and 2000? Explain.

S10.15* (a) Using the data in Exercise S10.9, and a base year of 1990, find the average of relative prices for:

(i) 1995 (ii) 2000

(b) Interpret the values found in (a).

S10.16 (a) Using the data in Exercise S10.9, and a base year of 1990, find the average of relative prices for 1995, except this time change the unit of coffee to 900 g.

(b) Compare your answer in (a) with that for Exercise S10.15(a)(i). What does this demonstrate about the average of relative prices?

S10.17* Suppose that you are now told that the quantities sold of each item in Exercise S10.9 during 1990 and 2000 were as shown below.

Item	Units (million) sold in 1990	Units (million) sold in 2000
Petrol (super)	76.4	93.5
T-bone steak	15.0	32.3
Coffee, instant	83.7	102.0

(a) Find the Laspeyres index in 2000 using 1990 as the base year.

(b) Find the Paasche index in 2000 using 1990 as the base year.

(c) Interpret the values found in (a) and (b).

S10.18 Use the values found in Exercise S10.17(a) and (b) to find the value of Fisher's ideal index in 2000 using 1990 as the base year. Interpret its value.

S10.19* Shown in the table below are the prices of 1 kg of onions, an 825 g can of peaches and a 1 kg leg of lamb in Brisbane for each of the September quarters 1980, 1990 and 2000.

		Price of one unit ($)		
Item	Unit	1980	1990	2000
Onions	1 kg	0.46	1.26	1.16
Peaches (can)	825 g	0.74	1.80	2.12
Leg of lamb	1 kg	4.15	5.71	5.85

Source: Australian Bureau of Statistics, *Average Retail Prices of Selected Items*, September Quarters 1980, 1990 and 2000, Cat. No. 6403.0

Using a base year of 1980, find the price relative for each item in:

(a) 1990 (b) 2000

S10.20 (a) Using the data in Exercise S10.19, and a base year of 1980, find the simple price index for each item in:

(i) 1990 (ii) 2000

(b) Use your answers in (a) to determine which item proportionately increased the most in price between:

(i) 1980 and 1990 (ii) 1990 and 2000 (iii) 1980 and 2000

S10.21* Using the data in Exercise S10.19, and a base year of 1980, find the simple aggregate index for:

(a) 1990 (b) 2000

S10.22 Interpret the values found in Exercise S10.21(a) and (b).

S10.23 (a) Using the data in Exercise S10.19, and a base year of 1980, find the simple aggregate index for 2000, except this time change the unit of onions to 10 kg.

(b) Compare your answer in (a) with that for Exercise S10.21(b). Why should it be different? Explain.

S10.24* (a) Using the data in Exercise S10.19, and a base year of 1980, find the average of relative prices for:

(i) 1990 (ii) 2000

(b) Interpret the values found in (a).

S10.25 Suppose that you are now told that the quantities sold of each item in Exercise S10.19 during 1990 and 2000 were as shown below.

Item	Units (million) sold in 1990	Units (million) sold in 2000
Onions	21.8	28.7
Peaches (can)	90.0	103.0
Leg of lamb	25.2	29.3

(a) Find the Laspeyres index in 2000 using 1990 as the base year.

(b) Find the Paasche index in 2000 using 1990 as the base year.

(c) Interpret the values found in (a) and (b).

(d) Do you feel that peaches have dominated these indexes? If so, is it to an unfair extent? Comment.

S10.26 Use the values found in Exercise S10.25(a) and (b) to find the value of Fisher's ideal index in 2000 using 1990 as the base year. Interpret its value.

S10.27* Tenants who live in a block of home units in Perth have agreed to pay weekly rent that is to be adjusted annually, based on the Perth CPI between the current year figure and the figure in the previous year.

Use the information in Table S10.12 to find the adjusted weekly rent (to the nearest dollar) in 1999/00 for tenants who had weekly rents in 1989/90 of:

(a) \$280 (b) \$360 (c) \$600

S10.28 One of the tenants in Exercise S10.27 paid a weekly rent of \$450 in 1994/95. If their rent was adjusted annually according to changes in the Perth CPI, what rent would they have paid in 1999/00?

S10.29 Another tenant in Exercise S10.27 paid a weekly rent of \$600 in 1997/98. If their rent was adjusted annually according to changes in the Perth CPI, what rent would they have paid in 1999/00?

S10.30* Use the information in Table S10.13 to determine which of the two categories 'Food' and 'Clothing' had the largest percentage increase in the weighted price during the 16 years between 1984/85 and 1999/00.

S10.31 Use the information in Table S10.13 to determine which of the two categories 'Transportation' and 'Tobacco and alcohol' had the largest percentage increase in the weighted price during the 10 years between 1989/90 and 1999/00.

S10.32 A divorced parent has agreed to make child-support payments to his former spouse by means of payments through the Child Support Agency (CSA). These payments are adjusted annually based on the CPI (weighted average of 8 capital cities, All Groups: for 1999/00 this figure was 124.7).

Use the information in Table S10.12 to find the adjusted weekly child-support payments (to the nearest dollar) in 1999/00 if the divorced parent made weekly payments in 1993/94 of:

(a) \$95 (b) \$180 (c) \$290

S10.33* You are now told that the divorced parent in Exercise S10.32 made weekly payments of \$350 in 1992/93. Assuming that annual adjustments were made as described, find the adjusted weekly child-support payments (to the nearest dollar) in 1999/00.

S10.34 For the faculty manager in Example S10.16, use Table S10.12 to find her 'real' annual salary (to the nearest dollar) in terms of 1994/95 dollars if her actual salary was:

(a) \$61 000 in 1996/97 (b) \$61 000 in 1997/98

(c) \$61 000 in 1999/00

S10.35 For the faculty manager in Example S10.16, use Table S10.12 to find what annual salary (to the nearest dollar) she would have had to receive in each of the following years in order for her purchasing power to be the same as it was in 1994/95, if she had started at that time on an annual salary of \$72 500:

(a) 1997/98 (b) 1998/99 (c) 1999/00

S10.36* A marketing assistant working in Canberra received a starting annual salary of \$42 500 in 1991/92. Assuming that he received annual salary increases in line with the Canberra CPI, use Table S10.12 to find his annual salary (to the nearest dollar) in:

(a) 1994/95 (b) 1996/97 (c) 1999/00

S10.37 (a) Use an appropriate publication of the Australian Bureau of Statistics to find the value of the CPI for each of Sydney and Canberra in 2000/01.

(b) Using the values found in (a) find the adjusted annual salary (to the nearest dollar) in 2000/01 for the people described in:

(i) Exercise S10.35 (ii) Exercise S10.36

S10.38 (a) Select a sample of any 12 household items, and go to appropriate stores and record their prices, weight or size, and brand name. These items can include, for example, groceries, meat, vegetables, takeaway food, clothing, hardware, electrical appliances, petrol, computer items, and so on.

(b) One month later, go to the same stores and record the prices of exactly the same items.

(c) Use the information in (a) and (b) to construct your own CPI to represent the change in the cost of household items from one month to the next.

(d) Interpret the value found in (c).

S10.39 Repeat Exercise S10.38, except now include only food products in the 12 items.

S10.40 Repeat Exercise S10.38, except now include only non-food products in the 12 items.

S10.41 Consider the two CPI values obtained in Exercises S10.39 and S10.40. Do you think that there is much difference in the change in the prices of food products as opposed to non-food products? Comment.

S10.42 Figure S10.2 illustrates an artist's display of the trend of 'real house prices' for various countries between 1979 and 1992. Note that the vertical axis is marked 'Index', so presumably the data represent some sort of index number.

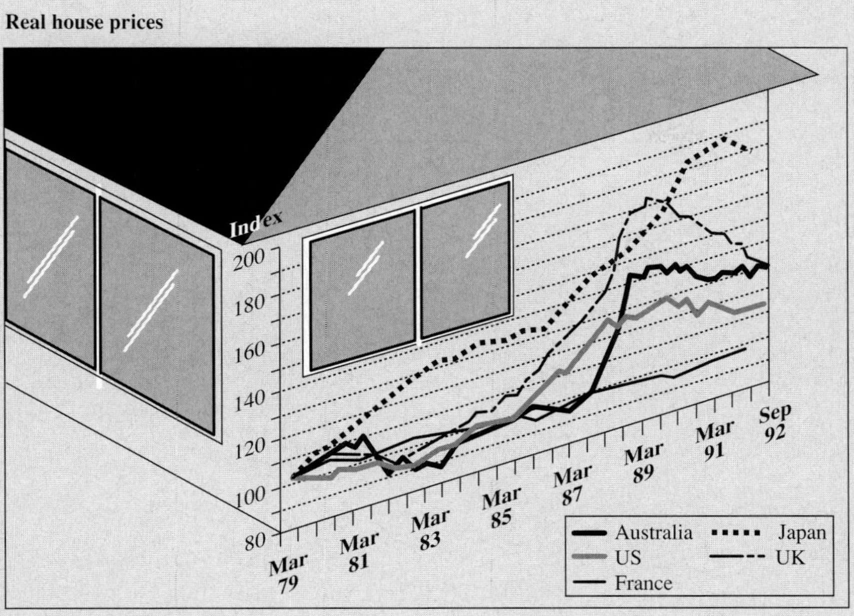

Figure S10.2 Real house prices for various countries, 1979–92

Source: *Sydney Morning Herald*, 9 January 1993, p. 57

(a) What base time period do you think is being used to calculate each index?

(b) List all the conclusions that you can draw from the display.

S10.43 List some of the features of the illustration in Figure S10.2. In particular, make note of any that you find attractive and any that you find unsatisfactory. Give reasons. Suggest any improvements that you might make.

Chapter S11

Time series and trend analysis

'Weather forecast for tonight: dark.'

(George Carlin, comedian, writer)

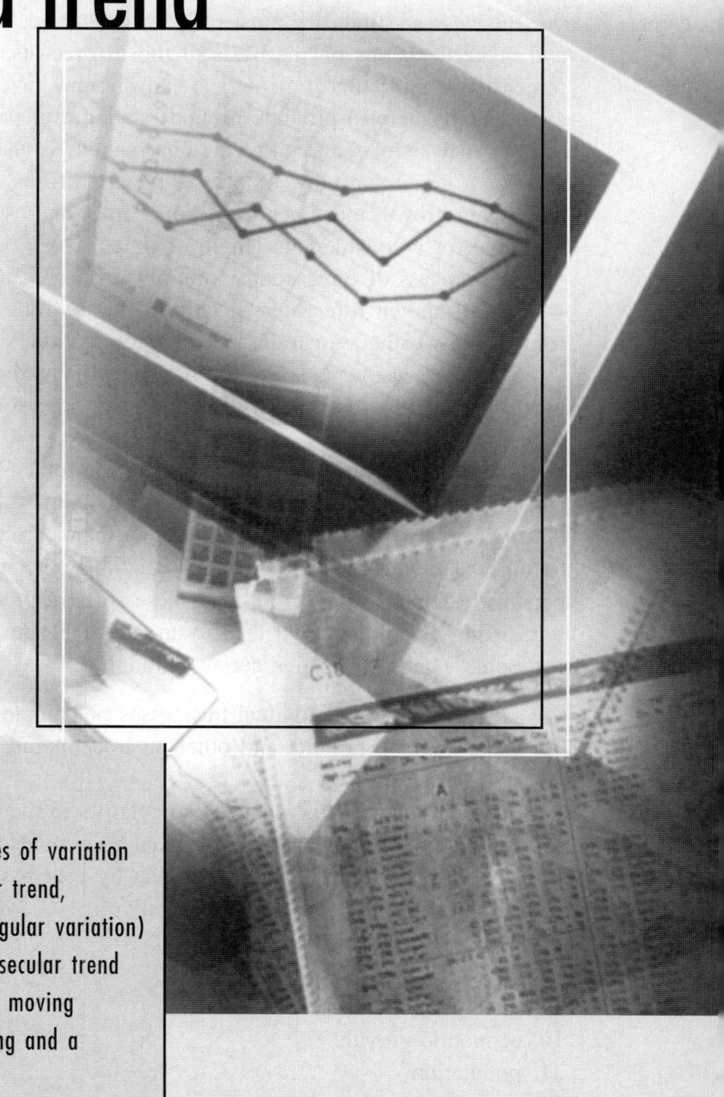

Learning objectives

In this chapter you will learn to:

- Describe a time series and explain its use
- Identify and interpret the four basic measures of variation that appear in a time series analysis (secular trend, seasonal variation, cyclical variation and irregular variation)
- Identify and use common methods of fitting secular trend lines to time series (including semi-averages, moving averages, least-squares, exponential smoothing and a growth model)
- Make forecasts

S11.1 INTRODUCTION

A **time series** consists of a set of observations that are measured at specified (usually equal) time intervals. Time series analysis attempts to identify those factors that exert an influence on the values in the series. Once these factors are identified, the time series may be projected into the future and used for both short-term and long-term forecasting.

Time series may be taken over very short periods (tiny fractions of a second in nuclear physics), over hours or days in some biological measurements, over years in economic contexts, or even over centuries in astronomy. The process is easiest to visualise in terms of years and months, as typically used in business and economics, and we shall therefore speak in those terms. You should, however, be prepared to translate the units to whatever is relevant in a given context.

Business and industry must plan for the future, so someone must forecast future activity levels and make decisions on the basis of those forecasts. Time series analysis is a basic tool for such forecasts. In particular, time series analysis may reveal a persistent pattern in the observations and consequently help to promote an understanding of past and current changes.

An analysis of the trend of the observations is also important for studying the progress of events and for a general review of prevailing conditions. The *trend* may be defined as the long-term underlying growth movement in a time series, and as such it can only be determined if the data are available for a considerable length of time.

The process of forecasting is not always easy, and it has often received adverse media publicity when well-known forecasters get things very wrong. This can include poor prediction of election results, the weather, Australia's economic position, sporting results, and so on.

Since forecasting is so tricky, it is rare for predictions to be *exactly* correct, regardless of how much information you have. The best we can hope for is to try to minimise the inevitable errors that will occur.

There are many techniques available for forecasting future events but in this chapter we will consider only a few. One of the better methods may be simply the 'gut feeling' or 'educated guess' that managers with a lot of experience in their particular business or industry may use. There may be no real scientific basis for their predictions other than what they, from their working knowledge, personally feel will happen. Often, these sorts of predictions, together with appropriate quantitative forecasting techniques, will lead to an acceptable result.

Some examples of time series include:

1. sales figures for individual businesses or complete industries
2. gross national product and other macroeconomic measures
3. share prices
4. the value of the Australian dollar relative to the US dollar
5. electricity consumption by individual households or by whole communities, on an hour-by-hour, day-by-day, month-by-month or year-by-year basis
6. number of employees on the payroll
7. number of staff working hours
8. costs of raw materials
9. unemployment rates
10. economic growth
11. population
12. foreign debt
13. trade figures

14. interest rates
15. working days lost

The list is endless, but many businesses keep a record of company information in their general bookkeeping activities so that the information is readily on hand if required. The time series data itself may be annual, semi-annual, quarterly, monthly, weekly or even daily: it depends on the type of business.

Example S11.1

A company is involved in the making of meat pies, which must be produced the night before and delivered fresh to shops the next day. In this case, the company would keep *daily* sales data, since the number of pies that may be sold has to be forecast every night.

Example S11.2

A service station sends out accounts to customers at the end of each month. The station keeps records of the amounts owed by such customers on a *monthly* basis, and so the accounts receivable from them may be summarised each month.

S11.2 TIME SERIES COMPONENTS

The four components usually identified are:

1. secular trend
2. seasonal variation
3. cyclical variation
4. irregular variation

While it is possible to break down a time series into these four components, the task is not always simple. For example, cycles are often difficult to recognise, let alone analyse, and they can vary in both intensity and length.

Secular trend

The **secular trend** is the long-term growth or decline of a series. In typical economic contexts, 'long-term' may mean 10 years or more. Essentially, the period should be long enough for any persistent pattern to emerge.

The secular trend therefore lets us describe the history of the time series, so that we could, for example, look at the success or otherwise of past management decisions. One situation may be where a company has opened a new branch or launched a new product, and the wisdom of that decision may be seen in the long-term trend of the time series.

Secular trends also allow us to look at past patterns or trends and use these to make some prediction about the future. In some situations it is also possible to isolate the effect of secular trends from the time series and hence make studies of the other components easier.

While trend estimates are often reliable, there are some instances where the usefulness of trend estimates is reduced by a high degree of irregularity in the original or seasonally adjusted series or by abrupt change in the time series characteristics of the original data.

Example S11.3

Figure S11.1 shows the secular trend of Australia's level of unemployed persons between November 1990 and November 2000. The trend estimate reached a peak of 946 800 in September 1993, before falling rapidly to 751 900 in June 1995. The trend then rose slowly for almost 2 years to 799 400 in March 1997 and then generally fell to 620 000 in November 2000. However, six months later in May 2001 it had risen again to 669 100.

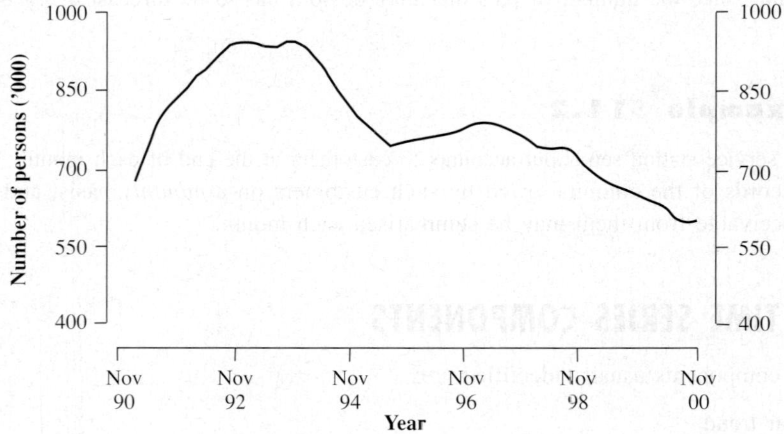

Figure S11.1 A graph illustrating the secular trend of the number of employed persons in Australia between November 1990 and November 2000

***Source*:** Australian Bureau of Statistics, *Labour Force*, November 2000, p. 3, Cat. No. 6202.0

Example S11.4

The graph in Figure S11.2 shows the amazing trend of the value Australian dollar versus the value of the US dollar during an 18-hour period on 8 November 2000. The occasion was the first day of the counting of the votes in the 2000 US Presidential election. In the early hours of the morning (Sydney time), as Democratic candidate Al Gore was seen to be doing well, the US dollar weakened (i.e. the Australian dollar rose), and when a major television network announced that he had won the state of Florida, the US dollar weakened further. However, when the network expressed some doubt about Gore's Florida victory, the US dollar rose sharply, especially when it announced that Republican candidate George W. Bush had in fact won the state. Finally, at around 6.00 p.m., when Al Gore retracted his concession of the election, the US dollar once again fell sharply.

The result of the US Presidential election was later battled out in the courts, with George W. Bush being declared the ultimate winner.

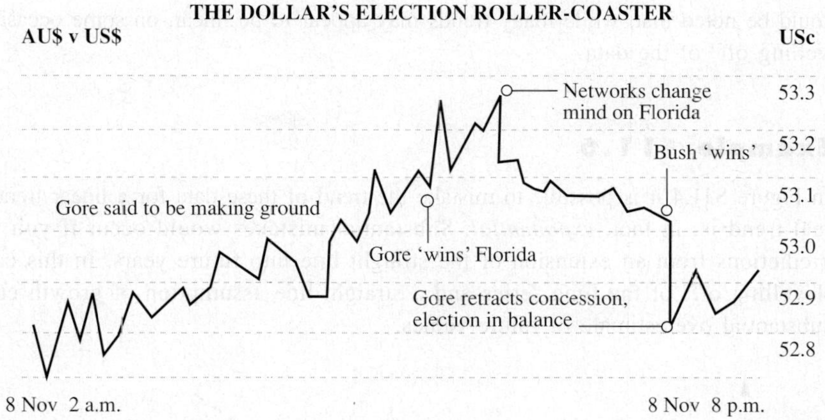

Figure S11.2 The changing value of the Australian dollar versus the US dollar during the first day of vote-counting in the 2000 US Presidential election

Source: *Sydney Morning Herald*, 10 November 2000, p. 26

Example S11.5

The Australian dollar is again highlighted in Figure S11.3, which shows the long-term trend of the exchange rate with the US dollar between 1970 and 2000. At its peak during the Whitlam government era in 1973–74, the Australian dollar was worth US$1.49, but 10 years later in March 1983, when Malcolm Fraser left office, it had dropped by over 35% to US$0.96.

The Hawke government let the dollar float in December 1983 when it was worth US$0.90 and it was still trading at US$0.89 in January 1989. It then fell to a trough of US$0.64 in September 1993, but peaked at US$0.81 in late 1996. From that point it has been largely downhill, with the value falling to below US$0.49 in March 2001.

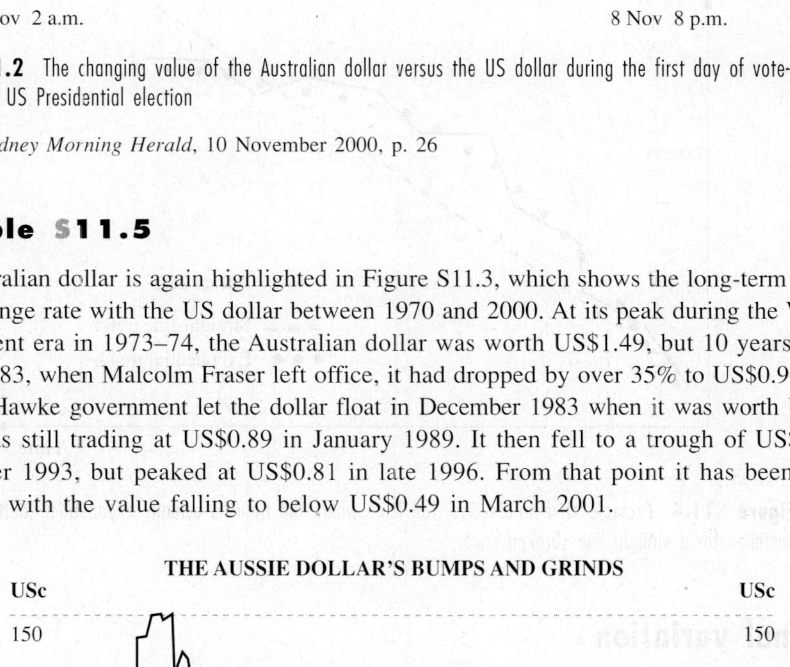

Figure S11.3 A graph illustrating the 30-year trend of the value of the AUS$ versus the value of the US$ between 1970 and 2000

Source: *Sydney Morning Herald*, 4 November 2000, p. 50

It should be noted that, while many trends may appear to be linear, on some occasions there may be a 'levelling off' of the data.

Example S11.6

In Figure S11.4 it is possible to mistake the trend of these data for a linear trend, whereas the real trend is, in fact, *exponential*. Substantial mistakes would occur if you tried to make predictions from an extension of the straight line into future years. In this case, there is a 'levelling off' of the time series and a straight-line assumption of growth could lead to a substantial overestimate of future values.

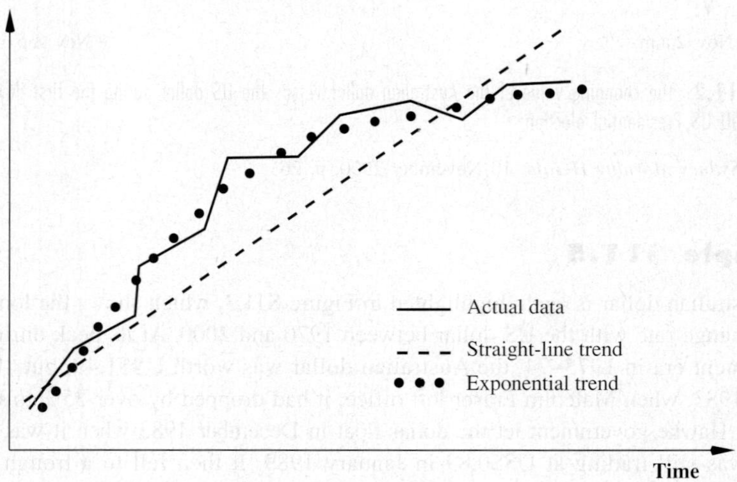

Figure S11.4 Example of a time series (full line) where the trend is actually exponential (dotted line) but might be mistaken for a straight line (broken line)

Seasonal variation

The **seasonal variation** of a time series is a pattern of change that recurs regularly over time. Seasonal patterns typically are one year long; that is, the pattern starts repeating itself at a fixed time each year.

Seasonal variations are usually due to the differences between seasons and to festive occasions such as Easter and Christmas. Examples include:

1. Airconditioner sales are greater in the summer months.
2. Heater sales are greater in the winter months.
3. The total number of people seeking work is large at the end of each year when students leave school.
4. Motels, hotels and camping grounds have a greater volume of customers in holiday seasons.
5. Medical practitioners report a substantial increase in the number of flu cases each winter.
6. Liquor outlets undergo increased sales during festive seasons.
7. Airline ticket sales (and prices!) increase during school holidays.

8. The Australian Consumer Price Index (CPI) is often at its highest during the December quarter and lowest during the March quarter.
9. The amount of electricity and water used varies within each 24-hour period.
10. The volume of work for tax agents increases dramatically around the time when income tax forms have to be filed.

While variations may recur every year, the concept of seasonal variation also extends to those patterns that occur monthly, weekly, daily or even hourly.

Time series graphs may be prepared using an adjustment for seasonal variations, although the other components of the series are not affected. Such graphs are said to be *seasonally adjusted* or *deseasonalised.*

The purpose of a seasonally adjusted series is to obtain a true picture of genuine movements in the time series after the seasonal effects have been removed.

Example S11.7

Figure S11.5 shows the trend of the unemployment rates for four countries between June 1988 and June 2000. These series have been seasonally adjusted to take into account such factors as the times of the year when school leavers are about to try to enter the workforce. Without this, there would be regular peaks and troughs to represent seasonal factors: the deseasonalising of the series smooths these out.

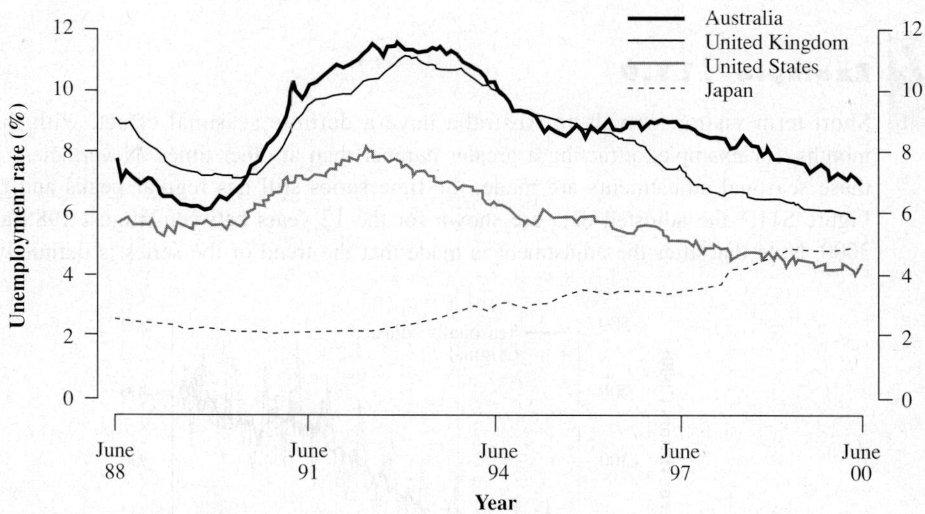

Figure S11.5 Seasonally adjusted unemployment rates for Australia, the United Kingdom, the United States and Japan between June 1988 and June 2000

Source: Australian Bureau of Statistics, *Australian Economic Indicators*, December 2000, Cat. No. 1350.0

Example S11.8

Figure S11.6 shows the balance of Australia's current account for the 12-year period between June 1988 and June 2000. This diagram shows a graph of both the seasonally adjusted balance and its trend line. If the balance were not seasonally adjusted, it could be very misleading by being either artificially too high or too low. The Australian Bureau of Statistics usually provides a graph of the seasonally adjusted figures in these cases, but also provides the actual (unadjusted) data in table form.

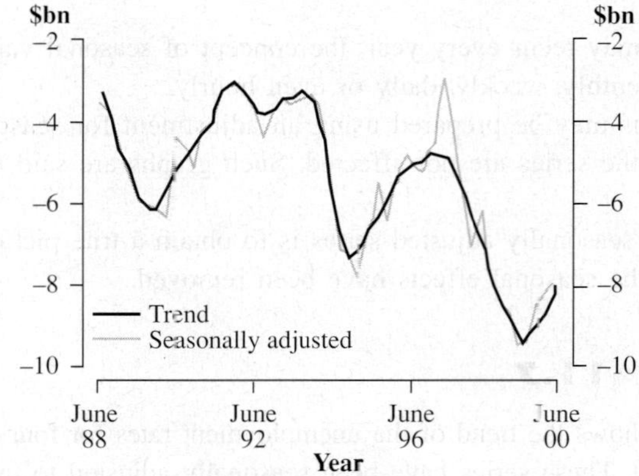

Figure S11.6 A graph showing the actual and seasonally adjusted trends of the balance of Australia's current account between June 1988 and June 2000

Source: Australian Bureau of Statistics, *Australian Economic Indicators*, December 2000, Cat. No. 1350.0

Example S11.9

Short-term visitor arrivals to Australia have a definite seasonal effect, with the summer months, for example, attracting a greater number than at other times. Nevertheless, even after these seasonal adjustments are made, the time series still has regular peaks and troughs. In Figure S11.7 the adjusted data are shown for the 12 years between August 1988 and August 2000. Note that after the adjustment is made that the trend of the series is definitely upwards.

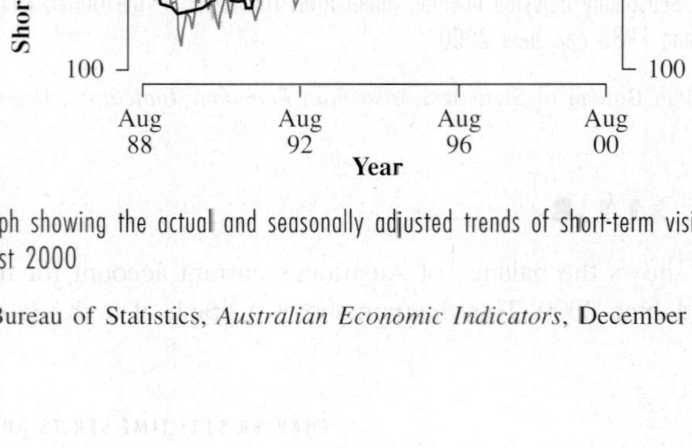

Figure S11.7 A graph showing the actual and seasonally adjusted trends of short-term visitors to Australia between August 1988 and August 2000

Source: Australian Bureau of Statistics, *Australian Economic Indicators*, December 2000, Cat. No. 1350.0

Example S11.10

Suppose there were some adverse publicity in December about ice-cream. It would be incorrect simply to compare the sales of ice-cream in June with those in December to determine the effect of the bad publicity. December would have higher sales in any case, because it is a warmer month. Comparisons of sales could only be made after seasonal variations had been allowed for. The true impact of the publicity could then be determined.

Cyclical variation

In a similar manner to seasonal variations, **cyclical variations** have recurring patterns, but have a longer and more erratic time scale. For instance, in economic contexts the term often describes the tendency for a particular industry to recover from a recession, become prosperous, decline into a recession and recover again. Unlike seasonal variation, the business cycle is extended over several years, and there is no guarantee that there will be any regularly recurring pattern.

Examples of causes of cyclical variation include:

1. floods
2. earthquakes/hurricanes
3. droughts
4. wars
5. changes in interest rates
6. major increases or decreases in the population
7. the opening of a new shopping complex
8. the building of a new airport
9. economic depressions or recessions
10. major sporting events, such as the Olympic Games
11. changes in consumer spending (i.e. lack of confidence)
12. changes in government monetary policy

In fact, the name 'cyclical variation' itself is an unfortunate one, since the word 'cyclical' often implies some sort of regularity. However, these cycles can be far from regular and it is usually impossible to predict just how long these periods of expansion and contraction will be. For example, cyclical variation can be caused by a war, although it is not clear that outbreaks of war occur in any regular pattern or have any predefined length.

There are techniques available for identifying the cyclical variation in a time series. One popular method is the **residual method**, but it is beyond the scope of this book.

Example S11.11

The ABS Experimental Composite Leading Indicator (XCLI) is a single time series designed to provide early signals of turning points in the Australian business cycle. Past performance of the XCLI shows it led turning points in the business cycle by between one-quarter and six-quarters, with the average being around two-quarters. Figure S11.8 shows the movements in the value of the XCLI between September 1988 and September 2000, along with the deviation of the gross domestic product (GDP). Note the trough in late 1990 and the peak in September 1994.

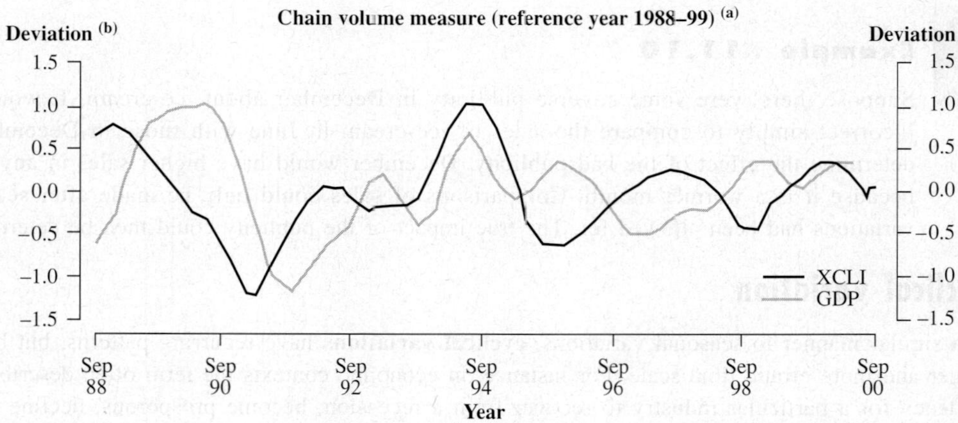

(a) The historical long-term trend growth rate of GDP is 1.03% in the June Quarter 2000 and the trend growth rate is 1.01%.
(b) Deviation is the unit of measure for the GDP series and it refers to the deviation of trend from its historical long-term trend. The XCLI series has no official unit of measure, i.e. it is dimensionless.

Figure S11.8 The value of the XCLI and its target, the business cycle in GDP, between September 1988 and September 2000

Source: Australian Bureau of Statistics, *Australian Economic Indicators*, December 2000, Cat. No. 1350.0

Example S11.12

A time series covering the period between 1988 and 2000 is illustrated in Figure S11.9, where the data represent the rate of inflation in Australia. The cyclical variation here is quite pronounced. From a high of 8.0% in 1990, the trend dropped markedly to around 1.2% in 1992, before rising to a peak of 4.8% in 1995. After another trough in 1997 it was rising again in 2000, partly due to the introduction of the GST.

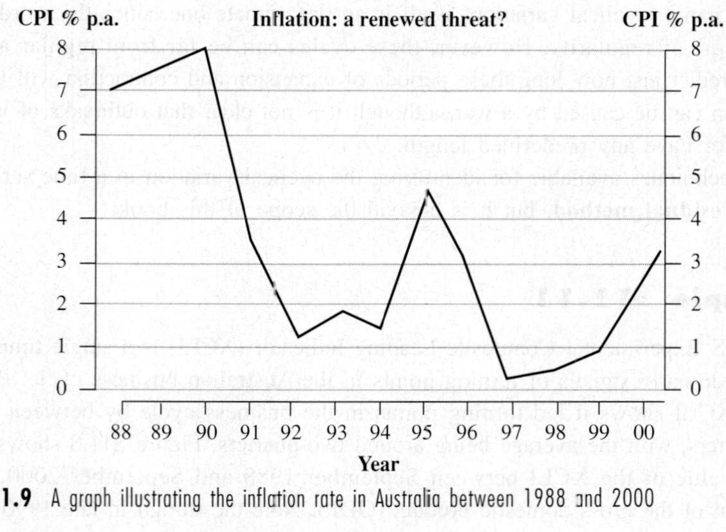

Figure S11.9 A graph illustrating the inflation rate in Australia between 1988 and 2000

Source: 'Labour Force', *The Bulletin*, 12 September 2000, p. 67

Irregular variation

Irregular variation in a time series occurs over varying (usually short) periods. It follows no regular pattern and is by nature unpredictable. It usually occurs randomly and may be linked to events that also occur randomly.

Examples of events that might cause irregular variation are:

1. the assassination (or disappearance) of a country's leader
2. short-term variations in the weather, such as unseasonably warm winters (they may affect sales of certain products)
3. sudden changes in interest rates
4. the collapse of large (or even small) companies
5. strikes (e.g. a strike by airline pilots affects many people working in the travel industry)
6. a government calling an unexpected election
7. sudden shifts in government policy
8. natural disasters
9. dramatic changes to the stock market
10. the effect of war in the Middle East on petrol prices around the world

Irregular variation cannot be explained mathematically. However, we can sometimes pinpoint its causes after the event (events such as those listed above), but it is not always easy. In general, if the variation in a time series cannot be accounted for by secular trend, or by seasonal or cyclical variation, then it is usually attributed to irregular variation.

Example S11.13

An example of irregular variation is illustrated in Figure S11.10, which shows the comparison of the quarterly growth rates for chain volume measures of Australia's GDP in both seasonally adjusted and trend terms between June 1986 and June 2000. Note the negative values in 1990 and 1991.

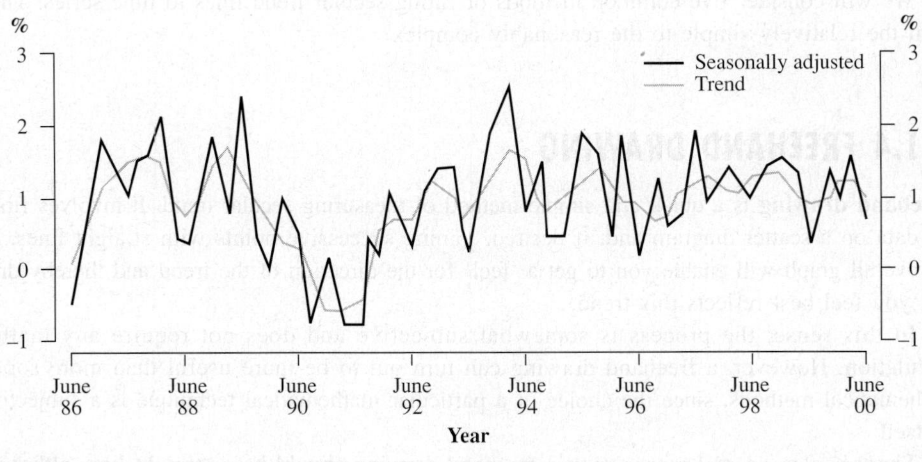

Figure S11.10 A graph illustrating quarterly growth rates for chain volume measures of Australia's GDP between June 1988 and June 2000

Source: Australian Bureau of Statistics, *Australian National Accounts*, June 2000, Cat. No. 5206.0

A further question to be decided is just how individual components interact to form a time series. In practice, we could assume that they all multiply or add together, but realistically the situation is more complex than this. Analysis of this problem is beyond the scope of this book.

S11.3 MEASUREMENT OF SECULAR TREND

The basic aim of estimating the trend of a time series is to describe the underlying movement of the series. When measuring a secular trend, a good starting point is to graph the data. In doing so, an obvious pattern may emerge, thus avoiding the need for mathematical calculations.

The trend itself may be represented by a smooth curve or a straight line. Keep in mind that we are looking at the trend over a long period and, in doing so, are 'averaging out' the highs and lows of the series. Also remember that measurement of secular trend can be somewhat subjective, depending on the method used to measure it. For example, to one person a straight line may appear reasonable, while to another a curve may seem more appropriate.

There are many reasons why we may want to examine the trend of data. These include:

1. For sales data, for example, the trend of the sales of one firm may be measured and then compared with the sales trend of competitive firms.
2. If a particular past trend can be reasonably expected to continue into the future, this may be used as a basis for future planning.
3. Performance of governments may be compared by examining the trend of appropriate economic indicators.
4. Consideration of the use of utilities such as gas, electricity and water will assist in planning for the future.
5. The state of the housing industry has shown marked trends over the past 15 years or so. An examination of the trend of median house prices gives a revealing picture of how the market has been progressing (and where it may be heading).

We will consider five common methods of fitting secular trend lines to time series. These range from the relatively simple to the reasonably complex.

S11.4 FREEHAND DRAWING

Freehand drawing is a quick and simple method of measuring secular trend. It involves first plotting the data on a scatter diagram and, if desired, joining successive points with straight lines. A look at the overall graph will enable you to get a 'feel' for the direction of the trend and thereby draw a line that you feel best reflects this trend.

In this sense, the process is somewhat subjective and does not require any mathematical calculation. However, a freehand drawing can turn out to be more useful than more sophisticated mathematical methods, since the choice of a particular mathematical technique is a subjective choice in itself.

There is also no real reason why a freehand drawing should be a straight line, although it may be easier to achieve.

Example S11.14

The graph in Figure S11.11 shows company profits before income tax in Australia between June 1988 and June 2000. A dotted straight line drawn in freehand has been superimposed and extended to June 2003 where it suggests that the figure will be about $12 000 million.

In Figure S11.12, the same graph as in Figure S11.11 has been used, but now a dotted freehand drawing of a curve, more of an exponential shape, has been superimposed. This takes into account the dip in the profits between 1994 and 1997 and takes a more conservative approach. This curve suggests a figure of about $9500 million in June 2003.

It is a matter of opinion which (if either) of the dotted lines in Figures S11.11 and S11.12 best describes the secular trend of this time series.

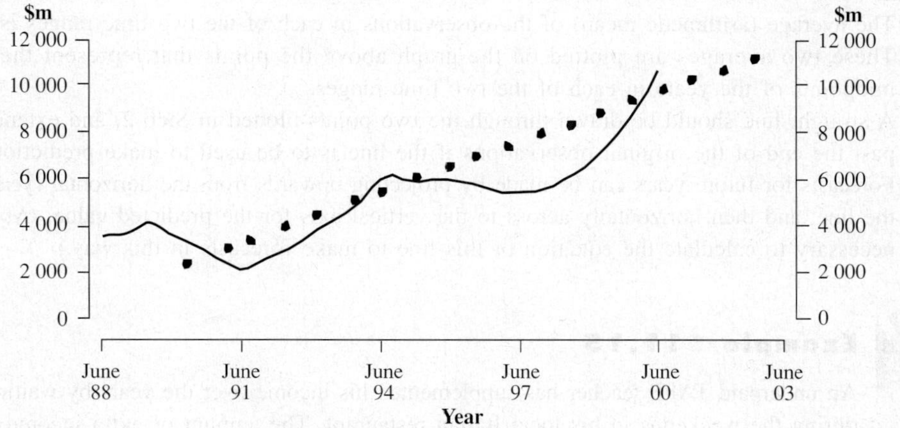

Figure S11.11 A straight-line trend curve fitted to company profits before income tax in Australia between June 1988 and June 2000

Source: Australian Bureau of Statistics, *Australian Economic Indicators*, December 2000, Cat. No. 1350.0

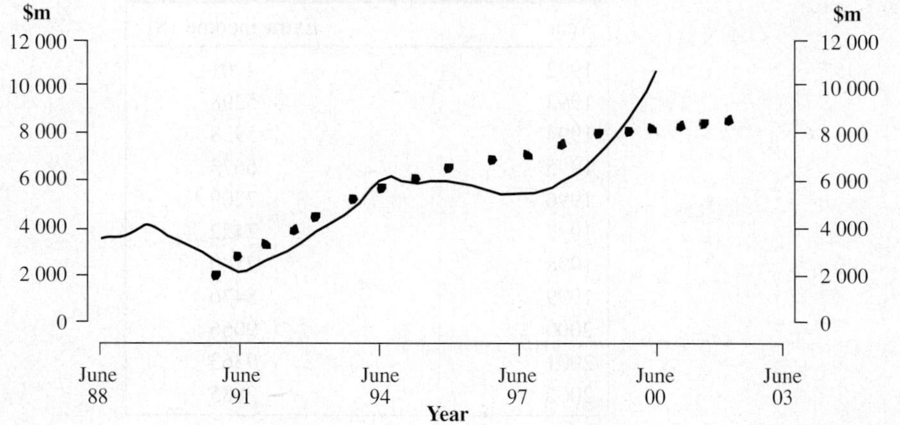

Figure S11.12 A freehand trend curve fitted to company profits before income tax in Australia between June 1988 and June 2000

Source: Australian Bureau of Statistics, *Australian Economic Indicators*, December 2000, Cat. No. 1350.0

S11.5 SEMI-AVERAGES

Of all the scientific methods of fitting a straight line to describe secular trend, the method of **semi-averages** is the easiest. However, the technique may not provide a very good outcome since it can be affected by outlying or unusual points in the series.

Nevertheless, in many cases it does provide a reasonable approximation of the trend of a time series if a straight line is to be used. The technique (assuming that the measurements are yearly) is as follows:

1. The data are divided into two equal time ranges. (If the number of years in the time series is *odd*, the middle year should be disregarded. For example, if the data comprised 25 annual data observations, then the first time range would consist of years 1–12 and the second of years 14–25. The observation for year number 13 would be ignored.)
2. The average (arithmetic mean) of the observations in each of the two time ranges is calculated. These two averages are plotted on the graph above the points that represent the respective midpoints of the years in each of the two time ranges.
3. A straight line should be drawn through the two points plotted in Step 2, and extended slightly past the end of the original observations if the line is to be used to make predictions.
4. Forecasts for future years can be made by projecting upwards from the horizontal (year) axis onto the line, and then horizontally across to the vertical axis for the predicted value. (*Note*: It is *not* necessary to calculate the equation of this line to make forecasts in this way.)

Example S11.15

An underpaid TAFE teacher has supplemented his income over the years by waiting on tables during the weekends in his local Italian restaurant. The amount of extra income that he has made by this activity is shown in Table S11.1 for the 11 years from 1992 to 2002.

Table S11.1 Extra income earned each year by the TAFE teacher in Example S11.15

Year	Extra income ($)
1992	4701
1993	5298
1994	5938
1995	6673
1996	7209
1997	7422
1998	7780
1999	8476
2000	9066
2001	9363
2002	9885

Find the secular trend of these data using the method of semi-averages, and draw it onto a graph of the data.

Solution

The solution is shown in Table S11.2. Note that, since we have 11 years (which is an odd number), we disregard the middle year of 1997.

Table S11.2 Calculations for Example S11.15

Year	Extra income ($)	Semi-totals ($)	Semi-averages ($)
1992	4701		
1993	5298		
1994	5938	29 819	5963.8
1995	6673		
1996	7209		
1997	7422	(disregard)	
1998	7780		
1999	8476		
2000	9066	44 570	8914.0
2001	9363		
2002	9885		

From Table S11.2, the *semi-total* for the 5 years 1992 to 1996 is $29 819. Hence, the *semi-average* for these years is $29 819 ÷ 5 = $5963.8.

The *semi-total* for the 5 years 1998 to 2002 is $44 570. Hence, the *semi-average* for these years is $44 570 ÷ 5 = $8914.0.

The semi-average line is drawn through the point 5963.8 directly above the year 1994 and the point 8914.0 directly above the year 2000.

Figure S11.13 shows the graph of the actual data in Table S11.1 as well as the semi-average line. Note that the line is quite a good fit.

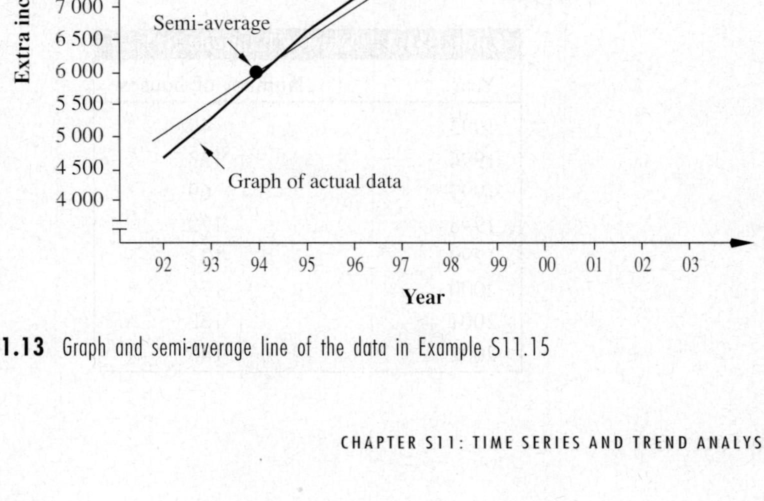

Figure S11.13 Graph and semi-average line of the data in Example S11.15

Example S11.16

Use the semi-average line in Figure S11.13 to predict what extra income the TAFE teacher might earn in 2003.

Solution

The dotted line in Figure S11.14 shows how we have projected upwards from 2003 onto the semi-average line and then moved horizontally across to the vertical axis. This leads to a prediction of about $10 400 for his extra income in 2003.

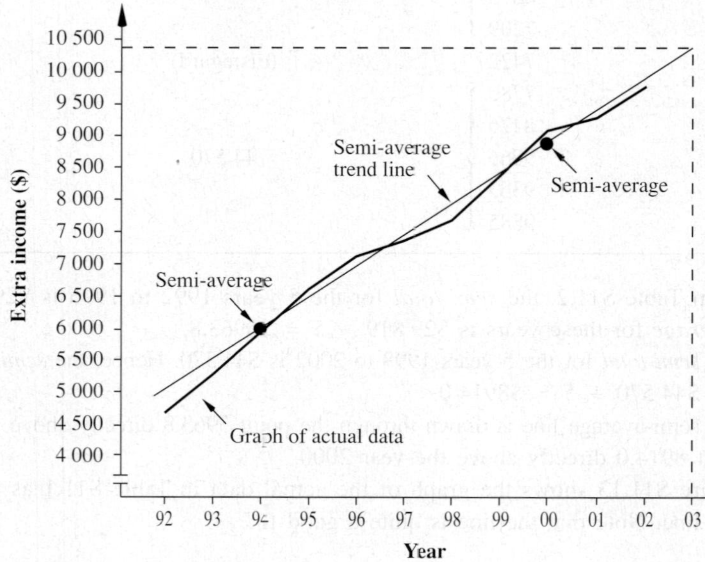

Figure S11.14 Graph, semi-average line and 2003 prediction for the data in Example S11.15

Example S11.17

The number of houses submitted for auction by a 'no sale–no charge' real-estate auction company over the period 1995 to 2002 is shown in Table S11.3.

Table S11.3 Data for Example S11.17

Year	Number of houses
1995	49
1996	133
1997	69
1998	170
1999	133
2000	175
2001	152
2002	185

Find the secular trend line using the method of semi-averages and draw it, along with a graph of the data.

Solution

This time we have an even number of years (8), so all the data can be used. The calculations are shown in Table S11.4.

Table S11.4 Calculations for Example S11.17

Year	Number of houses	Semi-totals ($)	Semi-averages ($)
1995	49		
1996	133	421	105.25
1997	69		
1998	170		
1999	133		
2000	175	645	161.25
2001	152		
2002	185		

From Table S11.4 the semi-total for the 4 years 1995–98 is 421. Hence, the semi-average for these years is $421 \div 4 = 105.25$.

The semi-total for the 4 years 1999–2002 is 645. Hence, the semi-average for these years is $645 \div 4 = 161.25$.

The semi-average line is drawn through the point 105.25 directly above the midpoint of the years 1996 and 1997, and the point 161.25 directly above the midpoint of the years 2000 and 2001.

Figure S11.15 shows the graph of the actual data in Table S11.3 as well as the semi-average line. Note that this line is also a reasonable fit of the trend.

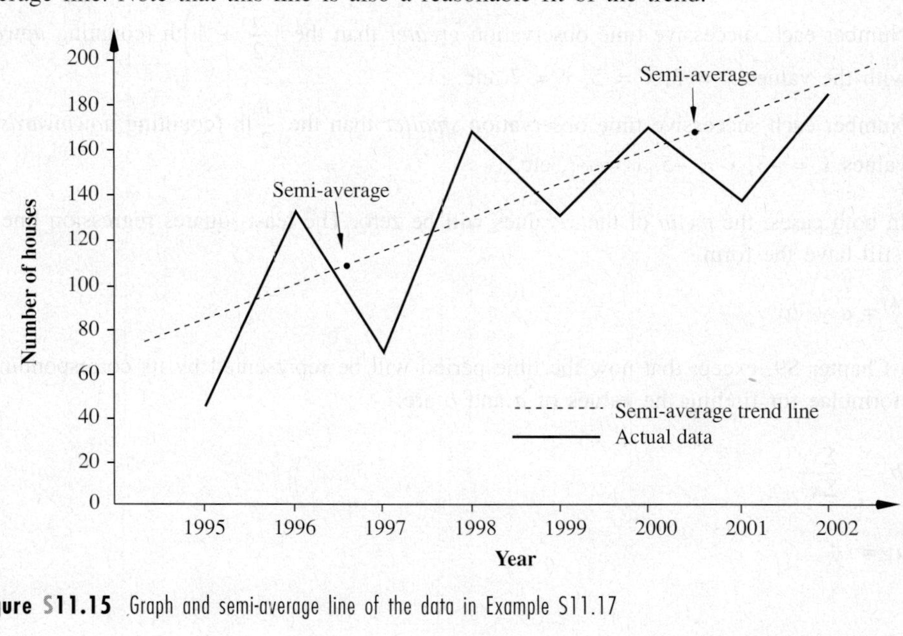

Figure S11.15 Graph and semi-average line of the data in Example S11.17

The method of semi-averages is simplistic, but it is at least a starting point. If the resulting straight line *does* appear to be a reasonable fit, it could be used in its own right; or perhaps a least-squares regression line could be tried, as in the next section.

S11.6 LEAST-SQUARES LINEAR REGRESSION

A more sophisticated way of fitting a straight line to a time series is to use the method of least-squares regression discussed in Chapter S9. In this case, the observations are the (dependent) y-variables and time is the (independent) x-variable.

Since in this case the x-variable is time units, the calculations may be simplified as follows.

Case I

Suppose that there are n time observations (x) where n is an *odd* number.

1. Let the middle time observation have the value $x = 0$.
2. Number each successive time observation *greater* than the middle one (counting *upwards*) with the values $x = 1$, $x = 2$, $x = 3$, and so on.
3. Number each successive time observation *smaller* than the middle one (counting *downwards*) with the values $x = -1$, $x = -2$, $x = -3$, and so on.

Case II

Suppose that there are n time observations (x) where n is an *even* number.

1. Let the $\frac{n}{2}$th observation have the value $x = -1$ and the $\left(\frac{n}{2} + 1\right)$th observation have the value $x = +1$.
2. Number each successive time observation *greater* than the $\left(\frac{n}{2} + 1\right)$th (counting *upwards by 2*) with the values $x = 3$, $x = 5$, $x = 7$, etc.
3. Number each successive time observation *smaller* than the $\frac{n}{2}$th (counting *downwards*) with the values $x = -3$, $x = -5$, $x = -7$, etc.

In both cases, the *mean* of the x-values will be zero. The least-squares regression line of y on x will still have the form

$$\hat{y} = a + bx \qquad \textbf{(S11.1)}$$

as in Chapter S9, except that now the time period will be represented by its corresponding x-value. The formulae for finding the values of a and b are:

$$b = \frac{\Sigma xy}{\Sigma x^2} \qquad \textbf{(S11.2)}$$

$$a = \bar{y} \qquad \textbf{(S11.3)}$$

Example S11.18

Find the least-squares regression line of extra income (in dollars) on year for the data in Example S11.15.

Solution

Since the number of time periods is *odd* (there are 11 years of data), this is Case I. Hence, we number the middle year (1997) by $x = 0$. The other years are as in Table S11.5, where the rest of the calculations are shown.

Table S11.5 Calculations for Example S11.18

Year	Value of x	Extra income ($) y	x^2	xy
1992	–5	4 701	25	–23 505
1993	–4	5 298	16	–21 192
1994	–3	5 938	9	–17 814
1995	–2	6 673	4	–13 346
1996	–1	7 209	1	–7 209
1997	0	7 422	0	0
1998	1	7 780	1	7 780
1999	2	8 476	4	16 592
2000	3	9 066	9	27 198
2001	4	9 363	16	37 452
2002	5	9 885	25	49 425
Total	0	81 811	110	55 381

Using Equation S11.2:

$$b = \frac{\Sigma xy}{\Sigma x^2} = \frac{55\,381}{110} = 503.46$$

and using Equation S11.3:

$$a = \bar{y} = \frac{81\,811}{11} = 7437.36$$

Hence, the least-squares regression line of y on x is:

$$\hat{y} = 7437.36 + 503.46x \qquad \textbf{(S11.4)}$$

Example S11.19

Use the least-squares regression line in Equation S11.4 in Example S11.18 to predict the extra income the TAFE teacher might earn in 2003 and 2004.

Solution

In this case, the year 2003 corresponds to $x = 6$. Using Equation S11.4, we get:

$$\hat{y} = 7437.36 + (503.46)(6) = 10\,458.12$$

Hence, we would predict extra income of about \$10 458 in 2003.

The year 2004 corresponds to $x = 7$. Using Equation S11.4, we get:

$$\hat{y} = 7437.36 + (503.46)(7) = 10\,961.58$$

Hence, we would predict extra income of about \$10 962 in 2004.

Note: The slope (coefficient of x) of the least-squares regression line (503.46) represents the amount of increase in extra income (in dollars) that the regression line predicts the TAFE teacher will receive each year.

Example S11.20

Find the least-squares regression line of number of houses submitted for auction on year for the data in Example S11.17.

Solution

Since the number of time periods is even (there are $n = 8$ years of data), this is Case II. Hence, we number the $\frac{n}{2}$th = 4th year (1998) by $x = -1$ and the $\left(\frac{n}{2} + 1\right)$th = 5th year (1999) by $x = +1$. The other years are as in Table S11.6, where the rest of the calculations are shown.

Table S11.6 Calculations for Example S11.20

Year	x	Number of houses y	x^2	xy
1995	−7	49	49	−343
1996	−5	133	25	−665
1997	−3	69	9	−207
1998	−1	170	1	−170
1999	1	133	1	133
2000	3	175	9	525
2001	5	152	25	760
2002	7	185	49	1295
Total	0	1066	168	1328

Using Equation S11.2:

$$b = \frac{\Sigma xy}{\Sigma x^2} = \frac{1328}{168} = 7.90$$

Using Equation S11.3:

$$a = \overline{y} = \frac{1066}{8} = 133.25$$

Hence, the least-squares regression line of y on x is:

$$\hat{y} = 133.25 + 7.90x \qquad \textbf{(S11.5)}$$

Example S11.21

Use the least-squares regression line given by Equation S11.5 in Example S11.20 to predict the number of houses that will be submitted for auction in 2003.

Solution

In this case, the year 2003 corresponds to $x = 9$. Using Equation S11.5, we get:

$$\hat{y} = 133.25 + (7.90)(9) = 204.35$$

Hence, we would predict that about 204 houses will be submitted for auction in 2003.

Despite the scientific nature of the least-squares regression technique, it must be reiterated that straight lines will still be inaccurate if the trend of the time series does *not* follow a straight line. In all cases, a scatter diagram of the observations should be drawn *first* to determine whether this is the case.

S11.7 MOVING AVERAGES

The method of **moving averages** is based on the premise that, if the values in a time series are *averaged* over a sufficient period, the effect of short-term variations will be reduced. That is, short-term cyclical, seasonal and irregular variations will be *smoothed out*, leaving an apparently smooth graph to show the overall trend.

By choosing the number of cases to be included in any one average, the degree of smoothing can be controlled. For example, for monthly data, a 12-month moving average smooths out seasonal effects, while for quarterly data a 4-quarter moving average smooths out quarterly variations.

The choice is not so simple if the data are *yearly*, since cyclical variations can be of different durations. One suggestion is to look at a graph of the data and obtain a rough indication of a 'typical' cycle length, and then use that number of years in each moving average. Note, however, that the larger the number of observations taken in the moving average, the greater the smoothing effect.

The technique for finding a moving average for a particular observation is to find the *average* of the m observations *before* the observation, the observation itself and the m observations *after* the observation.

Thus, a total of $(2m + 1)$ observations must be averaged each time a moving average is calculated. The value of m chosen should be a whole number (e.g. 1, 2, 3). The larger the value of m, the greater the smoothing effect on the data.

Example S11.22

If the data were yearly, for $m = 1$ we would find 3-year moving averages, for $m = 2$ we would find 5-year moving averages, and so on. We will first consider the case where we have an *odd* number of years used in the average.

Odd number of years used in the average

Steps in the calculation are:

1. Select the desired value of m.
2. For each observation x, calculate the *moving total*, where:
   Moving total = total of each group of m observations of which the middle one is observation x
3. For each observation x, calculate the *moving average*, where:

$$\text{Moving average} = \frac{\text{moving total}}{m}$$

Example S11.23

The 3-year *moving total* for an observation x would be the sum of the observation immediately *before* x, x itself and the observation immediately *after* x.

The 3-year *moving average* would be each of these moving totals divided by 3.

The 5-year *moving total* for an observation x would be the sum of the *two* observations immediately *before* x, x itself and the *two* observations immediately *after* x.

The 5-year *moving average* would be each of these moving totals divided by 5.

Note: It is not possible to calculate the moving totals and moving averages for *all* observations in the time series. For example, if a 3-year moving average is used, the moving average cannot be found for both the first and last observations. If a 5-year moving average is used, the moving average cannot be found for each of the first two and each of the last two observations.

Example S11.24

A white goods manufacturer has recorded the number of washing machines that have been sold each year through a national retailer between 1994 and 2002. The details are shown in Table S11.7.

Find the 3-year moving averages to obtain the trend of the data.

Solution

In this case, since we have 3-year moving averages, $m = 1$. The calculations are shown in Table S11.8, where the 3-year moving averages are given to the nearest integer. Note that the moving averages cannot be found for the first year (1994) and the last year (2002).

Table S11.7 Annual sales of washing machines between 1994 and 2002 for Example S11.24

Year	Number of sales
1994	1011
1995	1031
1996	976
1997	1020
1998	1195
1999	1174
2000	1261
2001	1330
2002	1384

Table S11.8 Calculation of the 3-year moving averages for the data in Example S11.24

Year	Number of sales	Three-year moving total	Three-year moving average
1994	1011	—	—
1995	1031	3018	1006
1996	976	3027	1009
1997	1020	3191	1064
1998	1195	3389	1130
1999	1174	3630	1210
2000	1261	3765	1255
2001	1330	3975	1325
2002	1384	—	—

An example of the calculation in Table S11.8 is for the year 1995, where:

$$\text{Moving total} = 1011 + 1031 + 976 = 3018$$

$$\text{Moving average} = \frac{3018}{3} = 1006$$

Example S11.25

Find the 5-year moving averages for the data in Example S11.24.

Solution

The calculations are shown in Table S11.9, where the 5-year moving averages are given to the nearest integer. Note that the moving averages cannot be found for the first 2 years (1994 and 1995) or the last 2 years (2001 and 2002).

Table S11.9 Calculation of the 5-year moving averages for the data in Example S11.24			
Year	Number of sales	Five-year moving total	Five-year moving average
1994	1011	—	—
1995	1031	—	—
1996	976	5233	1047
1997	1020	5396	1079
1998	1195	5626	1125
1999	1174	5980	1196
2000	1261	6344	1269
2001	1330	—	—
2002	1384	—	—

An example of the calculation in Table S11.9 is for the year 1997, where:

$$\text{Moving total} = 1031 + 976 + 1020 + 1195 + 1174 = 5396$$

$$\text{Moving average} = \frac{5396}{5} = 1079$$

Example S11.26

Use the calculations in Tables S11.8 and S11.9 to draw both the 3-year and 5-year moving-average trend curves for the data in Example S11.24. Comment.

Solution

The trend curves are shown in Figure S11.16, together with the original data. The 5-year moving average trend curve has smoothed out the variation the most.

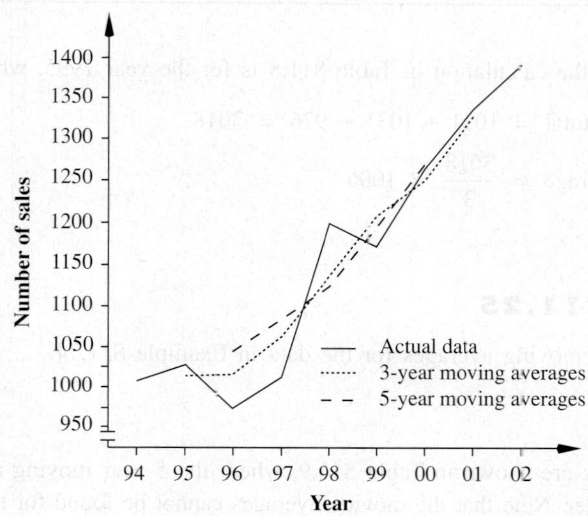

Figure S11.16 Three-year and five-year moving-average trend curves for the data in Example S11.24

Even number of years used in the average

Suppose now that we want to calculate moving averages where there is an *even* number of periods included in each average. Clearly, the situation is more complicated, since the same number of periods before and after a given period cannot be included.

The method used in this case is to calculate *two* moving averages for each period. These two averages are themselves averaged to obtain the final moving average.

Suppose, for example, that we wish to calculate a 4-year moving average for a set of observations in a time series. For each year x the steps are:

1. Calculate the 4-year *moving total* from year $(x - 2)$ to year $(x + 1)$. (Call this number y_1.)
2. Calculate the 4-year *moving average* from year $(x - 2)$ to year $(x + 1)$ by dividing y_1 by 4. (Call this number $\bar{y}_1$.)
3. Calculate the 4-year *moving total* from year $(x - 1)$ to year $(x + 2)$. (Call this number y_2.)
4. Calculate the 4-year *moving average* from year $(x - 1)$ to year $(x + 2)$ by dividing y_1 by 4. (Call this number $\bar{y}_2$.)
5. Calculate the *average* of $\bar{y}_1$ and $\bar{y}_2$. This is the (centred) 4-year moving average for year x.

Note: It is *not* possible to calculate the 4-year moving average for the first 2 years and the last 2 years in a time series.

To calculate, for example, the 6-year moving average for year x, replace $(x - 2)$ by $(x - 3)$, and $(x + 1)$ by $(x + 2)$ in Step 1. Replace $(x - 1)$ by $(x - 2)$, and $(x + 2)$ by $(x + 3)$ in Step 3. Replace the divisor 4 by 6 in Steps 2 and 4. Similar adjustments must be made for 8-year, 10-year, and so on, moving averages.

Example S11.27

The average daily sales of milk (in L) at a country store are shown in Table S11.10 for each of the years 1992 to 2002. Find the 4-year moving averages.

Table S11.10 Average daily milk sales for Example S11.27

Year	Average daily sales (L)
1992	47.6
1993	48.9
1994	51.5
1995	55.3
1996	57.9
1997	61.7
1998	65.3
1999	70.2
2000	76.1
2001	84.7
2002	93.2

Solution

The complete set of calculations is shown in Table S11.11.

Table S11.11 Calculations for Example S11.27

Year	Average daily sales (L)	Four-year total (Step 1)	Four-year average (Step 2)	Four-year total (Step 3)	Four-year average (Step 4)	Average of Step 2 and Step 4 (Step 5)
1992	47.6	—	—	—	—	—
1993	48.9	—	—	203.3	50.8	—
1994	51.5	203.3	50.8	213.6	53.4	52.1
1995	55.3	213.6	53.4	226.4	56.6	55.0
1996	57.9	226.4	56.6	240.2	60.0	58.3
1997	61.7	240.2	60.0	255.1	63.8	61.9
1998	65.3	255.1	63.8	273.3	68.3	66.0
1999	70.2	273.3	68.3	296.3	74.1	71.2
2000	76.1	296.3	74.1	324.2	81.0	77.6
2001	84.7	324.2	81.0	—	—	—
2002	93.2	—	—	—	—	—

For example, to calculate the 4-year moving average for 1994, the following steps are required:

1. Calculate the 4-year moving total from 1992 to 1995 (i.e. 203.3).
2. Hence, calculate the 4-year moving average from 1992 to 1995 (i.e. $\frac{203.3}{4} = 50.8$).
3. Calculate the 4-year moving total from 1993 to 1996 (i.e. 213.6).
4. Hence, calculate the 4-year moving average from 1993 to 1996 (i.e. $\frac{213.6}{4} = 53.4$).
5. Calculate the average of the two values found in Steps 2 and 4 (i.e. $\frac{50.8 + 53.4}{2} = 52.1$).

Example S11.28

Use the calculations in Table S11.11 to draw the 4-year moving average trend curve for the data in Example S11.27.

Solution

The trend curve is shown in Figure S11.17. Note that the moving averages follow the actual data fairly closely.

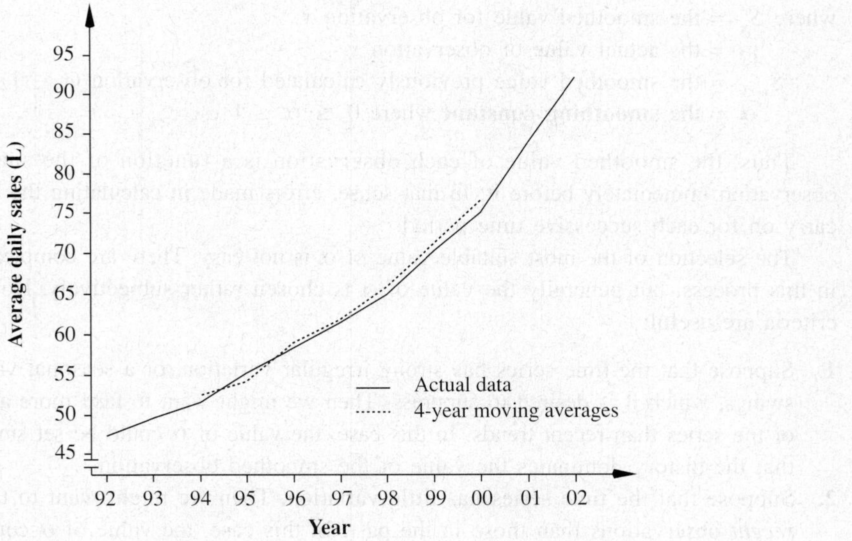

Figure S11.17 Four-year moving-average trend curve for the data in Example S11.27

Remarks

While the method of moving averages is easily understood and minimises the effect of extreme values, there are several drawbacks in its use. Among these are the fact that it cannot be brought up to date and may consequently be of little use for an analysis of the current period or for future predictions.

The choice of the number of periods to be included in the averaging process may also cause erratic behaviour. This is particularly so if there are strong seasonal effects that may be emphasised by some choices but suppressed by other choices.

S11.8 EXPONENTIAL SMOOTHING

Many of the objections about moving averages can be overcome if we use a kind of *continuous* moving average. In this case, the smoothed value for an observation x is a *weighted* average to two quantities, namely:

1. the actual value of observation x
2. a value derived from the history of the time series

This technique is known as **exponential smoothing** and is a method for continually revising an estimate in the light of more recent trends. It is based on averaging (or smoothing) the past values in a series in an exponential manner.

The smoothed value for an observation x is obtained from the *recurrence relation* in Equation S11.6:

$$S_x = \alpha y_x + (1 - \alpha)S_{x-1} \qquad \textbf{(S11.6)}$$

where S_x = the smoothed value for observation x

y_x = the actual value of observation x

S_{x-1} = the smoothed value previously calculated for observation $(x - 1)$

α = the **smoothing constant** where $0 \leq \alpha \leq 1$

Thus, the smoothed value of each observation is a function of the smoothed value of the observation immediately before it. In that sense, errors made in calculating the smoothed values will carry on for each successive time period.

The selection of the most suitable value of α is not easy. There are complex techniques to assist in this process, but generally the value of α is chosen rather subjectively. However, the following criteria are useful:

1. Suppose that the time series has strong irregular variation, or a seasonal variation causing wide swings, which it is desired to suppress. Then we might want to take more account of *past* trends of the series than recent trends. In this case, the value of α could be set small (say, $\alpha = 0.1$) so that the history dominates the value of the smoothed observation.
2. Suppose that the time series has little variation. Then we might want to take more account of *recent* observations than those in the past. In this case, the value of α could be set large (say, $\alpha = 0.9$). Recent observations will dominate the value of the smoothed observation, with previous values providing merely a kind of background stability.

On most occasions in practice the calculations are done using a computer. In addition, many time-series computer packages will also provide the most appropriate value of α.

The exponential smoothing approach has to have a starting point, so we choose the *first* smoothed value (i.e. S_1) to be the value of the first observation (y_1) in the time series.

Example S11.29

Consider the sales data in Example S11.27. Calculate the exponentially smoothed values for this time series using a smoothing constant of $\alpha = 0.40$. Plot the smoothed data on a graph with the original observations. Comment.

Solution

Since $\alpha = 0.40$, then $1 - \alpha = 0.60$. Using Equation S11.6, we have:

$$S_x = 0.40y_x + 0.60S_{x-1}$$

As stated above, we begin with S_1 = first observation = y_1 = 47.60. Then, if $x = 2$:

$$S_2 = 0.40y_2 + 0.60S_1 = (0.40)(48.90) + (0.60)(47.60) = 48.12$$

Then, if $x = 3$:

$$S_3 = 0.40y_3 + 0.60S_2 = (0.40)(51.50) + (0.60)(48.12) = 49.47$$

This process continues throughout the time series for each observation. The full set of calculations is shown in Table S11.12, and graphed in Figure S11.18.

In this example, the method of exponential smoothing has no particular merit since the time series values were always *increasing*. In such cases, the smoothed values will always be consistent with the actual values as seen in Figure S11.18.

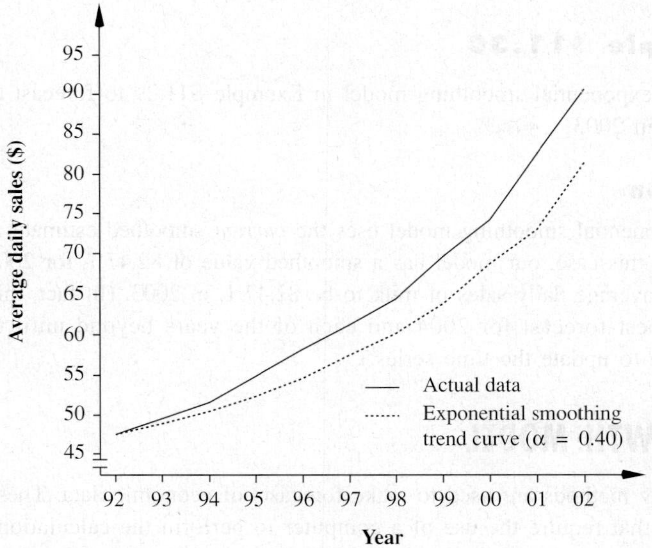

Figure S11.18 Graph of exponentially smoothed values for the data in Example S11.27

Table S11.12 Calculations for Example S11.29

Year	x	Observation (y_x)	S_{x-1}	$(1 - \alpha)\ S_{x-1}$	αy_x	S_x
1992	1	47.6				47.60
1993	2	48.9	47.60	28.56	19.56	48.12
1994	3	51.5	48.12	28.87	20.60	49.47
1995	4	55.3	49.47	29.68	22.12	51.80
1996	5	57.9	51.80	31.08	23.16	54.24
1997	6	61.7	54.24	32.54	24.68	57.22
1998	7	65.3	57.22	34.33	26.12	60.45
1999	8	70.2	60.45	36.27	28.08	64.35
2000	9	76.1	64.35	38.61	30.44	69.05
2001	10	84.7	69.05	41.43	33.88	75.31
2002	11	93.2	75.31	45.19	37.28	82.47

Note that the values of the observations in the time series in Example S11.29 are steadily *increasing*. In such cases, the smoothed values will be consistently *lower* than the actual values (see Figure S11.18). A remedy for this situation is to use an exponential smoothing model that involves *two* smoothing constants, but these models are beyond the scope of this book.

However, the method of exponential smoothing with a single smoothing constant can be very effective when a series has several components with wide variations.

Example S11.30

Use the exponential smoothing model in Example S11.29 to forecast the average daily sales of milk in 2003.

Solution

The exponential smoothing model uses the *current* smoothed estimate as a forecast for future years. In this case, our model has a smoothed value of 82.47 L for 2002. We would therefore forecast average daily sales of milk to be 82.47 L in 2003. (In fact, this would also currently be our best forecast for 2004 and each of the years beyond until more data were made available to update the time series.)

S11.9 GROWTH MODEL

In practice, many methods are used to make forecasts of economic data. These range from the very simple to those that require the use of a computer to perform the calculations.

The method of moving averages is useful for determining the secular trend of a time series, but cannot really be used for forecasting, as it cannot be brought up to date. We have already emphasised the dangers of assuming that all time series follow a straight line trend.

We now consider those series that appear to have little or no seasonal or cyclical variation. Suppose that we note from a graph of the data that the trend appears to be *exponential*. In this case, a **growth model** may be appropriate. A growth model is one that takes account of this exponential trend.

Suppose that we have a time series in which time is represented by the variable x and the corresponding observations are represented by the variable y. Further, suppose that we feel that the values of y are rising exponentially in relation to x. Then we may fit the model:

$$\hat{y} = ae^{bx} \qquad \textbf{(S11.7)}$$

where a and b are constants and $\hat{y}$ is the predicted value of y.

The aim now is to find the most appropriate values of a and b. The steps are:

1. Take the *natural logarithms* of the y-values (i.e. logarithms to the base e) to form a variable z such that:

$$z = \log_e y \qquad \textbf{(S11.8)}$$

2. Find the least-squares regression line of z on x, say:

$$z = c_1 + c_2 x$$

3. Then the values of a and b in Equation S11.7 are:

$$a = e^{c_1} \qquad \textbf{(S11.9)}$$

$$b = c_2 \qquad \textbf{(S11.10)}$$

Example S11.31

A company that sells large mainframe computers has worldwide sales for the years 1997 to 2002 as shown below.

Year	1997	1998	1999	2000	2001	2002
Sales	127	130	148	160	185	220

(a) Plot the data on a graph to determine whether a growth model might be appropriate.
(b) Construct the equation of the growth model.

Solution

(a) The graph of the data is shown in Figure S11.19. Since the trend seems to be increasing upwards in an exponential manner, it seems that a growth curve may be appropriate.

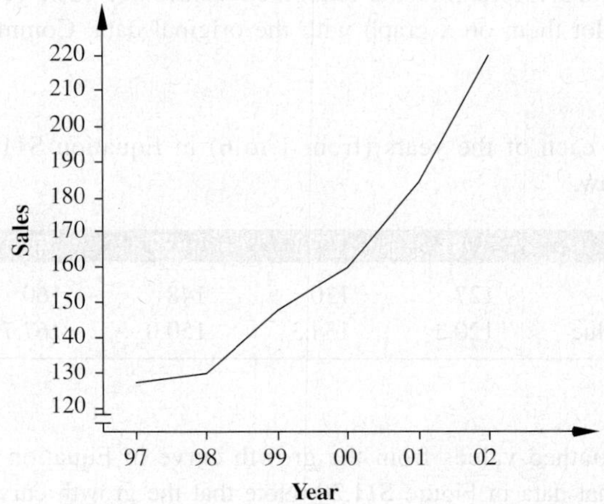

Figure S11.19 Graph of the data in Example S11.31

(b) Define the variables to be:

y = sales
x = year (define as 1, 2, 3, 4, 5, 6)
$z = \log_e y$

Hence, the values of the x and z-variables are:

x	1	2	3	4	5	6
z	4.844	4.868	4.997	5.075	5.220	5.394

Using the techniques of Chapter S9, it can be shown that the least-squares regression line of z on x is:

$$z = 4.678 + 0.111x$$

Therefore, $c_1 = 4.678$ and $c_2 = 0.111$.
From Equation S11.9:

$$a = e^{4.678} = 107.55$$

and from Equation S11.10:

$$b = 0.111$$

Hence, from Equation S11.7 the growth model is:

$$\hat{y} = 107.55\, e^{0.111x}$$

or, in terms of the original variables:

$$\text{Sales} = 107.55\, e^{0.111(\text{year})} \quad \textbf{(S11.11)}$$

Example S11.32

Use Equation S11.11 to find the smoothed values for each year for the data in Example S11.31, and plot them on a graph with the original data. Comment.

Solution

We substitute each of the years (from 1 to 6) in Equation S11.11 to obtain the smoothed values as below.

Year	1997	1998	1999	2000	2001	2002
Actual value	127	130	148	160	185	220
Smoothed value	120.2	134.3	150.0	167.7	187.3	209.3

These smoothed values from the growth curve in Equation S11.11 are plotted together with the original data in Figure S11.20. Note that the growth curve has a slight upward trend, although not as pronounced in recent years as that in the original data.

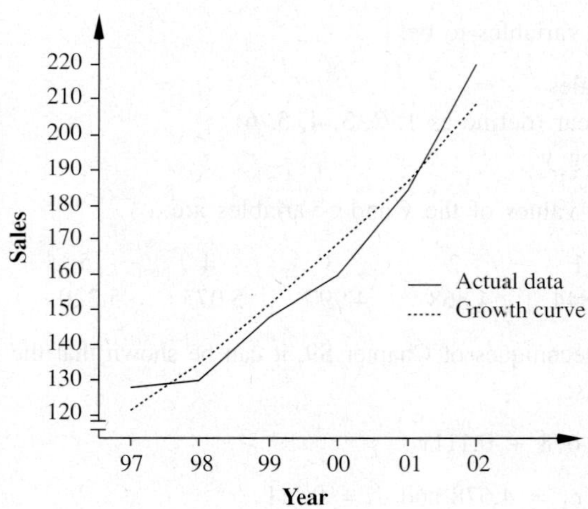

Figure S11.20 Graph of the smoothed values for the data in Example S11.31, calculated using the growth model

Example S11.33

Use the growth model in Example S11.31 to forecast the sales of computers in 2003.

Solution

To forecast the sales in 2003 we let *year* = 7 in Equation S11.11. This yields:

$$\begin{aligned} \text{Sales} &= 107.55\ e^{0.111(7)} \\ &= 107.55\ e^{0.777} \\ &= 233.9 \end{aligned}$$

Hence, we would forecast sales of about 234 computers in 2003. (Remember that in making this prediction we are assuming that past trends will continue into the future.)

S11.10 REMARKS

In this chapter we have considered only a handful of the models for describing the trend of a time series and only some of the techniques for predicting future values. There are many others that could be used, and statisticians are constantly striving for even better approaches.

However, while some remarkable achievements have been made in the area of time series analysis, there will never be a 'perfect' approach to the subject.

Quick Quiz

1. Give an example of a time series of Australian economic data that you feel would have a strong irregular component.
2. Explain in one or two sentences why government unemployment figures are usually presented as 'seasonally adjusted'.
3. Give an example of a time series consisting of 8 observations, and where the method of semi-averages provides a quite inaccurate measure of trend. Illustrate your answer using a graph.
4. Explain briefly why the method of moving averages cannot be used to make forecasts. What is the main purpose of moving averages?
5. For the method of exponential smoothing to be used, is it necessary for the time series data to have an exponential shape? Explain.
6. A time series consists of 12 yearly observations. Suppose you want to calculate a 6-year moving-average trend curve. For how many of the 12 observations could these moving averages be calculated? Explain.
7. Consider a time series of sales data where there is a constant increase each year in sales. Do you feel that a growth model fitted to such data would *overestimate* or *underestimate* the sales in future years? Explain.
8. If two successive 3-year moving averages are equal, what does this mean about the data?
9. Why do you think that economists often disagree with their forecasts of factors such as the inflation rate and the interest rate?
10. Give an example of data for which the time series might be recorded every hour. Explain why this time interval would be necessary.

Multiple Choice Questions

1. A difficulty with freehand drawing is that:
(a) it is subjective
(b) it is not scientific
(c) some people cannot draw well
(d) it depends on who draws the line
(e) all of the above may be difficulties

2. The least-squares regression line of weekly sales (in kg) on price (dollars per kg) for tomatoes at a certain fruit store is:

Weekly sales = 120 − 18.3 × price

If the price is $2.40 per kg in one week, this line predicts that the number of tomatoes (to the nearest kg) that will be sold will be:
(a) 76 (b) 84 (c) 102
(d) 164 (e) none of the above

3. The sales of bicycles at a discount store over a 6-year period are shown below:

Year	1	2	3	4	5	6
Sales	63	72	81	83	80	89

The 3-year moving total for year 3 is:
(a) 216 (b) 236 (c) 243
(d) 244 (e) none of the above

4. In Question 3, the 3-year moving average (to two decimal places) for year 5 is:
(a) 78.67 (b) 81.33 (c) 80.00
(d) 84.00 (e) none of the above

5. In Question 3, the 4-year moving average (to two decimal places) for year 5 is:
(a) 83.25 (b) 84.00 (c) 82.00
(d) not calculable (e) none of the above

6. A growth model may be useful when:
(a) all data points are roughly equal
(b) the trend is not exponential
(c) there is a seasonal or cyclic effect
(d) there is no seasonal or cyclic effect
(e) none of the above

7. A smoothing constant of $\alpha = 0.85$ might be used in exponential smoothing if:
(a) there were strong irregular variations
(b) there was little variation
(c) all the values were increasing
(d) there was no computer available to perform the calculations
(e) none of the above

8. The *secular trend* of a time series measures:
(a) the number of peaks and troughs
(b) the long-term growth or decline
(c) whether an appropriate measure of time has been chosen
(d) the cost of maintaining the series
(e) none of the above

9. *Linear regression* may be appropriate for a time series when:
(a) the unit of time is in years
(b) the trend of the data is exponential
(c) a growth model is also appropriate
(d) cyclical variation only is present
(e) none of the above

10. A time series may be *seasonally adjusted* only when:
(a) the data include the Christmas season
(b) patterns of change occur regularly over time
(c) the data are not increasing over time
(d) forecasts are not being made as a result
(e) none of the above

Exercises

S11.1 The number of divorces granted in Australia for each year between 1992 and 1999 is shown below.

Year	1992	1993	1994	1995	1996	1997	1998	1999
Divorces	45 729	48 363	48 312	49 712	52 466	51 300	51 400	52 600

Source: Australian Bureau of Statistics, *Marriages and Divorces*, Cat. No. 3310.0

(a) Plot the data on a graph and join the points.

(b) Draw a freehand curve on the graph in (a) and describe the trend of the data.

(c) Extend the curve in (b) to predict the number of divorces granted in Australia in 2000.

(d) Check with the relevant publication of the Australian Bureau of Statistics to see how accurate you were.

S11.2 (a) Calculate the 3-year moving averages for the time series data in Exercise S11.1.

(b) Plot the moving averages in (a) on a graph with the original data. Comment.

S11.3* The numbers of cars sold by a used-car dealer were recorded for the month of January in each of the years from 1995 to 2002. Sales were:

Year	1995	1996	1997	1998	1999	2000	2001	2002
Sales	43	52	49	55	60	68	74	79

(a) Plot the data on a graph and join the points.

(b) Draw the trend line on the graph in (a) using the method of semi-averages. What are these semi-averages?

S11.4 (a) Use the trend line in Exercise S11.3(b) to predict the number of used cars that will be sold in January 2003 and January 2004.

(b) Do you have much faith in these predictions? Give reasons.

S11.5 (a) Find the equation of the least-squares regression trend line for the data in Exercise S11.3. (Label the years as 1, 2, . . ., 8.)

(b) Use the line found in (a) to predict the number of used cars that will be sold in January 2003 and January 2004.

(c) Do you have much faith in these predictions? Give reasons.

S11.6* Calculate the 3-year moving averages for the data in Exercise S11.3.

S11.7 Calculate the 4-year moving averages for the data in Exercise S11.3.

S11.8* Calculate the exponentially smoothed values for the time series in Exercise S11.3. Choose $\alpha = 0.5$.

S11.9 Calculate the exponentially smoothed values for the time series in Exercise S11.3. Choose $\alpha = 0.1$.

S11.10 (a) Calculate the equation of a growth model for the time series in Exercise S11.3. (Label the years as 1, 2, ..., 8.)

(b) Do you think that a growth model is appropriate for this time series? Give reasons.

(c) Use the trend curve found in (a) to predict the number of used cars that will be sold in January 2003 and January 2004.

(d) Do you have much faith in these predictions? Give reasons.

S11.11 A Sydney-based company produces a line of wrapped cakes, which it sells to stores both in the city and country New South Wales. The company has recorded these sales for each of the 12 quarters between December 1998 and September 2001, as shown in the table below. The company is keen to produce a model that will enable it to predict its cake sales for each upcoming quarter.

Quarter	Number of cakes sold
December 1998	1874
March 1999	2142
June 1999	2605
September 1999	2767
December 1999	2924
March 2000	3270
June 2000	4020
September 2000	4764
December 2000	4137
March 2001	4016
June 2001	3886
September 2001	3770

(a) Plot the data on a graph and join the points.

(b) Can you detect any seasonal or cyclical variation in your graph in (a)? If so, clearly state the nature of the variation.

S11.12 (a) Calculate 3-quarterly moving averages for the data in Exercise S11.11.

(b) Calculate 4-quarterly moving averages for the data in Exercise S11.11.

(c) Plot the moving averages found in (a) and (b) onto a graph of the original data. Which set of moving averages best follows the trend of the data? Why do you think this might be the case?

S11.13 (a) Use the method of semi-averages to draw a trend line on the graph of the data in Exercise S11.11.

(b) Use the line found in (a) to predict the number of cakes sold in the December quarter 2001.

S11.14* (a) Use exponential smoothing (with $\alpha = 0.5$) to draw a trend line on the graph of the data in Exercise S11.11.

(b) Use the line found in (a) to predict the number of cakes sold in the December quarter 2001.

S11.15* (a) Use the method of least-squares regression to find the equation of the trend line for the data in Exercise S11.11.

(b) Use the line found in (a) to predict the number of cakes sold in the December quarter 2001.

S11.16 You are now told that the actual number of cakes sold in the December quarter 2001 was 3679. Which of the predictions in Exercises S11.13(b), S11.14(b) and S11.15(b) came the closest? Do you regard this best estimate as reasonably accurate? Give reasons.

S11.17* (a) Draw the least-squares regression line found in Equation S11.4 on a graph of the actual data in Table S11.1.

(b) Do you regard the line in (a) as a good fit? Give a reason.

S11.18 (a) Draw the least-squares regression line found in Equation S11.5 on a graph of the actual data in Table S11.3.

(b) Do you regard the line in (a) as a good fit? Give a reason.

S11.19 The table below shows the amount of private new capital expenditure (in $'000, current prices, not seasonally adjusted) for a large Australian manufacturing company for each of the 12 quarters between September 1999 and June 2002.

Quarter	Amount of private new capital expenditure ($'000)
September 1999	6 613
December 1999	7 685
March 2000	6 346
June 2000	8 117
September 2000	7 933
December 2000	9 331
March 2001	7 806
June 2001	9 377
September 2001	8 638
December 2001	9 938
March 2002	8 679
June 2002	11 347

(a) Plot the data on a graph using an appropriate vertical scale, and join the points.

(b) Can you detect any seasonal or cyclical variation in the graph found in (a)? If so, state the nature of this variation.

(c) How would you describe the secular trend of this series?

S11.20 (a) Use exponential smoothing (with $\alpha = 0.80$) to draw a trend line on the graph of the data in Exercise S11.19.

(b) Use the line found in (a) to predict the amount of private new capital expenditure (to the nearest $'000) in the September quarter:

(i) 2002 (ii) 2004

S11.21* Least-squares linear regression was used to find the equation of the trend line in Exercise S11.19. The 12 quarters of data were labelled (1, 2, 3, . . ., 12) and called the *x*-variable. The *y*-variable was the amount of new capital expenditure (in $'000). The least-squares regression equation of *y* on *x* was:

$$\hat{y} = 6495.8 + 296.16x$$

(a) Draw this line on the graph in Exercise S11.19(a). State whether you think that it is a good fit of the data.

(b) Use the regression equation above to predict the amount of private new capital expenditure (to the nearest \$'000) in the September quarter:

(i) 2002 (ii) 2004

S11.22 Consider the problem in Exercise S11.19 with the 12 quarters of data being labelled (1, 2, 3, . . .,12) and called the x-variable. Let $z = \log_e(y)$ where y = expenditure (in \$million). The least-squares regression equation of z on x is:

$$\hat{z} = 8.7976 + 0.03528x$$

(a) Use the above equation to construct the equation of the growth model.

(b) Plot the growth model on the graph in Exercise S11.19(a).

(c) Use the equation found in (a) to predict the amount of private new capital expenditure (to the nearest \$million) in the September quarter:

(i) 2002 (ii) 2004

S11.23 Suppose that the amount of private new capital expenditure by the company in Exercise S11.19 for the September quarter 2002 was in fact \$10 160 000. Compare this value with your estimates in Exercises S11.20(b)(i), S11.21(b)(i) and S11.22(c)(i). Which estimate came the closest? Do you regard this best estimate as reasonably accurate? Give reasons.

S11.24 Suppose that the amount of private new capital expenditure by the company in Exercise S11.19 for the September quarter 2002 was in fact \$12 250 000. Compare this value with your estimates in Exercises S11.20(b)(ii), S11.21(b)(ii) and S11.22(c)(ii). Which estimate came the closest? Do you regard this best estimate as reasonably accurate? Give reasons.

S11.25* A retail outlet reports the following sales (in \$'000) during the Christmas holiday season for each year between 1993 and 2001:

Year	1993	1994	1995	1996	1997	1998	1999	2000	2001
Sales (\$'000)	52	37	85	84	49	110	113	62	136

(a) Plot the data and join the points. Draw a freehand curve to describe the trend.

(b) Extend the freehand curve found in (a) to 2002, and use it to predict the sales for that year.

(c) Do you notice any particular variation or trend by looking at the plot of the original data? If so, state what you see.

S11.26 (a) Calculate the 3-year moving averages for the data in Exercise S11.25.

(b) Plot the moving averages found in (a) on a graph with the original data. Comment.

S11.27* (a) Calculate the 5-year moving averages for the data in Exercise S11.25.

(b) Plot the moving averages found in (a) on a graph with the original data. Comment.

S11.28* (a) Use the method of semi-averages to draw a trend line on a graph of the data in Exercise S11.25.

(b) Comment on how appropriate you feel the line found in (a) is in describing the trend of the data.

(c) In view of your answer to (b), do you think that it would be of much use to try to fit a least-squares regression line to the data in Exercise S11.25? Give a reason.

S11.29 (a) Use exponential smoothing (with $\alpha = 0.5$) to draw a trend line on the graph of the data in Exercise S11.25.

(b) Do you think that the smoothed values are a better fit than a straight line? Give a reason.

(c) Use the line found in (a) to predict the sales in 2002.

S11.30 Research the literature and briefly describe exponential smoothing with *two* smoothing constants. Discuss the role and effectiveness of the second constant.

S11.31* Determine which of the four time-series components correspond to each of the following events:

(a) a general decrease in the demand for slide rules

(b) an annual sale of the previous year's summer fashions

(c) a general recession in the economy

(d) strikes at a work plant resulting in orders not being filled

S11.32* Suppose that a time series has a seasonal component that cycles over a 12-month period, with one measurement each month. If we calculate the moving averages based on 12 successive measurements, explain why seasonal effects will be cancelled out.

S11.33 The numbers of visitors to an art gallery are shown below for each year frcm 1996 to 2001.

Year	1996	1997	1998	1999	2000	2001
Visitors	17 284	16 489	18 656	17 838	19 049	20 238

(a) Draw a scatter diagram of the data and join the successive points. (Make the scale on the vertical axis from 16 000 to 21 000.) Discuss whether you feel that a straight line would describe the trend of the data.

(b) You now have to estimate the number of visitors in both 2002 and 2003. Use any of the methods discussed in this chapter to come up with your best estimate.

S11.34 The number of new houses built in a large region of Australia is shown in the table below for the 6 quarters between June 2000 and September 2001.

Quarter ending	June 00	September 00	December 00	March 01	June 01	September 01
Number of new houses	26 610	25 679	25 831	20 207	20 734	19 876

(a) Plot the data on a graph and join the points.

(b) Draw the trend line on the graph in (a) using the method of semi-averages.

(c) Use the trend line found in (b) to predict the number of houses built in the region during the December 2001 quarter.

(d) The number of houses built in the region during the December 2001 quarter was in fact 21 619. Calculate both the absolute and percentage errors made by your prediction in (c).

S11.35* (a) For the data in Exercise S11.34, you are required to find the least-squares regression line of the number of new houses (y) on quarter (x). (Label the quarters 1, 2, 3, 4, 5, 6.) You may use the information that the product-moment correlation coefficient between these two variables is $r = -0.908$, with $s_x = 1.871$ and $s_y = 3187$.

(b) Use the regression line found in (a) to predict the number of houses built in the region during the December quarter 2001.

(c) Find the coefficient of determination. Comment.

(d) The number of houses built in the region during the December quarter 2001 was in fact 21 619. Calculate both the absolute and percentage errors made by your prediction in (c).

S11.36 (a) Calculate the 3-quarter moving averages for the data in Exercise S11.34.

(b) Plot the averages found in (a) on a graph against quarter. Comment on the trend.

S11.37 (a) Calculate the exponentially smoothed values for the time series in Exercise S11.34. Choose:

(i) $\alpha = 0.90$ (ii) $\alpha = 0.10$

(b) Plot the values found in (a) on a graph of the original data. Which of the values of α provided the best fit of the data?

S11.38 The average outside temperature in January for each of the 6 years between 1997 and 2002 was measured at a weather station in the northwest of Sydney. The results are shown below.

Year	1997	1998	1999	2000	2001	2002
Temperature (°C)	24.6	21.4	23.4	24.0	21.2	22.2

(a) Plot the data on a graph and join the points. (Make the vertical scale between 21.0 and 25.0.)

(b) Draw the trend line on the graph in (a) using the method of semi-averages. Does it seem to fit the data well?

(c) Use the trend line found in (b) to predict the average outside temperature at the same weather station in January 2003.

(d) The average outside temperature at the same weather station in January 2003 was in fact 20.5°C. What absolute and percentage errors did you make in your prediction in (c)? Comment.

S11.39 (a) For the data in Exercise S11.38, find the least-squares regression line of the average outside temperature at the same weather station in January on the last two digits of the year. Draw it on a scatter diagram. Does it seem to fit the data well?

(b) Use the line found in (a) to predict the average outside temperature at the same weather station in January 2003.

(c) The average outside temperature at the same weather station in January 2003 was in fact 20.5°C. What absolute and percentage errors did you make in your prediction in (c)? Comment.

S11.40 The average retail price of one dozen eggs in Hobart at 30 June is shown below at each of the 5-year intervals between 1971 and 1996.

Year	1971	1976	1981	1986	1991	1996
Price ($)	0.70	1.08	1.63	2.02	2.39	2.75

Source: Australian Bureau of Statistics, *Average Retail Prices of Selected Items*, June Quarters 1971, 1976, 1981, 1986, 1991 and 1996, Cat. No. 6403.0

(a) Plot the data on a graph and join the points.

(b) Try to predict the price of eggs (to the nearest cent) in Hobart at 30 June 2001 using:

(i) the method of semi-averages

(ii) least-squares linear regression

(iii) exponential smoothing with $\alpha = 0.20$

(iv) exponential smoothing with $\alpha = 0.80$

(c) In which of the predictions in (b) do you have the most faith? Why?

S11.41* (a) Using the data in Exercise S11.40, find the equation of the growth model. (Use the last two digits of the year, i.e. 71, 76, etc.)

(b) Use the growth model to predict the price of eggs (to the nearest cent) in Hobart on 30 June 2001.

(c) In view of the graph in Exercise S11.40(a), do you think that the growth model is appropriate for this problem? Why?

S11.42 The production of wheat in Australia (in '000 tonnes) is shown below for each of the 6 years (year ending 31 March) between 1989/90 and 1994/95.

Year	1989/90	1990/91	1991/92	1992/93	1993/94	1994/95
Production ('000 tonnes)	14 214	15 066	10 557	14 739	16 479	8 972

Source: Australian Bureau of Statistics, *Agricultural Statistics*, Cat. No. 7117.0

(a) Plot the data on a graph and join the points. (Place production on the vertical axis with a scale between 8000 and 28 000.)

(b) Find the secular trend using the method of semi-averages, and draw it onto the graph found in (a).

(c) Extend the line found in (b) to predict the production of wheat in Australia in 1999/2000.

(d) In view of the unusually low figure for 1994/95, do you think that your prediction in (c) might be fairly accurate? Explain.

(e) The production of wheat in Australia in 1999/2000 was in fact 25 000 000 tonnes. What actual and percentage errors did you make in your prediction in (c)? Were you correct in your assessment in (d)?

S11.43 Calculate the 3-year moving averages for the data in Exercise S11.42, plot them on a graph and join the points. Also plot on the same graph the original data and join the points. Comment.

Chapter S12

Hypothesis testing

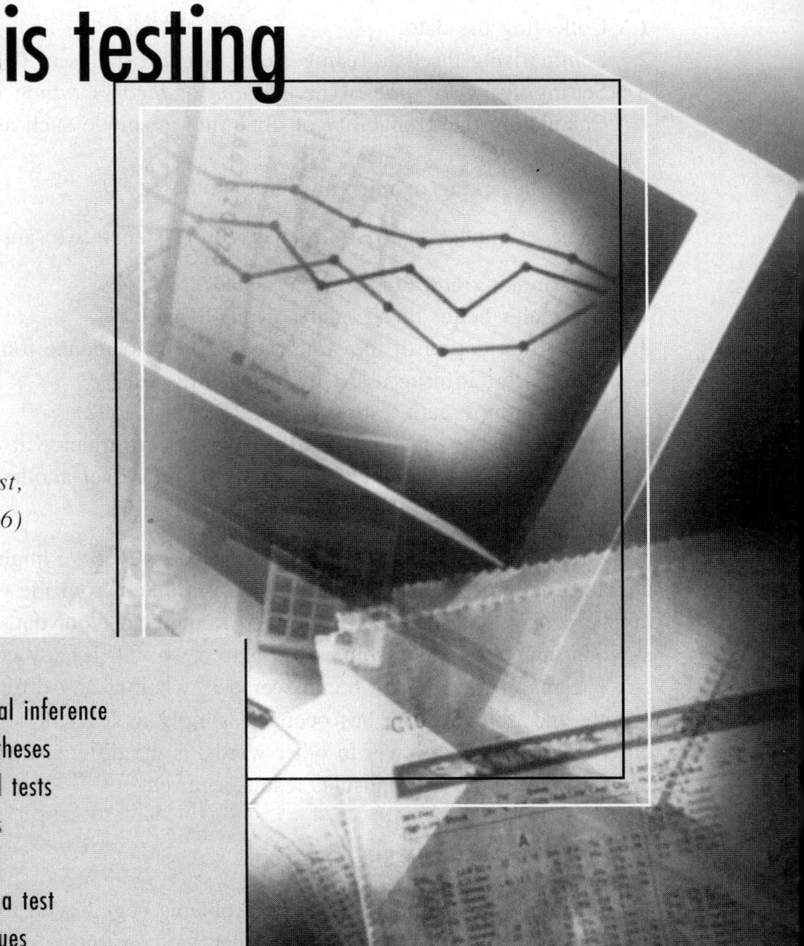

'It is a good morning exercise for a research scientist to discard a pet hypothesis every day before breakfast.'

(Konrad Lorenz, Austrian zoologist, On Aggression, *1966)*

Learning objectives

In this chapter you will learn to:

- Understand the principles of statistical inference
- Formulate null and alternative hypotheses
- Understand one-tailed and two-tailed tests
- Understand type I and type II errors
- Understand test statistics
- Understand the significance level of a test
- Understand and calculate critical values
- Understand the regions of acceptance and rejection
- Calculate and interpret a one-sample z-test statistic
- Calculate and interpret a one-sample t-test statistic
- Calculate and interpret a paired t-test statistic
- Calculate and interpret a two-sample t-test statistic
- Understand and calculate p-values

S12.1 STATISTICAL INFERENCE

One of the major roles of statisticians in practice is to draw conclusions from a set of data. This process is known as **statistical inference**, but it must always be borne in mind that, whatever conclusion is reached, it can always be wrong. However, in many circumstances we can put a probability on whether our conclusion is correct, and so we can make a decision that we could say is 'beyond reasonable doubt'.

In this decision-making process we usually undertake a series of steps, which can include some or all of the following:

1. Collecting the data.
2. Summarising the data (using either visual displays or descriptive statistics).
3. Setting up an *hypothesis* (i.e. a *claim* or *theory*), which is to be tested.
4. Calculating the probability of obtaining a sample such as the one we have if the hypothesis is *true*.
5. Either accepting or rejecting the hypothesis.

Some examples of questions that require statistics to answer are:

1. Are smoking and lung cancer related?
2. Are males better than females at mathematics?
3. Does fluoridation of the water supply help to reduce tooth decay?
4. Does a certain drug assist in curing a disease?
5. Has random breath-testing reduced the road toll?
6. Is performance at high school related to performance at university?
7. Do customers prefer one brand of a product over another?
8. Is profit related to the amount of advertising?

There are clearly a wide variety of questions that we might like to answer, and in most cases the conclusions will be made based on samples taken from the population. (Review the techniques for taking samples outlined in Chapter S5. Remember, your data and conclusions will only be as good as your samples. If the samples are biased, so will be any conclusions made from them.)

The major question to be answered is whether any difference between samples, or between a sample and a population, has occurred simply as a result of natural variation or because of a real difference between the two. In other words, is the difference encountered one that might 'reasonably' be expected or has an 'unusual' result occurred?

The null hypothesis

The technique for dealing with these problems begins with the formulation of an *hypothesis*, which is in the form of a statement. In particular, we construct a **null hypothesis**, which is a statement that nothing unusual has occurred. The notation for the null hypothesis is H_0.

In contrast, we also form an hypothesis that states that something unusual *has* occurred. This is called the **alternative hypothesis** and is denoted by H_1 or H_A.

The null and alternative hypotheses may be written together in the form:

H_0: (statement) v H_1 (alternative statement)

where the 'v' stands for *versus*.

Example S12.1

Some examples of general types of question and the corresponding hypotheses are shown below. The specific wording of an hypothesis for a question should be expressed in terms of the data for that question.

Question: Is this group different from the rest of the population?

Hypotheses:
H_0: The group is *not* different from the rest of the population.
v
H_1: The group *is* different from the rest of the population.

Question: Is the population mean μ equal to a specified value?
Hypotheses:
H_0: The population mean μ *is* equal to the specified value.
v
H_1: The population mean μ is *not* equal to the specified value.

Another way of expressing the null and alternative hypotheses is in the form of *symbols*. In the case where we have only a single sample, this takes the form in the null hypothesis of the population mean μ taking on a specified value μ_0. That is:

$H_0: \mu = \mu_0$

The alternative hypothesis

The alternative hypothesis may be classified as **two-tailed** or **one-tailed.**

Two-tailed test (two-sided alternative)
In this case, we do the test with no preconceived notion that the true value of μ is either above or below the hypothesised value of μ_0. The test is said to be *two-sided* (or two-tailed), and the alternative hypothesis is written:

$H_1: \mu \neq \mu_0$

Example S12.2

A paint manufacturer claims that its new plastic all-weather outdoor brand dries in an average of 20 minutes when used on timber houses. To test the claim, a scientist uses the paint on 30 different pieces of timber that might be used in the construction of houses.

In this case, there is *no* preconceived notion by the scientist about whether the claimed 20 minutes might be too long or too short. Therefore, a *two-sided* alternative might be appropriate. The hypotheses may be written:

$H_0: \mu = 20 \quad \text{v} \quad H_1: \mu \neq 20$

One-tailed test (one-sided alternative)

In this case, we do the test with a strong conviction that, if H_0 is not true (i.e. $\mu \neq \mu_0$), then it is clear that μ is either *greater than* μ_0 or *less than* μ_0.

If we feel that the only possible alternative is that μ is *greater* than μ_0, we write:

$$H_0: \mu > \mu_0$$

If we feel that the only possible alternative is that μ is *less* than μ_0, we write:

$$H_1: \mu < \mu_0$$

Example S12.3

The weights of full-term female babies born to mothers at a community hospital have been recorded over the years and found to have a mean of 3200 g.

A particular group of 50 mothers were identified as being heavy smokers throughout their pregnancy. It was felt by some medical researchers that heavy smoking usually results in the babies having a *lower* birth weight than other babies.

In this case, there is a preconceived notion that, if these babies do have an unusual birth weight, it will *only* be in the direction of *lower*. That is, it is felt there is *no chance* that they will have a *higher* birth weight. When setting up the alternative hypothesis, it may well be appropriate to use a *one-sided* alternative. The hypotheses in this case may be written:

$$H_0: \mu = 3200 \quad \text{v} \quad H_1: \mu < 3200$$

Although it is a matter of personal choice, you should be very careful in deciding to use a one-sided test, since it could be seen by some that you are prejudging the possible outcome before analysing the data, instead of being open-minded about the possible value of μ. On the other hand, if you genuinely feel that μ could possibly be in only one direction, a one-tailed test may be justified. It should be mentioned that many statisticians *always* use a two-sided alternative, even though there may be an argument for using a one-sided alternative.

Significance level

After the appropriate hypotheses have been formulated we must decide upon the *significance level* (or *α level*) of the test. This level represents the borderline probability between whether an event (or sample) has occurred by chance or whether an unusual event has taken place.

The most common significance level used is 0.05 (often called the 5% significance level), which is commonly written as $\alpha = 0.05$. Another widely used level is $\alpha = 0.01$ (or the 1% significance level). Although in theory *any* significance level may be used, these two are by far the most popular.

If we use, say, a 5% significance level, what we are saying in effect is that an event (or sample) that occurs less than 5% of the time is considered unusual. In this case, we will *reject* H_0 as being false if the probability of obtaining a sample like ours is *less* than 0.05, and *cannot reject* H_0 as being true if this probability is *more* than 0.05.

If we use, for example, a 1% significance level, we are saying that an event (or sample) that occurs less than 1% of the time is considered unusual. In this case, we will *reject* H_0 as being false if the probability of obtaining a sample like ours is *less* than 0.01, and *cannot reject* H_0 as being true if this probability is *more* than 0.01.

S12.2 ERRORS

In performing an hypothesis test, a statistician must be aware of the consequences of drawing the wrong conclusion. These consequences assist in deciding which significance level to use.

In effect, there are two possible errors in making a conclusion about a null hypothesis: type I errors and type II errors.

1. **Type I errors** occur when you *reject* H_0 as being false when H_0 is really *true*. The probability of making a type I error is, in fact, equal to α, the significance level of the test.
2. **Type II errors** occur when you *accept* H_0 as being true when H_0 is really *false*. The probability of making a type II error is denoted by the Greek letter β (beta). The calculation of the value of β is beyond the scope of this book.

Of course, we would like to avoid both errors as much as possible. Unfortunately, in trying to avoid one of them we increase the chances of making the other.

Example S12.4

A company has developed a drug that it feels may be a cure for certain types of cancer. It has collected vast amounts of data as a result of clinical trials and has asked you to examine the data and determine whether the drug actually works.

Solution

The null and alternative hypotheses (in words) are:

H_0: The drug *does not* work. v H_1: The drug *does* work.

At this point, we could decide on the significance level of the test. What might be the consequences of making a type I or a type II error?

Making a type I error: concluding the drug works when it really doesn't

1. Patients are given false hope.
2. Patients are taken off possibly more useful medication.
3. The company is put in the position of investing a great deal of money in a drug that eventually will be found out to be useless.
4. The company may eventually be open to legal action, which in turn may rebound onto you.

Making a type II error: concluding the drug doesn't work when it really does

1. The world (and cancer patients) are deprived of a cure.
2. Many people will die who could have been saved.
3. The company will have lost a great deal of money in lost profits when it does not market the drug.
4. If the company eventually discovers that the drug really does work, possible legal action against you may cite poor statistical advice.

A two-way table that summarises the relationship between rejecting/not rejecting H_0 and whether or not H_0 is true is given in Table S12.1 in terms of type I and type II errors.

Table S12.1 How type I and type II errors are made		
In reality / Decision	H_0 is true	H_0 is false
Reject H_0	Type I error	No error made
Not reject H_0	No error made	Type II error

Although Example S12.4 may seem somewhat extreme (since the relevant drug authorities would also have to be convinced that the drug works), it does highlight the problem faced when deciding which of the two types of error is the worse to make.

Since the probability of making a type I error is also the significance level of the test (α), if we particularly wanted to avoid making a type I error, we should choose a small value of α (e.g. $\alpha = 0.01$). However, in normal circumstances we might use $\alpha = 0.05$.

S12.3 ONE-SAMPLE *z*-TEST

In both this and the next section we will deal with the case of a single sample being chosen from a population and the question of whether that particular sample might be consistent with the rest of the population. To answer this question we construct a **test statistic** according to a particular formula. Exactly which test statistic is appropriate depends on the information available. However, it is very important that the correct one be used, since the use of an incorrect test statistic can lead to an erroneous conclusion.

In calculating the value of a test statistic, we will assume that the following information will always be available:

1. the *size* (n) of the sample
2. the *mean* ($\bar{x}$) of the sample
3. the *standard deviation* (s) of the sample.

Other information of interest to us might include:

4. Does the population have a normal distribution?
5. Is the population's standard deviation (σ) known?
6. Is the sample size (n) large? (By large, we mean that n is at least 25.)

We will classify the various types of test statistics according to the information available. In each case it is assumed that a random sample of size n with a mean of $\bar{x}$ and standard deviation of s is selected from the population. The hypothesised value of μ is μ_0. The first test statistic is known as a *one-sample z-test*.

The *z*-test statistic

Case I

Suppose that we have a situation where:

1. the population has a normal distribution and
2. the population standard deviation σ, is known

Case II

Suppose that we have a situation where:

1. the population has *any* distribution
2. the sample size n is large (i.e. at least 25) and
3. the value of σ is known

In both Case I and Case II we can use a *z-test statistic* defined by:

$$z = \frac{\bar{x} - \mu_0}{\frac{\sigma}{\sqrt{n}}} \quad \text{(S12.1)}$$

Case III

Suppose that we have a situation where:

1. the population has *any* distribution
2. the sample size n is large (i.e. at least 25) and
3. the value of σ is unknown (however, since n is large, the value of σ is approximated by the sample standard deviation s)

In Case III we can use a *z-test statistic* defined by:

$$z = \frac{\bar{x} - \mu_0}{\frac{s}{\sqrt{n}}} \quad \text{(S12.2)}$$

Once the value of the z-test statistic has been calculated, we must decide whether it is *significant.* By significant we mean a value that leads to *rejection* of the null hypothesis. There are several ways to approach this problem, but we will use a simple approach that uses *critical values.*

A critical value is one that represents the cut-off point for z-test statistics in deciding whether we do not reject or reject H_0. The particular critical value to use depends on two things:

1. whether we are using a one-sided or two-sided test, and
2. the significance level used

Rules for not rejecting or rejecting H_0 in a one-sample z-test

(a) Suppose $\alpha = 0.05$ and we are testing

H_0: $\mu = \mu_0$ v H_1: $\mu \neq \mu_0$ (i.e. two-sided)

Then *reject* H_0 if $|z| > 1.96$ and *not reject* H_0 if $|z| < 1.96$

(b) Suppose $\alpha = 0.01$ and we are testing

H_0: $\mu = \mu_0$ v H_1: $\mu \neq \mu_0$ (i.e. two-sided)

Then *reject* H_0 if $|z| > 2.58$ and *not reject* H_0 if $|z| < 2.58$

(c) Suppose $\alpha = 0.05$ and we are testing

H_0: $\mu = \mu_0$ v H_1: $\mu > \mu_0$ (i.e. one-sided)

Then *reject* H_0 if $z > 1.645$ and *not reject* H_0 if $z < 1.645$

(d) Suppose $\alpha = 0.05$ and we are testing

H_0: $\mu = \mu_0$ v H_1: $\mu < \mu_0$ (i.e. one-sided)

Then *reject* H_0 if $z < -1.645$ and *not reject* H_0 if $z > -1.645$

(e) Suppose $\alpha = 0.01$ and we are testing

H_0: $\mu = \mu_0$ v H_1: $\mu > \mu_0$ (i.e. one-sided)

Then *reject* H_0 if $z > 2.33$ and *not reject* H_0 if $z < 2.33$

(f) Suppose $\alpha = 0.01$ and we are testing

H_0: $\mu = \mu_0$ v H_1: $\mu < \mu_0$ (i.e. one-sided)

Then *reject* H_0 if $z < -2.33$ and *not reject* H_0 if $z > -2.33$

Note: The $|z|$ means the *absolute value* of z; that is, the value of z ignoring the sign. For example, $|-3.42| = 3.42$ and $|1.76| = 1.76$.

The critical values are, in fact, calculated from Table 1 on page S.453 (the normal distribution table). For example, in (a) the two-tailed critical value for z is obtained from finding the z value that cuts off the extreme 5% of values in total. This amounts to the extreme 2½% in each tail of the distribution. In other words, the area below the bottom cut-off point is 0.025 and the area above the top cut-off point is 0.025.

Figure S12.1 illustrates this point. Note also that the values of z for which H_0 is not rejected are called the *region of acceptance*, and the values of z for which H_0 are rejected are called the *region of rejection*.

S12.4 ONE-SAMPLE *t*-TEST

In this section we consider a test that has the same objective as a one-sample z-test. That is, to determine whether a sample is significantly different from the population as a whole. However, in this case the one-sample t-test has a different set of assumptions, as defined in Case IV.

CASE IV

Suppose that we have the situation where:

1. the population is normally distributed
2. the population standard deviation σ is *unknown* and
3. the sample size n is small

Then we can use a t-test statistic defined as:

$$t = \frac{\bar{x} - \mu_0}{\frac{s}{\sqrt{n}}} \quad \textbf{(S12.3)}$$

Unlike the z-test statistic, the t-test statistic has associated with it a quantity called **degrees of freedom**. In this case, the degrees of freedom are denoted by the Greek letter ν and are defined by $\nu = n - 1$.

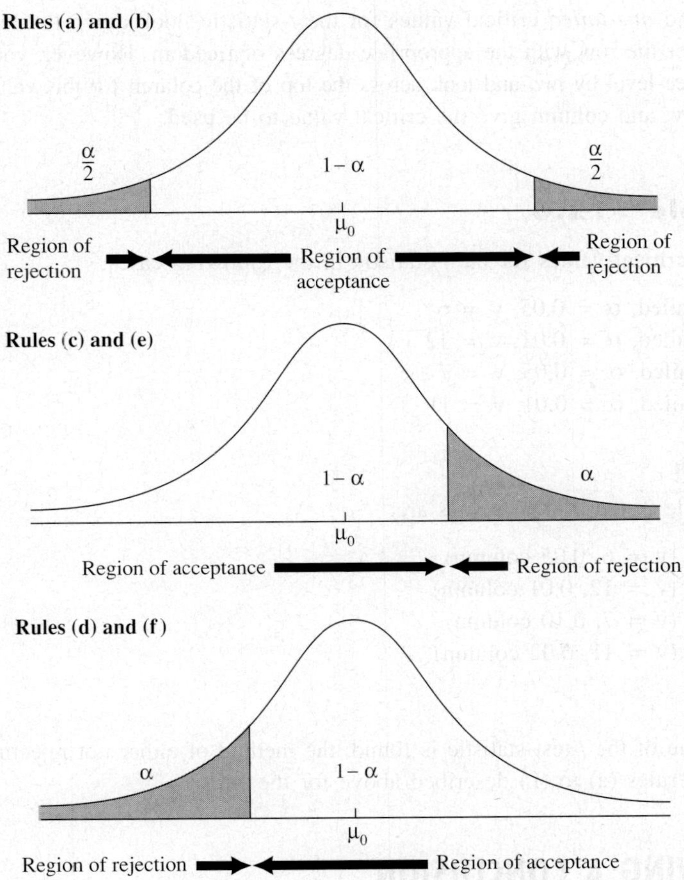

Figure S12.1 Regions of acceptance and rejection of H_0 for rules (a) to (f)

Degrees of freedom relate to the fact that the sum of *all* deviations in a sample of size n must add up to zero. Therefore, they cannot vary 'freely'. The constraint that their sum be zero eliminates one degree of freedom to $n - 1$. For our purpose, however, the sole function of the degrees of freedom is to identify the correct critical value to use when conducting a t-test.

When using a t-test statistic, the critical values for a z-test do *not* apply, and we must use special critical values in a table written especially for the t-test. These values depend on both the degrees of freedom and the significance level used in the test. This information is given in Table 2 on page S.454, and the critical values are given to three decimal places.

Example S12.5

To find *two-tailed* critical values for the t-statistic, look down the left-hand column of Table 2 for the row with the appropriate degrees of freedom and across the top of the column for the significance level used. The intersection of this row and column give the critical value to be used.

To find *one-tailed* critical values for the t-statistic, look down the left-hand column of Table 2 for the row with the appropriate degrees of freedom. However, you now *multiply* the significance level by *two* and look across the top of the column for this value. The intersection of this row and column give the critical value to be used.

Example S12.6

Find the critical values for the t-statistic in the following cases:

(a) two-tailed, $\alpha = 0.05$, $\nu = 6$
(b) two-tailed, $\alpha = 0.01$, $\nu = 12$
(c) one-tailed, $\alpha = 0.05$, $\nu = 7$
(d) one-tailed, $\alpha = 0.01$, $\nu = 11$

Solution

From Table 2, the critical values are:

(a) 2.447 ($\nu = 6$, 0.05 column)
(b) 3.055 ($\nu = 12$, 0.01 column)
(c) 1.895 ($\nu = 7$, 0.10 column)
(d) 2.718 ($\nu = 11$, 0.02 column)

After the value of the t-test statistic is found, the method of either not rejecting or rejecting H_0 is then similar to rules (a) to (f) described above for the z-test.

S12.5 DRAWING A CONCLUSION

Once a decision has been made to either accept or reject H_0, a conclusion should be written in terms of what the actual problem is. An example might be one of the form: 'Hence we conclude that the income of households in Perth is no different from that of households in other capital cities.'

To summarise, the steps that should be undertaken to perform a one-sample test are:

1. Set up the null and alternative hypotheses. This includes deciding whether you are using a one-sided or two-sided test.
2. Decide on the significance level you are using.
3. Write down the relevant data.
4. Decide on the test statistic to be used.
5. Calculate the value of the test statistic.
6. Find the relevant critical value and decide whether H_0 is to be *not rejected* or *rejected*.
7. Draw an appropriate conclusion.

The best way to become familiar with the principles of hypothesis testing is to do as many examples as possible. In many problems the hardest part is deciding on which is the correct test to use; the rest is simply following the appropriate formulae. However, be careful that you don't stumble at the last hurdle! Many students get everything correct except that their conclusions are backwards!

The following examples give a good cross-section of the possible combinations and rules for the various one-sample tests.

Example S12.7

Suppose that it is known that the weights of parcels delivered by a courier company follow a normal distribution with a mean of 6.50 kg and a standard deviation of 2.20 kg.

A random sample of 20 parcels sent by a particular law firm had a mean weight of 7.30 kg. Is there any evidence that this law firm sends parcels whose weight is different from that of other companies? Test at the 5% significance level.

Solution

Since there is no prior evidence that these parcels may be heavier or lighter, a *two-sided* alternative will be used. The hypotheses in *words* are:

H_0: The weights of parcels sent by this law firm are *not* different from those of other companies.

v

H_1: The weights of parcels sent by this law firm *are* different from those of other companies.

The hypotheses in *symbols* are:

$H_0: \mu = 6.50 \quad \text{v} \quad H_1: \mu \neq 6.50$

Data: normal population

$\sigma = 2.20$ is known
$n = 20$
$\bar{x} = 7.30$
$\mu_0 = 6.50$

We will use a two-sided test at $\alpha = 0.05$. These are the requirements for Case I and rule (a). Hence, a z-test statistic is appropriate. From Equation S12.1:

$$z = \frac{\bar{x} - \mu_0}{\frac{\sigma}{\sqrt{n}}} = \frac{7.30 - 6.50}{\frac{2.20}{\sqrt{20}}} = 1.62$$

The critical value at $\alpha = 0.05$ is 1.96. Since $|1.62| < 1.96$, *we cannot reject* H_0 at $\alpha = 0.05$.

We conclude that there is *no* evidence that this law company sends parcels of different weight from parcels sent by other companies.

Example S12.8

The manager of a nationwide discount store believes that there is always a heavier demand for power tools on Mondays compared to other days of the week. The sales of power tools each day in the store follow a normal distribution with a mean of 32.0 tools and a standard deviation of 6.3 tools.

To test his belief, a random sample of 16 Mondays was selected, and these had average sales of 35.0 tools. Use a one-sided test at $\alpha = 0.05$ to test the manager's belief.

Solution

Since the manager has a belief that the demand is *heavier* on Mondays, a *one-sided alternative* could be justified. The hypotheses in *words* are:

H_0: The demand for power tools on Mondays is *not* different from that on other days.

v

H_1: The demand for power tools on Mondays is *greater* than that on other days.

The hypotheses in *symbols* are:

$H_0: \mu = 32.0 \quad v \quad H_1: \mu > 32.0$

Data: normal population

$\sigma = 6.3$ is known

$n = 16$

$\bar{x} = 35.0$

$\mu_0 = 32.0$

We will use a one-sided test at $\alpha = 0.05$. These are the requirements for Case I and rule (c). Hence, a z-test statistic is appropriate. From Equation S12.1:

$$z = \frac{\bar{x} - \mu_0}{\frac{\sigma}{\sqrt{n}}} = \frac{35.0 - 32.0}{\frac{6.3}{\sqrt{16}}} = 1.90$$

The critical value at $\alpha = 0.05$ is 1.645. Since $1.90 > 1.645$, we *reject* H_0 at $\alpha = 0.05$.

We therefore conclude that the manager's belief may be correct. That is, the demand for power tools on Mondays is greater than that on other days.

Example S12.9

The personnel department of a company has been surveying employees and asking them how long it takes for them to travel from home to work each morning. It found that the distribution of times was skewed to the right with a mean of 21.6 minutes and a standard deviation of 7.2 minutes.

A random sample of 25 employees in the accounts section took an average of 24.1 minutes to travel to work. Are these employees different from other employees in their travel time?

Solution

Since there is no prior evidence that the time taken by these employees is longer or shorter than that of other employees, a *two-sided alternative* will be used. The hypotheses in *words* are:

H_0: The time taken to travel to work by employees in the accounts section is *not* different from that of other employees.

v

H_1: The time taken to travel to work by employees in the accounts section *is* different from that of other employees.

The hypotheses in *symbols* are:

H_0: $\mu = 21.6$ v H_1: $\mu \neq 21.6$

Data: non-normal population

$\sigma = 7.2$ is known
$n = 25$ (large)
$\bar{x} = 24.1$
$\mu_0 = 21.6$

We will use a two-sided test at $\alpha = 0.05$. These are the requirements for Case II and rule (a). Hence, a z-test statistic is appropriate. From Equation S12.1:

$$z = \frac{\bar{x} - \mu_0}{\frac{\sigma}{\sqrt{n}}} = \frac{24.1 - 21.6}{\frac{7.2}{\sqrt{25}}} = 1.74$$

The critical value at $\alpha = 0.05$ is 1.96. Since $|1.74| < 1.96$, we *cannot reject* H_0 at $\alpha = 0.05$.

We conclude that there is *no* evidence that employees in the accounts section take a different time to travel to work than other employees.

Example S12.10

Suppose now that the manager of the company in Example S12.9 had a preconceived idea that employees in the accounts section took *longer* than other employees to travel to work. Answer the problem now using a *one-sided* test at $\alpha = 0.05$.

Solution

The hypotheses in *words* are now:

H_0: The time taken to travel to work by employees in the accounts section is *not* different from that of other employees.

v

H_1: The time taken to travel to work by employees in the accounts section is *greater* than that of other employees.

The hypotheses in *symbols* are:

H_0: $\mu = 21.6$ v H_1: $\mu > 21.6$

Data: non-normal population

$\sigma = 7.2$ is known
$n = 25$ (large)

$\bar{x} = 24.1$
$\mu_0 = 21.6$

We will use a one-sided test at $\alpha = 0.05$. These are the requirements for Case II and rule (c). Hence, a z-test statistic is again appropriate. As before, Equation S12.1 is used, giving the same value of:

$z = 1.74$

The one-sided critical value at $\alpha = 0.05$ is 1.645. Since $1.74 > 1.645$, we *reject* H_0 at $\alpha = 0.05$.

We conclude now that there *is* evidence that employees in the accounts section take a *longer* time to travel to work than other employees.

Note that we obtained a different conclusion, using exactly the same data, from that in Example S12.9.

Example S12.11

Experience has shown that the number of matches in boxes follows a normal distribution. A manufacturer claims that the average number of matches in its boxes is 50.

A customer purchases a random sample of 9 boxes and counts the contents of each box. They were:

49 50 51 46 48 45 52 47 48

Based on this sample, should the customer believe the manufacturer's claim? Use a two-sided test at $\alpha = 0.05$.

Solution

Since there is no prior evidence that the average number of matches in the boxes is higher or lower than that claimed by the company, a *two-sided* alternative will be used. The hypotheses in *words* are:

H_0: The match company's claim *is* true.

v

H_1: The match company's claim is *not* true.

The hypotheses in *symbols* are:

H_0: $\mu = 50$ v H_1: $\mu \neq 50$

Data: normal population

σ = is unknown
$n = 9$ (small)
$\mu_0 = 50$

From the sample we can calculate $\bar{x} = 48.44$ and $s = 2.17$.

We will use a two-sided test at $\alpha = 0.05$. These are the requirements for Case III and rule (a). Hence, a t-test statistic is appropriate. Using Equation S12.3:

$$t = \frac{\bar{x} - \mu_0}{\frac{s}{\sqrt{n}}} = \frac{48.44 - 50}{\frac{2.17}{\sqrt{9}}} = -2.16$$

with $\nu = n - 1 = 8$

From Table 2, the two-tailed critical value at $\alpha = 0.05$ with degrees of freedom of $\nu = 8$ is 2.306. Since $|2.16| < 2.306$, we *cannot reject* H_0 at $\alpha = 0.05$.

We conclude that the claim made by the company that there is an average of 50 matches in its boxes may well be true.

Example S12.12

Suppose it was felt by a consumer organisation that there was *no chance* that the manufacturer in Example S12.11 would be placing *more* than an average of 50 matches in its boxes. That is, if its claim is *not* true, it can only be because the manufacturer is placing *less* than an average of 50 matches in its boxes.

In this case, we could conduct Example S12.11 using a *one-sided* test, where the alternative is that the manufacturer is placing *less* than an average of 50 matches in its boxes.

Solution

The hypotheses in *words* are:

H_0: The match company's claim *is* true.

v

H_1: The match company puts an average of *less* than 50 matches in its boxes.

The hypotheses in *symbols* are:

H_0: $\mu = 50$ v H_1: $\mu < 50$

Data: normal population

σ = is unknown

$n = 9$ (small)

$\mu_0 = 50$

We now have the requirements for Case III and rule (d). Hence, a *t*-test statistic is still appropriate.

The test statistic has the same value as in Example S12.11, that is $t = -2.16$ with degrees of freedom of $\nu = 8$. However, since we now have a *one-sided* alternative, to find the critical value at $\alpha = 0.05$ we look up the $\alpha = 0.10$ column with $\nu = 8$ and find a value of 1.860. Since the alternative is $\mu < 50$, the actual critical value is -1.860.

Since $-2.16 < -1.860$, we *reject* H_0 at $\alpha = 0.05$.

We conclude that the claim made by the company is false and that there is *less than* an average of 50 matches in the boxes.

Note that this disagrees with the conclusion made using the same data in Example S12.11. This is because in Example S12.12 it was assumed that the population average could not be *more than* 50; an assumption that might require a deal of justification, especially to the manufacturer!

Example S12.13

A taxi driver claims to take an average of \$12.00 on each fare, but the Taxation Office believes that the average is higher than that.

To test the driver's claim, the Taxation Office takes a random sample of 30 fares. The amounts that the taxi driver took on the fares in the sample had a mean of \$13.30 with a standard deviation of \$2.50. Test the driver's claim at $\alpha = 0.01$.

Solution

In this case, there is a prior notion that the taxi driver takes an average of *more than* \$12.00 on each fare. If it is felt there is *no chance* that the taxi driver is taking an average of *less than* \$12.00 on each fare, it may be appropriate to use a *one-sided alternative* on this occasion. The hypotheses in *words* are:

H_0: The driver *does* take an average of \$12.00 on each fare.

v

H_1: The driver takes an average of *more* than \$12.00 on each fare.

The hypotheses in *symbols* are:

$H_0: \mu = 12.00 \quad \text{v} \quad H_1: \mu > 12.00$

Data: non-normal population (since we do not know it *is* normal)

σ = is unknown
$n = 30$ (large)
$\bar{x} = 13.30$
$s = 2.50$
$\mu_0 = 12.00$

We will use a one-sided test at $\alpha = 0.01$. These are the requirements for Case III and rule (e). Hence, a z-test statistic is appropriate. Using Equation S12.2:

$$z = \frac{\bar{x} - \mu_0}{\frac{s}{\sqrt{n}}} = \frac{13.30 - 12.00}{\frac{2.50}{\sqrt{30}}} = 2.85$$

The one-sided critical value at $\alpha = 0.01$ is 2.33. Since $z = 2.85 > 2.33$, we *reject* H_0 at $\alpha = 0.01$.

We therefore do *not* believe the taxi driver's claim and conclude that there *is* evidence that the taxi driver takes an average of *more* than \$12.00 on each fare.

S12.6 PAIRED *t*-TEST

In the previous two sections we have dealt with problems in which an observed sample was compared with some population external to the sample. In both this and the following section we now consider problems in which there are two samples that are to be compared with each other. These are often referred to as **two-sample problems** and there are a number of test statistics available to investigate whether the two samples are indeed different from each other.

In some instances, the two samples have a structure such that the data are *paired*. That is, each member of one sample has a one-to-one linking with a corresponding member of the other sample in a way that is relevant to the question at hand. For this to be true, it follows that the samples must each contain the same number of observations.

Some examples of paired data are shown below.

Example S12.14

It is desired to see whether babies are capable of demonstrating left- or right-handedness. A natural experimental design might be to concentrate for a start on children of a specified age with right-handed parents, test each child for muscular coordination in each hand (with a score out of, say, 10) and record the results in a form such as Table S12.2, where 6 children are involved.

Table S12.2 Example of a table for the experiment in Example S12.14

Child	A	B	C	D	E	F
Right-handed scores	7	6	2	6	9	3
Left-handed scores	1	2	8	6	4	8

In Table S12.2 there are two samples, namely the right-handed scores (7, 6, 2, 6, 9, 3) and the left-handed scores (1, 2, 8, 6, 4, 8). To compare the values in these two samples requires the calculation of a *two-sample test statistic*.

In this case, the values in these two samples have a natural pairing. For example, the scores 7 and 1 may be considered 'paired' since they both belong to the same child (A). The same applies for the remaining scores, which may also be paired according to the same child.

In such cases, the most commonly used test statistic is a **paired *t*-test**, which takes into account this natural pairing. We will only consider the most common case where the alternative hypothesis is two-tailed. The steps involved in the analysis are as follows.

1. Construct the null and alternative (two-tailed) hypotheses. H_0 always states that there is *no difference* between the two samples, while H_1 always states that there is *some difference* between the two samples. In *symbol* form these are sometimes written as:

   $$H_0\text{: } \mu_d = 0 \quad \text{v} \quad H_1\text{: } \mu_d \neq 0$$

   where μ_d denotes the *population* mean difference between the pairs of observations

   Note that the value of μ_d cannot be calculated and will never be known. We are simply saying that under H_0 the assumed value of μ_d is zero, while under H_1 it is something other than zero.
2. Calculate the actual *differences* between the values in the two samples. It does not matter which sample values are subtracted from which, as long as you are consistent. It is important to retain the minus sign if the subtraction yields a *negative* number.

   For example, in Example S12.14, suppose we subtract the left-handed scores from the right-handed scores. The six differences (d) would be:

   d: 6 4 −6 0 5 −5

The only assumption required before using the paired *t*-test is that these differences come from a normal distribution. Fortunately, this is generally the case in practice and is usually taken for granted.

3. Calculate the mean ($\bar{x}_d$) and standard deviation (s_d) of the values of the differences *d*. In Example S12.14 it can be shown that:

$$\bar{x}_d = 0.67 \qquad s_d = 5.20$$

4. Calculate the value of the *paired t-test statistic*:

$$t = \frac{\bar{x}_d - \mu_d}{\frac{s_d}{\sqrt{n}}} \tag{S12.4}$$

where

n = the number of pairs of differences
ν = degrees of freedom = $n - 1$
μ_d = 0 (according to the null hypothesis)

In Example S12.14 we have $n = 6$ and so:

$$t = \frac{0.67 - 0}{\frac{5.20}{\sqrt{6}}} = 0.32$$

where $\nu = n - 1 = 6 - 1 = 5$ degrees of freedom

5. Look up the *critical value* in Table 2 for the desired significance level.

If |test statistic| > critical value, then H_0 is *rejected* and the conclusion is that H_1 applies. Hence, there is a significant difference between the two samples.

If |test statistic| < critical value, then H_0 *cannot be rejected* and the conclusion is that there is no evidence of a significant difference between the two samples.

In Example S12.14, from Table 2 using, say, $\alpha = 0.05$, the critical value is 2.571.

Since $|t| = |0.32| < 2.571$, we cannot reject H_0 at $\alpha = 0.05$.

We therefore conclude that there is no evidence of a difference between the right-handed and left-handed scores.

Example S12.15

The production output of a random sample of 10 factory workers is recorded on both Thursday and Friday of the same week. Table S12.3 shows the number of toy koalas produced by each worker on each day. Management wishes to determine whether there is any difference between the output of workers on these two days.

Investigate the situation using a significance level of $\alpha = 0.05$.

Table S12.3 Production data for Example S12.15

Worker	A	B	C	D	E	F	G	H	I	J
Thursday	273	221	186	271	234	253	281	183	177	200
Friday	243	225	188	270	219	233	271	178	187	190

Solution

Since the data are paired for this problem (with figures in each sample being linked as they belong to the same worker), we can use a paired *t*-test.

The null and alternative hypotheses may be written in *symbols* as:

$H_0: \mu_d = 0 \quad v \quad H_1: \mu_d \neq 0$

The hypotheses in *words* are:

H_0: There is *no* difference between the production on the two days

v

H_1: There *is* a difference between the production on the two days

Table S12.4 calculates the values of the differences between the two samples, where the Friday production has been subtracted from the Thursday production.

Table S12.4 Differences (*d*) for the production data for Example S12.15

Worker	A	B	C	D	E	F	G	H	I	J
Thursday	273	221	186	271	234	253	281	183	177	200
Friday	243	225	188	270	219	233	271	178	187	190
Differences (*d*)	30	−4	−2	1	15	20	10	5	−10	10

The statistics for these differences *d* are:

$n = 10 \qquad \bar{x}_d = 7.50 \qquad s_d = 12.06$

Using the formula for the paired *t*-test statistic in Equation S12.4:

$$t = \frac{7.50 - 0}{\frac{12.06}{\sqrt{10}}}$$

$$= 1.97$$

where $\nu = 10 - 1 = 9$ degrees of freedom.

From Table 2 using $\alpha = 0.05$, the critical value is 2.262. Since $|t| = 1.97 < 2.262$, we *cannot reject* H_0.

We therefore conclude that there is no evidence of a difference between production on the two days.

S12.7 TWO-SAMPLE *t*-TEST

As is the case with Section S12.6, once again we have two samples that we would like to compare with each other. However, this time suppose that the samples are *not* paired. That is, they are *independent*.

In this instance, the two samples need not contain the same number of observations, although in practice it is wise to try to make their sizes approximately the same, if possible.

The most common test statistic used in this situation is a *two-sample t-test* (also known as a *pooled t-test*). The assumption with this test is that the two samples come from normal distributions. Unless there is any reason to doubt that this is true, the usual practice is to take it for granted.

The calculation of the test statistic requires a little more work than that for the paired t-test. To simplify matters, it is useful to label the two samples as Sample 1 and Sample 2, although it doesn't matter which sample is which.

The steps for using a two-sample t-test are outlined below. Once again, only a two-sided alternative is demonstrated.

1. Construct the null and alternative (two-tailed) hypotheses. H_0 always states that there is *no difference* between the two samples, while H_1 always states that there is *some difference* between the two samples. In *symbol* form these may be written as:

$$H_0: \mu_1 = \mu_2 \quad \text{v} \quad H_1: \mu_1 \neq \mu_2$$

where μ_1 = the mean of the population from which Sample 1 was drawn
μ_2 = the mean of the population from which Sample 2 was drawn

We are saying here that under H_0 these two population means are equal, while under H_1 they are not equal.

2. Calculate the following statistics from the samples:

n_1 = the number of observations in Sample 1
n_2 = the number of observations in Sample 2
$\bar{x}_1$ = the mean of Sample 1
$\bar{x}_2$ = the mean of Sample 2
s_1 = the standard deviation of Sample 1
s_2 = the standard deviation of Sample 2

3. Calculate the value of the *pooled standard deviation* s_p according to Equation S12.5:

$$s_p = \sqrt{\frac{(n_1 - 1)s_1^2 + (n_2 - 1)s_2^2}{n_1 + n_2 - 2}} \qquad \textbf{(S12.5)}$$

Note that the value of s_p *always* lies between the values of s_1 and s_2.

4. Calculate the value of the *two-sample t-test statistic* as in Equation S12.6:

$$t = \frac{\bar{x}_1 - \bar{x}_2}{s_p\sqrt{\frac{1}{n_1} + \frac{1}{n_2}}} \qquad \textbf{(S12.6)}$$

where ν = degrees of freedom = $n_1 + n_2 - 1$

5. Look up the critical value in Table 2 for the desired significance level.

If |test statistic| > critical value, then H_0 is *rejected* and the conclusion is that H_1 applies. Hence, there is a significant difference between the two samples.

If |test statistic| < critical value, then H_0 *cannot be rejected* and the conclusion is that there is no evidence of a significant difference between the two samples.

Example S12.16

Members of two unions, A and B, are employed as tradespeople in a steelworks. Management is interested in whether there is a difference in the amount of overtime that members of each union work. Records for the number of hours worked in the previous two months are obtained for a random sample of 8 members from Union A and 12 members from Union B. This information is shown in Table S12.5.

Table S12.5 The number of hours of overtime worked for Example S12.16

Union A members	22.3	21.6	18.7	20.8	21.5	23.7	19.8	21.6
Union B members	25.8	24.7	26.4	25.0	20.7	24.8	22.4	23.1
	25.6	22.5	19.8	24.6				

Is this evidence of a difference between the amount of overtime worked between members of Union A and Union B? Test at a significance level of $\alpha = 0.05$.

Solution

In this case, the two samples are *independent* since there is no way of matching the values in each group. We can therefore use a two-sample *t*-test to perform the analysis.

Suppose that:

Sample 1 = the hours worked by Union A members
Sample 2 = the hours worked by Union B members

The null and alternative hypotheses in *symbols* are:

$$H_0: \mu_1 = \mu_2 \quad \text{v} \quad H_1: \mu_1 \neq \mu_2$$

The hypotheses in *words* are:

H_0: There is *no* difference between the amount of overtime worked by the members of Union A and Union B

v

H_1: There *is* a difference between the amount of overtime worked by the members of Union A and Union B

The following statistics can be calculated from the samples:

$n_1 = 8$
$n_2 = 12$
$\overline{x}_1 = 21.25$
$\overline{x}_2 = 23.78$
$s_1 = 1.52$
$s_2 = 2.08$

From Equation S12.5 the value of the pooled standard deviation s_p is:

$$\begin{aligned} s_p &= \sqrt{\frac{(n_1 - 1)s_1^2 + (n_2 - 1)s_2^2}{n_1 + n_2 - 2}} \\ &= \sqrt{\frac{(8 - 1)(1.52)^2 + (10 - 1)(2.08)^2}{8 + 12 - 2}} \\ &= \sqrt{3.54} \\ &= 1.88 \end{aligned}$$

From Equation S12.6 the two-sample t-test statistic is:

$$\begin{aligned} t &= \frac{\bar{x}_1 - \bar{x}_2}{s_p\sqrt{\frac{1}{n_1} + \frac{1}{n_2}}} \\ &= \frac{21.25 - 23.78}{1.88\sqrt{\frac{1}{8} + \frac{1}{12}}} \\ &= -3.81 \end{aligned}$$

where $\upsilon = 8 + 12 - 2 = 18$ degrees of freedom

From Table 2 using $\alpha = 0.05$, the critical value is 2.262. Since $|t| = |-3.81| = 3.81 > 2.262$, we *reject* H_0.

We therefore conclude that there is indeed evidence of a difference between the amount of overtime worked by the members of Union A and Union B.

S12.8 *p*-VALUES

Although we will not explore the concept in any depth, an alternative to using critical values for testing hypotheses is to use a ***p*-value** approach (where *p* stands for *probability*). In this case, the value of the test statistic is still calculated in the usual way.

However, in this case we now calculate the probability of obtaining a value *as extreme* as the value of the test statistic if H_0 were true. This probability is called the *p*-value of the test statistic. We then compare the *p*-value with α. Then:

1. If the *p*-value $> \alpha$, H_0 *cannot be rejected* at that α-level.
2. If the *p*-value $< \alpha$, H_0 *is rejected* at that α-level.

Note that whether the critical value or *p*-value approach is used in a particular problem, the conclusion will always be the same.

The idea is illustrated in Examples S12.17 and S12.18.

Example S12.17

Consider the problem in Example S12.7, where the value of z was found to be 1.62 and $\alpha = 0.05$ was used.

Since we used a *two-sided alternative*, the *p*-value is the probability that the *absolute value of z* is greater than 1.62. This area is shaded in Figure S12.2.

Using the techniques in Chapter S7, the shaded area yields:

$$p\text{-value} = 2 \times (0.5000 - 0.4474) = 0.1052$$

Since the p-value $= 0.1052 > 0.05$, we *cannot reject* H_0 (as concluded previously).

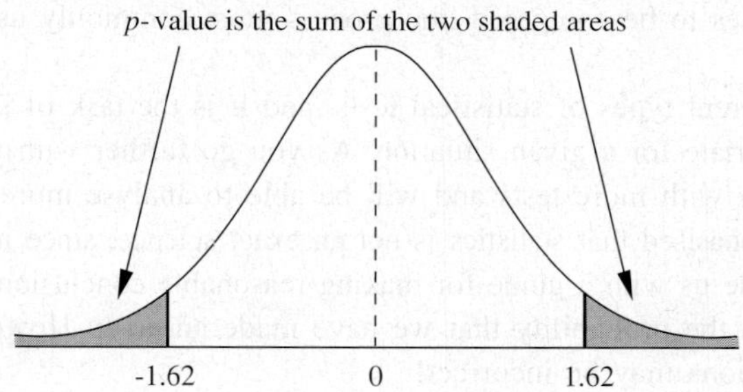

Figure S12.2 Calculation of the *p*-value for Example S12.17

Example S12.18

Consider the problem in Example S12.8, where the value of z was found to be 1.90 and $\alpha = 0.05$ was used.

Since we used a *one-sided alternative*, the p-value is the probability that the *value of z* is greater than 1.90. This area is shaded in Figure S12.3.

Using the techniques in Chapter S7, the shaded area yields:

$$p\text{-value} = 0.5000 - 0.4713 = 0.0287$$

Since the p-value $= 0.0287 < 0.05$, we *reject* H_0 (as concluded previously).

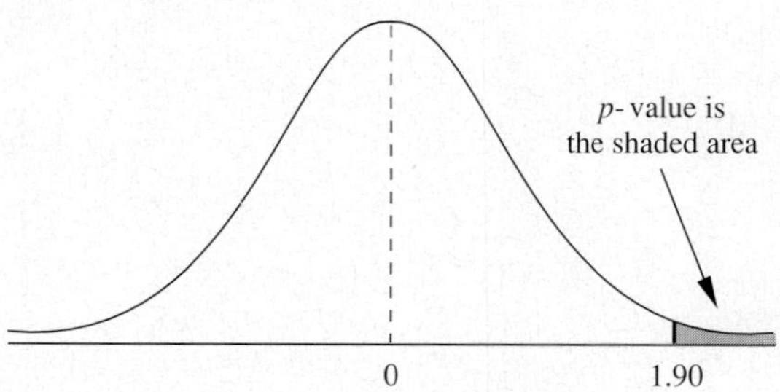

Figure S12.3 Calculation of the *p*-value for Example S12.18

S12.9 REMARKS

In this chapter we have used the term *cannot reject* H_0 whenever the absolute value of a calculated test statistic is less than the critical value from the tables. However, it should be pointed out that some people prefer to use the phrase *accept* H_0 in this case. This latter approach is somewhat misleading, since using the word 'accept' may imply, for example, that a hypothesised value of μ_0 is *absolutely correct*. In fact, their 'acceptance' of H_0 really implies that this value of μ_0 could *possibly* be correct.

Indeed, there are many other values of m_0 that could also possibly be correct and would also be found 'acceptable' if tested.

We have considered the most commonly used one-sample and two-sample tests, although the general principles apply to all hypothesis testing in statistics for problems involving larger numbers of samples and other instances where a conclusion is to be drawn from data collected. When there are more than two samples to be compared, the process most commonly used is called *Analysis of Variance (ANOVA)*.

There are many different types of statistical tests, and it is the task of statisticians to determine just which one is appropriate for a given situation. As you go further with your studies in statistics you will become familiar with more tests and will be able to analyse more sophisticated data sets.

It should also be emphasised that statistics is not an exact science, since it doesn't *prove* anything. What it does do is provide us with a guide for making reasonable conclusions based on the evidence before us, and even with the probability that we have made an error. However, the chance always remains that our conclusions may be incorrect!

Quick Quiz

1. Suppose that you want to test whether H_0: $\mu = 500$ v H_1: $\mu \neq 500$. You are told that the population is normally distributed and that a selected random sample of size 30 has a mean of 492 and a standard deviation of 20. State what type of test statistic you would use and find its value.
2. In Question 1, if you were testing at $\alpha = 0.01$, what would be the critical value?
3. Considering your answers to Questions 1 and 2, what conclusion would you draw?
4. In Question 3, would your conclusion have been any different if you were now told that the population is *not* normally distributed? Explain.
5. Suppose that you want to test whether H_0: $\mu = 25$ v H_1: $\mu < 25$. You are told that the population is normally distributed and that a selected random sample of size 10 has a mean of 22.5 and a standard deviation of 4.8. State what type of test statistic you would use and find its value.
6. In Question 5, if you were testing at $\alpha = 0.05$, what would be the critical value?
7. Considering your answers to Questions 5 and 6, what conclusion would you draw?
8. In Question 7, would your method of solving the problem have been any different if you were now told that the population is *not* normally distributed? Explain.
9. If when using a *z*-test you did not reject H_0 at $\alpha = 0.05$, explain why you would automatically also not reject H_0 at $\alpha = 0.01$.
10. You have a two-sample test to perform on some data where one sample is of size 10 and the other is of size 15. You feel tempted to use a paired *t*-test. Explain why this temptation would be clearly incorrect.

Multiple Choice Questions

1. A normal distribution has $\mu = 100$ and $\sigma = 40$. A random sample of size 25 is selected and found to have a mean of 90. The value of the *z*-test statistic for this mean is:
(a) 0.25 (b) −0.25 (c) 8
(d) −1.25 (e) none of the above

2. The significance level α of a test is:
(a) always equal to 0.05
(b) always equal to 0.50
(c) either 1.96 or 2.58
(d) the probability of making a type I error
(e) none of the above

3. A question involves testing whether international soccer players are any different in height from others in the population. A type II error in a two-tailed test for this problem would conclude that:
(a) soccer players are taller than others when they are really not
(b) soccer players are shorter than others when they are really not
(c) soccer players are the same height as others when they are really not
(d) soccer players are not the same height as others when they really are
(e) none of the above

4. In Question 3, a type I error in a two-tailed test in this problem would conclude that:
(a) soccer players are not taller than others when they really are
(b) soccer players are not shorter than others when they really are
(c) soccer players are not the same height as others when they really are
(d) soccer players are the same height as others when they are really not
(e) none of the above

5. In using a one-sample *t*-test, which of the following statements is true?
(a) the value of σ must be known
(b) *n* must be large
(c) the population must be normally distributed
(d) the degrees of freedom must be at least as large as the sample size
(e) none of the above

6. When using a one-sample one-sided *t*-test with a sample size of 15, if the alternative hypothesis is H_1: $\mu > 20$, the critical value to use at $\alpha = 0.05$ is:
(a) 2.145 (b) 1.761 (c) 1.753
(d) 1.328 (e) none of the above

7. You are about to perform a two-sample *t*-test and notice that the means of the two samples are identical although their standard deviations are different. Without actually doing the test, you could conclude straight away that:
(a) there is no difference between the two samples for any value of α
(b) there will be a difference between the two samples for any value of α
(c) the test will be inconclusive
(d) you cannot draw any conclusion without doing the test
(e) none of the above

8. The study of *ANOVA* deals with:
(a) the statistics of volcanoes
(b) which statistical test to use in a given situation
(c) comparing more than two samples
(d) a number of variable algorithms
(e) none of the above

9. The *pooled standard deviation*:

(a) involves the variation in the length of swimming pools
(b) is useful in calculating the two-sample *t*-test statistic
(c) is a combination of two standard deviations
(d) should only be used if the standard deviations are equal
(e) none of the above

10. Suppose that the *p*-value of a *z*-test statistic is 0.03. Then it follows that:

(a) H_0 cannot be rejected at $\alpha = 0.05$
(b) H_0 can be rejected at $\alpha = 0.01$
(c) the value of $|z|$ must be less than 1.96
(d) the value of $|z|$ must be greater than 2.58
(e) none of the above

Exercises

S12.1 The number of newspapers sold each Friday in a newsagency follows a normal distribution with $\mu = 120$ papers and $\sigma = 15$ papers. On the Friday before a Federal election, the newsagency sold 130 papers.

Is this a significant increase in the number of papers sold? Use a one-sided test at $\alpha = 0.05$ and draw appropriate conclusions.

S12.2 For Exercise S12.1, use a two-sided test at $\alpha = 0.05$ and draw appropriate conclusions.

S12.3 For Exercise S12.1, use a one-sided test at $\alpha = 0.01$ and draw appropriate conclusions.

S12.4* A hairdresser finds that the time taken to cut men's hair follows a bimodal distribution with a mean of 30 minutes and a standard deviation of 10 minutes. She notes that 36 males from a particular college take an average of 27 minutes to have their hair cut.

Does this group take less time than usual to have their hair cut? Use a one-sided test at $\alpha = 0.05$ and draw appropriate conclusions.

S12.5* For Exercise S12.4, use a two-sided test at $\alpha = 0.05$ and draw appropriate conclusions.

S12.6* For Exercise S12.4, use a one-sided test at $\alpha = 0.01$ and draw appropriate conclusions.

S12.7 The number of errors on the pages of a statistics textbook follows a normal distribution with a mean of 5.2 errors. In a chapter of 8 pages, the number of errors found on the pages was:

3 4 6 2 0 4 7 4

Does this chapter contain an unusual number of errors? Test at $\alpha = 0.05$.

S12.8 For Exercise S12.7, test at $\alpha = 0.01$.

S12.9 A company claims that the amount of soft drink in its bottles follows a normal distribution with an average of 750.0 mL. A random sample of 40 bottles revealed average contents of 748.6 mL with a standard deviation of 2.5 mL.

Based on this sample, do you believe the company's claim at the 5% significance level?

S12.10 For Exercise S12.9, do you believe the company's claim at the 1% significance level?

S12.11* Suppose that it can be assumed that the systolic blood pressure of 35-year-old males follows a normal distribution with $\mu = 120$ mm. A random sample of 9 such males in a stressful occupation had the following blood pressures (in mm):

125 123 120 119 136 142 118 143 147

Use a one-sided test at $\alpha = 0.05$ to determine whether this group has a significantly higher blood pressure than other such males.

S12.12* For Exercise S12.11, test at $\alpha = 0.05$ whether this group has a significantly *different* blood pressure from other such males.

S12.13 A company claims that its Bar-B-Q fire starters will burn for an average of 45 minutes. A random sample of 30 of these fire starters burned for an average of 44.0 minutes with a standard deviation of 1.8 minutes.

Use a two-sided test to decide whether you believe the company's claim at the 1% significance level.

S12.14 The ages (in years) of the 26 applicants for a university lectureship in accounting are:

28 35 25 29 30 27 33 48 32 39 26 31 32

52 34 29 33 37 26 25 28 40 36 30 35 30

Over the years, the average age of applicants has been 35.0 years.

Based on this sample, test at $\alpha = 0.05$ whether applicants are getting younger.

S12.15 Do the test in Exercise S12.14 at $\alpha = 0.01$.

S12.16 The marks in a statewide economics exam followed a distribution that was skewed to the right with a mean of 55 marks. A group of 25 students at a certain school obtained a mean mark of 53.0 with a standard deviation of 4.5 marks.

Test at $\alpha = 0.05$ whether this group did significantly worse than other students in the state.

S12.17* For Exercise S12.16, test at $\alpha = 0.05$ whether this group did significantly *differently* from other students in the state.

S12.18 The length of time that motorists wait for road service follows an unknown distribution with a mean of 50.0 minutes. On a particularly wet day, a random sample of 33 motorists had to wait an average of 52.6 minutes with a standard deviation of 8.4 minutes.

Did this group of motorists have to wait significantly *longer* than usual? Test at $\alpha = 0.05$.

S12.19 Loaves of bread are labelled as having an average net weight of 680 g. A random sample of 9 of these loaves had the following net weights (in g):

682 675 685 677 670 678 672 680 675

Based on this sample, test at $\alpha = 0.05$ whether the labelling is correct.

S12.20* An IQ test is standardised so that for Australian adults the mean is 100 and the standard deviation is 20. A random sample of 100 adults from a certain town obtained an average mark of 95 on this test.

What conclusion do you draw at $\alpha = 0.05$?

S12.21 A service station claims that it can give cars a grease and oil change in an average of 20 minutes 'while you wait'. A customer who observed the operation over a period of several weeks noted that for a random selection of 50 cars the mean waiting time was 23.5 minutes with a standard deviation of 4.5 minutes.

Do you believe the service station's claims? Use a two-sided test at $\alpha = 0.05$.

S12.22* A particular soldered component on a personal computer motherboard was tested for strength by measuring the amount of force required to break the solder. The manufacturer claims that the solder can withstand a large constant pressure for an average of 60 seconds before breaking. A random sample of 30 such components were tested using this pressure, and it was found that the solder broke after a mean of 58.4 seconds and a standard deviation of 3.6 seconds.

Is it possible that the manufacturer is telling the truth? Test at $\alpha = 0.01$.

S12.23 Police records indicate that, on a certain stretch of straight road, the amount by which speeding motorists exceed the speed limit follows a normal distribution with a mean of 16 km/h and a standard deviation of 6.5 km/h. On the Monday of a long weekend, a police officer recorded the amounts by which the 10 motorists booked for speeding exceeded the speed limit. They were (in km/h):

12 23 16 28 10 11 22 31 19 10

He believes that this group of motorists was travelling even faster than the usual speeding motorist.

Use an appropriate one-sided test at $\alpha = 0.05$ to see whether you agree with him.

S12.24 In Exercise S12.23, suppose that you are now told that the distribution is *not* normal. Would this alter the method that you used to solve the problem? If not, explain why. If so, explain how you would now proceed.

S12.25* A certain brand of cooking oil is advertised as containing an 'average of 10% saturated fats'. A random sample of 100 bottles revealed that their percentage of saturated fats had a mean of 10.9 with a standard deviation of 3.6.

Based on this evidence, do you believe the manufacturer's claim? Use a one-sided test at $\alpha = 0.05$.

S12.26 In Exercise S12.25, would you obtain a different conclusion if you used a two-sided test at $\alpha = 0.05$? Explain.

S12.27 A machine produces a metal part that is supposed to have a mean length of 25.00 mm. A random sample of 40 parts had a mean length of 24.30 mm with a standard deviation of 1.60 mm. Do you feel that the machine is producing parts with the correct mean length? Test at $\alpha = 0.05$.

S12.28 If H_0 is rejected at $\alpha = 0.01$, is it true that H_0 will *automatically* be rejected at $\alpha = 0.05$? Explain.

S12.29* If H_0 is rejected at $\alpha = 0.05$, is it true that H_0 will *automatically* be rejected at $\alpha = 0.01$? Explain.

S12.30 If H_0 is accepted at $\alpha = 0.01$, is it true that H_0 will *automatically* be accepted at $\alpha = 0.05$? Explain.

S12.31 If H_0 is accepted at $\alpha = 0.05$, is it true that H_0 will *automatically* be accepted at $\alpha = 0.01$? Explain.

S12.32* The number of French fries in a medium-sized packet sold at a fast-food outlet is usually known to follow a normal distribution with a mean of 18 fries. An inspector is investigating complaints that the number of fries in recent packets has been less than expected. To test this she selected a random sample of 9 packets and counted the number of fries in each. The results were:

14 12 20 16 15 15 17 18 13

Are the complaints justified? Use a one-sided test at $\alpha = 0.05$.

S12.33 It is claimed that the author of this textbook writes words with an average length of 4.5 characters.

Test this claim at $\alpha = 0.05$ by considering the length of the words in the first two sentences of Exercise S12.32.

S12.34 The birth weights of female babies at a major hospital is known to follow a normal distribution with $\mu = 3000$ g and a standard deviation of 500 g. A random sample of 50 babies born to mothers who had smoked throughout their pregnancy had a mean birth weight of 2780 g.

Is this evidence that babies born to mothers who smoked throughout their pregnancy had a lower birth weight than other babies? Use a one-sided test at $\alpha = 0.01$.

S12.35 In Exercise S12.34, if you had been told that the birth weights do *not* follow a normal distribution, would you have changed the way that you solved the problem? Explain.

S12.36* A charcoal chicken establishment advertises that its chickens have an average weight of 1.80 kg. A random sample of 25 chickens purchased by a customer over the past few weeks had a mean weight of 1.65 kg with a standard deviation of 0.60 kg.

Is it possible that the establishment is being truthful? Test at $\alpha = 0.05$.

S12.37 In previous seasons the number of goals scored in a match by a soccer team followed a normal distribution with a mean of 1.60 goals. In the first 8 matches of the current season they have scored the following number of goals:

2 1 3 2 4 2 5 3

The team coach believes that these results are evidence that the team is increasing the mean number of goals that it scores in its matches.

Do you agree? Use a one-sided test at $\alpha = 0.05$.

S12.38 In Exercise S12.37, suppose that in the ninth match of the season the team does not score any goals.

Do you agree with the coach now? Use a one-sided test at $\alpha = 0.05$.

S12.39 The weights of adults in a certain population follow a normal distribution with $\mu = 64.5$ kg and $\sigma = 4.8$ kg. There is a sign in the lift in an office block that states: 'This lift will not operate if the total weight of people in the lift exceeds 816 kg.' A manager complains that a random group of 12 adult workers entered the lift and it would not operate because of the excessive total weight of the passengers.

Is what the manager claims likely to be true? Use a one-sided test at $\alpha = 0.05$.

S12.40* The test scores of learner drivers over the years on a written driving test have followed a distribution that is skewed to the right with a mean of 65 and a standard deviation of 12.5. A random sample of 30 learner drivers from TAFE scored an average of 71.0.

Is this evidence that these TAFE students performed any differently from other learner drivers? Test at $\alpha = 0.01$.

S12.41 In Exercise S12.40, if you were now told that the test scores follow a *normal* distribution, would you change the way in which you solved the problem? Explain.

S12.42 Answer Exercise S12.36 using a *p-value* approach.

S12.43 Answer Exercise S12.39 using a *p-value* approach.

S12.44* Answer Exercise S12.40 using a *p-value* approach.

S12.45 Under the Australian justice system, an accused person is presumed innocent unless found guilty. Suppose that you are on a jury in which the accused person is charged with a crime for which the penalty is life in prison. Discuss what making a type I or type II error would mean in this case, and the possible consequences.

S12.46* A random sample of 16 females and 25 males were selected from TAFE business students and had their systolic blood pressures measured (in mm). For the female sample the mean was 113.7 mm and the standard deviation was 11.2 mm. For the male sample the mean was 119.6 mm and the standard deviation was 18.4 mm.

Assuming that blood pressures are normally distributed, test at $\alpha = 0.05$ whether there is any difference between the blood pressures of male and female TAFE business students.

S12.47 A group of 10 students sat for examinations in mathematics and English, both of which were marked out of 100. The student's individual marks were:

Student	1	2	3	4	5	6	7	8	9	10
Mathematics	83	72	91	62	80	43	59	36	82	75
English	54	92	64	88	59	76	62	71	42	50

Use an appropriate test to determine whether there is a significant difference between the two sets of marks. Test at $\alpha = 0.05$ and state any assumptions you are making.

S12.48 A company's marketing department is interested in comparing the weekly expenditure on food items of households in two localities A and B. A random sample of 9 households was chosen from locality A and another random sample of 9 was chosen from locality B. Their weekly expenditures (in dollars) were reported as follows.

Locality A	153	102	217	162	97	178	133	79	125
Locality B	147	93	165	105	115	127	142	83	113

Determine whether there is any difference between the amounts spent each week by households in the two localities. Use $\alpha = 0.01$.

S12.49 Six months later, the same 9 households in locality A in Exercise S12.48 were asked the same question. This time, their responses (in dollars and in the same order) were:

167 118 176 145 110 145 158 100 129

Do these figures indicate a change in the weekly expenditure of the households in locality A? Test at $\alpha = 0.05$.

S12.50 A cheese manufacturer is seeking to determine whether two different kinds of packaging of a certain kind of cheese differ in their ability to protect the cheese from mould. To do so, 20 equal-sized pieces of cheese are cut from the one block, and 10 are packaged by Method A and 10 are packaged by Method B. At 10 different locations, one Method A pack and one Method B pack are left until mould appears. The results (in days) before mould first appears are shown below.

Location	1	2	3	4	5	6	7	8	9	10
Method A	15	23	9	12	25	18	6	21	14	11
Method B	10	18	10	8	17	18	4	16	15	5

Does the evidence suggest that there is a difference between the ability of the two different kinds of packaging to protect the cheese from mould? Test at $\alpha = 0.05$.

S12.51 In Exercise S12.50, you now discover that the person you asked to carry out the experiment misunderstood your instructions, and in fact different locations were *not* used. That is, all 20 pieces of cheese were actually left in the one place.

(a) Would this information cause you to change your method of analysis in Exercise S12.50?

(b) If your answer to (a) is 'no', explain why not. If your answer to (a) is 'yes', perform the new analysis at $\alpha = 0.05$.

Chapter S13

Analysis of frequency data

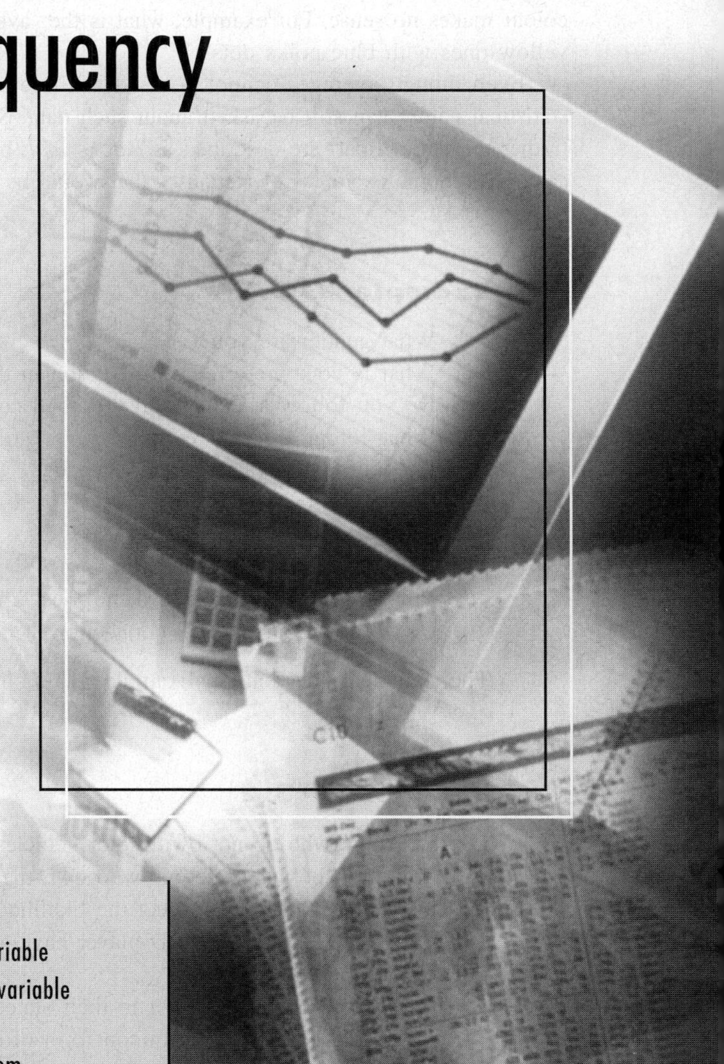

'I ain't lookin' to analyse you, categorise you, finalise you or advertise you.'

(Bob Dylan, American singer and songwriter, 'All I Really Want To Do', 1964)

Learning objectives

In this chapter you will learn to:

- Understand the meaning of a categorical variable
- Understand the difference between a single variable problem and a two variable problem
- Construct a table for a single variable problem
- Construct a contingency table for a two variable problem
- Analyse single variable data
- Analyse two variable data

S13.1 CATEGORICAL DATA

The test procedures in Chapter S12 concentrated heavily on what statisticians call 'location problems'; that is, testing whether something differs from something else 'on average'. Such problems do form a major part of statistical analyses, but there are many situations where a different type of test is appropriate.

In particular, data are often *non-numeric*, in the sense that each individual observation is a description rather than a number. For example, colours would conventionally be described verbally (red, green, yellow, etc.) rather than numerically. In such cases, the idea of calculating an 'average' colour makes no sense. For example, what is the 'average' of 2 black objects, 3 green ones and 15 yellow ones with blue polka dots?

Even though averages cannot be used in these circumstances, many important questions of a statistical nature can still be asked about such data. Systems where the observations are descriptive (rather than numerical) are described as *categorical*, because the individuals are being classified into categories. Some examples of responses that would be **categorical data** are shown in Example S13.1.

Example S13.1

1. What gender are you?
2. What is your preferred radio station in the mornings?
3. Do you follow a team in any football code? If so, what is the name of the team?
4. What colour are your eyes?
5. Do you have a valid driver's licence?
6. What suburb do you live in?
7. Have you ever travelled overseas?
8. Who is your favourite lecturer?
9. What magazine do you read the most often?
10. Do you have an Internet connection at home?

The statistical questions in Example S13.2 all involve categorical variables.

Example S13.2

1. Do people who do not own motor vehicles have a greater tendency to travel to work by public transport than people who own their own vehicles?
2. Does area of residence affect the likelihood of owning a motor vehicle?
3. Do people who own a DVD player have any different radio preferences from those who do not?
4. Do males and females differ in their stated interests in football teams? (Note that male/female is a particularly common form of categorical variable.)
5. Is there a significantly higher proportion of male wine-drinkers than female wine-drinkers?

Sometimes, questions such as 1–4 in Example S13.2 are typical of a great number of social science research problems, where questions of the possible *relationship* between categorical variables are of interest. These questions might also be expressed as in Example S13.3.

Example S13.3

1. Is there a relationship between car ownership and use of public transport to travel to work?
2. Is there a relationship between area of residence and car ownership?
3. Is there a relationship between owning a DVD player and preference for radio stations?
4. Is there a relationship between gender and interest in football teams?

On the other hand, these questions may also conveniently be expressed as questions about differences between *proportions*, such as:

1. Does the proportion of individuals travelling to work by public transport differ between vehicle owners and those who do not own a vehicle?
2. Does the proportion of people who own motor vehicles differ from one area of residence to another?
3. Does the proportion of people preferring various radio stations differ between those who own a DVD player and those who do not?
4. Is the proportion of males interested in certain football teams different from the corresponding proportion for females?

Although the above questions have been expressed in a different way, they are really asking the same thing. In one form the issue is whether a relationship between two categorical variables exists and the second asks whether there is a difference between two proportions. Since both questions are equivalent, the same technique of solution is applicable to each.

S13.2 THE PRESENTATION OF SINGLE VARIABLE CATEGORICAL DATA

Because we can express the same meaning in many different ways, it is common practice to have a standard form of presentation. From a theoretical point of view, it is convenient to work with *frequency* data; that is, data in which the *number* of occurrences of each category is recorded. In most cases, a table is the best option.

Examples of questions involving a single categorical variable are shown in Example S13.1, where the categories are the possible responses of that variable. A typical table for a single variable is illustrated in Examples S13.4 and S13.5. In these cases, a *one-dimensional* table showing all possible categorical outcomes of the variable along with the corresponding frequency is required.

Example S13.4

A fair six-sided die is rolled 60 times, with the outcomes shown in Table S13.1.

Table S13.1 Outcomes of 60 rolls of a fair six-sided die in Example S13.4

Category	1	2	3	4	5	6	*Total*
Frequency	8	7	12	13	5	15	60

Example S13.5

A random sample of 100 males from Melbourne was selected and asked about their interest in Australian Rules football. The results are shown in Table S13.2.

Table S13.2 Outcome of Australian Rules survey for Example S13.5

Category	Very interested	Moderate interest	Mild interest	No interest	*Total*
Frequency	42	25	18	15	100

S13.3 CONTINGENCY TABLES

Some problems involve two categorical variables and questions often arise about their relationship. Examples of these are shown in Examples S13.2 and S13.3. In this case, a *two-dimensional* table is appropriate, where one variable is presented along the rows and the other variable down the columns. The frequency of observations in each combination of the categorical outcomes of the variables is then entered into the appropriate box (called a *cell*) in the table. Such a table is often referred to as a **contingency table** and examples are shown in Examples S13.6 and S13.7.

Example S13.6

A survey was conducted in which 200 respondents were asked the following 2 questions:

1. In what part of Sydney do you live? (North, South, East, West)
2. Do you have an Internet connection in your place of residence? (Yes/No)

The responses are summarised in the contingency table in Table S13.3. The number of cells in a table can be found by multiplying the number of categories for each variable. In this case, there are $4 \times 2 = 8$ cells.

Table S13.3 A typical contingency table for the Internet survey in Example S13.6

Internet \ Live	North	South	East	West	*Total*
Yes	52	47	105	34	238
No	28	63	35	36	162
Total	80	110	140	70	400

Example S13.7

The managing director of a car manufacturing company is interested to know whether the colour of new cars purchased is associated with the gender of the buyer. Of a random sample of 800 new car buyers it was found that males purchased 42% of 300 white cars, 59% of 200 blue cars, 68% of 50 red cars and 38% of 250 other-coloured cars.

A typical contingency table for this data is shown in Table S13.4. This table includes all categories for each of the two variables 'gender' and 'colour of car chosen'. For the gender variable, in addition to the stated category 'male' there is also the implied category 'female'. For example, of the white car buyers, there were 42% of 300 = 126 males and therefore 300 – 126 = 174 females.

Table S13.4 A typical contingency table for the car purchase data in Example S13.7

Colour / Gender	White	Blue	Red	Other	*Total*
Male	126	118	34	95	373
Female	174	82	16	155	427
Total	300	200	50	250	800

The contingency tables in Tables S13.3 and S13.4 have characteristics that are common to all such tables. These include:

1. The final column is a total column.
2. The final row is a total row.
3. It generally does not matter which variable is along the columns and which is along the rows.
4. Frequencies must add up along each row.
5. Frequencies must add up down each column.
6. The value in the bottom right-hand corner of the table represents the total number of observations overall. It is often referred to as the *grand total* frequency.

S13.4 AN ANALYSIS OF SINGLE VARIABLE PROBLEMS

The simplest analysis of categorical data is if there is only a single variable. The question to be answered is whether an *observed* set of categorical data (with each observation being placed in exactly one category) is reasonably consistent with what was *expected* by some prior line of reasoning. The idea is expressed in Example S13.8.

Example S13.8

A statistician is presented with a six-sided die and asked to determine whether it is 'fair'. By this it is meant whether it is equally likely that the outcome will be a 1, 2, 3, 4, 5 or 6 when it is tossed.

To answer the question, the statistician rolls the die a total of 300 times. The outcomes are shown in Table S13.5.

Table S13.5 Outcomes of 300 die rolls for Example S13.8

Outcome	Frequency
1	48
2	57
3	60
4	42
5	44
6	49
Total	300

If the die is really fair, then there is a 1/6 probability that any given face will appear at any roll. Thus, in a loose sense, the 300 rolls would be 'expected' to yield 300 × 1/6 = 50 occurrences of each face.

On the other hand, a die has no memory and so no physical mechanism to cause it to give *exactly* equal frequencies for each face. It follows that *some* variation away from the 'expected' result could occur, even if the die were fair.

If the aim is to assess whether the die is really fair, then it is necessary to determine whether the discrepancy between the *observed* and *expected* frequencies is merely a natural one due to chance fluctuations. The alternative is that the discrepancy exists because the expectations are false. This decision is characteristic of the analysis of categorical data.

The steps involved in the analysis of a single variable problem are as follows. It is often called a **goodness-of-fit test**.

1. Construct the null hypothesis for the problem. This usually takes the general form of:

H_0: There is no difference between the observed frequencies and the expected frequencies

This should be modified for each individual problem. In Example S13.8 the null hypothesis would be:

H_0: The die *is* fair

The alternative hypothesis (using a two-sided alternative) would be:

H_1: The die is *not* fair

2. The *observed frequencies* come from the data of the problem. In Example S13.8, the observed frequencies are the actual values obtained for each category; that is, 48, 57, 60, 42, 44 and 49.

3. The *expected frequencies* are those we might 'expect' to occur if H_0 were true. In Example S13.8, since H_0 assumes that the die is fair, the expected frequency for each category is the same. That is, 300 × 1/6 = 50.

4. The measure of the discrepancy between the observed and expected frequencies is given by the **chi-square test** statistic. The formula is given in Equation S13.1.

$$\chi^2 = \sum_{\text{all categories}} \frac{(\text{observed frequency} - \text{expected frequency})^2}{\text{expected frequency}} \qquad \textbf{(S13.1)}$$

The symbol χ^2 is called 'chi-square', with the 'chi' being pronounced as 'ky'. Note that there is one term required in the calculation for each category of the variable. Also, since the square of a number can never be negative, the value of a χ^2-test statistic can also never be negative.

In Example S13.8, the calculations required for the χ^2-test statistic are:

$$\chi^2 = \frac{(48-50)^2}{50} + \frac{(57-50)^2}{50} + \frac{(60-50)^2}{50} + \frac{(42-50)^2}{50} + \frac{(44-50)^2}{50} + \frac{(49-50)^2}{50}$$

$$= 0.08 + 0.98 + 2.00 + 1.28 + 0.72 + 0.02$$

$$= 5.08$$

5. Associated with the test statistic are degrees of freedom. The degrees of freedom for a goodness-of-fit test are defined as:

$$\text{Degrees of freedom} = \text{number of categories} - 1 \qquad \textbf{(S13.2)}$$

In Example S13.8, since there are 6 categories, the degrees of freedom are $6 - 1 = 5$.

6. As is the case with the z-test statistic and the t-test statistic, the calculated value of the chi-square test statistic must be compared with a *critical value*. This critical value is obtained from Table 4 on page S.456. Two pieces of information are required: the **degrees of freedom** (down the left-hand column) and the **significance level** desired (across the top row).

7. In Example S13.8, if a significance level of, say, $\alpha = 0.05$ is desired, then we go to the degrees of freedom row 5 and column 0.05 to obtain a critical value of 11.07.

8. Compare the value of χ^2 that you calculated with the critical value from Table 4.

If $\chi^2 <$ the critical value, then H_0 *cannot be rejected.*
If $\chi^2 >$ the critical value, then H_0 *can be rejected.*

In Example S13.8 we have

$\chi^2 = 5.08 < 11.07$

Therefore, in this case H_0 cannot be rejected.

9. Based on the outcome of Step 8, an appropriate conclusion can be drawn. In Example S13.8, since H_0 cannot be rejected, the conclusion is that it is quite possible that the die may be fair. That is, the evidence of the outcomes of the rolls does not give us grounds to conclude that the die is not fair.

Example S13.9

The management of a company that manufactures a generic brand of soap powder wishes to know how purchasers rate its product. The sales manager claims that 50% would rate it as 'good', 30% as 'average' and 20% as 'poor'.

To test the claim, a market research team was engaged to sample the responses from a random sample of 200 users of the product. The results are summarised in Table S13.6.

Table S13.6 Outcome of a customer survey for Example S13.9	
Rating	Frequency
Good	83
Average	56
Poor	61
Total	200

Based on the sample responses, do you believe that the manager's claim could be true? Use $\alpha = 0.05$.

Solution

Since this is a problem involving frequencies and only a single variable (rating), a χ^2 goodness-of-fit test is appropriate. The null and alternative hypotheses are:

H_0: The manager's claim *is* true

H_1: The manager's claim is *not* true

From Table S13.6, the *observed frequencies* for each of the three categories (good, average and poor) are 83, 56 and 61.

The observed frequencies can be calculated from what would be expected if the claim was true. These are:

Good: $0.50 \times 200 = 100$
Average: $0.30 \times 200 = 60$
Poor: $0.20 \times 200 = 40$

The actual calculations required for the χ^2-test statistic are shown in Table S13.7.

Table S13.7 Calculation of the χ^2-test statistic for Example S13.9

Category	Observed	Expected	Observed − expected	(Obs. − exp.)^2	$\frac{(\text{Obs.} - \text{exp.})^2}{\text{exp.}}$
Good	83	100	−17	289	289/100 = 2.890
Average	56	60	−4	16	16/60 = 0.267
Poor	61	40	21	441	441/40 = 11.025
Total	200	200	0		$\chi^2 = 14.182$

From Table S13.7 the value of the χ^2-test statistic is 14.18. Since there are three categories, from Equation S13.2 the degrees of freedom are $3 - 1 = 2$. From Table 4, the critical value at $\alpha = 0.05$ is 5.991.

Since $\chi^2 = 14.182 > 5.991$, we *reject* H_0 at $\alpha = 0.05$.

We can therefore conclude that the manager's claimed proportions are *not* correct.

S13.5 ANALYSIS OF CONTINGENCY TABLES

The χ^2 technique used in the previous section can be generalised to the case where two variables are involved. In this case, the data will be in the form of a contingency table with any number of rows and columns. The steps involved in the analysis of contingency tables are shown below.

1. Construct the null hypothesis for the problem. This usually takes the general form that the two variables are *independent* or that there is *no relationship* between them.

   H_0: The two variables *are* independent

   or

   H_0: There is *no* relationship between the two variables

   This statement should be modified for each individual problem.

   The alternative hypothesis (using a two-sided alternative) would be:

   H_1: The two variables are *not* independent

   or

   H_1: There *is* a relationship between the two variables

2. The **observed frequencies** come from the data of the problem. There will be one observed frequency for each cell of the contingency table.
3. The **expected frequencies** are those that we might 'expect' to occur if H_0 were true. For each cell of the contingency table there will also be an expected frequency. The expected frequency for each cell can be found using the formula in Equation S13.3:

$$\text{Expected frequency for a cell} = \frac{\text{total frequency for that column} \times \text{total frequency for that row}}{\text{grand total frequency}} \quad \textbf{(S13.3)}$$

   where the *grand total frequency* can be found in the bottom right-hand corner of the table.

4. The measure of the discrepancy between the observed and expected frequencies is given by the *chi-square test statistic*. The formula is given in Equation S13.4:

$$\chi^2 = \sum_{\text{all cells}} \frac{(\text{observed frequency} - \text{expected frequency})^2}{\text{expected frequency}} \quad \textbf{(S13.4)}$$

   Note that there is one term required in the calculation for each cell of the table.

5. Associated with the test statistic are degrees of freedom. The degrees of freedom for a contingency table is given by:

$$\text{Degrees of freedom} = (\text{number of rows} - 1) \times (\text{number of columns} - 1) \quad \textbf{(S13.5)}$$

6. As is the case with the goodness-of-fit test, the calculated value of the chi-square test statistic must be compared with a critical value. This critical value is also obtained from Table 4 in the same way as before, using both the degrees of freedom and desired significance level.
7. Compare the value of χ^2 that you calculated with the critical value from Table 4.

   If $\chi^2 <$ the critical value, then H_0 *cannot be rejected.*
   If $\chi^2 >$ the critical value, then H_0 *can be rejected.*

8. Based on the outcome of Step 7, an appropriate conclusion can be drawn.

Example S13.10

The human resource manager of a major life insurance company has received complaints from the sales managers regarding the qualifications of the sales staff being employed. The sales managers want more sales staff who are university graduates. Of the 900 current sales staff, the human resource manager finds that only 300 are graduates. The human resource manager uses the sales records from the previous 12 months to classify the performance of all sales staff as either 'satisfactory' or 'unsatisfactory'. The frequencies in the various categories are shown in Table S13.8.

Table S13.8 Sales staff qualification and performance for Example S13.10

	Qualification		
Performance	Graduate	Non-graduate	*Total*
Satisfactory	170	310	480
Unsatisfactory	130	290	420
Total	300	600	900

Is there any relationship between sales performance and degree qualification? Test at $\alpha = 0.05$.

Solution

The null and alternative hypotheses for this problem are:

H_0: There is *no* relationship between sales performance and degree qualification

v

H_1: There *is* a relationship between sales performance and degree qualification

The expected frequency for each of the four cells can be found using Equation S13.3. For example, for the first row and first column we have:

$$\text{Expected frequency} = \frac{300 \times 480}{900} = 160$$

The remaining three expected frequencies can be obtained in the same way. Table S13.9 shows both the observed frequencies and the expected frequencies (in parentheses). Note that the expected frequencies also add up both along the rows and down the columns.

Table S13.9 Observed and expected frequencies for Example S13.10

	Qualification		
Performance	Graduate	Non-graduate	*Total*
Satisfactory	170 (160)	310 (320)	480
Unsatisfactory	130 (140)	290 (280)	420
Total	300	600	900

The value of the χ^2-test statistic can be found using Equation S13.4.

$$\chi^2 = \frac{(170 - 160)^2}{160} + \frac{(130 - 140)^2}{140} + \frac{(310 - 320)^2}{320} + \frac{(290 - 280)^2}{280}$$

$$= 0.625 + 0.714 + 0.313 + 0.357$$

$$= 2.009$$

Degrees of freedom $= (2 - 1) \times (2 - 1) = 1$

From Table 4, the critical value at $\alpha = 0.05$ is 3.841.

Since $\chi^2 = 2.009 < 3.841$, we *cannot reject* H_0 at $\alpha = 0.05$.

We therefore conclude that there does not appear to be any relationship between sales performance and whether the salesperson has a degree.

Example S13.11

A football enthusiast is interested in whether those people who follow a Rugby League team actually live in the same district as their team. She took a random sample of 200 Rugby League supporters from Sydney and classified their place of residence as West, North or elsewhere. She then asked them if they followed the Northern Eagles, one of Parramatta/Penrith/Wests Tigers or some other team. The results are shown in Table S13.10.

Table S13.10 Place of residence and team supported for Example S13.11

	Team			
Live	Northern Eagles	Parramatta/Penrith/Wests Tigers	Other	*Total*
West	30	45	15	90
North	42	19	19	80
Elsewhere	8	6	16	30
Total	80	70	50	200

Determine whether there is a relationship between place of residence and Rugby League team supported. Test at $\alpha = 0.01$.

Solution

The null and alternative hypotheses for this problem are:

H_0: There is *no* relationship between place of residence and Rugby League team supported

v

H_1: There *is* a relationship between place of residence and Rugby League team supported

The expected frequency for each of the 9 cells can be found using Equation S13.3. For example, for the first row and first column we have:

$$\text{Expected frequency} = \frac{80 \times 80}{200} = 32$$

The remaining 8 expected frequencies can be obtained in the same way. Table S13.11 shows both the observed frequencies and the expected frequencies (in parentheses). Note that the expected frequencies also add up both along the rows and down the columns.

Table S13.11 Place of residence and team supported for Example S13.11

	Team			
Live	Northern Eagles	Parramatta/Penrith/Wests Tigers	Other	*Total*
West	30 (32)	45 (28)	15 (20)	80
North	42 (32)	19 (28)	19 (20)	80
Elsewhere	8 (16)	6 (14)	16 (10)	40
Total	80	70	50	200

The value of the χ^2-test statistic can be found using Equation S13.4.

$$\begin{aligned}\chi^2 &= \frac{(30-32)^2}{32} + \frac{(45-28)^2}{28} + \frac{(15-20)^2}{20} + \frac{(42-32)^2}{32} + \frac{(19-28)^2}{28} \\ &\quad + \frac{(19-20)^2}{20} + \frac{(8-16)^2}{16} + \frac{(6-14)^2}{14} + \frac{(16-10)^2}{10} \\ &= 0.125 + 10.321 + 1.250 + 3.125 + 2.893 + 0.050 + 4.000 + 4.571 + 3.600 \\ &= 29.335\end{aligned}$$

Degrees of freedom = $(3 - 1) \times (3 - 1) = 4$
From Table 4, the critical value at $\alpha = 0.01$ is 13.277.
Since $\chi^2 = 29.335 > 13.277$, we *reject* H_0 at $\alpha = 0.01$.
We therefore conclude that there is a significant relationship between place of residence and Rugby League team supported.

S13.6 REMARKS

The chi-square procedures discussed in this chapter are based on *frequencies*. The numbers on which the formal arithmetic is done must be actual counts, *not* percentages or proportions. For example, if you have a problem in which the data are expressed as a percentage, these percentages must first be transformed into frequencies. Make sure that the counts are such that each individual is placed in exactly one category. In a contingency table, this means that each individual is in exactly one row and exactly one column; that is, in one cell of the table.

It should also be noted that chi-square analyses are technically only approximations. In particular, they may not be reliable when the expected frequency for any category or cell is less than 5. If this occurs, it may be overcome by combining similar categories into a larger category of similar meaning.

Quick Quiz

1. Under precisely what circumstances will the value of a χ^2-test statistic be exactly zero?
2. What would the conclusion be if the value of a χ^2-test statistic was exactly zero?
3. What is the χ^2 critical value for 10 degrees of freedom at $\alpha = 0.05$?
4. Convert 35% of 500 into a frequency.
5. Suppose you were doing a χ^2 goodness-of-fit with 7 categories. The value of the test statistic is 13.52. Would you *reject* or *not reject* H_0 at $\alpha = 0.05$? Why?
6. In Question 5, would you *reject* or *not reject* H_0 at $\alpha = 0.01$? Why?
7. A contingency table with 5 rows has 20 degrees of freedom. How many columns does the table have?
8. If a fair die were rolled 240 times, how many times would you expect the outcome to be either a 3 or a 4?
9. Describe briefly how you turn measurements of the numeric variable weight into a categorical variable.
10. What does it mean in a goodness-of-fit test if you find that the sum of the observed frequencies does not equal the sum of the expected frequencies?

Multiple Choice Questions

1. The *critical value* for a χ^2-test statistic with 15 degrees of freedom at $\alpha = 0.05$ is:
(a) 23.68 (b) 30.58 (c) 25.00
(d) 32.80 (e) none of the above

2. A contingency table has 6 rows and 8 columns. The number of *degrees of freedom* is:
(a) 48 (b) 35 (c) 13
(d) 40 (e) none of the above

3. Suppose when using a χ^2-test you rejected H_0 at $\alpha = 0.01$. It *automatically* follows that:
(a) H_0 would not be rejected at $\alpha = 0.05$
(b) H_0 would be rejected at $\alpha = 0.005$
(c) H_0 would be rejected at $\alpha = 0.05$
(d) H_0 would not be rejected at $\alpha = 0.10$
(e) none of the above

4. In a contingency table in which the grand total frequency is 400, the *expected frequency* for a cell where the row total is 50 and the column total is 30 is:
(a) 0.20 (b) 35 (c) 79
(d) 3.75 (e) none of the above

5. For a particular cell in a contingency table, the observed frequency is 10 and the expected frequency is 8. The contribution of this cell to the value of the χ^2-test statistic is:
(a) 80 (b) 63 (c) 0.40
(d) 0.20 (e) none of the above

6. A χ^2 goodness-of-fit test has 23 categories. The *critical value* at $\alpha = 0.05$ is approximately:
(a) 34.17 (b) 40.00 (c) 33.91
(d) 37.65 (e) none of the above

7. Suppose that H_0 states that there is no relationship between income level and donations to charity each year. Then H_1 would state that:
(a) income level and donations to charity have no association
(b) income level has nothing to do with charity donations
(c) if income levels rise, then donations to charity are unchanged
(d) income levels and donations to charity are related
(e) none of the above

8. If two variables are *independent*, then this means that:
(a) the value of one does not depend on the value of the other
(b) the value of one does depend on the value of the other
(c) the value of one can be used to predict the value of the other
(d) the variables are somehow related
(e) none of the above

9. If the sum of the first row of frequencies in a contingency table is equal to 250, it *automatically* follows that:
(a) the sum of the first column frequencies must also be 250
(b) the total sum of all column frequencies cannot be 250
(c) the grand total frequency must be greater than 250
(d) the degrees of freedom must be at least 10
(e) none of the above

10. If you calculated the value of a χ^2-test statistic to be *negative*, it *automatically* follows that:
(a) you must have made a mistake
(b) H_0 is rejected for all values of α
(c) H_0 is not rejected for all values of α
(d) the test is inconclusive
(e) none of the above

Exercises

S13.1 Sixty married women were chosen at random in a country town and asked whether they would prefer to live in a major city. Twenty-four answered 'yes'; the remainder 'no'. Test whether it is equally likely that half of all married women in the town would prefer to live in a major city. Use $\alpha = 0.05$.

S13.2 In a multinational glass products company 55% of the employees are female. The higher-level management of the company has 56 employees, 22 of whom are female. Is the proportion of females in higher-level management significantly different than for the company as a whole? Test using $\alpha = 0.05$.

S13.3 A union representative claimed that only 20% of employees in her industry were actively opposed to compulsory membership of a union. A management company undertook a private survey and found that 50 of a random sample of 200 employees opposed compulsory membership.

Do you accept the union representative's claim? Answer using $\alpha = 0.01$.

S13.4* The complete claim referred to in Exercise S13.3 was that 20% were actively opposed, 50% favoured compulsory membership and 30% were apathetic. In the sample, the results were:

50 actively opposed

60 in favour

90 apathetic

Are these results consistent with the claimed proportions? Test at $\alpha = 0.05$.

S13.5 In a Rugby Union season, three goal-kickers, Smith, Brown and Green, have the following success rates when taking kicks for penalty goals:

Smith: 34 goals from 50 attempts

Brown: 44 goals from 60 attempts

Green: 27 goals from 40 attempts

Is there any significant difference among the success rates for the three players? Test at $\alpha = 0.05$.

S13.6* In a survey, 57 out of a sample of 150 females and 88 out of a sample of 200 males thought that there should be a ban on gambling advertisements on television.

(a) Construct a contingency table for the data.

(b) Does this indicate a significant difference of opinion between the two sexes? Use $\alpha = 0.05$.

S13.7 A shopkeeper observes the number of cans of soft drink of various types sold during one week. The results were:

Drink \ Day	Monday	Tuesday	Wednesday	Thursday	Friday	Saturday
Cola drinks	145	145	130	285	305	80
Mineral water	75	40	30	115	115	245
Other 'fizzy' drinks	160	170	130	280	290	275

Does it appear that there is any relationship between the days of the week and the popularity of the various types of soft drinks? Use $\alpha = 0.05$.

S13.8 During a market research survey, a random sample of 400 smokers of all ages were asked to indicate their first preference among three brands of cigarettes. The results are set out below.

Brand preferred / Age of smoker	A	B	C	*Total*
Under 30	30	60	30	120
30–60	80	110	30	220
Over 60	30	10	20	60
Total	140	180	80	400

On considering these results, one of the manufacturers claimed that preference among brands is independent of age. Test at $\alpha = 0.05$ whether the evidence supports this claim.

S13.9 After looking at the data in Example S13.8, a manufacturing executive issued a press release that stated: 'We have reason to believe that 50% of the population prefer brand B rather than A or C.'

Test at $\alpha = 0.05$ whether the evidence supports the statement in the press release.

S13.10* Early in 2002 a drug company launched a new headache powder that was packaged in three different sizes, namely, small, medium and large. The company's market research indicated that public demand would be 20% for the small size, 50% for the medium size and 30% for the large size. A random sample of 500 customers subsequently showed that 120 bought the small size, 215 bought the medium size and 165 bought the large size. Do you feel that the market research was accurate? Test at $\alpha = 0.01$.

S13.11* Four machines are used to produce a component for videocassette recorders. Random samples of 200 components produced by each machine are selected and the number of defective components counted. The results are as follows:

Machine number	1	2	3	4	*Total*
Number of defectives	28	22	35	20	105

Do these results indicate whether the proportion of defectives produced varies from machine to machine? Test at $\alpha = 0.05$.

S13.12 A state government is interested in knowing the public's attitude to a continuance of the freeway system around Sydney. It surveyed 600 people of various age groups with the following results:

Opinion / Age	In favour	Opposed	*Total*
18–29	92	58	150
30–55	137	143	280
56 and over	121	49	170
Total	350	250	600

Do these results indicate any relationship between age and opinion on freeways? Test at $\alpha = 0.05$.

S13.13 In Exercise S13.12, combine the age categories '30–55' and '56 and over' into one category '30 and over'. Re-solve the question. Does your conclusion change?

S13.14* According to ratings as at 1 January 2002, a particular 7.00 p.m. television program had 20% of the viewing audience at that time. Before the next ratings period a private survey was taken because of the programming changes on the other channels, and this survey showed that from a random sample of 400 viewers, 64 tuned to that program. Does this indicate that the proportion of viewers tuned to the program had changed significantly? Test at $\alpha = 0.01$.

S13.15* The same 400 people in Exercise S13.14, categorised according to age, were asked to indicate the type of program they most enjoy. They gave the following results:

Program preferred \ Age	Under 20	20–40	Over 40	*Total*
News	46	28	26	100
Drama	32	59	29	120
Sport	82	73	25	180
Total	160	160	80	400

Do these results indicate any relationship between age and program preference? Test at $\alpha = 0.05$.

S13.16 In a major cricket study undertaken by John Croucher, one purpose was to determine if there was any relationship between the type of batsman and the location of the match for those 364 batsmen dismissed lbw in England/Australia Test matches played between 1946 and 1980. Batsmen were classified as good (batsmen or all-rounders) and poor (bowlers), and the location was either Australia or England. The relevant frequencies of lbw decisions are shown in the following table.

Batsman (category) \ Location	Australia	England	*Total*
Good	100	122	222
Poor	77	65	142
Total	177	187	364

Do these data indicate evidence of a relationship between the 'category of batsman' and 'the location of the match'? Test at $\alpha = 0.05$ and interpret your answer in the context of the question.

S13.17 The manager of a company that manufactures CD drives, CD-rewritable drives and DVD drives for notebook computers is interested to know whether there is a preference for particular types of drives among various customers. She classified customers into four categories: home, educational, small business and large business. A random sample of 200 customers was chosen with the following results:

Type of drive \ Customer type	Home	Educational	Small business	Large business	*Total*
CD	6	10	13	11	40
CD rewritable	21	33	20	16	90
DVD	13	17	17	23	70
Total	40	60	50	50	200

Is there any relationship between the type of drive selected and the type of customer? Test at $\alpha = 0.05$.

S13.18 A survey of 300 Australian-born males and 200 English-born males resident in New South Wales asked respondents to express their views on whether a member of British Royalty would be a suitable person to act as an Australian Governor-General. Of those surveyed, only 20% of the Australian-born respondents supported such an appointment, compared with 30% of the English-born respondents.

(a) Construct a contingency table of the data.

(b) Is there any significant difference between the proportions expressing support? Test at $\alpha = 0.01$.

S13.19* A wine manufacturer has 5 new varieties of wine (A, B, C, D and E), which are presented to visitors for tasting at the vineyards. Each visitor is asked to state their first preference. The manager claims that 40% will choose wine D, with the other wines being equally popular among the four of them. To test the claim, 80 visitors were chosen at random and their preferences recorded as below:

Wine	A	B	C	D	E
Number preferring that type	10	19	15	27	9

Do you accept the manager's claim? Test at $\alpha = 0.05$.

S13.20 A car magazine claims that 40% of new car buyers prefer a car that is Australian made. A survey of 200 new car buyers produced 62 who had bought Australian-made cars. Based on this evidence, do you accept the car magazine's claim? Test at $\alpha = 0.05$.

S13.21 Two years after the introduction of the GST in Australia, 500 managers of small businesses were asked whether they thought that it had been good for their organisation. Of these, 280 said 'yes', while the remainder said 'no'. Test at $\alpha = 0.05$ whether this is consistent with a government official's claim that 65% of small business managers would say that the GST had been good for their organisation.

Appendixes

Appendix A A NOTE ON SUMMATION NOTATION

In many areas of statistics there are problems in which the actual values of numerical observations are not known beforehand. Instead of describing these observations as '1st observation, 2nd observation, ..., 15th observation...', a more convenient method may be used. The most common technique is to denote the observations by a letter of the alphabet (e.g. x or y) and to label a particular observation in a sequence of observations by attaching a subscript to this letter. For instance, x_4, x_5 and y_{17} would mean, respectively, the 4th x-observation, the 5th x-observation and the 17th y-observation.

Example A.1

The ages of 10 randomly chosen customers in a restaurant are recorded below.

Customer	1	2	3	4	5	6	7	8	9	10
Age (years)	15	13	28	37	19	7	43	27	31	19

Suppose that this set of data is represented by the variable x. Then:

(a) $x_1 = 15$

(b) $x_5 = 19$

(c) $x_8 = 27$

(d) $x_3 + x_9 = 28 + 31 = 59$

(e) smallest x-value $= x_6 = 7$

(f) largest x-value $= x_7 = 43$

(g) $x_1 + x_2 + x_3 + x_4 + x_5 + x_6 + x_7 + x_8 + x_9 + x_{10} = 239$

Example A.1(g) involved the addition of 10 observations and, while the arithmetic was not difficult, it was tedious work to write out the full algebraic expression. If, for example, there were 4629 observations, it would be ridiculous to write out the expression in full. To avoid this problem, it is convenient to employ the 'sigma-operator' notation, where the Greek letter Σ (capital sigma) is used to correspond to the English word 'sum'. Thus, the expression 'Σx' means 'sum the x-values'.

In some instances, it may be desirable not to sum *all* the observations but only a certain *subset* of them. For example, you may be asked to sum all the values from x_2 to x_7 (both inclusive) in

Example A.1. This kind of expression may also be written using summation notation by indicating the *first* term to be included *below* the Σ sign and the *last* term to be included *above* the sigma sign.

Example A.2

(a) $\sum_{1}^{7} x$ means 'sum from x_1 to x_7, inclusive'.

(b) $\sum_{8}^{49} y$ means 'sum from y_8 to y_{49}, inclusive'.

If no figures appear above and below the Σ sign, it is assumed that *all* values are to be summed.

Example A.3

Below are some additions with their equivalent Σ notation.

Long version	Σ version
$x_1 + x_2 + x_3 + x_4 + x_5$	$\sum_{1}^{5} x$
$y_{12} + y_{13} + y_{14}$	$\sum_{12}^{14} y$
$z_{61} + z_{62} + z_{63} + z_{64}$	$\sum_{61}^{64} z$

Example A.4

Suppose $c_1 = 6$, $c_2 = 15$, $c_3 = 8$, $c_4 = 10$ and $c_5 = 7$. Then:

$$\sum_{1}^{5} c = 6 + 15 + 8 + 10 + 7 = 46$$

$$\sum_{1}^{3} c = 6 + 15 + 8 = 29$$

$$\sum_{2}^{4} c = 15 + 8 + 10 = 33$$

Sometimes, it is necessary to add some quantities that are *derived* from an original set of observations. In this case, write after the sigma sign the algebraic expression for the sort of quantity to be added. For instance, to add the squares of a set of *x*-observations, write 'Σx^2'. This means 'sum all the x^2-values'.

Example A.5

Suppose that the x-values are:

x_1	x_2	x_3	x_4	x_5	x_6
3	5	2	1	0	7

Then the corresponding x^2-values are:

9 25 4 1 0 49

and so:

$$\Sigma x^2 = 9 + 25 + 4 + 1 + 0 + 49 = 88$$

Note: The expression Σx^2 is *not* the same as $(\Sigma x)^2$. Σx^2 means 'calculate all the x^2-values, then add them up'. $(\Sigma x)^2$ means 'calculate Σx, then square the result'. Thus, in Example A.5, $\Sigma x = 3 + 5 + 2 + 1 + 0 + 7 = 18$, so that $(\Sigma x)^2 = 18 \times 18 = 324$, which is *not* the same as the value of 88 calculated for Σx^2.

Example A6

Suppose that a set of x-values is:

x_1	x_2	x_3	x_4	x_5	x_6
5	17	–1	6	0	–4

Then:

(a) $\Sigma x^2 = 5^2 + 17^2 + (-1)^2 + 6^2 + 0^2 + (-4)^2 = 367$

(b) $(\Sigma x)^2 = (5 + 17 - 1 + 6 + 0 - 4)^2 = (23)^2 = 529$

(c) Calculate $\Sigma(x - 3)$' means 'calculate all the values of $x - 3$ and add them up'. For this example the $(x - 3)$-values are:

2 14 –4 3 –3 –7

so that:

$$\Sigma(x - 3) = 2 + 14 - 4 + 3 - 3 - 7 = 5$$

(d) $[\Sigma(x - 3)]^2 = 5^2 = 25$

(e) To find $\Sigma(x - 3)^2$, calculate all the values of $(x - 3)^2$ and add them up. For this example the $(x - 3)^2$-values are:

$2^2 \quad 14^2 \quad (-4)^2 \quad 3^2 \quad (-3)^2 \quad (-7)^2$

so that:

$$\Sigma(x - 3)^2 = 2^2 + 14^2 + (-4)^2 + 3^2 + (-3)^2 + (-7)^2 = 283$$

The techniques used above also apply to multiple sets of observations. In the most simple case, suppose that there are two sets of observations, which are denoted by $x_1, x_2, x_3, \ldots$ and $y_1, y_2, y_3 \ldots$ For example, if both the height and weight were recorded for a group of people, with height denoted by x and weight by y, x_3 would be the height of the third person and y_3 would be that person's weight.

The expression 'Σxy' indicates that, for each pair of values, the product xy (i.e. x times y) should be calculated and the results added. Thus:

$$\Sigma xy = x_1y_1 + x_2y_2 + x_3y_3 + \ldots$$

Example A.7

Suppose that x-values and y-values are:

$x_1 = 6$ $\quad x_2 = 4$ $\quad x_3 = 9$ $\quad x_4 = 7$ $\quad x_5 = 3$

$y_1 = 2$ $\quad y_2 = 0$ $\quad y_3 = 15$ $\quad y_4 = 8$ $\quad y_5 = 6$

(a) $\Sigma x = 6 + 4 + 9 + 7 + 3 = 29$

(b) $\Sigma y = 2 + 0 + 15 + 8 + 6 = 31$

(c) $(\Sigma x)(\Sigma y) = 29 \times 31 = 899$

(d) $\Sigma xy = (6 \times 2) + (4 \times 0) + (9 \times 15) + (7 \times 8) + (3 \times 6) = 221$

Note: The expression Σxy is *not* the same as $(\Sigma x)(\Sigma y)$.

(e) Calculate the value of the expression:

$$E = \frac{1}{5}\Sigma xy - \left(\frac{\Sigma x}{5}\right)\left(\frac{\Sigma y}{5}\right)$$

From (a), $\Sigma x = 29$; from (b), $\Sigma y = 31$; from (d), $\Sigma xy = 221$.

Therefore, $E = \frac{1}{5}(221) - \left(\frac{29}{5}\right)\left(\frac{31}{5}\right) = 8.24.$

Example A.8

Suppose that a set of x-values is:

x_1	x_2	x_3	x_4	x_5
12	7	14	10	2

(a) $\Sigma x = 12 + 7 + 14 + 10 + 2 = 45$

(b) $(\Sigma x)^2 = 45^2 = 2025$

(c) $\Sigma x^2 = 12^2 + 7^2 + 14^2 + 10^2 + 2^2 = 493$

(d) $\Sigma(x - 9) = 3 - 2 + 5 + 1 - 7 = 0$

(e) $\Sigma |x - 9| = |3| + |-2| + |5| + |1| + |-7| = 18$

Appendix B A NOTE ON CALCULATORS

Introduction

There are many types of calculators on the market today, and it would be impossible to cover all the many features that they possess. Moreover, there are differences between brands as to how they label certain keys. The following sections will mention some of the more basic functions that many calculators can perform, but the list is by no means complete.

If you do not find some of the keys discussed here, consult the instruction booklet provided with your calculator.

Each reference to a specific key (apart from a numeric key) is enclosed in a box (□).

Clearing entries

If you mistakenly enter a number in the display during a calculation, you may press a *clear entry* key to erase the incorrect number only. (All other numbers that have been entered remain intact.) This key is usually labelled as:

[C] or [CE] or [CHX]

If you wish to clear *all* entries made in a calculation, press the *all clear* key. This key is usually labelled as [AC] : however, in *some* models, it is simply labelled as [C].

Basic operations

Most calculators have keys that perform the four basic operations, namely:

[+] [−] [×] [÷]

Depending on the type of calculator, numbers may be entered in one of two different ways:

1. If there is an [=] key or [EXE] key, the method used is *algebraic notation.*

2. If there is an [ENT] key, the method used is *reverse Polish notation.*

When using algebraic notation, enter numbers following elementary algebraic rules. For example, the calculation 3 + 2 would be performed by pressing the sequence:

3 [+] 2 [=]

and the calculator would give a display of 5. Note that pressing the [=] key instructs the calculator to complete the addition and display the answer.

When using reverse Polish notation, enter the two numbers *before* pressing the operation key. In the above example, this would be performed by the sequence:

3 [ENT] 2 [+]

and the calculator would give a display of 5. In this case, pressing the operation key [+] instructs the calculator to complete the addition and display the answer.

The remainder of the examples in this section are done assuming that the calculator used is one with an *algebraic* notation.

Care must be taken when dealing with *combined* operations; that is, calculations that include mixtures of operations. For example, suppose that we wanted to perform the calculation:

$6 \times (3 + 5)$

If we simply pressed the sequence:

6 [×] 3 [+] 5 [=]

the display would be 23. In this case, the calculator would have multiplied the 6 and the 3 to obtain 18, then added 5 to obtain 23. This is *not* what we intend. Instead, we should add what is in the brackets (i.e. the 3 and 5) first, obtain that answer on the display, then multiply by 6. That is, we should press the sequence:

3 [+] 5 [=] [×] 6 [=]

to obtain the correct answer of 48.

(Note that some calculators possess parenthesis keys of the type [[(] and [)]], which may also be used.)

Similarly, if we wanted to perform the calculation:

$(6 + 4) \times 3$

we would press the sequence:

6 [+] 4 [=] × 3 [=]

to obtain the correct answer of 30. If we pressed the sequence:

6 [+] 4 [×] 3 [=]

we would obtain the *incorrect* answer of 18.

Exponents

In some calculations we have to raise numbers to certain powers. In the special case where the power is 2, we are simply squaring the number; many calculators have a key labelled [x^2] that will do this directly. For example, if we wanted to find the square of 6.53, we would press the sequence:

6.53 [x^2]

and the calculator would perform the operation and display the correct answer of 42.6409.

In many calculators, if we enter a number and subsequently press the [×] key and then the [=] key, the calculator will display the square of the entered number. For example, pressing the sequence:

6.53 [×] [=]

would also display the correct answer of 42.6409.

Some calculators have a special power key labelled [x^y], which is used when raising numbers to higher powers. If this is the case, the answer to $(5.2)^4$ would be found from pressing the sequence:

5.2 [x^y] 4 [=]

and the correct answer of 731.1616 would be displayed. If a calculator does *not* have an [x^y] key, powers of numbers may be found by repeated multiplication if y is an integer.

Similarly, some calculators have a power key labelled [$x^{\frac{1}{y}}$], which is used when finding the yth root of x. For example, to find the 5th root of 28.7, the key sequence to press is:

28.7 [$x^{\frac{1}{y}}$] 5 [=]

to obtain the correct answer of 1.956935.

Memory

Most calculators have at least one *memory*. That is, they have the ability to store numbers while other calculations are performed. There are two different types of memory: the *storage-recall* and the *fully addressable*.

The simplest kind of memory is the *storage-recall* type. It is usually signified by an [STO] key or an [M in] key. If a number is entered into this type of memory, *any number that may have been in it previously is erased and replaced by the entered number*. For example, to enter the result of the calculation 2.5×3.4 into this memory we would press the sequence:

2.5 [×] 3.4 [=] [M in]

This would display the result of 8.5 and store the amount 8.5 in the memory. (Most calculators also give an indication that something is currently stored in the memory, usually 'M' somewhere on the display.)

To *recall* the amount in the memory, we can press a button labelled [MR] (for memory recall) or [RCL]. Note that memories are *not* cleared when these recall buttons are pressed.

A memory may be cleared in most cases by pressing the sequence:

0 [STO] or 0 [M in]

or by simply turning the calculator *off* and then *on*. However, most calculators retain their memory, even when turned off.

The *fully addressable* memory is one in which numbers can be added to (or subtracted from) numbers already in the memory without recalling the results to the display. The following keys are often available:

[M+] Enters a number from the display into the memory and *adds* the number in the display to the number that is already in the memory.

[M–] *Subtracts* the number in the display from the number in the memory.

[MR] *Recalls* the number currently in the memory to the display. (Pressing this key does *not* clear the memory.)

MC Clears the memory. (If the calculator does not have this key, the memory may be cleared by performing the sequences outlined for a storage-recall memory.)

For example, suppose we wished to perform the calculation:

$(21 \times 3) + (56 \times 4) - (32 \times 2)$

After first clearing the memory, we could press the sequence:

21 × 3 = M+ 56 × 4 = M+ 32 × 2 = M– MR

And the calculator would display the correct answer of 223. If we were to now press the AC key to clear the display, the figure of 223 would still remain in the memory until the memory itself was cleared.

If the calculator does *not* have an M– key, the amount to be subtracted could have its *sign* changed by pressing the +/– key and then the M+ key. For the above calculation, this would mean the sequence:

21 × 3 = M+ 56 × 4 = M+ 32 × 2 = +/– M+ MR

and the same result of 223 would be obtained.

Statistical calculators

Some calculators are able to perform basic statistical functions on a set of numbers. The calculator should first be set in its *statistical mode*. (Consult your calculator manual to see how this is done. For example, it may be set on some by moving a sliding switch and on others by pressing a sequence of keys.)

Also refer to your calculator manual to determine *how* data are to be entered. (In some cases, this is done via the key that has the word DATA or DT underneath it. This is *sometimes* the M+ key. When this is done, the letters SD may appear on the display. It is important to note that for many calculators, when the statistical mode is used, the *memory* is inoperative.)

The basic statistical functions provided usually include:

n the *number* of values entered

Σx the *sum* of the values entered

$\bar{x}$ the *mean* of the values entered

Σx^2 the *sum of the squares* of the values entered

σ_n the *standard deviation* of the values entered, using the formula in Equation S4.6

σ_{n-1} the *standard deviation* of the values entered, using the formula in Equations S4.7 or S4.8

For example, suppose we wish to find the basic statistics on the five numbers 34, 23, 56, 78, 45. First, we put the calculator in statistical mode. Suppose that values are entered via the M+ key. The sequence of keys to press is:

34 M+ 23 M+ 56 M+ 78 M+ 45 M+

At this stage, all values have been entered. To obtain the desired statistics, we now press the appropriate keys. (These statistics may be obtained in any order.) The result of pressing various keys is as follows:

Key pressed	Display
n	5
Σx	236
$\bar{x}$	47.2
Σx^2	12 930
σ_n	18.925 115 59
σ_{n-1}	21.158 922 47

More sophisticated statistical calculators may be able to perform advanced functions such as finding correlation and regression coefficients, factorials and the values of various test statistics.

Financial calculators

Financial calculators have been designed specifically to compute various financial functions. Examples include percentages, the amount of simple and compound interest accrued, the present and future value of an annuity, amortisation and cash flows. Such functions are available using inbuilt formulae that have been pre-programmed into the calculator. All the user has to do is to enter the data for the problem, and the result of the requested formula appears.

It is not unusual for financial/business calculators to contain several elementary statistical functions also, along with the capability to store text. Often, they are also *programmable*. That is, they allow users to write their own functions.

Other types of calculators

As well as being able to perform statistical or financial calculations with ease, most calculators (especially those designated as *scientific*) can also perform a wide range of mathematical functions, including those involving trigonometric and engineering data. Some of the more expensive also have the capability to draw graphs on the screen!

In any case, it would be unusual to find a student calculator that did not have a memory of keys for calculating the *square root* $\boxed{\sqrt{x}}$, *reciprocal* $\boxed{1/x}$, *logarithm* to the base 10 $\boxed{\log}$ or to the base *e* $\boxed{\ln}$ and a *power key* of the type $\boxed{x^y}$.

Indeed, today some calculators have become so sophisticated that they are virtually a handheld computer. Features such as coloured and multiline displays, memories of 1 MB or more, inbuilt spreadsheets and schedulers (to name but a few) have pushed the limits of their capabilities well beyond the basic level. Add to this the ability to beam messages and other information silently between similar model calculators, and these devices have become the ultimate cheating tool for students wanting to exchange answers during examinations.

For this reason, most universities, colleges and TAFE institutions have placed restrictions on the type of calculator that can be taken into an examination, usually banning those that are programmable or can store text information. Some institutions restrict students to using only one or two acceptable models, while others have spent tens of thousands of dollars purchasing vast quantities of calculators that are handed out to students during examinations in much the same way that tables used to be handed out many years ago.

Unfortunately, some students spend a great deal of money on a calculator that is packed with enticing features that they may never use. Moreover, these are often accompanied by intimidating lengthy instruction manuals that look all too complicated, and, to make matters worse, the calculator is one of the type that is prohibited from examinations.

Before purchasing any calculator, make sure that it has all the features that you are likely to need without going overboard. If in doubt, ask your lecturer for advice: it could save you a lot of time and trouble!

Appendix C A NOTE ON COMPUTER PACKAGES

Just as the electronic calculator became an indispensable tool for the student of the mid-1970s, so the computer has become a necessary device since the 1980s. On the market today there is a wide variety of brands of computers, all offering a dazzling range of features such as processor speeds of well over 1.5 GHz, something that would have been only dreamed about only 10 years earlier. During the initial period of change, most businesses and tertiary institutions moved from an operating system such as MS-DOS[1] (popular in the 1980s) to a Windows-type environment such as the early Microsoft Windows 3.1 or 3.11[1] and Windows NT.[1] In 1995 Microsoft Windows 95[1] was introduced with much fanfare, and in turn this was upgraded three years later to Windows 98.[1] At the end of the last century Windows 2000[1] appeared, quickly followed in early 2001 by Windows ME[1] (Millennium Edition). Most modern computer packages are also available for Apple Macintosh[2] computers.

In practice, most statisticians, and those working in finance, take advantage of one or more of the many software packages that are available to assist in solving their particular problems. Such packages may include a spreadsheet, a database, a word processor, a communications program, a statistical program or an easy connection to the Internet, or may perhaps be an *integrated* package that includes some or all of these functions.

For many offices the personal computer, often part of a network system, has replaced the mainframe as the source of computing power. A similar situation exists in most tertiary institutions, where students perform their computing work at a terminal or personal computer connected via a network to a file server containing the main software.

Computers may be operated by means of typing in commands via a keyboard or, in most cases, by means of a *mouse*. The mouse is manipulated by moving it to different positions on the desk (preferably on a *mouse pad*) such that the desk movements correspond to the movements of a pointer on the screen. When this pointer is on the desired feature of a program, an appropriate button on the mouse is clicked, once or twice, to operate that application.

A growing number of individuals are now connected to the *Internet* (World Wide Web), so that an enormous amount of information is literally only a click of a mouse button away. In turn, the proliferation of *email* enables communication with others anywhere in the world and many institutions now provide their students with at least limited access via personal accounts.

The major advances in computers include increases in clock speed, greater memory capacity, larger capacity hard disk drives and escalating graphics capabilities. The popularity of lightweight portable notebook computers with inbuilt modems enables the modern statistician to be in contact with colleagues and clients and carry out their work virtually anywhere in the world.

In the following sections we consider three of the more common computer packages used by students in Australia to carry out statistical analysis. A brief general description of each is provided, but for more in depth analysis you should of course refer to the appropriate manual.

---

1 Microsoft is a registered trademark, and Windows is a trademark of the Microsoft Corporation.
2 Macintosh is a registered trademark of Apple Computers.

Minitab®[3]

Minitab is a general-purpose data-analysis system that has been designed to organise and analyse data. It is useful for researchers, engineers, teachers, students, managers and statisticians. In fact, one of its outstanding features is that new users can learn many of its capabilities in a relatively short space of time, an attribute that makes it especially attractive for tertiary institutions, in which it enjoys worldwide popularity.

The versions of Minitab up to and including Release 8 were based on the MS-DOS operating system, while in late 1993 the first version for Windows, Release 9, appeared. Since then all versions have been Windows-based, including Releases 10, 10Xtra and 11, the latter appearing in 1996. Late 1997 saw the introduction of Release 12 and in 2001 Release 13 was still being used, which ran under the operating systems Windows 95/98/2000, Windows ME or Windows NT4.

A minimum of 16 MB of RAM is required to run Minitab, together with 35 MB of hard disk space for full installation. Also necessary is a CD-ROM drive and, for some capabilities, a mouse. A Pentium processor (or better) is also highly recommended.

Some of Minitab's data analysis features include:

1. Descriptive statistics, including means, medians, standard deviations and quartiles.
2. General graphics, including scatter plots, data displays, charts, histograms, three-dimensional graphs, pie charts and probability plots.
3. Regression analysis with one or more independent variables (residual analysis and stepwise procedures are also available), model selection, best subsets, prediction and confidence intervals, and a wide selection of graphs.
4. Analysis of Variance (ANOVA), including one-way and two-way multiple factor, general linear model (GLM), analysis of covariance (ANCOVA), multivariate analysis of variance (MANOVA) and a comprehensive summary of fitted models.
5. Time series analysis, including plots, trend analysis, decomposition analysis, single and double exponential smoothing, moving averages and ARIMA models.
6. Statistical process (quality) control, including run charts, Pareto charts, variable and attribute control charts and process capability analysis.
7. Reliability and survival analysis, including parametric distribution analysis, measures for how well a distribution fits the data and regression models for life data.
8. Designed experiments, including factorial designs for 3 to 15 factors, multiple response optimisation and sophisticated graphs.
9. Simulations and distributions, including random number generators and random sampling.

Depending on the version of Windows being used, the screen appearance of the dialogue boxes, windows and several software features of Minitab will vary slightly. However, the *contents* of these boxes and windows are the same for all versions of Windows.

When Minitab commences, the opening screen consists of a Session window (on the top part of the screen) and a Data window (on the lower part of the screen). This screen is shown in Figure C.1.

---

[3] Minitab is a registered trademark.

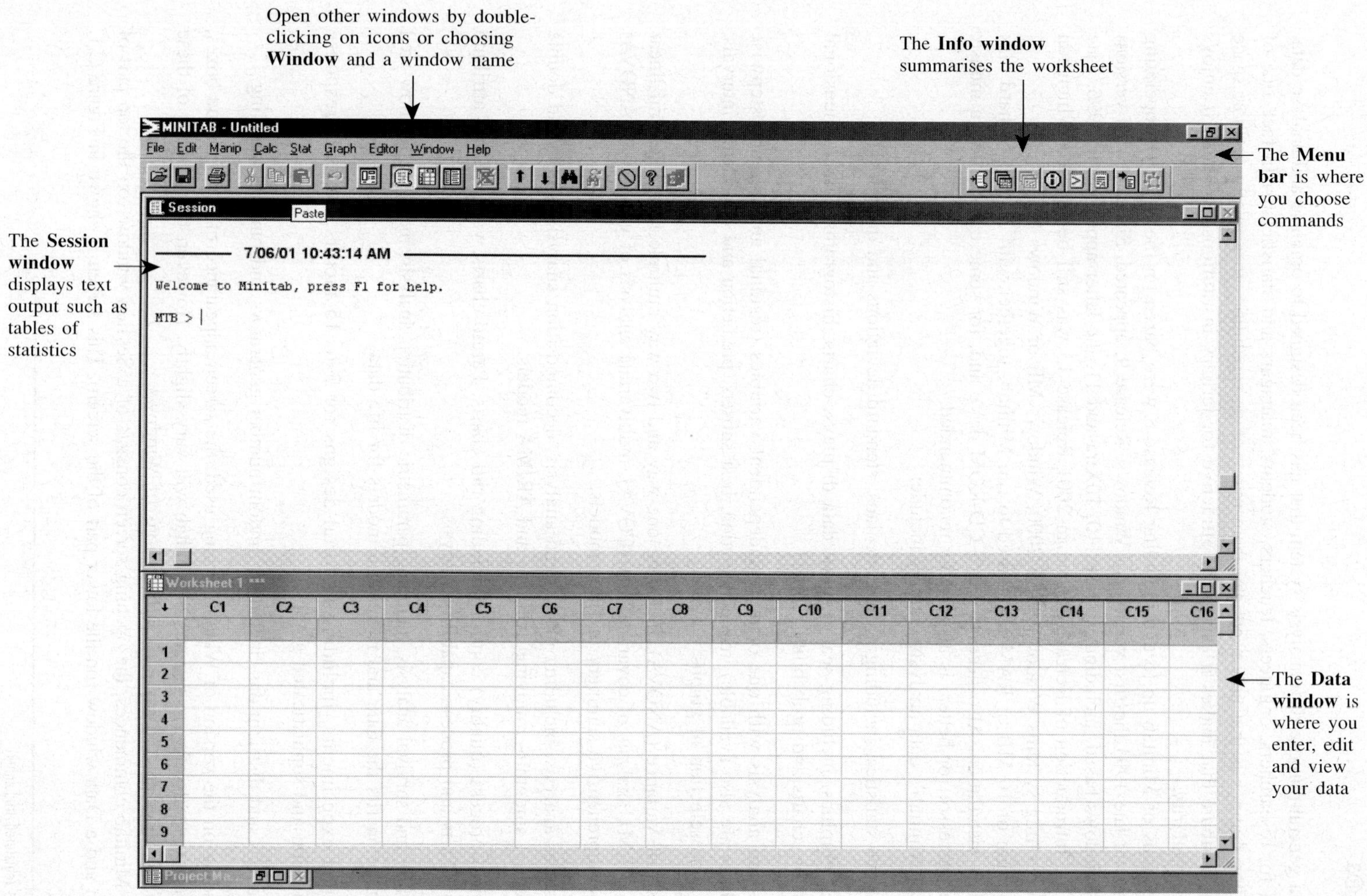

Figure C.1 An opening Minitab screen

***Source*:** Screen shot reprinted with permission from the Microsoft Corporation

There are three ways to access Minitab commands: with Menu commands; with toolbars; and with Session commands.

1. **Menu commands** In this case, click on an item in the Menu bar to open the Menu, then click on a Menu item to execute the command, open a submenu or open a dialogue box. (If a Menu item is dimmed, it is currently unavailable.) If you right-click in a Minitab window to open the shortcut menu, then click on a Menu item as in regular menus.

2. **Toolbars** When you click a toolbar button, Minitab performs an action or opens a dialogue box, exactly like the corresponding Menu command. Minitab's toolbars are detachable. You can drag them anywhere within the Minitab environment for easy access. However, detached toolbars always remain on top of any windows.

3. **Session commands** These are useful alternatives to Menu commands, especially when making macros to automate repetitive analyses. Most Session commands are simple, easy-to-remember words, such as PLOT, DESCRIBE or SORT. You can type commands in two places: the Command Line Editor and the Session window.

   (a) *To use the Command Line Editor*:
   Choose Edit ➔ Command Line Editor.
   Type the commands in the text box.
   Click Submit Commands.

   (b) *To use the Session window command prompt MTB>*:
   Make sure that the Session window is active.
   Choose Edit ➔ Enable Command Language.
   Type commands at the MTB> prompt.

In the Session window, the Minitab prompt MTB> will appear when the program is waiting for you to enter a command. There are numerous commands to read, edit and print data, draw plots and histograms, perform arithmetic calculations, undertake transformations, and carry out various statistical analyses and tests.

The data in Minitab are stored in columns, which are given the default names of C1, C2, C3, and so on until given specific names by the user (e.g. Cost, Profit, Year), that is, any name up to eight letters in length. Stored constants are denoted by K1, K2, K3, and so on. The total number of columns, stored constants and data points that can be used depends largely on the size of your computer's physical memory (RAM), virtual memory and the number of applications that you are running. This information is given when you start Minitab, with the default worksheet size being 100 000 data elements (cells).

Optional *subcommands* may also be used. These are often additional instructions as an adjunct to the main command. If a subcommand is used, the main command must end with a semi-colon and the final subcommand must end with a period. An example is:

```
MTB> DESCRIBE C1;
SUBC> BY C2.
```

The above sequence instructs Minitab to describe the data in column C1, but to separate the output into the various categories as given in column C2.

Another way of entering data into the worksheet is to enter it directly into the Data window. To do this, simply click into the appropriate cell and type the data point in the box. In this way, names

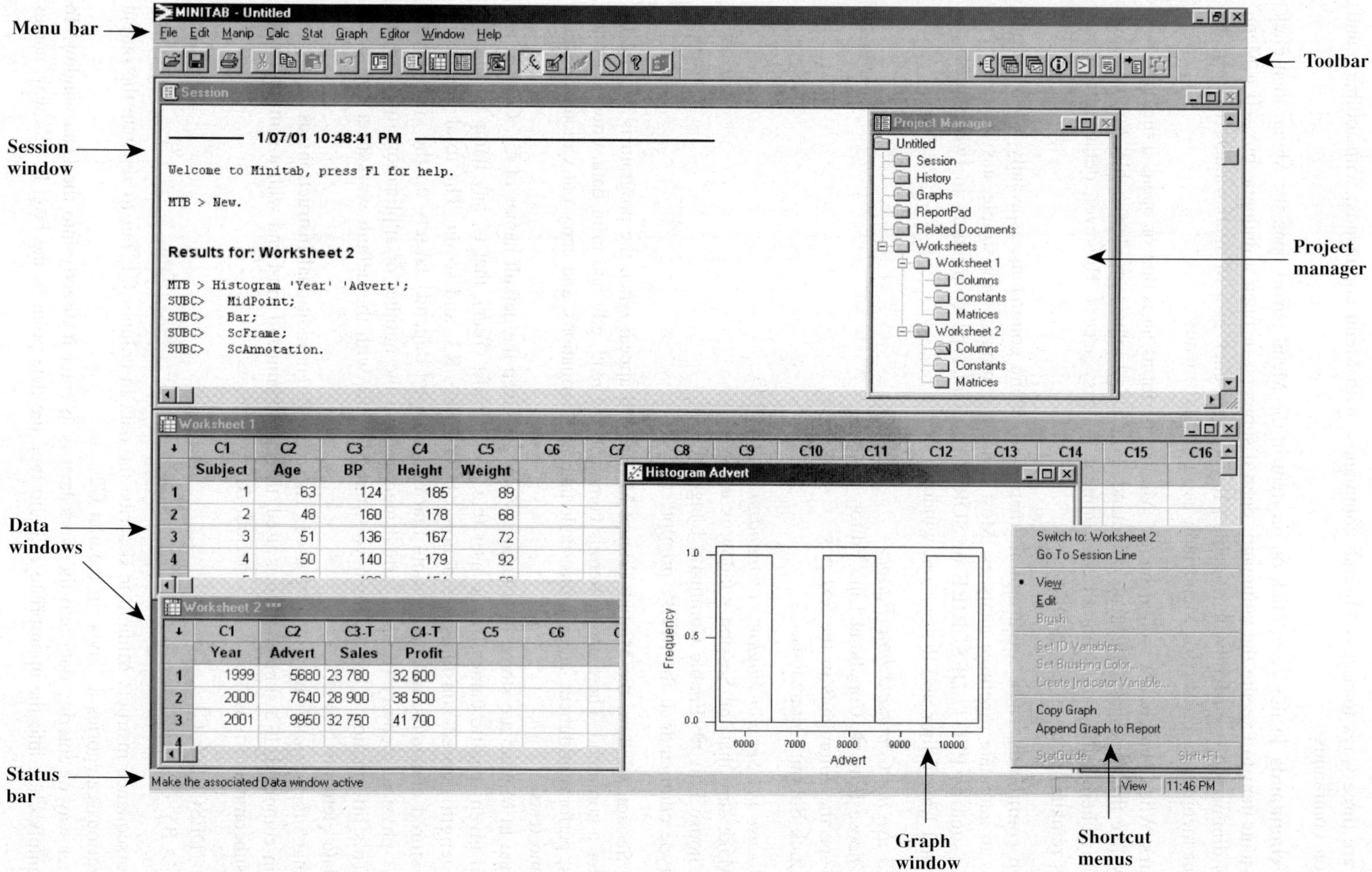

Figure C.2 An example of a worksheet for Minitab containing data, column names and graph

Source: Screen shot reprinted with permission from the Microsoft Corporation

of columns can also be entered by typing them in at the top of the columns. Other methods of entering data include reading the data in from a file or generating the data using a Minitab command. An example of a Minitab worksheet containing data and column names is shown in Figure C.2, along with a graph of some of the data.

As mentioned previously, one of the major assets of Minitab is its ease of use. Once the data are stored in appropriate columns, it is a simple matter of initiating the required command to obtain the desired output. For example, to obtain some basic statistics on data in, say, C4, we could enter at the Session window DESCRIBE C4 after the MTB> prompt. Alternatively, this could be done using the Command Line Editor by choosing Stat ➔ Basic Statistics ➔ Descriptive Statistics and then selecting the variable in column C4.

The resulting output would include:

1. the number of observations
2. the arithmetic mean
3. the median
4. the trimmed mean (i.e. the average of the observations after the largest 5% and smallest 5% are removed)
5. the standard deviation
6. the minimum and maximum values
7. the lower and upper quartiles

Minitab also has the ability to import data from, and export data to, an external spreadsheet such as *Microsoft Excel*. For full details of this and other features involving Minitab, refer to the *Minitab User's Guide*.

SPSS®[4]

SPSS (*Statistical Package for the Social Sciences*) is a powerful statistical package with wide-ranging capabilities. The latest version in 2001 was Version 10.1, which included an extensive help system, online tutorial and intuitive interface. The system requirements include a minimum of 32 MB of RAM with 32 MB of virtual memory. Microsoft Windows 95/98/2000 or ME or Microsoft Windows NT4.0, 68 MB of hard disk space, an SVGA monitor and Windows-compatible mouse are also essential.

Some of the statistics available on SPSS are as follows:

1. **Data summary:**
   frequency counts and measures
   descriptive statistics
   cross-tabulations
   case summary reports
   reports with summaries in rows and columns

[4] SPSS is a registered trademark of SPSS Inc. for its proprietary computer software.

2. **Multiple response:**
   summaries for data in which questions can have more than one answer
   frequency counts and tables

3. **Regression:**
   linear regression using weighted or unweighted least-squares
   several methods for variable entry and removal

4. **Compare means:**
   testing hypotheses about group differences
   descriptive statistics for subgroups
   independent, paired-sample and one-sample *t*-tests
   one-way analysis of variance (ANOVA)
   factorial ANOVA and covariance

5. **Non-parametric tests:**
   one-sample tests
   tests for two independent or related samples
   chi-square
   binomial
   improved data transformations
   interpolative values for missing values

6. **Graphics:**
   bar charts, including simple, clustered, stacked, dropped-shadow and hanging
   line drawings, including simple, multiple and projection
   simple and exploding pie charts
   histograms with or without normal curve superimposed
   various types of scatter plot, including simple, three-dimensional and animated rotating
   simple and clustered box plots
   time series plots
   probability plots

7. **Quality control charts:**
   Pareto
   X-bar
   range
   sigma
   individual chart

There are two main ways of opening SPSS for Windows:

1. SPSS Manager provides a standard Windows full-screen interface including online help about SPSS commands. Note that there are two windows. The SPSS Data Editor window displays data and allows them to be modified, while the Output window displays the results of the analysis. An example of a blank Data Editor window is shown in Figure C.3.

2. Production Mode (Batch Command Mode) is useful for experienced users who wish to execute a list of SPSS commands that have been saved to a command syntax file, by using either a text editor or SPSS to help you. This file must be named with a *.inc* extension.

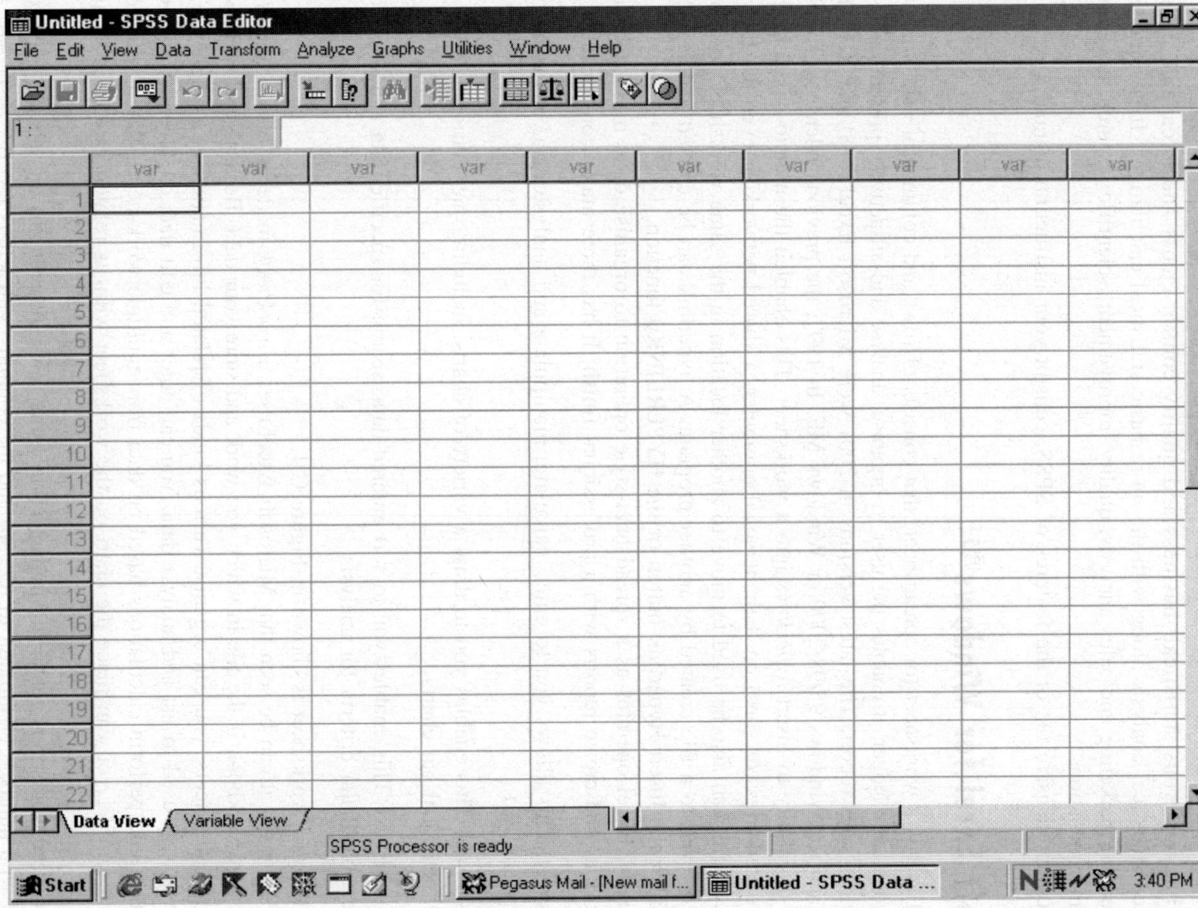

Figure C.3 A blank SPSS Data Editor window

Source: Screen shot reprinted with permission from SPSS Australia Pty Ltd and the Microsoft Corporation

A normal SPSS session consists of entering or loading a set of data and selecting the analysis to be performed from the Statistics Menu or one of the other pull-down menus. The output will then appear in the Output window, or may be saved to an output file if desired. Data can be in the form of an SPSS file, a spreadsheet file, a database file or an ASCII file. SPSS is able to produce quite sophisticated graphs, such as those that illustrate time series data.

SPSS has numerous statistical, business and quality control charts and is capable of quite sophisticated statistical analysis along with the more basic statistical calculations. It is suitable for survey research, marketing and sales analysis, quality improvement, scientific research, government reporting and education.

For a comprehensive list of the features of SPSS, consult your instruction manual.

Microsoft Excel for Windows®[5]

Microsoft Excel is a worksheet (or spreadsheet) that consists of rows and columns of data, somewhat like an accountant's ledger. It enables the user to organise, analyse and attractively present data such as a budget or sales report. The latest versions in 2001 were Microsoft Excel 2000 and XP, which were designed for Windows 95/98/2000 or Windows ME. In 1997, the previous release of Excel 97 introduced the ability to insert *hyperlinks* into a worksheet. This enabled the user to have one-click access to related files (or sites) on their personal computer, a shared network or even the Internet.

The hyperlink can also be used to move to another location in the same worksheet, to another worksheet or even to a file created by another program. A hyperlink can be created in Excel 2000 by using either the Insert Hyperlink button or the HYPERLINK0 function.

Essentially, Excel operates as a spreadsheet-style operation to organise data, make decisions, construct graphs and create reports with a professional finish. It has three major components:

1. **Worksheet** This allows you to enter, calculate, manipulate and analyse data, including both numbers and text.
2. **Charts Excel** This enables you to draw a variety of charts, including pie charts and two- and three-dimensional bar charts.
3. **Database Excel** This enables you to sort entered data, search for specific data and select data that meet particular criteria for analysis.

A blank Excel worksheet is shown in Figure C.4.

From Figure C.4 it can be seen that Microsoft Excel uses a *workbook* made up of *worksheets* (or spreadsheets). A workbook is the file in which you work and store your data. Because each workbook can contain many sheets, you can organise various kinds of related information in a single file.

Worksheets are used to list and analyse data. You can enter and edit data on several worksheets simultaneously and perform calculations based on data from multiple worksheets. When you create a chart, for example, you can place the chart on the worksheet with its related data on a separate chart sheet.

Each of these worksheets contains a frame in which the rows and columns are labelled. The rows are labelled with numbers (1 through to 65 536) and the columns with letters (A, B, …, Z, AA, AB, AC, … , IV), making 256 columns in all. The intersection of a row and a column is called a *cell*,

---

[5] Excel is a registered trademark of the Microsoft Corporation.

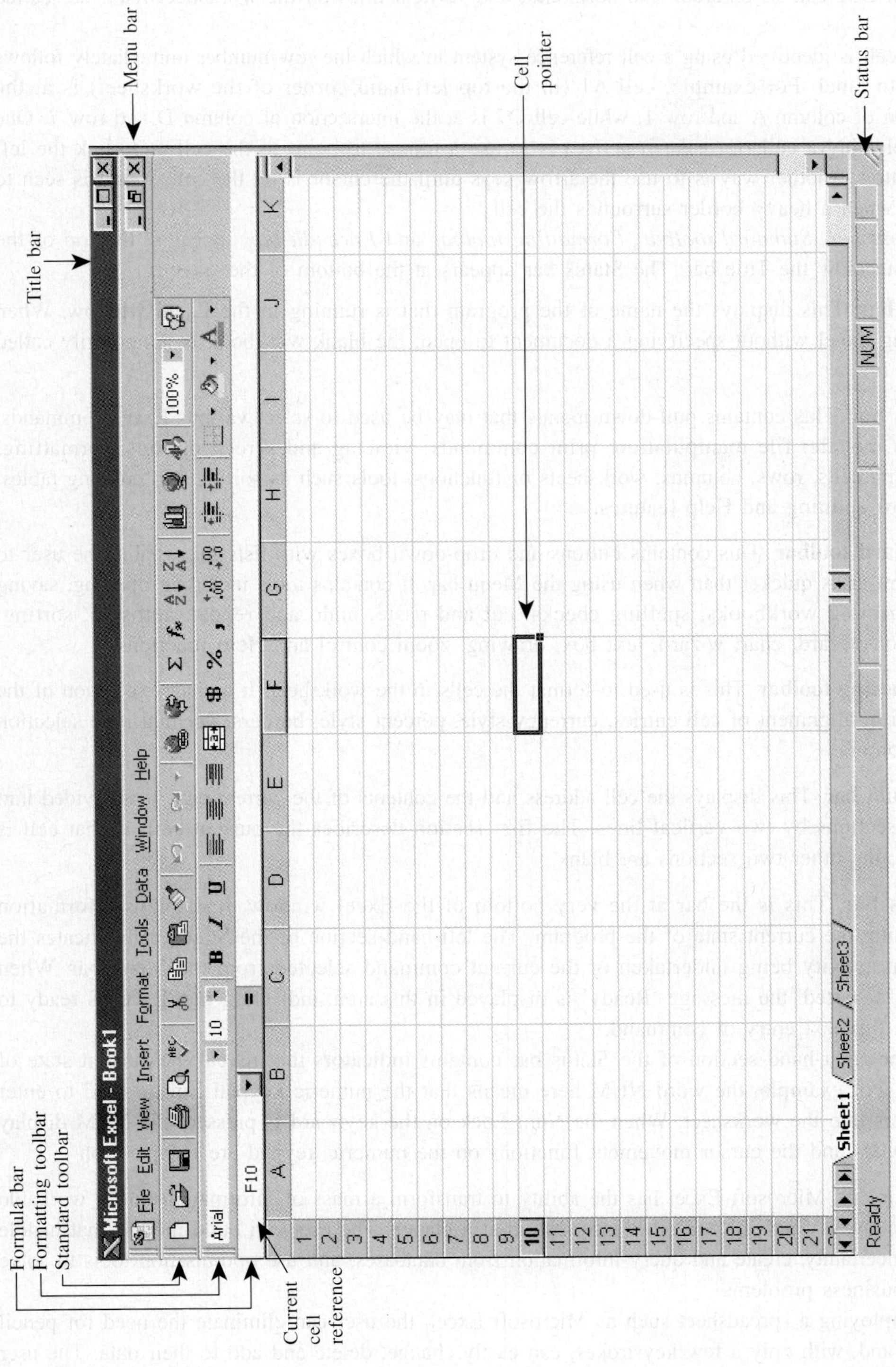

Figure C.4 A blank Excel worksheet

Source: Screen shot reprinted with permission from the Microsoft Corporation

into which data can be entered. The horizontal and vertical lines on the worksheet itself are called *gridlines*.

Each cell is identified using a cell reference system in which the row number immediately follows the column label. For example, cell A1 (in the top left-hand corner of the worksheet) is at the intersection of column A and row 1, while cell D7 is at the intersection of column D and row 7. One way of selecting a cell (to make it active) is to use a mouse to point to the cell and click the left mouse button. Another way is to use the arrow keys until the cursor is on the cell. A cell is seen to be active when a heavy border surrounds the cell.

A *Menu bar*, *Standard toolbar*, *Formatting toolbar* and *Formula bar* appear at the top of the screen just below the Title bar. The Status bar appears at the bottom of the screen.

1. **Title bar** This displays the name of the program that is running in the Excel window. When starting Excel without specifying a document to open, the blank workbook is temporarily called Book 1.

2. **Menu bar** This contains pull-down menus that may be used to select various Excel commands. These include: file manipulation; print commands; viewing and screen options; formatting; inserting cells, rows, columns, worksheets or functions; tools such as sorting and creating tables; window splitting and Help features.

3. **Standard toolbar** This contains buttons and drop-down boxes with lists that enable the user to perform tasks quicker than when using the Menu bar. It contains tools including opening, saving and printing workbooks, spelling checks, cut and paste, undo and repeat, autosum, sorting, function wizard, chart wizard, text box, drawing, zoom control and Help functions.

4. **Formatting toolbar** This is used to format the cells in the worksheet. It includes selection of the font size, alignment of cell entries, currency style, percent style, borders, decimal type selection and colours.

5. **Formula bar** This displays the cell address and the contents of the current cell. It is divided into three sections by two vertical lines. The first section describes the current cell. If that cell is empty, the other two sections are blank.

6. **Status bar** This is the bar at the very bottom of the Excel window. It displays information regarding the current state of the program. The left-hand section of the Status bar indicates the current activity being undertaken or the current command selected from the Menu bar. When Excel is started, the message 'Ready' is displayed in this area, indicating that Excel is ready to accept the next entry or command.

   The right-hand section of the Status bar contains indicators that reveal the current state of Excel. For example, the word NUM here means that the numeric keypad can be used to enter numbers into the worksheet. When the Num Lock on the keyboard is pressed, the NUM display disappears and the cursor movement functions on the numeric keypad are in operation.

In summary, Microsoft Excel has the ability to transform a mass of information into a workable form that conveys financial and other information at a glance. The user can build models that handle risk and uncertainty, create and query information from databases, and use optimisation tools to solve complex business problems.

By employing a spreadsheet such as Microsoft Excel, the user can eliminate the need for pencil and paper and, with only a few keystrokes, can easily change, delete and add to their data. The user can also perform calculations ranging from simple arithmetic calculations to more complex and mathematical formulae. The end result is the saving of an enormous amount of time, along with accountancy fees! Indeed, almost any job that uses rows and columns of numbers can be performed using Microsoft Excel. Its most common uses are for budgeting and financial planning in both business and the home.

For full details of how to use Microsoft Excel, refer to your instruction manual for the relevant version.

Appendix D TRIAL EXAMINATIONS

Paper A (Chapters S1–S6)

Time allowed: One hour
Number of questions: Three
Show all working
The questions are of equal value

1. The weight (in kg) of 10 packages sent by courier from an accounting company during one week is shown below.

1.5 2.3 0.8 2.8 5.4 4.5 3.7 5.8 1.5 1.7

(a) For these data calculate:

(i) the mean
(ii) the median
(iii) the mode
(iv) the lower quartile
(v) the upper quartile
(vi) the interquartile range

(b) Suppose the first five observations each had a weighting of 2 and the last five observations each had a weighting of 5. Calculate the weighted mean of the data.

2. A bank official is interested in how long customers typically spend when accessing automatic teller machines. A random sample of 12 customers was observed and the time they spent at the machine is shown below (in minutes):

4 5 2 7 5 7 9 4 3 5 5 4

(a) For these data calculate:

(i) the mean
(ii) the range
(iii) the standard deviation
(iv) the coefficient of variation

(b) Perform two checks to show that your answer for the standard deviation in (a)(iii) is reasonable.

(c) (i) Summarise the data in the form of a frequency distribution.
(ii) Calculate the mean from the frequency distribution in (c)(i). Verify that your answer agrees with that in (a)(i).

3. (a) A box contains 4 blue balls, 6 white balls and 2 brown balls. A ball is drawn at random from the box. Find the probability that the ball is:

(i) white
(ii) brown
(iii) white given it is not brown
(iv) blue given that it is either blue or brown

(b) In a particular used-car yard, 25% of the customers each day are female and 30% of customers have a mobile phone. If being female and having a mobile phone are independent, for an individual customer calculate the probability that:

(i) they are female and have a mobile phone
(ii) they are male and have a mobile phone
(iii) they are female or have a mobile phone

(c) A fair coin is tossed 4 times. Calculate the probability of obtaining at least 3 tails.

Paper B (Chapters S7–S13)

Time allowed: One hour
Number of questions: Three
Show all working
The questions are of equal value

1. The packaging process at a breakfast cereal company is arranged so that an average of 505 g is placed in each packet of cereal labelled as 'net weight 500 g'. Because of the variability of the process, the actual amount in the packets follows a normal distribution with a standard deviation of 8.0 grams.

(a) An individual packet is selected at random. Find the probability that the actual amount of cereal in this packet is:

(i) less than 500 g
(ii) more than 525 g
(iii) between 490 g and 510 g

(b) A random sample of 50 cereal packets is selected. Find the probability that the total weight of these packets is less than 25.2 kg.

(c) A random sample of 100 cereal packets is selected. How many of these packets would you expect to weigh less than 480 g?

2. A road transport company analyst wants to determine whether there is any relationship between the time it takes fully laden trucks to travel from Sydney to Melbourne and the weight of the load they are carrying. A random sample of 8 such trucks was selected with the following results:

Weight (tonnes) x	14	9	16	10	8	5	12	16
Time (hours) y	15	9	13	11	10	10	13	17

In answering the questions below you may use the following statistics for these data:

$\bar{x} = 11.25 \quad \bar{y} = 12.25 \quad s_x = 3.96 \quad s_y = 2.765 \quad S_{xy} = 64.61$

(a) (i) Calculate the correlation coefficient (to 3 decimal places) between *weight* and *time*.
(ii) Test the value in (i) for significance and draw an appropriate conclusion.

(b) (i) Calculate the least-squares regression line of *weight* on *time*.
(ii) Use your equation in (b)(i) to estimate the time the trip would take if the weight carried were 13 tonnes.

(c) Calculate the value of the coefficient of determination for your line in (b)(i) and interpret its value.

3. Shown in the table below are the prices of 1 L of unleaded petrol, 1 kg of rice and a 150 g jar of instant coffee in a suburb of inner-Sydney for each of the September quarters 1996 and 2001. Also shown is the quantity of each item sold in those quarters.

		Price of one unit ($)		Units sold ('000)	
Item	Unit	1996	2001	1996	2001
Petrol	1 L	0.67	0.98	125	286
Rice	1 kg	1.24	1.40	83	97
Coffee	150 g	5.69	6.50	12	14

(a) Using 1996 as the base year, calculate:

(i) the simple aggregate index for 2001
(ii) the average of relative prices for 2001
(iii) the Laspeyres index for 2001
(iv) the Paasche index for 2001
(v) Fisher's ideal index for 2001

(b) Interpret the value of the Laspeyres index found in (a)(iii).

Tables

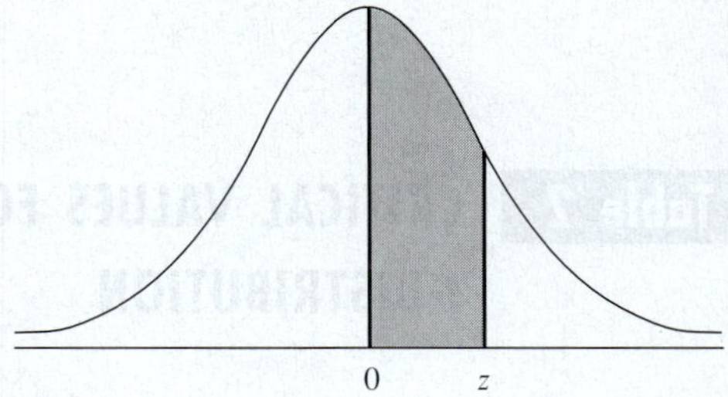

This table gives the area between 0 and $+z$.

Table 1 AREAS UNDER THE STANDARD NORMAL CURVE

z	.00	.01	.02	.03	.04	.05	.06	.07	.08	.09	z
0.0	.0000	.0040	.0080	.0120	.0160	.0199	.0239	.0279	.0319	.0359	**0.0**
0.1	.0398	.0438	.0478	.0517	.0557	.0596	.0636	.0675	.0714	.0753	0.1
0.2	.0793	.0832	.0871	.0910	.0948	.0987	.1026	.1064	.1103	.1141	0.2
0.3	.1179	.1217	.1255	.1293	.1331	.1368	.1406	.1443	.1480	.1517	0.3
0.4	.1554	.1591	.1628	.1664	.1700	.1736	.1772	.1808	.1844	.1879	0.4
0.5	.1915	.1950	.1985	.2019	.2054	.2088	.2123	.2157	.2190	.2224	**0.5**
0.6	.2257	.2291	.2324	.2357	.2389	.2422	.2454	.2486	.2517	.2549	0.6
0.7	.2580	.2611	.2642	.2673	.2703	.2734	.2764	.2793	.2823	.2852	0.7
0.8	.2881	.2910	.2939	.2967	.2995	.3023	.3051	.3078	.3106	.3133	0.8
0.9	.3159	.3186	.3212	.3238	.3264	.3289	.3315	.3340	.3365	.3389	0.9
1.0	.3413	.3438	.3461	.3485	.3508	.3531	.3554	.3577	.3599	.3621	**1.0**
1.1	.3643	.3665	.3686	.3708	.3729	.3749	.3770	.3790	.3810	.3830	1.1
1.2	.3849	.3869	.3888	.3907	.3925	.3944	.3962	.3980	.3997	.4015	1.2
1.3	.4032	.4049	.4066	.4082	.4099	.4115	.4131	.4147	.4162	.4177	1.3
1.4	.4192	.4207	.4222	.4236	.4251	.4265	.4279	.4192	.4306	.4319	1.4
1.5	.4332	.4345	.4357	.4370	.4382	.4394	.4406	.4418	.4429	.4441	**1.5**
1.6	.4452	.4463	.4474	.4484	.4495	.4505	.4515	.4525	.4535	.4545	1.6
1.7	.4554	.4564	.4573	.4582	.4591	.4599	.4608	.4616	.4625	.4633	1.7
1.8	.4641	.4649	.4656	.4664	.4671	.4678	.4686	.4693	.4699	.4706	1.8
1.9	.4713	.4719	.4726	.4732	.4738	.4744	.4750	.4756	.4761	.4767	1.9
2.0	.4772	.4778	.4783	.4788	.4793	.4798	.4803	.4808	.4812	.4817	**2.0**
2.1	.4821	.4826	.4830	.4834	.4838	.4842	.4846	.4850	.4854	.4857	2.1
2.2	.4861	.4864	.4868	.4871	.4875	.4878	.4881	.4884	.4887	.4890	2.2
2.3	.4893	.4896	.4898	.4901	.4904	.4906	.4909	.4911	.4913	.4916	2.3
2.4	.4918	.4920	.4922	.4925	.4927	.4929	.4931	.4932	.4934	.4936	2.4
2.5	.4938	.4940	.4941	.4943	.4945	.4946	.4948	.4949	.4951	.4952	**2.5**
2.6	.4953	.4955	.4956	.4957	.4959	.4960	.4961	.4962	.4963	.4964	2.6
2.7	.4965	.4966	.4967	.4968	.4969	.4970	.4971	.4972	.4973	.4974	2.7
2.8	.4974	.4975	.4976	.4977	.4977	.4978	.4979	.4979	.4980	.4981	2.8
2.9	.4981	.4982	.4982	.4983	.4984	.4984	.4985	.4985	.4986	.4986	2.9
3.0	.4987	.4987	.4987	.4988	.4988	.4989	.4989	.4989	.4990	.4990	**3.0**
3.1	.4900	.4991	.4991	.4991	.4992	.4992	.4992	.4992	.4993	.4993	3.1
3.2	.4993	.4993	.4994	.4994	.4994	.4994	.4994	.4995	.4995	.4995	3.2
3.3	.4995	.4995	.4995	.4996	.4996	.4996	.4996	.4996	.4996	.4997	3.3
3.4	.4997	.4997	.4997	.4997	.4997	.4997	.4997	.4997	.4997	.4998	3.4
3.5	.4998	.4998	.4998	.4998	.4998	.4998	.4998	.4998	.4998	.4998	**3.5**
3.6	.4998	.4998	.4999	.4999	.4999	.4999	.4999	.4999	.4999	.4999	3.6
3.7	.4999	.4999	.4999	.4999	.4999	.4999	.4999	.4999	.4999	.4999	3.7
3.8	.4999	.4999	.4999	.4999	.4999	.4999	.4999	.4999	.4999	.4999	3.8
3.9	.5000	.5000	.5000	.5000	.5000	.5000	.5000	.5000	.5000	.5000	3.9

Table 2 CRITICAL VALUES FOR THE *t*-DISTRIBUTION

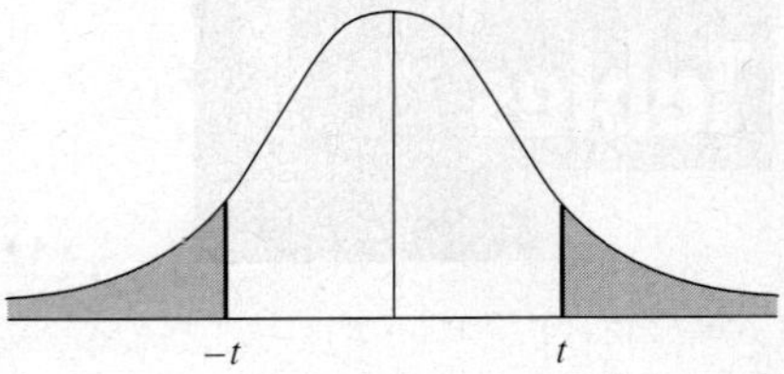

υ \ α	0.20	0.10	0.05	0.02	0.01	0.005	0.002	0.001	0.0005	α / υ
1	3.078	6.314	12.71	31.82	63.66	127.3	318.3	636.6	1273.	**1**
2	1.886	2.920	4.303	6.965	9.925	14.09	22.33	31.60	44.70	2
3	1.638	2.353	3.182	4.541	5.841	7.453	10.22	12.92	16.33	3
4	1.533	2.132	2.776	3.747	4.604	5.598	7.173	8.610	10.31	4
5	1.476	2.015	2.571	3.365	4.032	4.773	5.893	6.869	7.976	5
6	1.440	1.943	2.447	3.143	3.707	4.317	5.208	5.959	6.788	6
7	1.415	1.895	2.365	2.998	3.499	4.029	4.785	5.408	6.082	7
8	1.397	1.860	2.306	2.896	3.355	3.833	4.501	5.041	5.617	8
9	1.383	1.833	2.262	2.821	3.250	3.690	4.297	4.781	5.291	9
10	1.372	1.812	2.228	2.764	3.169	3.581	4.144	4.587	5.049	10
11	1.363	1.796	2.201	2.718	3.106	3.497	4.025	4.437	4.863	**11**
12	1.356	1.782	2.179	2.681	3.055	3.428	3.930	4.318	4.716	12
13	1.350	1.771	2.160	2.650	3.012	3.372	3.852	4.221	4.597	13
14	1.345	1.761	2.145	2.624	2.977	3.326	3.787	4.140	4.499	14
15	1.341	1.753	2.131	2.602	2.947	3.286	3.733	4.073	4.417	15
16	1.337	1.746	2.120	2.583	2.921	3.252	3.686	4.015	4.346	16
17	1.333	1.740	2.110	2.567	2.898	3.222	3.646	3.965	4.286	17
18	1.330	1.734	2.101	2.552	2.878	3.197	3.610	3.922	4.233	18
19	1.328	1.729	2.093	2.539	2.861	3.174	3.579	3.883	4.187	19
20	1.325	1.725	2.086	2.528	2.845	3.153	3.552	3.850	4.146	20
21	1.323	1.721	2.080	2.518	2.831	3.135	3.527	3.819	4.110	**21**
22	1.321	1.717	2.074	2.508	2.819	3.119	3.505	3.792	4.077	22
23	1.319	1.714	2.069	2.500	2.807	3.104	3.485	3.768	4.047	23
24	1.318	1.711	2.064	2.492	2.797	3.091	3.467	3.745	4.021	24
25	1.316	1.708	2.060	2.485	2.787	3.078	3.450	3.725	3.996	25
26	1.315	1.706	2.056	2.479	2.779	3.067	3.435	3.707	3.974	26
27	1.314	1.703	2.052	2.473	2.771	3.057	3.421	3.690	3.954	27
28	1.313	1.701	2.048	2.467	2.763	3.047	3.408	3.674	3.935	28
29	1.311	1.699	2.045	2.462	2.756	3.038	3.396	3.659	3.918	29
30	1.310	1.697	2.042	2.457	2.750	3.030	3.385	3.646	3.902	30
35	1.306	1.690	2.030	2.438	2.724	2.996	3.340	3.591	3.836	**35**
40	1.303	1.684	2.021	2.423	2.704	2.971	3.307	3.551	3.788	40
45	1.301	1.679	2.014	2.412	2.690	2.952	3.281	3.520	3.752	45
50	1.299	1.676	2.009	2.403	2.678	2.937	3.261	3.496	3.723	50
60	1.296	1.671	2.000	2.390	2.660	2.915	3.232	3.460	3.681	60
70	1.294	1.667	1.994	2.381	2.648	2.899	3.211	3.435	3.651	70
80	1.292	1.664	1.990	2.374	2.639	2.887	3.195	3.416	3.629	80
90	1.291	1.662	1.987	2.368	2.632	2.878	3.183	3.402	3.612	90
∞	1.282	1.645	1.960	2.326	2.576	2.807	3.090	3.291	3.481	∞

Table 3 CRITICAL VALUES FOR THE RANK CORRELATION COEFFICIENT

n	Critical value
5	0.9000
6	0.8286
7	0.7450
8	0.7143
9	0.6833
10	0.6364
11	0.6091
12	0.5804
13	0.5549
14	0.5341
15	0.5179
16	0.5000
17	0.4853
18	0.4716
19	0.4579
20	0.4451
22	0.4241
24	0.4061
26	0.3894
28	0.3749
30	0.3620

Source: Adapted from W. J. Conover, *Practical Nonparametric Statistics*, New York, Wiley, 1971, p. 390

Table 4 CRITICAL VALUES FOR THE CHI-SQUARE DISTRIBUTION

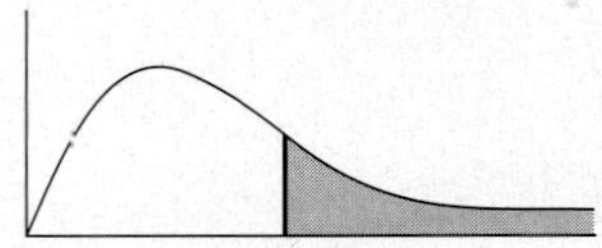

This table gives the critical values for various degrees of freedom v and significance levels α.

v \ α	0.10	0.05	0.025	0.01	0.005	0.001	α / v
1	2.706	3.841	5.024	6.635	7.879	10.827	**1**
2	4.605	5.991	7.378	9.210	10.597	13.815	2
3	6.151	7.815	9.348	11.345	12.838	16.266	3
4	7.779	9.488	11.143	13.277	14.860	18.466	4
5	9.236	11.070	12.832	15.086	16.750	20.515	5
6	10.645	12.592	14.449	16.812	18.548	22.457	**6**
7	12.017	14.067	16.013	18.475	20.278	24.321	7
8	13.362	15.507	17.535	20.090	21.955	26.124	8
9	14.684	16.919	19.023	21.666	23.589	27.877	9
10	15.987	18.307	20.483	23.209	25.188	29.588	10
11	17.275	19.675	21.920	24.725	26.757	31.264	**11**
12	18.549	21.026	23.337	26.217	28.300	32.909	12
13	19.812	22.362	24.736	27.688	29.819	34.527	13
14	21.064	23.685	26.119	29.141	31.319	36.124	14
15	22.307	24.996	27.488	30.578	32.801	37.698	15
16	23.542	26.296	28.845	32.000	34.267	39.252	**16**
17	24.769	27.587	30.191	33.409	35.718	40.791	17
18	25.989	28.869	31.526	34.805	37.156	42.312	18
19	27.204	30.144	32.852	36.191	38.582	43.819	19
20	28.412	31.410	34.170	37.566	39.997	45.314	20
21	29.615	32.671	35.479	38.932	41.401	46.796	**21**
22	30.813	33.924	36.781	40.289	42.796	48.268	22
23	32.007	35.172	38.076	41.638	44.181	49.728	23
24	33.196	36.415	39.364	42.980	45.558	51.179	24
25	34.382	37.652	40.646	44.314	46.928	52.619	25
26	35.563	38.885	41.923	45.642	48.290	54.051	**26**
27	36.741	40.113	43.195	46.963	49.645	55.475	27
28	37.916	41.337	44.461	48.278	50.994	56.892	28
29	39.087	42.557	45.722	49.588	52.335	58.301	29
30	40.256	43.773	46.979	50.892	53.672	59.702	30
40	51.805	55.758	59.342	63.691	66.766	73.403	**40**
50	63.167	67.505	71.420	76.154	79.490	86.660	50
60	74.397	79.082	83.298	88.379	91.952	99.608	60
70	85.527	90.531	95.023	100.425	104.215	112.317	70
80	96.578	101.979	106.629	112.329	116.321	124.839	80
90	107.565	113.145	118.136	124.116	128.299	137.208	90
100	118.498	124.342	129.561	135.807	140.170	149.449	100

Summary of useful statistical formulae

Some of the more commonly used statistical formulae are listed below, together with their corresponding equation number in the text.

1. The estimate of the *mode* (Mo) from a grouped frequency distribution is:

$$\text{Mo} = L + \frac{d_1}{d_1 + d_2}(i) \qquad \textbf{(S3.1)}$$

2. The estimate of the *median* ($\tilde{x}$) from a grouped frequency distribution is:

$$\tilde{x} = L + \frac{\left(\frac{n}{2} - C\right)}{f}(i) \qquad \textbf{(S3.2)}$$

3. The *arithmetic mean* ($\bar{x}$) of a set of n observations is:

$$\bar{x} = \frac{\Sigma x}{n} \qquad \textbf{(S3.3)}$$

4. If an arbitrary origin is used, the *arithmetic mean* ($\bar{x}$) is:

$$\bar{x} = \text{arbitrary origin} + \frac{\Sigma d}{n} \qquad \textbf{(S3.4)}$$

5. If a set of x-values is arranged in frequency table form, the *arithmetic mean* ($\bar{x}$) is:

$$\bar{x} = \frac{\Sigma fx}{\Sigma f} \qquad \textbf{(S3.5)}$$

6. The estimate for the *weighted arithmetic mean* from a grouped frequency distribution is:

$$\bar{x} \doteqdot \frac{\Sigma fm}{\Sigma f} \qquad \textbf{(S3.6)}$$

where m = midpoint of the class interval

7. If each x-value in a set of observations is assigned a corresponding weight (w), the *weighted arithmetic mean* is:

$$\bar{x} = \frac{\Sigma xw}{\Sigma w} \qquad \textbf{(S3.7)}$$

8. The estimate of the *lower quartile* (Q_1) from a grouped frequency distribution is:

$$Q_1 = L + \frac{\left(\frac{n}{4} - C\right)}{f}(i) \qquad \text{(S3.8)}$$

9. The estimate of the *upper quartile* (Q_3) from a grouped frequency distribution is:

$$Q_3 = L + \frac{\left(\frac{3n}{4} - C\right)}{f}(i) \qquad \text{(S3.9)}$$

10. The *geometric mean* of $x_1, x_2, \ldots x_n$ is:

$$\text{Geometric mean} = \sqrt[n]{x_1 x_2 x_3 \ldots x_n} \qquad \text{(S3.10)}$$

11. Range = largest observation – smallest observation **(S4.1)**

12. Interquartile range = upper quartile – lower quartile **(S4.2)**
$= Q_3 - Q_1$

13. Residual $= x - \bar{x}$ **(S4.3)**

14. The *mean deviation* of a set of n observations is:

$$\text{Mean deviation} = \frac{\Sigma|x - \bar{x}|}{n} \qquad \text{(S4.4)}$$

15. If the x-values are arranged in the form of a frequency distribution, the *mean deviation* is:

$$\text{Mean deviation} = \frac{\Sigma f|x - \bar{x}|}{\Sigma f} \qquad \text{(S4.5)}$$

16. The *standard deviation* (σ) of a population of size N is:

$$\sigma = \sqrt{\frac{\Sigma(x - \bar{x})^2}{N}} \qquad \text{(S4.6)}$$

17. The *standard deviation* (s) of a sample of n observations is:

$$s = \sqrt{\frac{\Sigma(x - \bar{x})^2}{n - 1}} \qquad \text{(S4.7)}$$

18. An alternative formula for the *standard deviation* (s) of a sample of n observations is:

$$s = \sqrt{\frac{\Sigma x^2 - n\bar{x}^2}{n - 1}} \qquad \text{(S4.8)}$$

19. If each x-value occurs with frequency f, the *standard deviation* (s) is:

$$s = \sqrt{\frac{\Sigma f(x - \bar{x})^2}{\Sigma f - 1}} \qquad \text{(S4.9)}$$

20. The formula for estimating the *standard deviation* (s) from a grouped frequency distribution is:

$$s \doteqdot \sqrt{\frac{\Sigma f(m - \bar{x})^2}{\Sigma f - 1}} \qquad \text{(S4.10)}$$

21. The *coefficient of variation* (V) is:

$$V = 100\left(\frac{s}{\bar{x}}\right)\% \qquad \text{(S4.11)}$$

22. If two events A and B are *mutually exclusive*, the following relationship holds:

$$\text{P}(A \text{ or } B) = \text{P}(A) + \text{P}(B) \qquad \text{(S6.1)}$$

23. Suppose that $A_1, A_2, A_3 \ldots A_n$ are n mutually exclusive events. Then:

$$\text{P}(A_1 \text{ or } A_2 \text{ or } A_3 \ldots \text{ or } A_n) = \text{P}(A_1) + \text{P}(A_2) + \text{P}(A_3) + \ldots + \text{P}(A_n) \qquad \text{(S6.2)}$$

24. If two events A and B are *independent*, the following relationship holds:

$$\text{P}(A \text{ and } B) = \text{P}(A) \times \text{P}(B) \qquad \text{(S6.3)}$$

25. Suppose that $A_1, A_2, A_3, \ldots, A_n$ are n independent events. Then:

$$\text{P}(A_1 \text{ and } A_2 \text{ and } A_3 \ldots \text{ and } A_n) = \text{P}(A_1) \times \text{P}(A_2) \times \text{P}(A_3) \times \ldots \times \text{P}(A_n) \qquad \text{(S6.4)}$$

26. If two events A and B are *mutually exclusive* and are *complements* of each other, the following relationship holds:

$$\text{P}(A) + \text{P}(B) = 1 \qquad \text{(S6.5)}$$

27. P(*at least one* person has the characteristic) = 1 − P(*no* person has the characteristic) (S6.6)

28. For *any* two events, A and B, the following relationship holds:

$$\text{P}(A|B) = \frac{\text{P}(A \text{ and } B)}{\text{P}(B)} \qquad \text{(S6.7)}$$

29. When two events are *not* mutually exclusive, the following *general addition law* should be used:

$$\text{P}(A \text{ or } B) = \text{P}(A) + \text{P}(B) - \text{P}(A \text{ and } B) \qquad \text{(S6.8)}$$

30. If two events A and B are *not* independent, the following relationship holds:

$$P(B) = P(B|A)P(A) + P(B|not\ A)P(not\ A) \qquad \textbf{(S6.9)}$$

31. A *standard score* is defined as:

$$\text{Standard score} = z = \frac{\text{observed value} - \text{mean}}{\text{standard deviation}} \qquad \textbf{(S7.1)}$$

32. The *standard score* (z) of an observation x from a population with mean μ and standard deviation σ is:

$$z = \frac{x - \mu}{\sigma} \qquad \textbf{(S7.2)}$$

33. To convert a standard score z to a *raw score* x, for a distribution with mean μ and standard deviation σ, use:

$$x = \mu + z\sigma \qquad \textbf{(S7.3)}$$

34. To convert a mean of $\bar{x}$ to a z-score use:

$$z = \frac{\bar{x} - \mu}{\frac{\sigma}{\sqrt{n}}} \qquad \textbf{(S7.4)}$$

35. If σ is known:

$$\text{Standard error of the mean} = \frac{\sigma}{\sqrt{n}} \qquad \textbf{(S7.5)}$$

36. If σ is unknown and n is large:

$$\text{Standard error of the mean} \doteqdot \frac{s}{\sqrt{n}} \qquad \textbf{(S7.6)}$$

37. If σ is known and the population is normal, a *95% confidence interval* for μ is:

$$\left(\bar{x} - 1.96\frac{\sigma}{\sqrt{n}}, \bar{x} + 1.96\frac{\sigma}{\sqrt{n}}\right) \qquad \textbf{(S7.7)}$$

38. If σ is unknown and n is large, a *95% confidence interval* for μ is:

$$\left(\bar{x} - 1.96\frac{s}{\sqrt{n}}, \bar{x} + 1.96\frac{s}{\sqrt{n}}\right) \qquad \textbf{(S7.8)}$$

39. To calculate the *product-moment correlation coefficient* (r), use:

$$r = \frac{S_{xy}}{\sqrt{(S_{xx})(S_{yy})}} \qquad \textbf{(S8.1)}$$

40. $S_{xx} = \Sigma(x - \bar{x})^2 = (n - 1)s_x^2$ **(S8.2)**

41. $S_{yy} = \Sigma(y - \bar{y})^2 = (n - 1)s_y^2$ **(S8.3)**

42. $S_{xy} = \Sigma(x - \bar{x})(y - \bar{y})$ **(S8.4)**

43. An alternative formula for r is:

$$r = \frac{S_{xy}}{(n - 1)s_x s_y} \qquad \textbf{(S8.5)}$$

44. Another alternative formula for r is:

$$r = \frac{\Sigma xy - n\bar{x}\bar{y}}{(n - 1)s_x s_y} \qquad \textbf{(S8.6)}$$

45. The *rank correlation coefficient* (r_s) is defined as:

$$r_s = 1 - \frac{6\Sigma d^2}{n^3 - n} \qquad \textbf{(S8.7)}$$

where d = the difference between the corresponding rankings
n = the number of pairs of observations

46. The *test statistic* (z) for testing a value of r for significance is:

$$z = r\sqrt{n - 1} \qquad \textbf{(S8.8)}$$

47. The *population* least-squares regression line is denoted by:

$$\hat{y} = \alpha + \beta x \qquad \textbf{(S9.1)}$$

where α and β are constants

48. The *sample* least-squares regression line is denoted by:

$$\hat{y} = a + bx \qquad \textbf{(S9.2)}$$

where a and b are constants

49. The *regression coefficient b* may be found from:

$$b = \frac{S_{xy}}{S_{xx}} \tag{S9.3}$$

50. The *regression coefficient a* may be found from:

$$a = \bar{y} - b\bar{x} \tag{S9.4}$$

51. An alternative expression for the *regression coefficient b* is:

$$b = r\left(\frac{s_y}{s_x}\right) \tag{S9.5}$$

52. The *coefficient of determination* (r^2) may be found from the calculating procedures for r, or directly from:

$$r^2 = \frac{(S_{xy})^2}{S_{xx}S_{yy}} \tag{S9.6}$$

53. The *price relative* of an item in period x where y is the base period is:

$$\text{Price relative} = \frac{p_n}{p_o} \tag{S10.1}$$

where p_n = price in period x
p_o = price in base period y

54. Simple price index = price relative × 100

$$= \frac{p_n}{p_o} \times 100 \tag{S10.2}$$

55. Simple aggregate index $= \dfrac{\Sigma p_n}{\Sigma p_o} \times 100$ **(S10.3)**

where Σp_n = sum of the current period prices
Σp_o = sum of the base period prices

56. Average of relative prices $= \dfrac{\sum\left(\dfrac{p_n}{p_o} \times 100\right)}{k}$ **(S10.4)**

where $\dfrac{p_n}{p_o} \times 100$ = the simple price index of an item
k = the number of items

57. Laspeyres index $= \dfrac{\Sigma p_n q_o}{\Sigma p_o q_o} \times 100$ **(S10.5)**

where $\Sigma p_n \; q_o$ = equivalent expenditure incurred in the current period
$\Sigma p_o \; q_o$ = expenditure incurred in the base period

58. Paasche index $= \dfrac{\Sigma p_n q_n}{\Sigma p_o q_n} \times 100$ **(S10.6)**

where $\Sigma p_n q_n$ = expenditure incurred in the current period
$\Sigma p_o q_n$ = equivalent expenditure incurred in the base period

59. Fisher's ideal index $= \sqrt{\text{(Laspeyres index) (Paasche index)}}$

$$= \sqrt{\frac{\Sigma p_n q_o \Sigma p_n q_n}{\Sigma p_o q_o \Sigma p_o q_n}} \times 100$$ **(S10.7)**

60. To *deflate* an actual salary and express it in terms of '*real*' *salary* (of the base year), use:

$$\text{'Real salary'} = \text{actual salary in current year} \times \frac{\text{CPI in base year}}{\text{CPI in current year}}$$ **(S10.8)**

61. To calculate the '*adjusted*' *annual salary*, use:

$$\text{'Adjusted' annual salary} = \text{base year salary} \times \frac{\text{CPI in current year}}{\text{CPI in base year}}$$ **(S10.9)**

62. The *least-squares regression line* of y on x is:

$$\hat{y} = a + bx$$ **(S11.1)**

63. The value of b in Equation S11.1 is:

$$b = \frac{\Sigma xy}{\Sigma x^2}$$ **(S11.2)**

64. The value of a in Equation S11.1 is:

$$a = \bar{y}$$ **(S11.3)**

65. The recurrence relation that defines *exponential smoothing* is:

$$S_x = \alpha y_x + (1 - \alpha)S_{x-1}$$ **(S11.6)**

where S_x = smoothed value for observation x
S_{x-1} = immediately previous smoothed value
y_x = actual value of observation x

66. A *growth* model between two variables x and y is of the form:

$$\hat{y} = ae^{bx} \qquad \textbf{(S11.7)}$$

where a and b are constants

67. To transform y-values to z-values for exponential smoothing, use:

$$z = \log_e y \qquad \textbf{(S11.8)}$$

68. The value of a in Equation S11.7 is:

$$a = e^{c_1} \qquad \textbf{(S11.9)}$$

69. The value of b in Equation S11.7 is:

$$b = c_2 \qquad \textbf{(S11.10)}$$

70. The formula for a one-sample *z-test statistic* where σ is *known* is:

$$z = \frac{\bar{x} - \mu_0}{\frac{\sigma}{\sqrt{n}}} \qquad \textbf{(S12.1)}$$

71. The formula for a one-sample *z-test statistic* where σ is *unknown* is:

$$z = \frac{\bar{x} - \mu_0}{\frac{s}{\sqrt{n}}} \qquad \textbf{(S12.2)}$$

72. The formula for a one-sample *t-test statistic* is:

$$t = \frac{\bar{x} - \mu_0}{\frac{s}{\sqrt{n}}} \qquad \textbf{(S12.3)}$$

with $\nu = n - 1$ degrees of freedom.

73. The formula for a *paired t-test statistic* is:

$$t = \frac{\bar{x}_d - \mu_d}{\frac{s_d}{\sqrt{n}}} \qquad \textbf{(S12.4)}$$

where n = number of pairs of differences
υ = degrees of freedom = $n - 1$
μ_d = 0 (according to the null hypothesis)

74. The formula for the *pooled standard deviation* s_p is:

$$s_p = \sqrt{\frac{(n_1 - 1)s_1^2 + (n_2 - 1)s_2^2}{n_1 + n_2 - 2}} \qquad \textbf{(S12.5)}$$

75. The formula for the *two-sample t-test statistic* is:

$$t = \frac{\bar{x}_1 - \bar{x}_2}{s_p\sqrt{\frac{1}{n_1} + \frac{1}{n_2}}} \qquad \textbf{(S12.6)}$$

76. The formula for the χ^2*-test statistic* for a *goodness-of-fit test* is:

$$\chi^2 = \sum_{\text{all categories}} \frac{(\text{observed frequency} - \text{expected frequency})^2}{\text{expected frequency}} \qquad \textbf{(S13.1)}$$

77. *Degrees of freedom* for a χ^2 goodness-of-fit test are:

Degrees of freedom = number of categories – 1 **(S13.2)**

78. The *expected frequency* for a cell in a contingency table is:

$$\text{Expected frequency for a cell} = \frac{\text{total frequency for that column} \times \text{total frequency for that row}}{\text{grand total frequency}} \qquad \textbf{(S13.3)}$$

79. The formula for the χ^2*-test statistic* for a *contingency table test* is:

$$\chi^2 = \sum_{\text{all cells}} \frac{(\text{observed frequency} - \text{expected frequency})^2}{\text{expected frequency}} \qquad \textbf{(S13.4)}$$

80. *Degrees of freedom* for a contingency table are:

Degrees of freedom = (number of rows – 1) × (number of columns – 1) **(S13.5)**

Solutions to selected exercises

Chapter S1

S1.3 (a) Examples include those that require tables, graphs, charts, etc. (e.g. company reports).

(b) Examples include those that require a conclusion (e.g. market research, clinical trials).

(c) Examples include a report that requires both visual displays and a conclusion.

S1.6 One definition given by the Operational Research Society is: 'Operations research is the application of the methods of science to complex problems arising in the direction and management of large systems of people, machines, materials and money in industry, business, government and defence. The approach is to develop a scientific model of the system incorporating factors such as chance and risk, with which to predict and compare the outcomes of alternative decisions, strategies or controls. The purpose is to help management determine its policy and actions scientifically.'

S1.7 Briefly, you would have to select a random sample of males and females from your institution and ask them whether they are smokers or non-smokers. The samples should be reasonably large (say, at least 25 each) and roughly of the same size. You could then calculate the proportion of each group who are smokers. This would give you a rough idea of whether there is a difference, but an appropriate statistical test would have to be used for a proper determination.

S1.10 (a) No influence.

(b) Possibly increases as males live longer (e.g. medical advances).

(c) Influenced by the economy (e.g. inflation).

(d) Possibly influenced by unemployment.

(e) Influenced by the economy (e.g. interest rates).

(f) Possibly influenced by the economy (maybe less in times of recession).

S1.16 Before the advent of calculators they would have used slide rules or logarithm tables.

S1.20 Conclusion is wrong. To draw any conclusion you would have to know the proportion of the population who are *not* convicted bank robbers but like chocolate ice-cream.

S1.21 *Demography* is the study of vital and social statistics such as births, deaths, diseases, marriages, etc. of populations.

S1.27 *Quantitative analysis* pertains to the describing or measuring of numerical quantities, while *qualitative analysis* is concerned mainly with non-numerical descriptions.

S1.28 A *clinical trial* is an experiment concerned with the observation and treatment of disease in patients.

S1.32 The idea is to convey that the product has been tested at the highest level by experts. However, it is meaningless since statistical tests do not *prove* anything; they merely give an indication of the effectiveness. Also, it does not state exactly *who* conducted the university tests: it could have been first-year students or administrative staff with little or no expertise in the field.

S1.35 An *actuary* is a statistician who calculates risks, rates, etc. according to probabilities derived from population statistics. Actuaries are often employed by insurance companies in the field of general insurance or life assurance.

Chapter S2

S2.1 (a)

18	18	18	19	19	19	19	19	19	19
19	20	20	20	21	21	21	21	22	22
22	22	23	23	23	23	23	24	24	24
25	25	25	26	26	26	27	27	28	28
28	29	30	30	31	32	32	32	33	35

(b) and (c)

Time (seconds)	Frequency	Cumulative frequency
18	3	3
19	8	11
20	3	14
21	4	18
22	4	22
23	5	27
24	3	30
25	3	33
26	3	36
27	2	38
28	3	41
29	1	42
30	2	44
31	1	45
32	3	48
33	1	49
35	1	50
	$\Sigma f = 50$	

S2.2 (a)

Class interval (seconds)	Frequency
18–20	14
21–23	13
24–26	9
27–29	6
30–32	6
33–35	2
	$\Sigma f = 50$

(b) and (c)

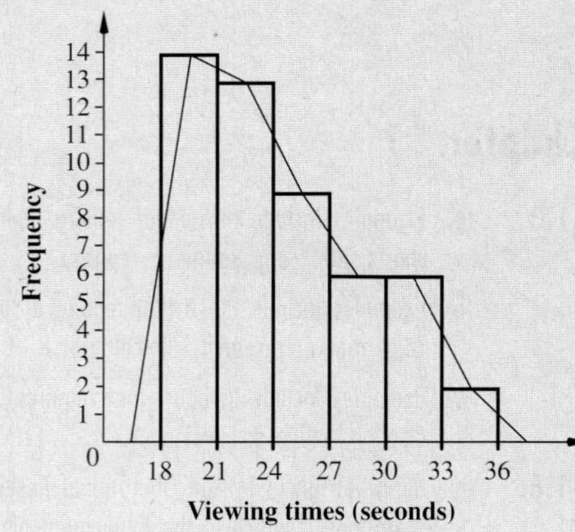

(d) exponential

(e)

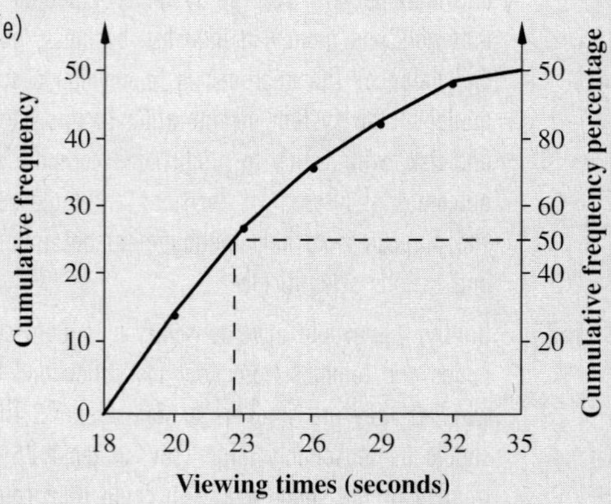

(f) 22 seconds

S2.6 (a) and (b)

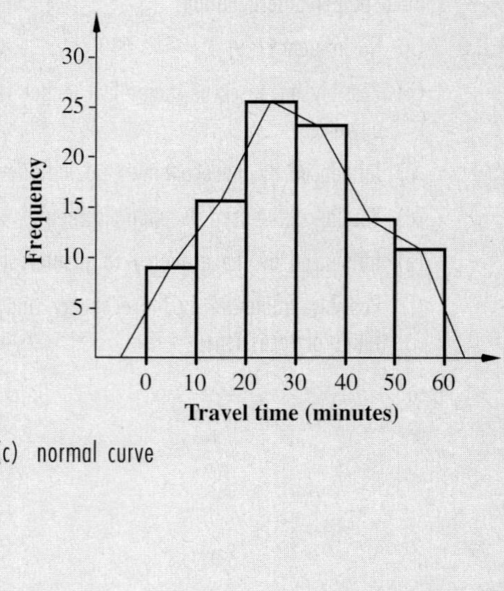

(c) normal curve

S2.8 (a)

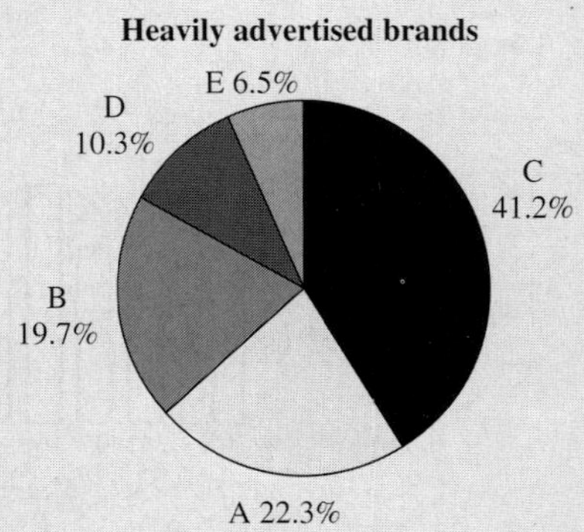

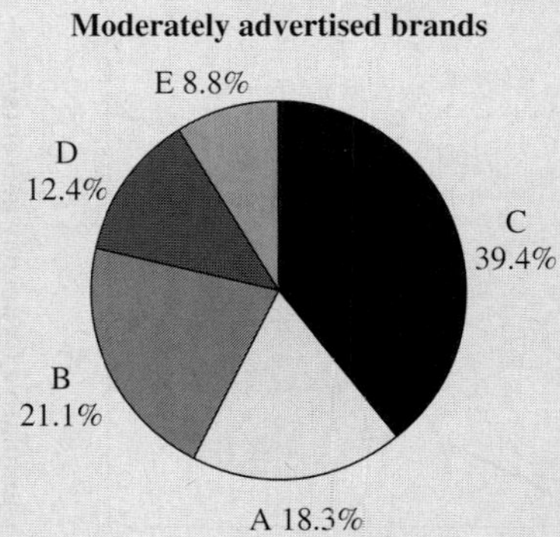

(b)

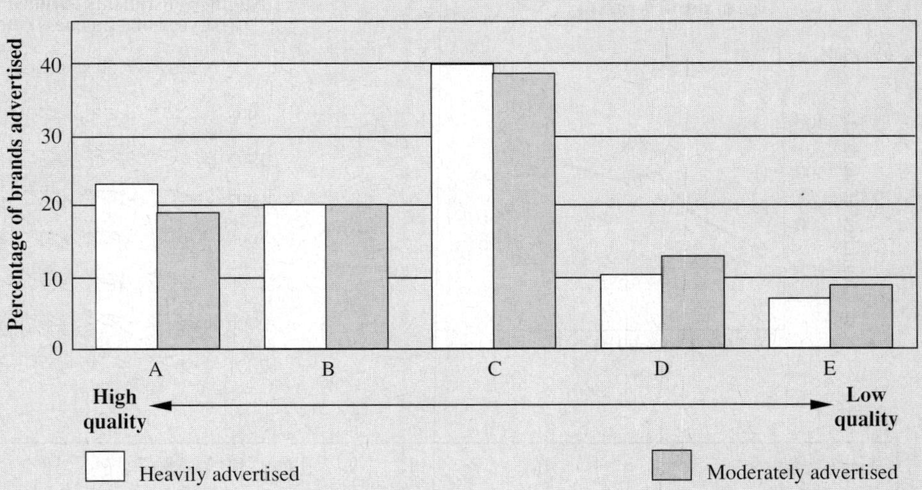

S2.10 (a)

Class interval (years)	Frequency
40–49	2
50–59	5
60–69	12
70–79	10
80–89	5
90–99	2
	$\Sigma f = 36$

(b) and (c)

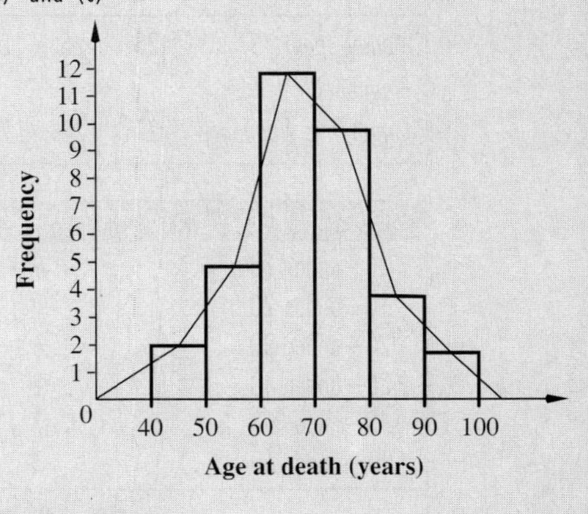

(d) normal curve

(e)

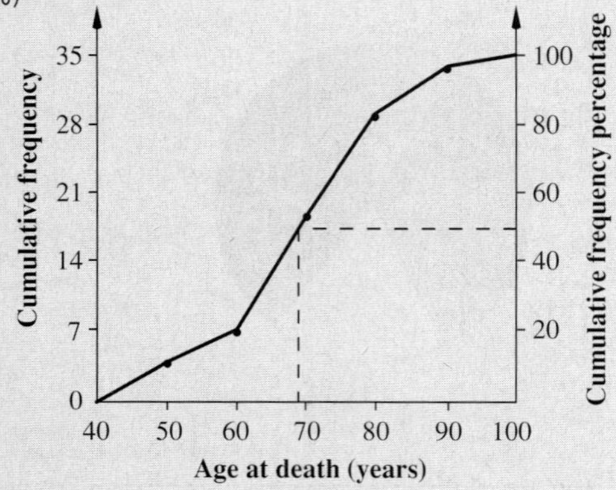

(b)

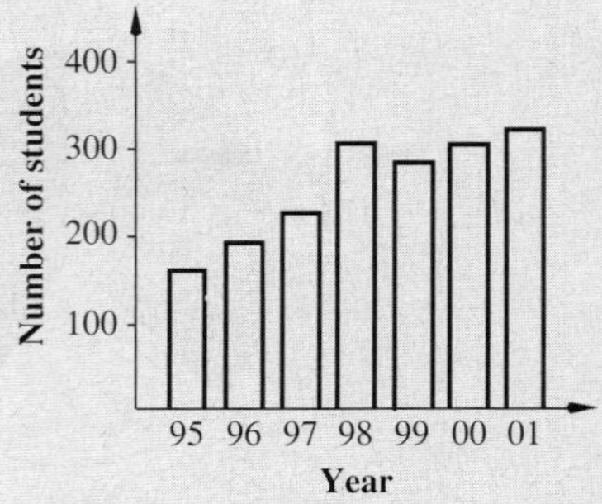

(f) 69 years

S2.11 (a)

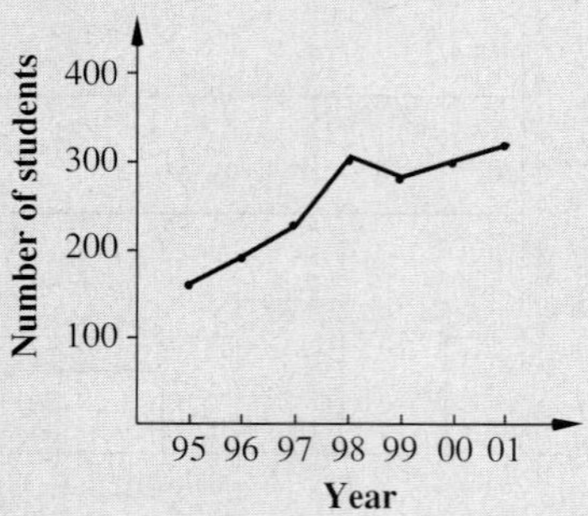

(c) **Number of students enrolled in the first year of a business course**

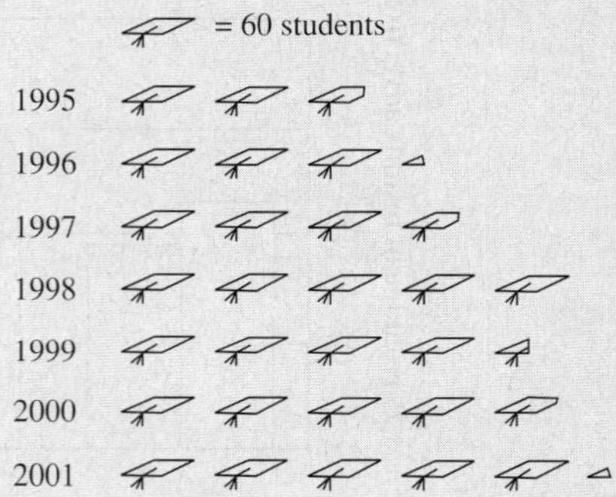

S2.13 (a) and (b)

Money spent ($)	3	4	6	7	8	9	12	14	15	16	18	19	20	21
Frequency	1	1	2	2	2	1	1	1	2	1	3	1	1	1
Cumulative frequency	1	2	4	6	8	9	10	11	13	14	17	18	19	20

Money spent ($)	23	25	29	33	36	42	47	48	50	51	52	54	56
Frequency	1	2	1	1	1	2	1	4	1	1	3	1	1
Cumulative frequency	21	23	24	25	26	28	29	33	34	35	38	39	40

(c)

Money spent ($)	Frequency
0–under 10	9
10–under 20	9
20–under 30	6
30–under 40	2
40–under 50	7
50–under 60	7
	$\Sigma f = 40$

(d) and (e)

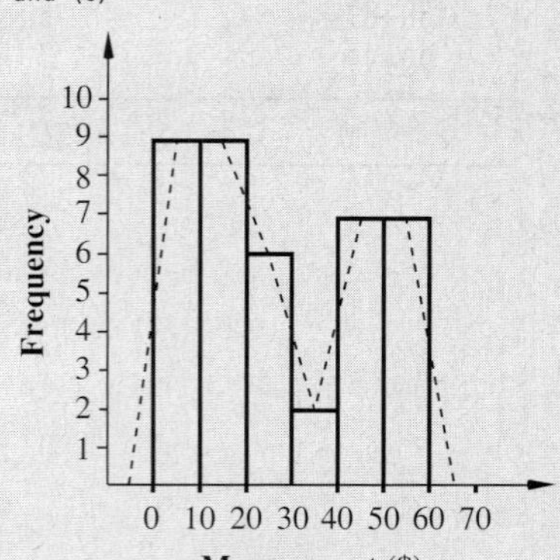

S2.19 **Responses to the question: 'Are you satisfied or dissatisfied with the way John Howard is doing his job as Prime Minister?'**

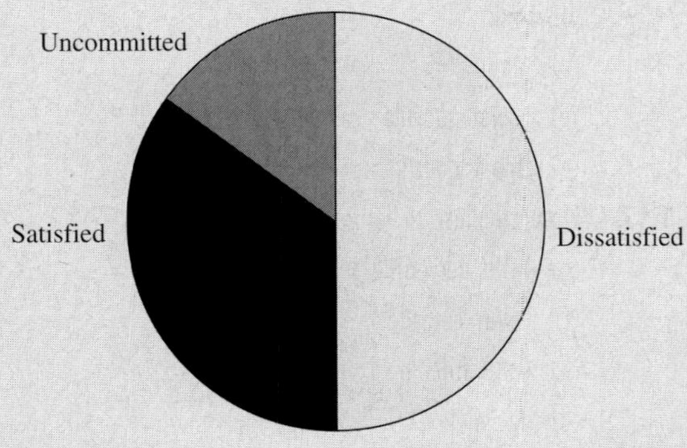

S2.20 The display is a graphical representation incorporating a multiple bar chart and pictogram. Overall, it is clear and user-friendly, various lengths giving a quick comparison. Absolute values, however, are not explicit and require extra calculations. The reader may also be confused by lengths not starting at zero.

S2.23

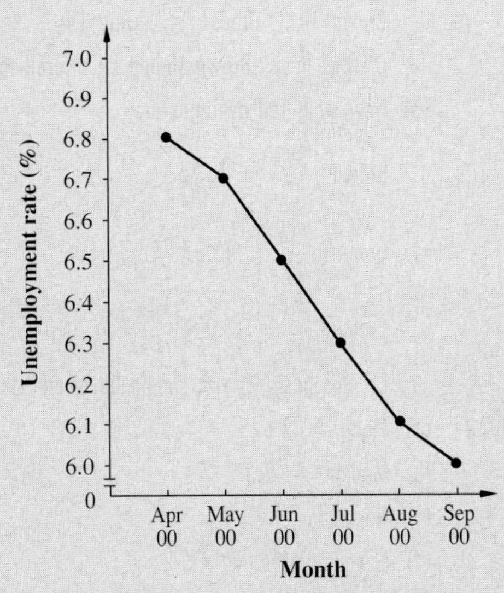

This graph is probably not a fair representation since zero is suppressed on the vertical axis. This has made the drop in the unemployment rate more dramatic than it really is.

S2.29 (a) 8 8 16 16 17 17 19 19 26 26
26 26 27 32 36 36 38 41 41 54

(b)

x	8	16	17	19	26	27	32	36	38	41	54
f	2	2	2	2	4	1	1	2	1	2	1

S2.30

Number of burglaries	Frequency
0–under 10	2
10–under 20	6
20–under 30	5
30–under 40	4
40–under 50	2
50–under 60	1
	$\Sigma f = 20$

Chapter S3

S3.1 (a) Mean $= \dfrac{\$600.00}{12} = \$50.00$

(b) If 40 is subtracted from each item, the new data are −5.72, 12.76, −1.74, and so on. The sum of these is 120.00. Hence, the mean is

$$\$40 + \frac{\$120.00}{12} = \$50.00$$

(c) Median $= \dfrac{\$50.21 + \$51.63}{2} = \$50.92$

S3.2 (a) Mode = 7

(b) Median = 7

(c) Mean $= \dfrac{55}{8} = 6.875$

(d) Geometric mean $= \sqrt[8]{1\,881\,600} = 6.086$

S3.3 (a) Mean = 4.44, mode = 2, median = 3

(b) Consider the data:

0 1 2 2 4 4 4 23

Mean = 5, mode = 4, median = 3

S3.5 Both the mean and the median are multiplied by that fixed number. In this case, the mean goes from 53.25 to 10 × 53.25 = 532.5, while the median goes from 53 to 10 × 53 = 530.

S3.7 The data have not been arranged in order of magnitude. The correct median is 5.

S3.8 (a) Mean salary = \$21 929

(b) Median salary = \$16 000

(c) Modal salary = \$15 000

(d) The mean salary is clearly not representative, because it is dominated by the two extremely high values. The mode is extremely close to the lower end of the range. Only the median is 'central'.

S3.10 (a)

	Season		
Cricketer	99/00	00/01	01/02
Amos	42.83	39.27	36.38
Taylor	28.16	25.00	21.43

(b) Combined averages:

99/00: $\dfrac{1858}{54} = 34.41$

00/01: $\dfrac{1489}{47} = 31.68$

01/02: $\dfrac{1225}{46} = 26.63$

(c) Three seasons combined:

Amos: $\dfrac{2431}{61} = 39.85$

Taylor: $\dfrac{2141}{86} = 24.90$

S3.11 (a) Time taken from A to B = $\frac{1}{2}$ hour

Time taken from B to A = $\frac{1}{4}$ hour

Average speed

$$= \frac{\text{total distance}}{\text{total time taken}}$$

$$= \frac{20}{\frac{1}{2} + \frac{1}{4}}$$

$= 26.67$ km/h

(b) Let x = time taken (in hours) on return trip. Then

$$30 = \frac{20}{\frac{1}{2} + x}$$

which yields $x = \frac{1}{6}$ h = 10 min. To travel 10 km in 10 min, the motorist must average 60 km/h.

S3.15 (a) Mode = 10

(b) Mean = $\dfrac{292}{32} = 9.125$

(c) Median = 10

(d) $Q_1 = 7$

(e) $Q_3 = 12$

S3.16 (a) Mean = $\dfrac{\Sigma fx}{\Sigma f} = \dfrac{105}{52} = 2.02$

(b) Median = average of 26th and 27th observations = 2

(c) Mode = 1

(d) Lower quartile = 1

(e) Upper quartile = 3

S3.17 From smallest to largest:

(g) 20th percentile

(f) lower quartile

(h) 0.38 fractile

(b) decile 4

(d) median

(e) decile 6

(c) 70th percentile

(a) upper quartile

S3.20 (a) Unweighted averages are:

Brown 14.2, Green 13.2, Gray 14.0.

On this basis, Brown should be offered the position.

(b) New weighted averages are:

Brown: $\dfrac{289}{21} = 13.76$

Green: $\dfrac{284}{21} = 13.52$

Gray: $\dfrac{273}{21} = 13.00$

On this basis, Brown should be offered the position.

S3.22 (a) Mean = 9.625

(b) Median = 9.5

(c) No mode

(d) Lower quartile = 4.5

(e) Upper quartile = 14.5

S3.25 (a) Unweighted mean score = $\dfrac{23}{5} = 4.60$

(b) Weighted mean score

$= (0.3 \times 5) + (0.25 \times 7) + (0.25 \times 3) + (0.15 \times 6) + (0.05 \times 2)$

$= 5$

S3.28 (a) $\bar{x} = \frac{4780}{8} = 597.5$ points

(b) $\tilde{x} = \frac{573 + 601}{2} = 587$ points

(c) $Q_1 = \frac{506 + 570}{2} = 538$ points

(d) $Q_3 = \frac{672 + 686}{2} = 679$ points

S3.32 (a) Mean $= \frac{171}{64} = 2.67$ goals

(b) Median $= 3$ goals

(c) Mode $= 3$ goals

(d) Lower quartile $= 1.5$ goals

(e) Upper quartile $= 4$ goals

S3.36 (a) Mean $= 100 + \frac{561}{7} = 180.14$ cm

(b) Mean $= 180 + \frac{1}{7} = 180.14$ cm

S3.38 (a) (i) Mean = \$220

(ii) Median = \$228

(b) Type I: \$152, type II: \$114, type III: \$93.33

(c) Mean $= \frac{\$660}{6} = \$110$

Chapter S4

S4.1 (a) Mean $= \frac{179}{10} = 17.9$

(b) Median $= \frac{18 + 19}{2} = 18.5$

(c) Range $= 28 - 5 = 23$

(d) Lower quartile = 3rd (sorted) observation from the lower end = 15

Upper quartile = 3rd (sorted) observation from the upper end = 21

(e) Interquartile range $= 21 - 15 = 6$

(f) Mean deviation $= \frac{49.2}{10} = 4.92$

(g) Standard deviation = 6.76

Range × 30% = 6.90, so value for *s* seems reasonable.

(h) Coefficient of variation $= 100\left(\frac{6.76}{17.9}\right) = 37.8\%$

S4.2 (a) Range = \$78.00 − \$18.80 = \$59.20

(b) Mean deviation $= \frac{\$144.8}{10} = \$14.48$

(c) Lower quartile = \$32.50

Upper quartile = \$56.00

Interquartile range = \$56.00 − \$32.50 = \$23.50

(d) Quartile deviation $= \frac{\$23.5}{2} = \$11.75$

(e) Mean = \$43.90

Standard deviation = \$18.76

(f) Coefficient of variation $= 100\left(\frac{18.76}{43.90}\right) = 42.7\%$

S4.3 (a) Mean $= \frac{\Sigma fx}{\Sigma f} = \frac{48}{24} = 2$

(b) Median = average of 12th and 13th observations = 2

(c) Mode = 1

(d) Range $= 4 - 0 = 4$

(e) Lower quartile = 1

Upper quartile = 3

Interquartile range $= 3 - 1 = 2$

(f) Mean deviation $= \frac{22}{24} = 0.92$

(g) Standard deviation = 1.18

S4.5 (a)

x	0	1	2	3	4	5	*Total*
f	5	5	4	4	3	5	26

(b) Range $= 5 - 0 = 5$

(c) Mean $= \frac{62}{26} = 2.385$

(d) Mean deviation $= \frac{40.769}{26} = 1.568$

(e) Standard deviation = 1.81

S4.6 (a) Mean $= \frac{191}{10} = 19.1$ minutes

(b) Median $= \frac{15 + 20}{2} = 17.5$ minutes

(c) Range $= 43 - 2 = 41$ minutes

(d) Mean deviation $= \frac{105}{10} = 10.5$ minutes

(e) Standard deviation = 13.20 minutes

(f) Coefficient of variation $= 100\left(\dfrac{13.20}{19.1}\right) = 69.1\%$

S4.8 (a) Average (mean) $\doteqdot$ \$391

(b) Range $\doteqdot$ \$600 − \$0 = \$600

(c) Lower quartile $\doteqdot 350 + \dfrac{(12.5 - 12)50}{9}$

$= \$352.78$

Upper quartile $\doteqdot 450 + \dfrac{(37.5 - 31)50}{8}$

$= \$490.62$

Interquartile range $\doteqdot$ \$490.62 − \$352.78

= \$137.84

(d) Mean deviation = \$109.44

(e) Standard deviation $\doteqdot$ \$136.07

S4.10 Midpoints of class intervals = 45, 55, 65, 75, 85, 95

(a) Estimate of mean $= \dfrac{\Sigma fm}{\Sigma f} = \dfrac{4060}{60} = 67.67$

(b) Estimate of standard deviation = 13.13

(c)

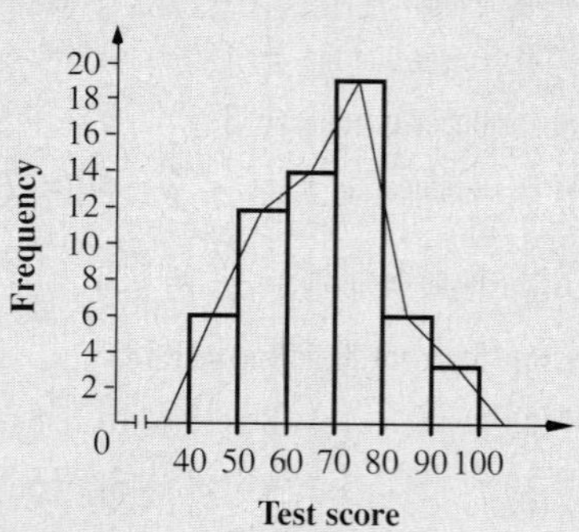

(d) One standard deviation either side of the estimated mean approximately covers the interval (54.5, 80.8). This interval seems to include about 40 of the students, so our percentage estimate would be about 67%.

S4.11 (a)

x	f
0	2
1	2
2	4
3	3
4	2
5	2
6	1
	$\Sigma f = 16$

(b) Mean $= \dfrac{\Sigma fx}{\Sigma f} = \dfrac{43}{16} = 2.69$

(c) Median = average of 8th and 9th observations

$= \dfrac{2 + 3}{2} = 2.5$

(d) Mode = 2

(e) Range = 6 − 0 = 6

(f) Lower quartile = 1.5

Upper quartile = 4

Interquartile range = 4 − 1.5 = 2.5

(g) Mean deviation $= \dfrac{23}{16} = 1.44$

(h) Standard deviation = 1.78

(i) Quartile deviation $= \dfrac{2.5}{2} = 1.25$

S4.13 Midpoints = 2.5, 7.5, 12.5, 17.5, 22.5

(a) Mean = 11.5 days

(b) Standard deviation = 5.98 days

(c) Range = 25

Range × 30% = 7.50, so value for s is reasonable.

S4.14 (a) Mean $= \dfrac{1172}{12} = 97.67$

(*Note*: Bradman went on to play a further 48 completed Test innings. However, his mean score (including not-outs) did not change all that much in his total of 80 Test innings. All told, he finished with an average of 99.94.)

Median $= \dfrac{58 + 79}{2} = 68.5$

No mode

(b) Range = 334 − 1 = 333

(c) Lower quartile $= \dfrac{14 + 18}{2} = 16$

Upper quartile $= \dfrac{123 + 131}{2} = 127$

Interquartile range = 127 − 16 = 111

(d) Mean deviation $= \dfrac{931.33}{12} = 77.61$

(e) Standard deviation = 103.67

S4.16 *One* such example is the two numbers 0 and 1. The range is 1 and the standard deviation is 0.71, i.e. 71% of the range.

S4.19
$$\begin{aligned}\sum(x-\bar{x}) &= \sum x - \sum \bar{x}\\ &= \sum x - n\bar{x}\\ &= \sum x - n\left(\frac{\sum x}{n}\right)\\ &= \sum x - \sum x\\ &= 0\end{aligned}$$

S4.22 An example is 1, 2, 2, 5, 5.
Mean = interquartile range = 3

S4.24 (b) (i) $\bar{x}$ = 295 500
(ii) $\tilde{x}$ = 202 000
(iii) Range = 702 000
(iv) Mean deviation = 192 200
(v) s = 239 341

S4.26 (a) Range = 11
(b) Mean deviation = 2.75
(c) Standard deviation = 3.55

S4.28 Standard deviation = 1.65

S4.30 (a) to (d) Each increases by 10%.

S4.32 They sold the same number:
Parlour A: 47.6 − 2(4.8) = 38 pizzas
Parlour B: 51.2 − 2(6.6) = 38 pizzas

S4.35 $\Sigma(x - \bar{x})^2 = (n - 1)s^2$
$= 6(12.8)^2 = 983.04$

Chapter S5

S5.1

(a) Discrete	(b) Continuous
(c) Discrete	(d) Continuous
(e) Discrete	(f) Discrete
(g) Discrete	(h) Discrete
(i) Discrete	(j) Continuous

Note, however, that (h) and (i) are rather debatable; for instance, interest rates are usually quoted as discrete values, such as 14% or 14.5%, but there is nothing in principle to prevent an interest rate of, say, 14.138 241%.

S5.2 (a) Estimate the mean aggregate UAI score of all New South Wales students in that year using a random sample of 100 students.
(b) Estimate the mean aggregate UAI score of all students in Australia who did an HSC in that year using the marks of the New South Wales students as a sample.

S5.5 Two of the many possible methods of obtaining a random sample are:
(a) Assign each student a unique number (0001 to 1500) and use a table of random numbers to select 50.
(b) Write the names of the students on identical tags, one per tag, place the tags in a barrel and withdraw 50 tags (spinning the barrel between each draw).

S5.7 Yes, provided some number could be assigned to each person on the electoral roll. (Consumers *not* on the roll would be excluded.)

S5.9 (a) 50 20 80 100 40 70 80 10 40 50
(b) Actual total = 538
Total of rounded data = 540
Absolute error = 540 − 538 = 2
(c) Relative error $= \dfrac{2}{538} \doteqdot 0.0037$
(d) Relative error percentage = 100 × 0.0037
= 0.37%

S5.11 (a)

\$2.20	\$2.90	\$2.40	\$2.00	\$1.80
\$2.10	\$2.40	\$2.90	\$1.60	\$1.50

(b) Actual total = \$21.81
Total of rounded data = \$21.80
Absolute error = \$0.01
Relative error $= \dfrac{\$0.01}{\$21.80} \doteqdot 0.000\,46$
Relative error percentage = 100 × 0.000 46
= 0.046%
(c) \$2 \$3 \$2 \$2 \$2 \$2 \$2 \$3 \$2 \$2
(d) Total of rounded data = \$22.00
Absolute error = \$0.19
Relative error $= \dfrac{\$0.19}{\$21.80} \doteqdot 0.0087$

Relative error percentage $= 100 \times 0.0087\%$
$= 0.87\%$

S5.12 Assign each of the 92 people a number from 01 through to 92. Then select a random sample of 10 numbers using the table. One method would be to select the first 2 digits of each number, working vertically downwards in the first column. This yields:

35 88 54 77 53 89 03 82 69 31

Then select those people who have assigned numbers corresponding to the above.

S5.14 Suppose we take the first 3 digits, starting at column 1 and working our way down the columns. Ignore any number greater than 486. We get:

359 031 311 403 466 309 294 307 304 260

Workers with these identification numbers should be chosen.

S5.15 Best estimate $= (0.6 \times 0.5) + (0.75 \times 0.5)$
$= 0.675$ or 67.5%

S5.16 *Step 1*: Calculate the proportion of the population in each suburb relative to the total population of the city,

e.g. $A = \frac{250}{800} = 0.3125$

Step 2: Determine the amount of the random sample who would come from each suburb,

e.g. $A = 200 \times 0.3125$
$= 62.5$ (say, 63)

Step 3: Select the relevant number of people from each suburb by using a random sampling technique. Electoral rolls would assist in this process.

S5.20 (a) \$0.40 \$1.50 \$1.00 \$0.80 \$1.40
\$1.60 \$0.80 \$0.40 \$0.70 \$0.90
\$1.70 \$0.20

(b) Absolute error = actual total − rounded total
= \$11.09 − \$11.40
= \$0.31 (ignore the sign)

Relative error $= \frac{\$0.31}{\$11.09} = 0.027\,95$

Relative error percentage $= 0.027\,95 \times 100$
$= 2.795\%$

S5.27 (a) Number of males = 30
Number of females = 130

(b) *Step 1*: Determine the size of the 4 categories, i.e. female/executive, female/clerical, male/executive, male/clerical.

Step 2: Determine the proportion of the staff in each category, e.g. female/clerical
$= \frac{130}{200} = 0.65$.

Step 3: Calculate the number of staff in each category required in the sample, e.g. female/clerical $= 40 \times 0.65 = 26$.

Step 4: Allocate the staff with individual numbers 001–200. Use a random number technique to select staff according to the strata numbers previously calculated in Step 3.

S5.29 *Advantage*: Anonymity of phone interview may result in more objective answers.

Disadvantages: Bias towards certain sections of the population depending on time of call (e.g. retired, elderly, unemployed). Also, you may get a low response rate.

S5.34 (a) Toss the coin twice. Possible (equally likely) outcomes are:

HH HT TH TT

Assign one of these outcomes to each person.

(b) Assign the 4 staff the numbers 1, 2, 3, 4. Roll the die once and select the person whose number appears. If a 5 or 6 is rolled, roll the die again until a 1, 2, 3 or 4 is rolled.

S5.37 (a) 6446 4754 3566 1498 1884 470 196 311
(b) 6450 4750 3570 1500 1880 470 200 310
(c) 6400 4800 3600 1500 1900 500 200 300

S5.39 Search the relevant court and other records, and find listings of those people convicted of armed bank robbery, along with their sentences and the actual time they spent in prison. Select a random sample (of at least 25) of these, and calculate the mean time they spent in prison.

S5.42 Select a random sample of groceries (at least 25 items) of various types from each of the stores, and write down the price each time. (Each store must sell *exactly* the same item.) Find the total cost of the items in each store and compare.

S5.44 Use random number tables to select a random sample of, say, 50 pages. Use the tables again to select one line number on each of these pages. Then use the tables once more to select one word from each of these lines. Find the mean length of these 50 words and use this as your estimate.

Chapter S6

S6.1 (a) P(all heads)

$= P(HHH) = \left(\frac{1}{2}\right)^3 = \frac{1}{8}$

(b) P(all tails)

$= P(TTT) = \left(\frac{1}{2}\right)^3 = \frac{1}{8}$

(c) P(two heads and one tail)

$= P(HHT \text{ or } HTH \text{ or } THH)$

$= P(HHT) + P(HTH) + P(THH)$

$= \frac{1}{8} + \frac{1}{8} + \frac{1}{8} = \frac{3}{8}$

S6.5 (a) $P(\text{red}) = \frac{3}{16}$

(b) $P(\text{black}) = \frac{5}{16}$

(c) $P(\text{blue}) = \frac{8}{16} = \frac{1}{2}$

(d) P(blue or red)

$= P(\text{blue}) + P(\text{red})$

$= \frac{8}{16} + \frac{3}{16} = \frac{11}{16}$

S6.7 (a) $P(\text{neither}) = \left(\frac{12}{20}\right)^2 = 0.36$

(b) $P(\text{both}) = \left(\frac{8}{20}\right)^2 = 0.16$

(c) P(at least one) = 1 − P(none)

= 1 − 0.36

= 0.64

S6.9 (a) P(male but not personal) = 0.18

(b) P(personal but not male) = 0.30

(c) P(both female and corporate) = 0.15

(d) P(both personal and small business) = 0

(e) P(not small business) = 0.80

(f) P(not male and not corporate) = 0.42

S6.12 P(none owns a video camera) = $(0.85)^4$ = 0.522

S6.15 Number of integers containing digit 1 = 20

P(integer selected contains digit 1) $= \frac{20}{100} = 0.20$

S6.18 P(both same colour)

= P(both red or both white or both black)

= P(both red) + P(both white) + P(both black)

$= \left(\frac{5}{20} \times \frac{4}{19}\right) + \left(\frac{6}{20} \times \frac{5}{19}\right) + \left(\frac{9}{20} \times \frac{8}{19}\right)$

$= \frac{122}{380} = 0.321$

S6.19 (a) $P(\text{male}) = \frac{41}{80}$

(b) $P(\text{female}) = \frac{39}{80}$

(c) $P(\text{very angry}) = \frac{16}{80} = \frac{1}{5}$

(d) $P(\text{angry}) = \frac{38}{80} = \frac{19}{40}$

(e) $P(\text{annoyed}) = \frac{26}{80} = \frac{13}{40}$

S6.20 (a) P(male but not very angry) $= \frac{31}{80}$

(b) P(very angry but not male) $= \frac{3}{40}$

(c) P(both female and annoyed) $= \frac{1}{8}$

(d) P(both male and very angry) $= \frac{1}{8}$

(e) P(not annoyed) $= \frac{27}{40}$

(f) P(not male and not very angry) $= \frac{33}{80}$

S6.24 Assume that the subjects are independent. In this case:

P(passes all 12 subjects) = $(0.90)^{12}$ = 0.282

S6.26 Number of possible outcomes

= 2 × 2 × 2 × 2 = 16

Complete list is:

HHHH	HTHT
HHHT	THTH
HHTH	THHT
HTHH	HTTT
THHH	THTT
HHTT	TTHT
HTTH	TTTH
TTHH	TTTT

P(3 or 4 heads)

= P(3 heads) + P(4 heads)

$= \frac{4}{16} + \frac{1}{16} = \frac{5}{16}$

S6.28 P(at least two odd) $= \frac{3}{5}$

S6.30 (a) P(6 smokers) = $(0.34)^6$ = 0.00154

(b) P(no smokers) = $(0.66)^6$ = 0.0827

(c) P(at least 1 smoker) = 1 − 0.0827 = 0.9173

S6.32 P(arrest and conviction) = 0.25 × 0.40 = 0.10

Assume the two events are independent.

S6.35 P(both aces) = $\frac{1}{221}$

S6.36 P(both aces) = $\left(\frac{1}{13}\right)^2 = \frac{1}{169}$

S6.40 (a) Both absent:

0.10 × 0.05 × 200 = 1 day

(b) Both present:

0.90 × 0.95 × 200 = 171 days

(c) One or both absent:

(1 − 0.855) × 200 = 29 days

(d) One or both present:

(1 − 0.005) × 200 = 199 days

S6.46 (a) Computer and no modem: 462 households

(b) Computer and modem: 138 households

(c) No computer and no modem: 1400 households

S6.49 Define the events:

E = student enrols in electronics

A = student enrols in astronomy

P = student passes the elective

(a)

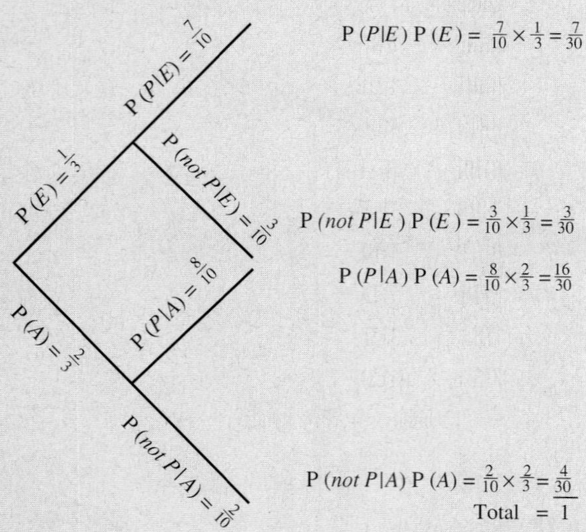

(b) (i) P(*E* and *P*)

$= P(P \mid E)P(E)$

$= \frac{1}{3} \times \frac{7}{10} = \frac{7}{30}$

(b) (ii) P(*A* and *not P*)

$= P(not\ P \mid A)P(A)$

$= \frac{2}{10} \times \frac{2}{3} = \frac{4}{30} = \frac{2}{15}$

(iii) P(*P*)

$= P(P \mid A)P(A) + P(P \mid E)P(E)$

$= \left(\frac{8}{10} \times \frac{2}{3}\right) + \left(\frac{7}{10} \times \frac{1}{3}\right) = \frac{23}{30}$

S6.51 (a) Let *H* = weather is hot

A = Anna wins match

S = Sarah wins match

Probability tree is shown below.

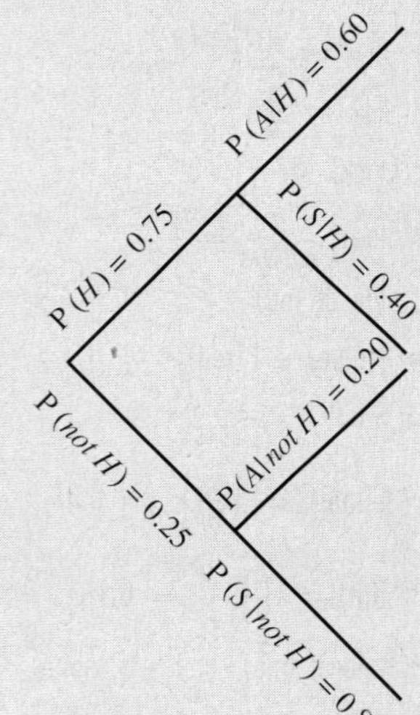

(b) (i) P(*S* and *H*)

$= P(S \mid H)P(H)$

= 0.40 × 0.75

= 0.30

(ii) P(*A* and *not H*)

$= P(A \mid not\ H)P(not\ H)$

= 0.20 × 0.25

= 0.05

(iii) P(A)

$= P(A \mid H)P(H) + P(A \mid not\ H)P(not\ H)$

$= (0.60 \times 0.75) + (0.20 \times 0.25)$

$= 0.50$

(iv) P(S)

$= P(S \mid H)P(H) + P(S \mid not\ H)P(not\ H)$

$= (0.40 \times 0.75) + (0.80 \times 0.25)$

$= 0.50$

Chapter S7

S7.2 (a)

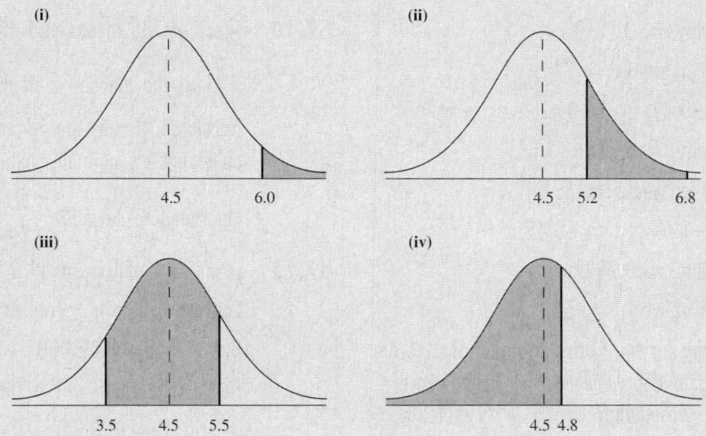

(b) (i) $z = 1.5$;
proportion
$= 0.5000 - 0.4332$
$= 0.0668$

(ii) $z_1 = 0.7$, $z_2 = 2.3$;
proportion
$= 0.4893 - 0.2580$
$= 0.2313$

(iii) $z_1 = -1.0$, $z_2 = 1.0$;
proportion
$= 0.3413 + 0.3413$
$= 0.6826$

(iv) $z = 0.3$;
proportion
$= 0.5000 + 0.1179$
$= 0.6179$

S7.4 (a) $z = 0.7$;
proportion
$= 0.5000 + 0.2580 = 0.7580$

(b) $z = -0.5$;
proportion
$= 0.5000 + 0.1915 = 0.6915$

(c) $z = -0.3$;
proportion
$= 0.5000 - 0.1179 = 0.3821$

(d) $z = 1.7$;
proportion
$= 0.5000 - 0.4554 = 0.0446$

(e) $z_1 = -2.3$, $z_2 = 2.7$;
proportion
$= 0.4893 + 0.4965 = 0.9858$

(f) $z_1 = -1.1$, $z_2 = 0.5$;
proportion
$= 1 - (0.3643 + 0.1915) = 0.4442$

S7.5 $\mu = 75$ and $\sigma = 12$

(a) (i) $z = 1.17$;
Probability $= 0.8790$
Number of weeks $= 176$

(ii) $z = -0.42$;
Probability = 0.6628
Number of weeks = 133

(iii) $z = 0.42$;
Probability = 0.3372
Number of weeks = 67

(iv) $z = -1.17$;
Probability = 0.1210
Number of weeks = 24

(v) $z_1 = -0.42$, $z_2 = 0.42$;
Probability = 0.3256
Number of weeks = 65

(vi) $z_1 = -1.67$, $z_2 = -0.42$;
Probability = 0.2897
Number of weeks = 58

(b) We are looking for the 90th percentile. The *z*-score corresponding to this value is $z = 1.28$. Hence, required fuel consumption = 75 + (12)(1.28) = 90.36 L

S7.7 (a) and (b)

$\mu = 100$ and $\sigma = 25$

Grade 1: $z = 1.28$;
Cut-off mark = 100 + (25)(1.28) = 132
Expected number = 200

Grade 2: $z = 0.52$;
Cut-off mark = 100 + (25)(0.52) = 113
Expected number = 400

Grade 3: $z = -0.52$;
Cut-off mark = 100 + (25)(−0.52) = 87
Expected number = 800

Grade 4: $z = -1.28$;
Cut-off mark = 100 + (25)(−1.28) = 68
Expected number = 400

Grade 5: cut-off mark is below 68
Expected number = 200

S7.8 (a) Median = 0.80 m

(b) Upper quartile corresponds to a *z*-score of 0.675.
Upper quartile
= 0.80 + (0.675 × 0.25) = 0.969 m

(c) Lower quartile corresponds to a *z*-score of −0.675.
Lower quartile
= 0.80 − (0.675 × 0.25) = 0.631 m

S7.9 To find $P(x \leq 112)$: $z = 1.83$; required probability = 0.5000 + 0.4664 = 0.9664

S7.10 These are 38 means each based on 25 items. The '*n*' in $\frac{\sigma}{\sqrt{n}}$ is the number of items from which each mean is calculated. Hence, the 38 means may be expected to be approximately normally distributed with mean $\mu = 126$ and standard deviation $\frac{\sigma}{\sqrt{n}} = \frac{27}{\sqrt{25}} = 5.4$

S7.13 $\mu = 83.4$ and $\sigma = 11.2$

Convert 100 to a *z*-score and get $z = 1.48$.

$P(z > 1.48) = 0.5000 - 0.4306 = 0.0694$

Hence, proportion with a sodium content > 100 g = 0.0694.

So expected number in a random sample of size 500 with a sodium content > 100 g is 0.0694 × 500 = 34.70, i.e. about 34 or 35 males.

S7.14 4.00 corresponds to a *z*-score of 0.52.

Hence, $0.52 = \frac{4.00 - 3.50}{\sigma}$

So $\sigma = 0.96$ L

S7.17 If she places the bet within 12 minutes, the average service time of each of the 9 previous customers was less than

$\frac{12}{9} = 1\frac{1}{3}$ minutes = 80 seconds.

So the required probability is:

P(average of 9 items < 80 if $\mu = 90$, $\sigma = 25$)

$$= P\left(z < \frac{80 - 90}{25/\sqrt{9}}\right)$$

$= P(z < -1.20)$
$= 0.5000 - 0.3849 = 0.1151$

S7.19 $\mu = 70$, $\sigma = 20$ and $n = 50$ (large)

$P(\bar{x} < 68) = P(z < -0.71)$

$= 0.5000 - 0.2611 = 0.2389$

Hence, required probability = 0.2389.

S7.21 $\mu = 5.2$ and $\sigma = 1.2$

$P(\bar{x} < 5.5) = P(z < 0.25)$

$= 0.5000 + 0.0987 = 0.5987$

Hence, required percentage is about 60%.

S7.22 $\mu = 103.3$ and $\sigma = 20.0$

z-score of 95th percentile $= 1.645$

Hence, amount of material recycled

$= \mu + \sigma z$

$= 103.3 + (20.0)(1.645)$

$= 136.2$ kg

S7.23 $n = 25$(large), $\bar{x} = 5.68$ and $s = 2.53$

(a) Point estimate of μ is $\bar{x} = 5.68$

(b) Standard error of the estimate

$$= \frac{s}{\sqrt{n}} = \frac{2.53}{\sqrt{25}} = 0.506$$

S7.24 Required 95% confidence interval for μ is:

$$\left(\bar{x} - \frac{1.96s}{\sqrt{n}}, \bar{x} + \frac{1.96s}{\sqrt{n}}\right)$$

$= (5.68 - (1.96)(0.506),$
$5.68 + (1.96)(0.506))$
$= (4.69, 6.67)$

S7.29 $n = 16$, $\sigma = 12.0$ and $\bar{x} = 84.6$

Required 95% confidence interval for μ is:

$$\left(\bar{x} - \frac{1.96\sigma}{\sqrt{n}}, \bar{x} + \frac{1.96\sigma}{\sqrt{n}}\right)$$

$$= \left(84.6 - \frac{(1.96)(12.0)}{\sqrt{16}}, 84.6 + \frac{(1.96)(12.0)}{\sqrt{16}}\right)$$

$= (78.72, 90.48)$

S7.32 $n = 16$, $\sigma = 12.0$ and $\bar{x} = 84.6$

Required 90% confidence interval for μ is:

$$\left(\bar{x} - \frac{1.645\sigma}{\sqrt{n}}, \bar{x} + \frac{1.645\sigma}{\sqrt{n}}\right)$$

$$= \left(84.6 - \frac{(1.645)(12.0)}{\sqrt{16}},\right.$$

$$\left.84.6 + \frac{(1.645)(12.0)}{\sqrt{16}}\right)$$

$= (79.66, 89.54)$

S7.34 (a) 95% confidence interval is:

$$\left(7.20 - (1.96)\frac{(2.50)}{\sqrt{60}}, 7.20 + (1.96)\frac{(2.50)}{\sqrt{60}}\right)$$

$= (6.57, 7.83)$

99% confidence interval is:

$$\left(7.20 - (2.58)\frac{(2.50)}{\sqrt{60}}, 7.20 + (2.58)\frac{(2.50)}{\sqrt{60}}\right)$$

$= (6.37, 8.03)$

(b) No. The population average waiting time is unlikely to be 6.00 minutes because 6.00 does not lie in either of the two confidence intervals in (a).

S7.38 $\sigma = \dfrac{103.5 - 90}{0.675} = 20$

S7.41 No. The sample size is too small for the Central Limit Theorem to hold.

S7.46 95% confidence interval for μ is (158.12, 161.88)

99% confidence interval for μ is (157.52, 162.48)

S7.50 $n = \left(\dfrac{2 \times 1.96 \times 17}{8.33}\right)^2 = 64$ cans

S7.53 (a) $P(\bar{x} > 37.0) = 0.1635$

(b) $P(\bar{x} < 39.0) = 0.9934$

(c) $P(32.0 \le \bar{x} \le 38.0) = 0.9555$

Chapter S8

S8.1 (a)

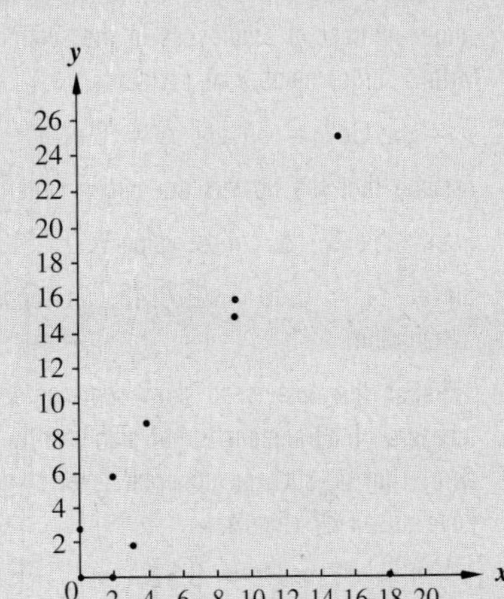

(b) $n = 10$ and $r = 0.487$.

(c) Assume that beer consumption per week and number of cigarettes smoked per day by male adults are uncorrelated.

$z = 0.487\sqrt{9} = 1.461$

Since $|z| = 1.461 < 1.96$, conclude that the value of r is *not* significant.

Based on this sample, the assumption is *correct*. That is, there is *no* correlation between beer consumption per week and number of cigarettes smoked per day by male adults.

S8.3 $n = 22$ and $r = -0.42$.

(a) Assume that weights and times are uncorrelated.

$$z = -0.42\sqrt{21} = -1.92$$

Since $|z| = 1.92 < 1.96$, conclude that the value of r is *not* significant.

(b) Based on this sample, the assumption made in (a) is *correct*. That is, there is *no* correlation between weight and time taken to run 100 m for females.

S8.4 $n = 12$ and $r_s = 0.63$.

Assume that the number of industrial accidents and number of employees are uncorrelated. From Table 3, the critical value is 0.5804.

Since $|r_s| = 0.63 > 0.5804$, we *reject* the assumption. It appears that a significant correlation exists between the number of industrial accidents and number of employees. (In particular, it appears that a larger number of employees in the section is associated with a larger number of accidents.)

S8.5 $n = 8$, $\Sigma d^2 = 34$ and $r_s = 0.60$.

Assume that the tipsters are *not* in general agreement.

From Table 3, the critical value is 0.7143.

Since $|r_s| = 0.60 < 0.7143$, we *cannot reject* the assumption.

Although the tipsters do show *some* agreement, the measure of agreement is not significantly different from zero. That is, such an apparent level of agreement could have arisen by chance.

S8.9 (a) $n = 12$ and $r = 0.52$.

Assume that sales volume of petrol in Melbourne and maximum daily temperature are uncorrelated.

$z = 0.52\sqrt{11} = 1.72$.

Since $|z| = 1.72 < 1.96$, we *cannot reject* the assumption. Conclude that this value of r is *not* significant.

$n = 25$ and $r = 0.42$.

Assume that sales volume of petrol in Adelaide and maximum daily temperature are uncorrelated.

$z = 0.42\sqrt{24} = 2.06$.

Since $|z| = 2.06 > 1.96$, we *reject* the assumption. Conclude that this value of r is significant.

(b) In (a) the higher value of r was not significant, while the smaller value of r was. This demonstrates that the value of r alone is *not* sufficient to test for significance. The sample size also must be considered.

S8.11 $n = 50$ and $r = -0.38$.

Assume that the length of time the agent has been with the company and the amount of new business written are uncorrelated.

$z = -0.38\sqrt{49} = -2.66$

Since $|z| = 2.66 > 1.96$, we *reject* the assumption. Hence, this value of r is significant. Conclude that the length of time the agent has been with the company and the amount of new business written are *negatively* related.

S8.14 (a) Speed score

81 90 62 80 43 76 58 82 90 36

Rank

7 9.5 4 6 2 5 3 8 9.5 1

Accuracy score

27 38 37 60 65 52 82 47 56 18

Rank

2 4 3 8 9 6 10 5 7 1

Differences (d)

5 5.5 1 −2 −7 −1 −7 3 2.5 0

(b) $\Sigma d^2 = 174.5$

$r_s = -0.058$

(c) Assume that speed and accuracy are uncorrelated. From Table 3, the critical value for $n = 10$ is 0.6364.

Since $|r_s| = 0.058 < 0.6364$, we *cannot reject* the assumption. Conclude that, based on this sample, speed and accuracy are *not* significantly related.

S8.16 (a)

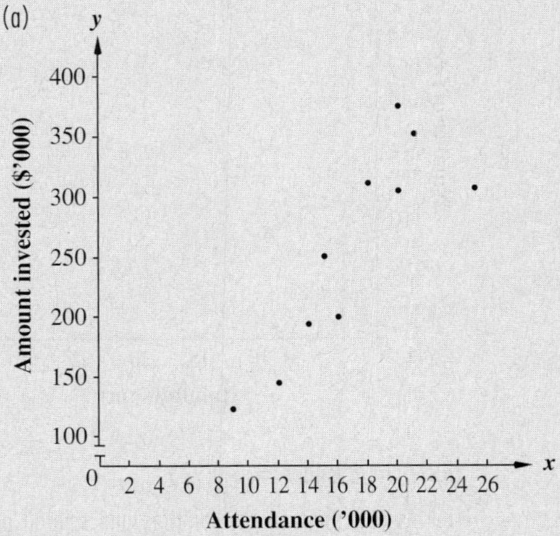

(b) $n = 10$ and $r = 0.821$.

(c) Assume that attendance at race meetings and amount invested on the TAB are uncorrelated.

$z = 0.821\sqrt{9} = 2.463$.

Since $|z| = 2.463 > 1.96$, we *reject* the assumption. Conclude that there is a significant *positive* relationship between attendance at race meetings and the amount invested on the TAB.

S8.18 (a)

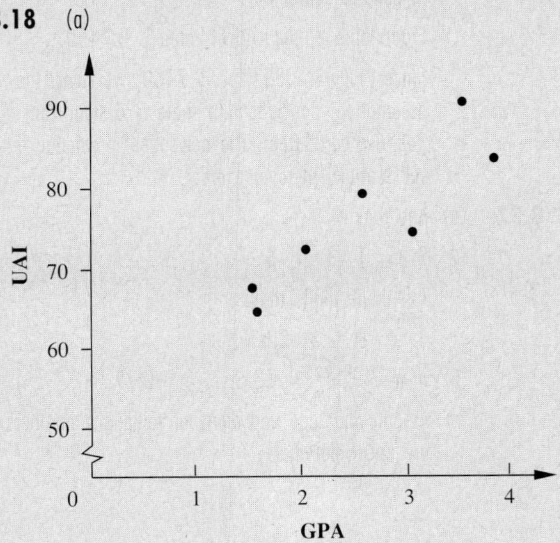

(b) $n = 7$ and $r = 0.904$.

(c) Assume that UAI and GPA of students are uncorrelated.

$z = 0.904\sqrt{6} = 2.21$.

Since $|z| = 2.21 > 1.96$, we *reject* the assumption. Conclude that there is a significant *positive* relationship between UAI and GPA of students.

S8.20 $n = 20$ and $r = 0.925$.

Assume that total beer consumption in Victoria and average weekly wage of schoolteachers in that state are uncorrelated.

$z = 0.925\sqrt{19} = 4.03$.

Since $|z| = 4.03 > 1.96$, the value of r is significant.

It would be tempting to conclude here that total beer consumption in Victoria and average weekly wage of schoolteachers in that state *are* correlated. However, the correlation is plainly spurious. Beer consumption has risen because there are more people and schoolteachers' salaries have risen because of inflation. The two variables are not directly related to each other, although they have both risen together over the years.

S8.21 (a)

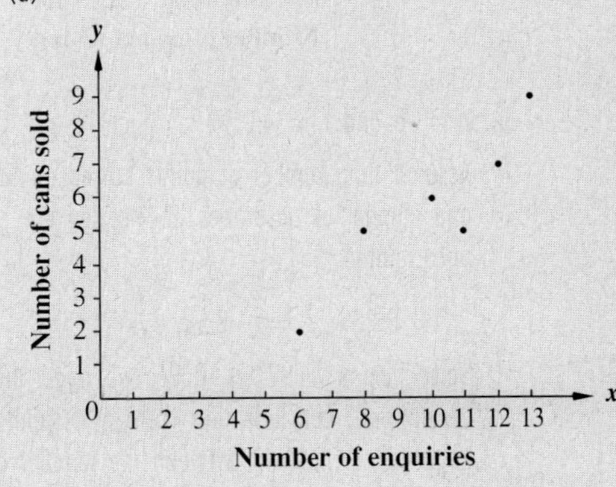

(b) $n = 6$ and $r = 0.875$.

(c) Assume that number of enquiries and number of cans sold per day are uncorrelated.

$z = 0.875\sqrt{5} = 1.96$.

Since $|z| = 1.96$, the result is *inconclusive*.

Based on this sample, we are unable to draw any firm conclusion as to whether a relationship exists. To test whether a correlation exists in such cases, the manager should try to gather more data, if possible.

S8.23 The situation arises frequently when the values of variables rise as time passes, and this dominates any more subtle effects. For instance, the number of people in employment and the amounts paid out in unemployment benefits are notionally negatively correlated, since increasing employment should cause reduced unemployment. As time passes, however, the number of people in the population increases and also general wage levels increase, so that it would appear (over a period of years) that high numbers of employed people are associated with high levels of unemployment benefits.

S8.24 (a)

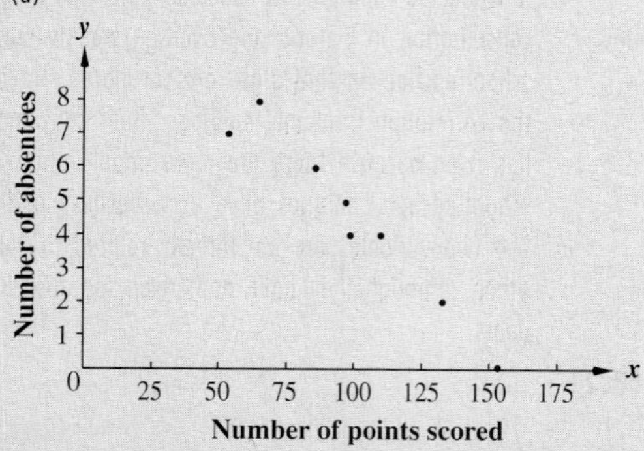

(b) $n = 8$ and $r = -0.979$.

(c) Assume that number of points scored on Sunday and number of absentees on Monday are uncorrelated.

$z = 0.979\sqrt{7} = -2.59$.

Since $|z| = 2.59 > 1.96$, we *reject* the assumption. Conclude that there is a significant *negative* relationship between the number of points scored on Sunday and the number of absentees on Monday.

S8.27 (a) From the scatter diagram we would estimate the value of r to be very close to -1.

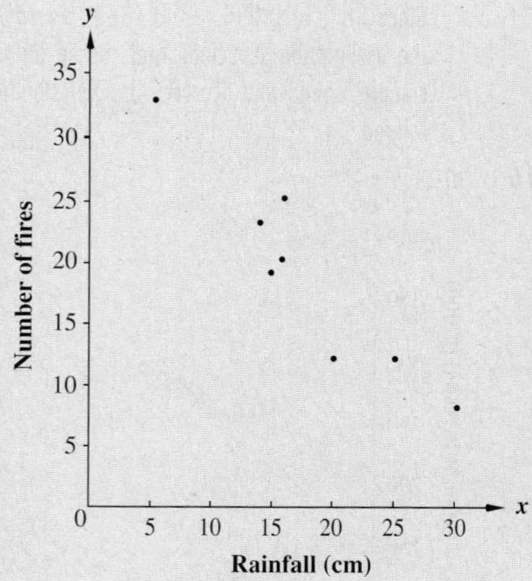

(b) $n = 8$ and $r = -0.929$.

(c) Assume that number of fires and rainfall are uncorrelated.

$z = 0.929\sqrt{7} = -2.46$.

Since $|z| = 2.46 > 1.96$, we *reject* the assumption. Conclude that there is a significant *negative* relationship between the number of fires and rainfall.

S8.30 (a) $n = 7$, $\Sigma d^2 = 104$ and $r_s = -0.857$.

(b) Assume that the predicted and actual results are *not* in general agreement.

From Table 3, the critical value is 0.7450.

Since $|r_s| = 0.857 > 0.7450$, we *reject* the assumption. Conclude that there is a significant *negative* correlation. The newspaper prediction is substantially in reverse order.

S8.32 (a) Age rank:

2 5 1 3 4 6

Telephone digits rank:

5 4 6 2.5 2.5 1

(b) $n = 6$, $\Sigma d^2 = 62.5$, $r_s = -0.79$.

(c) Assume that age and digits of telephone numbers are uncorrelated.

From Table 3, the critical value is 0.8286.

Since $|r_s| = 0.79 < 0.8286$, we *cannot reject* the assumption. Conclude that, based on this sample, there is *no* significant correlation between age and digits of telephone numbers.

S8.35 For example:

x	1	2	3	4	5	6	7	8
y	3	5	10	17	20	33	46	78

In this case, $\Sigma d^2 = 0$ and so $r_s = 1$.

Chapter S9

S9.1 (a) The number of new policies is the *y*-variable and the number of years of experience is *x*.

$\Sigma(x - \bar{x})(y - \bar{y}) = 849.42$

$\Sigma(x - \bar{x})^2 = 384.92$

$\bar{y} = 34.92,\ \bar{x} = 9.42,\ n = 12$

$$b = \frac{849.42}{384.92} = 2.207$$

$$a = 34.92 - (2.207 \times 9.42)$$
$$= 14.13$$

The regression line is:

$\hat{y} = 14.13 + 2.21x$

or:

estimated number of new policies
= 14.13 + 2.21 × (number of years of experience)

(b) $\Sigma(y - \bar{y})^2 = 2436.9$ and $r^2 = 0.77$.

This implies that 77% of the variation in the number of new policies written annually is attributable to the number of years of experience.

(c) From the regression line using $x = 15$,

$\hat{y} = 14.13 + (2.21)(15) = 47.28$

Hence, we estimate the number of policies to be about 47.

S9.2 From Equation S9.5, if $r = 0$ then the regression line has a slope of $b = 0$. The regression line then becomes the horizontal line $\hat{y} = \bar{y}$.

S9.4 The slope of the regression line is 2.21. This means that, for every extra year of experience, agents will write an average of 2.21 new policies annually.

S9.5 They are the same when *x* and *y* are *perfectly* correlated, i.e. either $r = 1$ or $r = -1$.

S9.7 (a) $r = 0.949$ and so the coefficient of determination is:

$r^2 = (0.949)^2 = 0.901$.

(b) This means that 90.1% of the variation in council expenditure is due to the number of children in the electorate. The remaining 9.9% of variation is due to other factors.

S9.8 (a) and (c)

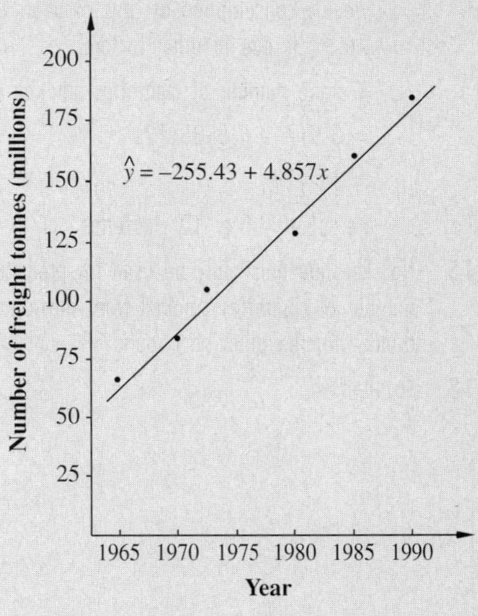

(b) Number of freight tonnes (millions) = −255.43 + 4.857(year)

(c) The line is a good fit of the data.

(d) If year = 95, then number of freight tonnes predicted = 205 985 000.

(e) $$\%\text{ error} = \left(\frac{215 - 206}{215}\right) \times 100$$
$$= 4.2\%$$

S9.13 (a) Any set of data that lie on a straight line (i.e. $r = -1$ or $+1$) is correct.

One example is:

x	1	2	3	4	5
y	2	4	6	8	10

(b) The characteristic is that the points should all lie on a straight line.

S9.14 (a) The least-squares regression line of y on x is:

$$\hat{y} = 3.517 + 0.6585x$$

or:

average number of cigarettes per day

= 3.517 + 0.6585 × (average weekly consumption of cans of beer)

(b) Since $r = 0.487$, then the coefficient of determination is $r^2 = (0.487)^2 = 0.237$.

Hence, 23.7% of the variation in the average number of cigarettes per day is due to the average weekly consumption of cans of beer. The remaining 76.3% is due to other factors.

(c) Average number of cigarettes smoked per day

= 3.517 + 0.6585(12)

= 11.419

i.e. about 11 or 12 cigarettes

S9.15 No. The line must only be used for predicting the number of cigarettes smoked from the amount of beer drunk, *not* the other way round.

S9.18 (a) and (c)

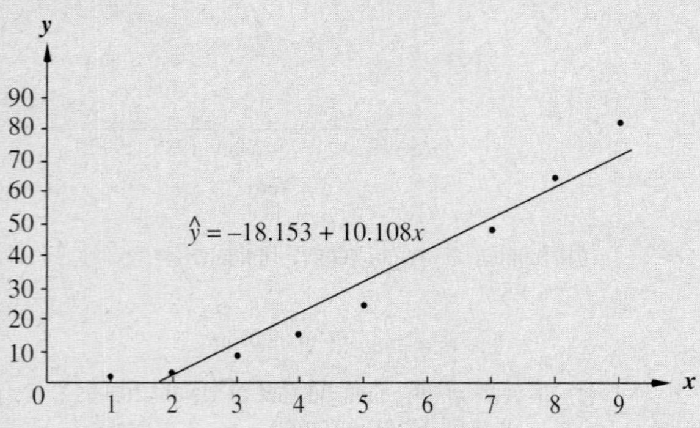

(b) The least-squares regression line of y on x is

$$\hat{y} = -18.153 + 10.108x$$

(d) If $x = 6$, then the line predicts

$$\hat{y} = -18.153 + 10.108(6)$$

$$= 42.5.$$

However, there is clearly a *quadratic* relationship between x and y (i.e. in the sample $y = x^2$ for each data point). Hence, a better prediction would be $\hat{y} = (6)^2 = 36$.

S9.21 (a) and (c)

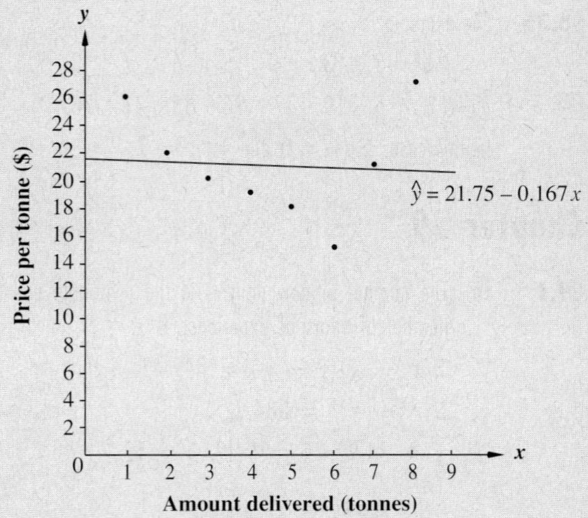

(b) The least-squares regression line is:

$$\hat{y} = 21.75 - 0.167x$$

or:

price per tonne ($)

= 21.75 − 0.167(number of tonnes)

(d) We have $n = 8$ and $r = -0.102$.

Hence, the coefficient of determination is $r^2 = (-0.102)^2 = 0.010$.

Therefore, only 1% of the variation in price per tonne is due to the number of tonnes. The remainder of the variation is due to other factors. Thus, a straight line provides a very poor fit of these data.

S9.22 The slope of the regression line, given by S_{xy}/S_{xx}, must always have the same sign as S_{xy}, since S_{xx} cannot be negative. Similarly, the sign of the correlation coefficient r must always be the same as that of S_{xy}, since the denominator in its formula cannot be negative.

S9.24 (a) $y = 5, 4, 3, 2, 1$

(b)

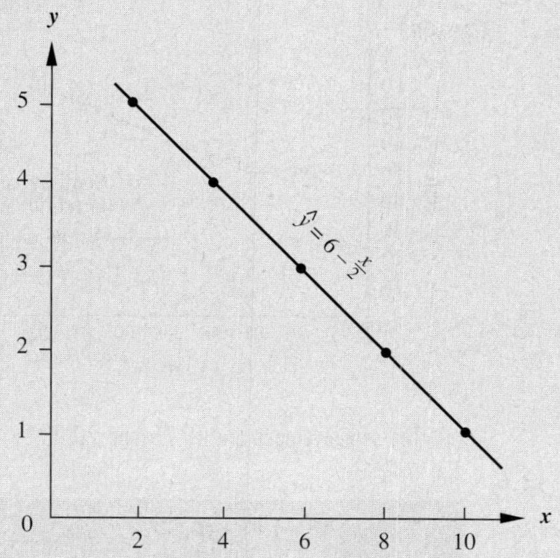

(c) $r = -1$

(d) $\hat{y} = 6.00 - 0.50x$

This is the same as the original equation.

S9.29 (a)

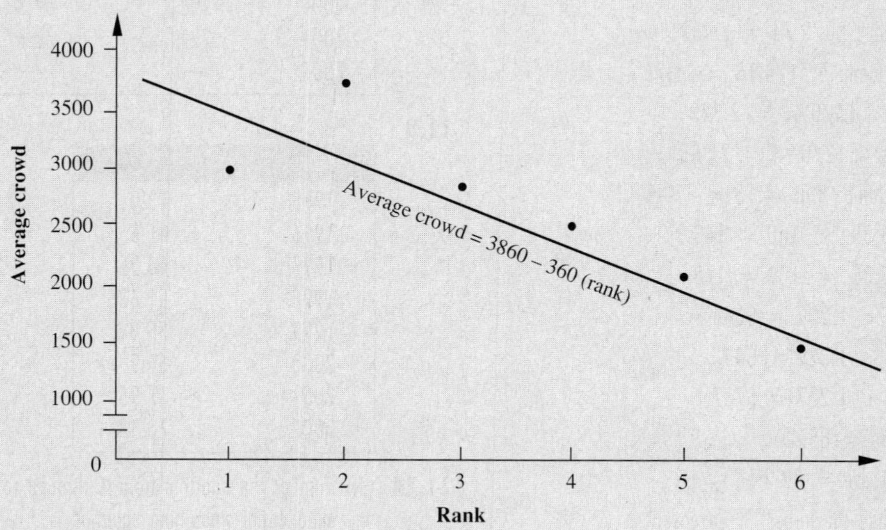

(b) Average crowd = 3860 − 360(rank)

(c) $r^2 = 78.7\%$

Line is a fairly good fit.

S9.36 (a)

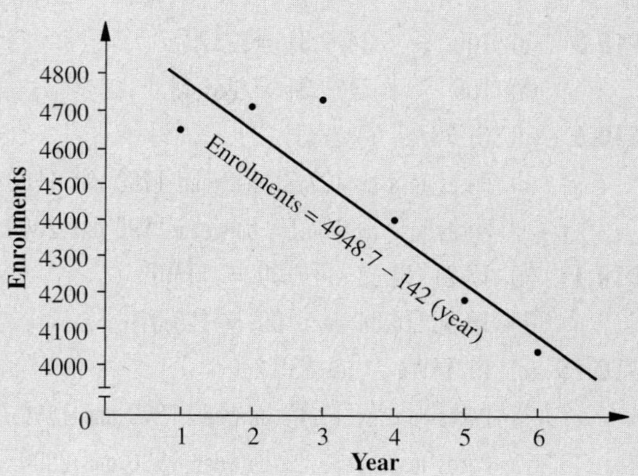

(b) Enrolments = 4948.7 − 142(year)

(c) $r^2 = 82.1\%$

Line is a good fit.

Chapter S10

S10.3 (a) 100 × 5.34/2.31 = 231.2

(b) 100 × 6.14/2.31 = 265.8

S10.6 (a) (i) 237.2 (ii) 276.5

(b) Prices rose by 137.2% between 1980 and 1990.

Prices rose by 176.5% between 1980 and 2000.

S10.11 (a) 17.21/14.84 × 100 = 116.0

(b) 18.55/14.84 × 100 = 125.0

S10.15 (a) (i) 119.4 (ii) 135.2

(b) Prices rose by 19.4% between 1990 and 1995.

Prices rose by 35.2% between 1990 and 2000.

S10.17 (a) Laspeyres index = 762.74/558.07 × 100 = 136.7

(b) Paasche index = 1089.7/818.99 × 100 = 133.1

(c) In (a) prices rose by 36.7% between 1990 and 2000.

In (b) prices rose by 33.1% between 1990 and 2000.

S10.19 (a) Onions: 1.26/0.46 = 2.739

Peaches: 1.80/0.74 = 2.432

Leg of lamb: 5.71/4.15 = 1.376

(b) Onions: 1.16/0.46 = 2.522

Peaches: 2.12/0.74 = 2.865

Leg of lamb: 5.85/4.15 = 1.410

S10.21 (a) 8.77/5.35 × 100 = 163.9

(b) 9.13/5.35 × 100 = 170.7

S10.27 (a) \$280 × 1.229 = \$344

(b) \$360 × 1.229 = \$442

(c) \$600 × 1.229 = \$737

S10.30 Food increase = 85.4%

Clothing increase = 52.2%

Thus, food had the largest % increase.

S10.33 \$350 × 124.7/108.4 = \$403

S10.36 (a) \$42 500 × 115.1/107.8 = \$45 378

(b) \$42 500 × 121.2/107.8 = \$47 783

(c) \$42 500 × 124.2/107.8 = \$48 966

Chapter S11

S11.3 (a) and (b)

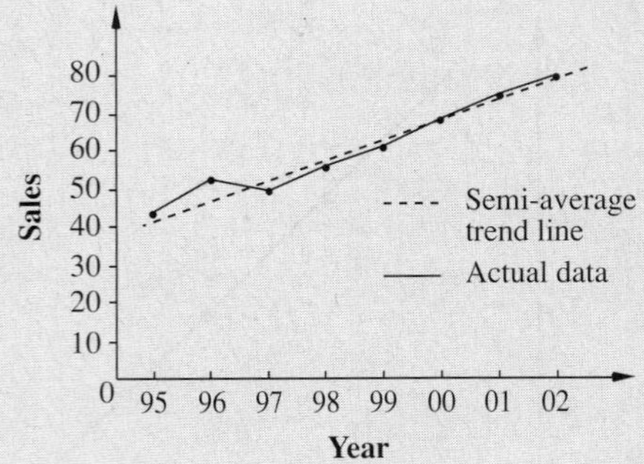

(b) The semi-averages are 49.75 and 70.25.

S11.6

Year	3-year moving total	3-year moving average
1995	—	—
1996	144	48.00
1997	156	52.00
1998	164	54.67
1999	183	61.00
2000	202	67.33
2001	221	73.67
2002	—	—

S11.8

Year	S_x
1995	43.0
1996	47.5
1997	48.25
1998	51.625
1999	55.81
2000	61.9
2001	67.95
2002	73.48

S11.14 (a) Number the quarters from 1 through to 12. The exponential smoothing equation is

$$S_x = 0.5y_x + 0.5S_{x-1}$$

Quarter (x)	S_x
1	1874
2	2008
3	2373.5
4	2686
5	2845.5
6	3097
7	3645
8	4392
9	4450.5
10	4076.5
11	3951
12	3826.5

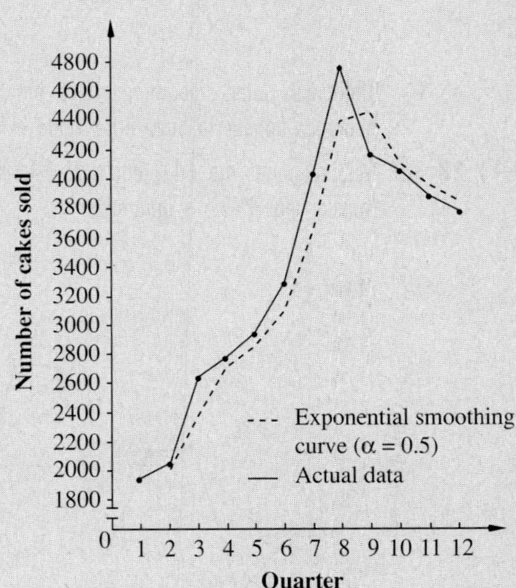

(b) The exponential smoothing model would use the current period forecast of about 3826 as the forecast number of cakes sold in the December quarter 2001. However, if we extend the curve in (a), the predicted number of cakes sold in the December quarter 2001 is about 3700.

S11.15 (a) Number the quarters (x) from 1 through to 12.

Let y = number of cakes sold.

Then the least-squares regression line of y on x is

$\hat{y} = 1997.3 + 208.02x$

(b) If $x = 13$,

$\hat{y} = 1997.3 + (208.02)(13) = 4701.56$

Hence, we would predict about 4702 cakes sold in the December quarter 2001.

S11.17 (a)

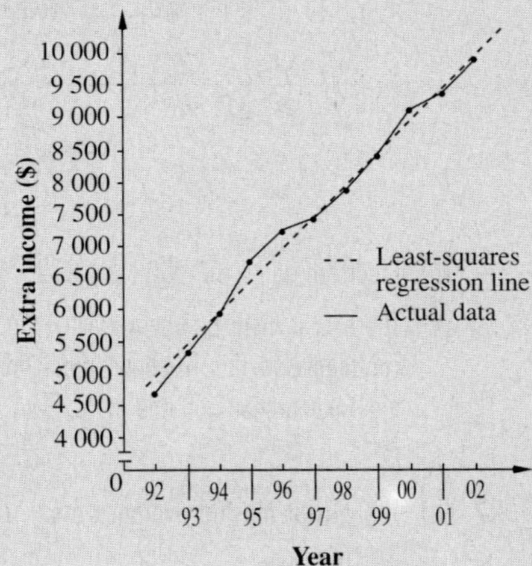

(b) It is a reasonable fit. (In fact, $r^2 = 71.2\%$.)

S11.21 (a)

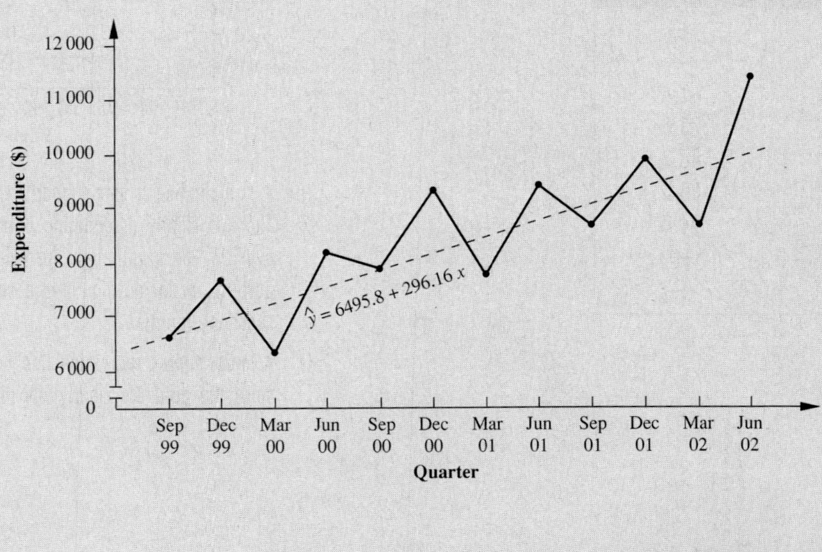

(b) (i) \$10 346 million

(ii) \$12 715 million

S11.25 (a)

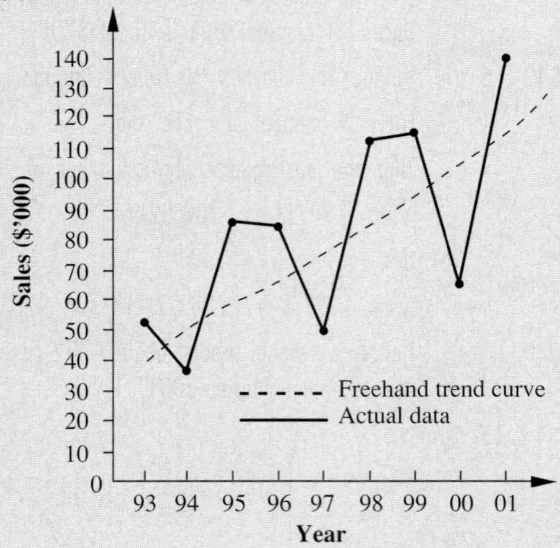

(b) Predicted sales for 2002 = \$125 000

(c) There is a distinct 3-year cycle. You could *not* have confidence in this freehand prediction, which does not take account of this cycle.

S11.27 (a) The 5-year moving averages are:

Year	Moving average
1993	—
1994	—
1995	61.4
1996	73.0
1997	88.2
1998	83.6
1999	94.0
2000	—
2001	—

(b)

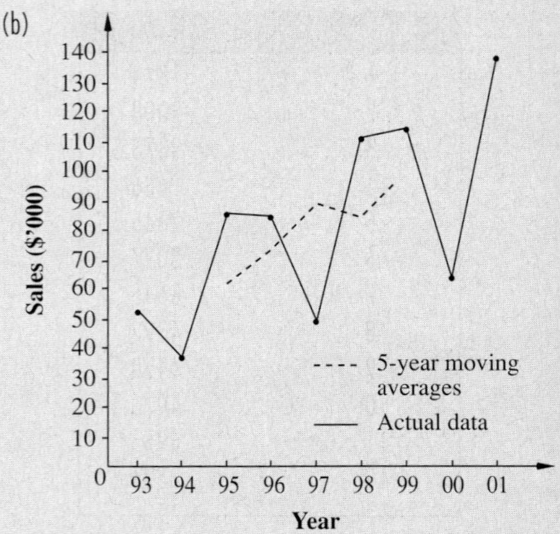

There was some smoothing effect, but the moving averages appear to have little value in this case.

S11.28 (a) Semi-averages are 64.5 and 105.25. (Note that the middle year 1997 is ignored.)

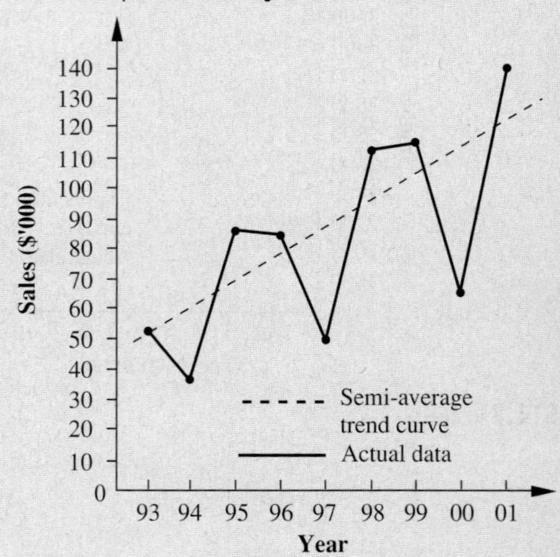

(b) A straight line is *not* a good fit of this time series. There is clearly a repeating pattern in the series every 3 years, although the values within the pattern are increasing. It is a similar effect to seasonal variation.

(c) A least-squares regression line would be of little use since the data are clearly *not* linear.

S11.31 (a) Secular trend.

(b) Seasonal variations.

(c) Cyclical fluctuations.

(d) Probably irregular movements, but possibly cyclical fluctuations if the stakes are major and prolonged, and possibly seasonal variations in cases such as mail and brewery strikes at Christmas time.

S11.32 Each average will include one measurement from each month. Thus, the seasonal effect from any one month appears exactly once with exactly the same weighting as that for any other month, and so any high seasonal effects in some months will be balanced out by the low ones in other months.

S11.35 (a) $b = -0.908 \times \frac{3186.1}{1.871} = -1546.6$

$a = 23\,156.2 - (-1546.6)(3.5)$

$= 28\,569.3$

$\hat{y} = 28\,569.3 - 1546.6x$

(b) $\hat{y} = 28\,569.3 - 1546.6(7)$

$= 17\,743.1$

This line predicts that about 17 743 new houses will be built in the private sector during the December quarter 2001.

(c) $r^2 = (-0.908)^2 = 82.4\%$, which indicates that a straight line is a reasonable fit of the data.

(d) Absolute error = 3876 houses

Percentage error = 17.9%

S11.41 (a) $c_1 = -4.038$, $c_2 = 0.053\,94$

$\log_e y = z$

$= -4.038 + 0.053\,94(\text{year})$

$a = e^{-4.038} = 0.017\,63$

$\hat{y} = 0.017\,63e^{0.05394(\text{year})}$

(b) Let year = 101 in (a).

$\hat{y} = 0.017\,63e^{0.05394(101)} = \$4.10$

(c) No. Growth is *not* exponential. It is more linear. Therefore, the prediction in Exercise 11.40(b)(ii) is likely to be more accurate.

Chapter S12

S12.4 $H_0: \mu = 30$ v $H_1: \mu < 30$

$n = 36$ (large), $\bar{x} = 27$, $\sigma = 10$

Test statistic is $z = -1.80$.

Critical value at $\alpha = 0.05$ is -1.645.

Since $z = -1.80 < -1.645$, H_0 is *rejected* at $\alpha = 0.05$.

We therefore conclude that this group did take *less* time.

S12.5 $H_0: \mu = 30$ v $H_1: \mu \neq 30$

$n = 36$ (large), $\bar{x} = 27$, $\sigma = 10$

Test statistic is $z = -1.80$.

Critical value at $\alpha = 0.05$ is 1.96.

Since $|z| = 1.80 < 1.96$, H_0 *cannot be rejected* at $\alpha = 0.05$.

We therefore conclude that this group takes the *same* time as other males.

S12.6 $H_0: \mu = 30$ v $H_1: \mu < 30$

$n = 36$ (large), $\bar{x} = 27$, $\sigma = 10$

Test statistic is $z = -1.80$.

Critical value at $\alpha = 0.01$ is -2.33.

Since $z = -1.80 > -2.33$, H_0 *cannot be rejected* at $\alpha = 0.01$.

We therefore conclude that this group takes the *same* time as other males.

S12.11 $H_0: \mu = 120$ v $H_1: \mu > 120$

$n = 9$, $\bar{x} = 130.33$, $s = 10.93$, population normal

Test statistic is $t = 2.67$ with $\nu = 8$.

Critical value at $\alpha = 0.05$ is 1.86.

Since $t = 2.67 > 1.86$, H_0 is *rejected*.

We therefore conclude that this group does have a significantly *higher* blood pressure than other such males.

S12.12 $H_0: \mu = 120$ v $H_1: \mu \neq 20$

$n = 9$, $\bar{x} = 130.33$, $s = 10.93$, population normal

Test statistic is $t = 2.67$ with $\nu = 8$.

Critical value at $\alpha = 0.05$ is 2.31.

Since $|t| = 2.67 > 2.31$, H_0 is *rejected* at $\alpha = 0.05$.

We therefore conclude that this group has a significantly *different* blood pressure from other such males.

S12.17 $H_0: \mu = 55.0$ v $H_1: \mu \neq 55.0$

$n = 25$ (large), $\bar{x} = 53.0$, $s = 4.5$

Test statistic is $z = -2.22$.

Critical value at $\alpha = 0.05$ is 1.96.

Since $|z| = 2.22 > 1.96$, H_0 is *rejected* at $\alpha = 0.05$.

We therefore conclude that this group did significantly *differently* from other students in the state.

S12.20 $H_0: \mu = 100$ v $H_1: \mu \neq 100$

$n = 100$ (large), $\bar{x} = 95$, $\sigma = 20$

Test statistic is $z = 2.50$.

Critical value at $\alpha = 0.05$ is 1.96.

Since $|z| = 2.50 > 1.96$, H_0 is *rejected.*

We therefore conclude that the adults in this town performed *differently* in this IQ test.

Note: If you had some *prior* notion that these adults were going to do significantly better or worse, perhaps a one-sided test could be used.

S12.22 Use a *one-sided test.*

$H_0: \mu = 60$ v $H_1: \mu < 60$

$n = 30$, $\bar{x} = 58.4$, $s = 3.6$

Test statistic is $z = -2.43$.

Since $z = -2.43 < -2.33$, *reject* H_0 at $\alpha = 0.01$.

No, they are *not* telling the truth. Conclude that the solder breaks at an average of *less than* 60 seconds.

S12.25 Use a *one-sided test.*

$H_0: \mu = 10$ v $H_1: \mu > 10$

$n = 100$, $\bar{x} = 10.9$, $s = 3.6$

Test statistic is $z = 2.5$.

Since $z = 2.5 > 1.645$, *reject* H_0 at $\alpha = 0.05$.

No, do *not* believe the claim. Conclude that the average fat content is *more than* 10%.

S12.29 No.

Two-sided: If $|z| > 1.96$, it does *not* always follow that $|z| > 2.58$.

One-sided: If $z > 1.645$, it does *not* always follow that $z > 2.33$.

S12.32 Use a *one-sided test.*

$H_0: \mu = 18$ v $H_1: \mu < 18$

$n = 9$, $\bar{x} = 15.56$, $s = 2.51$

Test statistic is $t = 2.92$ with $\nu = 8$.

One-sided critical value at $\alpha = 0.05$ is 1.860.

Since $t = 2.92 > 1.860$, *reject* H_0 at $\alpha = 0.05$.

The complaints *are* justified. Conclude that the mean number of fries is *less than* 18.

S12.36 Use a *two-sided test.*

$H_0: \mu = 1.80$ v $H_1: \mu \neq 1.80$

$n = 25$, $\bar{x} = 1.65$, $s = 0.60$

Test statistic is $z = 1.25$.

Since $|z| = 1.25 < 1.96$, *cannot reject* H_0 at $\alpha = 0.05$.

Yes, it is *possible* that the establishment is being truthful.

S12.40 Use a *two-sided test.*

$H_0: \mu = 65$ v $H_1: \mu \neq 65$

$n = 30$, $\bar{x} = 71$, $\sigma = 12.5$

Test statistic is $z = 2.63$.

Since $|z| = 2.63 > 2.58$, *reject* H_0 at $\alpha = 0.01$.

Yes, it appears that these TAFE students *did* perform differently from other learner drivers.

S12.44 Use a *two-sided test.*

H_0: $\mu = 65$ v H_1: $\mu \neq 65$

$n = 30$, $\bar{x} = 71$, $\sigma = 12.5$

Test statistic is $z = 2.63$.

p-value = 2(0.5000 − 0.4957) = 0.0086

Since 0.0086 < 0.01, *reject* H_0 at $\alpha = 0.01$.

Yes, it appears that these TAFE students *did* perform differently from other learner drivers. (Note that this conclusion is identical to that in Exercise S12.40, as it should be.)

S12.46 Let Sample 1 = females, Sample 2 = males.

Since blood pressures are normally distributed, a two-sample *t*-test can be used.

The null and alternative hypotheses in symbols are:

H_0: $\mu_1 = \mu_2$ v H_1: $\mu_1 \neq \mu_2$

Data: $x_1 = 113.7$, $s_1 = 11.2$, $x_2 = 119.6$, $s_2 = 18.4$, $n_1 = 16$, $n_2 = 25$

The *pool standard deviation* is $s_p = 16.02$.

The *t-test statistic* is $t = 1.15$ with $\nu = 39$ degrees of freedom.

From Table 2, the *critical value* at $\alpha = 0.05$ is 2.021.

Since $|t| = 1.15 < 2.021$, H_0 *cannot be rejected* at $\alpha = 0.05$.

We conclude that there appears to be no significant difference between the blood pressures of male and female TAFE business students.

Chapter S13

S13.4 H_0: The results are consistent with the claimed proportions

v

H_1: The results are not consistent with the claimed proportions

The goodness-of-fit table is shown below.

Category	Opposed	In favour	Apathetic	*Total*
Observed frequency	50	60	90	200
Expected frequency	40	100	60	200

Therefore, $\chi^2 = 2.5 + 16 + 15 = 33.5$ with $\nu = 2$ degrees of freedom.

From Table 4, the critical value at $\alpha = 0.05$ is 5.991.

Since $\chi^2 = 33.5 > 5.991$, we *reject* H_0 at $\alpha = 0.05$.

We therefore conclude that the results are not consistent with the claimed proportions.

S13.6 (a) The contingency table is shown below.

Sex \ Opinion	Ban	Not ban	*Total*
Male	88(82.9)	112(117.1)	200
Female	57(62.1)	93(87.9)	150
Total	145	205	350

(b) H_0: There is no difference of opinion between the two sexes

v

H_1: There is a difference of opinion between the two sexes

The expected frequencies are shown in parentheses in the contingency table in 13.6(a). The value of the χ^2 test-statistic is 1.25 with $\nu = 1$ degree of freedom.

From Table 4, the critical value at $\alpha = 0.05$ is 3.841.

Since $\chi^2 = 1.25 < 3.841$, we *cannot reject* H_0 at $\alpha = 0.05$.

We therefore conclude that there does not appear to be a difference between the two sexes on the issue of banning gambling advertising on television.

S13.10 H_0: The market research is accurate

v

H_1: The market research is not accurate

The goodness-of-fit table is shown below.

Category	Small	Medium	Large	*Total*
Observed frequency	120	215	165	500
Expected frequency	100	250	150	500

Therefore, $\chi^2 = (120 - 100)^2/100 + (215 - 250)^2/250 + (165 - 150)^2/150$

$= 4.000 + 4.900 + 1.500$

$= 10.400$

with $\nu = 2$ degrees of freedom.

From Table 4, the critical value at $\alpha = 0.01$ is 9.210.

Since $\chi^2 = 10.400 > 9.210$, we *reject* H_0 at $\alpha = 0.01$.

We therefore conclude that the results are not consistent with the market research and that the market research was not accurate.

S13.11 In this problem there is an 'implied' category of *not defective*. The complete contingency table, with the expected frequencies shown in parentheses, is shown below.

Category \ Machine no.	1	2	3	4	*Total*
Defective	28(26.25)	22(26.25)	35(26.25)	20(26.25)	105
Not defective	172(173.75)	178(173.75)	165(173.75)	180(173.75)	695
Total	200	200	200	200	800

H_0: The proportion of defectives does not vary from machine to machine

v

H_1: The proportion of defectives does vary from machine to machine

The value of the χ^2 test-statistic is 6.00 with $\nu = 3$ degrees of freedom.

From Table 4, the critical value at $\alpha = 0.05$ is 7.815.

Since $\chi^2 = 6.00 < 7.815$, we *cannot reject* H_0 at $\alpha = 0.05$.

We therefore conclude that there is no evidence of the proportion of defectives varying from machine to machine.

S13.14 H_0: There has been no change in the proportion of viewers

v

H_1: There has been a change in the proportion of viewers

The goodness-of-fit table is shown below.

Category	Tuned to program	Not tuned to program	*Total*
Observed frequency	64	336	400
Expected frequency	80	320	400

Therefore, $\chi^2 = (64 - 80)^2/80 + (336 - 320)^2/320$

$= 3.200 + 0.800$

$= 4.000$

with $\nu = 1$ degree of freedom.

From Table 4, the critical value at $\alpha = 0.05$ is 3.841.

Since $\chi^2 = 4.000 > 3.841$, we *reject* H_0 at $\alpha = 0.05$.

We therefore conclude that there has been a difference in the proportion of viewers who tuned to the program. It has increased significantly.

S13.19 H_0: The manager's claim is true

v

H_1: The manager's claim is not true

The goodness-of-fit table is shown below.

Category	A	B	C	D	E	*Total*
Observed frequency	10	19	15	27	9	80
Expected frequency	12	12	12	32	12	80

Therefore,
$$\chi^2 = (10 - 12)^2/12 + (19 - 12)^2/12 + (15 - 12)^2/12 + (27 - 32)^2/32 + (9 - 12)^2/12$$
$$= 0.333 + 4.083 + 0.750 + 0.781 + 0.750$$
$$= 6.697$$

with $\nu = 4$ degrees of freedom.

From Table 4, the critical value at $\alpha = 0.05$ is 9.488.

Since $\chi^2 = 6.697 < 9.488$, we *cannot reject* H_0 at $\alpha = 0.05$.

We therefore conclude that the manager's claim may well be true.

Glossary

A

ABS *see* **Australian Bureau of Statistics**

absolute error the magnitude of the difference between the approximated value of a quantity and its actual value; found by ignoring the sign of the difference

accumulated value of an annuity *see* **future value of an annuity**

acid test ratio a measure of liquidity obtained by excluding the less liquid current assets and the less pressing current liabilities

alternative hypothesis (H_1) the hypothesis that is accepted if the null hypothesis (H_0) is rejected

amortisation the payment of a debt by making regular payments to cover the principal and interest on the balance of the principal. When the last payment is made, the debt is extinguished

annuities periodic payments of equal amounts of money made over a period

annuity due an annuity where the payments are made at the beginning of each period

antilogarithm the value of the number that corresponds to a given logarithm

arithmetic mean *see* mean

array data that have been sorted in order of magnitude

asset a resource of value controlled by a business (e.g. cash, bank deposits, inventory, receivables)

asset turnover the ratio of sales assets to total assets

ATO Australian Taxation Office

Australian Bureau of Statistics (ABS) the main source of official Australian statistical data

average *see* **mean**

average of relative prices a composite index number that is the average of simple price indexes. It is of limited appeal since it is unweighted

average of weighted relative prices *see* **Laspeyres index**

B

balance sheet a statement that lists the resources of value (assets) and liabilities; a statement of the financial position at a specific date

bar chart a visual display of categorical data, consisting of vertical bars where the heights of the bars represent frequencies. Types of bar charts include simple, multiple and compound

base *see* logarithm

base period the period to which other periods are compared. It should be a 'normal' period with respect to the other periods and should not be chosen too far in the past

bias tendency for certain individuals or objects to have a larger or smaller chance of being selected in a sample

bimodal for discrete data, where there are two modes; for continuous data, where there are two peaks in the curve

bivariate data data that consist of two variables

book value the amount remaining of the original cost of an asset that has not been charged as an expense. Also called (current) written-down value

break-even analysis calculates how many items must be produced and sold in order that the income gained exactly equals the cost of manufacturing them

brokerage the commission charged by a stockbroker who acts on your behalf to buy or sell stocks and shares

C

cash discount a reduction in the price of goods when the purchaser pays in the form of cash or cheque, or within a specified period

categorical data data in which the observations are placed in categories

CATI Computer-Assisted Telephone Interview

cell a square representing a combination of outcomes in a contingency table

Central Limit Theorem the statement that if large samples are taken from a population, the sample means will follow a normal distribution, regardless of the shape of the population. It is the reason why the normal distribution is so important

chain discount a discount that is more than a single discount on an item or service. The purpose is often to add a further inducement for the item to be bought or the service to be used

characteristic the integer part of a logarithm

Chi-square test statistical procedure for testing hypotheses on frequency data

class interval a range of values for data; used in histograms

cluster sampling a sampling method in which the population is divided into groups (or clusters) where those in each group have similar characteristics. Some of those clusters are randomly selected for inclusion in the overall sample

coefficient a factor by which the rest of an algebraic term is multiplied

coefficient of determination the square of the correlation coefficient. It also indicates how well a straight line fits a set of bivariate data. Its value lies between 0 and 1

coefficient of variation a measure of the relative variability of a set of data; the ratio (expressed as a percentage) of the mean to the standard deviation

cold calling a survey technique in which a respondent is telephoned without warning

commission the income received as a result of selling goods and/or services; often expressed as a percentage of the value of the good/service

common logarithm *see* **logarithm**

complementary events events that are complements of each other; together they form the whole sample space

composite index number an index number that is constructed from changes in a number of different items

compound interest interest gained when the interest itself earns interest from period to period

conditional probability the probability that an event occurs given that another event occurs

confidence interval a range of values within which we can say that a population parameter will lie with a specified probability

constant a term whose value does not change

Consumer Price Index (CPI) the index most commonly used in Australia as a general indicator of the rate of price change for consumer goods and services

contingency table two-way table in which one variable is placed down the rows and the other variable is placed along the columns

continuous data data that are not made up of naturally distinct units. These data can never be measured with complete accuracy

contribution margin per unit the amount available per unit to contribute to the meeting of fixed costs

correlation coefficient a measure of the degree of association between two variables. It has a value between −1 and +1

CPI *see* **Consumer Price Index**

critical value a value, obtained from tables, with which the calculated value of a test statistic is compared to decide whether H_0 should be rejected or not

cubic equation an equation containing an x^3 term as the term with the highest power

cumulative frequency distribution a frequency distribution in which the frequencies for each observation (or class interval) are progressively added

cumulative frequency percentage cumulative frequencies expressed as a percentage of the total frequency

current asset ratio the ratio of current assets to current liabilities. Also known as *working-capital ratio*

current period the period for which an index number is to be found

current written-down value *see* **book value**

cyclical variation a time series component that indicates the expansion and contractions in the series. It measures the ability of a business (or economy) to recover from a recession, prosper in good times, contract again, etc.

D

debtor turnover a measure of the operating efficiency of the credit policy of a business expressed as a ratio of gross credit sales to average gross accounts receivables

deciles divisions of a distribution into 10 equal parts where each part contains the same number of observations

deferred annuity an annuity where the payments do not commence until a period of time has elapsed

degree of an expression the highest exponent of any variable contained in the expression

degrees of freedom a parameter of the t-distribution that measures how many values in a sample of a fixed size are free to vary

dependent events events whose outcomes depend on each other

dependent variable the variable whose value is to be predicted

depreciable asset an asset that depreciates in value over time

depreciation the amount for taxation purposes that is allowed for the decreasing value of equipment used for the purpose of generating income. It represents the allocation of part of the cost of an asset as an expense over its estimated useful life. Methods of calculating depreciation include **prime cost method**, **diminishing value method**, **units-of-production method** and **sum-of-the-years' digits method**

descriptive statistics the collection, condensation and display of information

deseasonalised *see* **seasonally adjusted**

diminishing value method a depreciation method; assumes that the depreciation of an asset is highest in the earliest years of its life and calculates it as a fixed percentage of the preceding book value. Also called the **reducing-balance method**

discount a reduction in price given usually to induce a customer to purchase a good or service. Types of discounts include sales, trade and cash

discounted value of an annuity *see* **present value of an annuity**

discrete data data in which the values are distinct and isolated

dispersion the extent to which the observations in a set of data are spread about the mean. Also called **variation**

effective life of an asset the period for which the asset can reasonably be expected to be used in producing assessable income if it is maintained in good condition

E

effective rate the real rate of interest that is actually paid by the borrower when a flat rate of interest is quoted

efficiency ratio the ability of a company to achieve the maximum return for the lowest possible level of assets

equation a statement that two expressions are equal

event some subset of a sample space

exponent a number that is written as a superscript to another number called a base. Also called **power** or **index**

exponential smoothing a technique for smoothing time series in which the values are replaced by estimates calculated using a combination of old and recent trends

expression a combination of constants and variables by the use of arithmetic operations

F

face value the value of a loan or investment before any interest is added on or any discount deducted

financial calculator calculator with specific keys designed to calculate specific financial formulae

Fisher's ideal index an index that is the square root of the product of the **Laspeyres index** and the **Paasche index**. It may be considered as a compromise between these two indexes

fixed cost a cost that is independent of the number of items produced

flat rate the interest rate that is based on the original principal, even though the amount of the loan outstanding is reducing

formula a rule or principle that is expressed in terms of algebraic symbols

fractile a measure that gives the fraction of a population below a given value

fraction an expression of the form $\frac{a}{b}$

freehand drawing a technique for drawing a line to indicate the secular trend of a time series. There are no rules for drawing such lines, so the technique is very subjective

frequency the number of times that an observation occurs

frequency distribution a listing of each observation together with its frequency

frequency polygon a line graph of a grouped frequency distribution; usually drawn on a histogram

future value of an annuity the amount due at the end of the term, which includes all the payments that have been made up to and including the final payment period. Also known as accumulated value of an annuity

G

Gaussian distribution an alternative name for the **normal distribution**

general addition law a relationship used in the calculation of probability for events that are *not* mutually exclusive

geometric mean the *n*th root of a product of *n* observations; used to average rates, ratios and growth rates

goodness-of-fit a measure of how well a straight line fits a set of bivariate data. *See* **coefficient of determination**

Goods and Services Tax *see* **GST**

gradient of a line the rate at which the variable *y* changes with respect to the variable *x*. Also called the *slope of a line*

graph a visual display of the relationship between variables

gross income the total amounts received or accrued

grouped frequency distribution data that have been grouped into class intervals, with the frequency of observations in each class interval given

growth model the fitting of an exponential trend to a time series; usually used in series where there is little or no seasonal or cyclical variation

GST a Goods and Services Tax introduced into Australia on 1 July 2000. This is a broad-based tax of 10 per cent on most supplies of goods and services consumed in Australia

H

harmonic mean a measure of central tendency used to average time rates

histogram a visual display of numerical data (in particular, a grouped frequency distribution). It is a simple vertical bar chart in which the bars are adjacent and the height of each bar represents the frequency of the corresponding class interval

I

impossible event an event that contains no outcomes. Also known as the empty set

independent events events where the occurrence of one event does not affect the likelihood of the other event occurring

independent variable a variable that is used to make a prediction of the dependent variable

index *see* **exponent**

index number a measure that compares two figures. It is expressed as a ratio of the figure in the current period to the figure in the base period

inflation the persistent increase in the general level of prices

inflationary spiral an effect where inflation and interest rates continually push each other higher

input tax credit this is the amount allowed to offset the GST included in the price paid for an acquisition or the GST paid on importations used in the enterprise

integer a whole number (e.g. 1, 2, 3)

interest any income derived from capital that has been invested. Also the amount of money that is paid for the use of borrowed money

interfractile range the range of values between two specified fractiles

interquartile range a measure of dispersion that is the difference between the upper and lower quartiles. Also called **midspread**

intersection of events the set of outcomes that are in all events

interval estimate *see* **confidence interval**

irregular variation a time series component that represents the residual variation after the components of secular trend, seasonal variation and cyclical variation have been eliminated. It represents unexplained variation

K

kurtosis a measure of skewness indicating the extent to which values are concentrated around the mode

L

Laspeyres index a commonly used weighted index where the weights are the quantities of each item bought in the base period. Also known as the **average of weighted relative prices**

least-squares linear regression the fitting of a straight line to a set of data. It is based on minimising the sum of the squares of the errors that the line makes

least-squares linear regression line this is the line of best fit according to the criterion of least-squares regression

liability the claim on a company's assets by those who have lent money or extended credit to it

like terms terms that differ only in their numerical coefficient

linear equation an equation in which the exponent of all variables is 1

linear interpolation a method for approximating values between two tabulated values

linear regression *see* **least-squares linear regression**

liquidity ratio the ability of a company to pay its immediate debts

logarithm the logarithm of a number N to a base b is the power to which b must be raised to obtain N. If the base is 10, it is called a **common logarithm**

lottery method a method of sampling where each population member is assigned a number and a selection of numbers is drawn randomly

M

mantissa the decimal part of a logarithm

marginal tax rate the percentage tax that must be paid on each dollar in a particular income range

mean a measure of central tendency in which the sum of all observations is divided by the number of observations. Also known as the **arithmetic mean** or **average**

mean deviation a measure of dispersion that is the average of the absolute value of the residuals

median the middle observation of a set of data. That is, 50% of values are less than the median and 50% are more than the median

Medicare levy federal tax paid by certain taxpayers in Australia to assist in paying the cost of free healthcare

method of moving averages a technique for smoothing time series data that relies on replacing each observation with a moving average

midspread *see* **interquartile range**

mixed congruential generator a method of generating sequences of random numbers

mode for discrete data, the observation(s) with the largest frequency; for continuous data, the values of the variable where the smooth curve has peaks

moving average an average assigned to a time series observation, consisting of the average of one (or more) points both before and after it; used to smooth data

multiple regression regression in which there is one dependent (y) variable but more than one independent (x) variable

multistage sampling a sampling method in which the population is broken into groups from which a number of subgroups are selected randomly

mutually exclusive events events that cannot both occur at the same time

N

natural logarithm a logarithm to the base e

net profit margin the ratio of net profit to sales

normal curve a smooth curve describing the shape of the normal distribution

normal distribution a symmetrical statistical distribution that is bell-shaped. Its importance is due to the **Central Limit Theorem**

null hypothesis (H_0) a statement that nothing unusual has occurred. It will be either be rejected or not rejected by using a statistical test

O

observed frequency the actual data that arise from an experiment

ogive curve a graph of a cumulative frequency distribution. It can be used to estimate the median or percentiles but may not be particularly accurate

one-tailed test the statistical test used when the alternative hypothesis (H_1) is in one direction only

open-ended class interval a class interval that has no lower/upper bound (e.g. above 100, below 20)

ordered pairs pairs of observations that are presented in a specified order

ordinary annuity an annuity where the payments are made at the end of each period

origin the point where the two variables x and y both have the value of zero

outliers observations that are unusual or inconsistent with the rest of the observations

P

Paasche index a weighted index where the weights used are the quantities of each item bought in the *current* period

Pearson product-moment correlation coefficient *see* **correlation coefficient**

percentage the relative value of one number to another, expressed as the numerator of the equivalent fraction where the denominator is 100. The symbol used is %

percentiles measures that divide a distribution into 100 equal parts. The xth percentile has x% of the observations below it

perpetuity an annuity in which the payments continue indefinitely

pictogram a visual display of data that uses pictures

plant articles or assets that are used to produce assessable income, or are installed ready for use to produce income and are held in reserve

point estimate a single estimate of a population parameter

pooled standard deviation a combination of two sample standard deviations that is used in the two-sample t-test statistic

population the complete set of data about which one seeks information

population parameter a value calculated from a population (e.g. mean, median, standard deviation)

power *see* **exponent**

present value of an annuity the value of the annuity at the beginning of the initial rent period. Also called the **discounted value**

price relative the ratio of the price in the current period to the price of the same item in the base period

primary data data that are collected by the person or organisation that will be using the information

prime cost method depreciation method; assumes that the value of an asset depreciates by a fixed sum each period over the life of the asset. Also called the **straight-line method**

principal the original amount of money that is lent or invested, or the current balance of the original amount. Also called capital

probability the proportion of possible outcomes in a sample space that yield an event

product-moment correlation coefficient a measure of the strength of the relationship between two variables. Denoted by r

profit and loss statement a statement for a business that produces the net profit of sales ratio. Also known as an *income statement*

profit ratio the relationship of profit (gross and/or net) to some financial quantity (e.g. total assets or sales)

proportion a ratio that expresses the value of one quantity relative to the value of the other

pseudo random numbers numbers that for all purposes look random, but are in fact generated by a mathematical formula

Q

quadratic equation an equation containing an x^2 term as the term with the highest power

quartic equation an equation containing an x^4 term as the term with the highest power

quartiles measures that divide a distribution into four equal parts. The top 25% is cut off by the upper (3rd) quartile and the bottom 25% is cut off by the lower (1st) quartile; the 2nd quartile is the median

quartile class the class interval that contains the relevant quartile

quartile deviation one-half of the value of the interquartile range

quota sampling a non-random sampling method that involves the selection of quotas of individuals of certain types

R

random numbers a set of digits (often in 5 × 5 blocks) that have no pattern; used to assist in selecting a random sample

random sample a sample in which every member of the population has an equal chance of being selected

range a measure of dispersion that is the difference between the largest and smallest observations in a set of data

rank correlation coefficient a measure of the degree of association between two sets of rankings

rate a ratio expressed with a specified base

ratio a method of comparing two numbers or rates. It is usually expressed as a fraction, percentage or decimal and represents the relative value of one quantity to another quantity

raw data data in the form in which they are collected

reducing-balance method *see* **diminishing value method**

reference position (for the decimal point) the position immediately to the right of the first non-zero digit

relative error the ratio of the absolute value of the error made in estimating a quantity to the actual value of the quantity

relative error percentage the absolute error expressed as a percentage of the actual value of the quantity

residual the distance an observation is away from the mean

residual method a technique for removing the cyclical variation in a time series

return on investment the ratio of net profit to total assets

S

sales tax a wholesale (federal) tax on goods, administered and collected by the Australian

Taxation Office. As from 1 July 2000 it was replaced by the GST

salvage value of an asset the estimated value of the asset, if any, at the end of its effective life. Also known as **scrap value**

sample some part of a population

sample space the set of all possible outcomes of an experiment

sample statistic a value calculated from a sample

sampling error the random discrepancy between a measurement from a sample and the true population value

scatter diagram a visual display of bivariate data in which ordered pairs of data are plotted on a coordinate axis system. In a time series, these points are sometimes joined with straight lines. Also called a **scatter plot**

scatter plot *see* **scatter diagram**

scrap value of an asset *see* **salvage value of an asset**

scientific notation a shorthand way of writing very large or very small numbers

seasonal variation a time series component that indicates the repeating pattern of the series. Possible causes are festive occasions and climatic conditions (e.g. summer and winter)

seasonally adjusted when the factor of seasonal variation has been eliminated from a time series. Also called **deseasonalised**

secondary data information that has already been collected by someone else

secular trend a time series component that measures the long-term movement in the series

semi-averages a crude method of fitting a straight line to a time series. It is affected by outlying observations

semi-interquartile range *see* **quartile deviation**

sigma notation *see* **summation notation**

significance level the probability of making a type I error

simple aggregate index a composite index number that is the ratio (expressed as a percentage) of the price of a basket of goods in the current year to the price of exactly the same basket of goods in the base year. This index has serious drawbacks

simple index number an index number that is constructed from changes in the price of a single item

simple interest the interest earned per annum calculated on the amount of the principal only

simple price index the **price relative** expressed as a percentage

simultaneous equations equations that must be solved at the same time in order to find the values of the variables that will solve *both* equations

sinking fund a fund into which periodic payments are made so as to accumulate a specified nominated amount of money at the end of a specified period

skewed distribution a distribution that is not symmetrical. That is, the mean, the median and the mode are not all equal

skewness a measure of the symmetry (or lack of symmetry) of a distribution

slope of a line *see* **gradient of a line**

smoothing the use of techniques to eliminate the 'bumps' (or small variations) in data (especially time series data)

smoothing constant a constant (denoted by α) used in the exponential smoothing of a time series

Spearman rank correlation coefficient *see* **rank correlation coefficient**

spurious correlation correlation that seems to occur when two variables appear to be correlated but in fact are not. This often happens when the two variables are each related to a third variable, but not to each other

stamp duty a state duty levied on some legal, commercial and financial transactions. The rate of duty depends on the state involved

standard deviation a measure of dispersion that uses all observations and measures the 'average' distance of observations away from the mean

standard error of the mean a measure of accuracy of the point estimate of the population mean. If the sample is *large*, it is the sample standard deviation divided by the square root of the size of the sample

standard normal distribution a normal distribution that has a mean of 0 and a standard deviation of 1

standard score *see* **z-score**

statistical inference the process of drawing conclusions from available information and making decisions

statistician a person who can collect data, present them, analyse them and draw conclusions from them

stock turnover the ratio of the cost of goods sold to the average stock held; a measure of the efficiency of a company's stock

straight-line method *see* **prime cost method**

stratified sampling a non-random sampling method that involves dividing the population into strata (subsets) and taking samples of equal size from each of the strata

sum-of-the-years' digits method a depreciation method; assumes that the depreciation of an asset is highest in the earliest years of its life and calculates it as a varying fraction of the preceding book value

summation notation a shorthand method of indicating that values of a variable are to be added; denoted by the Greek letter Σ. Also known as **sigma notation**

T

t-distribution a statistical distribution, of similar shape to the normal distribution, that is useful for estimations involving small samples

taxable income assessable income minus allowable deductions

term part of an algebraic expression

test statistic a quantity that is calculated to determine whether the null hypothesis (H_0) is to be rejected or not

time series a set of observations measured at specified (usually equal) time intervals

total asset turnover the ratio of sales to total assets

trade discount a reduction in the price of goods when purchases are made from one business by another business

transposition the moving of terms from one side of an equation to the other

trend curve a curve that follows the (secular) trend of the data

turning point the point where a graph reaches a maximum or minimum

two-sample problem problem in which two samples are to be compared with each other

two-tailed test the statistical test used when the alternative hypothesis (H_1) is in two directions

type I error rejecting the null hypothesis (H_0) when it really is true

type II error not rejecting the null hypothesis (H_0) when it really is false

U

unbiased sample a sample that is genuinely representative of the population

undeducted cost of an asset this is the cost of the asset less the sum of the depreciation deductions and any further amounts that could not be deducted because the asset was used other than for producing assessable income (such as private use) or because a deduction is prevented by the income tax law (such as use for leisure facilities)

unimodal data data that have a single mode

union of events the set of outcomes that are in *any* of the events

units-of-production method a depreciation method; assumes that the depreciation of an asset depends on the number of items produced and calculates it as a fixed cost per unit of production

V

variable a term representing a quantity that may have a number of different values

variable cost a cost that changes with the number of items produced

variation *see* **dispersion**

Venn diagram a visual display of sample spaces and events. Such diagrams consist mainly of circles or rectangles

W

weighted index number an index number that allows greater importance (or weightings) to be attached to some items. Weightings can include such factors as quantities sold or quantities consumed of each item

weighted mean mean that assigns weights (or measures of relative importance) to the observations to be averaged; often used in time series measurements

wholesale sales tax *see* **sales tax**

written-down value *see* **book value**

Y

yield the amount of return on an investment expressed as a percentage of the price paid

Z

***z*-score** the transformation of a value in terms of the number of standard deviations it is from the mean

Index

D

E

F

G

H

Q

R

S

T

U

V

W

X

Y

Z